Open Problems in Mathematics

John Forbes Nash, Jr. • Michael Th. Rassias
Editors

Open Problems in Mathematics

Springer

Editors
John Forbes Nash, Jr. (Deceased)
Department of Mathematics
Princeton University
Princeton, NJ, USA

Michael Th. Rassias
Department of Mathematics
Princeton University
Princeton, NJ, USA

ETH-Zürich
Department of Mathematics
Zürich, Switzerland

ISBN 978-3-319-32160-8 ISBN 978-3-319-32162-2 (eBook)
DOI 10.1007/978-3-319-32162-2

Library of Congress Control Number: 2016941333

Printed on acid-free paper

This Springer imprint is published by Springer Nature
The registered company is Springer International Publishing AG Switzerland

Preface

John Forbes Nash, Jr. and Michael Th. Rassias

Learn from yesterday, live for today, hope for tomorrow.
The important thing is not to stop questioning.
– Albert Einstein (1879–1955)

It has become clear to the modern working mathematician that no single researcher, regardless of his knowledge, experience, and talent, is capable anymore of overviewing the major open problems and trends of mathematics in its entirety. The breadth and diversity of mathematics during the last century has witnessed an unprecedented expansion.

In 1900, when David Hilbert began his celebrated lecture delivered before the International Congress of Mathematicians in Paris, he stoically said:

> Who of us would not be glad to lift the veil behind which the future lies hidden; to cast a glance at the next advances of our science and at the secrets of its development during future centuries? What particular goals will there be toward which the leading mathematical spirits of coming generations will strive? What new methods and new facts in the wide and rich field of mathematical thought will the new centuries disclose?

Perhaps Hilbert was among the last great mathematicians who could talk about mathematics as a whole, presenting problems which covered most of its range at the time. One can claim this, not because there will be no other mathematicians of Hilbert's caliber, but because life is probably too short for one to have the opportunity to expose himself to the allness of the realm of modern mathematics. Melancholic as this thought may sound, it simultaneously creates the necessity and aspiration for intense collaboration between researchers of different disciplines. Thus, overviewing open problems in mathematics has nowadays become a task which can only be accomplished by collective efforts.

The scope of this volume is to publish invited survey papers presenting the status of some essential open problems in pure and applied mathematics, including old and new results as well as methods and techniques used toward their solution. One expository paper is devoted to each problem or constellation of related problems. The present anthology of open problems, notwithstanding the fact that it ranges over a variety of mathematical areas, does not claim by any means to be complete, as such a goal would be impossible to achieve. It is, rather, a collection of beautiful mathematical questions which were chosen for a variety of reasons. Some were chosen for their undoubtable importance and applicability, others because they constitute intriguing curiosities which remain unexplained mysteries on the basis of current knowledge and techniques, and some for more emotional reasons. Additionally, the attribute of a problem having a somewhat *vintage flavor* was also influential in our decision process.

The book chapters have been contributed by leading experts in the corresponding fields. We would like to express our deepest thanks to all of them for participating in this effort.

Princeton, NJ, USA
April, 2015

John F. Nash, Jr.

Michael Th. Rassias

Contents

A Farewell to "A Beautiful Mind and a Beautiful Person"

Michael Th. Rassias

Having found it very hard to resign myself to John F. Nash's sudden and so tragic passing, I postponed writing my commemorative addendum to our jointly composed preface until this compilation of papers on open problems was almost fully ready for publication. Now that I have finally built up my courage for coming to terms with John Nash's demise, my name, which joyfully adjoins his at the end of the above preface, now also stands sadly alone below the following bit of reminiscence from my privileged year as his collaborator and frequent companion.

It all started in September 2014, in one of the afternoon coffee/tea meetings that take place on a daily basis in the common room of Fine Hall, the building housing the Mathematics Department of Princeton University. John Nash silently entered the room, poured himself a cup of decaf coffee and then sat alone in a chair close by. That was when I first approached him and had a really pleasant chat about problems in the interplay of game theory and number theory. From that day onwards, our discussions became ever more frequent, and we eventually decided to prepare this volume *Open Problems in Mathematics*. The day we made this decision, he turned to me and said with his gentle voice, "I don't want to be *just a name* on the cover though. I want to be really involved." After that, we met almost daily and discussed for several hours at a time, examining a vast number of open problems in mathematics ranging over several areas. During these discussions, it became even clearer to me that his way of thinking was very different from that of almost all other mathematicians I have ever met. He was thinking in an unconventional, most creative way. His quick and distinctive mind was still shining bright in his later years.

This volume was practically almost ready before John and Alicia Nash left in May for Oslo, where he was awarded the 2015 Abel Prize from the Norwegian Academy of Science and Letters. We had even prepared the preface of this volume, which he was so much looking forward to see published. Our decision to include handwritten signatures, as well, was along the lines of the somewhat vintage flavor and style that he liked.

John Nash was planning to write a brief article on an open problem in game theory, which was the only problem we had not discussed yet. He was planning to prepare it and discuss about it after his trip to Oslo. Thus, he never got the opportunity to write it. On this note, and notwithstanding my ‘last-minute’ invitation, Professor Eric Maskin generously accepted to contribute a paper presenting an important open problem in cooperative game theory.

With this opportunity, I would also like to say just a few words about the man behind the mathematician. In the famous movie *A Beautiful Mind*, which portrayed his life, he was presented as a really combative person. It is true that in his early years he might have been, having also to battle with the demons of his illness. Being almost 60 years younger than him, I had the chance to get acquainted with his personality in his senior years. All the people who were around him, including myself, can avow that he was a truly wonderful person. Very kind and disarmingly simple, as well as modest. This is the reason why, among friends at Princeton, I used to humorously say that the movie should have been called *A Beautiful Mind and a Beautiful Person*. What was certainly true though was the dear love between John and Alicia Nash, who together faced and overcame the tremendous challenges of John Nash’s life. It is somehow a romantic tragedy that fate bound them to even leave this life together.

In history, one can say that among the mathematicians who have reached greatness, there are some—a selected few—who have gone beyond greatness to become legends. John Nash was one such legend.

The contributors of papers and myself cordially dedicate this volume to the memory and rich mathematical legacy of John F. Nash, Jr.

Princeton, NJ, USA Michael Th. Rassias

Introduction
John Nash: Theorems and Ideas

Misha Gromov

Nash was not building big theories, he did not attempt to dislodge old concepts and to promote new ones, he didn't try to be paradoxical.

Nash was solving classical mathematical problems, difficult problems, something that nobody else was able to do, not even to imagine how to do it.

His landmark theorem of 1956—one of the main achievements of mathematics of the twentieth century–reads:

> All Riemannian manifolds X can be realised as smooth submanifolds in Euclidean spaces $\mathbb{R}^q$, such that the smoothness class of the submanifold realising an X in $\mathbb{R}^q$ equals that of the Riemannian metric g on X and where the dimension q of the ambient Euclidean space can be universally bounded in terms of the dimension of X.[1]

And as far as C^1-*smooth* isometric embeddings $f : X \to \mathbb{R}^q$ are concerned, there is no constraint on the dimension of the Euclidean space except for what is dictated by the topology of X:

> Every C^1-smooth n-dimensional submanifold X_0 in $\mathbb{R}^q$ for $q \geq n+1$ can be deformed (by a C^1-isotopy) to a new C^1-position such that the induced Riemannian metric on X_0 becomes equal to a given g.[2]

At first sight, these are natural classically looking theorems. But what Nash has discovered in the course of his constructions of isomeric embeddings is far from "classical"—it is something that brings about a dramatic alteration of our understanding of the basic logic of analysis and differential geometry. Judging from

M. Gromov
IHÉS, 36 route de Chartres, 41990 Bures-sur-Yvette, France

e-mail: gromov@ihes.fr

[1]This was proven in the 1956 paper for C^r-smooth metrics, $r = 3, 4, \ldots, \infty$; the existence of *real analytic* isometric embeddings of *compact* manifolds with *real analytic* Riemannian metrics to Euclidean spaces was proven by Nash in 1966.

[2]Nash proved this in his 1954 paper for $q \geq n+2$, where he indicated that a modification of his method would allow $q = n+1$ as well. This was implemented in a 1955 paper by Nico Kuiper.

the classical perspective, what Nash has achieved in his papers is as impossible as the story of his life.

Prior to Nash, the following two heuristic principles, vaguely similar to the first and the second laws of thermodynamics, have been (almost?) unquestionably accepted by analysts:

1. Conservation of Regularity. *The smoothness of solutions f of a "natural" functional, in particular a differential, equation $\mathscr{D}(f) = g$ is determined by the equation itself but not by a particular class of functions f used for the proof of the existence of solutions.*
2. Increase of Irregularity. *If some amount of regularity of potential solutions f of our equations has been lost, it cannot be recaptured by any "external means," such as artificial smoothing of functions.*

Instances of the first principle can be traced to the following three Hilbert's problems:

5th: *Continuous groups are infinitely differentiable, in fact, real analytic.*
19th: *Solutions of "natural" elliptic PDE are real analytic.*
Also Hilbert's formulation of his **13th** problem on

non-representability of "interesting" functions in many variables
by superpositions of **continuous** *functions in fewer variables*

is motivated by this principle:

$$\textit{continuous} \Leftrightarrow \textit{real analytic}$$

as far as superpositions of functions are concerned.

Nash C^1-isometric embedding theorem shattered the conservation of regularity idea: the system of differential equations that describes isometric immersions $f : X \to \mathbb{R}^q$ may have no analytic or not even C^2-smooth solution f.

But, according to Nash's 1954 theorem, *if $q > dim(X)$, and if X is diffeomorphic, to $\mathbb{R}^n$, $n < q$, or to the n-sphere, then, no matter what Riemannian metric g you are given on this X, there are lots of isometric C^1-embeddings $X \to \mathbb{R}^q$.*[3]

Now, look at an equally incredible Nash's approach to *more regular, say C^∞-smooth, isometric embeddings*. The main lemma used by Nash, his *implicit (or inverse) function theorem*, may seem "classical" unless you read the small print:

Let $\mathscr{D} : F \to G$ be a C^∞-smooth non-linear differential operator between spaces F and G of C^∞-sections of two vector bundles over a manifold X.

---

[3]In the spirit of Nash but probably independently, the *continuous* $\Leftrightarrow$ *real analytic* equivalence for superpositions of functions was disproved by Kolmogorov in 1956; yet, in essence, Hilbert's **13th** problem remains unsolved: *are there algebraic (or other natural) functions in many variables that are not superpositions of* **real analytic** *functions in two variables?*

Also, despite an enormous progress, "true" Hilbert's 19th problem remains widely open: *what are possible singularities of solutions of elliptic PDE systems, such as minimal subvarieties and Einstein manifolds.*

If the linearization $\mathcal{L} = \mathcal{L}_{f_0}(f)$ *of* $\mathcal{D}$ *at a point* $f_0 \in F$ *is invertible at* $g_0 = \mathcal{D}(f_0) \in G$ *by a* **differential** *operator linear in* g*, say* $\mathcal{M} = \mathcal{M}_{f_0}(g)$*, then* $\mathcal{D}$ *is also invertible (by a nonlinear non-differential operator) in a (small fine)* C^∞*-neighborhood of* $\mathcal{D}(f_0) \in G$*.*

You must be a novice in analysis or a genius like Nash to believe anything like that can be ever true and/or to have a single nontrivial application.

First of all, who has ever seen inversions of differential operators again by *differential* ones?

And second of all, how on earth can $\mathcal{D}$ be inverted by means of $\mathcal{M}$ when both operators, being differential, *increase* irregularity?

But Nash writes down a simple formula for a linearized inversion $\mathcal{M}$ for the metric inducing operator $\mathcal{D}$, and he suggests a compensation for the loss of regularity by the use of *smoothing operators*.

The latter may strike you as realistic as a successful performance of perpetuum mobile with a mechanical implementation of Maxwell's demon... unless you start following Nash's computation and realize to your immense surprise that the smoothing does work in the hands of John Nash.

This, combined with a few ingenious geometric constructions, leads to C^∞-smooth isometric embeddings $f : X \to \mathbb{R}^q$ for $q = 3n^3/2 + O(n^2)$, $n = dim(X)$.

Besides the above, Nash has proved a few other great theorems, but it is his work on isometric immersions that opened a new world of mathematics that stretches in front of our eyes in yet unknown directions and still waits to be explored.

$P \stackrel{?}{=} NP$

Scott Aaronson

Abstract In 1950, John Nash sent a remarkable letter to the National Security Agency, in which—seeking to build theoretical foundations for cryptography—he all but formulated what today we call the $\mathsf{P} \stackrel{?}{=} \mathsf{NP}$ problem, and consider one of the great open problems of science. Here I survey the status of this problem in 2016, for a broad audience of mathematicians, scientists, and engineers. I offer a personal perspective on what it's about, why it's important, why it's reasonable to conjecture that $\mathsf{P} \neq \mathsf{NP}$ is both true and provable, why proving it is so hard, the landscape of related problems, and crucially, what progress has been made in the last half-century toward solving those problems. The discussion of progress includes diagonalization and circuit lower bounds; the relativization, algebrization, and natural proofs barriers; and the recent works of Ryan Williams and Ketan Mulmuley, which (in different ways) hint at a duality between impossibility proofs and algorithms.

1 Introduction

> Now my general conjecture is as follows: for almost all sufficiently complex types of enciphering, especially where the instructions given by different portions of the key interact complexly with each other in the determination of their ultimate effects on the enciphering, the mean key computation length increases exponentially with the length of the key, or in other words, the information content of the key ... The nature of this conjecture is such that I cannot *prove* it, even for a special type of ciphers. Nor do I expect it to be proven.—John Nash, 1950 [171]

In 1900, David Hilbert challenged mathematicians to design a "purely mechanical procedure" to determine the truth or falsehood of any mathematical statement.

For a more recent version of this survey, with numerous corrections and revisions as well as new material, the reader is invited to visit http://www.scottaaronson.com/papers/pvsnp.pdf

S. Aaronson (✉)
Department of Computer Science, University of Texas, Austin, USA
e-mail: scott@scottaaronson.com

J.F. Nash, Jr., M.Th. Rassias (eds.), *Open Problems in Mathematics*,
DOI 10.1007/978-3-319-32162-2_1

That goal turned out to be impossible. But the *question*—does such a procedure exist, and why or why not?—helped launch three related revolutions that shaped the twentieth century: one in math and science, as reasoning itself became a subject of mathematical analysis; one in philosophy, as the results of Gödel, Church, Turing, and Post showed the limits of formal systems and the power of self-reference; and one in technology, as the electronic computer achieved, not all of Hilbert's dream, but enough of it to change the daily experience of most people on earth.

The $\mathsf{P} \stackrel{?}{=} \mathsf{NP}$ problem is a modern refinement of Hilbert's 1900 question. The problem was explicitly posed in the early 1970s in the works of Cook and Levin, though versions were stated earlier—including by Gödel in 1956, and as we see above, by John Nash in 1950. In plain language, $\mathsf{P} \stackrel{?}{=} \mathsf{NP}$ asks whether there's a *fast* procedure to answer all questions *that have short answers that are easy to verify mechanically*. Here one should think of a large jigsaw puzzle with (say) 10^{1000} possible ways of arranging the pieces, or an encrypted message with a similarly huge number of possible decrypts, or an airline with astronomically many ways of scheduling its flights, or a neural network with millions of weights that can be set independently. All of these examples share two key features:

(1) a finite but exponentially-large space of possible solutions; and
(2) a fast, mechanical way to check whether any claimed solution is "valid." (For example, do the puzzle pieces now fit together in a rectangle? Does the proposed airline schedule achieve the desired profit? Does the neural network correctly classify the images in a test suite?)

The $\mathsf{P} \stackrel{?}{=} \mathsf{NP}$ question asks whether, under the above conditions, there's a general method to *find* a valid solution whenever one exists, and which is enormously faster than just trying all the possibilities one by one, from now till the end of the universe, like in Jorge Luis Borges' Library of Babel.

Notice that Hilbert's goal has been amended in two ways. On the one hand, the new task is "easier" because we've restricted ourselves to questions with only finitely many possible answers, each of which is easy to verify or rule out. On the other hand, the task is "harder" because we now insist on a *fast* procedure: one that avoids the exponential explosion inherent in the brute-force approach.

Of course, to discuss such things mathematically, we need to pin down the meanings of "fast" and "mechanical" and "easily checked." As we'll see, the $\mathsf{P} \stackrel{?}{=} \mathsf{NP}$ question corresponds to one natural choice for how to define these concepts, albeit not the only imaginable choice. For the impatient, P stands for "Polynomial Time," and is the class of all decision problems (that is, infinite sets of yes-or-no questions) solvable by a standard digital computer—or for concreteness, a Turing machine—using a polynomial amount of time. By this, we mean a number of elementary logical operations that is upper-bounded by the bit-length of the input question raised to some fixed power. Meanwhile, NP stands for "Nondeterministic Polynomial Time," and is the class of all decision problems for which, if the answer is "yes," then there is a polynomial-size proof that a Turing machine can *verify* in polynomial time. It's immediate that $\mathsf{P} \subseteq \mathsf{NP}$, so the question is whether this containment is proper (and hence $\mathsf{P} \neq \mathsf{NP}$), or whether $\mathsf{NP} \subseteq \mathsf{P}$ (and hence $\mathsf{P} = \mathsf{NP}$).

1.1 *The Importance of* $P \stackrel{?}{=} NP$

Before getting formal, it seems appropriate to say something about the significance of the $P \stackrel{?}{=} NP$ question. $P \stackrel{?}{=} NP$, we might say, shares with Hilbert's original question the character of a "math problem that's more than a math problem": a question that reaches inward to ask about mathematical reasoning itself, and also outward to everything from philosophy to natural science to practical computation.

To start with the obvious, essentially all the cryptography that we currently use on the Internet—for example, for sending credit card numbers—would be broken if $P = NP$ (and if, moreover, the algorithm were efficient in practice, a caveat we'll return to later). Though he was writing 21 years before $P \stackrel{?}{=} NP$ was explicitly posed, this is the point Nash was making in the passage with which we began.

The reason is that, in most cryptography, the problem of finding the decryption key is an NP search problem: that is, we know mathematically how to *check* whether a valid key has been found. The only exceptions are cryptosystems like the one-time pad and quantum key distribution, which don't rely on any computational assumptions (but have other disadvantages, such as the need for huge pre-shared keys or for special communication hardware).

The metamathematical import of $P \stackrel{?}{=} NP$ was also recognized early. It was articulated, for example, in Kurt Gödel's now-famous 1956 letter to John von Neumann, which sets out what we now call the $P \stackrel{?}{=} NP$ question. Gödel wrote:

> If there actually were a machine with [running time] $\sim Kn$ (or even only with $\sim Kn^2$) [for some constant K independent of n], this would have consequences of the greatest magnitude. That is to say, it would clearly indicate that, despite the unsolvability of the Entscheidungsproblem, the mental effort of the mathematician in the case of yes-or-no questions could be completely [added in a footnote: apart from the postulation of axioms] replaced by machines. One would indeed have to simply select an n so large that, if the machine yields no result, there would then also be no reason to think further about the problem.

Expanding on Gödel's observation, some modern commentators have explained the importance of $P \stackrel{?}{=} NP$ as follows. It's well-known that $P \stackrel{?}{=} NP$ is one of the seven Clay Millennium Problems (alongside the Riemann Hypothesis, the Yang-Mills mass gap, etc.), for which a solution commands a million-dollar prize [66]. But even among those problems, $P \stackrel{?}{=} NP$ has a special status. For if someone discovered that $P = NP$, and if moreover the algorithm was efficient in practice, that person could solve not merely one Millennium Problem but *all seven* of them—for she'd simply need to program her computer to search for formal proofs of the other six conjectures.[1] Of course, if (as most computer scientists believe) $P \neq NP$,

[1]Here we're using the observation that, once we fix a formal system (say, first-order logic plus the axioms of ZF set theory), deciding whether a given statement has a proof at most n symbols long in that system is an NP problem, which can therefore be solved in time polynomial in n assuming

a proof of *that* would have no such world-changing implications, but even the fact that such a proof could *rule out* those implications underscores the enormity of what we're asking.

I should be honest about the caveats. While theoretical computer scientists (including me!) have not always been above poetic flourish, $\mathsf{P} \stackrel{?}{=} \mathsf{NP}$ is not quite equivalent to the questions of "whether human creativity can be automated," or "whether anyone who can appreciate a symphony is Mozart, anyone who can recognize a great novel is Jane Austen." Apart from the obvious point that *no* purely mathematical question could fully capture these imponderables, there are also more specific issues.

For one thing, while $\mathsf{P} \stackrel{?}{=} \mathsf{NP}$ has tremendous relevance to artificial intelligence, it says nothing about the *differences*, or lack thereof, between humans and machines. Indeed, $\mathsf{P} \neq \mathsf{NP}$ would represent a limitation on *all* classical digital computation, one that might plausibly apply to human brains just as well as to electronic computers. Nor does $\mathsf{P} \neq \mathsf{NP}$ rule out the possibility of robots taking over the world. To defeat humanity, presumably the robots wouldn't need to solve arbitrary NP problems in polynomial time: they'd merely need to be smarter than *us*, and to have imperfect heuristics better than the imperfect heuristics that *we* picked up from a billion years of evolution! Conversely, while a proof of $\mathsf{P} = \mathsf{NP}$ might hasten a robot uprising, it wouldn't guarantee one. For again, what $\mathsf{P} \stackrel{?}{=} \mathsf{NP}$ asks is not whether *all* creativity can be automated, but only *that creativity whose fruits can be quickly checked by computer programs that we know how to write.*

To illustrate, suppose we wanted to program a computer to create new Mozart-quality symphonies and Shakespeare-quality plays. If $\mathsf{P} = \mathsf{NP}$, and the algorithm were efficient in practice, then that really would imply that these feats could be reduced to a seemingly-easier problem, of programming a computer to *recognize* such symphonies and plays when given them. And interestingly, $\mathsf{P} = \mathsf{NP}$ might *also* help with the recognition problem: for example, by letting us train a neural network that reverse-engineered the expressed aesthetic preferences of hundreds of human experts. But how well that neural network would perform is an empirical question outside the scope of mathematics.

1.2 *Objections to* $\mathsf{P} \stackrel{?}{=} \mathsf{NP}$

After modest exposure to the $\mathsf{P} \stackrel{?}{=} \mathsf{NP}$ problem, many people come up with what they consider an irrefutable objection to its phrasing or importance. Since the same objections tend to recur, in this section I'll collect the most frequent ones and make some comments about them.

$\mathsf{P} = \mathsf{NP}$. We're also assuming that the other six Clay conjectures have ZF proofs that are not too enormous: say, 10^{12} symbols or fewer, depending on the exact running time of the assumed algorithm. In the case of the Poincaré Conjecture, this can almost be taken to be a fact, modulo the translation of Perelman's proof [179] into the language of ZF.

1.2.1 The Asymptotic Objection

Objection $P \stackrel{?}{=} NP$ talks only about asymptotics—i.e., whether the running time of an algorithm grows polynomially or exponentially with the size n of the question that was asked, as n goes to infinity. It says nothing about the number of steps needed for concrete values of n (say, a thousand or a million), which is all anyone would ever care about in practice.

Response It was realized early in the history of computer science that "number of steps" is not a robust measure of hardness, because it varies too wildly from one machine model to the next (from Macs to PCs and so forth), and also depends heavily on low-level details of how the problem is encoded. The asymptotic complexity of a problem could be seen as *that contribution to its hardness that is clean and mathematical*, and that survives the vicissitudes of technology. Of course, real-world software design requires thinking about many non-asymptotic contributions to a program's efficiency, from compiler overhead to the layout of the cache (as well as many considerations that have nothing to do with efficiency at all). But any good programmer knows that the asymptotics matter as well.

More specifically, many people object to theoretical computer science's equation of "polynomial" with "efficient" and "exponential" with "inefficient," given that for any practical value of n, an algorithm that takes 1.0000001^n steps is clearly preferable to an algorithm that takes n^{1000} steps. This would be a strong objection, if such algorithms were everyday phenomena. Empirically, however, computer scientists found that there *is* a strong correlation between "solvable in polynomial time" and "solvable efficiently in practice," with most (but not all) problems in P that they care about solvable in linear or quadratic or cubic time, and most (but not all) problems outside P that they care about requiring c^n time via any known algorithm, for some c significantly larger than 1. Furthermore, even when the first polynomial-time algorithm discovered for some problem takes (say) n^6 or n^{10} time, it often happens that later advances lower the exponent, or that the algorithm runs much faster in practice than it can be guaranteed to run in theory. This is what happened, for example, with linear programming, primality testing, and Markov Chain Monte Carlo algorithms.

Having said that, *of course* the goal is not just to answer some specific question like $P \stackrel{?}{=} NP$, but to learn the truth about efficient computation, whatever it might be. If practically-important NP problems turn out to be solvable in n^{1000} time but not in n^{999} time, or in 1.0000001^n time, then so be it. From this perspective, one could argue that $P \stackrel{?}{=} NP$ simply serves as a marker of ignorance: in effect we are saying, "if we can't even answer *this*, then surely we can't answer the more refined questions either."

1.2.2 The Polynomial-Time Objection

Objection But why should we draw the border of efficiency at the polynomial functions, as opposed to any other class of functions—for example, functions upper-bounded by n^2, or functions of the form $n^{\log^c n}$ (called *quasipolynomial* functions)?

Response There is a good theoretical answer to this: it's because polynomials are the smallest class of functions that contain the linear functions, and that are closed under basic operations like addition, multiplication, and composition. For this reason, they are the smallest class that ensures that we can compose "efficient algorithms" a constant number of times, and still get an algorithm that is efficient overall. For the same reason, polynomials are *also* the smallest class that ensures that our "set of efficiently solvable problems" is independent of the low-level details of the machine model.

Having said that, much of algorithms research *is* about lowering the order of the polynomial, for problems already known to be in P; and theoretical computer scientists *do* use looser notions like quasipolynomial time whenever they are needed.

1.2.3 The Kitchen-Sink Objection

Objection $\mathsf{P} \stackrel{?}{=} \mathsf{NP}$ is limited, because it talks only about discrete, deterministic algorithms that find exact solutions in the worst case—and also, because it ignores the possibility of natural processes that might exceed the limits of Turing machines, such as analog computers, biological computers, or quantum computers.

Response For every assumption mentioned above, there is now a major branch of theoretical computer science that studies what happens when one relaxes the assumption: for example, randomized algorithms, approximation algorithms, average-case complexity, and quantum computing. I'll discuss some of these branches in Sect. 5. Briefly, though, there are deep reasons why many of these ideas are thought to leave the original $\mathsf{P} \stackrel{?}{=} \mathsf{NP}$ problem in place. For example, according to the $\mathsf{P} = \mathsf{BPP}$ conjecture (see Sect. 5.4.1), randomized algorithms yield no more power than P, while careful analyses of noise, energy expenditure, and the like suggest that the same is true for analog computers (see [3]). Meanwhile, the famous *PCP Theorem* and its offshoots (see Sect. 3) have shown that, for many NP problems, there cannot even be a polynomial-time algorithm to *approximate* the answer to within a reasonable factor, unless $\mathsf{P} = \mathsf{NP}$.

In other cases, new ideas have led to major, substantive *strengthenings* of the $\mathsf{P} \neq \mathsf{NP}$ conjecture (see Sect. 5): for example, that there exist NP problems that are hard even on random inputs, or hard even for a quantum computer. Of course, proving $\mathsf{P} \neq \mathsf{NP}$ itself is a prerequisite to proving any of these strengthened versions.

There's one part of this objection that's so common that it requires some separate comments. Namely, people will say that even if $\mathsf{P} \neq \mathsf{NP}$, *in practice* we can find almost always find good enough solutions to the problems we care about, for example by using heuristics like simulated annealing or genetic algorithms, or by using special structure or symmetries in real-life problem instances.

Certainly there are cases where this assumption is true. But there are also cases where it's false: indeed, the entire field of cryptography is about *making* the assumption false! In addition, I believe our practical experience is biased by the fact that we don't even *try* to solve search problems that we "know" are hopeless—yet that wouldn't be hopeless in a world where $P = NP$ (and where the algorithm was efficient in practice). For example, presumably no one would try using brute-force search to look for a formal proof of the Riemann Hypothesis one billion lines long or shorter, or a 10-megabyte program that reproduced most of the content of Wikipedia within a reasonable time (possibly needing to encode many of the principles of human intelligence in order to do so). Yet both of these are "merely" NP search problems, and things one could seriously contemplate in a world where $P = NP$.

1.2.4 The Mathematical Snobbery Objection

Objection $P \stackrel{?}{=} NP$ is not a "real" math problem, because it talks about Turing machines, which are arbitrary human creations, rather than about "natural" mathematical objects like integers or manifolds.

Response The simplest reply is that $P \stackrel{?}{=} NP$ is not about Turing machines at all, but about *algorithms*, which seem every bit as central to mathematics as integers or manifolds. Turing machines are just one particular formalism for expressing algorithms, as the Arabic numerals are one particular formalism for integers. And crucially, just like the Riemann Hypothesis is still the Riemann Hypothesis in base-17 arithmetic, so essentially *every* formalism for deterministic digital computation ever proposed gives rise to the same complexity classes P and NP, and the same question about whether they are equal. (This observation is known as the *Extended Church-Turing Thesis*.)

This objection might also reflect lack of familiarity with recent progress in complexity theory, which has drawn on Fourier analysis, arithmetic combinatorics, representation theory, algebraic geometry, and dozens of other subjects about which yellow books are written. Furthermore, in Sect. 6.6, we'll see Geometric Complexity Theory (GCT), a breathtakingly ambitious program for proving $P \neq NP$ that throws almost the entire arsenal of modern mathematics at the problem, including geometric invariant theory, plethysms, quantum groups, and Langlands-type correspondences. Regardless of whether GCT's specific conjectures pan out, they illustrate in detail how progress toward proving $P \neq NP$ will plausibly involve deep insights from many parts of mathematics.

1.2.5 The Sour Grapes Objection

Objection $P \stackrel{?}{=} NP$ is *so* hard that it's impossible to make anything resembling progress on it, at least at this stage in human history—and for that reason, it's unworthy of serious effort or attention. Indeed, we might as well treat such questions as if their answers were formally independent of set theory, as for all we know they are (a possibility discussed further in Sect. 3.1).

Response One of the main purposes of this survey is to explain what we know now, relevant to the $\mathsf{P} \stackrel{?}{=} \mathsf{NP}$ problem, that we didn't know 10 or 20 or 30 years ago. It's true that, if "progress" entails having a solution already in sight, or being able to estimate the time to a solution, I know of no progress of *that* kind! But by the same standard, one would have to say there was no "progress" toward Fermat's Last Theorem in 1900—even as mathematicians, partly motivated by Fermat's problem, were laying foundations of algebraic number theory that *did* eventually lead to Wiles's proof. In this survey, I'll try to convey how, over the last few decades, insights about circuit lower bounds, relativization and arithmetization, pseudorandomness and natural proofs, the "duality" between lower bounds and algorithms, the permanent and determinant manifolds, and more have transformed our understanding of what a $\mathsf{P} \neq \mathsf{NP}$ proof could look like.

I should point out that, even supposing $\mathsf{P} \stackrel{?}{=} \mathsf{NP}$ is *never* solved, it's already been remarkably fruitful as an "aspirational" or "flagship" question, helping to shape research in algorithms, cryptography, learning theory, derandomization, quantum computing, and other things that theoretical computer scientists work on. Furthermore, later we'll see examples of how seemingly-unrelated progress in some of those other areas, unexpectedly ended up tying back to the quest to prove $\mathsf{P} \neq \mathsf{NP}$.

1.2.6 The Obviousness Objection

Objection It is intuitively obvious that $\mathsf{P} \neq \mathsf{NP}$. For that reason, a proof of $\mathsf{P} \neq \mathsf{NP}$—confirming that indeed, we can't do something that no reasonable person would ever have imagined we could do—gives almost no useful information.

Response This objection is perhaps less common among mathematicians than others, since were it upheld, it would generalize to *almost all* of mathematics! Like with most famous unsolved math problems, the quest to prove $\mathsf{P} \neq \mathsf{NP}$ is "less about the destination than the journey": there might or might not be surprises in the answer itself, but there will *certainly* be huge surprises (indeed, there already have been huge surprises) along the way. More concretely: to make a sweeping statement like $\mathsf{P} \neq \mathsf{NP}$, about what polynomial-time algorithms *can't* do, will require an unprecedented understanding of what they *can* do. This will almost certainly entail the discovery of many new polynomial-time algorithms, some of which could have practical relevance. In Sect. 6, we will see many more subtle examples of the "duality" between algorithms and impossibility proofs, with progress on each informing the other.

Of course, to whatever extent you regard $\mathsf{P} = \mathsf{NP}$ as a live possibility, the Obviousness Objection is not open to you.

1.2.7 The Constructivity Objection

Objection Even if P = NP, the proof could be nonconstructive—in which case it wouldn't have any of the amazing implications discussed in Sect. 1.1, because we wouldn't know the algorithm.

Response A nonconstructive proof that an algorithm exists is indeed a theoretical possibility, though one that has reared its head only a few times in the history of computer science.[2] Even then, however, once we knew that an algorithm *existed*, we would have a massive inducement to try to find it. The same is true if, for example, the first proof of P = NP only gave an n^{1000} algorithm, but we suspected that an n^2 algorithm existed.[3]

1.3 Further Reading

There were at least four previous major survey articles about $P \stackrel{?}{=} NP$: Michael Sipser's 1992 "The History and Status of the P versus NP Question" [207]; Stephen Cook's 2000 "The P versus NPProblem" [66], which was written for the announcement of the Clay Millennium Prize; Avi Wigderson's 2006 "P,

---

[2]The most celebrated examples of nonconstructive proofs that algorithms exist all come from the *Robertson-Seymour graph minors theory*, one of the great achievements of twentieth-century combinatorics (for an accessible introduction, see for example Fellows [75]). The Robertson-Seymour theory typically deals with *parameterized* problems: for example, "given a graph G, decide whether G can be embedded on a sphere with k handles." In those cases, typically a fast algorithm A_k can be abstractly shown to exist for every value of k. The central problem is that each A_k requires hard-coded data—in the above example, a finite list of obstructions to the desired embedding—that no one knows how to find given k, and whose size might also grow astronomically as a function of k. On the other hand, once the finite obstruction set for a given k was known, one could then use it to solve the problem for any graph G in time $O\left(|G|^3\right)$, where the constant hidden by the big-O depended on k.

Robertson-Seymour theory also provides a few examples of non-parameterized problems that are abstractly proved to be in P but with no bound on the exponent, or abstractly proved to be $O(n^3)$ or $O(n^2)$ but with no bound on the constant. Thus, one cannot rule out the possibility that the same would happen with an NP-complete problem, and Donald Knuth [131] has explicitly speculated that P = NP will be proven in that way. To me, however, it is unclear whether he speculates this because there is a positive reason for thinking it true, or just because it would be cool and interesting if it *was* true.

[3]As an amusing side note, there is a trick called *Levin's universal search* [141], in which one "dovetails" over all Turing machines $M_1, M_2, \ldots$ (that is, for all t, runs $M_1, \ldots, M_t$ for t steps each), halting when and if any M_i has outputs a valid solution to one's NP search problem. If we know P = NP, then we know this particular algorithm will find a valid solution, whenever one exists, in polynomial time—because clearly *some* M_i does so, and all the machines other than M_i increase the total running time by "only" a polynomial factor! With more work, one can even decrease this to a constant factor. Admittedly, however, the polynomial or constant factor will be so enormous as to negate this algorithm's practical use.

NP, and Mathematics—A Computational Complexity Perspective"[232]; and Eric Allender's 2009 "A Status Report on the P versus NP Question" [21]. All four are excellent, so it's only with trepidation that I add another entry to the crowded arena. I hope that, if nothing else, this survey shows how much has continued to occur. I cover several major topics that either didn't exist a decade ago, or existed only in much more rudimentary form: for example, the algebrization barrier, "ironic complexity theory" (including Ryan Williams's NEXP $\not\subset$ ACC result), the "chasm at depth three" for the permanent, and the Mulmuley-Sohoni Geometric Complexity Theory program.

The seminal papers that set up the intellectual framework for $\mathsf{P} \stackrel{?}{=} \mathsf{NP}$, posed it, and demonstrated its importance include those of Edmonds [74], Cobham [65], Cook [67], Karp [122], and Levin [141]. See also Trakhtenbrot [223] for a survey of Soviet thought about *perebor*, as brute-force search was referred to in Russian in the 1950s and 60s.

The classic text that introduced the wider world to P, NP, and NP-completeness, and that gave a canonical (and still-useful) list of hundreds of NP-complete problems, is Garey and Johnson [86]. Some recommended computational complexity theory textbooks—in rough order from earliest to most recent, in the material they cover—are Sipser [208], Papadimitriou [175], Schöning [199], Moore and Mertens [155], and Arora and Barak [27]. Surveys on particular aspects of complexity theory will be recommended where relevant throughout the survey.

Those seeking a nontechnical introduction to $\mathsf{P} \stackrel{?}{=} \mathsf{NP}$ might enjoy Lance Fortnow's charming book *The Golden Ticket* [80], or his 2009 popular article for *Communications of the ACM* [79]. My own *Quantum Computing Since Democritus* [6] gives something between a popular and a technical treatment.

2 Formalizing $\mathsf{P} \stackrel{?}{=} \mathsf{NP}$ and Central Related Concepts

The $\mathsf{P} \stackrel{?}{=} \mathsf{NP}$ problem is normally phrased in terms of *Turing machines*: a theoretical model of computation proposed by Alan Turing in 1936, which involves a one-dimensional tape divided into discrete squares, and a finite control that moves back and forth on the tape, reading and writing symbols. For a formal definition, see, e.g., Sipser [208] or Cook [66].

In this survey, I won't define Turing machines, for the simple reason that *if you know any programming language—C, Java, Python, etc.—then you already know something that's equivalent to Turing machines for our purposes.* More precisely, the *Church-Turing Thesis* holds that virtually any model of digital computation one can define will be equivalent to Turing machines, in the sense that Turing machines can simulate that model and vice versa. A modern refinement, the *Extended Church-Turing Thesis*, says that moreover, these simulations will incur at most a polynomial overhead in time and memory. If we accept this, then there's a well-defined notion of "solvable in polynomial time by a digital computer," which is independent of the low-level details of the computer's architecture: the instruction set, the rules for accessing memory, etc. This licenses us to ignore those details. The main caveats here are that

(1) the computer must be classical, discrete, and deterministic (it's not a quantum computer, an analog device, etc., nor can it call a random-number generator or any other external resource), and
(2) there must be no *a-priori* limit on how much memory the computer can address, even though any program that runs for finite time will only address a finite amount of memory.[4,5]

We can now define P and NP, in terms of Turing machines for concreteness—but, because of the Extended Church-Turing Thesis, the reader is free to substitute other computing formalisms such as Lisp programs, λ-calculus, stylized assembly language, or cellular automata.

A *language* is a set of binary strings, $L \subseteq \{0, 1\}^*$, where $\{0, 1\}^*$ is the set of all binary strings of all (finite) lengths. Of course a language can be infinite, even though every string in the language is finite. One example is the language consisting of all palindromes: for instance, 00, 11, 0110, 11011, etc., but not 001 or 1100. A more interesting example is the language consisting of all binary encodings of prime numbers: for instance, 10, 11, 101, and 111, but not 100.

A binary string $x \in \{0, 1\}^*$, for which we want to know whether $x \in L$, is called an *instance* of the general problem of deciding membership in L. Given a Turing machine M and an instance x, we let $M(x)$ denote M run on input x (say, on a tape initialized to $\cdots 0\#x\#0 \cdots$, or x surrounded by delimiters and blank or 0 symbols). We say that $M(x)$ *accepts* if it eventually halts and enters an "accept" state, and we say that M *decides* the language L if for all $x \in \{0, 1\}^*$,

$$x \in L \Longleftrightarrow M(x) \text{ accepts}.$$

The machine M may also contain a "reject" state, which M enters to signify that it has halted without accepting. Let $|x|$ be the length of x (i.e., the number of bits). Then we say M is *polynomial-time* if there exists a polynomial p such that $M(x)$ halts, either accepting or rejecting, after at most $p(|x|)$ steps, for all $x \in \{0, 1\}^*$.

[4]The reason for this caveat is that, if a programming language were inherently limited to (say) 64K of memory, there would be only finitely many possible program behaviors, so in principle we could just cache everything in a giant lookup table. Many programming languages do impose a finite upper bound on the addressable memory, but they could easily be generalized to remove this restriction (or one could consider programs that store information on external I/O devices).

[5]I should stress that, once we specify which computational models we have in mind—Turing machines, Intel machine code, etc.—the polynomial-time equivalence of those models is typically a *theorem*, though a rather tedious one. The "thesis" of the Extended Church-Turing Thesis, the part not susceptible to proof, is that all *other* "reasonable" models of digital computation will also be equivalent to those models.

Now, P, or Polynomial-Time, is the class of all languages L for which there exists a Turing machine M that decides L in polynomial time. Also, NP, or Nondeterministic Polynomial-Time, is the class of languages L for which there exists a Turing machine M, and a polynomial p, such that for all $x \in \{0,1\}^*$,

$$x \in L \iff \exists w \in \{0,1\}^{p(|x|)} \; M(x,w) \text{ accepts.}$$

In other words, NP is the class of languages L for which, whenever $x \in L$, there exists a polynomial-size "witness string" w, which enables a polynomial-time "verifier" M to recognize that indeed $x \in L$. Conversely, whenever $x \notin L$, there must be no w that causes $M(x,w)$ to accept.

There is an earlier definition of NP, which explains its ungainly name. Namely, we can define a *nondeterministic Turing machine* as a Turing machine that "when it sees a fork in the road, takes it": that is, that is allowed to transition from a single state at time t to multiple possible states at time $t+1$. We say that a machine "accepts" its input x, if there *exists* a list of valid transitions between states, $s_1 \rightarrow s_2 \rightarrow s_3 \rightarrow \cdots$, that the machine could make on input x that terminates in an accepting state s_{Accept}. The machine "rejects" if there is no such accepting path. The "running time" of such a machine is the maximum number of steps taken along *any* path, until the machine either accepts or rejects. We can then define NP as the class of all languages L for which there exists a nondeterministic Turing machine that decides L in polynomial time. It is clear that NP, so defined, is equivalent to the more intuitive verifier definition that we gave earlier. In one direction, if we have a polynomial-time verifier M, then a nondeterministic Turing machine can create paths corresponding to all possible witness strings w, and accept if and only if there exists a w such that $M(x,w)$ accepts. In the other direction, if we have a nondeterministic Turing machine M', then a verifier can take as its witness string w a description of a claimed path that causes $M'(x)$ to accept, then check that the path indeed does so.

Clearly $\mathsf{P} \subseteq \mathsf{NP}$, since an NP verifier M can just ignore its witness w, and try to decide in polynomial time whether $x \in L$ itself. The central conjecture is that this containment is strict.

Conjecture 1. $\mathsf{P} \neq \mathsf{NP}$.

2.1 NP*-Completeness*

A further concept, not part of the statement of $\mathsf{P} \stackrel{?}{=} \mathsf{NP}$ but central to any discussion of it, is NP-*completeness*. To explain this requires a few more definitions. An *oracle Turing machine* is a Turing machine that, at any time, can submit an instance x to an "oracle": a device that, in a single time step, returns a bit indicating whether x belongs to some given language L. An oracle that answers all queries consistently with L is called an *L-oracle*, and we write M^L to denote the (oracle) Turing machine

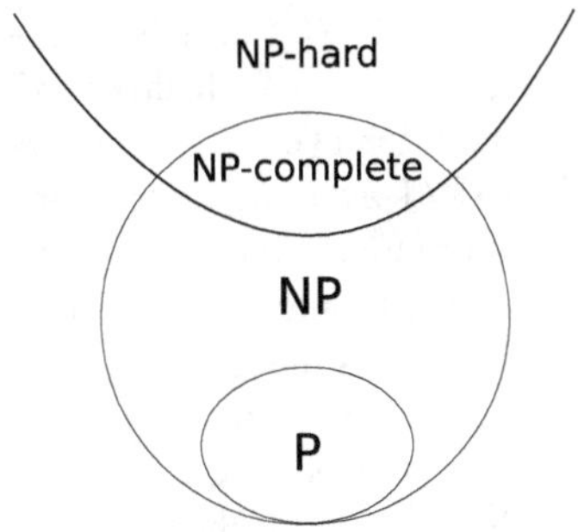

Fig. 1 P, NP, NP-hard, and NP-complete

M with L-oracle. We can then define P^L, or P *relative to L*, as the class of all languages L' for which there exists an oracle machine M such that M^L decides L' in polynomial time. If $L' \in \mathsf{P}^L$, then we also write $L' \leq_{\mathsf{P}}^{T} L$, which means "$L'$ is polynomial-time Turing-reducible to L." Note that polynomial-time Turing-reducibility is indeed a partial order relation (i.e., it is transitive and reflexive).

A language L is NP-*hard* (technically, NP-hard under Turing reductions[6]) if $\mathsf{NP} \subseteq \mathsf{P}^L$. Informally, NP-hard means "at least as hard as any NP problem": if we had a black box for an NP-hard problem, we could use it to solve all NP problems in polynomial time. Also, L is NP-*complete* if L is NP-hard *and* $L \in \mathsf{NP}$. Informally, NP-complete problems are the hardest problems in NP. (See Fig. 1.)

A priori, it is not completely obvious that NP-hard or NP-complete problems even exist. The great discovery of theoretical computer science in the 1970s was that hundreds of problems of practical importance fall into these classes: indeed, what is unusual is to find a hard NP problem that is *not* NP-complete.

More concretely, consider the following languages:

- 3SAT is the language consisting of all encodings of Boolean formulas φ over n variables, which consist of ANDs of "3-clauses" (i.e., ORs of up to three variables or their negations), such that there exists at least one assignment that satisfies φ. Here is an example, for which one can check that there's *no* satisfying assignment:

$$(x \vee y \vee z) \wedge (\overline{x} \vee \overline{y} \vee \overline{z}) \wedge (x \vee \overline{y}) \wedge (\overline{x} \vee y) \wedge (y \vee \overline{z}) \wedge (\overline{y} \vee z)$$

[6]In practice, often one only needs a special kind of Turing reduction called a *many-one reduction* or *Karp reduction*, which is a polynomial-time algorithm that maps every yes-instance of L' to a yes-instance of L, and every no-instance of L' to a no-instance of L. The additional power of Turing reductions—to make multiple queries to the L-oracle (with later queries depending on the outcomes of earlier ones), post-process the results of those queries, etc.—is needed only in a minority of cases. Nevertheless, for conceptual simplicity, throughout this survey I'll talk in terms of Turing reductions.

- HAMILTONCYCLE is the language consisting of all encodings of undirected graphs, for which there exists a cycle that visits each vertex exactly once (a Hamilton cycle).
- TSP (Traveling Salesperson Problem) is the language consisting of all encodings of ordered pairs $\langle G, k\rangle$, such that G is a graph with positive integer weights, k is a positive integer, and G has a Hamilton cycle of total weight at most k.
- CLIQUE is the language consisting of all encodings of ordered pairs $\langle G, k\rangle$, such that G is an undirected graph, k is a positive integer, and G contains a clique with at least k vertices.
- SUBSETSUM is the language consisting of all encodings of positive integer tuples $\langle a_1, \ldots, a_k, b\rangle$, for which there exists a subset of the a_i's that sums to b.
- 3COL is the language consisting of all encodings of undirected graphs G that are 3-*colorable* (that is, the vertices of G can be colored red, green, or blue, so that no two adjacent vertices are colored the same)

All of these languages are easily seen to be in NP. The famous *Cook-Levin Theorem* says that one of them—3SAT—is also NP-hard, and hence NP-complete.

Theorem 2 (Cook-Levin Theorem [67, 141]). 3SAT *is* NP-*complete.*

A proof of Theorem 2 can be found in any theory of computing textbook (for example, [208]). Here I'll confine myself to saying that Theorem 2 can be proved in three steps, each of them routine from today's standpoint:

(1) One constructs an artificial language that is "NP-complete essentially by definition": for example,

$$L = \left\{ \left(\langle M\rangle, x, 0^s, 0^t \right) : \exists w \in \{0,1\}^s \text{ suchthat} M(x,w) \text{ acceptsin} \leq t \text{ steps} \right\},$$

where $\langle M\rangle$ is a description of the Turing machine M.

(2) One then reduces L to the CIRCUITSAT problem, where we are given as input a description of a Boolean circuit C built of AND, OR, and NOT gates, and asked whether there exists an assignment $x \in \{0,1\}^n$ for the input bits such that $C(x) = 1$. To do that, in turn, is more like electrical engineering than mathematics: given a Turing machine M, one simply builds up a Boolean logic circuit that simulates the action of M on the input (x, w) for t time steps, whose size is polynomial in the parameters $|\langle M\rangle|$, $|x|$, s, and t, and which outputs 1 if and only if M ever enters its accept state.

(3) Finally, one reduces CIRCUITSAT to 3SAT, by creating a new variable for each gate in the Boolean circuit C, and then creating clauses to enforce that the variable for each gate G equals the AND, OR, or NOT (as appropriate) of the variables for G's inputs. For example, one can express the constraint $a \wedge b = c$ by

$$(a \vee \overline{c}) \wedge (b \vee \overline{c}) \wedge \left(\overline{a} \vee \overline{b} \vee c\right).$$

One then constrains the variable for the final output gate to be 1, yielding a 3SAT instance φ that is satisfiable if and only if the CIRCUITSAT instance was (i.e., iff there existed an x such that $C(x) = 1$).

Note that the algorithms to reduce L to CIRCUITSAT and to 3SAT—i.e., to convert M to C and C to φ—run in polynomial time (actually linear time), so we do indeed preserve NP-hardness. Also, the reason for the 3 in 3SAT is simply that a Boolean AND or OR gate has one output bit and two input bits, so it relates three bits in total. The analogous 2SAT problem turns out to be in P.

Once one knows that 3SAT is NP-complete, "the floodgates are open." One can then prove that countless other NP problems are NP-complete by reducing 3SAT to them, and then reducing those problems to others, and so on. The first indication of how pervasiveness NP-completeness really was came from Karp [122] in 1972. He showed, among many other results:

Theorem 3 (Karp [122]). HAMILTONCYCLE, TSP, CLIQUE, SUBSETSUM, *and* 3COL *are all* NP-*complete.*

Today, so many combinatorial search problems have been proven NP-complete that, whenever one encounters a new such problem, a useful rule of thumb is that it's "NP-complete unless it has a good reason not to be"!

Note that, if any NP-complete problem is in P, then all of them are, and P = NP. Conversely, if any NP-complete problem is not in P, then none of them are, and P $\neq$ NP.

One application of NP-completeness is to reduce the number of logical quantifiers needed to state the P $\neq$ NP conjecture. Let $\mathcal{PT}$ be the set of all polynomial-time Turing machines, and given a language L, let $L(x) = 1$ if $x \in L$ and $L(x) = 0$ otherwise. Then a "naïve" statement of P $\neq$ NP would be

$$\exists L \in \mathsf{NP}\ \forall M \in \mathcal{PT}\ \exists x\ M(x) \neq L(x).$$

(Here, by quantifying over all languages in NP, we really mean quantifying over all verification algorithms that define such languages.) Once we know that 3SAT (for example) is NP-complete, we can state P $\neq$ NP as simply:

$$\forall M \in \mathcal{PT}\ \exists x\ M(x) \neq 3\text{Sat}(x).$$

In words, we can pick any NP-complete problem we like; then P $\neq$ NP is equivalent to the statement that *that* problem is not in P.

2.2 *Other Core Concepts*

A few more concepts give a fuller picture of the $\mathsf{P} \stackrel{?}{=} \mathsf{NP}$ question, and will be referred to later in the survey. In this section, we restrict ourselves to concepts that were explored in the 1970s, around the same time as $\mathsf{P} \stackrel{?}{=} \mathsf{NP}$ itself was formulated, and that are covered alongside $\mathsf{P} \stackrel{?}{=} \mathsf{NP}$ in any undergraduate textbook. Other important concepts, such as nonuniformity, randomness, and one-way functions, will be explained as needed in Sect. 5.

2.2.1 Search, Decision, and Optimization

For technical convenience, P and NP are defined in terms of languages or "decision problems," which have a single yes-or-no bit as the desired output (i.e., given an input x, is $x \in L$?). To put practical problems into this decision format, typically we ask something like: *does there exist* a solution that satisfies the following list of constraints? But of course, in real life we don't merely want to know whether a solution exists; we want to *find* a solution whenever there is one! And given the many examples in mathematics where explicitly finding an object is harder than proving its existence, one might worry that this would also occur here. Fortunately, though, shifting our focus from decision problems to search problems doesn't change the $\mathsf{P} \stackrel{?}{=} \mathsf{NP}$ question at all, because of the following classic observation.

Proposition 4. *If* $\mathsf{P} = \mathsf{NP}$*, then for every language* $L \in \mathsf{NP}$ *(defined by a verifier* M*), there is a polynomial-time algorithm that actually finds a witness* $w \in \{0,1\}^{p(n)}$ *such that* $M(x,w)$ *accepts, for all* $x \in L$*.*

Proof. The idea is to learn the bits of an accepting witness $w = w_1 \cdots w_{p(n)}$ one by one, by asking a series of NP decision questions. For example:

- Does there exist a w such that $M(x,w)$ accepts and $w_1 = 0$?

  If the answer is "yes," then next ask:

- Does there exist a w such that $M(x,w)$ accepts, $w_1 = 0$, and $w_2 = 0$?

  Otherwise, next ask:

- Does there exist a w such that $M(x,w)$ accepts, $w_1 = 1$, and $w_2 = 0$?

Continue in this manner until all $p(n)$ bits of w have been set. (This can also be seen as a binary search on the set of all $2^{p(n)}$ possible witnesses.) ∎

Note that there *are* problems for which finding a solution is believed to be much harder than deciding whether one exists. A classic example, as it happens, is the problem of finding a Nash equilibrium of a matrix game. Here Nash's theorem guarantees that an equilibrium always exists, but an important 2006 result of Daskalakis et al. [71] gave evidence that there is no polynomial-time algorithm to *find* an equilibrium.[7] The upshot of Proposition 4 is just that search and decision are equivalent for the NP-*complete* problems.

---

[7]Technically, Daskalakis et al. showed that the search problem of finding a Nash equilibrium is complete for a complexity class called PPAD. This could be loosely interpreted as saying that the

In practice, perhaps even more common than search problems are *optimization problems*, where we have some efficiently-computable cost function, say $C : \{0,1\}^n \to \{0,1,\ldots,2^{p(n)}\}$, and the goal is to find a solution $x \in \{0,1\}^n$ that maximizes $C(x)$. Fortunately, we can always reduce optimization problems to search and decision problems, by simply asking to find a solution x such that $C(x) \geq K$, and doing a binary search to find the largest K for which such an x still exists. So again, if $\mathsf{P} = \mathsf{NP}$ then all NP optimization problems are solvable in polynomial time. On the other hand, it is important to remember that, while "is there an x such that $C(x) \geq K$?" is an NP question, "does $\max_x C(x) = K$?" and "does x^* maximize $C(x)$?" are presumably *not* NP questions, because no single x is a witness to a yes-answer.

More generally, the fact that decision, search, and optimization all hinge on the same $\mathsf{P} \stackrel{?}{=} \mathsf{NP}$ question has meant that many people—including experts—freely abuse language by referring to search and optimization problems as "NP-complete." Strictly they should call such problems NP-hard, while reserving "NP-complete" for suitable associated decision problems.

2.2.2 The Twilight Zone: Between P and NP-complete

We say a language L is NP-*intermediate* if $L \in \mathsf{NP}$, but L is neither in P nor NP-complete. One might hope, not only that $\mathsf{P} \neq \mathsf{NP}$, but that there would be a dichotomy, with all NP problems either in P or else NP-complete. However, a classic result by Ladner [135] rules that possibility out.

Theorem 5 (Ladner [135]). *If* $\mathsf{P} \neq \mathsf{NP}$, *then there exist* NP-*intermediate languages.*

While Theorem 5 is theoretically important, the NP-intermediate problems that it yields are extremely artificial (requiring diagonalization to construct). On the other hand, as we'll see, there are also problems of real-world importance—particularly in cryptography and number theory—that are believed to be NP-intermediate, and a proof of $\mathsf{P} \neq \mathsf{NP}$ could leave the status of those problems open. (Of course, a proof of $\mathsf{P} = \mathsf{NP}$ would mean there were *no* NP-intermediate problems, since every NP problem would then be both NP-complete and in P.)

2.2.3 coNP and the Polynomial Hierarchy

Let $\overline{L} = \{0,1\}^* \setminus L$ be the *complement* of L: that is, the set of strings not in L. Then the complexity class

problem is "as close to NP-hard as it could possibly be, subject to Nash's theorem showing why the decision version is trivial."

$$\mathsf{coNP} := \{\overline{L} : L \in \mathsf{NP}\}$$

consists of the complements of all languages in NP. Note that this is *not* the same as $\overline{\mathsf{NP}}$, the set of all non-NP languages! Rather, $L \in \mathsf{coNP}$ means that whenever $x \notin L$, there's a short proof of non-membership that can be efficiently verified.

A natural question is whether NP is *closed under complement*: that is, whether $\mathsf{NP} = \mathsf{coNP}$. If $\mathsf{P} = \mathsf{NP}$, then certainly $\mathsf{P} = \mathsf{coNP}$, and hence $\mathsf{NP} = \mathsf{coNP}$ also. On the other hand, we could imagine a world where $\mathsf{NP} = \mathsf{coNP}$ even though $\mathsf{P} \neq \mathsf{NP}$. In that world, there would always be short proofs of *un*satisfiability (or of the *non*existence of cliques, Hamilton cycles, etc.), but those proofs could be intractable to find. A generalization of the $\mathsf{P} \neq \mathsf{NP}$ conjecture says that this doesn't happen:

Conjecture 6. $\mathsf{NP} \neq \mathsf{coNP}$.

A further generalization of P, NP, and coNP is the *polynomial hierarchy* PH. Defined by analogy with the *arithmetic hierarchy* in computability theory, PH is an infinite sequence of classes whose zeroth level equals P, and whose kth level (for $k \geq 1$) consists of all problems that are in P^L or NP^L or coNP^L, for some language L in the $(k-1)$st level. More succinctly, we write $\Sigma_0^{\mathsf{P}} = \mathsf{P}$, and

$$\Delta_k^{\mathsf{P}} = \mathsf{P}^{\Sigma_{k-1}^{\mathsf{P}}}, \qquad \Sigma_k^{\mathsf{P}} = \mathsf{NP}^{\Sigma_{k-1}^{\mathsf{P}}}, \qquad \Pi_k^{\mathsf{P}} = \mathsf{coNP}^{\Sigma_{k-1}^{\mathsf{P}}}$$

for all $k \geq 1$.[8] A more intuitive definition of PH is as the class of languages that are definable using a polynomial-time predicate with a constant number of alternating universal and existential quantifiers: for example, $L \in \Pi_2^{\mathsf{P}}$ if and only if there exists a polynomial-time machine M and polynomial p such that for all x,

$$x \in L \iff \forall w \in \{0,1\}^{p(|x|)}\, \exists z \in \{0,1\}^{p(|x|)}\ M(x, w, z) \text{ accepts.}$$

NP is then the special case with just one existential quantifier, over witness strings w.

If $\mathsf{P} = \mathsf{NP}$, then the entire PH "recursively unwinds" down to P: for example,

$$\Sigma_2^{\mathsf{P}} = \mathsf{NP}^{\mathsf{NP}} = \mathsf{NP}^{\mathsf{P}} = \mathsf{NP} = \mathsf{P}.$$

Moreover, one can show that if $\Sigma_k^{\mathsf{P}} = \Pi_k^{\mathsf{P}}$ or $\Sigma_k^{\mathsf{P}} = \Sigma_{k+1}^{\mathsf{P}}$ for any k, then all the levels above the $k$thcome "crashing down" to $\Sigma_k^{\mathsf{P}} = \Pi_k^{\mathsf{P}}$. On the other hand, a collapse at the kth level isn't known to imply a collapse at any *lower* level. Thus, we get an infinite sequence of stronger and stronger conjectures: first $\mathsf{P} \neq \mathsf{NP}$, then $\mathsf{NP} \neq \mathsf{coNP}$, then $\Sigma_2^{\mathsf{P}} \neq \Pi_2^{\mathsf{P}}$, and so on. In the limit, we can conjecture the following:

[8] In defining the kth level of the hierarchy, we could also have given oracles for Π_{k-1}^{P} rather than $\Sigma_{k-1}^{\mathsf{P}}$: it doesn't matter. Note also that "an oracle for complexity class $\mathcal{C}$" should be read as "an oracle for any $\mathcal{C}$-complete language L."

Conjecture 7. *All the levels of* PH *are distinct—i.e., the infinite hierarchy is strict.*

This is a generalization of $\mathsf{P} \neq \mathsf{NP}$ that many computer scientists believe, and that has many useful consequences that aren't known to follow from $\mathsf{P} \neq \mathsf{NP}$ itself.

It's also interesting to consider $\mathsf{NP} \cap \mathsf{coNP}$, which is the class of languages that admit short, easily-checkable proofs for both membership *and* non-membership. Here is yet another strengthening of the $\mathsf{P} \neq \mathsf{NP}$ conjecture:

Conjecture 8. $\mathsf{P} \neq \mathsf{NP} \cap \mathsf{coNP}$.

Of course, if $\mathsf{NP} = \mathsf{coNP}$, then the $\mathsf{P} \stackrel{?}{=} \mathsf{NP} \cap \mathsf{coNP}$ question becomes equivalent to the original $\mathsf{P} \stackrel{?}{=} \mathsf{NP}$ question. But it's conceivable that $\mathsf{P} = \mathsf{NP} \cap \mathsf{coNP}$ even if $\mathsf{NP} \neq \mathsf{coNP}$ (Fig. 2).

2.2.4 Factoring and Graph Isomorphism

As an application of these concepts, let's consider two languages that are suspected to be NP-intermediate. First, FAC—a language variant of the factoring problem—consists of all ordered pairs of positive integers $\langle N, k \rangle$ such that N has a nontrivial divisor at most k. Clearly a polynomial-time algorithm for FAC can be converted into a polynomial-time algorithm to output the prime factorization (by repeatedly doing binary search to peel off N's smallest divisor), and vice versa. Second, GRAPHISO—that is, graph isomorphism—consists of all encodings of pairs of undirected graphs $\langle G, H \rangle$, such that $G \cong H$. It's easy to see to see that FAC and GRAPHISO are both in NP.

More interestingly, FAC is actually in $\mathsf{NP} \cap \mathsf{coNP}$. For one can prove that $\langle N, k \rangle \notin$FAC by exhibiting the unique prime factorization of N, and showing that it only involves primes greater than k.[9] But this has the striking consequence that *factoring cannot be* NP*-complete unless* $\mathsf{NP} = \mathsf{coNP}$. The reason is the following.

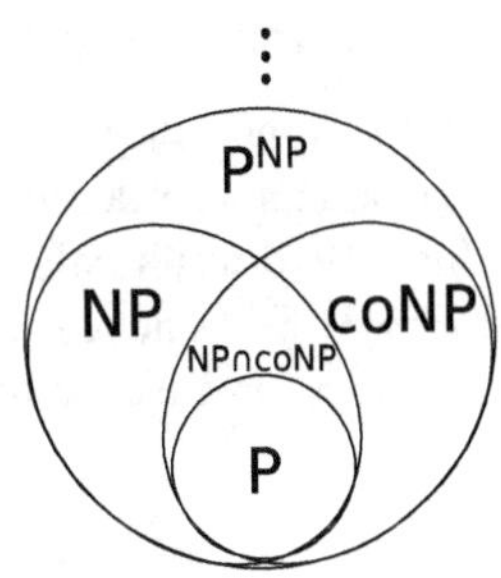

Fig. 2 The polynomial hierarchy

[9]This requires one nontrivial result, that every prime number has a succinct certificate—or in other words, that primality testing is in NP [180]. Since 2002, it is even known that primality testing is in P [14].

Proposition 9. *If any* $\mathsf{NP} \cap \mathsf{coNP}$ *language is* NP*-complete, then* $\mathsf{NP} = \mathsf{coNP}$*, and hence* PH *collapses to* NP.

Proof. Suppose $L \in \mathsf{NP} \cap \mathsf{coNP}$. Then $\mathsf{P}^L \subseteq \mathsf{NP} \cap \mathsf{coNP}$, since one can prove the validity of every answer to every query to the L-oracle (whether the answer is 'yes' or 'no'). So if $\mathsf{NP} \subseteq \mathsf{P}^L$, then $\mathsf{NP} \subseteq \mathsf{NP} \cap \mathsf{coNP}$ and hence $\mathsf{NP} = \mathsf{coNP}$. ■

GRAPHISO is not quite known to be in $\mathsf{NP} \cap \mathsf{coNP}$. However, it has been proven to be in $\mathsf{NP} \cap \mathsf{coNP}$ under a plausible assumption about pseudorandom generators [130]—and even with no assumptions, Boppana, Håstad, Zachos [49] proved the following.

Theorem 10 ([49]). *If* GRAPHISO *is* NP*-complete, then* PH *collapses to* Σ_2^{P}.

As this survey was being written, Babai [32] announced the following breakthrough result.

Theorem 11 (Babai [32]). GRAPHISO *is solvable in* $n^{\text{polylog}\, n}$ *time.*

Of course, this gives even more dramatic evidence that GRAPHISO is not NP-complete: if it was, then *all* NP problems would be solvable in $n^{\text{polylog}\, n}$ time as well.

2.2.5 Space Complexity

PSPACE is the class of languages L decidable by a Turing machine that uses a polynomial number of bits of *space* or *memory*, with no restriction on the number of time steps. Certainly $\mathsf{P} \subseteq \mathsf{PSPACE}$, since in t time steps, a serial algorithm can access at most t memory cells. More generally, it is not hard to see that $\mathsf{P} \subseteq \mathsf{NP} \subseteq \mathsf{PH} \subseteq \mathsf{PSPACE}$, but *none* of these containments have been proved to be strict. The following conjecture—asserting that polynomial space is strictly stronger than polynomial time—is perhaps second only to $\mathsf{P} \neq \mathsf{NP}$ itself in notoriety.

Conjecture 12. $\mathsf{P} \neq \mathsf{PSPACE}$.

If $\mathsf{P} \neq \mathsf{NP}$, then certainly $\mathsf{P} \neq \mathsf{PSPACE}$ as well, but the converse is not known.

One can also define a nondeterministic variant of PSPACE, called $\mathsf{NPSPACE}$. But a 1970 result called *Savitch's Theorem* [197] shows that actually $\mathsf{PSPACE} = \mathsf{NPSPACE}$.[10] The reasons for this are extremely specific to space, and do not seem to suggest any avenue to proving $\mathsf{P} = \mathsf{NP}$, the analogous statement for time.

[10] A further surprising result from 1987, called the *Immerman-Szelepcsényi Theorem* [110, 218], says that $\mathsf{NSPACE}(f(n)) = \mathsf{coNSPACE}(f(n))$ for every "reasonable" memory bound $f(n)$. (By contrast, Savitch's Theorem produces a quadratic blowup when simulating nondeterministic space by deterministic space, and it remains open whether that blowup can be removed.) This further illustrates how space complexity behaves differently than we expect time complexity to behave.

2.2.6 Counting Complexity

Given an NP search problem, besides asking whether a solution exists, it is also natural to ask how many solutions there are. To capture this, in 1979 Valiant [226] defined the class #P (pronounced "sharp-P") of combinatorial counting problems. Formally, a function $f : \{0,1\}^* \to \mathbb{N}$ is in #P if and only if there is a polynomial-time Turing machine M, and a polynomial p, such that for all $x \in \{0,1\}^*$,

$$f(x) = \left|\left\{w \in \{0,1\}^{p(|x|)} : M(x,w) \text{ accepts}\right\}\right|.$$

Note that, unlike P, NP, and so on, #P is not a class of languages (i.e., decision problems). However, there are two ways we can compare #P to language classes. The first is by considering $\mathsf{P}^{\#\mathsf{P}}$: that is, P with a #P oracle. We then have $\mathsf{NP} \subseteq \mathsf{P}^{\#\mathsf{P}} \subseteq \mathsf{PSPACE}$, as well as the following highly non-obvious inclusion, called *Toda's Theorem*.

Theorem 13 (Toda [222]). $\mathsf{PH} \subseteq \mathsf{P}^{\#\mathsf{P}}$.

The second way is by considering a complexity class called PP (Probabilistic Polynomial-Time). PP can be defined as the class of languages $L \subseteq \{0,1\}^*$ for which there exist #P functions f and g such that for all inputs $x \in \{0,1\}^*$,

$$x \in L \Longleftrightarrow f(x) \geq g(x).$$

It is not hard to see that $\mathsf{NP} \subseteq \mathsf{PP} \subseteq \mathsf{P}^{\#\mathsf{P}}$. More interestingly, one can use binary search to show that $\mathsf{P}^{\mathsf{PP}} = \mathsf{P}^{\#\mathsf{P}}$, so in that sense PP is "almost as strong as #P."

In practice, given any known NP-complete problem (3SAT, CLIQUE, SUBSETSUM, etc.), the counting version of that problem (denoted #3SAT, #CLIQUE, #SUBSETSUM, etc.) will be #P-complete. Indeed, it is open whether there is any NP-complete problem that violates that rule. However, the converse is false: for example, the problem of deciding whether a graph has a perfect matching is in P, but Valiant [226] showed that counting the *number* of perfect matchings is #P-complete.

The #P-complete problems are believed to be "genuinely much harder" than the NP-complete problems, in the sense that—in contrast to the situation with PH—even if $\mathsf{P} = \mathsf{NP}$ we would still have no idea how to prove $\mathsf{P} = \mathsf{P}^{\#\mathsf{P}}$. On the other hand, we do have the following nontrivial result.

Theorem 14 (Stockmeyer [212]). *Suppose* $\mathsf{P} = \mathsf{NP}$. *Then in polynomial time, we could approximate any* #P *function to within a factor of* $1 \pm \varepsilon$, *for any* $\varepsilon = 1/n^{O(1)}$.

2.2.7 Beyond Polynomial Resources

Of course, one can consider many other time and space bounds besides polynomial. Before entering into this, I should offer a brief digression on the use of asymptotic notation in theoretical computer science, since such notation will also be used later in the survey.

- $f(n)$ is $O(g(n))$ if there exist nonnegative constants A, B such that $f(n) \leq Ag(n) + B$ for all n (i.e., g is an asymptotic upper bound on f).
- $f(n)$ is $\Omega(g(n))$ if $g(n)$ is $O(f(n))$ (i.e., g is an asymptotic *lower* bound on f).
- $f(n)$ is $\Theta(g(n))$ if $f(n)$ is $O(g(n))$ and $g(n)$ is $O(f(n))$ (i.e., f and g grow at the same asymptotic rate).
- $f(n)$ is $o(g(n))$ if for all positive A, there exists a B such that $f(n) \leq Ag(n) + B$ for all n (i.e., g is a *strict* asymptotic upper bound on f).

Now let $\mathsf{TIME}(f(n))$ be the class of languages decidable in $O(f(n))$ time, let $\mathsf{NTIME}(f(n))$ be the class decidable in nondeterministic $O(f(n))$ time—that is, with a witness of size $O(f(n))$ that is verified in $O(f(n))$ time—and let $\mathsf{SPACE}(f(n))$ be the class decidable in $O(f(n))$ space.[11] We can then write $\mathsf{P} = \bigcup_k \mathsf{TIME}(n^k)$ and $\mathsf{NP} = \bigcup_k \mathsf{NTIME}(n^k)$ and $\mathsf{PSPACE} = \bigcup_k \mathsf{SPACE}(n^k)$. It is also interesting to study the exponential versions of these classes:

$$\mathsf{EXP} = \bigcup_k \mathsf{TIME}\left(2^{n^k}\right),$$

$$\mathsf{NEXP} = \bigcup_k \mathsf{NTIME}\left(2^{n^k}\right),$$

$$\mathsf{EXPSPACE} = \bigcup_k \mathsf{SPACE}\left(2^{n^k}\right).$$

Note that by "exponential," here we mean not just $2^{O(n)}$, but $2^{p(n)}$ for any polynomial p.

Along with $\mathsf{P} \subseteq \mathsf{PSPACE}$, there is another fundamental relation between time and space classes:

Proposition 15. $\mathsf{PSPACE} \subseteq \mathsf{EXP}$.

Proof. Consider a deterministic machine whose state can be fully described by $p(n)$ bits of information (e.g., the contents of a polynomial-size Turing machine tape, plus a few extra bits for the location and internal state of tape head). Clearly such a machine has at most $2^{p(n)}$ possible states. Thus, after $2^{p(n)}$ steps, either the machine has halted, or else it has entered an infinite loop and will never accept. So to decide whether the machine accepts, it suffices to simulate it for $2^{p(n)}$ steps. ■

[11] Unlike P or PSPACE, classes like $\mathsf{TIME}(n^2)$, $\mathsf{SPACE}(n^3)$, etc. can be sensitive to whether we are talking about Turing machines, RAM machines, or some other model of computation. But in any case, one can simply fix one of those models any time the classes are mentioned in this survey, and nothing will go wrong.

More generally, we get an infinite interleaved hierarchy of deterministic, nondeterministic, and space classes:

$$\mathsf{P} \subseteq \mathsf{NP} \subseteq \mathsf{PSPACE} \subseteq \mathsf{EXP} \subseteq \mathsf{NEXP} \subseteq \mathsf{EXPSPACE} \subseteq \cdots$$

There is also a "higher-up" variant of the $\mathsf{P} \neq \mathsf{NP}$ conjecture, which not surprisingly is *also* open:

Conjecture 16. $\mathsf{EXP} \neq \mathsf{NEXP}$.

We can at least prove a close relationship between the $\mathsf{P} \stackrel{?}{=} \mathsf{NP}$ and $\mathsf{EXP} \stackrel{?}{=} \mathsf{NEXP}$ problems, via a trick called "padding" or "upward translation":

Proposition 17. *If* $\mathsf{P} = \mathsf{NP}$, *then* $\mathsf{EXP} = \mathsf{NEXP}$.

Proof. Let $L \in \mathsf{NEXP}$, and let its verifier run in $2^{p(n)}$ time for some polynomial p. Then consider the language

$$L' = \left\{ x0^{2^{p(|x|)}} : x \in L \right\},$$

which consists of the inputs in L, but "padded out with an exponential number of trailing zeroes." Then $L' \in \mathsf{NP}$, since verifying that $x \in \{0,1\}^n$ is in L takes $2^{p(n)}$ time, which is linear in $n + 2^{p(n)}$ (the length of $x0^{2^{p(|x|)}}$). So by assumption, $L' \in \mathsf{P}$ as well. But this means that $L \in \mathsf{EXP}$, since given $x \in \{0,1\}^n$, we can simply pad x out with $2^{p(n)}$ trailing zeroes ourselves, then run the algorithm that takes time polynomial in $n + 2^{p(n)}$. ■

For the same reason, if $\mathsf{P} = \mathsf{PSPACE}$, then $\mathsf{EXP} = \mathsf{EXPSPACE}$. On the other hand, padding only works in one direction: as far as anyone knows today, we could have $\mathsf{P} \neq \mathsf{NP}$ even if $\mathsf{EXP} = \mathsf{NEXP}$.

To summarize, $\mathsf{P} \stackrel{?}{=} \mathsf{NP}$ is just the tip of an iceberg; there seems to be an extremely rich structure both below and above the NP-complete problems. Until we can prove $\mathsf{P} \neq \mathsf{NP}$, however, most of that structure will remain conjectural.

3 Beliefs About $\mathsf{P} \stackrel{?}{=} \mathsf{NP}$

Just as Hilbert's question turned out to have a negative answer, so too in this case, most computer scientists conjecture that $\mathsf{P} \neq \mathsf{NP}$: that there exist rapidly checkable problems that are *not* rapidly solvable, and for which brute-force search is close to the best that one can do. This is not a unanimous opinion. At least one famous computer scientist, Donald Knuth [131], has professed a belief that $\mathsf{P} = \mathsf{NP}$, while another, Richard Lipton [148], professes agnosticism. Also, in a poll of mathematicians and theoretical computer scientists conducted by William Gasarch [87] in 2002, there were 61 respondents who said $\mathsf{P} \neq \mathsf{NP}$, but also 9 who said $\mathsf{P} = \mathsf{NP}$. Admittedly, it can be hard to tell whether declarations that $\mathsf{P} = \mathsf{NP}$ are meant seriously, or are merely attempts to be contrarian. However, we can surely agree with Knuth and Lipton that we are far from understanding the limits of efficient computation, and that there are further surprises in store.

In this section, I'd like to explain why, *despite* our limited understanding, many of us feel roughly as confident about $\mathsf{P} \neq \mathsf{NP}$ as we do about (say) the Riemann Hypothesis, or other conjectures in math—not to mention empirical sciences—that most experts believe without proof.[12]

The first point is that, when we ask whether $\mathsf{P} = \mathsf{NP}$, we are not asking whether heuristic optimization methods (such as SAT-solvers) can *sometimes* do well in practice; or whether there are *sometimes* clever ways to avoid exponential search. If you believe, for example, that there is *any* cryptographic one-way function—that is, any transformation of inputs $x \to f(x)$ that is easy to compute but hard to invert—then that is enough for $\mathsf{P} \neq \mathsf{NP}$. Such an f need not have any "nice" mathematical structure (like the discrete logarithm function); it could simply be, say, the evolution function of some arbitrary cellular automaton.

It is sometimes claimed that, when we consider $\mathsf{P} \stackrel{?}{=} \mathsf{NP}$, there is a "symmetry of ignorance": yes, we have no idea how to solve NP-complete problems in polynomial time, but we *also* have no idea how to prove that impossible, and therefore anyone is free to believe whatever they like. In my view, however, what breaks the symmetry is the *immense, well-known difficulty of proving lower bounds*. Simply put: even if we suppose $\mathsf{P} \neq \mathsf{NP}$, I don't believe there's any great mystery about why a proof has remained elusive. A rigorous impossibility proof is often a tall order, and many times in history—e.g., with Fermat's Last Theorem, the Kepler Conjecture, or the problem of squaring the circle—such a proof was requested *centuries* before mathematical understanding had advanced to the point where it became a realistic possibility! And as we'll see in Sects. 4 and 6, today we know something about the difficulty of proving even "baby" versions of $\mathsf{P} \neq \mathsf{NP}$; about the barriers that have been overcome and the others that remain to be.

By contrast, if $\mathsf{P} = \mathsf{NP}$, then there is, at least, a puzzle about why the whole software industry, over half a century, has failed to uncover any promising leads for, say, a fast algorithm to invert arbitrary one-way functions (just the algorithm itself, not necessarily a proof that it works). The puzzle is heightened when we realize that, in many real-world cases—such as linear programming, primality testing, and network routing—fast methods to handle a problem in practice *did* come decades before a full theoretical understanding of why the methods worked.

Another reason to believe $\mathsf{P} \neq \mathsf{NP}$ comes from the hierarchy theorems, which we'll meet in Sect. 6.1. Roughly speaking, these theorems imply that "most" pairs of complexity classes are unequal; the problem, in most cases, is simply that we

[12] I like to joke that, if computer scientists had been physicists, we'd simply have declared $\mathsf{P} \neq \mathsf{NP}$ to be an observed law of Nature, analogous to the laws of thermodynamics. A Nobel Prize would even be given for the discovery of that law. (And in the unlikely event that someone later proved $\mathsf{P} = \mathsf{NP}$, a second Nobel Prize would be awarded for the law's overthrow.)

can't prove this for *specific* pairs! For example, in the chain of complexity classes $\mathsf{P} \subseteq \mathsf{NP} \subseteq \mathsf{PSPACE} \subseteq \mathsf{EXP}$, we know that $\mathsf{P} \neq \mathsf{EXP}$, which implies that *at least one* of $\mathsf{P} \neq \mathsf{NP}$, $\mathsf{NP} \neq \mathsf{PSPACE}$, and $\mathsf{PSPACE} \neq \mathsf{EXP}$ must hold. So we might say: given the provable reality of a rich lattice of unequal complexity classes, one needs to offer a special argument if one thinks two classes collapse, but not necessarily if one thinks they're different.

To my mind, however, the strongest argument for $\mathsf{P} \neq \mathsf{NP}$ involves the thousands of problems that have been shown to be NP-complete, and the thousands of other problems that have been shown to be in P. If just one of these problems had turned out to be both NP-complete *and* in P, that would have immediately implied $\mathsf{P} = \mathsf{NP}$. Thus, one could argue, the $\mathsf{P} \neq \mathsf{NP}$ hypothesis has had thousands of opportunities to be "falsified by observation." Yet somehow, in every case, the NP-completeness reductions and the polynomial-time algorithms conspicuously avoid meeting each other—a phenomenon that I once described as the "invisible fence" [7].

This phenomenon becomes particularly striking when we consider *approximation algorithms* for NP-hard problems, which return not necessarily an optimal solution but a solution within some factor of optimal. To illustrate, there is a simple polynomial-time algorithm that, given a 3SAT instance φ, finds an assignment that satisfies at least a $7/8$ fraction of the clauses.[13] Conversely, in 1997 Johan Håstad [105] proved the following striking result.

Theorem 18 (Håstad [105]). *Suppose there is a polynomial-time algorithm that, given as input a satisfiable* 3SAT *instance* φ, *outputs an assignment that satisfies at least a* $7/8 + \varepsilon$ *fraction of the clauses, where* $\varepsilon > 0$ *is any constant. Then* $\mathsf{P} = \mathsf{NP}$.

Theorem 18 is one (strong) version of the *PCP Theorem* [29, 30], which is considered one of the crowning achievements of theoretical computer science. The PCP Theorem yields many other examples of "sharp NP-completeness thresholds," where as we numerically adjust the required solution quality, an optimization problem undergoes a sudden "phase transition" from being in P to being NP-complete. Other times there is a gap between the region of parameter space known to be in P and the region known to be NP-complete. One of the major aims of contemporary research is to close those gaps, for example by proving the so-called *Unique Games Conjecture* [127].

We see a similar "invisible fence" if we shift our attention from approximation algorithms to Leslie Valiant's program of "accidental algorithms" [227]. The latter are polynomial-time algorithms, often for planar graph problems, that exist for

[13]Strictly speaking, this is for the variant of 3SAT in which every clause must have *exactly* three literals, rather than at most three.

Also note that, if we allow the use of randomness, then we can satisfy a $7/8$ fraction of the clauses *in expectation* by just setting each of the n variables uniformly at random! This is because a clause with three literals has $2^3 - 1 = 7$ ways to be satisfied, and only one way to be unsatisfied. A deterministic polynomial-time algorithm that's *guaranteed* to satisfy at least $7/8$ of the clauses requires only a little more work.

certain parameter values but not for others, for reasons that are utterly opaque if one doesn't understand the strange cancellations that the algorithms exploit. A prototypical result is the following:

Theorem 19 (Valiant [227]). *Let* PLANAR3SAT *be a special case of* 3SAT *in which the bipartite graph of clauses and variables is a planar graph. Now consider the following problem: given an instance of* PLANAR3SAT *which is monotone (i.e., has no negations), and in which each variable occurs twice, count the number of satisfying assignments mod k. This problem is in* P *for* $k = 7$*, but is* NP*-hard under randomized reductions for* $k = 2$.[14]

Needless to say (because otherwise you would have heard!), in not one of these examples have the "P region" and the "NP-complete region" of parameter space been discovered to overlap. For example, in Theorem 19, the NP-hardness proof just happens to fail if we ask about the number of solutions mod 7, the very case for which an algorithm is known. If $\mathsf{P} = \mathsf{NP}$ then this is, at the least, an unexplained coincidence. If $\mathsf{P} \neq \mathsf{NP}$, on the other hand, then it makes perfect sense.

3.1 Independent of Set Theory?

Since the 1970s, there has been speculation that $\mathsf{P} \neq \mathsf{NP}$ might be independent (that is, neither provable or disprovable) from the standard axiom systems for mathematics, such as Zermelo-Fraenkel set theory. To be clear, this would mean that either

(1) $\mathsf{P} \neq \mathsf{NP}$, but that fact could never be proved (at least not in our usual formal systems), or else
(2) a polynomial-time algorithm for NP-complete problems *does* exist, but it can never be proven to work, or to halt in polynomial time.

Because $\mathsf{P} \neq \mathsf{NP}$ is a purely arithmetical statement (a Π_2-sentence), it can't simply be excised from mathematics, as some formalists would do with (say) the Continuum Hypothesis or the Axiom of Choice. A polynomial-time algorithm for 3Sat either exists or it doesn't! But that doesn't imply that we can prove which.

In 2003, I wrote a survey article [1] about whether $\mathsf{P} \stackrel{?}{=} \mathsf{NP}$ is formally independent, which somehow never got around to offering any opinion about the *likelihood* of that eventuality! So for the record: I regard the independence of $\mathsf{P} = \mathsf{NP}$ as a farfetched possibility, as I do for the Riemann hypothesis, Goldbach's conjecture, and other unsolved problems of "ordinary" mathematics. At the least, I'd say that the independence of $\mathsf{P} \stackrel{?}{=} \mathsf{NP}$ has the status right now of a "free-floating speculation" with little or no support from past mathematical experience.

---

[14] Indeed, a natural conjecture would be that the problem is NP-hard under randomized reductions for *all* $k \neq 7$, but this remains open (Valiant, personal communication).

There have been celebrated independence results over the past century, but as far as I know they all fall into four classes:

(1) Independence of statements that are themselves about formal systems: for example, that assert their own unprovability in ZF set theory, or ZF's consistency. This is the class produced by Gödel's incompleteness theorems.
(2) Independence of statements in transfinite set theory, such as the Axiom of Choice (AC) and the Continuum Hypothesis (CH). Unlike "ordinary" mathematical statements—$P \neq NP$, the Riemann hypothesis, etc.—the set-theoretic ones can't be rephrased in the language of elementary arithmetic; only questions about their *provability* from various axiom systems are arithmetical. For that reason, one can question whether AC, CH, and so on need to have definite truth-values at all, independent of the axiom system. In any case, the independence of set-theoretic principles seems different in kind, and less "threatening," than the independence of arithmetical statements.[15]
(3) Independence from "weak" systems, which don't encompass all accepted mathematical reasoning. Goodstein's Theorem [91], and the non-losability of the Kirby-Paris hydra game [129] are two examples of interesting arithmetical statements that can be proved using small amounts of set theory (or ordinal induction), but not within Peano arithmetic.
(4) Independence from ZF of strange combinatorial statements, which (alas) would never have been studied if not for their independence. Harvey Friedman [84] has produced striking examples of such statements.

Of course, it's possible that $P \neq NP$ is unprovable, but that that fact *itself* will forever elude proof: indeed, maybe the question of the independence of $P \neq NP$ is itself independent of set theory, and so on ad infinitum! But one can at least say that, if $P \neq NP$ (or for that matter, the Riemann hypothesis, Goldbach's conjecture, etc.) were *proven* independent of ZF, it would be an unprecedented development: probably history's first example of an independence result that didn't fall into one of the four classes above.[16]

The proof of independence would also have to be unlike any known independence proof. Ben-David and Halevi [39] noticed that the techniques used to prove statements such as Goodstein's Theorem independent of Peano arithmetic, actually prove independence from the stronger theory $PA + \Pi_1$: that is, Peano arithmetic plus the set of all true arithmetical Π_1-sentences (sentences with a single

[15] Note also that, by the *Shoenfield absoluteness theorem* [202], one's beliefs about the Axiom of Choice, the Continuum Hypothesis, or other statements proven independent of ZF via forcing can have no effect on the provability of arithmetical statements such as $P \neq NP$.

[16] If a Π_1-sentence like the Goldbach Conjecture or the Riemann Hypothesis were known to be independent of ZF, then it would also be known to be *true*, since any counterexample would have a trivial finite proof! On the other hand, we could also imagine, say, the Goldbach Conjecture being proven equivalent to the consistency of ZF, in which case we could say only that *either* ZF is consistent and Goldbach is true but ZF doesn't prove either, or *else* ZF proves anything. In any case, none of this directly applies to $P \neq NP$, which is a Π_2-sentence.

universal quantifier and no existential quantifiers). However, if $\mathsf{P} \neq \mathsf{NP}$ could be proven independent of $\mathrm{PA} + \Pi_1$, that would mean that no Π_1-sentence implying $\mathsf{P} \neq \mathsf{NP}$ could hold. And thus, for example, NP-complete problems would have to be solvable in $n^{f(n)}$ time for all computable functions f, no matter how slowly growing. We would "almost" have $\mathsf{P} = \mathsf{NP}$.

As Sect. 6 will discuss, there are various formal *barriers*—including the relativization, algebrization, and natural proofs barriers—that explain why certain existing techniques cannot be powerful enough to prove $\mathsf{P} \neq \mathsf{NP}$. These barriers can be interpreted, literally, as proofs that $\mathsf{P} \neq \mathsf{NP}$ is formally unprovable from certain sets of axioms: namely, axioms that capture, or at least come close to capturing, the power of the techniques in question (relativizing, algebrizing, or naturalizing techniques) [28, 112, 190]. In all these cases, however, the axiom sets are known not to capture all techniques in complexity theory: there are existing results that go beyond these axiom sets (albeit, only a few that go beyond all of them at once). Thus, these barriers indicate gaps in our current techniques, rather than in the foundations of mathematics.

4 Why Is Proving $\mathsf{P} \neq \mathsf{NP}$ Difficult?

Let's suppose that $\mathsf{P} \neq \mathsf{NP}$. Then given the disarming simplicity of the statement, why is proving it so difficult? As mentioned above, complexity theorists have identified three technical barriers, called *relativization* [34], *natural proofs* [191], and *algebrization* [10], that any proof of $\mathsf{P} \neq \mathsf{NP}$ will need to overcome. They've also shown that it's possible to surmount each of these barriers, though there are few results that surmount all of them simultaneously. The barriers will be discussed alongside progress toward proving $\mathsf{P} \neq \mathsf{NP}$ in Sect. 6.

However, one can also say something more conceptual, and possibly more illuminating, about the meta-question of why it's so hard to prove hardness. In my view, the central reason why proving $\mathsf{P} \neq \mathsf{NP}$ is hard is simply that, in case after case, there *are* amazingly clever ways to avoid brute-force search, and the diversity of those ways rivals the diversity of mathematics itself. And even if, as I said in Sect. 3, there seems to be an "invisible electric fence" separating the NP-complete problems from the slight variants of those problems that are in P—still, almost any *argument* anyone can imagine for why the NP-complete problems are hard would, if it worked, also apply to the variants in P.

To illustrate, we saw in Sect. 2.1 that 3SAT is NP-complete. We also saw that 2SAT, which is like 3SAT except with two variables per clause rather than three, is in P: indeed, 2SAT is solvable in linear time. Other variants of satisfiability that are in P include HORNSAT (where each clause is an OR of arbitrary many non-negated variables and at most one negated variable), and XORSAT (where each clause is a linear equation mod 2, such as $x_2 \oplus x_7 \oplus x_9 \equiv 1 \pmod 2$).

Likewise, even though it's NP-complete to decide whether a given graph is 3-colorable, one can decide in linear time whether a graph is 2-colorable. Also, even though SUBSETSUM is NP-complete, one can easily decide whether there's a subset of $a_1, \ldots, a_k$ summing to b in time that's nearly linear in $a_1 + \cdots + a_k$. In other words, if each a_i is required to be encoded in "unary" notation (that is, as a list of a_i ones) rather than in binary, then SUBSETSUM is in P.

As a more interesting example, finding the maximum clique in a graph is NP-complete, as are finding the minimum vertex cover, the chromatic number, and so on. Yet in the 1960s, Edmonds [74] famously showed that the *maximum matching*—that is, the largest set of edges no two of which share a vertex—can be found in P. To a casual observer, matching doesn't look terribly different from the other graph optimization problems, but it *is* different.

Or consider linear, semidefinite, and convex programming. These techniques yield hundreds of optimization problems that "seem at first like they should be NP-complete," yet are solvable in P. A few examples are finding maximum flows, finding equilibria of two-player zero-sum games, training linear classifiers, and optimizing over quantum states and unitary transformations.[17]

We can also give examples of "shocking" algorithms for problems that are clearly in P. Most famously, the problem of multiplying two $n \times n$ matrices, $C = AB$, seems like it should "obviously" require $\sim n^3$ steps: $\sim n$ steps for each of the n^2 entries of the product matrix C. But in 1968, Strassen [214] discovered an algorithm that takes only $O\left(n^{\log_2 7}\right)$ steps. There has since been a long sequence of further improvements, culminating in the $O\left(n^{2.376}\right)$ algorithm by Coppersmith and Winograd [70], and its recent improvements to $O\left(n^{2.374}\right)$ by Stothers [213] and to $O\left(n^{2.373}\right)$ by Vassilevska Williams [230]. Thus, letting ω be the *matrix multiplication exponent* (i.e., the least

---

[17] I won't have much to say about linear or semidefinite programming in this survey, so perhaps this is as good a place as any to mention that today, we know a great deal about the impossibility of solving NP-complete problems in polynomial time by formulating them as "natural" linear programs. This story starts in 1987, with a preprint by Swart [217] that claimed to prove P $=$ NP by reducing the Traveling Salesperson Problem to a linear program with n^8 variables and constraints. Swart's preprint inspired a landmark paper by Yannakakis [244] (making it possibly the most productive failed P $=$ NP proof in history!), in which Yannakakis showed that there is no "symmetric" linear program with $n^{o(n)}$ variables and constraints that has the "Traveling Salesperson Polytope" as its projection onto a subset of the variables. This ruled out Swart's approach. Yannakakis also showed that the polytope corresponding to the maximum matching problem has no symmetric LP of subexponential size, but the polytope for the minimum spanning tree problem *does* have a polynomial-size LP. In general, expressibility by such an LP is sufficient for a problem to be in P, but not necessary.

Later, in 2012, Fiorini et al. [76] substantially improved Yannakakis's result, getting rid of the symmetry requirement. There have since been other major results in this direction: in 2014, Rothvoß[194] showed that the perfect matching polytope requires exponentially-large LPs (again with no symmetry requirement), while in 2015, Lee, Raghavendra, and Steurer [140] extended many of these lower bounds from linear to semidefinite programs.

Collectively, these results rule out one "natural" approach to proving P $=$ NP: namely, to start from famous NP-hard optimization problems like TSP, and then find a polynomial-size linear or semidefinite program that projects onto the polytope whose extreme points are the valid solutions. Of course, we can't yet rule out the possibility that linear or semidefinite programs could help prove P $=$ NP in some more indirect way (or via some NP-hard problem other than the specific ones that were studied); ruling *that* out seems essentially tantamount to proving P $\neq$ NP itself.

ω such that $n \times n$ matrices can be multiplied in $n^{\omega+o(1)}$ time), all we know today is that $\omega \in [2, 2.373]$. Some computer scientists conjecture that $\omega = 2$; but in any case, just like with attempts to prove $\mathsf{P} \neq \mathsf{NP}$, an obvious obstruction to proving $\omega > 2$ is that the proof had better *not* yield $\omega = 3$, or even a "natural-looking" bound like $\omega \geq 2.5$.

The breadth of clever polynomial-time algorithms might seem like a trite observation, incommensurate with the challenge of explaining why it's so hard to prove $\mathsf{P} \neq \mathsf{NP}$. Yet we have evidence to the contrary. Over the decades, there have been hundreds of flawed 'proofs' announced for $\mathsf{P} \neq \mathsf{NP}$. The announcement that received the most attention, including coverage in *The New York Times* and other major media outlets, was that of Deolalikar [73] in 2010. But in every such case that I'm aware of, *the proof could ultimately be rejected on the ground that, if it worked, then it would also yield superpolynomial lower bounds for problems known to be in* P.

With some flawed $\mathsf{P} \neq \mathsf{NP}$ proofs, this is easy to see: for example, perhaps the author proves that 3SAT must take exponential time, by some argument that's fearsome in technical details, but that ultimately boils down to "there are 2^n possible assignments to the variables, and clearly any algorithm must spend at least one step rejecting each of them." A general-purpose refutation of such arguments is simply that, if they worked, then they'd work equally well for 2SAT. Alternatively, one could point out that, as we'll see in Sect. 5.1, it's known how to solve 3SAT in 1.3^n time. So a $\mathsf{P} \neq \mathsf{NP}$ proof had *better* not imply a $\Omega(2^n)$ lower bound for 3SAT.

In the case of Deolalikar's $\mathsf{P} \neq \mathsf{NP}$ attempt [73], the details were more complicated, but the bottom line ended up being similar. Deolalikar appealed to certain statistical properties of the set of satisfying assignments of a *random* 3SAT instance. The claim was that, for reasons having to do with logical definability, those statistical properties precluded 3SAT from having a polynomial-time algorithm. During an intense online discussion, however, skeptics pointed out that random XORSAT—which we previously mentioned as a satisfiability variant in P—gives rise to solution sets indistinguishable from those of random 3SAT, with respect to the properties Deolalikar was using (see for example [229]). This implied that there must be one or more bugs in the proof, though it still left the task of finding them (which was also done).

None of this means that proving $\mathsf{P} \neq \mathsf{NP}$ is impossible. *A priori*, it might also have been hard to imagine a proof of the unsolvability of the halting problem, but of course we know that such a proof exists. As we'll see in Sect. 6.1, a central difference between the two cases is that methods from logic—namely, diagonalization and self-reference—worked to prove the unsolvability of the halting problem, but there's a precise sense in which these tricks *can't* work (at least not by themselves) to prove $\mathsf{P} \neq \mathsf{NP}$. A related difference comes from the *quantitative* character of $\mathsf{P} \neq \mathsf{NP}$: somehow, any proof will need to explain why (say) a 1.00001^n or $2^{\sqrt{n}}$ algorithm for 3SAT is impossible, even though a 1.3^n algorithm exists. In some sense, this need to make fine quantitative distinctions—to say that, yes, brute-force search *can* be beaten, but only by this much for this problem and by that much for that one—puts a lower bound on the sophistication of any $P \neq NP$ proof.

5 Strengthenings of the $P \neq NP$ Conjecture

I'll now survey various strengthenings of the $P \neq NP$ conjecture, which are often needed for applications to cryptography, quantum computing, fine-grained complexity, and elsewhere. Some of these strengthenings will play a role when, in Sect. 6, we discuss the main approaches to proving $P \neq NP$ that have been tried.

5.1 *Different Running Times*

There has been a great deal of progress on beating brute-force search for many NP-complete problems, even if the resulting algorithms still take exponential time. For example, the following was shown by Schöning.

Theorem 20 (Schöning [198]). *There is a randomized algorithm that solves* 3SAT *in* $O(1.3^n)$ *time.*

For many NP-complete problems like HAMILTONCYCLE, for which the obvious brute-force algorithm takes $\sim n!$ time, it is also possible to reduce the running time to $O(2^n)$, or sometimes even to $O(c^n)$ for some $c < 2$, through clever dynamic programming.

How far can these algorithms be pushed? For example, is it possible that 3SAT could be solvable in $2^{O(\sqrt{n})}$ time, as various NP-intermediate problems like FACTORING are known to be? An important conjecture called the Exponential Time Hypothesis, or ETH, asserts that the answer is no:

Conjecture 21 (Exponential Time Hypothesis). *Any deterministic algorithm for* 3SAT *takes* $\Omega(c^n)$ *steps, for some constant* $c > 1$.

The ETH is an ambitious strengthening of $P \neq NP$, one that has found many applications in recent years. Often, for example, assuming the ETH, it can be shown that some particular polynomial-time or quasipolynomial-time algorithm is optimal (i.e., that its exponent can't be improved), whereas nothing similar is known assuming only $P \neq NP$. One example is the problem of computing the *edit distance* between two strings, or the minimum number of insertions, deletions, and replacements needed to transform one string to the other. Here a quadratic-time algorithm has long been known, while in a recent breakthrough, Backurs and Indyk [33] proved that algorithm to be essentially optimal assuming the ETH. A second example is the problem of approximating the value of a two-prover "free game":

here Aaronson et al. [9] gave an $n^{O(\log n)}$ algorithm, and also proved that algorithm essentially optimal assuming the ETH.

5.2 Nonuniform Algorithms and Circuits

$\mathsf{P} \stackrel{?}{=} \mathsf{NP}$ asks whether there is a *single* algorithm that, for each input size n, solves an NP-complete problem like 3SAT in time polynomial in n. But one could also allow a different algorithm for each input size: for example, imagine a clever approach that yielded a speedup for $n = 1000$ but was swamped by constant factors for $n = 100$, another approach that worked for $n = 10,000$ but not $n = 1000$, and so on. To formalize this notion, let P/poly be the class of languages L for which there exists a polynomial-time Turing machine M, as well as an infinite set of "advice strings" $a_1, a_2, \ldots$, where a_n is $p(n)$ bits long for some polynomial p, such that for all n and all $x \in \{0,1\}^n$, we have

$$M(x, a_n) \text{ accepts} \iff x \in L.$$

An equivalent way to define P/poly is as the class of languages recognized by a family of *polynomial-size circuits*: that is, networks of Boolean logic gates (such as AND, OR, NOT), with the input bits $x_1, \ldots, x_n$ at the bottom and with a bit determining whether $x \in L$ at the top, where the network can be different for each input length n.[18] The *size* of such a circuit is simply the number of logic gates in it. P/poly is a nonuniform generalization of P: certainly $\mathsf{P} \subset \mathsf{P/poly}$, but there is no containment in the other direction.[19]

Now, the nonuniform version of the $\mathsf{P} \neq \mathsf{NP}$ conjecture is the following.

Conjecture 22. $\mathsf{NP} \not\subset \mathsf{P/poly}$.

If $\mathsf{P} = \mathsf{NP}$, then certainly $\mathsf{NP} \subset \mathsf{P/poly}$, but the converse need not hold. About the closest we have to a converse is the *Karp-Lipton Theorem* [123]:

Theorem 23. *If* $\mathsf{NP} \subset \mathsf{P/poly}$, *then* PH *collapses to* Σ_2^{P}.

Proof. Consider a complete problem for Π_2^{P}: say, "for all $x \in \{0,1\}^{p(n)}$, does there exist a $y \in \{0,1\}^{p(n)}$ such that $A(x,y)$ accepts?", for some polynomial p and polynomial-time algorithm A. Assuming $\mathsf{NP} \subset \mathsf{P/poly}$, we can solve that problem in Σ_2^{P} as follows:

[18] Despite the term "circuit," which comes from electrical engineering, circuits in theoretical computer science are always *free* of cycles; they proceed from the inputs to the output via layers of logic gates.

[19] This is a rare instance where the non-containment can actually be *proved*: for example, any unary language (i.e., language of the form $\{0^n : n \in S\}$) is clearly in P/poly, but there is an uncountable infinity of such languages, so almost all of them cannot be in P.

- "Does there exist a circuit C such that for all x, the algorithm $A(x, C(x))$ accepts?"

For if $\mathsf{NP} \subset \mathsf{P/poly}$ and $\forall x \exists y A(x, y)$ is true, then clearly there exists a polynomial-size circuit C that takes x as input, and outputs a y such that $A(x, y)$ accepts. So we can simply use the existential quantifier in our Σ_2^{P} algorithm to guess a description of that circuit.

We conclude that, if $\mathsf{NP} \subset \mathsf{P/poly}$, then $\Pi_2^{\mathsf{P}} \subseteq \Sigma_2^{\mathsf{P}}$ (and by symmetry, $\Sigma_2^{\mathsf{P}} \subseteq \Pi_2^{\mathsf{P}}$). But this is known to cause a collapse of the entire polynomial hierarchy to Σ_2^{P}. ■

In summary, while most complexity theorists conjecture that $\mathsf{NP} \not\subset \mathsf{P/poly}$, as far as we know it is a stronger conjecture than $\mathsf{P} \neq \mathsf{NP}$. Indeed, it is even plausible that future techniques could prove $\mathsf{P} \neq \mathsf{NP}$ without proving $\mathsf{NP} \not\subset \mathsf{P/poly}$: for example, as we will discuss in Sect. 6.1, we can currently prove $\mathsf{P} \neq \mathsf{EXP}$, but cannot currently prove $\mathsf{EXP} \not\subset \mathsf{P/poly}$, or even $\mathsf{NEXP} \not\subset \mathsf{P/poly}$. Despite this, as we will see in Sect. 6, *most* of the techniques that have been explored for proving $\mathsf{P} \neq \mathsf{NP}$, would actually yield the stronger result $\mathsf{NP} \not\subset \mathsf{P/poly}$ if they worked. For that reason, $\mathsf{P/poly}$ plays a central role in work on the $\mathsf{P} \stackrel{?}{=} \mathsf{NP}$ question.

There is one other aspect of circuit complexity that will play a role later in this survey, and that is *depth*. The depth of a circuit simply means the length of the longest path from an input bit to the output bit—or, if we think of the logic gates as organized into layers, then the number of layers. There is a subclass of $\mathsf{P/poly}$ called NC^1 (the NC stands for "Nick's Class," after Nick Pippenger) which consists of all languages that are decided by a family of circuits that have polynomial size and *also* depth $O(\log n)$.[20] One can also think of NC^1 as the class of problems solvable in logarithmic time (nonuniformly) using a polynomial number of parallel processors. It is conjectured that $\mathsf{P} \not\subset \mathsf{NC}^1$ (that is, not all efficient algorithms can be parallelized), but alas, even showing that $\mathsf{NEXP} \not\subset \mathsf{NC}^1$ remains open at present.

Another way to define NC^1 is as the class of languages decidable by a family of polynomial-size Boolean *formulas*. In theoretical computer science, a formula just means a circuit where every gate has fanout 1 (that is, where a gate cannot have its output fed as input to multiple other gates). To see the equivalence: in one direction, by replicating subcircuits wherever necessary, clearly any circuit of depth d and size s can be "unraveled" into a formula of depth d and size at most $2^d s$, which is still polynomial in n if $d = O(\log n)$ and $s = n^{O(1)}$. In the other direction, there is an extremely useful fact proved by Brent [52], called "depth reduction."

Proposition 24 (Brent [52]). *Given any Boolean formula of size S, there is an equivalent formula of size $S^{O(1)}$ and depth $O(\log S)$.*[21]

[20] Note that, if each logic gate depends on at most 2 inputs, then $\log_2 n$ is the smallest depth that allows the output to depend on all n input bits.

[21] Bshouty, Cleve, and Eberly [53] showed that the size of the depth-reduced formula can even be taken to be $O(S^{1+\varepsilon})$, for any constant $\varepsilon > 0$.

Because of Proposition 24, the minimum depth D of any formula for a Boolean function f is simply $\Theta(\log S)$, where S is the minimum size of any formula for f. For circuits, by contrast, size and depth are two independent variables, which might in general be related only by $D \leq S \leq 2^D$.

5.3 Average-Case Complexity

If $\mathsf{P} \neq \mathsf{NP}$, that means that there are NP problems for which no Turing machine succeeds at solving *all* instances in polynomial time. But often, especially in cryptography, we need more than that. It would be laughable to advertise a cryptosystem on the grounds that there *exist* messages that are hard to decode! Thus, it is natural to ask whether there are NP problems that are hard "in the average case" or "on random instances," rather than merely in the worst case. More pointedly, does the existence of such problems follow from $\mathsf{P} \neq \mathsf{NP}$, or is it a different, stronger assumption?

The first step is to clarify what we mean by a "random instance." For some NP-complete problems, it makes sense to ask about a *uniform* random instance: for example, we can consider 3SAT with n variables and $m = \alpha n$ uniformly-random clauses (for some constant α), or 3COLORING on an Erdös-Rényi random graph. In those cases, the difficulty tends to vary wildly with the problem and the precise distribution. With 3SAT, for example, if the clause/variable ratio α is too small, then random instances are trivially satisfiable, while if α is too large, then they are trivially unsatisfiable. But there is a "sweet spot," $\alpha \approx 4.25$, where random 3SAT undergoes a phase transition from satisfiable to unsatisfiable, and where the difficulty seems to blow up accordingly. Even at the threshold, however, random 3SAT might still be much easier than worst-case 3SAT: the breakthrough "survey propagation algorithm" [50] can solve random 3SAT quickly, even for α extremely close to the threshold.[22] More generally, there has been a great deal of work on understanding particular distributions over instances, often using tools from statistical physics (for an accessible introduction, see for example Moore and Mertens [155]). Unfortunately, there are almost no known reductions among these sorts of distributional problems, which would let us say that if one of them is hard then so is another. The reason is that almost any imaginable reduction from problem A to problem B will map a random instance of A to an extremely special, *non-*random instance of B.

This means that, if we want to pick random instances of NP-complete problems and be confident they are hard, then we might need carefully-tailored distributions. Levin [142], and Li and Vitányi [143], observed that there exists a "universal distribution" $\mathcal{D}$ with the remarkable property that *any algorithm that fails on any instance, will also fail with high probability with respect to instances drawn from* $\mathcal{D}$.

[22] But making matters more complicated still, survey propagation fails badly on random 4SAT.

Briefly, one constructs $\mathcal{D}$ by giving each string $x \in \{0,1\}^*$ a probability proportional to $2^{-K(x)}$, where $K(x)$ is the *Kolmogorov complexity* of x: that is, the number of bits in the shortest computer program whose output is x. One then argues that, given any algorithm A, one can design a short computer program that brute-force searches for the first instances on which A fails—and for that reason, if there are any such instances, then $\mathcal{D}$ will assign them a high probability.

In this construction, the catch is that there is no feasible way actually to *sample* instances from the magical distribution $\mathcal{D}$. Thus, given a family of distributions $\mathcal{D} = \{\mathcal{D}_n\}_n$, where $\mathcal{D}_n$ is over $\{0,1\}^n$, call $\mathcal{D}$ *efficiently samplable* if there exists a Turing machine that takes as input a positive integer n and a uniformly random string $r \in \{0,1\}^{p(n)}$ (for some polynomial p), and that outputs a sample from $\mathcal{D}_n$ in time polynomial in n.[23] Then the real question, one might say, is whether any NP-complete problems are hard on average with respect to efficiently samplable distributions. More formally, does the following conjecture hold?

Conjecture 25 (NP **Hard on Average).** *There exists a language $L \in$ NP, as well as an efficiently samplable family of distributions $\mathcal{D} = \{\mathcal{D}_n\}_n$, such that for all polynomial-time algorithms A, there exists an n such that*

$$\Pr_{x \sim \mathcal{D}_n} [A(x) = L(x)] < 0.51.$$

Here $L(x) \in \{0,1\}$ denotes the characteristic function of L.

It is a longstanding open problem whether $P \neq NP$ implies Conjecture 25. There are NP-intermediate problems—one famous example being the discrete logarithm problem—that are known to have the remarkable property of *worst-case/average-case equivalence*. That is, any polynomial-time algorithm for these problems that works on (say) 10% of instances implies a polynomial-time algorithm for *all* instances; and conversely, if the problem is hard at all then it is hard on average. However, despite decades of work, no one has been able to show worst-case/average-case equivalence for any NP-complete problem (with respect to any efficiently samplable distribution), and there are known obstacles to such a result. For details, see for example the survey by Bogdanov and Trevisan [47].

5.3.1 Cryptography and One-Way Functions

One might hope that, even if we cannot base secure cryptography solely on the assumption that $P \neq NP$, at least we could base it on Conjecture 25. But there is one more obstacle. In cryptography, we do not merely need NP problems for which it is easy to generate hard instances: rather, we need NP problems for which it is easy to generate hard instances, *along with secret solutions to those instances*. This

---

[23] We could also allow sampling from some distribution *close* to $\mathcal{D}_n$, but we will ignore that complication here.

motivates the definition of a *one-way function (OWF)*, perhaps the central concept of modern cryptography. Let $f = \{f_n\}_n$ be a family of functions, with $f_n : \{0,1\}^n \to \{0,1\}^{p(n)}$ for some polynomial p. Then we call f a one-way function family if

(1) f_n is computable in time polynomial in n, but
(2) f_n is hard to invert: that is, for all polynomial-time algorithms A and all polynomials q,

$$\Pr_{x \sim \{0,1\}^n} [f_n (A (f_n (x))) = f_n (x)] < \frac{1}{q(n)}.$$

We then make the following conjecture.

Conjecture 26. *There exists a one-way function family.*

Conjecture 26 is stronger than Conjecture 25, which in turn is stronger than $\mathsf{P} \neq \mathsf{NP}$. Indeed, it is not hard to show the following.

Proposition 27. *Conjecture 26 holds if and only if there exists a fast way to generate hard random* 3SAT *instances with "planted solutions": that is, an efficiently samplable family of distributions* $\mathcal{D} = \{\mathcal{D}_n\}_n$ *over* (φ, x) *pairs, where* φ *is a satisfiable* 3SAT *instance and* x *is a satisfying assignment to* φ*, such that for all polynomial-time algorithms A and all polynomials q,*

$$\Pr_{\varphi \sim \mathcal{D}_n} [A(\varphi) \text{ finds a satisfying assignment to } \varphi] < \frac{1}{q(n)}.$$

Proof. Given a one-way function family f, we can generate a hard random 3SAT instance with a planted solution by choosing $x \in \{0,1\}^n$ uniformly at random, computing $f_n(x)$, and then using the Cook-Levin Theorem (Theorem 2) to construct a 3SAT instance that encodes the problem of finding a preimage of $f_n(x)$. Conversely, given a polynomial-time algorithm that takes as input a positive integer n and a random string $r \in \{0,1\}^{p(n)}$ (for some polynomial p), and that outputs a hard 3SAT instance φ_r together with a planted solution x_r to φ_r, the function $f_n(r) := \varphi_r$ will necessarily be one-way (since inverting f_n would let us find a satisfying assignment to φ_r). ■

Conjecture 26 turns out to suffice for building most of the ingredients of private-key cryptography, notably including pseudorandom generators [106] and pseudorandom functions [88]. Furthermore, while Conjecture 26 is formally stronger than $\mathsf{P} \neq \mathsf{NP}$, Proposition 27 suggests that the two conjectures are conceptually similar: "all we are asking for" is a hard NP problem, together with a fast way to generate hard solved instances of it.

This contrasts with the situation for *public*-key cryptography—i.e., the kind of cryptography that does not require any secrets to be shared in advance, and which is used for sending credit-card numbers over the web. To create a secure public-key cryptosystem, one needs something even stronger than Conjecture 26: for example, a *trapdoor* OWF,[24] which is an OWF with the additional property that it becomes easy to invert if one is given a secret "trapdoor" string generated along with the function. We do, of course, have candidates for secure public-key cryptosystems, which are based on problems such as factoring, discrete logarithms (over both multiplicative groups and elliptic curves), and finding short nonzero vectors in lattices. To date, however, all public-key cryptosystems require "sticking one's neck out," and conjecturing the hardness of some specific NP-intermediate problem, something with much more structure than any known NP-complete problem. In other words, for public-key cryptography, today one needs additional conjectures that go fundamentally beyond $\mathsf{P} \neq \mathsf{NP}$, or even the existence of OWFs.

5.4 Randomized Algorithms

Even assuming $\mathsf{P} \neq \mathsf{NP}$, we can still ask whether NP-complete problems can be solved in polynomial time with help from random bits. This is a different question than whether NP is hard on average: whereas before we were asking about algorithms that solve *most* instances (with respect to some distribution), now we are asking about algorithms that solve *all* instances, for *most* choices of some auxiliary random numbers.

Historically, algorithm designers have often resorted to randomness, to deal with situations where *most* choices that an algorithm could make are fine, but any *specific* choice will lead to terrible behavior on certain inputs. For example, in Monte Carlo simulation, used throughout science and engineering, one estimates the volume of a high-dimensional object by just sampling random points, and then checking what fraction of them lie inside.

A second example concerns primality testing: that is, deciding the language

$$\text{PRIMES} = \{N : N \text{ is a binary encoding of a prime number}\}.$$

In modern cryptosystems such as RSA, it is just as important that primality testing be *easy* as that the related factoring problem be hard. In the 1970s, Rabin [181] and Solovay and Strassen [210] showed how to decide PRIMES in time polynomial in $\log N$ (i.e., the number of digits of N). The small catch was that their algorithms were randomized: in addition to N, they required a second input r; and for each N, the algorithms were guaranteed to succeed for most r's but not all of them. Miller [154] also proposed a deterministic polynomial-time algorithm for PRIMES, but could only prove the algorithm correct assuming the Extended Riemann Hypothesis. Finally, after decades of work on the problem, in 2002 Agrawal, Kayal, and Saxena [14] gave an unconditional proof that PRIMES is in P. In other words, if one only

[24]There are closely-related objects, such as "lossy" trapdoor OWFs (see [178]), that also suffice for building public-key cryptosystems.

cares about testing primality in polynomial time, and not about the degree of the polynomial, then randomness was never needed after all.

A third example concerns the problem of *polynomial identity testing* (PIT). Here we are given as input a circuit or formula, composed of addition and multiplication gates, that computes a polynomial $p : \mathbb{F} \to \mathbb{F}$ over a finite field $\mathbb{F}$. The question is whether p is the identically-zero polynomial—that is, whether the identity $p(x) = 0$ holds. If $\deg(p) \ll |\mathbb{F}|$, then the Fundamental Theorem of Algebra immediately suggests a way to solve this problem: simply pick an $x \in \mathbb{F}$ uniformly at random and check whether $p(x) = 0$. Since a nonzero polynomial p can vanish on at most $\deg(p)$ points, the probability that we will "get unlucky" and choose one of those points is at most $\deg(p) / |\mathbb{F}|$. To this day, no one knows of any deterministic approach that achieves similar performance,[25] and derandomizing PIT is considered one of the frontier problems of theoretical computer science. For details, see for example the survey of Shpilka and Yehudayoff [205].

5.4.1 BPP and Derandomization

What is the power of randomness more generally? Can *every* randomized algorithm be derandomized, as ultimately happened with PRIMES? To explore these issues, complexity theorists study several randomized generalizations of the class P. We will consider just one of them: *Bounded-Error Probabilistic Polynomial-Time*, or BPP, is the class of languages $L \subseteq \{0,1\}^*$ for which there exists a polynomial-time Turing machine M, as well as a polynomial p, such that for all inputs $x \in \{0,1\}^n$,

$$\Pr_{r \in \{0,1\}^{p(n)}} [M(x,r) = L(x)] \geq \frac{2}{3}.$$

In other words, for every x, the machine M must correctly decide whether $x \in L$ "most of the time" (that is, for most choices of r). Crucially, here we can easily replace the constant $2/3$ by any other number between $1/2$ and 1, or even by a function like $1 - 2^{-n}$. So for example, if we wanted to know $x \in L$ with 0.999999 confidence, then we'd simply run M several times, with different independent values of r, and then output the majority vote among the results.

It is clear that $\mathsf{P} \subseteq \mathsf{BPP} \subseteq \mathsf{PSPACE}$; more interestingly, Sipser [206] and Lautemann [139] proved that BPP is contained in $\Sigma_2^{\mathsf{P}} \cap \Pi_2^{\mathsf{P}}$ (that is, the second level of PH). The Rabin-Miller and Solovay-Strassen algorithms imply that PRIMES $\in$ BPP.

---

[25] At least, not for *arbitrary* polynomials computed by small formulas or circuits. A great deal of progress has been made derandomizing PIT for restricted classes of polynomials. In fact, the deterministic primality test of Agrawal et al. [14] was based on a derandomization of one extremely special case of PIT.

Today, most complexity theorists conjecture that what happened to PRIMES can happen to all of BPP:

Conjecture 28. P = BPP.

The reason for this conjecture is that it follows from the existence of good enough *pseudorandom generators*, which we could use to replace the random string r in any BPP algorithm M by deterministic strings that "look random, as far as M can tell." Furthermore, work in the 1990s showed that, if we grant certain extremely plausible lower bounds on circuit size, then these pseudorandom generators exist. Perhaps the most striking result along these lines is that of Impagliazzo and Wigderson [115]:

Theorem 29 (Impagliazzo-Wigderson [115]). *Suppose there exists a language decidable in 2^n time, which requires nonuniform circuits of size $2^{\Omega(n)}$. Then* P = BPP.

Of course, if P = BPP, then the question of whether randomized algorithms can efficiently solve NP-complete problems is just the original $\mathsf{P} \stackrel{?}{=} \mathsf{NP}$ question in a different guise. Ironically, however, the "obvious" approach to proving P = BPP is to prove a strong circuit lower bound—and if one knew how to do that, perhaps one could prove P ≠ NP as well!

Even if we don't assume P = BPP, it's easy to show that deterministic *nonuniform* algorithms (see Sect. 5.2) can simulate randomized algorithms:

Proposition 30 (Adleman [11]). BPP ⊂ P/poly.

Proof. Let the language L be decided by a BPP algorithm that uses $p(n)$ random bits. Then by using $q(n) = O(n \cdot p(n))$ random bits, running the algorithm $O(n)$ times with independent random bits each time, and outputting the majority answer, we can push the probability of error on any given input $x \in \{0,1\}^n$ from $1/3$ down to (say) 2^{-2n}. Thus, the probability that there *exists* an $x \in \{0,1\}^n$ on which the algorithm errs is at most $2^n\left(2^{-2n}\right) = 2^{-n}$. This means, in particular, that there must be a fixed choice for the random string $r \in \{0,1\}^{q(n)}$ that causes the algorithm to succeed on all $x \in \{0,1\}^n$. So to decide L in P/poly, we simply "hardwire" that r as the advice. ∎

By combining Theorem 23 with Proposition 30, we immediately obtain that if NP ⊆ BPP, then the polynomial hierarchy collapses to the second level. So the bottom line is that the NP ⊆ BPP question is likely identical to the $\mathsf{P} \stackrel{?}{=} \mathsf{NP}$ question, but is extremely tightly related even if not.

5.5 *Quantum Algorithms*

The class BPP might not exhaust what the physical world lets us efficiently compute, with quantum computing an obvious contender for going further. In 1993, Bernstein and Vazirani [42] defined the complexity class BQP, or Bounded-Error Quantum Polynomial-Time, as a quantum-mechanical generalization of BPP.

(Details of quantum computing and BQP are beyond the scope of this survey, but see [6, 172].) Bernstein and Vazirani, along with Adleman et al. [12], also showed some basic containments:

$$\mathsf{P} \subseteq \mathsf{BPP} \subseteq \mathsf{BQP} \subseteq \mathsf{PP} \subseteq \mathsf{P}^{\#\mathsf{P}} \subseteq \mathsf{PSPACE}.$$

In 1994, Shor [203] famously showed that the factoring and discrete logarithm problems are in BQP—and hence, that a scalable quantum computer, if built, could break almost all currently-used public-key cryptography. To design his quantum algorithms, Shor had to exploit extremely special properties of factoring and discrete logarithm, which are not known to hold for NP-complete problems.

The quantum analogue of the $\mathsf{P} \stackrel{?}{=} \mathsf{NP}$ question is the question of whether $\mathsf{NP} \subseteq \mathsf{BQP}$: that is, *can quantum computers solve* NP*-complete problems in polynomial time?*[26] Most quantum computing researchers conjecture that the answer is no:

Conjecture 31. $\mathsf{NP} \not\subset \mathsf{BQP}$.

Naturally, there is little hope of proving Conjecture 31 at present, since any proof would imply $\mathsf{P} \neq \mathsf{NP}$! We do not even know today how to prove conditional statements (analogous to what we have for BPP and P/poly): for example, that if $\mathsf{NP} \subseteq \mathsf{BQP}$ then PH collapses. On the other hand, it *is* known that, if a fast quantum algorithm for NP-complete problems exists, then in some sense it will have to be extremely different from Shor's or any other known quantum algorithm. For example, if we ignore the structure of NP-complete problems, and just consider the abstract task of searching an unordered list, then quantum computers can provide at most a square-root speedup over the classical running time [40]. This implies that there exists an oracle A such that $\mathsf{NP}^A \not\subset \mathsf{BQP}^A$. Note that the square-root speedup is actually achievable, using *Grover's algorithm* [97]. For most NP-complete problems, however, the fastest known quantum algorithm will be obtained by simply layering Grover's algorithm on top of the fastest known classical algorithm, yielding a quadratic speedup but no more.[27] So for example, as far as anyone knows today, even a quantum computer would need $2^{\Omega(n)}$ time to solve 3SAT.

[26] One can also consider the QMA-complete problems, which are a quantum generalization of the NP-complete problems themselves (see [48]), but we will not pursue that here.

[27] One can artificially design an NP-complete problem with a superpolynomial quantum speedup over the best known classical algorithm by, for example, taking the language

$$L = \left\{0\varphi 0 \cdots 0 \mid \varphi \text{ is a satisfiable 3Sat instance of size } n^{0.01}\right\} \cup$$
$$\left\{1x \mid x \text{ is a binary encoding of a positive integer with an odd number of distinct prime factors}\right\}.$$

Clearly L is NP-complete, and a quantum algorithm can decide L in $O\left(c^{n^{0.01}}\right)$ time for some c, whereas the best known classical algorithm will take $\exp(n)$ time.

Conversely, there are also NP-complete problems with no significant quantum speedup known—say, because the best known classical algorithm is based on dynamic programming, and it's unknown how to combine that with Grover's algorithm. A candidate example is the Traveling Salesman Problem, which is solvable in $O(2^n \operatorname{poly}(n))$ time using the Held-Karp

Of course, one can also wonder whether the physical world might provide computational resources even *beyond* quantum computing (based on black holes? closed timelike curves? modifications to quantum mechanics?), and if so, whether *those* resources might enable the polynomial-time solution of NP-complete problems. Such speculations are beyond the scope of this article, but see for example [3].

6 Progress

One common view among mathematicians is that questions like $\mathsf{P} \stackrel{?}{=} \mathsf{NP}$, while undoubtedly important, are just too hard to make any progress on in the present state of mathematics. It's true that we seem to be nowhere close to a solution, but in this section, I'll build a case that the extreme pessimistic view is unwarranted. I'll explain what genuine knowledge I think we have, relevant to proving $\mathsf{P} \neq \mathsf{NP}$, that we didn't have thirty years ago or in some cases ten years ago. One could argue that, if $\mathsf{P} \neq \mathsf{NP}$ is a distant peak, then all the progress has remained in the foothills. On the other hand, scaling the foothills has *already* been highly nontrivial, so anyone who wants to work on $\mathsf{P} \neq \mathsf{NP}$ had better get acquainted with what's been done.

More concretely, I'll tell a story here of the interaction between *lower bounds* and *barriers*: on the one hand, actual successes in proving superpolynomial or exponential lower bounds in interesting models of computation; but on the other, explanations for why the techniques used to achieve those successes don't extend to prove $\mathsf{P} \neq \mathsf{NP}$. We'll see how the barriers influence the next generation of lower bound techniques, which are sometimes specifically designed to evade the barriers, or evaluated on their potential to do so.

With a single exception—namely, the Mulmuley-Sohoni Geometric Complexity Theory program—I'll restrict my narrative to ideas that have already had definite successes in proving limits on computation, beyond what had previously been known. The drawback of this choice is that in many cases, the ideas that are concrete enough to have worked for *something*, are also concrete enough that we understand why they can't work for $\mathsf{P} \neq \mathsf{NP}$! My defense is that this section would be unmanageably long, if it had to cover *every* idea about how $\mathsf{P} \neq \mathsf{NP}$ might someday be proved.

I should, however, at least mention some important approaches to lower bounds that will be missing from my subsequent narrative. The first is *descriptive complexity theory*; see for example the book of Immerman [111] for a good introduction. Descriptive complexity characterizes many complexity classes in terms of their logical expressive power: for example, P corresponds to sentences expressible in first-order logic with linear order and a least fixed point; NP to sentences expressible in existential second-order logic; PSPACE to sentences expressible in second-order

---

dynamic programming algorithm [107], whereas Grover's algorithm seems to yield only the worse bound $O(\sqrt{n!})$.

logic with transitive closure; and EXP to sentences expressible in second-order logic with a least fixed point. The hope is that characterizing complexity classes in this way, with no explicit mention of resource bounds, will make it easier to see which are equal and which different. There is one major piece of evidence for this hope: namely, descriptive complexity played an important role in the proof by Immerman [110] that nondeterministic space is closed under complement (though the independent proof by Szelepcsényi [218] of the same result did not use these ideas). Descriptive complexity theory has not yet led to new separations.

The second approach is *lower bounds via communication complexity*. Given a Boolean function $f : \{0,1\}^n \to \{0,1\}$, consider the following communication game: Alice receives an x such that $f(x) = 0$, Bob receives an input x such that $f(x) = 1$, and their goal is to agree on an index $i \in \{1, \ldots, n\}$ such that $x_i \neq y_i$. Let C_f be the communication complexity of this game: that is, the minimum number of bits that Alice and Bob need to exchange to win the game, if they use an optimal protocol (and where the communication cost is maximized over all x, y pairs). Then in 1990, Karchmer and Wigderson [121] showed the following remarkable connection.

Theorem 32 (Karchmer-Wigderson [121]). *For any f, the minimum depth of any Boolean circuit for f is equal to C_f.*

Combined with Proposition 24 (depth-reduction for formulas), Theorem 32 implies that every Boolean function f requires formulas of size at least 2^{C_f}: in other words, *communication lower bounds imply formula-size lower bounds*. Since communication complexity is a well-established area of theoretical computer science with many strong lower bounds (see for example the book by Kushilevitz and Nisan [134]), one might therefore hope that lower-bounding the communication cost of the "Karchmer-Wigderson game" would be a viable approach to proving $\mathsf{NP} \not\subset \mathsf{NC}^1$ or $\mathsf{P} \not\subset \mathsf{NC}^1$, either of which would constitute a huge step toward $\mathsf{P} \neq \mathsf{NP}$.

See Sect. 6.2.2 for Karchmer and Wigderson's applications of a similar connection to *monotone* formula-size lower bounds. Also see Aaronson and Wigderson [10] for further connections between communication complexity and computational complexity—including even a "communication complexity lower bound" that if true would imply $\mathsf{P} \neq \mathsf{NP}$. Of course, the question is whether these translations merely shift the difficulty of complexity class separations to a superficially different setting, or whether they set the stage for genuinely new insights.

The third approach is *lower bounds via derandomization*. In Sect. 5.4.1, we discussed the discovery in the 1990s that, if sufficiently strong circuit lower bounds hold, then $\mathsf{P} = \mathsf{BPP}$: that is, every randomized algorithm can be made deterministic with only a polynomial slowdown. In the early 2000s, it was discovered that converse statements often hold as well: that is, *derandomizations of randomized algorithms imply circuit lower bounds*. Probably the most-cited result along these lines is that of Kabanets and Impagliazzo [118]:

Theorem 33 ([118]). *Suppose the polynomial identity testing problem from Sect. 5.4 is in* P. *Then either* NEXP $\not\subset$ P/poly, *or else the permanent function has no polynomial-size arithmetic circuits (see Sect. 6.5.1).*

As usual, the issue is that it is not clear whether we should interpret this result as giving a plausible path toward proving circuit lower bounds (namely, by derandomizing PIT), or simply as explaining why derandomizing PIT will be hard (namely, because doing so will imply circuit lower bounds)!

The fourth approach could be called *lower bounds via "innocent-looking" combinatorics problems*. Here is an example: given an $n \times n$ matrix A, say over the finite field $\mathbb{F}_2$, call A *rigid* if not only does A have rank $\Omega(n)$, but any matrix obtained by changing $O(n^{1/10})$ entries in each row of A also has rank $\Omega(n)$. It is easy to show, via a counting argument, that almost all matrices $A \in \mathbb{F}_2^{n \times n}$ are rigid. On the other hand, Valiant [224] made the following striking observation in 1977: if we manage to find any *explicit example* of a rigid matrix, then we also get an explicit example of a Boolean function that cannot be computed by any circuit of linear size and logarithmic depth.

For another connection in the same spirit, given a 3-dimensional tensor $A \in \mathbb{F}_2^{n \times n \times n}$, let the *rank* of A be the smallest r such that A can be written as the sum of r rank-one tensors (that is, tensors of the form $t_{ijk} = x_i y_j z_k$). Then it is easy to show, via a counting argument, that almost all tensors $A \in \mathbb{F}_2^{n \times n \times n}$ have rank $\Omega(n^2)$. On the other hand, Strassen [215] observed in 1973 that, if we find any explicit example of a 3-dimensional tensor with rank r, then we also get an explicit example of a Boolean function with circuit complexity $\Omega(r)$. [28] Alas, proving that any explicit matrix is rigid, or that any explicit tensor has superlinear rank, have turned out to be staggeringly hard problems—as perhaps shouldn't surprise us, given the implications for circuit lower bounds!

The rest of the section is organized as follows:

- Section 6.1 covers logical techniques, which typically fall prey to the relativization barrier.
- Section 6.2 covers combinatorial techniques, which typically fall prey to the natural proofs barrier.
- Section 6.3 covers "hybrid" techniques (logic plus arithmetization), many of which fall prey to the algebrization barrier.

---

[28] Going even further, Raz [184] proved in 2010 that, if we manage to show that any explicit d-dimensional tensor $A : [n]^d \to \mathbb{F}$ has rank at least $n^{d(1-o(1))}$, then we've also shown that the $n \times n$ permanent function has no polynomial-size arithmetic formulas. It's easy to construct explicit d-dimensional tensors with rank $n^{\lfloor d/2 \rfloor}$, but the current record is an explicit d-dimensional tensor with rank at least $2n^{\lfloor d/2 \rfloor} + n - O(d \log n)$ [18].

Note that, if we could show that the permanent had no $n^{O(\log n)}$-size arithmetic formulas, that would imply Valiant's famous Conjecture 66: that the permanent has no polynomial-size arithmetic *circuits*. However, Raz's technique seems incapable of proving formula-size lower bounds better than $n^{\Omega(\log \log n)}$

- Sections 6.4 covers "ironic complexity theory" (as exemplified by the recent work of Ryan Williams), or the use of nontrivial algorithms to prove circuit lower bounds.
- Section 6.5 covers arithmetic circuit lower bounds, which *probably* fall prey to arithmetic variants of the natural proofs barrier (though this remains disputed).
- Section 6.6 covers Mulmuley and Sohoni's Geometric Complexity Theory (GCT), an audacious program to tackle $\mathsf{P} \stackrel{?}{=} \mathsf{NP}$ and related problems by reducing them to questions in algebraic geometry and representation theory (and which is also an example of "ironic complexity theory").

Note that, for the approaches covered in Sects. 6.4 and 6.6, no formal barriers are yet known.

6.1 Logical Techniques

In the 1960s, Hartmanis and Stearns [102] realized that, by simply "scaling down" Turing's diagonalization proof of the undecidability of the halting problem, one can at least prove *some* separations between complexity classes. In particular, one can generally show that more of the same resource (time, memory, etc.) lets one decide more languages than less of that resource. Here is a special case of their so-called *Time Hierarchy Theorem.*

Theorem 34 (Hartmanis-Stearns [102]). $\mathsf{P} \neq \mathsf{EXP}$.

Proof. Let

$$L = \{(\langle M \rangle, x, 0^n) : M(x) \text{ halts in at most } 2^n \text{ steps}\}.$$

Clearly $L \in \mathsf{EXP}$. On the other hand, suppose by contradiction that $L \in \mathsf{P}$. Then there is some polynomial-time Turing machine A such that $A(z)$ accepts if and only if $z \in L$. Let A run in $p(n + |\langle M \rangle| + |x|)$ time. Then using A, we can easily produce another machine B that does the following:

- Takes input $(\langle M \rangle, 0^n)$.
- Runs forever if $M(\langle M \rangle, 0^n)$ halts in at most 2^n steps; otherwise halts.

Note that, if B halts at all, then it halts after only $p(2n + 2|\langle M \rangle|) = n^{O(1)}$ steps.

Now consider what happens when B is run on input $(\langle B \rangle, 0^n)$. If $B(\langle B \rangle, 0^n)$ runs forever, then $B(\langle B \rangle, 0^n)$ halts. Conversely, if $B(\langle B \rangle, 0^n)$ halts, then for all sufficiently large n, it halts in fewer than 2^n steps, but that means that $B(\langle B \rangle, 0^n)$ runs forever. So we conclude that B, and hence A, cannot have existed. ∎

More broadly, the same argument shows that there are languages decidable in $O(n^2)$ time but not in $O(n)$ time, in $O(n^3)$ time but not in $O(n^2)$ time, and so on for almost every natural pair of runtime bounds. (Technically, we have $\mathsf{TIME}(f(n)) \neq$

$\mathsf{TIME}(g(n))$ for every f, g that are *time-constructible*—that is, there exist Turing machines that run for $f(n)$ and $g(n)$ steps given n as input—and that are separated by more than a $\log n$ multiplicative factor.) Likewise, the *Space Hierarchy Theorem* shows that there are languages decidable in $O(f(n))$ space but not in $O(g(n))$ space, for all natural $f(n) \gg g(n)$. Cook [68] also proved a hierarchy theorem for the nondeterministic time classes, which will play an important role in Sect. 6.4:

Theorem 35 (Nondeterministic Time Hierarchy Theorem [68]). *For all time-constructible f, g such that $f(n+1) = o(g(n))$, we have* $\mathsf{NTIME}(f(n)) \neq \mathsf{NTIME}(g(n))$.

One amusing consequence of the hierarchy theorems is that we know, for example, that $\mathsf{P} \neq \mathsf{SPACE}(n)$, even though we don't know either that $\mathsf{P} \not\subset \mathsf{SPACE}(n)$ or that $\mathsf{SPACE}(n) \not\subset \mathsf{P}$! For suppose by contradiction that $\mathsf{P} = \mathsf{SPACE}(n)$. Then by a padding argument (cf. Proposition 17), P would also contain $\mathsf{SPACE}(n^2)$, and therefore equal $\mathsf{SPACE}(n^2)$. But then we'd have $\mathsf{SPACE}(n) = \mathsf{SPACE}(n^2)$, violating the Space Hierarchy Theorem.

In summary, there really is an infinite hierarchy of harder and harder computable problems. Complexity classes don't collapse in the most extreme ways imaginable, with (say) everything solvable in linear time.

6.1.1 Circuit Lower Bounds Based on Counting

A related idea—not exactly "diagonalization," but counting arguments made explicit—can also be used to show that certain problems cannot be solved by polynomial-size *circuits*. This story starts with Claude Shannon [201], who made the following fundamental observation in 1949.

Proposition 36 (Shannon [201]). *There exists a Boolean function $f : \{0,1\}^n \to \{0,1\}$, on n variables, such that any circuit to compute f requires at least $\Omega(2^n/n)$ logic gates. Indeed, almost all Boolean functions on n variables (that is, a $1 - o(1)$ fraction of them) have this property.*[29]

Proof. There are 2^{2^n} different Boolean functions f on n variables, but only

$$\sum_{t=1}^{T} \binom{n}{2}\binom{n+1}{2}\cdots\binom{n+t-1}{2} < (n+T)^{2T}$$

different Boolean circuits with n inputs and at most T NAND gates. Since each circuit can only represent one function, and since $(n+T)^{2T} = o\left(2^{2^n}\right)$ when $T = o(2^n/n)$, it follows by a counting argument (i.e., the pigeonhole principle) that *some*

---

[29] With some effort, Shannon's lower bound can be shown to be tight: that is, every n-variable Boolean function *can* be represented by a circuit of size $O(2^n/n)$. (The obvious upper bound is $O(n2^n)$.)

f must require a circuit with $T = \Omega(2^n/n)$ NAND gates—and indeed, that almost all of the 2^{2^n} possible f's must have this property. The number of AND, OR, and NOT gates required is related to the number of NAND gates by a constant factor, so is also $\Omega(2^n/n)$. ■

Famously, Proposition 36 shows that there *exist* Boolean functions that require exponentially large circuits—in fact, that almost all of them do—yet it fails to produce a single example of such a function! It tells us nothing whatsoever about 3SAT or CLIQUE or any other particular function that might interest us. In that respect, it is similar to Shannon's celebrated proof that almost all codes are good error-correcting codes, which also fails to produce a single example of such a code. Just like, in the decades after Shannon, the central research agenda of coding theory was to "make Shannon's argument explicit" by finding *specific* good error-correcting codes, so too the agenda of circuit complexity has been to "make Proposition 36 explicit" by finding specific functions that provably require large circuits.

In some cases, the mere fact that we know, from Proposition 36, that hard functions *exist* lets us "bootstrap" to show that particular complexity classes must contain hard functions. Here is an example of this.

Theorem 37. $\mathsf{EXPSPACE} \not\subset \mathsf{P/poly}$.

Proof. Let n be sufficiently large. Then by Proposition 36, there exist functions $f : \{0,1\}^n \to \{0,1\}$ with circuit complexity at least $c2^n/n$, for some constant $c > 0$. Thus, if we list all the 2^{2^n} functions in lexicographic order by their truth tables, there must be a first function in the list, call it f_n, with circuit complexity at least $c2^n/n$. We now define

$$L := \bigcup_{n \geq 1} \{x \in \{0,1\}^n : f_n(x) = 1\}.$$

Then by construction, $L \notin \mathsf{P/poly}$. On the other hand, enumerating all n-variable Boolean functions, calculating the circuit complexity of each, and finding the first one with circuit complexity at least $c2^n/n$ can all be done in exponential space. Hence $L \in \mathsf{EXPSPACE}$. ■

There is also a "scaled-down version" of Theorem 37, proved in the same way:

Theorem 38. *For every fixed k, there is a language in* PSPACE *that does not have circuits of size n^k.*[30]

By being a bit more clever, Kannan [120] lowered the complexity class in Theorem 37 from EXPSPACE to $\mathsf{NEXP}^{\mathsf{NP}}$.

Theorem 39 (Kannan [120]). $\mathsf{NEXP}^{\mathsf{NP}} \not\subset \mathsf{P/poly}$.

[30]Crucially, this will be a different language for each k; otherwise we would get $\mathsf{PSPACE} \not\subset \mathsf{P/poly}$, which is far beyond our current ability to prove.

Proof. First, we claim that $\mathsf{EXP}^{\mathsf{NP}^{\mathsf{NP}}} \not\subset \mathsf{P/poly}$. The reason is simply a more careful version of the proof of Theorem 37: in $\mathsf{EXP}^{\mathsf{NP}^{\mathsf{NP}}}$, we can do an explicit binary search for the lexicographically first Boolean function $f_n : \{0,1\}^n \to \{0,1\}$ such that every circuit of size at most (say) $c2^n/n$ disagrees with f_n on some input x. (Such an f_n must exist by a counting argument.)

Next, suppose by contradiction that $\mathsf{NEXP}^{\mathsf{NP}} \subset \mathsf{P/poly}$. Then certainly $\mathsf{NP} \subset \mathsf{P/poly}$. By the Karp-Lipton Theorem (Theorem 23), this implies that $\mathsf{PH} = \Sigma_2^{\mathsf{P}}$, so in particular $\mathsf{P}^{\mathsf{NP}^{\mathsf{NP}}} = \mathsf{NP}^{\mathsf{NP}}$. By upward translation (as in Proposition 17), this in turn means that $\mathsf{EXP}^{\mathsf{NP}^{\mathsf{NP}}} = \mathsf{NEXP}^{\mathsf{NP}}$. But we already know that $\mathsf{EXP}^{\mathsf{NP}^{\mathsf{NP}}}$ does not have polynomial-size circuits, and therefore neither does $\mathsf{NEXP}^{\mathsf{NP}}$. ∎

Amusingly, if one works out the best possible lower bound that one can get from Theorem 39 on the circuit complexity of a language in $\mathsf{NEXP}^{\mathsf{NP}}$, it turns out to be *half-exponential*: that is, a function f such that $f(f(n))$ grows exponentially. Such functions exist, but have no closed-form expressions.

Directly analogous to Theorem 38, a "scaled-down" version of the proof of Theorem 39 shows that, for every fixed k, there is a language in $\Sigma_2^{\mathsf{P}} = \mathsf{NP}^{\mathsf{NP}}$ that does not have circuits of size n^k.

In Sect. 6.3, we will discuss slight improvements to these results that can be achieved with algebraic methods. Nevertheless, it (sadly) remains open even to show that $\mathsf{NEXP} \not\subset \mathsf{P/poly}$, or that there is a language in NP that does not have linear-sized circuits.

6.1.2 The Relativization Barrier

The magic of diagonalization, self-reference, and counting arguments is how abstract and general they are: they never require us to "get our hands dirty" by understanding the inner workings of algorithms or circuits. But as was recognized early in the history of complexity theory, the price of generality is that the logical techniques are extremely limited in scope.

Often the best way to understand the limits of a proposed approach for proving a statement S, is to examine *what else besides* S the approach would prove if it worked—i.e., which stronger statements S' the approach "fails to differentiate" from S. If any of the stronger statements are false, then the approach cannot prove S either.

That is exactly what Baker et al. [34] did for diagonalization in 1975, when they articulated the *relativization barrier*. Their central insight was that almost all the techniques we have for proving statements in complexity theory—such as $\mathcal{C} \subseteq \mathcal{D}$ or $\mathcal{C} \not\subset \mathcal{D}$, where $\mathcal{C}$ and $\mathcal{D}$ are two complexity classes—are so general that, if they work at all, then they actually prove $\mathcal{C}^A \subseteq \mathcal{D}^A$ or $\mathcal{C}^A \not\subset \mathcal{D}^A$ *for all possible oracles* A. In other words: if all the machines that appear in the proof are enhanced in the same way, by being given access to the same oracle, the proof is completely oblivious to that change, and goes through just as before. A proof with this property is said to "relativize," or to hold "in all possible relativized worlds" or "relative to any oracle."

Why do so many proofs relativize? Intuitively, because the proofs only do things like using one Turing machine M_1 to simulate a second Turing machine M_2 step-by-step, without examining either machine's internal structure. In that case, if M_2 is given access to an oracle A, then M_1 can still simulate M_2 just fine, provided that M_1 is *also* given access to A, in order to simulate M_2's oracle calls.

To illustrate, the reader might want to check that the proofs of Theorems 37, 38, and 39 can be straightforwardly modified to show that, more generally:

- $\mathsf{P}^A \neq \mathsf{EXP}^A$ for all oracles A.
- $\mathsf{EXPSPACE}^A \not\subset \mathsf{P}^A/\mathsf{poly}$.

Alas, Baker, Gill, and Solovay then observed that no relativizing technique can possibly resolve the $\mathsf{P} \stackrel{?}{=} \mathsf{NP}$ question. For, unlike (say) $\mathsf{P} \stackrel{?}{=} \mathsf{EXP}$ or the unsolvability of the halting problem, $\mathsf{P} \stackrel{?}{=} \mathsf{NP}$ admits "contradictory relativizations": there are some oracle worlds where $\mathsf{P} = \mathsf{NP}$, and others where $\mathsf{P} \neq \mathsf{NP}$. For that reason, any proof of $\mathsf{P} \neq \mathsf{NP}$ will need to "notice," at some point, that there are no oracles in "our" world: it will have to use techniques that *fail* relative to certain oracles.

Theorem 40 (Baker-Gill-Solovay [34]). *There exists an oracle A such that* $\mathsf{P}^A = \mathsf{NP}^A$, *and another oracle B such that* $\mathsf{P}^B \neq \mathsf{NP}^B$.

Proof Sketch. To make $\mathsf{P}^A = \mathsf{NP}^A$, we can just let A be any PSPACE-complete language. Then it is not hard to see that $\mathsf{P}^A = \mathsf{NP}^A = \mathsf{PSPACE}$.

To make $\mathsf{P}^B \neq \mathsf{NP}^B$, we can (for example) let B be a random oracle, as observed by Bennett and Gill [41]. We can then, for example, define

$$L = \left\{0^n : \text{the first } 2^n \text{ bits of } B \text{ contain a run of } n \text{ consecutive } 1'\text{s}\right\}.$$

Clearly $L \in \mathsf{NP}^B$. By contrast, one can easily that $L \notin \mathsf{P}^B$ with probability 1 over B: in this case, there *really is* nothing for a deterministic Turing machine to do but brute-force search, requiring exponentially many queries to the B oracle. ■

We also have the following somewhat harder result.

Theorem 41 (Wilson [243]). *There exists an oracle A such that* $\mathsf{NEXP}^A \subset \mathsf{P}^A/\mathsf{poly}$, *and such that every language in* NP^A *has linear-sized circuits with A-oracle gates (that is, gates that query A).*

In other words, any proof even of $\mathsf{NEXP} \not\subset \mathsf{P}/\mathsf{poly}$—that is, of a circuit lower bound just "slightly" beyond those that have already been proven—will require non-relativizing techniques. One can likewise show that non-relativizing techniques will be needed to make real progress on many of the other open problems of complexity theory (such as proving $\mathsf{P} = \mathsf{BPP}$).

If the relativization barrier seems too banal, the way to appreciate it is to try to invent techniques, for proving inclusions or separations among complexity classes, that *fail* to relativize. It's harder than it sounds! A partial explanation for this was given by Arora et al. [28], who reinterpreted the relativization barrier in logical terms. From their perspective, a relativizing proof is simply any proof that "knows" about complexity classes, only through axioms that assert the classes' closure properties, as well as languages that they *do* contain (for example, P contains the empty language; if L_1 and L_2 are both in P, then so are Boolean combinations like $\overline{L_1}$ and $L_1 \cap L_2$). These axioms can be shown to imply statements such as $\mathsf{P} \neq \mathsf{EXP}$. But other statements, like $\mathsf{P} \neq \mathsf{NP}$, can be shown to be independent of the axioms, by constructing models of the axioms where those statements are false. One constructs those models by using oracles to "force in" additional languages—such as PSPACE-complete languages, if one wants a world where $\mathsf{P} = \mathsf{NP}$—which the axioms might not *require* to be contained in complexity classes like P and NP, but which they don't prohibit from being contained, either. The conclusion is that any proof of $\mathsf{P} \neq \mathsf{NP}$ will need to appeal to deeper properties of the classes P and NP, properties that don't follow from these closure axioms.

6.2 *Combinatorial Lower Bounds*

Partly because of the relativization barrier, in the 1980s attention shifted to combinatorial approaches: that is, approaches where one tries to prove superpolynomial lower bounds on the number of operations of *some* kind needed to do *something*, by actually "rolling up one's sleeves" and delving into the messy details of what the operations do (rather than making abstract diagonalization arguments). These combinatorial approaches enjoyed some spectacular successes, some of which seemed at the time like they were within striking distance of proving $\mathsf{P} \neq \mathsf{NP}$. Let us discuss some examples.

6.2.1 Proof Complexity

Suppose we are given a 3SAT formula φ, and we want to prove that φ has no satisfying assignments. One natural approach to this is called *resolution*: we repeatedly pick two clauses of φ, and then "resolve" the clauses (or "smash them together") to derive a new clause that logically follows from the first two. This is most useful when one of the clauses contains a non-negated literal x, and the other contains the corresponding negated literal $\overline{x}$. For example, from the clauses $(x \vee y)$ and $(\overline{x} \vee z)$, it is easily seen that we can derive $(y \vee z)$. The new derived clause can then be added to the list of clauses, and used as an input to future resolution steps.

Now, if we ever derive the empty clause ( )—say, by smashing together (x) and $(\overline{x})$—then we can conclude that our original 3SAT formula φ must have been unsatisfiable. For in that case, φ entails a clause that's not satisfied by *any* setting of variables. Another way to say this is that resolution is a *sound* proof system. By doing an induction on the number of variables in φ, it's not hard to show that resolution is also *complete*:

Proposition 42. *Resolution is a complete proof system for the unsatisfiability of* kSAT. *In other words, given any unsatisfiable* kSAT *formula* φ, *there exists some sequence of resolution steps that produces the empty clause.*

So the key question about resolution is just *how many* resolution steps are needed to derive the empty clause, starting from an unsatisfiable formula φ. If that number could be upper-bounded by a polynomial in the size of φ, it would follow that $\mathsf{NP} = \mathsf{coNP}$. If, moreover, an appropriate sequence of resolutions could actually be *found* in polynomial time, it would follow that $\mathsf{P} = \mathsf{NP}$.

On the other hand, when one proves completeness by induction on the number of variables n, the only upper bound one gets on the number of resolution steps is 2^n. And indeed, in 1985, Haken proved the following celebrated result.

Theorem 43 (Haken [101]). *There exist* kSAT *formulas, involving* $n^{O(1)}$ *variables and clauses, for which any resolution proof of unsatisfiability requires at least* $2^{\Omega(n)}$ *resolution steps. An example is a* kSAT *formula that explicitly encodes the "nth Pigeonhole Principle": that is, the statement that is no way to map* $n+1$ *pigeons into* n *holes, without mapping two pigeons to the same hole.*

Haken's nontrivial proof formalized the intuition that any resolution proof of the Pigeonhole Principle will ultimately be stuck "reasoning locally": "let's see, if I put this pigeon there, and that one there ... darn, it *still* doesn't work!" Such a proof has no ability to engage in higher-level reasoning about the total *number* of pigeons.

Since Haken's breakthrough, there have been many other exponential lower bounds on the sizes of unsatisfiability proofs, typically for proof systems that generalize resolution in some way (see Beame and Pitassi [36] for a good survey). These, in turn, often let us prove exponential lower bounds on the running times of certain kinds of algorithms. For example, there is a widely-used class of kSAT algorithms called DPLL (Davis-Putnam-Logemann-Loveland) algorithms [72], which are based on pruning the search tree of possible satisfying assignments. DPLL algorithms have the property that, if one looks at the search tree of their execution on an unsatisfiable kSAT formula φ, one can *read off* a resolution proof that φ is unsatisfiable. From that fact, together with Theorem 43, it follows that there exist kSAT formulas (for example, the Pigeonhole Principle formulas) for which any DPLL algorithm requires exponential time.

In principle, if one could prove superpolynomial lower bounds for *arbitrary* proof systems (constrained only by the proofs being checkable in polynomial time), one would get $\mathsf{P} \neq \mathsf{NP}$, and even $\mathsf{NP} \neq \mathsf{coNP}$! However, perhaps this motivates turning our attention to lower bounds on circuit size, which tend to be somewhat easier than the analogous proof complexity lower bounds, and which—if generalized to arbitrary Boolean circuits—would "merely" imply $\mathsf{P} \neq \mathsf{NP}$ and $\mathsf{NP} \not\subset \mathsf{P/poly}$, rather than $\mathsf{NP} \neq \mathsf{coNP}$.

6.2.2 Monotone Circuit Lower Bounds

Recall, from Sect. 5.2, that if we could merely prove that any family of *Boolean circuits* to solve some NP problem required a superpolynomial number of AND, OR, and NOT gates, then that would imply $P \neq NP$, and even the stronger result $NP \not\subset P/poly$ (that is, NP-complete problems are not efficiently solvable by nonuniform algorithms).

Now, some NP-complete languages L have the interesting property being *monotone*: that is, changing an input bit from 0 to 1 can change the answer from $x \notin L$ to $x \in L$, but never from $x \in L$ to $x \notin L$. An example is the CLIQUE language: say, the set of all encodings of n-vertex graphs G, as adjacency matrices of 0s and 1s, such that G contains a clique on at least $\sqrt{n}$ vertices. It's not hard to see that one can decide any such language using a *monotone circuit*: that is, a Boolean circuit of AND and OR gates only, no NOT gates. For the CLIQUE language, for example, a circuit could simply consist of an OR of $\binom{n}{\sqrt{n}}$ ANDs, one for each possible clique. It thus becomes interesting to ask what are the *smallest* monotone circuits for monotone NP-complete languages.

In 1985, Alexander Razborov, then a graduate student, astonished the complexity theory world with the following result.

Theorem 44 (Razborov [187]). *Any monotone circuit for* CLIQUE *requires at least* $n^{\Omega(\log n)}$ *gates.*

Subsequently, Alon and Boppana [25] improved this, to show that any monotone circuit to detect a clique of size $\sim (n/\log n)^{2/3}$ must have size $\exp\left(\Omega\left((n/\log n)^{1/3}\right)\right)$. I won't go into the proof of Theorem 44 here, but it uses beautiful combinatorial techniques, including (in modern versions) the Erdös-Rado sunflower lemma.

The significance of Theorem 44 is this: if we could now merely prove that *any circuit for a monotone language can be made into a monotone circuit without much increasing its size*, then we'd immediately get $P \neq NP$ and even $NP \not\subset P/poly$. And indeed, this was considered a potentially-viable approach to proving $P \neq NP$ for some months. Alas, the approach turned out to be a dead end, for the following reason.

Theorem 45 ([188, 220]). *There are monotone languages even in* P *that require exponentially-large monotone circuits. An example is the* MATCHING *language, consisting of all adjacency-matrix encodings of n-vertex graphs that admit a matching on at least* $\sim (n/\log n)^{2/3}$ *vertices. This language requires monotone circuits of size* $\exp\left(\Omega\left((n/\log n)^{1/3}\right)\right)$.

Thus, while Theorem 44 stands as a striking example of the power of combinatorics to prove circuit lower bounds, ultimately it tells us not about the hardness of NP-complete problems, but only about the weakness of monotone circuits. Theorem 45 implies that, even if we are trying to compute a monotone Boolean function (such as the MATCHING function), allowing ourselves the non-monotone

NOT gate can yield an exponential reduction in circuit size. Alas, Razborov's techniques break down completely as soon as a few NOT gates are available.[31]

I should also mention lower bounds on monotone *depth*. In the STCON (s, t-connectivity) problem, we're given as input the adjacency matrix of an undirected graph, and asked whether or not there is a path between two designated vertices s and t. By using their connection between circuit depth and communication complexity (see Sect. 6), Karchmer and Wigderson [121] were able to prove that any monotone circuit for STCON requires $\Omega\left(\log^2 n\right)$ depth—and as a consequence, that any monotone *formula* for STCON requires $n^{\Omega(\log n)}$ size. Since STCON is known to have monotone circuits of polynomial size, this implies in particular that monotone formula size and monotone circuit size are not polynomially related.

6.2.3 Small-Depth Circuits and the Random Restriction Method

Besides restricting the allowed gates (say, to AND and OR only), there's a second natural way to "hobble" a circuit, and thereby potentially make it easier to prove lower bounds on circuit size. Namely, we can restrict the circuit's *depth*, the number of layers of gates between input and output. If the allowed gates all have a "fanin" of 1 or 2 (that is, they all take only 1 or 2 input bits), then clearly any circuit that depends nontrivially on all n of the input bits must have depth at least $\log_2 n$. On the other hand, if we allow gates of *unbounded fanin*—for example, ANDs or XORs or MAJORITYs on unlimited numbers of inputs—then it makes sense to ask what can be computed even by circuits of *constant* depth. Constant-depth circuits are very closely related to *neural networks*, which also consist of a small number of layers of "logic gates" (i.e., the neurons), with each neuron allowed to have very large "fanin"—i.e., to accept input from many or all of the neurons in the previous layer.

If we don't also restrict the number of gates or neurons, then it turns out that *every* function can be computed in small depth:

Proposition 46. *Every Boolean function $f : \{0,1\}^n \to \{0,1\}$ can be computed by an unbounded-fanin, depth-3 circuit of size $O(n2^n)$: namely, by an OR of ANDs of input bits and their negations.*

Proof. We simply need to check whether the input, $x \in \{0,1\}^n$, is one of the z's such that $f(z) = 1$:

[31] Note that, if we encode the input string using the so-called *dual-rail representation*—in which every 0 is represented by the 2-bit string 01, and every 1 by 10—then the monotone circuit complexities of CLIQUE, MATCHING, and so on *do* become essentially equivalent to their non-monotone circuit complexities, since we can push all the NOT gates to the bottom layer of the circuit using de Morgan's laws. Unfortunately, Razborov's lower bound techniques also break down under dual-rail encoding.

$$f(x) = \bigvee_{z=z_1\cdots z_n\,:\,f(z)=1} \left(\left(\bigwedge_{i\in\{1,\ldots,n\}\,:\,z_i=1} x_i \right) \wedge \left(\bigwedge_{i\in\{1,\ldots,n\}\,:\,z_i=0} \overline{x}_i \right) \right).$$

■

Similarly, in typical neural network models, every Boolean function can be computed by a network with $\sim 2^n$ neurons arranged into just two layers.

So the interesting question is what happens if we restrict both the depth *and* the number of gates or neurons. More formally, let AC^0 be class of languages $L \subseteq \{0,1\}^*$ for which there exists a family of circuits $\{C_n\}_{n\geq 1}$, one for each input size n, such that:

(1) $C_n(x)$ outputs 1 if $x \in L$ and 0 if $x \notin L$, for all n and $x \in \{0,1\}^n$.
(2) Each C_n consists of unbounded-fanin AND and OR gates, as well as NOT gates.
(3) There is a polynomial p such that each C_n has at most $p(n)$ gates.
(4) There is a constant d such that each C_n has depth at most d.

Clearly AC^0 is a subclass of $\mathsf{P/poly}$; indeed we recover $\mathsf{P/poly}$ by omitting condition (4). Now, one of the major triumphs of complexity theory in the 1980s was to understand AC^0, as we still only dream of understanding $\mathsf{P/poly}$. It's not just that we know $\mathsf{NP} \not\subset \mathsf{AC}^0$; rather, it's that we know in detail which problems are and aren't in AC^0 (even problems within P), and exactly how many gates are needed for each given depth d. As the most famous example, let PARITY be the language consisting of all strings with an odd number of '1' bits. Then:

Theorem 47 (Ajtai [16], Furst-Saxe-Sipser [85]). PARITY *is not in* AC^0.

While the original lower bounds on the size of AC^0 circuits for PARITY were only slightly superpolynomial, Theorem 47 was subsequently improved by Yao [245] and then by Håstad [103], the latter of whom gave an essentially optimal result: namely, any AC^0 circuit for PARITY of depth d requires at least $2^{\Omega\left(n^{1/(d-1)}\right)}$ gates.

The first proofs of Theorem 47 used what is called the *method of random restrictions*. In this method, we assume by contradiction that we have a size-s, depth-d, unbounded-fanin circuit C for our Boolean function—say, the PARITY function. We then randomly fix most of the input bits to 0 or 1, while leaving a few input bits unfixed. What we hope to find is that the random restriction "kills off" an entire layer of gates—because any AND gate that takes even one constant 0 bit as input can be replaced by the constant 0 function, and likewise, any OR gate that takes even one 1 bit as input can be replaced by the constant 1 function. Thus, any AND or OR gate with a large fanin is extremely likely to be killed off; gates with small fanin might not be killed off, but can be left around to be dealt with later. We then repeat this procedure, randomly restricting most of the remaining unfixed bits, in order to kill off the next higher layer of AND and OR gates, and so on through all d layers. By the time we are done, we have reduced C to a shadow of its former self: specifically, to a circuit that depends on only a constant number of input bits. Meanwhile, even though only a tiny fraction of the input bits (say, $n^{1/d}$ of them) remain unfixed, we still have a nontrivial Boolean function on those bits: indeed, it is easy to see that any restriction of the PARITY function to a subset S of bits will either be PARITY itself, or else NOT(PARITY). But a circuit of constant size clearly can't compute a Boolean function that depends on $\sim n^{1/d}$ input bits. This yields our desired contradiction.

At a high level, there were three ingredients needed for the random restriction method to work. First, the circuit needed to built out of AND and OR gates, which are likely to get killed off by random restrictions. The method would *not* have worked if the circuit contained unbounded-fanin MAJORITY gates (as a neural network does), or even unbounded-fanin XOR gates. Second, it was crucial that the circuit depth d was small, since we needed to shrink the number of unfixed input variables by a large factor d times, and then still have unfixed variables left over. It turns out that random restriction arguments can yield *some* lower bound whenever $d = o\left(\frac{\log n}{\log\log n}\right)$, but not beyond that. Third, we needed to consider a function, such as PARITY, that remains nontrivial even after the overwhelming majority of input bits have been randomly fixed to 0 or 1. The method wouldn't have worked, for example, for the n-bit AND function (which is unsurprising, since the AND function *does* have a depth-1 circuit, consisting of a single AND gate!).

The original proofs for PARITY $\notin \mathsf{AC}^0$ have been generalized and improved on in many ways. For example, Linial et al. [144] examined the weakness of AC^0 circuits from a different angle: "turning lemons into lemonade," they gave a quasipolynomial-time algorithm to *learn* arbitrary AC^0 circuits with respect to the uniform distribution over inputs. Also, proving a conjecture put forward by Linial and Nisan [145] (and independently Babai), Braverman [51] showed that AC^0 circuits cannot distinguish the outputs of a large range of pseudorandom generators from truly random strings.

Meanwhile, Håstad [103] showed that for every d, there are functions computable by AC^0 circuits of depth d that require exponentially many gates for AC^0 circuits of depth $d-1$. This implies that there exists an oracle relative to which PH is infinite. Improving that result, Rossman, Servedio, and Tan [193] very recently showed that the same functions Håstad had considered require exponentially many gates even to *approximate* using AC^0 circuits of depth $d-1$. This implies that PH is infinite relative to a *random* oracle with probability 1, resolving a 30-year-old open problem.

The random restriction method has also had other applications in complexity theory, besides to AC^0. Most notably, it's been used to prove polynomial lower bounds on *formula size*. The story of formula-size lower bounds starts in 1961 with Subbotovskaya [216], who used random restrictions to show that the n-bit PARITY function requires formulas of size $\Omega\left(n^{1.5}\right)$. Later Khrapchenko [128] improved this to $\Omega\left(n^2\right)$, which is tight.[32] Next, in 1987, Andreev [26] constructed a different

[32] Assume for simplicity that n is a power of 2. Then $x_1 \oplus \cdots \oplus x_n$ can be written as $y \oplus z$, where $y := x_1 \oplus \cdots \oplus x_{n/2}$ and $z := x_{n/2+1} \oplus \cdots \oplus x_n$. This in turn can be written as $(y \wedge \overline{z}) \vee (\overline{y} \wedge z)$. Expanding recursively now yields a size-n^2 formula for PARITY, made of AND, OR, and NOT gates.

Boolean function in P that could be shown, again using random restrictions, to require formulas of size $n^{2.5-o(1)}$. This was subsequently improved to $n^{2.55-o(1)}$ by Impagliazzo and Nisan [114], to $n^{2.63-o(1)}$ by Paterson and Zwick [176], and finally to $n^{3-o(1)}$ by Håstad [104] and to $\Omega\left(\frac{n^3}{(\log n)^2(\log\log n)^3}\right)$ by Tal [219]. Unfortunately, the random restriction method seems fundamentally incapable of going beyond $\Omega\left(n^3\right)$. On the other hand, for Boolean circuits rather than formulas, we still have no lower bound better than *linear* for any function in P (or for that matter, in NP).!

6.2.4 Small-Depth Circuits and the Polynomial Method

For our purposes, the most important extension of Theorem 47 was achieved by Smolensky [209] and Razborov [189] in 1987. Let $\mathsf{AC}^0[m]$ be the class of languages decidable by a family of constant-depth, polynomial-size, unbounded-fanin circuits with AND, OR, NOT, and MOD-m gates (gates output 1 if their number of '1' input bits is divisible by m, and 0 otherwise). Then Smolensky and Razborov extended the class of circuits for which lower bounds can be proven from AC^0 to $\mathsf{AC}^0[p]$, whenever p is prime.

Theorem 48 (Smolensky [209], Razborov [189]). *Let p and q be distinct primes. Then* MOD_q*, the set of all strings with Hamming weight divisible by q, is not in* $\mathsf{AC}^0[p]$*. Indeed, any* $\mathsf{AC}^0[p]$ *circuit for* MOD_q *of depth d requires* $2^{\Omega\left(n^{1/2d}\right)}$ *gates. As a corollary, the* MAJORITY *function is also not in* $\mathsf{AC}^0[p]$*, and also requires* $2^{\Omega\left(n^{1/2d}\right)}$ *gates to compute using* $\mathsf{AC}^0[p]$ *circuits of depth d.*

It is not hard to show that $\mathsf{AC}^0[p] = \mathsf{AC}^0\left[p^k\right]$ for any $k \geq 1$, and thus, one also gets lower bounds against $\mathsf{AC}^0[m]$, whenever m is a prime power.

The proof of Theorem 48 uses the so-called *polynomial method.* Here one argues that, if a function f can be computed by a constant-depth circuit with AND, OR, NOT, and MOD-p gates, then f can also be approximated by a low-degree polynomial over the finite field $\mathbb{F}_p$. One then shows that a function of interest, such as the MOD_q function (for $q \neq p$), *can't* be approximated by any such low-degree polynomial. This provides the desired contradiction.

The polynomial method is famously specific in scope: it's still not known how to generalize the method even to $\mathsf{AC}^0[m]$ circuits, where m is not a prime power. The reason why it breaks down there is simply that there are no finite fields of non-prime-power order. And thus, to take an example, it's still open whether the n-bit MAJORITY function has a constant-depth, polynomial-size, unbounded-fanin circuit consisting of AND, OR, NOT, and MOD-6 gates (or even entirely of MOD-6 gates)!

Stepping back, it's interesting to ask whether the constant-depth circuit lower bounds evade the relativization barrier explained in Sect. 6.1.2. There's some disagreement about whether it's even sensible to feed oracles to tiny complexity classes such as AC^0 (see Allender and Gore [22] for example). However, to whatever extent it *is* sensible, the answer is that these lower bounds do evade relativization. For example, if by $\left(\mathsf{AC}^0\right)^A$, we mean AC^0 extended by "oracle gates" that query A, then it's easy to construct an A such that $\left(\mathsf{AC}^0\right)^A = \mathsf{P}^A$: for example, any A that is P-complete under AC^0-reductions will work. On the other hand, we know from Theorem 47 that $\mathsf{AC}^0 \neq \mathsf{P}$ in the "real," unrelativized world.

6.2.5 The Natural Proofs Barrier

Despite the weakness of AC^0 and $\mathsf{AC}^0[m]$ circuits, the progress on lower bounds for them suggested what seemed to many researchers like a plausible path to proving $\mathsf{NP} \not\subset \mathsf{P/poly}$, and hence $\mathsf{P} \neq \mathsf{NP}$. That path is simply to generalize the random restriction and polynomial methods further and further, to get lower bounds for more and more powerful classes of circuits. The first step, of course, would be to generalize the polynomial method to handle $\mathsf{AC}^0[m]$ circuits, where m is not a prime power. Then one could handle what are called TC^0 circuits: that is, constant-depth, polynomial-size, unbounded-fanin circuits with MAJORITY gates (or, as in a neural network, *threshold gates*, which output 1 if a certain weighted affine combination of the input bits exceeds 0, and 0 otherwise). Next, one could aim for polynomial-size circuits of logarithmic depth: that is, the class NC^1. Finally, one could push all the way to polynomial-depth circuits: that is, the class $\mathsf{P/poly}$.

Unfortunately, we now know that this path hits a profound barrier at TC^0, if not earlier—a barrier that explains why the random restriction and polynomial methods haven't taken us further toward a proof of $\mathsf{P} \neq \mathsf{NP}$. Apparently this barrier was known to some experts in the 1980s, but it was first articulated in print in 1993 by Razborov and Rudich [191], who called it the *natural proofs barrier*.

The basic insight is that combinatorial techniques, such as the method of random restrictions, do more than advertised: in some sense, they do too much for their own good. In particular, not only do they let us show that certain specific functions, like PARITY, are hard for AC^0; they even let us certify that a *random* function is hard for AC^0. Indeed, such techniques give rise to an *algorithm*, which takes as input the truth table of a Boolean function $f : \{0,1\}^n \to \{0,1\}$, and which has the following two properties.

(1) **"Constructivity."** The algorithm runs in time polynomial in the size of f's truth table (that is, polynomial in 2^n).
(2) **"Largeness."** If f is chosen uniformly at random, then with probability at least $1/n^{O(1)}$ over f, the algorithm certifies that f is hard (i.e., that f is not in some circuit class $\mathcal{C}$, such as AC^0 in the case of the random restriction method).

If a lower bound proof gives rise to an algorithm satisfying (1) and (2), then Razborov and Rudich call it a *natural proof*. In many cases, it's not entirely obvious that a lower bound proof is natural, but with some work one can show that it is. To illustrate, in the case of the random restriction method, the algorithm could check that f has a large fraction of its Fourier mass on high-degree Fourier coefficients, or that f has high "average sensitivity" (that is, if x and y are random inputs that differ only in a single bit, then with high probability $f(x) \neq f(y)$). These tests have the following three properties:

- They are easy to perform, in time polynomial in the truth table size 2^n.
- A random function f will pass these tests with overwhelming probability (that is, such an f will "look like PARITY" in the relevant respects).
- The results of Linial, Mansour, and Nisan [144] show that any f that passes these tests remains nontrivial under most random restrictions, and for that reason, cannot be in AC^0.

But now, twisting the knife, Razborov and Rudich point out that any natural lower bound proof is self-defeating, in that *it yields an efficient algorithm to solve some of the same problems that we'd set out to prove were hard.* More concretely, suppose we have a natural lower bound proof against the circuit class $\mathcal{C}$. Then by definition, we also have an efficient algorithm A that, given a random Boolean function $f : \{0,1\}^n \to \{0,1\}$, certifies that $f \notin \mathcal{C}$ with at least $1/n^{O(1)}$ probability over f. But this means that $\mathcal{C}$ cannot contain very strong families of *pseudorandom functions*: namely, functions $f : \{0,1\}^n \to \{0,1\}$ that are indistinguishable from "truly" random functions, even by algorithms that can examine their entire truth tables and use time polynomial in 2^n.

Why not? Because A can *never* certify $f \notin \mathcal{C}$ if f is a pseudorandom function, computable in $\mathcal{C}$. But A certifies $f \notin \mathcal{C}$ with $1/n^{O(1)}$ probability over a truly random f. Thus, A serves to distinguish random from pseudorandom functions with non-negligible[33] bias—so the latter were never really pseudorandom at all.

To recap, we've shown that, if there's any natural proof that any function is not in $\mathcal{C}$, then all Boolean functions computable in $\mathcal{C}$ can be distinguished from random functions by $2^{O(n)}$-time algorithms. That might not sound so impressive, since $2^{O(n)}$ is a lot of time. But a key observation is that, for most of the circuit classes $\mathcal{C}$ that we care about, there are families of pseudorandom functions $\{f_s\}_s$ on n bits that are conjectured to require $2^{p(n)}$ time to distinguish from truly random functions, where $p(n)$ is as large a polynomial as we like (related to the length of the random "seed" s). It follows from results of Naor and Reingold [170] that in TC^0 (constant-depth, polynomial-size threshold circuits), there are functions that can't be distinguished from random functions in $2^{O(n)}$ time, *unless* the factoring and discrete logarithm problems are solvable in $O\left(2^{n^\varepsilon}\right)$ time for every $\varepsilon > 0$. (For comparison, the best *known* algorithms for these problems take roughly $2^{n^{1/3}}$ time.) Likewise, Banerjee et al. [35] showed that in TC^0, there are functions that can't be distinguished from random in $2^{O(n)}$ time, unless noisy systems of linear equations can be solved in $O\left(2^{n^\varepsilon}\right)$ time for every $\varepsilon > 0$.

It's worth pausing to let the irony sink in. Razborov and Rudich are pointing out that, as we showed certain problems (factoring and discrete logarithm) to be harder and harder via a natural proof, we'd simultaneously show those same problems to be easier and easier! Indeed, any natural proof showing that these problems took

[33] In theoretical computer science, the term *non-negligible* means lower-bounded by $1/n^{O(1)}$.

at least $t(n)$ time, would also show that they took *at most* roughly $2^{t^{-1}(n)}$ time. As a result, no natural proof could possibly show these problems take more than half-exponential time: that is, time $t(n)$ such that $t(t(n))$ grows exponentially.

Here, perhaps, we are finally face-to-face with a central conceptual difficulty of the $\mathsf{P} \stackrel{?}{=} \mathsf{NP}$ question: namely, we're trying to prove that certain functions are hard, but the problem of deciding whether a function is hard is *itself* hard, according to the very sorts of conjectures that we're trying to prove.[34]

Of course, the natural proofs barrier didn't prevent complexity theorists from proving strong lower bounds against AC^0. But the result of Linial et al. [144] can be interpreted as saying that this is *because* AC^0 is not yet powerful enough to express pseudorandom functions. When we move just slightly higher, to TC^0 (constant-depth threshold circuits), we *do* have pseudorandom functions under plausible hardness assumptions, and—not at all coincidentally, according to Razborov and Rudich—we no longer have strong circuit lower bounds. In that sense, natural proofs explains almost precisely why the progress toward proving $\mathsf{P} \neq \mathsf{NP}$ via circuit complexity stalled where it did. The one complication in the story is the $\mathsf{AC}^0[m]$ classes, for which we don't yet have strong lower bounds (though see Sect. 6.4), but *also* don't have pseudorandom function candidates. For those classes, it's still possible that natural proofs could succeed.

As Razborov and Rudich themselves stressed, the take-home message is *not* that we should give up on proving $\mathsf{P} \neq \mathsf{NP}$. In fact, since the beginning of complexity theory, we've had at least one technique that easily evades the natural proofs barrier: namely, diagonalization (the technique used to prove $\mathsf{P} \neq \mathsf{EXP}$; see Sect. 6.1)! The reason why diagonalization evades the barrier is that it zeroes in on a specific property of the function f being lower-bounded—namely, the fact that f is EXP-complete, and thus able to simulate all P machines—and thereby avoids the trap of arguing that "f is hard because it looks like a random function." Of course, diagonalization is subject to the relativization barrier (see Sect. 6.1.2), so the question still stands of how to evade relativization and natural proofs simultaneously; we'll return to that question in Sect. 6.3.

More broadly, there are many cases in mathematics where we can prove that some object O of interest to us has a property P, even though we have no hope of finding a general polynomial-time algorithm to decide whether *any* given object has property P, or even to certify a large fraction of objects as having property P. In such cases, often we prove that O has property P by exploiting special symmetries in O—symmetries that have little to do with why O has property P, but everything to do with why we can *prove* it has the property. As an example, a random graph is an *expander graph* (that is, a graph on which a random walk mixes rapidly) with overwhelming probability. But since the general problem of deciding whether a

[34]Technically, the problem of distinguishing random from pseudorandom functions is equivalent to the problem of inverting one-way functions, which is not *quite* as strong as solving NP-complete problems in polynomial time—only solving average-case NP-complete problems with planted solutions. For more see Sect. 5.3.

graph is an expander is NP-hard, if we want a *specific* graph G that's provably an expander, typically we need to construct G with a large amount of symmetry: for example, by taking it to be the Cayley graph of a finite group. Similarly, even though we expect that there's no general efficient algorithm to decide if a Boolean function f is hard,[35] given as input f's truth table, we might be able to prove that certain *specific* f's (for example, NP- or #P-complete ones) are hard by exploiting their symmetries. Geometric Complexity Theory (see Sect. 6.6) is the best-known development of that particular hope for escaping the natural proofs barrier.

But GCT is not the only way to use symmetry to evade natural proofs. As a vastly smaller example, I [2, Appendix 10] proved an exponential lower bound on the so-called *manifestly orthogonal formula size* of a function $f : \{0,1\}^n \rightarrow \{0,1\}$ that outputs 1 if the input x is a codeword of a linear error-correcting code, and 0 otherwise. Here a *manifestly orthogonal formula* is a formula over $x_1, \ldots, x_n, \overline{x}_1, \ldots, \overline{x}_n$ consisting of OR and AND gates, where every OR must be of two subformulas over the same set of variables, and every AND must be of two subformulas over disjoint sets of variables. My lower bound wasn't especially difficult, but what's notable about it took crucial advantage of a *symmetry* of linear error-correcting codes: namely, the fact that any such code can be recursively decomposed as a disjoint union of Cartesian products of smaller linear error-correcting codes. My proof thus gives no apparent insight into how to certify that a *random* Boolean function has manifestly orthogonal formula size $\exp(n)$, and possibly evades the natural proofs barrier (if there *is* such a barrier in the first place for manifestly orthogonal formulas).

Another proposal for how one could use symmetry to evade the natural proofs barrier comes from a beautiful 2010 paper by Allender and Koucký [24] (see also Allender's survey [20]). These authors show that, if one wanted to prove that certain specific NC^1 problems were not in TC^0, thereby establishing the breakthrough separation $TC^0 \neq NC^1$, it would suffice to show that those problems had no TC^0 circuits of size $n^{1+\varepsilon}$, for any constant $\varepsilon > 0$. To achieve this striking "bootstrap," from an $n^{1+\varepsilon}$ lower bound to a superpolynomial one, Allender and Koucký exploit the *self-reducibility* of the NC^1 problems in question, the fact that they can be reduced to smaller instances of themselves. Crucially, this self-reducibility would *not* hold for a random function. For this reason, the proposed lower bound method has at least the potential to evade the natural proofs barrier. Indeed, it's not even totally implausible that a natural proof could yield an $n^{1+\varepsilon}$ lower bound for TC^0

[35]The problem, given as input the truth table of a Boolean function $f : \{0,1\}^n \rightarrow \{0,1\}$, of computing or approximating the circuit complexity of f is called the Minimum Circuit Size Problem (MCSP). It is a longstanding open problem whether or not MCSP is NP-hard; at any rate, there are major obstructions to *proving* it NP-hard with existing techniques (see Kabanets and Cai [117] and Murray and Williams [169]). On the other hand, MCSP cannot be in P (or BPP) unless there are no cryptographically-secure pseudorandom generators. At any rate, what is relevant to natural proofs is just whether there is an efficient algorithm to *certify a large fraction* of Boolean functions as being hard, which is a weaker requirement than solving MCSP.

circuits, with the bootstrapping from $n^{1+\varepsilon}$ to superpolynomial the only non-natural part of the argument.[36]

6.3 *Arithmetization*

In the previous sections, we saw that there are logic-based techniques (like diagonalization) that suffice to prove $\mathsf{P} \neq \mathsf{EXP}$ and $\mathsf{NEXP}^{\mathsf{NP}} \not\subset \mathsf{P/poly}$, and that indeed evade the natural proofs barrier, but that are blocked from proving $\mathsf{P} \neq \mathsf{NP}$ by the relativization barrier. Meanwhile, there are combinatorial techniques (like random restrictions) that suffice to prove circuit lower bounds against AC^0 and $\mathsf{AC}^0[p]$, and that evade the relativization barrier, but that are blocked from proving lower bounds against $\mathsf{P/poly}$ (and hence, from proving $\mathsf{P} \neq \mathsf{NP}$) by the natural proofs barrier.

This situation raises a question: couldn't we simply *combine* techniques that evade relativization but not natural proofs, with techniques that evade natural proofs but not relativization, in order to evade both? As it turns out, we can.

6.3.1 IP = PSPACE

The story starts with a dramatic development in complexity theory around 1990, though not one that obviously bore on $\mathsf{P} \neq \mathsf{NP}$ or circuit lower bounds. In the 1980s, theoretical cryptographers became interested in so-called *interactive proof systems*, which are protocols where a computationally-unbounded but untrustworthy prover (traditionally named Merlin) tries to convince a skeptical polynomial-time verifier (traditionally named Arthur) that some mathematical statement is true, via a two-way conversation, in which Arthur can randomly generate challenges and then evaluate Merlin's answers to them.

More formally, let IP (Interactive Proof) be the class of all languages $L \subseteq \{0,1\}^*$ for which there exists a probabilistic polynomial-time algorithm for Arthur with the following properties. Arthur receives an input string $x \in \{0,1\}^n$ (which Merlin also knows), and then generates up to $n^{O(1)}$ challenges to send to Merlin. Each challenge is a string of up to $n^{O(1)}$ bits, and each can depend on x, on Arthur's internal random bits, *and* on Merlin's responses to the previous challenges. (We also allow Arthur, if he likes, to keep some random bits hidden, without sending them to Merlin—though

---

[36] Allender and Koucký's paper partly builds on 2003 work by Srinivasan [211], who showed that, to prove $\mathsf{P} \neq \mathsf{NP}$, one would "merely" need to show that any algorithm to compute weak approximations for the MAXCLIQUE problem takes $\Omega(n^{1+\varepsilon})$ time, for some constant $\varepsilon > 0$. The way Srinivasan proved this striking statement was, again, by using a sort of self-reduciblity: he showed that, if there's a polynomial-time algorithm for MAXCLIQUE, then by running that algorithm on smaller graphs sampled from the original graph, one can solve approximate versions of MAXCLIQUE in $n^{1+o(1)}$ time.

surprisingly, this turns out not to make any difference [90].) We think of Merlin as trying his best to persuade Arthur that $x \in L$; at the end, Arthur decides whether to accept or reject Merlin's claim. We require that for all inputs x:

- If $x \in L$, then there is some strategy for Merlin (i.e., some function determining which message to send next, given x and the sequence of challenges so far[37]) that causes Arthur to accept with probability at least $2/3$ over his internal randomness.
- If $x \notin L$, then regardless of what strategy Merlin uses, Arthur rejects with probability at least $2/3$ over his internal randomness.

Clearly IP generalizes NP: indeed, we recover NP if we get rid of the interaction and randomness aspects, and just allow a single message from Merlin, which Arthur either accepts or rejects. In the other direction, it's not hard to show that $\mathsf{IP} \subseteq \mathsf{PSPACE}$.[38]

The question asked in the 1980s was: *does interaction help? how much bigger is* IP *than* NP*?* It was observed that IP contains at least a few languages that aren't known to be in NP, such as graph non-isomorphism. This is so because of a simple, famous, and elegant protocol [89]: given two n-vertex graphs G and H, Arthur can pick one of the two uniformly at random, randomly permute its vertices, then send the result to Merlin. He then challenges Merlin: *which graph did I start from, G or H?* If $G \ncong H$, then Merlin, being computationally unbounded, can easily answer this challenge by solving graph isomorphism. If, on the other hand, $G \cong H$, then Merlin sees the same distribution over graphs regardless of whether Arthur started from G or H, so he must guess wrongly with probability $1/2$.

Despite such protocols, the feeling in the late 1980s was that IP should be only a "slight" extension of NP. This feeling was buttressed by a result of Fortnow and Sipser [83], which said that there exists an oracle A such that $\mathsf{coNP}^A \not\subset \mathsf{IP}^A$, and hence, any interactive protocol even for coNP (e.g., for proving Boolean formulas unsatisfiable) would require non-relativizing techniques.

Yet in the teeth of that oracle result, Lund et al. [151] showed nevertheless that $\mathsf{coNP} \subseteq \mathsf{IP}$ "in the real world"—and not only that, but $\mathsf{P}^{\#\mathsf{P}} \subseteq \mathsf{IP}$. This was quickly improved by Shamir [200] to the following striking statement:

Theorem 49 ([151, 200]). $\mathsf{IP} = \mathsf{PSPACE}$.

Theorem 49 means, for example, that if a computationally-unbounded alien came to Earth, it could not merely beat us in games of strategy like chess—rather, the

[37]We can assume without loss of generality that Merlin's strategy is deterministic, since Merlin is computationally unbounded, and any convex combination of strategies must contain a deterministic strategy that causes Arthur to accept with at least as great a probability as the convex combination does.

[38]This is so because a polynomial-space Turing machine can treat the entire interaction between Merlin and Arthur as a game, in which Merlin is trying to get Arthur to accept with the largest possible probability. The machine can then evaluate the exponentially large game tree using depth-first recursion.

alien could mathematically prove to us, via a short conversation and to statistical certainty, that it knew how to play *perfect* chess. Theorem 49 has been hugely influential in complexity theory for several reasons, but one reason was that it illustrated, dramatically and indisputably, that the relativization barrier need not inhibit progress.

So how was this amazing result achieved, and why does the proof *fail* relative to certain oracles? The trick is what we now call *arithmetization*. This means that we take a Boolean formula or circuit—involving, for example, AND, OR, and NOT gates—and then reinterpret the Boolean gates as arithmetic operations over some larger finite field $\mathbb{F}_p$. More concretely, the Boolean AND ($x \wedge y$) becomes multiplication (xy), the Boolean NOT becomes the function $1 - x$, and the Boolean OR ($x \vee y$) becomes $x + y - xy$. Note that if $x, y \in \{0, 1\}$, then we recover the original Boolean operations. But the new operations make sense even if $x, y \notin \{0, 1\}$, and they have the effect of lifting our Boolean formula or circuit to a multivariate polynomial over $\mathbb{F}_p$. Furthermore, the degree of the polynomial can be upper-bounded in terms of the size of the formula or circuit.

The advantage of this lifting is that polynomials, at least over large finite fields, have powerful error-correcting properties that are unavailable in the Boolean case. These properties ultimately derive from the Fundamental Theorem of Algebra: a nonzero, degree-d univariate polynomial has at most d roots. As a consequence, if $q, q' : \mathbb{F}_p \to \mathbb{F}_p$ are two degree-d polynomials that are unequal (and $d \ll p$), then with high probability, their inequality can be seen by querying them at a random point:

$$\Pr_{x \in \mathbb{F}_p} \left[q(x) = q'(x) \right] \leq \frac{d}{p}.$$

Let me now give a brief impression of how one proves Theorem 49, or at least the simpler result $\mathsf{coNP} \subseteq \mathsf{IP}$. Let $\varphi(x_1, \ldots, x_n)$ be, say, a 3SAT formula that Merlin wants to convince Arthur is unsatisfiable. Then Arthur first lifts φ to a multivariate polynomial $q : \mathbb{F}_p^n \to \mathbb{F}_p$, of degree $d \leq |\varphi|$ (where $|\varphi|$ is the size of φ), over the finite field $\mathbb{F}_p$, for some $p \gg 2^n$. Merlin's task is equivalent to convincing Arthur of the following equation:

$$\sum_{x_1, \ldots, x_n \in \{0,1\}} q(x_1, \ldots, x_n) = 0.$$

To achieve this, Merlin first sends Arthur the coefficients of a univariate polynomial $q_1 : \mathbb{F}_p \to \mathbb{F}_p$. Merlin claims that q_1 satisfies

$$q_1(x_1) = \sum_{x_2, \ldots, x_n \in \{0,1\}} q(x_1, x_2, \ldots, x_n), \tag{1}$$

and also satisfies $q_1(0) + q_1(1) = 0$. Arthur can easily check the latter equation for himself. To check Eq. (1), Arthur picks a random value $r_1 \in \mathbb{F}_p$ for x_1 and sends it to Merlin. Then Merlin replies with a univariate polynomial q_2, for which he claims that

$$q_2(x_2) = \sum_{x_3,\ldots,x_n \in \{0,1\}} q(r_1, x_2, x_3, \ldots, x_n).$$

Arthur checks that $q_2(0) + q_2(1) = q_1(r_1)$, then picks a random value $r_2 \in \mathbb{F}_p$ for x_2 and sends it to Merlin, and so on. Finally, Arthur checks that q_n is indeed the univariate polynomial obtained by starting from the arithmetization of φ, then fixing $x_1, \ldots, x_{n-1}$ to $r_1, \ldots, r_{n-1}$ respectively. The Fundamental Theorem of Algebra ensures that, if Merlin lied at any point in the protocol, then with high probability at least one of Arthur's checks will fail.

Now, to return to the question that interests us: why does this protocol escape the relativization barrier? The short answer is: because if the Boolean formula φ involved oracle gates, then we wouldn't have been able to arithmetize φ. By arithmetizing φ, we did something "deeper" with it, more dependent on its structure, than simply evaluating φ on various Boolean inputs (which would have continued to work fine had an oracle been involved). Arithmetization made sense because φ was built out of AND and OR and NOT gates, which we were able to reinterpret arithmetically. But how would we arithmetically reinterpret an oracle gate?

6.3.2 Hybrid Circuit Lower Bounds

To recap, $\mathsf{PSPACE} \subseteq \mathsf{IP}$ is a non-relativizing inclusion of complexity classes. But can we leverage that achievement to prove non-relativizing *separations* between complexity classes, with an eye toward $\mathsf{P} \neq \mathsf{NP}$? Certainly, by combining $\mathsf{IP} = \mathsf{PSPACE}$ with the Space Hierarchy Theorem (which implies $\mathsf{SPACE}(n^k) \neq \mathsf{PSPACE}$ for every fixed k), we get that $\mathsf{IP} \not\subset \mathsf{SPACE}(n^k)$ for every fixed k. Likewise, by combining $\mathsf{IP} = \mathsf{PSPACE}$ with Theorem 38 (that PSPACE does not have circuits of size n^k for fixed k), we get that IP doesn't have circuits of size n^k either. Furthermore, both of these separations can be shown to be non-relativizing, using techniques from [56]. But can we get more interesting separations?

The key to doing so turns out to be a beautiful corollary of the $\mathsf{IP} = \mathsf{PSPACE}$ theorem. To state the corollary, we need one more complexity class: MA (Merlin-Arthur) is a probabilistic generalization of NP. It's defined as the class of languages $L \subseteq \{0,1\}^*$ for which there exists a probabilistic polynomial-time verifier M, and a polynomial p, such that for all inputs $x \in \{0,1\}^*$:

- If $x \in L$ then there exists a witness string $w \in \{0,1\}^{p(|x|)}$ such that $M(x, w)$ accepts with probability at least $2/3$ over its internal randomness.
- If $x \notin L$, then $M(x, w)$ rejects with probability at least $2/3$ over its internal randomness, for all w.

Clearly MA contains NP and BPP. It can also be shown that $\mathsf{MA} \subseteq \Sigma_2^{\mathsf{P}} \cap \Pi_2^{\mathsf{P}}$ and that $\mathsf{MA} \subseteq \mathsf{PP}$, where PP is the counting class from Sect. 2.2.6. Now, here's the corollary of Theorem 49:

Corollary 50. *If* $\mathsf{PSPACE} \subset \mathsf{P/poly}$, *then* $\mathsf{PSPACE} = \mathsf{MA}$.

Proof. Suppose $\mathsf{PSPACE} \subset \mathsf{P/poly}$, let $L \in \mathsf{PSPACE}$, and let x be an input in L. Then as an MA witness proving that $x \in L$, Merlin simply sends Arthur a description of a polynomial-size circuit C that simulates the PSPACE prover, in an interactive protocol that convinces Arthur that $x \in L$. (Here we use one additional fact about Theorem 49, beyond the mere fact that $\mathsf{IP} = \mathsf{PSPACE}$: that, in the protocol, Merlin can run a PSPACE algorithm to decide which message to send next.) Then Arthur simulates the protocol, using C to compute Merlin's responses to his random challenges, and accepts if and only if the protocol does. Hence $L \in \mathsf{MA}$. ■

Likewise:

Corollary 51. *If* $\mathsf{P}^{\#\mathsf{P}} \subset \mathsf{P/poly}$, *then* $\mathsf{P}^{\#\mathsf{P}} = \mathsf{MA}$.

(Again, here we use the observation that, in the protocol proving that $\mathsf{P}^{\#\mathsf{P}} \subseteq \mathsf{IP}$, Merlin can run a $\mathsf{P}^{\#\mathsf{P}}$ algorithm to decide which message to send next.)

Let's now see how we can use these corollaries of $\mathsf{IP} = \mathsf{PSPACE}$ to prove new circuit lower bounds. Let $\mathsf{MA}_{\mathsf{EXP}}$ be "the exponential-time version of MA," with a $2^{p(n)}$-size witness that can be probabilistically verified in $2^{p(n)}$ time: in other words, the class that is to MA as NEXP is to NP. Then:

Theorem 52 (Buhrman-Fortnow-Thierauf [56]). $\mathsf{MA}_{\mathsf{EXP}} \not\subset \mathsf{P/poly}$.

Proof. Suppose by contradiction that $\mathsf{MA}_{\mathsf{EXP}} \subset \mathsf{P/poly}$. Then certainly $\mathsf{PSPACE} \subset \mathsf{P/poly}$, which means that $\mathsf{PSPACE} = \mathsf{MA}$ by Corollary 50. By a padding argument (see Proposition 17), this means that $\mathsf{EXPSPACE} = \mathsf{MA}_{\mathsf{EXP}}$. But we already saw in Theorem 37 that $\mathsf{EXPSPACE} \not\subset \mathsf{P/poly}$, and therefore $\mathsf{MA}_{\mathsf{EXP}} \not\subset \mathsf{P/poly}$ as well. ■

Note in particular that if we could prove $\mathsf{MA} = \mathsf{NP}$, then we would also have $\mathsf{MA}_{\mathsf{EXP}} = \mathsf{NEXP}$ by padding, and hence $\mathsf{NEXP} \not\subset \mathsf{P/poly}$ by Theorem 52. This provides another example of how derandomization can lead to circuit lower bounds, a theme mentioned in Sect. 6.

A second example involves the class PP.

Theorem 53 (Vinodchandran [231]). *For every fixed k, there is a language in* PP *that does not have circuits of size* n^k.

Proof. Fix k, and suppose by contradiction that PP has circuits of size n^k. Then in particular, $\mathsf{PP} \subset \mathsf{P/poly}$, so $\mathsf{P}^{\mathsf{PP}} = \mathsf{P}^{\#\mathsf{P}} \subset \mathsf{P/poly}$, so $\mathsf{P}^{\#\mathsf{P}} = \mathsf{PP} = \mathsf{MA}$ by Corollary 51. But we noted in Sect. 6.1 that Σ_2^{P} does not have circuits of size n^k.

And $\Sigma_2^P \subseteq P^{\#P}$ by Toda's Theorem (Theorem 13), so $P^{\#P}$ doesn't have circuits of size n^k either. Therefore neither does PP.[39] ■

As a final example, Santhanam [195] showed the following (we omit the proof).

Theorem 54 (Santhanam [195]). *For every fixed k, there is an* MA *"promise problem"*[40] *that does not have circuits of size n^k.*

The above results clearly evade the natural proofs barrier, because they give lower bounds against strong circuit classes such as P/poly, or the set of all size-n^k circuits for fixed k. This is not so surprising when we observe that the proofs build on the simpler results from Sect. 6.1, which already used diagonalization to evade the natural proofs barrier.

What is more interesting is that these results *also* evade the relativization barrier. Of course, one might guess as much, after noticing that the proofs use the non-relativizing IP = PSPACE theorem. But to show rigorously that the circuit lower bounds *themselves* fail to relativize, one needs to construct oracles relative to which the circuit lower bounds are false. This is done by the following results, whose somewhat elaborate proofs we omit:

Theorem 55 (Buhrman-Fortnow-Thierauf [56]). *There exists an oracle A such that* $MA_{EXP}^A \subset P^A/poly$.

Theorem 56 (Aaronson [4]). *There exists an oracle A relative to which all languages in* PP *have linear-sized circuits.*

The proofs of both of these results also easily imply that there exists an oracle relative to which all MA promise problems have linear-sized circuits.

The bottom line is that, by combining non-relativizing results like IP = PSPACE with non-naturalizing results like EXPSPACE $\not\subset$ P/poly, we can prove interesting circuit lower bounds that neither relativize *nor* naturalize. So then why couldn't we keep going, and use similar techniques to prove NEXP $\not\subset$ P/poly, or even P $\neq$ NP? Is there a third barrier, to which even the arithmetization-based lower bounds are subject?

---

[39] Actually, for this proof one does not really need either Toda's Theorem, *or* the slightly-nontrivial result that Σ_2^P does not have circuits of size n^k. Instead, one can just argue directly that at any rate, $P^{\#P}$ does not have circuits of size n^k, using a slightly more careful version of the argument of Theorem 37. For details see Aaronson [4].

[40] In complexity theory, a *promise problem* is a pair of subsets $\Pi_{YES}, \Pi_{NO} \subseteq \{0,1\}^*$ with $\Pi_{YES} \cap \Pi_{NO} = \varnothing$. An algorithm solves the problem if it accepts all inputs in Π_{YES} and rejects all inputs in Π_{NO}. Its behavior on inputs neither in Π_{YES} nor Π_{NO} (i.e., inputs that "violate the promise") can be arbitrary. A typical example of a promise problem is: given a Boolean circuit C, decide whether C accepts at least $2/3$ of all inputs $x \in \{0,1\}^n$ or at most $1/3$ of them, promised that one of those is true. This problem is in BPP (or technically, PromiseBPP). The role of the promise here is to get rid of those inputs for which random sampling would accept with probability between $1/3$ and $2/3$, violating the definition of BPP.

6.3.3 The Algebrization Barrier

In 2008, Avi Wigderson and I [10] showed that, alas, there is a third barrier. In particular, while the arithmetic techniques used to prove $\mathsf{IP} = \mathsf{PSPACE}$ do evade relativization, they crash up against a modified version of relativization that is "wise" to those techniques. We called this modified barrier the *algebraic relativization* or *algebrization* barrier. We then showed that, in order to prove $\mathsf{P} \neq \mathsf{NP}$—or for that matter, even to prove $\mathsf{NEXP} \not\subset \mathsf{P/poly}$, or otherwise go even slightly beyond the results of Sect. 6.3.2—one would need techniques that evade the algebrization barrier (and *also*, of course, evade natural proofs).

In more detail, we can think of an oracle as just an infinite collection of Boolean functions, $f_n : \{0,1\}^n \to \{0,1\}$ for each n. Now, by an *algebraic oracle*, we mean an oracle that provides access not only to f_n for each n, but also to a low-degree extension $\widetilde{f_n} : \mathbb{F}^n \to \mathbb{F}$ of f_n over some large finite field $\mathbb{F}$. This extension must have the property that $\widetilde{f_n}(x) = f_n(x)$ for all $x \in \{0,1\}^n$, and it must be a polynomial of low degree—say, at most $2n$. But such extensions always exist, and querying them *outside* the Boolean cube $\{0,1\}^n$ might help even for learning about the Boolean part f_n.

The point of algebraic oracles is that they capture what we could do if we had a formula or circuit for f_n, and were willing to evaluate it not only on Boolean inputs, but on non-Boolean ones as well, in the manner of $\mathsf{IP} = \mathsf{PSPACE}$. In particular, we saw in Sect. 6.3.1 that, given (say) a 3SAT formula φ, we can "lift" φ to a low-degree polynomial $\tilde{\varphi}$ over a finite field $\mathbb{F}$ by reinterpreting the AND, OR, and NOT gates in terms of field addition and multiplication. So if we're trying to capture to power of arithmetization relative to an oracle function f_n, then it stands to reason that we should also be allowed to lift f_n.

Once we do so, we find that the non-relativizing results based on arithmetization, such as $\mathsf{IP} = \mathsf{PSPACE}$, *relativize with respect to algebraic oracles* (or "algebrize"). That is:

Theorem 57. $\mathsf{IP}^{\widetilde{A}} = \mathsf{PSPACE}^{\widetilde{A}}$ *for all algebraic oracles* $\widetilde{A}$*. Likewise,* $\mathsf{PSPACE}^{\widetilde{A}} \subset \mathsf{P}^{\widetilde{A}}/\mathsf{poly}$ *implies* $\mathsf{PSPACE}^{\widetilde{A}} = \mathsf{MA}^{\widetilde{A}}$ *for all algebraic oracles* $\widetilde{A}$*, and so on for all the interactive proof results.*

The intuitive reason is that, any time (say) Arthur needs to arithmetize a formula φ containing A-oracle gates in an interactive protocol, he can handle non-Boolean inputs to the A-oracle gates by calling $\widetilde{A}$.

As a consequence of Theorem 57, the circuit lower bounds of Sect. 6.3.2 are algebrizing as well: for example, for all algebraic oracles $\widetilde{A}$, we have $\mathsf{MA}_{\mathsf{EXP}}^{\widetilde{A}} \not\subset \mathsf{P}^{\widetilde{A}}/\mathsf{poly}$, and $\mathsf{PP}^{\widetilde{A}}$ does not have size-n^k circuits with $\widetilde{A}$-oracle gates.

Admittedly, the original paper of Aaronson and Wigderson [10] only managed to prove a weaker version of Theorem 57. It showed, for example, that for all algebraic oracles $\widetilde{A}$, we have $\mathsf{PSPACE}^{A} \subseteq \mathsf{IP}^{\widetilde{A}}$, and $\mathsf{MA}_{\mathsf{EXP}}^{\widetilde{A}} \not\subset \mathsf{P}^{A}/\mathsf{poly}$. As a result, it had to define algebrization in a convoluted way, where some complexity classes received the algebraic oracle $\widetilde{A}$ while others received only the "original" oracle A, and which received which depended on what kind of result one was talking about (e.g., an inclusion or a separation). Shortly afterward, Impagliazzo, Kabanets, and Kolokolova [112] managed to fix this defect of algebrization, proving Theorem 57 even when all classes receive the same algebraic oracle $\widetilde{A}$, but only at the cost of jettisoning Aaronson and Wigderson's conclusion that any proof of $\mathsf{NEXP} \not\subset \mathsf{P/poly}$ will require non-algebrizing techniques. Very recently, Aydınlıoğlu and Bach [174] showed how to get the best of both worlds, with a uniform definition of algebrization *and* the conclusion about NEXP vs. P/poly.

In any case, the main point of [10] was that to prove $\mathsf{P} \neq \mathsf{NP}$, or otherwise go further than the circuit lower bounds of Sect. 6.3.2, one will need non-algebrizing techniques: techniques that fail to relativize in a "deeper" way than $\mathsf{IP} = \mathsf{PSPACE}$ fails to relativize. Let us see why this is true for $\mathsf{P} \neq \mathsf{NP}$.

Theorem 58 (Aaronson-Wigderson [10]). *There exists an algebraic oracle* $\widetilde{A}$ *such that* $\mathsf{P}^{\widetilde{A}} = \mathsf{NP}^{\widetilde{A}}$. *As a consequence, any proof of* $\mathsf{P} \neq \mathsf{NP}$ *will require non-algebrizing techniques.*

Proof. We can just let A be any PSPACE-complete language, and then let $\widetilde{A}$ be its unique extension to a collection of multilinear polynomials over $\mathbb{F}$ (that is, polynomials in which no variable is ever raised to a higher power than 1). The key observation is that the multilinear extensions are themselves computable in PSPACE. So we get a PSPACE-complete oracle $\widetilde{A}$, which collapses P and NP for the same reason as in the original argument of Baker et al. [34] (see Theorem 40). ■

Likewise, Aaronson and Wigderson [10] showed that any proof of $\mathsf{P} = \mathsf{NP}$, or even $\mathsf{P} = \mathsf{BPP}$, would need non-algebrizing techniques. They also proved the following somewhat harder result, whose proof we omit.

Theorem 59 ([10]). *There exists an algebraic oracle* $\widetilde{A}$ *such that* $\mathsf{NEXP}^{\widetilde{A}} \subset \mathsf{P}^{\widetilde{A}}/\mathsf{poly}$. *As a consequence, any proof of* $\mathsf{NEXP} \not\subset \mathsf{P/poly}$ *will require non-algebrizing techniques.*

Note that this explains almost exactly why progress stopped where it did: $\mathsf{MA}_{\mathsf{EXP}} \not\subset \mathsf{P/poly}$ can be proved with algebrizing techniques, but $\mathsf{NEXP} \not\subset \mathsf{P/poly}$ cannot be.

I should mention that Impagliazzo et al. [112] gave a logical interpretation of algebrization, extending the logical interpretation of relativization given by Arora et al. [28]. In particular, Impagliazzo et al. show that the algebrizing statements can be seen as all those statements that follow from "algebrizing axioms for computation," which include basic closure properties, *and also* the ability to lift any Boolean computation to a larger finite field. Statements like $\mathsf{P} \neq \mathsf{NP}$ are then provably independent of the algebrizing axioms.

6.4 *Ironic Complexity Theory*

There is one technique that has had some striking recent successes in proving circuit lower bounds, and that bypasses the natural proofs, relativization, *and* algebrization barriers. This technique might be called "ironic complexity theory." It uses the existence of surprising algorithms in one setting to show the *non*existence of algorithms in another setting. It thus reveals a "duality" between upper and lower bounds, and reduces the problem of proving impossibility theorems to the much better-understood task of designing efficient algorithms.[41]

At a conceptual level, it is not hard to see how algorithms can lead to lower bounds. For example, suppose someone discovered a way to verify arbitrary exponential-time computations efficiently, thereby proving $\mathsf{NP} = \mathsf{EXP}$. Then as an immediate consequence of the Time Hierarchy Theorem ($\mathsf{P} \neq \mathsf{EXP}$), we would get $\mathsf{P} \neq \mathsf{NP}$. As another example, suppose someone discovered that every language in P had linear-size circuits. Then $\mathsf{P} = \mathsf{NP}$ would imply that every language in PH had linear-size circuits—but since we know that is not the case (see Sect. 6.1), we could again conclude that $\mathsf{P} \neq \mathsf{NP}$. Conversely, if someone proved $\mathsf{P} = \mathsf{NP}$, that wouldn't be a total disaster for lower bounds research: at least it would immediately imply $\mathsf{EXP} \not\subset \mathsf{P/poly}$ (via $\mathsf{EXP} = \mathsf{EXP}^{\mathsf{NP}^{\mathsf{NP}}}$), and the existence of languages of P and NP that don't have linear-size circuits!

Examples like this can be multiplied, but there is an obvious problem with them: they each show a separation, but only assuming a collapse that is considered extremely unlikely to happen. However, recently researchers have managed to use surprising algorithms that *do* exist, and collapses that *do* happen, to achieve new lower bounds. In this section I'll give two examples.

6.4.1 Time-Space Tradeoffs

At the moment, no one can prove that solving 3SAT requires more than linear time (let alone exponential time!), on realistic models of computation like random-access machines.[42] Nor can anyone prove that solving 3SAT requires more than $O(\log n)$ bits of memory. But the situation is not completely hopeless: at least we can prove there's no algorithm for 3SAT that uses both linear time *and* logarithmic memory! Indeed, we can do better than that.

A bit of background: just as one can scale PSPACE up to $\mathsf{EXPSPACE}$ and so on, one can also scale PSPACE down to $\mathsf{LOGSPACE}$, which the class of languages L decidable by a Turing machine that uses only $O(\log n)$ bits of read/write

---

[41] Indeed, the hybrid circuit lower bounds of Sect. 6.3.2 could already be considered examples of ironic complexity theory. In this section, we discuss other examples.

[42] On unrealistic models such as one-tape Turing machines, one can prove up to $\Omega(n^2)$ lower bounds for 3SAT and many other problems (even recognizing palindromes), but only by exploiting the fact that the tape head needs to waste a lot of time moving back and forth across the input.

memory, in addition to a read-only memory that stores the n-bit input itself. We have $\mathsf{LOGSPACE} \subseteq \mathsf{P}$, for the same simple reason why $\mathsf{PSPACE} \subseteq \mathsf{EXP}$ (see Proposition 15). We also have $\mathsf{LOGSPACE} \neq \mathsf{PSPACE}$ by the Space Hierarchy Theorem. On the other hand, no one has proven even that $\mathsf{LOGSPACE} \neq \mathsf{NP}$.

Now, a "time-space tradeoff theorem" shows that any algorithm to solve some problem must use *either* more than T time or else more than S space. Let me state perhaps the canonical time-space tradeoff theorem for 3SAT (though it's since been improved):

Theorem 60 (Lipton-Viglas [149]). *No random-access machine can solve* 3SAT *simultaneously in* $n^{\sqrt{2}-\varepsilon}$ *time and* $n^{o(1)}$ *space, for any* $\varepsilon > 0$.

Here, a "random-access machine" means a machine that can access an arbitrary memory location in $O(1)$ time, as usual in practical programming. This makes Theorem 60 *stronger* than one might have assumed: it holds not merely for unrealistically weak models such as Turing machines, but for realistic models as well. Also, again, "$n^{o(1)}$ space" means that we get the n-bit 3SAT instance itself in a read-only memory, and also get $n^{o(1)}$ bits of auxiliary read/write memory.

While Theorem 60 is obviously a far cry from $\mathsf{P} \neq \mathsf{NP}$, it does rely essentially on 3SAT being NP-complete: we don't yet know how to prove analogous results for matching, linear programming, or other natural problems in P.[43] This makes Theorem 60 fundamentally different from (say) the PARITY $\notin \mathsf{AC}^0$ result of Sect. 6.2.3.

Let $\mathsf{DTISP}(T, S)$ be the class of languages decidable by an algorithm, running on a RAM machine, that uses $O(T)$ time and $O(S)$ space. Then Theorem 60 can be stated more succinctly as

$$\text{3Sat} \notin \mathsf{DTISP}\left(n^{\sqrt{2}-\varepsilon}, n^{o(1)}\right)$$

for all $\varepsilon > 0$.

At a high level, Theorem 60 is proved by assuming the opposite, and then deriving stranger and stranger consequences until one ultimately gets a contradiction with the Nondeterministic Time Hierarchy Theorem (Theorem 35). There are three main ideas that go into how one does this. The first idea is a tight version of the Cook-Levin Theorem (Theorem 2). In particular, one can show, not merely that 3SAT is NP-complete, but that 3SAT is complete for $\mathsf{NTIME}(n)$ (that is, nondeterministic

[43] On the other hand, by proving size-depth tradeoffs for so-called *branching programs*, researchers have been able to obtain time-space tradeoffs for certain special problems in P. Unlike the 3SAT tradeoffs, the branching program tradeoffs involve only *slightly* superlinear time bounds; on the other hand, they really do represent a fundamentally different way to prove time-space tradeoffs, one that makes no appeal to NP-completeness, diagonalization, or hierarchy theorems. As one example, in 2000 Beame et al. [37], building on earlier work by Ajtai [17], used branching programs to prove the following: there exists a problem in P, based on binary quadratic forms, for which any RAM algorithm (even a nonuniform one) that uses $n^{1-\Omega(1)}$ space must also use $\Omega\left(n \cdot \sqrt{\log n / \log\log n}\right)$ time.

linear-time on a RAM machine) under nearly linear-time reductions. That means that, to prove Theorem 60, it suffices to prove a non-containment of complexity classes:

$$\mathsf{NTIME}(n) \not\subset \mathsf{DTISP}\left(n^{\sqrt{2}-\varepsilon}, n^{o(1)}\right)$$

for all $\varepsilon > 0$.

The second idea is called "trading time for alternations." Consider a deterministic computation that runs for T steps and uses S bits of memory. Then we can "chop the computation up" into k blocks, $B_1, \ldots, B_k$, of T/k steps each. The statement that the computation accepts is then equivalent to the statement that *there exist* S-bit strings $x_0, \ldots, x_k$, such that

(i) x_0 is the computation's initial state,
(ii) *for all* $i \in \{1, \ldots, k\}$, the result of starting in state x_{i-1} and then running for T/k steps is x_i, and
(iii) x_k is an accepting state.

We can summarize this as

$$\mathsf{DTISP}(T, S) \subseteq \Sigma_2\mathsf{TIME}\left(Sk + \frac{T}{k}\right),$$

where the Σ_2 means that we have two alternating quantifiers: an existential quantifier over $x_1, \ldots, x_k$, followed by a universal quantifier over i. Choosing $k := \sqrt{T/S}$ to optimize the bound then gives us

$$\mathsf{DTISP}(T, S) \subseteq \Sigma_2\mathsf{TIME}\left(\sqrt{TS}\right).$$

So in particular,

$$\mathsf{DTISP}\left(n^c, n^{o(1)}\right) \subseteq \Sigma_2\mathsf{TIME}\left(n^{c/2+o(1)}\right).$$

The third idea is called "trading alternations for time." If we assume by way of contradiction that

$$\mathsf{NTIME}(n) \subseteq \mathsf{DTISP}\left(n^c, n^{o(1)}\right) \subseteq \mathsf{TIME}(n^c),$$

then in particular, for all $b \geq 1$, we can remove the inner quantifier to get

$$\Sigma_2\mathsf{TIME}\left(n^b\right) \subseteq \mathsf{NTIME}\left(n^{bc}\right).$$

So putting everything together, if we consider a constant $c > 1$, and use padding (as in Proposition 17) to talk about $\mathsf{NTIME}(n^2)$ rather than $\mathsf{NTIME}(n)$, then the starting assumption that 3SAT is solvable in $n^{c-\varepsilon}$ time and $n^{o(1)}$ space implies that

$$\begin{aligned}\mathsf{NTIME}\left(n^2\right) &\subseteq \mathsf{DTISP}\left(n^{2c}, n^{o(1)}\right) \\ &\subseteq \Sigma_2\mathsf{TIME}\left(n^{c+o(1)}\right) \\ &\subseteq \mathsf{NTIME}\left(n^{c^2+o(1)}\right).\end{aligned}$$

But if $c^2 < 2$, then this contradicts the Nondeterministic Time Hierarchy Theorem (Theorem 35). This completes the proof of Theorem 60. Notice that the starting hypothesis about 3SAT was applied not once but twice, which was how the final running time became n^{c^2}.

Later, using a more involved argument, Fortnow and van Melkebeek [81] improved Theorem 60, to show that 3SAT can't be solved by a RAM machine using $n^{\phi-\varepsilon}$ time and $n^{o(1)}$ space, where $\phi = \frac{1+\sqrt{5}}{2} \approx 1.618$ is the golden ratio. Subsequently Williams [234] improved the time bound in this result to $n^{\sqrt{3}-\varepsilon}$, and then [235] to $n^{2\cos\pi/7-\varepsilon}$, with the help of computer search for the optimal sequence of "moves" in an alternation-trading argument. In 2012, however, Buss and Williams [63] showed that no alternation-trading proof can possibly improve that exponent beyond the peculiar constant $2\cos\pi/7 \approx 1.801$. There have been many related time-space tradeoff results, including for #P-complete and PSPACE-complete problems, but I won't cover them here (see van Melkebeek [152] for a survey).

Alternation-trading has had other applications in complexity theory, other than to time-space tradeoffs. In particular, it played a key role in a celebrated 1983 result of Paul et al. [177], whose statement is tantalizingly similar to $\mathsf{P} \neq \mathsf{NP}$.

Theorem 61 (Paul et al. [177]). $\mathsf{TIME}(n) \neq \mathsf{NTIME}(n)$, *if we define these classes using multiple-tape Turing machines.*

In this case, the key step was to show, via a clever combinatorial argument involving "pebble games," that for multi-tape Turing machines, deterministic linear time can be simulated in $\Sigma_4\mathsf{TIME}(f(n))$, for some f that's *slightly* sublinear. This, combined with the assumption $\mathsf{TIME}(n) = \mathsf{NTIME}(n)$, is then enough to produce a contradiction with a time hierarchy theorem.

What can we say about barriers? All the results mentioned above clearly evade the natural proofs barrier, because they ultimately rely on diagonalization, and (more to the point) because classes like $\mathsf{TIME}(n)$ and $\mathsf{DTISP}\left(n^{\sqrt{2}}, n^{o(1)}\right)$ contain plausible pseudorandom function candidates. Whether they evade the relativization barrier (let alone algebrization) is a trickier question; it depends on subtle details of the oracle access mechanism. There are some definitions of the classes $\mathsf{TIME}(n)^A$, $\mathsf{DTISP}(T, S)^A$, and so on under which these results relativize, and others under which they don't: for details, see for example Moran [156].

On the definitions that cause these results *not* to relativize, the explanation for how is that the proofs "look inside" the operations of a RAM machine or a multi-tape Turing machine *just enough* for something to break down if certain kinds of oracle calls are present. To illustrate, in the proof of Theorem 60 above, we nondeterministically guessed the complete state of the machine at various steps in its execution, taking advantage of the fact that the state was an $n^{o(1)}$-bit string. This wouldn't have worked had there been an n-bit query written onto an oracle tape (even if the oracle tape were write-only). Likewise, in the proof of Theorem 61, the combinatorial pebble arguments use specific properties of multi-tape Turing machines that might fail for RAM machines, let alone for oracle machines.

Because their reasons for failing to relativize have nothing to do with lifting to large finite fields, I conjecture that, with a suitable oracle access mechanism, Theorems 60 and 61 would also be non-algebrizing. But this remains to be shown.

6.4.2 $\mathsf{NEXP} \not\subset \mathsf{ACC}$

In Sect. 6.2.4, we saw how Smolensky [209] and Razborov [189] managed to prove strong lower bounds against the class $\mathsf{AC}^0[p]$, or constant-depth, polynomial-size circuits of AND, OR, NOT, and MOD p gates, where p is a prime. This left the frontier of circuit lower bounds as $\mathsf{AC}^0[m]$, where m is a composite. Slightly more ambitiously, we could hope for lower bounds against a complexity class called ACC, which consists of all languages decidable by constant-depth, polynomial-size circuits with AND, OR, NOT, and MOD m gates for *any* m (where we can mix multiple m's in the same circuit).

Meanwhile, we saw in Sect. 6.3 how Buhrman et al. [56] proved that $\mathsf{MA_{EXP}} \not\subset \mathsf{P/poly}$, but how this cannot be extended even to $\mathsf{NEXP} \not\subset \mathsf{P/poly}$ using algebrizing techniques. Indeed, it remains open even to prove $\mathsf{NEXP} \not\subset \mathsf{TC}^0$.

This state of affairs—and its continuation for decades—helps to explain why many theoretical computer scientists were electrified when Ryan Williams proved the following in 2011.

Theorem 62 (Williams [242]). $\mathsf{NEXP} \not\subset \mathsf{ACC}$.

If we compare it against the ultimate goal of proving $\mathsf{NP} \not\subset \mathsf{P/poly}$, Theorem 62 looks almost laughably weak: it shows only that Nondeterministic Exponential Time, a class vastly larger than NP, is not in ACC, a circuit class vastly smaller than P/poly. But a better comparison is against where we were before. The proof of Theorem 62 was noteworthy not only because it defeats all the known barriers (relativization, algebrization, and natural proofs), but also because it brings together almost *all* known techniques in Boolean circuit lower bounds, including diagonalization, the polynomial method, interactive proof results, and ironic complexity theory. So it is worth at least sketching the elaborate proof, so we can see how a lower bound at the current frontier operates. (For further details, I recommend two excellent expository articles by Williams himself [236, 240].)

At a stratospherically high level, the proof of Theorem 62 is built around the Nondeterministic Time Hierarchy Theorem, following a program that Williams had previously laid out in [238]. More concretely, we assume that $\mathsf{NTIME}(2^n) \subset \mathsf{ACC}$.

We then use that assumption to show that $\mathsf{NTIME}(2^n) = \mathsf{NTIME}(2^n/n^k)$ for some positive k: a slight speedup of nondeterministic machines, but enough to achieve a contradiction with Theorem 35.

How do we use the assumption $\mathsf{NTIME}(2^n) \subset \mathsf{ACC}$ to violate the Nondeterministic Time Hierarchy Theorem? The key to this—and this is where "ironic complexity theory" enters the story—is a faster-than-brute-force algorithm for a problem called ACCSAT. Here we are given as input a description of ACC circuit C, and want to decide whether there exists an input $x \in \{0,1\}^n$ such that $C(x) = 1$. The core of Williams's proof is the following straightforwardly algorithmic result.

Lemma 63 (Williams [242]). *There is a deterministic algorithm that solves* ACCSAT, *for* ACC *circuits of depth d with n inputs, in $2^{n-\Omega(n^\delta)}$ time, for some constant $\delta > 0$ that depends on d.*

The proof of Lemma 63 is itself a combination of several ideas. First, one appeals to a powerful structural result of Yao [246], Beigel-Tarui [38], and Allender-Gore [23] from the 1990s, which shows that functions in ACC are representable in terms of low-degree polynomials.

Lemma 64 ([23, 38, 246]). *Let $f : \{0,1\}^n \to \{0,1\}$ be computable by an* ACC *circuit of size s and depth d. Then $f(x)$ can be expressed as $g(p(x))$, where $p : \{0,1\}^n \to \mathbb{N}$ is a polynomial of degree $\log^{O(1)} s$ that is a sum of $\exp\left(\log^{O(1)} s\right)$ monomials with coefficients of* 1, *and $g : \mathbb{N} \to \{0,1\}$ is some efficiently computable function. (Here the constant in the big-O depends on d.)*

The proof of Lemma 64 uses some elementary number theory, and is closely related to the polynomial method from Sect. 6.2.4, by which one shows that any $\mathsf{AC}^0[p]$ function can be approximated by a low-degree polynomial over the finite field $\mathbb{F}_p$.[44]

Next, one devises a faster-than-brute-force algorithm that, given a function $g(p(x))$ as above, decides whether there exists an $x \in \{0,1\}^n$ such that $g(p(x)) = 1$. The first step is to give an algorithm that constructs a table of all 2^n values of $p(x)$, for all the 2^n possible values of x, in $\left(2^n + s^{O(1)}\right) n^{O(1)}$ time, rather than the $O(2^n s)$ time that one would need naïvely. (In other words, this algorithm uses only $n^{O(1)}$ time on average per entry in the table, rather than $O(s)$ time—an improvement if s is superpolynomial.) Here there are several ways to go: one can use a fast rectangular matrix multiplication algorithm due to Coppersmith [69], but one can also just use a dynamic programming algorithm reminiscent of the Fast Fourier Transform.

Now, by combining this table-constructing algorithm with Lemma 64, we can immediately solve ACCSAT, for an ACC circuit of size $s = 2^{n^{o(1)}}$, in $2^n n^{O(1)}$ time, which is better than the $O(2^n s)$ time that we would need naïvely. However, this still isn't good enough to prove Lemma 63, which demands a $2^{n-\Omega(n^\delta)}$ algorithm. So

[44]Interestingly, both the polynomial method and the proof of Lemma 64 are also closely related to the proof of Toda's Theorem (Theorem 13), that $\mathsf{PH} \subseteq \mathsf{P}^{\#\mathsf{P}}$.

there is a further trick: given an ACC circuit C of size $n^{O(1)}$, one first "shaves off" n^{δ} of the n variables, building a new ACC circuit C' that takes as input the $n - n^{\delta}$ remaining variables, and that computes the OR over all $2^{n^{\delta}}$ possible assignments to the n^{δ} shaved variables.[45] The new circuit C' has size $2^{O(n^{\delta})}$, so one can construct the table, and thereby solve ACCSAT for C' (and hence for C), in time $2^{n-\Omega(n^{\delta})}$.

Given Lemma 63, as well as the starting assumption $\mathsf{NTIME}(2^n) \subset \mathsf{ACC}$, there is still a lot of work to do to prove that $\mathsf{NTIME}(2^n) = \mathsf{NTIME}(2^n/n^k)$. Let me summarize the four main steps:

(1) One first uses a careful, quantitative version of the Cook-Levin Theorem (Theorem 2), to reduce the problem of simulating an $\mathsf{NTIME}(2^n)$ machine to a problem called SUCCINCT3SAT. In that problem, one is given a circuit C whose truth table encodes an exponentially large 3SAT instance φ, and the problem is to decide whether or not φ is satisfiable.
(2) One next appeals to a result of Impagliazzo, Kabanets, and Wigderson [113], which says that if $\mathsf{NEXP} \subset \mathsf{P/poly}$, then the satisfying assignments for satisfiable SUCCINCT3SAT instances can themselves be constructed by polynomial-size circuits.
(3) One massages the result (2) to get a conclusion about ACC: very roughly speaking, if $\mathsf{NEXP} \subset \mathsf{ACC}$, then given any satisfiable SUCCINCT3SAT instance Φ, there is an equivalent instance Φ' in which the circuit C is an ACC circuit. Furthermore, a satisfying assignment for Φ' can itself be constructed by an ACC circuit W; and the problems of verifying that Φ and Φ' are equivalent, and that W does indeed encode a satisfying assignment for Φ', can be solved in slightly less than 2^n time nondeterministically, if we use the fact (Lemma 63) that ACCSAT is solvable in $2^{n-\Omega(n^{\delta})}$ time. Note that in this argument (which is the most complicated part of the proof), one uses the assumption $\mathsf{NEXP} \subset \mathsf{ACC}$ not just once but several times.
(4) Putting everything together, we get that $\mathsf{NTIME}(2^n)$ machines can be reduced to SUCCINCT3SAT instances, which can then (assuming $\mathsf{NEXP} \subset \mathsf{ACC}$, and using the ACCSAT algorithm) be decided in $\mathsf{NTIME}(2^n/n^k)$ for some positive k. But that contradicts the Nondeterministic Time Hierarchy Theorem (Theorem 35).

Let me mention some improvements and variants of Theorem 62. Already in his original paper [242], Williams noted that the proof actually yields a stronger result, that $\mathsf{NTIME}(2^n)$ has no ACC circuits of "third-exponential" size: that is, size $f(n)$ where $f(f(f(n)))$ grows exponentially. He also gave a second result, that $\mathsf{TIME}(2^n)^{\mathsf{NP}}$—that is, deterministic exponential time with an NP oracle—has no ACC circuits of size $2^{n^{o(1)}}$. More recently, Williams has extended Theorem 62 to show that $\mathsf{NTIME}(2^n)/1 \cap \mathsf{coNTIME}(2^n)/1$ (where the $/1$ denotes 1 bit of

[45]Curiously, this step can only be applied to the ACC circuits themselves, which of course allow OR gates. It cannot be applied to the Boolean functions of low-degree polynomials that one derives from the ACC circuits.

nonuniform advice) does not have ACC circuits of size $n^{\log n}$ [239], and also to show that even ACC circuits of size $n^{\log n}$ with threshold gates at the bottom layer cannot compute all languages in NEXP [241].

At this point, it's appropriate to make some general remarks about the proof of Theorem 62 and the prospects for pushing it further. First of all, why did this proof only yield lower bounds for functions in the huge complexity class NEXP, rather than EXP or NP or even P? The short answer is that, in order to prove that a class $\mathcal{C}$ is not in ACC via this approach, we need to use the assumption $\mathcal{C} \subset \mathsf{ACC}$ to violate a hierarchy theorem for $\mathcal{C}$-like classes. However, there's a bootstrapping problem: the mere fact that $\mathcal{C}$ *has* small ACC circuits doesn't imply that we can *find* those circuits in a $\mathcal{C}$-like class, in order to obtain the desired contradiction. When $\mathcal{C} = \mathsf{NEXP}$, we can use the nondeterministic guessing power of the NTIME classes simply to *guess* the small ACC circuits for NEXP, but even when $\mathcal{C} = \mathsf{EXP}$ this approach seems to break down.

A second question is: what in Williams's proof was specific to ACC? Here the answer is that the proof used special properties of ACC in one place only: namely, in the improved algorithm for ACCSAT (Lemma 63). This immediately suggests a possible program to prove $\mathsf{NEXP} \not\subset \mathcal{C}$ for larger and larger circuit classes $\mathcal{C}$. For example, let TC^0SAT be the problem where we are given as input a TC^0 circuit C (that is, a neural network, or constant-depth circuit of threshold gates), and we want to decide whether there exists an $x \in \{0,1\}^n$ such that $C(x) = 1$. Then if we could solve TC^0SAT even slightly faster than brute force—say, in $O(2^n/n^k)$ time for some positive k—Williams's results would immediately imply $\mathsf{NEXP} \not\subset \mathsf{TC}^0$.[46] Likewise, recall from Sect. 2.1 that CIRCUITSAT is the satisfiability problem for *arbitrary* Boolean circuits. If we had an $O(2^n/n^k)$ algorithm for CIRCUITSAT, then Williams's results would imply the long-sought $\mathsf{NEXP} \not\subset \mathsf{P/poly}$.

A third question is: how does the proof of Theorem 62 evade the known barriers? Because of the way the algorithm for ACCSAT exploits the structure of ACC circuits, we shouldn't be surprised if the proof evades the relativization and algebrization barriers. And indeed, using the techniques of Wilson [243] and of Aaronson and Wigderson [10], one can easily construct an oracle A such that $\mathsf{NEXP}^A \subset \mathsf{ACC}^A$, and even an algebraic oracle $\widetilde{A}$ such that $\mathsf{NEXP}^{\widetilde{A}} \subset \mathsf{ACC}^{\widetilde{A}}$, thereby showing that $\mathsf{NEXP} \not\subset \mathsf{ACC}$ is a non-relativizing and non-algebrizing result. Meanwhile, because it uses diagonalization (in the form of the Nondeterministic Time Hierarchy Theorem), we might say that the proof of Theorem 62 has the "capacity" to evade natural proofs. On the other hand, as we alluded to in Sect. 6.2.5, it's not yet clear whether ACC is powerful enough to compute pseudorandom functions—and thus, whether it even *has* a natural proofs barrier to evade! The most we can say is that *if* ACC has a natural proofs barrier, *then* Theorem 62 evades it.

---

[46] Very recently, Kane and Williams [119] managed to give an explicit Boolean function that requires depth-2 threshold circuits with $\Omega\left(n^{3/2}/\log^3 n\right)$ gates. However, their argument does not proceed via a better-than-brute-force algorithm for depth-2 TC^0SAT; the latter problem appears to remain open.

Given everything we saw in the previous sections, a final question arises: is there some fourth barrier, beyond relativization, algebrization, and natural proofs, which will inherently prevent even Williams's techniques from proving $\mathsf{P} \neq \mathsf{NP}$, or even (say) $\mathsf{NEXP} \not\subset \mathsf{TC}^0$? One reasonable answer is that this question is premature: in order to identify the barriers to a given set of techniques, we first need to know formally what the techniques *are*—i.e., what properties all the theorems using those techniques have in common—but we can't know that until the techniques have had a decade or more to solidify, and there are at least three or four successful examples of their use. Another answer is that yes, there is (or might be) an obvious "barrier" to the continuation of Williams's program. This barrier is that a faster-than-brute-force algorithm for TC^0SAT, let alone for more general problems like CIRCUITSAT, might simply not exist. If there's some threshold of expressive power in a circuit, beyond which brute-force search really *does* become the fastest possible algorithm for circuit satisfiability, then ironic complexity theory (or at least this incarnation of it) will run out of the irony that it needs as fuel.

Even if so, though, I see Theorem 62 as having contributed something important to the quest to prove $\mathsf{P} \neq \mathsf{NP}$, by demonstrating just how much nontrivial work can get done, and how many barriers can be overcome, *along the way* to applying a 1960s-style hierarchy theorem. Williams's result makes it possible to imagine that, in the far future, $\mathsf{P} \neq \mathsf{NP}$ might be proved by assuming the opposite, then deriving stranger and stranger consequences using thousands of pages of mathematics barely comprehensible to anyone alive today—and yet still, the *coup de grâce* will be a diagonalization argument, barely different from what Turing did in 1936.

6.5 Arithmetic Complexity Theory

Besides Turing machines and Boolean circuits acting on bits, there's another kind of computation that has enormous relevance to the attempt to prove $\mathsf{P} \neq \mathsf{NP}$. Namely, we can consider computer programs that operate directly on elements of a field, such as the reals or complexity. Perhaps the easiest way to do this is via *arithmetic circuits*, which take as input a collection of elements $x_1, \ldots, x_n$ of a field $\mathbb{F}$,[47] and whose operations consist of adding or multiplying any two previous elements—or any previous element and any scalar from $\mathbb{F}$—to produce a new $\mathbb{F}$-element. We then consider the minimum number of operations needed to compute some polynomial $g : \mathbb{F}^n \to \mathbb{F}$, as a function of n. For concreteness, one can think of $\mathbb{F}$ as the reals $\mathbb{R}$, although we are most interested in algorithms that work over any $\mathbb{F}$, and that compute g *as a formal polynomial*, rather than as just a function over a particular $\mathbb{F}$.[48]

---

[47] I'll restrict to fields here for simplicity, but one can also consider (e.g.) rings.

[48] To clarify, 0 and $2x$ are equal as functions over the finite field $\mathbb{F}_2$, but not equal as formal polynomials.

At first glance, arithmetic circuits seem more powerful than Boolean circuits, because they have no limit of finite precision: for example, an arithmetic circuit could multiply π and e in a single time step. From another angle, however, arithmetic circuits are weaker, because they have no facility (for example) to extract individual bits from the binary representations of the $\mathbb{F}$ elements: they can *only* manipulate them as $\mathbb{F}$ elements. In general, the most we can say is that, *if* an input has helpfully been encoded using the elements $0, 1 \in \mathbb{F}$ only, *then* an arithmetic circuit can simulate a Boolean one, by using $x \to 1 - x$ to simulate NOT, multiplication to simulate Boolean AND, and so on. But for arbitrary inputs, such a simulation might be impossible.

Thus, arithmetic circuits represent a different kind of computation: or rather, a generalization of the usual kind, since we can recover ordinary Boolean computation by setting $\mathbb{F} = \mathbb{F}_2$. A major reason to focus on arithmetic circuits is that it often seems easier—or better, less absurdly hard!—to understand circuit size in the arithmetic setting than in the Boolean one. The usual explanation given for this is the so-called "yellow books argument": arithmetic complexity brings us closer to continuous mathematics, about which we have centuries' worth of deep knowledge (e.g., algebraic geometry and representation theory) that's harder to apply in the Boolean case.

One remark: in the rest of the section, I'll talk exclusively about arithmetic *circuit* complexity: that is, about nonuniform arithmetic computations, and the arithmetic analogues of questions such as NP versus $\mathsf{P/poly}$ (see Sect. 5.2). But it's also possible to develop a theory of *arithmetic Turing machines*, which (roughly speaking) are like arithmetic circuits except that they're uniform, and therefore need loops, conditionals, memory registers, and so on. See the book of Blum, Cucker, Shub, and Smale (BCSS) [46] for a beautiful exposition of this theory. In the BCSS framework, one can ask precise analogues of the $\mathsf{P} \stackrel{?}{=} \mathsf{NP}$ question for Turing machines over arbitrary fields $\mathbb{F}$, such as $\mathbb{R}$ or $\mathbb{C}$, recovering the "ordinary, Boolean" $\mathsf{P} \stackrel{?}{=} \mathsf{NP}$ question exactly when $\mathbb{F}$ is finite. At present, no implications are known among the $\mathsf{P} \stackrel{?}{=} \mathsf{NP}$, the $\mathsf{P}_{\mathbb{R}} \stackrel{?}{=} \mathsf{NP}_{\mathbb{R}}$, and the $\mathsf{P}_{\mathbb{C}} \stackrel{?}{=} \mathsf{NP}_{\mathbb{C}}$ questions, although it's known that $\mathsf{P}_{\mathbb{C}} \neq \mathsf{NP}_{\mathbb{C}}$ implies $\mathsf{NP} \not\subset \mathsf{P/poly}$ (see for example Bürgisser [57, Chap. 8]).[49]

The problems of proving $\mathsf{P}_{\mathbb{R}} \neq \mathsf{NP}_{\mathbb{R}}$ and $\mathsf{P}_{\mathbb{C}} \neq \mathsf{NP}_{\mathbb{C}}$ are known to be closely related to the problem of proving arithmetic circuit lower bounds, which we'll discuss in the following sections. I can't resist giving one example of a connection, due to BCSS [46]. Given a positive integer n, let $\tau(n)$ be the number of operations in the smallest arithmetic circuit that takes the constant 1 as its sole input, and that computes n using additions, subtractions, and multiplications. For example, we have

[49]The central difference between the $\mathsf{P}_{\mathbb{R}} \stackrel{?}{=} \mathsf{NP}_{\mathbb{R}}$ and $\mathsf{P}_{\mathbb{C}} \stackrel{?}{=} \mathsf{NP}_{\mathbb{C}}$ questions is simply that, because $\mathbb{R}$ is an ordered field, one defines Turing machines over $\mathbb{R}$ to allow comparisons ($<, \leq$) and branching on their results.

- $\tau(2) = 1$ via $1 + 1$,
- $\tau(3) = 2$ via $1 + 1 + 1$,
- $\tau(4) = 2$ via $(1 + 1)^2, \ldots$

Also, let $\tau^*(n)$ be the minimum of $\tau(kn)$ over all positive integers k.

Theorem 65 (BCSS [46]). *Suppose $\tau^*(n!)$ grows faster than $(\log n)^{O(1)}$. Then $\mathsf{P}_{\mathbb{C}} \neq \mathsf{NP}_{\mathbb{C}}$.*[50]

6.5.1 Permanent Versus Determinant

The central problem studied in arithmetic complexity theory—if you like, the arithmetic analogue of the NP vs. P/poly problem—is the *permanent versus determinant problem*. The problem concerns the following two functions of an $n \times n$ matrix $X \in \mathbb{F}^{n \times n}$:

$$\text{Per}(X) = \sum_{\sigma \in S_n} \prod_{i=1}^{n} x_{i,\sigma(i)},$$

$$\text{Det}(X) = \sum_{\sigma \in S_n} (-1)^{\text{sgn}(\sigma)} \prod_{i=1}^{n} x_{i,\sigma(i)}.$$

Despite the similarity of their definitions—they are identical apart from the $(-1)^{\text{sgn}(\sigma)}$—the permanent and determinant have some dramatic differences. The determinant is computable in polynomial time, for example by using Gaussian elimination. (Indeed, the determinant is computable in $O(n^\omega)$ time, where $\omega \in [2, 2.373]$ is the matrix multiplication exponent; see Sect. 4.) The determinant has many other interpretations—for example, the product of X's eigenvalues, and the volume of the parallelepiped spanned by its row vectors—giving it a central role in linear algebra and geometry.

By contrast, Valiant [226] proved in 1979 that the permanent is #P-complete. Thus, a polynomial-time algorithm for the permanent would imply even more than $\mathsf{P} = \mathsf{NP}$: it would yield an efficient algorithm, not merely to solve NP-complete problems, but even to count how many solutions they have. In some sense, the #P-completeness of the permanent *explains* why Per, unlike Det, has no simple geometric or linear-algebraic interpretations: if such interpretations existed, then presumably they would imply $\mathsf{P} = \mathsf{P}^{\#\mathsf{P}}$.

In the arithmetic model, there exist arithmetic circuits of size $O(n^3)$, and even size $O(n^\omega)$, that compute $\text{Det}(X)$ as a formal polynomial in the entries of X, and that work over an arbitrary field $\mathbb{F}$. By contrast, Valiant conjectured the following.

[50] In later work, Bürgisser [59] showed that the same conjecture about $\tau^*(n!)$ (called the *τ-conjecture*) would also imply Valiant's Conjecture 66, that the permanent has no polynomial-size arithmetic circuits.

Conjecture 66 (Valiant's Conjecture). *Any arithmetic circuit for* Per (X) *requires size superpolynomial in n, over any field of characteristic other than* 2.[51,52]

Bürgisser [58] showed that, if Conjecture 66 fails over any field of positive characteristic, or if it fails over any field of characteristic zero *and the Generalized Riemann Hypothesis holds*, then $\mathsf{P}^{\#\mathsf{P}} \subset \mathsf{P/poly}$, and hence $\mathsf{NP} \subset \mathsf{P/poly}$.[53] (The main difficulty in proving this result is just that an arithmetic circuit might have very large constants hardwired into it.) On the other hand, no converses to this result is currently known. It's conceivable, for example, that we could have $\mathsf{P} = \mathsf{P}^{\#\mathsf{P}}$ for some "inherently Boolean" reason, even if the permanent required arithmetic circuits of exponential size. To put it another way, proving Conjecture 66 could serve as an "arithmetic warmup"—some would even say an "arithmetic prerequisite"—to proving Boolean separations such as $\mathsf{P}^{\#\mathsf{P}} \not\subset \mathsf{P/poly}$ and $\mathsf{P} \neq \mathsf{NP}$.

Better yet, Conjecture 66 turns out to be implied by (and nearly equivalent to) an appealing mathematical conjecture, which makes no direct reference to computation or circuits. Let's say that the $n \times n$ permanent *linearly embeds* into the $m \times m$ determinant, if it is possible to express Per (X) (for an $n \times n$ matrix $X \in \mathbb{F}^{n \times n}$) as Det $(L(X))$, where $L(X)$ is an $m \times m$ matrix each of whose entries is an affine combination of the entries of X. Then let $D(n)$ be the smallest m such that the $n \times n$ permanent linearly embeds into the $m \times m$ determinant.

Grenet [92] showed that $D(n) \leq 2^n$. By contrast, the best current lower bound on $D(n)$ is quadratic, and was proved by Mignon and Ressayre [153] in 2004, following a long sequence of linear lower bounds:

Theorem 67 (Mignon and Ressayre [153]). $D(n) \geq n^2/2$.

(Actually, Mignon and Ressayre proved Theorem 67 only for fields of characteristic 0. Their result was then extended to all fields of characteristic other than 2 by Cai et al. [64] in 2008.)

The basic idea of the proof of Theorem 67 is to consider the *Hessian matrix* of a polynomial $p : \mathbb{F}^N \to \mathbb{F}$, or the matrix of second partial derivatives, evaluated at some particular point $X_0 \in \mathbb{F}^N$:

$$H_p(X_0) := \begin{pmatrix} \frac{\partial^2 p}{\partial x_1^2}(X_0) & \cdots & \frac{\partial^2 p}{\partial x_1 \partial x_N}(X_0) \\ \vdots & \ddots & \vdots \\ \frac{\partial^2 p}{\partial x_N \partial x_1}(X_0) & \cdots & \frac{\partial^2 p}{\partial x_N^2}(X_0) \end{pmatrix}.$$

---

[51] Fields of characteristic 2, such as $\mathbb{F}_2$, are a special case: there, the permanent and determinant are *equivalent*, so in particular Per (X) has polynomial-size arithmetic circuits.

[52] In the literature, Conjecture 66 is often called the $\mathsf{VP} \neq \mathsf{VNP}$ conjecture, with VP and VNP being arithmetic analogues of P and NP respectively. I won't use that terminology in this survey, for several reasons: (1) VP is arguably more analogous to NC than to P, (2) VNP is arguably more analogous to #P than to NP, and (3) Conjecture 66 is almost always studied as a nonuniform conjecture, more analogous to $\mathsf{NP} \not\subset \mathsf{P/poly}$ than to $\mathsf{P} \neq \mathsf{NP}$.

[53] Indeed, #P would even have polynomial-size circuits of depth $\log^{O(1)} n$.

Here we mean the "formal" partial derivatives of p: even if $\mathbb{F}$ is a finite field, one can still symbolically differentiate a polynomial over $\mathbb{F}$, to produce new polynomials over smaller sets of variables. In general, when we're trying to lower-bound the difficulty of computing a polynomial p, a common technique in arithmetic complexity is to look at various partial derivatives $\frac{\partial^k p}{\partial x_{i_1} \cdots \partial x_{i_k}}$—and in particular, at the dimensions of vector spaces spanned by those partial derivatives, or the ranks of matrices formed from them—and then argue that, if p had a small circuit (or formula, or whatever), then those dimensions or ranks couldn't possibly be as high as they are.

In the case of Theorem 67, one proves the following two statements:

(1) If p is the permanent, of an $n \times n$ matrix of $N = n^2$ indeterminates, then there exists a point $X_0 \in \mathbb{F}^N$ such that $\operatorname{rank}\left(H_p\left(X_0\right)\right) = N$.
(2) If p is the determinant of an $m \times m$ matrix of affine functions in the N indeterminates, then $\operatorname{rank}\left(H_p\left(X\right)\right) \leq 2m$ for every X.

Combining these, one gets $m \geq n^2/2$, if p is both the $n \times n$ permanent and an $m \times m$ determinant.

So to summarize, the "blowup" $D(n)$ in embedding the permanent into the determinant is known to be at least quadratic and at most exponential. The huge gap here becomes a bit less surprising, once one knows that $D(n)$ is tightly connected to the arithmetic circuit complexity of the permanent. In particular, recall that a *formula* is just a circuit in which every gate has a fanout of 1. Then Valiant [225] showed the following:

Theorem 68 (Valiant [225]). *$D(n) \leq F(n) + 1$, where $F(n)$ is the size of the smallest arithmetic formula for the $n \times n$ permanent.*

Thus, if we could prove that $D(n)$ grew faster than any polynomial, we'd have shown that the permanent has no polynomial-size formulas. But heightening the interest still further, Valiant et al. [228] showed that in the arithmetic world, there's a surprisingly tight connection between formulas and circuits:

Theorem 69 (Valiant et al. [228]). *If a degree-d polynomial has an arithmetic circuit of size s, then it also has an arithmetic formula of size $(sd)^{O(\log d)}$.*

Theorem 69 implies that $C(n) \leq D(n)^{O(\log n)}$, where $C(n)$ is the size of the smallest arithmetic *circuit* for the $n \times n$ permanent. This means that, if we could prove that $D(n)$ grew not only superpolynomially but faster than $n^{O(\log n)}$, we'd also have shown that $C(n)$ grew superpolynomially, thereby establishing Valiant's Conjecture 66.

But lower-bounding $D(n)$ is not merely sufficient for proving Valiant's Conjecture; it's also necessary! For recall that the $n \times n$ determinant has an arithmetic circuit of size $O\left(n^3\right)$, and even $O\left(n^\omega\right)$. So we get the following chain of implications:

$$D(n) > n^{O(\log n)} \implies F(n) > n^{O(\log n)} \text{ (by Theorem 68)}$$

$$\Longrightarrow C(n) > n^{O(1)} \text{ (by Theorem 69; this is Valiant's Conjecture 66)}$$

$$\Longrightarrow D(n) > n^{O(1)} \text{(by the} n^{O(1)} \text{ arithmetic circuit for determinant)}$$

$$\Longrightarrow F(n) > n^{O(1)} \text{ (by Theorem 68).}$$

Today, a large fraction of the research aimed at proving $\mathsf{P} \neq \mathsf{NP}$ is aimed, more immediately, at proving Valiant's Conjecture 66 (see Agrawal [13] for a survey focusing on that goal). The hope is that, on the one hand, powerful tools from algebraic geometry and other fields can be brought to bear on Valiant's problem, but on the other, that solving it could provide insight about the original $\mathsf{P} \stackrel{?}{=} \mathsf{NP}$ problem.

6.5.2 Arithmetic Circuit Lower Bounds

I won't do justice in this survey to the now-impressive body of work motivated by Conjecture 66; in particular, I'll say little about proof techniques. Readers who want to learn more about arithmetic circuit lower bounds should consult Shpilka and Yehudayoff [205, Chap. 3] for an excellent survey circa 2010, or Saraf [196] for a 2014 update. Briefly, though, computer scientists have tried to approach Conjecture 66 much as they've approached $\mathsf{NP} \not\subset \mathsf{P/poly}$, by proving lower bounds against more and more powerful arithmetic circuit classes. In that quest, they've had some notable successes (paralleling the Boolean successes), but have also run up against some major differences from the Boolean case.

For starters, just as Razborov [187] and others considered monotone Boolean circuits, one can also consider *monotone arithmetic circuits* (over fields such as $\mathbb{R}$ or $\mathbb{Q}$), in which all coefficients need to be positive. Since the determinant involves -1 coefficients, it doesn't make sense to ask about monotone circuits for $\text{Det}(X)$, but one can certainly ask about the monotone circuit complexity of $\text{Per}(X)$. And already in 1982, Jerrum and Snir [116] proved the following arithmetic counterpart of Razborov's Theorem 44:

Theorem 70 (Jerrum and Snir [116]). *Any monotone circuit for* $\text{Per}(X)$ *requires size* $2^{\Omega(n)}$.

As another example, just as computer scientists considered constant-depth Boolean circuits (the classes AC^0, ACC, TC^0, and so on), so we can also consider *constant-depth arithmetic circuits*, which are conventionally denoted $\Sigma\Pi$, $\Sigma\Pi\Sigma$, etc. to indicate whether they represent a multivariate polynomial as a sum of products, a sum of product of sums, etc. It's trivial to prove exponential lower bounds on the sizes of depth-two ($\Sigma\Pi$) circuits: that just amounts to lower-bounding the number of monomials in a polynomial. More interesting is the following result:

Theorem 71 (Grigoriev and Karpinski [93], Grigoriev and Razborov [94]). *Over a finite field, any* $\Sigma\Pi\Sigma$ *circuit for* $\operatorname{Det}(X)$ *requires size* $2^{\Omega(n)}$. *(Indeed, this is true even for circuits representing* $\operatorname{Det}(X)$ *as a function.)*

Curiously, over *infinite* fields, the best lower bound that we have is still a much weaker one, due to Shpilka and Wigderson [204]:

Theorem 72 (Shpilka and Wigderson [204]). *Over infinite fields, any* $\Sigma\Pi\Sigma$ *circuit for* $\operatorname{Det}(X)$ *requires size* $\Omega\left(n^4/\log n\right)$.

Theorems 71 and 72 are stated for the determinant, although they have analogues for the permanent. In any case, these results certainly don't succeed in showing that the permanent is *harder* than the determinant.

The situation is better when we restrict the fanin of the multiplication gates. In particular, by a $\Sigma\Pi^{[a]}\Sigma\Pi^{[b]}$ circuit, let's mean a depth-4 circuit where every inner multiplication gate has fanin at most a, and every bottom multiplication gate has fanin at most b. Then in 2013, Gupta et al. [99] proved the following.

Theorem 73 (Gupta et al. [99]). *Any* $\Sigma\Pi^{[O(\sqrt{n})]}\Sigma\Pi^{[\sqrt{n}]}$ *circuit for* $\operatorname{Per}(X)$ *or* $\operatorname{Det}(X)$ *requires size* $2^{\Omega(\sqrt{n})}$.

Subsequently, Kayal et al. [126] proved a size lower bound of $n^{\Omega(\sqrt{n})}$ for such circuits, though not for the permanent or determinant but for a different explicit polynomial.

The situation is also better when we restrict to *homogeneous* arithmetic circuits. These are circuits where every gate is required to compute a homogeneous polynomial: that is, one where all the monomials have the same degree. Here Nisan and Wigderson [173] established the following in 1997.

Theorem 74 (Nisan and Wigderson [173]). *Over any field, any homogeneous* $\Sigma\Pi\Sigma$ *circuit for* $\operatorname{Det}(X)$ *requires size* $2^{\Omega(n)}$.

Going further, in 2014 Kayal et al. [125] gave an explicit polynomial for which any homogeneous $\Sigma\Pi\Sigma\Pi$ circuit requires size $n^{\Omega(\sqrt{n})}$.

It's natural to wonder: why are we stuck talking about depth-3 and depth-4 arithmetic circuits? Why couldn't we show that the permanent and determinant have no *constant-depth* arithmetic circuits of subexponential size, just like Theorem 47 and its successors showed that PARITY has no constant-depth Boolean circuits of subexponential size? After all, wasn't the whole point of arithmetic complexity that it was supposed to be *easier* than Boolean complexity?

In 2008, Agrawal and Vinay [15] gave a striking answer to these questions; they called their answer "the chasm at depth four." In particular, building on the earlier work of Valiant et al. [228] (Theorem 69), Agrawal and Vinay showed that, if we managed to prove strong enough lower bounds for depth-4 arithmetic circuits, then we'd also get superpolynomial lower bounds for *arbitrary* arithmetic circuits! Here's one special case of their result:

Theorem 75 (Agrawal and Vinay [15]). *Suppose that* Per(X) *requires depth-*4 *arithmetic circuits (even homogeneous ones) of size* $2^{\Omega(n)}$. *Then* Per(X) *requires arithmetic circuits of superpolynomial size, and Valiant's Conjecture 66 holds.*

Subsequently, Koiran [133] and Tavenas [221] showed that Valiant's Conjecture would follow, not merely from a $2^{\Omega(n)}$ size lower bound for homogeneous depth-4 circuits computing the permanent, but from *any* size lower bound better than $n^{\Omega(\sqrt{n})}$. In an even more exciting development, Gupta et al. [98] reduced the depth from four to three (though only for fields of characteristic 0, and no longer allowing homogeneity):

Theorem 76 (Gupta et al. [98]). *Suppose that* Per(X) *requires depth-*3 *arithmetic circuits of size more than* $n^{\Omega(\sqrt{n})}$, *over fields of characteristic* 0. *Then* Per(X) *requires arithmetic circuits of superpolynomial size, and Valiant's Conjecture 66 holds.*

These results can be considered extreme versions of the depth reduction of Brent [52] (see Proposition 24). I should mention that all of these results hold, not just for the permanent, but for *any* homogeneous polynomial of degree $n^{O(1)}$. In particular, by applying their depth reduction "in the opposite direction" for the determinant, Gupta et al. [98] were able to show that there *exist* depth-3 arithmetic circuits of size $n^{O(\sqrt{n})}$ for Det(X). This provides an interesting counterpoint to the result of Nisan and Wigderson [173] (Theorem 74), which showed that size $2^{\Omega(n)}$ is needed for the determinant if we restrict to depth-3 *homogeneous* circuits.

There are yet other results in this vein, which give yet other tradeoffs. But perhaps we should step back from the flurry of theorems and try to summarize the situation. After decades of research in arithmetic circuit complexity, we now have lower bounds of the form $n^{\Omega(\sqrt{n})}$ on the sizes of depth-3 and depth-4 arithmetic circuits computing explicit polynomials (subject to various technical restrictions). On the other hand, we also have a deep explanation for why the progress has stopped at the specific bound $n^{\Omega(\sqrt{n})}$: because *any lower bound even slightly better than that would already prove Valiant's Conjecture, that the permanent is superpolynomially harder than the determinant!* It's as if, in arithmetic complexity, we reach a terrifying precipice—beyond which we can no longer walk but need to fly—sooner than we do in the Boolean case. And around 2014, we learned exactly where that precipice is and walked right up to it, but we still haven't jumped.

In this connection, it's worth pointing out that, with the exception of Theorem 67 by Mignon and Ressayre [153], none of the results in this section actually *differentiate* the permanent from the determinant: that is, none of them prove a lower bound for Per(X) better than the analogous lower bound known for Det(X). Eventually, of course, any proof of Valiant's Conjecture *will* need to explain why the permanent is harder than the determinant, which is one of the main motivations for the Mulmuley-Sohoni program (see Sect. 6.6).

Let me end this section by discussing two striking results of Ran Raz that didn't quite fit into the narrative above. The first result is a superpolynomial lower bound on the sizes of *multilinear formulas*. An arithmetic formula is called *multilinear* if the polynomial computed by each gate is a multilinear polynomial (that is, no variable is raised to a higher power than 1). Notice that the permanent and determinant are both multilinear polynomials. For that reason, they can be computed by multilinear formulas, and it makes sense to ask about the size of the smallest such formulas.

In a 2004 breakthrough, Raz [182] proved the following.

Theorem 77 (Raz [182]). *Any multilinear formula for* $\mathrm{Per}(X)$ *or* $\mathrm{Det}(X)$ *requires size* $n^{\Omega(\log n)}$.[54]

What made Theorem 77 a breakthrough was that there was no restriction on the formula's depth. The proof was via the random restriction method from Sect. 6.2.3, combined with the idea (common in arithmetic complexity) of using matrix rank as a progress measure. In more detail, let $p : \{0,1\}^n \to \mathbb{R}$ be a polynomial computed by a small multilinear formula: for simplicity, we'll take p's inputs to be Boolean. Then basically, one randomly partitions p's input variables into two small sets $X = \{x_1, \ldots, x_k\}$ and $Y = \{y_1, \ldots, y_k\}$, and a large set Z of size $n - 2k$. (Here one should imagine, say, $k = n^{1/3}$.) One then randomly fixes the variables in Z to 0's or 1's, while leaving the variables in X and Y unfixed. Next, one defines a matrix $M \in \mathbb{R}^{2^k \times 2^k}$, whose rows are indexed by the 2^k possible assignments to X, whose columns are indexed by the 2^k possible assignments to Y, and whose (X, Y) entry equals $p(X, Y, Z)$. Finally, one proves the following two statements:

- With high probability, M has rank much smaller than 2^k. This is the hard part of the proof: one uses the assumption that p has a small multilinear formula, and then argues by induction on the formula.
- If p represents the function f of interest to us (say, the permanent or determinant), then $\mathrm{rank}(M) = 2^k$ with certainty.

Together, these yield the desired contradiction, showing that f can't have had a small multilinear formula after all.

It seems likely that the lower bound in Theorem 77 could be improved from $n^{\Omega(\log n)}$ all the way up to $2^{\Omega(n)}$, but this remains open. Raz and Yehudayoff [185] did manage to prove an exponential lower bound for *constant-depth* multilinear formulas computing the permanent or determinant; and in a separate work [186], they also proved a $2^{\Omega(n)}$ lower bound for "non-cancelling" multilinear formulas computing an explicit polynomial f (not the permanent or determinant). Here "non-cancelling"—a notion that I defined in [2]—basically means that nowhere in the formula are we allowed to add two polynomials that "almost perfectly" cancel each other out, leaving only a tiny residue.

[54] An immediate corollary is that any multilinear *circuit* for $\mathrm{Per}(X)$ or $\mathrm{Det}(X)$ requires *depth* $\Omega(\log^2 n)$.

Of course, just like with the arithmetic circuit lower bounds discussed earlier, so far all the known multilinear formula lower bounds fail to distinguish the permanent from the determinant.

The second result of Raz's concerns so-called *elusive functions*. Given a polynomial curve $f : \mathbb{C} \to \mathbb{C}^n$, Raz calls f *elusive* if f is not contained in the image of any polynomial mapping $g : \mathbb{C}^{n-1} \to \mathbb{C}^n$ of degree 2. He then proves the following striking theorem.

Theorem 78 (Raz [183]). *Suppose there exists an elusive function whose coefficients can be computed in polynomial time. Then* Per (X) *requires arithmetic circuits of superpolynomial size, and Valiant's Conjecture 66 holds.*

Arguably, this makes Valiant's Conjecture look *even more* like a question of pure algebraic geometry than it did before! As evidence that the "elusive function" approach to circuit lower bounds is viable, Raz then constructed an explicit f that was elusive in a weak sense, which was already enough to imply the following new lower bound:

Theorem 79 (Raz [183]). *For every r, there is an explicit polynomial p with n variables and degree $O(r)$, such that any depth-r arithmetic circuit for p (over any field) requires size $n^{1+\Omega(1/r)}$.*

6.5.3 Arithmetic Natural Proofs?

In Sect. 6.5.2, we saw arithmetic circuit lower bounds that, again and again, seem to go "right up to the brink" of proving Valiant's Conjecture, but then stop short. Given this, it's natural to wonder what the barriers are to further progress in arithmetic complexity, and how they relate to the barriers in the Boolean case.

We've already discussed one obvious barrier, which is that eventually we need techniques that work for the permanent but *fail* for the determinant. It might also be interesting to define an arithmetic analogue of the relativization barrier (Sect. 6.1.2). To my knowledge, this hasn't been done, but my guess is that in the arithmetic setting, the natural choices for oracles would look a lot like the algebraic oracles studied by Aaronson and Wigderson [10] (see Sect. 6.3.3). With a notion of "oracle" in hand, one could probably show that most arithmetic circuit lower bounds require arithmetically non-relativizing techniques. On the other hand, this wouldn't be much of an obstruction, since even the results discussed in Sect. 6.5.2 should *already* evade the relativization barrier, for the same reason as those of Sects. 6.2.3 and 6.2.4.

In the rest of this section, I'd like to discuss the contentious question of whether or not arithmetic circuit complexity faces a natural proofs barrier, in the sense of Razborov and Rudich [191]. Recall from Sect. 6.2.5 that a circuit lower bound proof is called *natural* if, besides proving that the specific function f of interest to us is not in a circuit class $\mathcal{C}$, the proof also provides a *polynomial-time algorithm* A that takes as input a function's truth table, and that certifies a $1/n^{O(1)}$ fraction of all functions as not belonging to $\mathcal{C}$. Such an A can be used to distinguish functions in $\mathcal{C}$ from random functions with non-negligible bias. Meanwhile, the class $\mathcal{C}$ has a *natural proofs barrier* if $\mathcal{C}$ contains *pseudorandom function families*, which cannot be so distinguished from random functions, and whose existence is therefore incompatible with the existence of A.

In the arithmetic setting, presumably we'd call a proof *natural* if it yields a polynomial-time algorithm[55] that takes as input, say, the complete output table of a homogeneous degree-d polynomial $p : \mathbb{F}^n \to \mathbb{F}$ over a finite field $\mathbb{F}$, and that certifies a $1/n^{O(1)}$ fraction of all such polynomials as not belonging to the arithmetic circuit class $\mathcal{C}$. Also, we'd say that $\mathcal{C}$ has a *natural proofs barrier* if $\mathcal{C}$ contains *pseudorandom polynomial families*. By this, we mean families of homogeneous degree-d polynomials, $p_s : \mathbb{F}^n \to \mathbb{F}$, that no $|\mathbb{F}|^{O(n)}$-time algorithm can distinguish from uniformly-random homogeneous degree-d polynomials with non-negligible bias. (We can no longer talk about uniformly-random *functions*, since an algorithm can easily ascertain, for example, that p_s is a degree-d polynomial.) By exactly the same logic as in the Boolean case, if $\mathcal{C}$ is powerful enough to compute pseudorandom polynomials, then no natural proof can show that a polynomial is not in $\mathcal{C}$.

Now, one point that's *not* disputed is that all the arithmetic circuit lower bounds discussed in Sect. 6.5.2 are natural in the above sense. I didn't say much about how the lower bounds are proved, but as mentioned in Sect. 6.5.1, arithmetic circuit lower bounds generally proceed by finding some parameter $\alpha(p)$ associated with a polynomial p—say, the rank of its Hessian matrix, or the dimension of a vector space spanned by p's partial derivatives—to use as a "progress measure." The proof then argues that

(1) $\alpha(p)$ is large for the specific polynomial p of interest to us (say, the permanent or determinant), but
(2) every gate added to our circuit or formula can only increase $\alpha(p)$ by so much,

thereby implying that p requires many gates. Furthermore, virtually any progress measure α that's a plausible choice for such an argument—and certainly the ones used in the existing result—will be computable in $|\mathbb{F}|^{O(n)}$ time, and will be maximized by a *random* polynomial p of the appropriate degree. Alas, this implies that the argument is natural! If the circuit class $\mathcal{C}$ has a natural proofs barrier, then no such argument can possibly prove $p \notin \mathcal{C}$.

The only part that's controversial is whether arithmetic circuit classes *do* have a natural proofs barrier. To show that they did, we'd need plausible candidates for pseudorandom polynomials—e.g., homogeneous degree-d polynomials $p : \mathbb{F}^n \to \mathbb{F}$ that actually have small arithmetic circuits, but that look to any efficient test just like random homogeneous polynomials of degree d. The trouble is that, while cryptographers know a great deal about how to construct pseudorandom functions,

[55] For simplicity, here I'll assume that we mean an "ordinary" (Boolean) polynomial-time algorithm, though one could also require polynomial-time algorithms in the arithmetic model.

the accepted constructions are all "inherently Boolean"; they don't work in the setting of low-degree polynomials over a finite field.

Thus, to take one example, the work of Goldreich, Goldwasser, and Micali (GGM) [88], combined with that of Håstad et al. [106], shows how to build a pseudorandom function family starting from any *one-way function* (see Sect. 5.3.1). And indeed, Razborov and Rudich [191] used a variant of the GGM construction in their original paper on natural proofs. However, if we try to implement the GGM construction using arithmetic circuits—say, using multiplication for the AND gates, $1 - x$ for the NOT gates, etc.—we'll find that we've produced an arithmetic circuit of $n^{O(1)}$ depth, which computes a polynomial of $\exp\left(n^{O(1)}\right)$ degree: far too large.

As I mentioned in Sect. 6.2.5, if we're willing to assume the hardness of specific cryptographic problems, then there are also much more direct constructions of pseudorandom functions, which produce circuits of much lower depth. In particular, there's the construction of Naor and Reingold [170], which is based on factoring and discrete logarithm; and that of Banerjee et al. [35], which is based on noisy systems of linear equations. Unfortunately, examination of these constructions reveals that they, too, require treating the input as a string of bits rather than of finite field elements. So for example, the Naor-Reingold construction involves modular exponentiation, which of course goes outside the arithmetic circuit model, where only addition and multiplication are allowed.

At this point I can't resist stating my own opinion, which is that the issue here is partly technical but also partly social. Simply put: Naor-Reingold and Banerjee et al. are taken to be relevant to natural proofs, because factoring, discrete logarithm, and solving noisy systems of linear equations have become *accepted by the community of cryptographers* as plausibly hard problems. Since real computers use Boolean circuits, and since in practice one normally needs pseudorandom *functions* rather than polynomials, cryptographers have had extremely little reason to study pseudorandom low-degree polynomials that are computed by small arithmetic circuits over finite fields. If they *had* studied that, though, it seems entirely plausible that they would've found decent candidates for such polynomials, and formed a social consensus that they indeed seem hard to distinguish from random polynomials.

Motivated by that thought, in a 2008 blog post [5], I offered my own candidate for a pseudorandom family of polynomials, $p_s : \mathbb{F}^n \to \mathbb{F}$, which are homogeneous of degree $d = n^{O(1)}$. My candidate was simply this: motivated by Valiant's result [225] that the determinant can express any arithmetic formula (Theorem 68), take the random seed s to encode d^2 uniformly-random linear functions, $L_{i,j} : \mathbb{F}^n \to \mathbb{F}$ for all $i, j \in \{1, \ldots, d\}$. Then set

$$p_s(x_1, \ldots, x_n) := \operatorname{Det} \begin{pmatrix} L_{1,1}(x_1, \ldots, x_n) & \cdots & L_{1,d}(x_1, \ldots, x_n) \\ \vdots & \ddots & \vdots \\ L_{d,1}(x_1, \ldots, x_n) & \cdots & L_{d,d}(x_1, \ldots, x_n) \end{pmatrix}.$$

My conjecture is that, at least when d is sufficiently large, a random p_s drawn from this family should require $\exp\left(d^{\Omega(1)}\right)$ time to distinguish from a random homogeneous polynomial of degree d, if we're given the polynomial $p : \mathbb{F}^n \to \mathbb{F}$ by a table of $|\mathbb{F}|^n$ values. If d is a large enough polynomial in n, then $\exp\left(d^{\Omega(1)}\right)$ is greater than $|\mathbb{F}|^{O(n)}$, so the natural proofs barrier would apply.

So far there's been little study of this conjecture, with the exception of a 2012 paper by Kayal [124], which proved the following.

Theorem 80 (Kayal [124]). *Let $d \leq \sqrt{n}$, and suppose we're given black-box access to a degree-d polynomial $p : \mathbb{F}^n \to \mathbb{F}$, which is promised to be the permanent or determinant of a $d \times d$ matrix of linear forms. Then there exists a randomized, $n^{O(1)}$-time algorithm to reconstruct the linear forms.*

If we don't need to reconstruct the linear forms, but only break my pseudorandom polynomial candidate, then the results of Mignon and Ressayre [153] (Theorem 67) let us push the range where the algorithm works all the way up to $d \leq n/\sqrt{2}$. It's not known what happens for larger d.

Of course, if our goal is to prove $\mathsf{P} \neq \mathsf{NP}$ or $\mathsf{NP} \not\subset \mathsf{P/poly}$, then perhaps the whole question of arithmetic natural proofs is ultimately beside the point. For to prove $\mathsf{NP} \not\subset \mathsf{P/poly}$, we'll need to become experts at overcoming the natural proofs barrier *in any case*: either in the arithmetic world, or if not, then when we move from the arithmetic world back to the Boolean one.

6.6 Geometric Complexity Theory

I'll end this survey with some *extremely* high-level remarks about Geometric Complexity Theory (GCT): a breathtakingly ambitious program to prove $\mathsf{P} \neq \mathsf{NP}$ and related conjectures using algebraic geometry and representation theory. This program has been pursued since the late 1990s primarily by Ketan Mulmuley, though with important contributions by Milind Sohoni and others.

I like to describe GCT as "the string theory of computer science." Like string theory, GCT has the aura of an intricate theoretical superstructure from the far future, impatiently being worked on today despite our inability to test some key premises. Both theories have attracted interest partly because of "miraculous coincidences" (for string theory, these include anomaly cancellations and the prediction of gravitons; for GCT, exceptional properties of the permanent and determinant, and remarkable algorithms to compute the multiplicities of irreps). Both have been described as beautiful, deep, compelling, and even "the only game in town" (not surprisingly, a claim disputed by the fans of rival ideas!). And like with string theory, there's scarcely any part of modern mathematics that's *not* known or believed to be relevant to GCT.[56]

For both string theory and GCT, however, the central problem has been to say something novel and verifiable about the real-world phenomena that motivated the

[56]Although so far, I haven't seen a conjectured role for topology or logic in GCT.

theory in the first place, and to do so in a way that depends essentially on the theory's core tenets (rather than on inspirations or analogies, or on incidental fragments of the theory). For string theory, that would mean making confirmed predictions or (e.g.) explaining the masses and generations of the elementary particles; for GCT, it would mean proving new circuit lower bounds. Indeed, proponents of both theories freely admit that one might spend one's entire career on the theory, without living to see a payoff of that kind.[57]

GCT is not an easy subject, and I don't pretend to be an expert. Part of the difficulty is inherent, while part of it is that the primary literature on GCT is not optimized for beginners: it contains a profusion of new ideas, extended analogies, speculations, theorems, speculations given acronyms and then referred to *as if* they were theorems, and advanced mathematics assumed as background, all in papers and manuscripts that cite each other in spaghetti fashion, making it difficult to track down where a given claim is proved.

Mulmuley regards the beginning of GCT as a 1997 paper of his [158], which used algebraic geometry to prove a circuit lower bound for the classic MAXFLOW problem. In MAXFLOW, we're given as input a description of an n-vertex directed graph G, with nonnegative integer weights on the edges (called *capacities*), along with designated source and sink vertices s and t, and a nonnegative integer w (the *target*). The problem is to decide whether w units of a liquid can be routed from s to t, with the amount of liquid flowing along an edge e never exceeding e's capacity. The MAXFLOW problem is famously in P,[58] but the known algorithms are inherently serial, and it remains open whether they can be parallelized. More concretely, is MAXFLOW in NC^1, or some other small-depth circuit class? Note that any proof of MAXFLOW $\notin NC^1$ would imply the spectacular separation $NC^1 \neq P$. Nevertheless, Mulmuley [158] managed to prove a strong lower bound for a restricted class of circuits, which captures almost all the MAXFLOW algorithms known in practice.

Theorem 81 (Mulmuley [158]). *Consider an arithmetic circuit for* MAXFLOW, *which takes as input the integer capacities and the integer target w.*[59] *The gates, which all have fanin* 2, *can perform integer addition, subtraction, and multiplication* $(+, -, \times)$ *as well as integer comparisons* $(=, \leq, <)$. *A comparison gate returns the integer* 1 *if it evaluates to "true" and* 0 *if it evaluates to "false." The circuit's final*

---

[57]Furthermore, just as string theory didn't predict what new data there *has* been in fundamental physics in recent decades (e.g., the dark energy), so GCT played no role in, e.g., the proof of $NEXP \not\subset ACC$ (Sect. 6.4.2), or the breakthrough lower bounds for small-depth circuits computing the permanent (Sect. 6.5.2). In both cases, to point this out is to hold the theory to a high and probably unfair standard, but also, *ipso facto*, to pay the theory a compliment.

[58]MAXFLOW can easily be reduced to linear programming, but Ford and Fulkerson [78] also gave a much faster and direct way to solve it, which can be found in any undergraduate algorithms textbook. There have been further improvements since.

[59]We can assume, if we like, that G, s, and t are hardwired into the circuit. We can also allow constants such as 0 and 1 as inputs, but this is not necessary, as we can also generate constants ourselves using comparison gates and arithmetic.

output must be 1 *if the answer is "yes" and* 0 *if the answer is "no." Direct access to the bit-representations of the integers is not allowed, nor (for example) are the floor and ceiling functions.*

Any such circuit must have depth $\Omega\left(\sqrt{n}\right)$*. Moreover, this holds even if the capacities are restricted to be* $O\left(n^2\right)$*-bit integers.*

Similar to previous arithmetic complexity lower bounds, the proof of Theorem 81 proceeds by noticing that any small-depth arithmetic circuit separate the yes-instances from the no-instances via a small number of low-degree algebraic surfaces. It then appeals to results from algebraic geometry (e.g., the Milnor-Thom Theorem) to show that, for problems of interest (such as MAXFLOW), no such separation by low-degree surfaces is possible. As Mulmuley already observed in [158], the second part of the argument would work just as well for a *random* problem. So we get a natural proof in the Razborov-Rudich sense (Sect. 6.2.5), meaning that the same technique can't possibly work to prove $\mathsf{NC}^1 \neq \mathsf{P}$: it must break down when arbitrary bit operations are allowed. Nevertheless, for Mulmuley, Theorem 81 was strong evidence that algebraic geometry provides a route to separating complexity classes.

The central papers on GCT, written between 2001 and 2013, are as follows:

- GCT1 [167] by Mulmuley and Sohoni, which already contains almost all of the main ideas, and should be read (perhaps along with GCT6) before attempting the others.
- GCT2 [168], by Mulmuley and Sohoni, which has further technical results about representation-theoretic obstructions, and which conjectures that the orbit closures relevant to GCT are essentially *captured* by their representation-theoretic data.
- GCT3 [166], by Mulmuley, Narayanan, and Sohoni, which gives an efficient algorithm, based on linear programming, to decide the positivity of Littlewood-Richardson coefficients. Bürgisser and Ikenmeyer [60] later gave a combinatorial algorithm for the same problem.
- GCT4 [45], by Blasiak, Mulmuley, and Sohoni, which gives a "positive formula" for certain Kronecker coefficients (putting the computation of those coefficients in #P).
- GCT5 [165] by Mulmuley, which gives a conditional derandomization of Noether's Normalization Lemma, a subject of perhaps tangential relevance to the rest of the GCT program. Forbes and Shpilka [77] later gave an unconditional version of the same result.
- GCT6 [162] by Mulmuley, which steps back and revisits the basics of GCT in a somewhat more philosophical way, explaining what Mulmuley calls "the Flip" (and what this survey called "ironic complexity theory"), and arguing the need for explicit obstructions.
- GCT7 [159] and GCT8 [160] by Mulmuley, which give conjectures about quantum groups that, if true, would yield positive formulas for certain plethysm problems (again, putting those problems in #P).

Readers seeking a less overwhelming introduction can try Mulmuley's overviews in *Communications of the ACM* [164] or in *Journal of the ACM* [163]; or lecture notes and other materials on Mulmuley's GCT website [157]; or surveys by Regan [192] or Landsberg [137] (the latter is for geometers, not computer scientists). In my view, however, perhaps the most beginner-friendly exposition of GCT yet written is contained in Joshua Grochow's Ph.D. thesis [95, Chap. 3]. Indeed, some readers might want to put my survey down at this point and read Grochow's thesis instead.

6.6.1 From Complexity to Algebraic Geometry

So, what *is* GCT? It's easiest to understand GCT as a program to prove Valiant's Conjecture 66—that is, to show that any affine embedding of the $n \times n$ permanent over $\mathbb{C}$ into the $m \times m$ determinant over $\mathbb{C}$ requires (say) $m = 2^{n^{\Omega(1)}}$, and hence, that the permanent requires exponential-size arithmetic circuits. GCT also includes an even more speculative program to prove Boolean lower bounds, such as $\mathsf{NP} \not\subset \mathsf{P/poly}$ and hence $\mathsf{P} \neq \mathsf{NP}$. However, if one accepts the premises of GCT in the first place (e.g., the primacy of algebraic geometry for circuit lower bounds), then one might as well start with the permanent versus determinant problem, since that's where the core ideas of GCT come through the most clearly.

The first observation Mulmuley and Sohoni make is that Valiant's Conjecture 66 can be translated into an algebraic-geometry conjecture about *orbit closures*. In more detail, consider a group G that acts on a set V: in the examples that interest us, G will be a group of complex matrices, while V will be a high-dimensional vector space over $\mathbb{C}$ (e.g., the space of all homogeneous degree-n polynomials over some set of variables). Then the *orbit* of a point $v \in V$, denoted Gv, is the set $\{g \cdot v : g \in G\}$. Also, the *orbit closure* of x, denoted $\overline{Gv}$, is the closure of Gv in the usual complex topology. It contains all the points that can be arbitrarily well approximated by points in Gv.

To be more concrete, let $G = \mathrm{GL}_{m^2}(\mathbb{C})$ be the general linear group: the group of all invertible $m^2 \times m^2$ complex matrices. Also, let V be the vector space of all homogeneous degree-m complex polynomials over m^2 variables; the variables are labeled $x_{1,1}, \ldots, x_{m,m}$, and we'll think of them as the entries of an $m \times m$ matrix X. (Recall that a *homogeneous* polynomial is one where all the monomials have the same degree m; restricting to them lets GCT talk about linear maps rather than affine ones.) Then G acts on V in an obvious way: for all matrices $A \in G$ and polynomials $p \in V$, we can set $(A \cdot p)(x) := p(Ax)$. Indeed, this action is a *group representation*: each $A \in G$ acts linearly on the coefficients of p, so we get a homomorphism from G to the linear transformations on V.

Now, we'd like to interpret both the $m \times m$ determinant and the $n \times n$ permanent ($n \ll m$) as points in V, in order to phrase the permanent versus determinant problem in terms of orbit closures. For the determinant, this is trivial: Det_m *is* a degree-m homogeneous polynomial over the x_{ij}'s. The $n \times n$ permanent, on the other hand, is a lower-degree polynomial over a smaller set of variables. GCT solves that problem by considering the so-called *padded permanent*,

$$\mathrm{Per}^*_{m,n}(X) := x_{m,m}^{m-n}\,\mathrm{Per}_n(X|_n),$$

where $X|_n$ denotes the top-left $n \times n$ submatrix of X, and $x_{m,m}$ is just some entry of X that's not in $X|_n$. This is a homogeneous polynomial of degree m.

Let $\chi_{\mathrm{Det},m} := \overline{G \cdot \mathrm{Det}_m}$ be the orbit closure of the $m \times m$ determinant, and let $\chi_{\mathrm{Per},m,n} := \overline{G \cdot \mathrm{Per}^*_{m,n}}$ be the orbit closure of the padded $n \times n$ permanent. I can now state the central conjecture that GCT, in its current incarnation, seeks to prove.

Conjecture 82 (Mulmuley-Sohoni Conjecture). *If $m = 2^{n^{o(1)}}$, then for all sufficiently large n we have $\mathrm{Per}^*_{m,n} \notin \chi_{\mathrm{Det},m}$, or equivalently $\chi_{\mathrm{Per},m,n} \not\subset \chi_{\mathrm{Det},m}$. In other words, the padded permanent is not in the orbit closure of the determinant.*

Mulmuley and Sohoni's first major observation is that a proof of Conjecture 82 (or indeed, *any* lower bound on m better than $n^{\Omega(\log n)}$) would imply a proof of Valiant's Conjecture:

Proposition 83 ([167]). *Suppose there's an affine embedding of the $n \times n$ permanent into the $m \times m$ determinant: i.e., $D(n) \leq m$, in the notation of Sect. 6.5.1. Then $\mathrm{Per}^*_{m,n} \in \chi_{\mathrm{Det},m}$.*

Let me make two remarks about Proposition 83. First, for the proposition to hold, it's crucial that we're talking about the orbit *closure*, not just about the orbit. It's easy to see that $\mathrm{Per}^*_{m,n} \notin G \cdot \mathrm{Det}_m$—for example, because every element of $G \cdot \mathrm{Det}_m$ is *irreducible* (can't be factored into lower-degree polynomials), whereas the padded permanent is clearly reducible. But that tells us only that there's no *invertible* linear transformation of the variables that turns the determinant into the padded permanent, not that there's no linear transformation at all. In GCT, linear changes of variable play the role of reductions—so, while the orbit of f plays the role of the "f-complete problems," it's the orbit closure that plays the role of the complexity class of functions reducible to f.[60]

[60] More broadly, I've found, there are many confusing points about GCT whose resolutions require reminding ourselves that we're talking about orbit *closures*, and not only about orbits. For example, the plan of GCT is ultimately to show, roughly speaking, that the $n \times n$ permanent has "too little symmetry" to be embedded into the $m \times m$ determinant, unless m is much larger than n. But in that case, what about (say) a random arithmetic formula f of size n, which has *no* nontrivial symmetries, but which clearly *can* be embedded into the $(n + 1) \times (n + 1)$ determinant, by Theorem 68? Even though this f clearly isn't characterized by its symmetries, mustn't the embedding obstructions for f be a strict superset of the embedding obstructions for the permanent—since f's symmetries are a strict subset of the permanent's symmetries—and doesn't that give rise to a contradiction? According to Mulmuley (personal communication), the solution to the apparent paradox is that this argument *would* be valid if we were talking only about orbits, but it's not valid for orbit closures. With orbit closures, the set of obstructions doesn't depend in a simple way on the symmetries of the original function, so that it's possible that an obstruction for the permanent would fail to be an obstruction for f.

The second remark is that, despite their similarity, it's unknown whether Conjecture 82 is *equivalent* to Valiant's conjecture that the permanent requires affine embeddings into the determinant of size $D(n) > 2^{n^{o(1)}}$. The reason is that, as far as anyone knows, there might be points in the orbit closure of the determinant that aren't in its "endomorphism orbit" (that is, the set of polynomials that have not-necessarily-invertible linear embeddings into the determinant) In complexity terms, these points would be homogeneous degree-m polynomials that can be arbitrarily well approximated by determinants of $m \times m$ matrices of linear functions, but not represented exactly.

See Grochow [95] for further discussion of both issues.

6.6.2 Characterization by Symmetries

So far, it seems like all we've done is restated Valiant's Conjecture in a more abstract language and slightly generalized it. But now we come to the central insight of GCT, which is that the permanent and determinant are both special, highly symmetric functions, and we can leverage that fact to learn more about their orbit closures than we could if they were arbitrary functions. For starters, Per (X) is symmetric under permuting X's rows or columns, transposing X, and multiplying the rows or columns by scalars that multiply to 1. That is, we have

$$\text{Per}(X) = \text{Per}\left(X^T\right) = \text{Per}(PXQ) = \text{Per}(AXB) \tag{2}$$

for all permutation matrices P and Q, and all diagonal matrices A and B such that Per (A) Per $(B) = 1$. The determinant has an even larger symmetry group: we have

$$\text{Det}(X) = \text{Det}\left(X^T\right) = \text{Det}(AXB) \tag{3}$$

for all matrices A and B such that Det (A) Det $(B) = 1$.

But there's a further point: it turns out that the permanent and determinant are both *uniquely characterized* (up to a constant factor) by their symmetries, among all homogeneous polynomials of the same degree. More precisely:

Theorem 84. *Let p be any degree-m homogeneous polynomial in the entries of $X \in \mathbb{C}^{m \times m}$ that satisfies $p(X) = p(PXQ) = p(AXB)$ for all permutation matrices P, Q and diagonal A, B with* Per (A) Per $(B) = 1$. *Then $p(X) = \alpha$* Per (X) *for some $\alpha \in \mathbb{C}$. Likewise, let p be any degree-m homogeneous polynomial in the entries of $X \in \mathbb{C}^{m \times m}$ that satisfies $p(X) = p(AXB)$ for all A, B with* Det (A) Det $(B) = 1$. *Then $p(X) = \alpha$* Det (X) *for some $\alpha \in \mathbb{C}$.*[61]

[61]Note that we don't even need to assume the symmetry $p(X) = p\left(X^T\right)$; that comes as a free byproduct. Also, it might seem like "cheating" that we use the permanent to state the symmetries that characterize the permanent, and likewise for the determinant. But we're just using the permanent and determinant as convenient ways to specify which matrices A, B we want, and could

Theorem 84 is fairly well-known in representation theory; see Grochow [95, Propositions 3.4.3 and 3.4.5] for an elementary proof, using Gaussian elimination for the determinant and even simpler considerations for the permanent. Notice that we're not merely saying that any polynomial p with the same symmetry group as the permanent is a multiple of the permanent (and similarly for the determinant), but rather that any p whose symmetry group *contains* the permanent's is a multiple of the permanent.

In a sense, Theorem 84 is the linchpin of the entire GCT program. Among other things, it's GCT's answer to the question of how it will overcome the natural proofs barrier. For notice that, if we picked a degree-m homogeneous polynomial at random, it almost certainly *wouldn't* be uniquely characterized by its symmetries, as the permanent and determinant are.[62] Thus, if a proof that the permanent is hard relies on symmetry-characterization, we need not fear that the same proof would work for a random homogeneous polynomial, and thereby give us a way to break arithmetic pseudorandom functions (Sect. 6.5.3). While this isn't mentioned as often, Theorem 84 should also let GCT overcome the relativization and algebrization barriers, since (for example) a polynomial that was $\#\mathsf{P}^A$-complete for some oracle A, rather than #P-complete like the permanent was, would not have the same symmetries as the permanent itself.

6.6.3 The Quest for Obstructions

Because the permanent and determinant are characterized by their symmetries, and because they satisfy another technical property called "partial stability," Mulmuley and Sohoni observe that a field called *geometric invariant theory* can be used to get a handle on their orbit closures. I won't explain the details of how this works (which involve something called Luna's Étale Slice Theorem [150]), but will just state the punchline.

Given a set $S \subseteq \mathbb{C}^N$, define $R[S]$, or the *coordinate ring of* S, to be the vector space of all complex polynomials $q : \mathbb{C}^N \to \mathbb{C}$, with two polynomials identified if they agree on all points $x \in S$. Then we'll be interested in $R_{\text{Det}} := R[\chi_{\text{Det},m}]$ and $R_{\text{Per}} := R[\chi_{\text{Per},m,n}]$: the coordinate rings of the orbit closures of the determinant and the padded permanent. In this case, $N = \binom{m^2+m-1}{m}$ is the dimension of the vector space of homogeneous degree-m polynomials over m^2 variables. So the coordinate rings are vector spaces of polynomials over N variables: truly enormous objects.

Next, let $q : \mathbb{C}^N \to \mathbb{C}$ be one of these "big" polynomials, whose inputs are the coefficients of a "small" polynomial p (such as the permanent or determinant). Then we can define an action of the general linear group, $G = \text{GL}_{m^2}(\mathbb{C})$, on q, via $(A \cdot q)(p(x)) := q(p(Ax))$ for all $A \in G$. In other words, we take the action of G on

give slightly more awkward symmetry conditions that avoided them. (This is especially clear for the permanent, since if A is diagonal, then $\text{Per}(A)$ is just the product of the diagonal entries.)

[62] See Grochow [95, Proposition 3.4.9] for a simple proof of this, via a dimension argument.

the "small" polynomials p that we previously defined, and use it to induce an action on the "big" polynomials q. Notice that this action fixes the coordinate rings R_{Det} and R_{Per} (i.e., just shifts their points around), simply because the action of G fixes the orbit closures $\chi_{\text{Det},m}$ and $\chi_{\text{Per},m,n}$ themselves. As a consequence, the actions on G on R_{Det} and R_{Per} give us two representations of the group G: that is, homomorphisms that map the elements of G to linear transformations on the vector spaces R_{Det} and R_{Per} respectively. Call these representations ρ_{Det} and ρ_{Per} respectively.

Like most representations, ρ_{Det} and ρ_{Per} can be decomposed uniquely into direct sums of *irreducible representations*, or "irreps" (which are not further decomposable). In particular, let $\rho : G \to \mathbb{C}^{k \times k}$ be any irrep of G. Then ρ occurs with some nonnegative integer *multiplicity*, call it $\lambda_{\text{Det}}(\rho)$, in ρ_{Det}, and with some possibly different multiplicity, call it $\lambda_{\text{Per}}(\rho)$, in ρ_{Per}. We're now ready for the theorem that sets the stage for the rest of GCT.

Theorem 85 (Mulmuley-Sohoni [167]). *Suppose there exists an irrep ρ such that $\lambda_{\text{Per}}(\rho) > \lambda_{\text{Det}}(\rho)$. Then $\text{Per}^*_{m,n} \notin \chi_{\text{Det},m}$: that is, the padded permanent is not in the orbit closure of the determinant.*

Note that Theorem 85 is not an "if and only if": even if $\text{Per}^*_{m,n} \notin \chi_{\text{Det},m}$, there's no result saying that the reason must be representation-theoretic. In GCT2 [168], Mulmuley and Sohoni *conjecture* that the algebraic geometry of $\chi_{\text{Det},m}$ is in some sense completely determined by its representation theory, but if true, that would have to be for reasons rather specific to $\chi_{\text{Det},m}$ (or other "complexity-theoretic" orbit closures).

If $\lambda_{\text{Per}}(\rho) > \lambda_{\text{Det}}(\rho)$, then Mulmuley and Sohoni call ρ an *obstruction* to embedding the permanent into the determinant. Any obstruction would be a *witness* to the permanent's hardness: if one likes, it would prove that the $m \times m$ determinant has "too much symmetry" to express the padded $n \times n$ permanent, unless m is much larger than n. From this point forward, GCT is focused entirely on the hunt for such an obstruction.

A priori, one could imagine proving nonconstructively that an obstruction ρ must exist, without actually finding the obstruction. However, Mulmuley and Sohoni emphatically reject that approach. They want not merely any proof of Conjecture 82, but an "explicit" proof: that is, one that yields an algorithm that actually *finds* an obstruction ρ witnessing $\text{Per}^*_{m,n} \notin \chi_{\text{Det},m}$, in time polynomial in m and n. Alas, as you might have gathered, the representations ρ_{Det} and ρ_{Per} are fearsomely complicated objects—so even if we accept for argument's sake that obstructions should exist, we seem a very long way from algorithms to find them in less than astronomical time.[63]

For now, therefore, Mulmuley and Sohoni argue that the best way to make progress toward Conjecture 82 is to work on *more and more efficient algorithms*

[63] In principle, ρ_{Det} and ρ_{Per} are infinite-dimensional representations, so an algorithm could search them forever for obstructions without halting. On the other hand, if we impose some upper bound on the degrees of the polynomials in the coordinate ring, we get an algorithm that takes "merely" doubly- or triply-exponential time.

to compute the multiplicities of irreps in complicated representations like ρ_{Det} and ρ_{Per}. The hope is that, in order to design those algorithms, we'll be forced to acquire such a deep understanding that we'll then know exactly where to look for a ρ such that $\lambda_{\mathrm{Per}}(\rho) > \lambda_{\mathrm{Det}}(\rho)$. So that's the program that they've pursued for the last decade, for example in GCT 3, 4, 7, and 8 [45, 159, 160, 166].

The central idea here—that the path to proving $\mathsf{P} \neq \mathsf{NP}$ will go through *discovering new algorithms*, rather than through ruling them out—is GCT's version of "ironic complexity theory," discussed in Sect. 6.4. What I've been calling "irony" in this survey, Mulmuley calls "The Flip" [161]: that is, flipping lower-bound problems into upper-bound problems, which we have a far better chance of solving.

Stepping back from the specifics of the GCT program, Mulmuley's view is that, before we prove (say) $\mathsf{NP} \not\subset \mathsf{P/poly}$, a natural intermediate goal is to find an algorithm A that takes a positive integer n as input, runs for $n^{O(1)}$ time (or even $\exp\left(n^{O(1)}\right)$ time), and then outputs a proof that 3SAT instances of size n have no circuits of size m, for some superpolynomial function m. Such an algorithm wouldn't immediately prove $\mathsf{NP} \not\subset \mathsf{P/poly}$, because we might still not know how to prove that A succeeded for every n. Even so, it would clearly be a titanic step forward, since we could run A and check that it *did* succeed for every n we chose, perhaps even for n's in the billions. At that point, we could say either that $\mathsf{NP} \not\subset \mathsf{P/poly}$, or else that $\mathsf{NP} \subset \mathsf{P/poly}$ only "kicks in" at such large values of n as to have few or no practical consequences. Furthermore, Mulmuley argues, we'd then be in a much better position to prove $\mathsf{NP} \not\subset \mathsf{P/poly}$ outright, since we'd "merely" have to analyze A, whose very existence would obviously encode enormous insight about the problem, and prove that it worked for all n.[64]

Mulmuley and others have indeed made progress in discovering efficient algorithms to the compute multiplicities of irreps, though the state-of-the-art is still extremely far from what Mulmuley conjectures is possible. To give an example, a *Littlewood-Richardson coefficient* is the multiplicity of a given irrep in a tensor product of two irreps of the general linear group $\mathrm{GL}_n(\mathbb{C})$. In GCT3 [166], Mulmuley et al. observed that a result of Knutson and Tao [132] implies that one can use linear programming, not necessarily to compute Littlewood-Richardson coefficients in polynomial time, but at least to decide whether they're positive or zero.[65] Bürgisser and Ikenmeyer [60] later gave a faster polynomial-time algorithm for the same problem; theirs was purely combinatorial and based on maximum flow. Note that, even if we only had efficient algorithms for positivity, those could still be useful for finding so-called *occurrence obstructions*: that is, irreps ρ such that

[64] A very loose analogy: well before Andrew Wiles proved Fermat's Last Theorem [233], that $x^n + y^n = z^n$ has no nontrivial integer solutions for any $n \geq 3$, number theorists knew a reasonably efficient algorithm that took an exponent n as input, and that (in practice, in all the cases that were tried) proved FLT for that n. Using that algorithm, in 1993—just before Wiles announced his proof—Buhler et al. [55] proved FLT for all n up to 4 million.

[65] As pointed out in Sect. 2.2.6, there are other cases in complexity theory where deciding positivity is much easier than exact counting: for example, deciding whether a graph has at least one perfect matching (counting the *number* of perfect matchings is #P-complete).

$\lambda_{\mathrm{Per}}(\rho) > 0$ even though $\lambda_{\mathrm{Det}}(\rho) = 0$. Thus, the "best-case scenario" for GCT is that the permanent's hardness can be witnessed not just by any obstructions, but by occurrence obstructions.

In many cases, we don't even know yet how to represent the multiplicity of an irrep as a #P function: at best, we can represent it as a *difference* between two #P functions. In those cases, the current research effort in GCT aims to give "positive formulas" for the multiplicities: in other words, to represent them as sums of exponentially many *nonnegative* terms, and thereby place their computation in #P itself. The hope is that a #P formula could be a first step toward a polynomial-time algorithm to decide positivity. To illustrate, Blasiak et al. achieved this for "two-row Kronecker coefficients" in GCT4 [45]. For other kinds of irreps, the quest for positive formulas has led to *extremely* deep waters, involving (for example) quantum groups and conjectured generalizations of the Riemann Hypothesis over finite fields. I won't go into this here, but see GCT7 [159] and GCT8 [160] for this part of the story.

6.6.4 GCT and $\mathsf{P} \stackrel{?}{=} \mathsf{NP}$

Suppose—let's dream—that everything above worked out perfectly. That is, suppose GCT led to the discovery of explicit obstructions for embedding the padded permanent into the determinant, and thence to a proof of Valiant's Conjecture 66. How would GCT go even further, to prove $\mathsf{P} \neq \mathsf{NP}$?

The short answer is that Mulmuley and Sohoni [167] defined an NP function called E, as well as a P-complete[66] function called H, and showed them to be characterized by symmetries in only a slightly weaker sense than the permanent and determinant are. The E and H functions aren't nearly as natural as the permanent and determinant, but they suffice to show that $\mathsf{P} \neq \mathsf{NP}$ could in principle be proven by finding explicit representation-theoretic obstructions, which in this case would be representations associated with the orbit of E but not with the orbit of H. Alas, because E and H are functions over a *finite* field $\mathbb{F}_q$ rather than over $\mathbb{C}$, the relevant algebraic geometry and representation theory would all be over finite fields as well. This leads to mathematical questions even less well-understood (!) than the ones discussed earlier, providing some additional support for the intuition that proving $\mathsf{P} \neq \mathsf{NP}$ should be "even harder" than proving Valiant's Conjecture.

For illustration, let me now define Mulmuley and Sohoni's E function. Let X_0 and X_1 be two $n \times n$ matrices over the finite field $\mathbb{F}_q$. Also, given a binary string $s = s_1 \cdots s_n$, let X_s be the $n \times n$ matrix obtained by choosing the ith column from X_0 if $s_i = 0$ or from X_1 if $s_i = 1$ for all $i \in \{1, \ldots, n\}$. We then set

[66] A language L is called P-*complete* if (1) $L \in \mathsf{P}$, and (2) every $L' \in \mathsf{P}$ can be reduced to L by some form of reduction weaker than arbitrary polynomial-time ones (LOGSPACE reductions are often used for this purpose).

$$F(X_0, X_1) := \prod_{s \in \{0,1\}^n} \mathrm{Det}(X_s),$$

and finally set

$$E(X_0, X_1) := 1 - F(X_0, X_1)^q$$

to obtain a function in $\{0, 1\}$. Testing whether $E(X_0, X_1) = 1$ is clearly an NP problem, since an NP witness is a single s such that $\mathrm{Det}(X_s) = 0$. Furthermore, Gurvits [100] showed that, at least over $\mathbb{Z}$, testing whether $F(X_0, X_1) = 0$ is NP-complete. Interestingly, it's not known whether E itself is NP-complete—though of course, to prove P $\neq$ NP, it would suffice to put any NP problem outside P, not necessarily an NP-complete one.

The main result about E (or rather F) is the following:

Theorem 86 (see Grochow [95, Proposition 3.4.6]). *Let $p : \mathbb{F}_q^{2n^2} \to \mathbb{F}_q$ be any homogeneous, degree-$n2^n$ polynomial in the entries of X_0 and X_1 that's divisible by* $\mathrm{Det}(X_0)\,\mathrm{Det}(X_1)$*, and suppose every linear symmetry of F is also a linear symmetry of p. Then $p(X_0, X_1) = \alpha F(X_0, X_1)$ for some $\alpha \in \mathbb{F}_q$.*

The proof of Theorem 86 involves some basic algebraic geometry. Even setting aside GCT, it would be interesting to know whether the existence of a plausibly-hard NP problem that's characterized by its symmetries had direct complexity-theoretic applications.

One last remark: for some complexity classes, such as BQP, we currently lack candidate problems characterized by their symmetries, so even by the speculative standards of this section, it's unclear how GCT could be used to prove (say) P $\neq$ BQP or NP $\not\subset$ BQP.

6.6.5 Reports from the Trenches

Within the past few years, there's been surprisingly rapid progress on connecting GCT to mainstream complexity theory; reproving known lower bounds using GCT or "GCT-like" methods; and investigating the truth or falsehood of some of GCT's main hypotheses. Let me mention a few highlights.

First, it's now known that the central property of the permanent and determinant that GCT seeks to exploit—namely, their characterization by symmetries—does indeed have complexity-theoretic applications. In particular, Mulmuley [161] observed that one can use the symmetry-characterization of the permanent to give a different, and in many ways nicer, proof of the classic result of Lipton [146] that the permanent is *self-testable*: that is, given a circuit C that's alleged to compute $\mathrm{Per}(X)$, in randomized polynomial time one can either verify that $C(X) = \mathrm{Per}(X)$

for most matrices X, or else find a counterexample where $C(X) \neq \text{Per}(X)$.[67] Subsequently, Kayal [124] took Mulmuley's observation further by exploiting symmetry-characterization more heavily. Recall Kayal's Theorem 80: that when $d \leq \sqrt{n}$, there's a randomized polynomial-time algorithm to decide whether a given degree-d polynomial p (to which we have black-box access) is the permanent or determinant of a $d \times d$ matrix of affine functions f_{ij}, and if so, to find suitable f_{ij}'s. Now, in the same paper [124], Kayal also showed that if q is an *arbitrary* polynomial, then deciding whether there exist A and b such that $p(x) = q(Ax + b)$ is NP-hard. So what is it about the permanent and determinant that made the problem so much easier? The answer is their symmetry-characterization, along with more specific properties of the Lie algebras of the stabilizer groups of Per and Det.

For more about the algorithmic consequences of symmetry-characterization, see for example Grochow [95, Chap. 4], who also partially derandomized Kayal's algorithm and applied it to other problems such as matrix multiplication. In my view, an exciting challenge right now is to use the symmetry-characterization of the permanent, or perhaps of the E function from Sect. 6.6.4, to prove other new complexity results—*not* necessarily circuit lower bounds—that hopefully evade the relativization and algebrization barriers.

In a second recent development, we now know that nontrivial circuit lower bounds—albeit, not state-of-the-art ones—can indeed be proved by finding representation-theoretic obstructions, just as GCT proposed. Recall from Sect. 6 that the *rank* of a 3-dimensional tensor $A \in \mathbb{F}^{n \times n \times n}$ is the smallest r such that A can be written as the sum of r rank-one tensors, $t_{ijk} = x_i y_j z_k$. If $\mathbb{F}$ is a continuous field like $\mathbb{C}$, then we can also define the *border rank* of A to be the smallest r such that A can be written as the *limit* of rank-r tensors. It's known that border rank can be strictly less than rank.[68] Border rank was introduced in 1980 by Bini et al. [44] to study matrix multiplication algorithms: indeed, one could call that the first appearance of orbit closures in computational complexity, decades before GCT.

More concretely, the *$n \times n$ matrix multiplication tensor*, say over $\mathbb{C}$, is defined to be the 3-dimensional tensor $M_n \in \mathbb{C}^{n^2 \times n^2 \times n^2}$ whose $(i,j), (j,k), (i,k)$ entries are all 1, and whose remaining $n^6 - n^3$ entries are all 0. In 1973, Strassen [215] proved a key result connecting the rank of M_n to the complexity of matrix multiplication:

Theorem 87 (Strassen [215]). *The rank of M_n equals the minimum number of nonscalar multiplications in any arithmetic circuit that multiplies two $n \times n$ matrices.*

---

[67] This result of Lipton's provided the germ of the proof that IP = PSPACE; see Sect. 6.3.1.
Mulmuley's test improves over Lipton's by, for example, requiring only nonadaptive queries to C rather than adaptive ones.

[68] A standard example is the $2 \times 2 \times 2$ tensor whose $(2, 1, 1)$, $(1, 2, 1)$, and $(1, 1, 2)$ entries are all 1, and whose 5 remaining entries are all 0. One can check that this tensor has a rank of 3 but border rank of 2.

As an immediate corollary, the border rank of M_n lower-bounds the arithmetic circuit complexity of $n \times n$ matrix multiplication. Bini [43] showed, moreover, that the exponent ω of matrix multiplication is the same if calculated using rank or border rank—so in that sense, the two are asymptotically equal.

In a tour-de-force, in 2004 Landsberg [136] proved that the border rank of M_2 is 7, using non-GCT differential geometry methods. A corollary was that any procedure to multiply two 2×2 matrices requires at least 7 nonscalar multiplications, precisely matching the upper bound discovered by Strassen [214] in 1969. (The trivial upper bound is 8 multiplications.)

I can now state what Bürgisser and Ikenmeyer [61] used GCT to achieve in 2013.

Theorem 88 (Bürgisser and Ikenmeyer [61]). *There are representation-theoretic (GCT) occurrence obstructions that witness that the border rank of M_n is at least $\frac{3}{2}n^2 - 2$.*

By comparison, the state-of-the-art lower bound on the border rank of matrix multiplication, obtained using non-GCT methods, is the following.

Theorem 89 (Landsberg and Ottaviani [138]). *The border rank of M_n is at least $2n^2 - n$.*

Of course, since both lower bounds are still quadratic, neither of them shows that matrix multiplication requires more than $\sim n^2$ time, but it's interesting to see how the best current GCT and non-GCT bounds compare.

Meanwhile, another recent insight is that most known circuit lower bounds can be put into a "broadly GCT-like" format. In an elegant 2014 paper, Grochow [96] showed that the AC^0 and $\mathsf{AC}^0[p]$ lower bounds of Sects. 6.2.3 and 6.2.4, the embedding lower bound of Mignon and Ressayre (Theorem 67), the lower bounds for small-depth arithmetic circuits and multilinear formulas of Sect. 6.5.2, and many other results can each be seen as constructing a *separating module*: that is, a "big polynomial" that takes as input the coefficients of an input polynomial, that vanishes for all polynomials in some complexity class $\mathcal{C}$, but that doesn't vanish for a polynomial q for which we're proving that $q \notin \mathcal{C}$. Interestingly, the lower bounds that *don't* fit into this format—such as $\mathsf{MA}_{\mathsf{EXP}} \not\subset \mathsf{P/poly}$ (Theorem 52) and $\mathsf{NEXP} \not\subset \mathsf{ACC}$ (Theorem 62)—essentially all use diagonalization as a key ingredient. Grochow's result is an important unification, and it *sounds* like an impressive success for GCT. On the other hand, it's important to understand that Grochow didn't show that the known circuit lower bounds yield *representation-theoretic* obstructions: only that they yield separating modules. In other words, we might say, Grochow showed that the known circuit lower bounds fit into an abstract construal of the "GCT program," but not that they fit into its really meaty part, into the part that makes GCT a *program*.

A third development, just in 2015, is that we now have some initial results from actual searches for irreps that could serve as obstructions to embedding the permanent into the determinant. Unfortunately, not all the news is encouraging. For example, Ikenmeyer and Panova [109] have shown that "rectangular Kronecker coefficients" can't be used to give occurrence obstructions separating $\chi_{\mathrm{Per},m,n}$ from

$\chi_{\text{Det},m}$. In other words, there's a large class of irreps such that, *if* they occur at least once in the padded permanent representation ρ_{Per}, then they also occur at least once in the determinant representation ρ_{Det}. Likewise, Bürgisser, Ikenmeyer, and Hüttenhain [62] have shown that any occurrence obstructions separating $\chi_{\text{Per},m,n}$ from $\chi_{\text{Det},m}$ must be "holes" of a certain monoid, further limiting where to look. It's entirely possible that these negative results are artifacts of "looking under the lamp-post"—i.e., only at those irreps ρ for which one can actually calculate today whether $\lambda_{\text{Per}}(\rho)$ and $\lambda_{\text{Det}}(\rho)$ are nonzero—but the worry is that irreps ρ such that $\lambda_{\text{Det}}(\rho) = 0$ might be incredibly rare, in which case occurrence obstructions could be hard to find if they even exist.

On the positive side, Ikenmeyer, Mulmuley, and Walter [108] have shown that there are superpolynomially many Kronecker coefficients that *do* vanish, thereby raising hope that occurrence obstructions might exist after all. Notably, they proved this result by first giving a #P formula for the relevant class of Kronecker coefficients, thereby illustrating the GCT strategy of first looking for algorithms and only later looking for the obstructions themselves.

6.6.6 The Lessons of GCT

Unsurprisingly, expert opinion is divided about GCT's prospects. Some feel that GCT does little more than take complicated questions and make them even more complicated. Others feel it's a perfectly natural and reasonable approach—maybe even the *only* extant approach that stands a chance—and that the complication is an inevitable byproduct of finally grappling with the real issues. Of course, one can also "cheer GCT from the sidelines" without feeling prepared to work on it oneself, particularly given the unclear prospects for any computer-science payoff in the foreseeable future. (Mulmuley once told me he thought it would take a hundred years until GCT led to major complexity class separations, and he's the *optimist*!)

Personally, I'd call myself a qualified fan of GCT, in much the same way and for the same reasons that I'm a qualified fan of string theory. I think all complexity theorists should learn about GCT—for one thing, because it has deep general lessons for the quest to prove $\mathsf{P} \neq \mathsf{NP}$, *even if* it ends up not succeeding, or not succeeding as Mulmuley envisioned. This section is devoted to what I believe those lessons are.

A first lesson is that we can in principle evade the relativization, algebrization, and natural proofs barriers by using the existence of complete problems with special properties: as a beautiful example, the property of being "characterized by symmetries," which the permanent and determinant both enjoy. A second lesson is that "ironic complexity theory" has even further reach than one might have thought: one could use the existence of surprisingly fast algorithms, not merely to show that certain complexity collapses would violate hierarchy theorems, but also to help *find certificates* that problems are hard. A third lesson is that there's at least one comprehensible route by which a circuit lower bound proof would need to know about huge swathes of "traditional, continuous" mathematics, as many computer scientists have long suspected (or feared!).

But none of those lessons really gets at how GCT changed my own thinking about $\mathsf{P} \stackrel{?}{=} \mathsf{NP}$. One of the most striking features of GCT is that, even as the approach stands today, it "knows" about various nontrivial problems in P, such as maximum flow and linear programming (because they're involved in deciding whether the multiplicities of irreps are nonzero). We knew, of course, that any proof of $\mathsf{P} \neq \mathsf{NP}$ would need to "know" that linear programming, matching, and so on are in P and are therefore different from 3SAT. (Indeed, that's one way to express the main difficulty of the $\mathsf{P} \stackrel{?}{=} \mathsf{NP}$ problem.) So the fact that GCT knows about all these polynomial-time algorithms seems both impressive and reassuring. But what's strange is that GCT seems to know the *upper* bounds, not the lower bounds—the power of the algorithms, but not their limitations! In other words, consider a hypothetical proof of $\mathsf{P} \neq \mathsf{NP}$ using GCT. If we ignore the details, and look from a distance of a thousand miles, the proof seems to be telling us not: "You see how weak P is? You see all these problems it can't solve?" but rather, "You see how strong P is? You see all these amazing, nontrivial problems it *can* solve?" The proof would seem to building up an impressive case for the wrong side of the argument!

One response would be to point out that this is math, not a political debate, and leave it at that. But perhaps one can do better. Let A be a hypothetical problem, like matching or linear programming, that's in P for a nontrivial reason, and that's also definable purely in terms of its symmetries, as the permanent and determinant are. Then we can define an orbit closure χ_A, which captures all problems reducible to A. By assumption, χ_A must be contained in χ_{P}, the orbit closure corresponding to a P-complete problem, such as Mulmuley and Sohoni's H function (see Sect. 6.6.4). And hence, there must *not* be any representation-theoretic obstruction to such a containment. In other words, if we were to compute the multiplicities $m_1, m_2, \ldots$ of all the irreps in the representation associated with χ_A, as well as the multiplicities $n_1, n_2, \ldots$ of the irreps associated with χ_{P}, we'd necessarily find that $m_i \leq n_i$ for all i.

Furthermore, by the general philosophy of GCT, once we had our long-sought positive formulas for these multiplicities (based on nonstandard quantum groups, the Riemann hypothesis, silly putty, herbal shampoo, or whatever), we might well be able to use those formulas to *prove* the above inequalities.

Now let's further conjecture—following GCT2 [168]—that the orbit closures χ_A and χ_{P} are completely captured by their representation-theoretic data. In that case, by showing that there's no representation-theoretic obstruction to $\chi_A \subseteq \chi_{\mathsf{P}}$, we would have proved, nonconstructively, that there *exists* a polynomial-time algorithm for A! And for that reason, we shouldn't be surprised if the algorithmic techniques that are used to solve A (matching, linear programming, or whatever) have already implicitly appeared in getting to this point. Indeed, we should be worried if they didn't appear.[69]

[69]It would be interesting to find a function in P, more natural than the P-complete H function, that's completely characterized by its symmetries, and then try to understand explicitly why there's

More broadly, Mulmuley has repeatedly stressed that $P \neq NP$ is a "universal mathematical statement": it says there's no polynomial-time algorithm for 3SAT, no matter which area of math we tried to use to construct such an algorithm. And therefore, after we find that an easy "opening gambit"—like the diagonalization used to proved the unsolvability of the halting problem—doesn't work, we shouldn't be shocked if nearly every area of math ends up playing some role in the proof.

Now, a natural reaction to this observation would be, not awe at the profundity of $P \stackrel{?}{=} NP$, but rather complete despair. Since math is infinite, and since the possible "ideas for polynomial-time algorithms" are presumably unbounded, why doesn't the "universality" of $P \stackrel{?}{=} NP$ mean that the task of proving could go on forever? At what point, according to the GCT philosophy, can we ever say "enough! we've discovered enough polynomial-time algorithms; now we're ready to flip things around and proceed to proving $P \neq NP$"?

The point I want to make is that the earlier considerations about χ_A and χ_P immediately suggest an answer to this question. Namely, we're ready to stop when we've discovered nontrivial polynomial-time algorithms, not for all problems in P, but for *all problems in* P *that are characterized by their symmetries*. For let B be a problem in P that isn't characterized by symmetries. Then the orbit closure χ_B is contained in χ_P, and if we could *prove* that $\chi_B \subseteq \chi_P$, then we would've nonconstructively shown the existence of a polynomial-time algorithm for B. But our hypothetical $P \neq NP$ proof doesn't need to know about that. For since B isn't characterized by symmetries, the GCT arguments aren't going to be able to prove $\chi_B \subseteq \chi_P$ anyway.

The above would seem to motivate a thorough investigation of which functions in P (or NC^1, etc.) can be characterized by their symmetries. If GCT is going to work at all, then the set of such functions, while presumably infinite, ought to be classifiable into a finite number of families. The speculation suggests itself that these families might roughly correspond to the different algorithmic techniques: Gaussian elimination, matching, linear programming, polynomial factorization, etc., and of course whatever other techniques haven't yet been discovered. As a concrete first step toward these lofty visions, it would be interesting to find some example of a function in P that's characterized by its symmetries, like the determinant is, and that's in P only because of the existence of nontrivial polynomial-time algorithms for (say) matching or linear programming.

6.6.7 Reservations

The reader might have gotten the impression, from the last section, that I think GCT is the most important advance on $P \stackrel{?}{=} NP$ ever made. In this section I'll set out a few of my reservations.

---

no representation-theoretic obstruction to that function's orbit closure being contained in χ_P—something we already know must be true.

First, it remains unclear whether we'll be able to use representation theory, in the foreseeable future, not merely to prove Valiant's Conjecture or $\mathsf{P} \neq \mathsf{NP}$, but to prove *any* new lower bound: for example, $\mathsf{EXP} \not\subset \mathsf{ACC}$, or the permanent requiring arithmetic formulas of size $\Omega(n^4)$, or really *anything* that would convince skeptics of GCT's power. I regard this as possible, and Grochow's demonstration [96] that most existing circuit lower bounds can be phrased in vaguely GCT-like ways is evidence for it. On the other hand, if we want "full GCT"—meaning not just separating modules, but specifically *representation-theoretic* obstructions—then it's also possible that the program is so challenging that getting it to work for anything is as hard as getting it to work for Valiant's Conjecture. In practice, I predict that there will be a rush among theoretical computer scientists to learn GCT when, and only when, it's used to prove *some* impressive new lower bound that had foiled other approaches.

Second, the story of the border rank of the matrix multiplication tensor, set out in Sect. 6.6.5, gives me pause, because it raises the possibility that even if representation-theoretic obstructions *do* exist, proving their existence will be *even harder* than proving complexity class separations in some more direct way. One possible "failure mode" for GCT is that, after centuries of struggle, mathematicians and computer scientists finally prove Valiant's Conjecture and $\mathsf{P} \neq \mathsf{NP}$—and then, after *further* centuries of struggle, it's shown that GCT could've proven these results as well (but only with quantitatively weaker bounds).

Third, there are major aspects of complexity theory that GCT seems not to capture. For example, we saw in Sect. 6.6.5 that *diagonalization*—which despite reports of its demise, has reliably shown up in one lower bound after another over the decades, from the Time Hierarchy Theorem to time-space tradeoffs to $\mathsf{NEXP} \not\subset \mathsf{ACC}$—seems to elude the GCT framework. A related point is that GCT, as it stands, has no way to take advantage of *uniformity*: for example, no way to prove $\mathsf{P} \neq \mathsf{NP}$, without also proving the stronger result $\mathsf{NP} \not\subset \mathsf{P/poly}$. However, given that we can prove $\mathsf{P} \neq \mathsf{EXP}$ but can't even prove $\mathsf{NEXP} \not\subset \mathsf{TC}^0$, it seems conceivable that uniformity could help in proving $\mathsf{P} \neq \mathsf{NP}$.

But perhaps my most important reservation is with a central argument that Mulmuley has offered for GCT in recent years [161, 163, 164]. His argument is that, even if GCT isn't literally the *only* way forward on $\mathsf{P} \stackrel{?}{=} \mathsf{NP}$, still, the choice of GCT to go after explicit obstructions is in some sense *provably unavoidable*—and furthermore, GCT is the "simplest" approach to finding the explicit obstructions, so Occam's Razor all but forces us to try GCT first. I completely agree that GCT represents a natural attack plan. It's something to try. But I don't think we have any theorem that can support the interpretation that GCT's choices are inevitable, or "basically" inevitable. And I can easily envision that progress will come by freely "mixing and matching" GCT ideas with non-GCT ones.

In more detail, we can identify at least four major successive decisions that GCT makes:

(1) To prove $\mathsf{P} \neq \mathsf{NP}$, we should start by proving Valiant's Conjecture 66.

(2) To prove Valiant's Conjecture, the natural approach is to prove Conjecture 82, about orbit closures.
(3) To prove Conjecture 82, the natural approach is to find explicit representation-theoretic embedding obstructions.
(4) To find those obstructions, we should start by finding faster algorithms (or algorithms in lower complexity classes) to learn about the multiplicities of irreps.

All four decisions are reasonable, but not one is obvious. And of course, even if every proposition in a list has high probability individually (or high probability conditioned on its predecessors), their conjunction could have probability close to zero!

As we saw in Sect. 6.5, decision (1) predates GCT by decades, so there's no need to revisit it here. Meanwhile, decision (2) seems to involve only a small strengthening of what needs to be proved, in return for a large gain in elegance. But there's plenty to question about decisions (3) and (4).

Regarding (3): we saw, in Sect. 6.6.5, that even if a given embedding of orbit closures is impossible, the reason might simply not be reflected in representation theory—and even if it is, it might be *harder* to prove that there's a representation-theoretic obstruction than that there's some other obstruction, and one might get only a weaker lower bound that way. At least, that's what seems to be true so far with the border rank of the matrix multiplication tensor. This would especially be an issue if it turned out that occurrence obstructions were rare or nonexistent, so that finding representation-theoretic obstructions would require counting the multiplicities of irreps and comparing them (rather than just checking whether the multiplicities are zero).

But let me concentrate on (4). Is it clear that we must, in Mulmuley's words, "go for explicitness": that is, look for an efficient algorithm that takes a specific n and m as input, and tries to find a witness that it's impossible to embed the padded $n \times n$ permanent into the $m \times m$ determinant? Why not just look directly for a *proof*, which (if we found it) would work for arbitrary n and $m = n^{O(1)}$?

Mulmuley's argument for explicitness rests on what he calls the "flip theorems" [161]. These theorems, in his interpretation, assert that *any* successful approach to circuit lower bounds (not just GCT) will yield explicit obstructions as a byproduct. And thus, all GCT is doing is bringing into the open what any proof of Valiant's Conjecture or $\mathsf{NP} \not\subset \mathsf{P/poly}$ will eventually need to confront anyway.

Let me now state some of the flip theorems. First, building on a 1996 learning algorithm of Bshouty et al. [54], in 2003 Fortnow, Pavan, and Sengupta showed that, if NP-complete problems are hard at all, then there must be short lists of instances that cause all small circuits to fail.

Theorem 90 (Fortnow et al. [82]). *Suppose* $\mathsf{NP} \not\subset \mathsf{P/poly}$. *Then for every n and k, there's a list of* 3SAT *instances $\varphi_1, \ldots, \varphi_\ell$, of length at most $\ell = n^{O(1)}$, such* *that every circuit C of size at most n^k fails to decide at least one φ_i in the list. Furthermore, such a list can be found in the class* $\mathsf{BPP}^{\mathsf{NP}}$.[70]

Atserias [31] showed that, in the statement of Theorem 90, we can also swap the NP oracle for an oracle for the circuit C itself: in other words, the list $\varphi_1, \ldots, \varphi_\ell$ can also be found in the class BPP^C.

Likewise, if the permanent is hard, then there must be short lists of matrices that cause all small arithmetic circuits to fail—and here the lists are much easier to find than they are in the Boolean case.

Theorem 91 (Mulmuley [161, 163]). *Suppose Valiant's Conjecture 66 holds (i.e., the permanent has no polynomial-size arithmetic circuits, over finite fields $\mathbb{F}$ with $|\mathbb{F}| \gg n$). Then for every n and k, there's a list of matrices $A_1, \ldots, A_\ell \in \mathbb{F}^{n \times n}$, of length at most $\ell = n^{O(1)}$, such that for every arithmetic circuit C of size at most n^k, there exists an i such that $C(A_i) \neq \operatorname{Per}(A_i)$. Indeed, a random list $A_1, \ldots, A_\ell$ will have that property with $1 - o(1)$ probability. Furthermore, if polynomial identity testing has a black-box derandomization,[71] then such a list can be found in deterministic $n^{O(1)}$ time.*

While not hard to prove, Theorems 90 and 91 are conceptually interesting: they show that the entire hardness of 3SAT and of the permanent can be "concentrated" into a small number of instances. My difficulty is that *this* sort of "explicit obstruction" to computing 3SAT or the permanent, seems barely related to the sorts of explicit obstructions that GCT is seeking. The obstructions of Theorems 90 and 91 aren't representation-theoretic; they're simply lists of hard instances. Furthermore, a list like $\varphi_1, \ldots, \varphi_\ell$ or $A_1, \ldots, A_\ell$ is *not* an easy-to-verify witness that 3SAT or the permanent is hard, because we'd still need to check that the list worked against all of the exponentially many $n^{O(1)}$-sized circuits. Having such a list reduces a two-quantifier (Π_2^{P}) problem to a one-quantifier (NP) problem, but it still doesn't put the problem in P—and we have no result saying that if, for example, the permanent is hard, then there must be obstructions that can be verified in $n^{O(1)}$ time. Perhaps the best we can say is that, if we *proved* the permanent was hard, then we'd immediately get, for every n, an "obstruction" that could be both found and verified in 0 time steps! But for all we know, the complexity of finding provable obstructions could jump from $\exp\left(n^{O(1)}\right)$ to 0 as our understanding improved, without ever passing through $n^{O(1)}$.

Thus, I find, GCT's suggestion to look for faster obstruction-finding (or obstruction-recognizing) algorithms is a useful *guide*, a heuristic, a way to organize

---

[70] In fact, the list can be found in the class $\mathsf{ZPP}^{\mathsf{NP}}$, where ZPP stands for Zero-Error Probabilistic Polynomial-Time. This means that, whenever the randomized algorithm succeeds in constructing the list, it's *certain* that it's done so.

[71] The polynomial identity testing problem was defined in Sect. 5.4. Also, by a "black-box derandomization," we mean a deterministic polynomial-time algorithm that outputs a *hitting set*: that is, a list of points $x_1, \ldots, x_\ell$ such that, for all small arithmetic circuits C that don't compute the identically-zero polynomial, there exists an i such that $C(x_i) \neq 0$. What makes the derandomization "black-box" is that the choice of $x_1, \ldots, x_\ell$ doesn't depend on C.

our thoughts about how on earth we're going to find, say, an irrep ρ that blocks the embedding of the permanent into the determinant. But it's not the only possible way forward.

7 Conclusions

Some will say that this survey's very length, the bewildering zoo of approaches and variations and results and barriers that it covered, is a sign that no one has any real clue about the $\mathsf{P} \stackrel{?}{=} \mathsf{NP}$ problem—or at least, that *I* don't. Among those who think that, perhaps someone will write a shorter survey that points unambiguously to the right way forward!

But it's also possible that the business of proving brute-force search unavoidable seems complicated because it *is* complicated—because the reasons why brute-force search *can* be avoided in specific cases are as diverse as mathematics itself, and because here (unlike, say, for the halting problem), an argument ruling out all those reasons must be similarly deep and wide-ranging. I confess to limited sympathy for the idea that someone will just set aside everything that's already known, think hard about the structural properties of the sets of languages accepted by deterministic and nondeterministic polynomial-time Turing machines, and find a property that holds for one set but not the other, thereby proving $\mathsf{P} \neq \mathsf{NP}$. For I keep coming back to the question: if a hands-off, aprioristic approach sufficed for $\mathsf{P} \neq \mathsf{NP}$, then why did it apparently *not* suffice for all the weaker separations that we've surveyed here?

At the same time, I hope our tour of the progress in lower bounds has made the case that there's no reason (yet!) to elevate $\mathsf{P} \stackrel{?}{=} \mathsf{NP}$ to some plane of metaphysical unanswerability, or assume it to be independent of the axioms of set theory, or anything like that. The experience of complexity theory, including the superpolynomial lower bounds that people *did* manage to prove after struggle and effort, is consistent with $\mathsf{P} \stackrel{?}{=} \mathsf{NP}$ being "merely a math problem"—albeit, a math problem that happens to be well beyond the current abilities of civilization, much like the solvability of the quintic in the 1500s, or Fermat's Last Theorem in the 1700s. When we're faced with such a problem, a natural response is to want to deepen our understanding of the entire subject (in this case, algorithms and computation) surrounding the problem—not merely because that's a prerequisite to someday capturing the famous beast, but because *regardless*, the new knowledge gained along the way will hopefully find uses elsewhere. In our case, modern cryptography, quantum computing, and parts of machine learning could all be seen as flowers that bloomed in the garden of $\mathsf{P} \stackrel{?}{=} \mathsf{NP}$.

Obviously I don't know how $\mathsf{P} \neq \mathsf{NP}$ will ultimately be proved—if I did, this would be a very different survey! It seems plausible that a successful approach would yield the stronger result $\mathsf{NP} \not\subset \mathsf{P/poly}$ (i.e., that it wouldn't take advantage of uniformity); that it would start by proving Valiant's Conjecture (i.e., that the algebraic case would precede the Boolean one); and that it would draw on many core areas of mathematics, but none of these things are even close to certain. The only prediction I feel confident in making is that the idea of "ironic complexity theory"—i.e., of a profound duality between upper and lower bounds, where the way to prove that there's no fast algorithm for problem A is to *discover* fast algorithms for problems B, C, and D—is here to stay. As we saw, ironic complexity theory is at the core of the GCT program, but it's *also* at the core of Williams's proof of NEXP $\not\subset$ ACC, which in other respects is about as far from GCT as possible. The natural proofs barrier also provides a sort of contrapositive to ironic complexity, showing how the *nonexistence* of efficient algorithms is often what *prevents* us from proving lower bounds. If 3SAT is ever placed outside P, I'm willing to bet that the proof will place many other problems *inside* P—or at any rate, in smaller complexity classes than previously known.

So, if we take an optimistic attitude (optimistic about proving intractability!), then which breakthroughs should we seek next? What's on the current horizon? There are hundreds of possible answers to that question—we've already encountered some in this survey—but if I had to highlight a few:

- Prove lower bounds against nonuniform TC^0—for example, by finding a better-than-brute-force algorithm for the neural network satisfiability problem.[72]
- Prove a lower bound better than $n^{\lfloor d/2 \rfloor}$ on the rank of an explicit d-dimensional tensor, or construct one of the many other algebraic or combinatorial objects (rigid matrices, elusive functions, etc.) that are known to imply new circuit lower bounds.
- Advance proof complexity to the point where we could, for example, prove a superpolynomial lower bound on the number of steps needed to convert some n-input, polynomial-size Boolean circuit C into some equivalent circuit C', via moves that each swap out a size-$O(1)$ subcircuit for a different size-$O(1)$ subcircuit with identical input/output behavior.[73]
- Prove a superpolynomial lower bound on the number of 2-qubit quantum gates needed to implement some explicit n-qubit unitary transformation U. (Remarkably, as far as anyone knows today, one could succeed at this without needing to prove *any* new classical circuit lower bound. On the other hand, it's plausible that one would need to overcome a unitary version of the natural proofs barrier.)

[72]In 1999, Allender [19] showed that the permanent, and various other natural #P-complete problems, can't be solved by LOGTIME-*uniform* TC^0 circuits: in other words, constant-depth threshold circuits for which there's an $O(\log n)$-time algorithm to output the ith bit of their description, for any i. Indeed, these problems can't be solved by LOGTIME-uniform TC^0 circuits of size $f(n)$, where f is any function that can yield an exponential when iterated a constant number of times. The proof uses a hierarchy theorem, it would be interesting to know whether it relativizes.

[73]Examples are deleting two successive NOT gates, or applying de Morgan's laws. By the completeness of Boolean algebra, one can give local transformation rules that suffice to convert any n-input Boolean circuit into any equivalent circuit using at most $\exp(n)$ moves.

From talking to experts, this problem seems closely related to the problem of proving superpolynomial lower bounds for so-called *Frege proofs*, but is possibly easier.

- Clarify whether ACC has a natural proofs barrier, and prove ACC lower bounds for problems in EXP or below.
- Clarify whether there's an arithmetic natural proofs barrier, or something else preventing us from crossing the "chasm at depth three."
- Prove any lower bound on $D(n)$, the determinantal complexity of the $n \times n$ permanent, better than Mignon and Ressayre's $\Omega\left(n^2/2\right)$.
- Derandomize polynomial identity testing—or failing that, prove *some* derandomization theorem that implies a strong new circuit lower bound.
- Discover a new class of polynomial-time algorithms, especially but not only for computing the multiplicities of irreps.
- Pinpoint additional special properties of 3SAT or the permanent that could be used to evade the natural proofs barrier, besides the few that are already known, such as symmetry-characterization and the ability to simulate machines in a hierarchy theorem.
- Prove any interesting new lower bound by finding a representation-theoretic obstruction to an embedding of orbit closures.
- Perhaps my favorite challenge: *find new, semantically-interesting ways to "hobble" the classes of polynomial-time algorithms and polynomial-size circuits*, besides the ways that have already been studied, such as restricted memory, restricted circuit depth, monotone gates only, arithmetic operations only, and restricted families of algorithms (such as DPLL and certain linear and semidefinite programming relaxations). Any such restriction that one discovers is effectively a new slope that one can try to ascend up the $\mathsf{P} \neq \mathsf{NP}$ mountain, possibly gentler than the previous slopes.

Going even further on a limb, I've long wondered whether massive computer search could give any insight into complexity lower bound questions, beyond the relatively incidental ways it's been used for this purpose already (e.g., [237]). For example, could we feasibly discover the smallest arithmetic circuits to compute the permanents of 4×4 and 5×5 and 6×6 matrices? And if we did, would examination of those circuits yield any clues about what to prove for general n? The conventional wisdom has always been "no" to both questions. For firstly, as far as anyone knows today, the computational complexity of such a search will grow not merely exponentially but *doubly* exponentially with n (assuming the optimal circuits that we're trying to find grow exponentially themselves), and examination of the constants suggests that $n = 4$ might already be out of reach. And secondly, even if we knew the optimal circuits, they'd tell us nothing about the existence or nonexistence of clever algorithms that start to win only at much larger values of n. After all, many of the theoretically efficient algorithms that we know today only overtake the naïve algorithms for n's in the thousands or millions.[74]

[74]There are even what Richard Lipton has termed "galactic algorithms" [147], which beat their asymptotically-worse competitors, but only for values of n that likely exceed the information storage capacity of the galaxy or the observable universe. The currently-fastest matrix multiplication

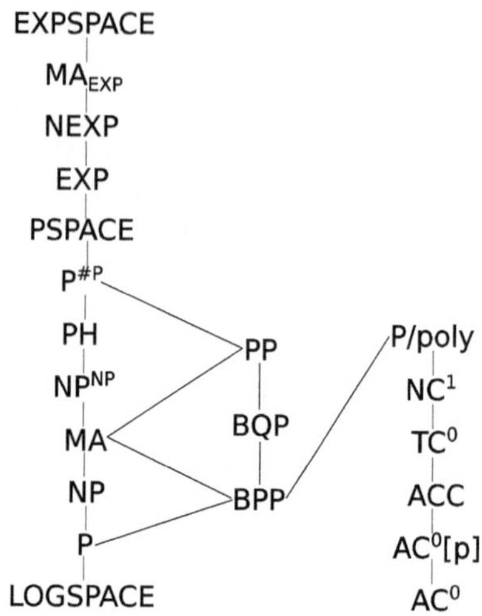

Fig. 3 Known inclusion relations among 21 of the complexity classes that appear in this survey

These arguments have force. Even so, I think we should be open to the possibility that someday, advances in raw computer power, and especially theoretical advances that decrease the effective size of the search space, might change the calculus and open up the field of "experimental complexity theory." One way that could happen would be if breakthroughs in GCT, or some other such program, brought the quest for "explicit obstructions" (say, to the 1000×1000 permanent having size-10^{20} circuits) within the range of computer search, if still not within the range of the human mind.

Or of course, the next great advance might come from some direction that I didn't even mention here, whether because no one grasps its importance yet or simply because *I* don't. But regardless of what happens, the best fate this survey could possibly enjoy would be to contribute, in some tiny way, to making itself outdated.

Appendix: Glossary of Complexity Classes

To help you remember all the supporting characters in the ongoing soap opera of which P and NP are the stars, this appendix contains short definitions of the complexity classes that appear in this survey, with references to the sections where the classes are discussed in more detail. For a fuller list, containing over 500 classes, see for example my Complexity Zoo [8]. All the classes below are classes of decision problems—that is, languages $L \subseteq \{0, 1\}^*$. The known inclusion relations among many of the classes are also depicted in Fig. 3.

---

algorithms *might* fall into this category, although the constants don't seem to be known in enough detail to say for sure.

Π_2^{P}: $\mathsf{coNP}^{\mathsf{NP}}$, the second level of the polynomial hierarchy (with universal quantifier in front). See Sect. 2.2.3.

Σ_2^{P}: $\mathsf{NP}^{\mathsf{NP}}$, the second level of the polynomial hierarchy (with existential quantifier in front). See Sect. 2.2.3.

AC^0: The class decidable by a nonuniform family of polynomial-size, constant-depth, unbounded-fanin circuits of AND, OR, and NOT gates. See Sect. 6.2.3.

$\mathsf{AC}^0\,[m]$: AC^0 enhanced by MOD m gates, for some specific value of m (the case of m a prime power versus a non-prime-power are dramatically different). See Sect. 6.2.4.

ACC: AC^0 enhanced by MOD m gates, for every m simultaneously. See Sect. 6.4.2.

BPP: Bounded-Error Probabilistic Polynomial-Time. The class decidable by a polynomial-time randomized algorithm that errs with probability at most $1/3$ on each input. See Sect. 5.4.1.

BQP: Bounded-Error Quantum Polynomial-Time. The same as BPP except that we now allow quantum algorithms. See Sect. 5.5.

coNP: The class consisting of the complements of all languages in NP. Complete problems include unsatisfiability, graph non-3-colorability, etc. See Sect. 2.2.3.

$\mathsf{DTISP}\,(f\,(n)\,,g\,(n))$: See Sect. 6.4.1.

EXP: Exponential-Time, or $\bigcup_k \mathsf{TIME}\left(2^{n^k}\right)$. Note the permissive definition of "exponential," which allows any polynomial in the exponent. See Sect. 2.2.7.

EXPSPACE: Exponential-Space, or $\bigcup_k \mathsf{SPACE}\left(2^{n^k}\right)$. See Sect. 2.2.7.

IP: Interactive Proofs, the class for which a "yes" answer can be proven (to statistical certainty) via an interactive protocol in which a polynomial-time verifier Arthur exchanges a polynomial number of bits with a computationally-unbounded prover Merlin. Turns out to equal PSPACE [200]. See Sect. 6.3.1.

LOGSPACE: Logarithmic-Space, or $\mathsf{SPACE}\,(\log n)$. Note that only read/write memory is restricted to $O\,(\log n)$ bits; the n-bit input itself is stored in a read-only memory. See Sect. 6.4.1.

MA: Merlin-Arthur, the class for which a "yes" answer can be proven to statistical certainty via a polynomial-size message from a prover ("Merlin"), which the verifier ("Arthur") then verifies in probabilistic polynomial time. Same as NP except that the verification can be probabilistic. See Sect. 6.3.2.

$\mathsf{MA}_{\mathsf{EXP}}$: The exponential-time analogue of MA, where now Merlin's proof can be $2^{n^{O(1)}}$ bits long, and Arthur's probabilistic verification can also take $2^{n^{O(1)}}$ time. See Sect. 6.3.2.

NC^1: The class decidable by a nonuniform family of polynomial-size Boolean formulas—or equivalently, polynomial-size Boolean circuits of fanin 2 and depth $O\,(\log n)$. The subclass of P/poly that is "highly parallelizable." See Sect. 5.2.

NEXP: Nondeterministic Exponential-Time, or $\bigcup_k \mathsf{NTIME}\left(2^{n^k}\right)$. The exponential-time analogue of NP. See Sect. 2.2.7.

NP: Nondeterministic Polynomial-Time, or $\bigcup_k \mathsf{NTIME}(n^k)$. The class for which a "yes" answer can be proven via a polynomial-size witness, which is verified by a deterministic polynomial-time algorithm. See Sect. 2.

NTIME $(f(n))$: Nondeterministic $f(n)$-Time. The class for which a "yes" answer can be proven via an $O(f(n))$-bit witness, which is verified by a deterministic $O(f(n))$-time algorithm. Equivalently, the class solvable by a nondeterministic $O(f(n))$-time algorithm. See Sect. 2.2.7.

P: Polynomial-Time, or $\bigcup_k \mathsf{TIME}(n^k)$. The class solvable by a deterministic polynomial-time algorithm. See Sect. 2.

$\mathsf{P}^{\#\mathsf{P}}$: P with an oracle for #P problems (i.e., for counting the exact number of accepting witnesses for any problem in NP). See Sect. 2.2.6.

P/poly: P enhanced by polynomial-size "advice strings" $\{a_n\}_n$, which depend only on the input size n but can otherwise be chosen to help the algorithm as much as possible. Equivalently, the class solvable by a nonuniform family of polynomial-size Boolean circuits (i.e., a different circuit is allowed for each input size n). See Sect. 5.2.

PH: The Polynomial Hierarchy. The class expressible via a polynomial-time predicate with a constant number of alternating universal and existential quantifiers over polynomial-size strings. Equivalently, the union of $\Sigma_1^{\mathsf{P}} = \mathsf{NP}$, $\Pi_1^{\mathsf{P}} = \mathsf{coNP}$, $\Sigma_2^{\mathsf{P}} = \mathsf{NP}^{\mathsf{NP}}$, $\Pi_2^{\mathsf{P}} = \mathsf{coNP}^{\mathsf{NP}}$, and so on. See Sect. 2.2.3.

PP: Probabilistic Polynomial-Time. The class decidable by a polynomial-time randomized algorithm that, for each input x, guesses the correct answer with probability greater than $1/2$. Like BPP but without the bounded-error ($1/3$ versus $2/3$) requirement, and accordingly believed to be much more powerful. See Sect. 2.2.6.

PSPACE: Polynomial-Space, or $\bigcup_k \mathsf{SPACE}(n^k)$. See Sect. 2.2.5.

SPACE $(f(n))$: The class decidable by a serial, deterministic algorithm that uses $O(f(n))$ bits of memory (and possibly up to $2^{O(f(n))}$ time). See Sect. 2.2.7.

TC^0: AC^0 enhanced by MAJORITY gates. Also corresponds to "neural networks" (polynomial-size, constant-depth circuits of threshold gates). See Sect. 6.2.5.

TIME $(f(n))$: The class decidable by a serial, deterministic algorithm that uses $O(f(n))$ time steps (and therefore, $O(f(n))$ bits of memory). See Sect. 2.2.7.

Acknowledgements I thank Andy Drucker, Adam Klivans, Ashley Montanaro, Leslie Valiant, and Ryan Williams for helpful observations and for answering questions. I especially thank Michail Rassias for his exponential patience with my delays completing this article. Supported by an NSF Waterman Award, under grant no. 1249349.

References

1. S. Aaronson. Is P versus NP formally independent? *Bulletin of the EATCS*, (81), October 2003.
2. S. Aaronson. Multilinear formulas and skepticism of quantum computing. In *Proc. ACM STOC*, pages 118–127, 2004. quant-ph/0311039, www.scottaaronson.com/papers/mlinsiam.

pdf.
3. S. Aaronson. NP-complete problems and physical reality. *SIGACT News*, March 2005. quant-ph/0502072.
4. S. Aaronson. Oracles are subtle but not malicious. In *Proc. Conference on Computational Complexity*, pages 340–354, 2006. ECCC TR05-040.
5. S. Aaronson. Arithmetic natural proofs theory is sought, 2008. www.scottaaronson.com/blog/?p=336.
6. S. Aaronson. *Quantum Computing Since Democritus*. Cambridge University Press, 2013.
7. S. Aaronson. The scientific case for $P \neq NP$, 2014. www.scottaaronson.com/blog/?p=1720.
8. S. Aaronson et al. The Complexity Zoo. www.complexityzoo.com.
9. S. Aaronson, R. Impagliazzo, and D. Moshkovitz. AM with multiple Merlins. In *Proc. Conference on Computational Complexity*, pages 44–55, 2014. arXiv:1401.6848.
10. S. Aaronson and A. Wigderson. Algebrization: a new barrier in complexity theory. *ACM Trans. on Computation Theory*, 1(1), 2009. Earlier version in Proc. ACM STOC'2008.
11. L. Adleman. Two theorems on random polynomial time. In *Proc. IEEE FOCS*, pages 75–83, 1978.
12. L. Adleman, J. DeMarrais, and M.-D. Huang. Quantum computability. *SIAM J. Comput.*, 26(5):1524–1540, 1997.
13. M. Agrawal. Determinant versus permanent. In *Proceedings of the International Congress of Mathematicians*, 2006.
14. M. Agrawal, N. Kayal, and N. Saxena. PRIMES is in P. *Annals of Mathematics*, 160(2):781–793, 2004. Preprint released in 2002.
15. M. Agrawal and V. Vinay. Arithmetic circuits: a chasm at depth four. In *Proc. IEEE FOCS*, pages 67–75, 2008.
16. M. Ajtai. Σ_1^1-formulae on finite structures. *Annals of Pure and Applied Logic*, 24(1):1–48, 1983.
17. M. Ajtai. A non-linear time lower bound for Boolean branching programs. *Theory of Computing*, 1(1):149–176, 2005. Earlier version in Proc. IEEE FOCS'1999, pp. 60–70.
18. B. Alexeev, M. A. Forbes, and J. Tsimerman. Tensor rank: some lower and upper bounds. In *Proc. Conference on Computational Complexity*, pages 283–291, 2011.
19. E. Allender. The permanent requires large uniform threshold circuits. *Chicago Journal of Theoretical Computer Science*, 7:19, 1999.
20. E. Allender. Cracks in the defenses: scouting out approaches on circuit lower bounds. In *Computer Science in Russia*, pages 3–10, 2008.
21. E. Allender. A status report on the P versus NP question. *Advances in Computers*, 77:117–147, 2009.
22. E. Allender and V. Gore. On strong separations from AC^0. In *Fundamentals of Computation Theory*, pages 1–15. Springer Berlin Heidelberg, 1991.
23. E. Allender and V. Gore. A uniform circuit lower bound for the permanent. *SIAM J. Comput.*, 23(5):1026–1049, 1994.
24. E. Allender and M. Koucký. Amplifying lower bounds by means of self-reducibility. *J. of the ACM*, 57(3):1–36, 2010. Earlier version in Proc. IEEE Complexity'2008, pp. 31–40.
25. N. Alon and R. B. Boppana. The monotone circuit complexity of Boolean functions. *Combinatorica*, 7(1):1–22, 1987.
26. A. E. Andreev. On a method for obtaining more than quadratic effective lower bounds for the complexity of π-schemes. *Moscow Univ. Math. Bull.*, 42:63–66, 1987. In Russian.
27. S. Arora and B. Barak. *Complexity Theory: A Modern Approach*. Cambridge University Press, 2009. Online draft at www.cs.princeton.edu/theory/complexity/.
28. S. Arora, R. Impagliazzo, and U. Vazirani. Relativizing versus nonrelativizing techniques: the role of local checkability. Manuscript, 1992.
29. S. Arora, C. Lund, R. Motwani, M. Sudan, and M. Szegedy. Proof verification and the hardness of approximation problems. *J. of the ACM*, 45(3):501–555, 1998. Earlier version in Proc. IEEE FOCS'1992, pp. 14–23.

30. S. Arora and S. Safra. Probabilistic checking of proofs: a new characterization of NP. *J. of the ACM*, 45(1):70–122, 1998. Earlier version in Proc. IEEE FOCS'1992, pp. 2–13.
31. A. Atserias. Distinguishing SAT from polynomial-size circuits, through black-box queries. In *Proc. Conference on Computational Complexity*, pages 88–95, 2006.
32. L. Babai. Graph isomorphism in quasipolynomial time. arXiv:1512.03547, 2015.
33. A. Backurs and P. Indyk. Edit distance cannot be computed in strongly subquadratic time (unless SETH is false). In *Proc. ACM STOC*, pages 51–58, 2015.
34. T. Baker, J. Gill, and R. Solovay. Relativizations of the P=?NP question. *SIAM J. Comput.*, 4:431–442, 1975.
35. A. Banerjee, C. Peikert, and A. Rosen. Pseudorandom functions and lattices. In *Proc. of EUROCRYPT*, pages 719–737, 2012.
36. P. Beame and T. Pitassi. Propositional proof complexity: past, present, and future. *Current Trends in Theoretical Computer Science*, pages 42–70, 2001.
37. P. Beame, M. E. Saks, X. Sun, and E. Vee. Time-space trade-off lower bounds for randomized computation of decision problems. *J. of the ACM*, 50(2):154–195, 2003. Earlier version in Proc. IEEE FOCS'2000, pp. 169–179.
38. R. Beigel and J. Tarui. On ACC. *Computational Complexity*, 4:350–366, 1994. Earlier version in Proc. IEEE FOCS'1991, pp. 783–792.
39. S. Ben-David and S. Halevi. On the independence of P versus NP. Technical Report TR714, Technion, 1992.
40. C. Bennett, E. Bernstein, G. Brassard, and U. Vazirani. Strengths and weaknesses of quantum computing. *SIAM J. Comput.*, 26(5):1510–1523, 1997. quant-ph/9701001.
41. C. H. Bennett and J. Gill. Relative to a random oracle A, $P^A \neq NP^A \neq coNP^A$ with probability 1. *SIAM J. Comput.*, 10(1):96–113, 1981.
42. E. Bernstein and U. Vazirani. Quantum complexity theory. *SIAM J. Comput.*, 26(5):1411–1473, 1997. Earlier version in Proc. ACM STOC'1993.
43. D. Bini. Relations between exact and approximate bilinear algorithms. applications. *Calcolo*, 17(1):87–97, 1980.
44. D. Bini, G. Lotti, and F. Romani. Approximate solutions for the bilinear form computational problem. *SIAM J. Comput.*, 9(4):692–697, 1980.
45. J. Blasiak, K. Mulmuley, and M. Sohoni. *Geometric complexity theory IV: nonstandard quantum group for the Kronecker problem*, volume 235 of *Memoirs of the American Mathematical Society*. 2015. arXiv:cs.CC/0703110.
46. L. Blum, F. Cucker, M. Shub, and S. Smale. *Complexity and Real Computation*. Springer-Verlag, 1997.
47. A. Bogdanov and L. Trevisan. Average-case complexity. *Foundations and Trends in Theoretical Computer Science*, 2(1), 2006. ECCC TR06-073.
48. A. D. Bookatz. QMA-complete problems. *Quantum Information and Computation*, 14(5-6):361–383, 2014. arXiv:1212.6312.
49. R. B. Boppana, J. Håstad, and S. Zachos. Does co-NP have short interactive proofs? *Inform. Proc. Lett.*, 25:127–132, 1987.
50. A. Braunstein, M. Mézard, and R. Zecchina. Survey propagation: An algorithm for satisfiability. *Random Structures and Algorithms*, 27(2):201–226, 2005.
51. M. Braverman. Poly-logarithmic independence fools AC^0 circuits. In *Proc. Conference on Computational Complexity*, pages 3–8, 2009. ECCC TR09-011.
52. R. P. Brent. The parallel evaluation of general arithmetic expressions. *J. of the ACM*, 21:201–206, 1974.
53. N. H. Bshouty, R. Cleve, and W. Eberly. Size-depth tradeoffs for algebraic formulae. *SIAM J. Comput.*, 24(4):682–705, 1995. Earlier version in Proc. IEEE FOCS'1991, pp. 334–341.
54. N. H. Bshouty, R. Cleve, R. Gavaldà, S. Kannan, and C. Tamon. Oracles and queries that are sufficient for exact learning. *J. Comput. Sys. Sci.*, 52(3):421–433, 1996.
55. J. Buhler, R. Crandall, R. Ernvall, and T. Metsänkylä. Irregular primes and cyclotomic invariants to four million. *Mathematics of Computation*, 61(203):151–153, 1993.

56. H. Buhrman, L. Fortnow, and T. Thierauf. Nonrelativizing separations. In *Proc. Conference on Computational Complexity*, pages 8–12, 1998.
57. P. Bürgisser. Completeness and reduction in algebraic complexity theory. 2000. Available at math-www.uni-paderborn.de/agpb/work/habil.ps.
58. P. Bürgisser. Cook's versus Valiant's hypothesis. *Theoretical Comput. Sci.*, 235(1):71–88, 2000.
59. P. Bürgisser. On defining integers and proving arithmetic circuit lower bounds. *Computational Complexity*, 18(1):81–103, 2009. Earlier version in Proc. STACS'2007, pp. 133–144.
60. P. Bürgisser and C. Ikenmeyer. Deciding positivity of Littlewood-Richardson coefficients. *SIAM J. Discrete Math.*, 27(4):1639–1681, 2013.
61. P. Bürgisser and C. Ikenmeyer. Explicit lower bounds via geometric complexity theory. In *Proc. ACM STOC*, pages 141–150, 2013. arXiv:1210.8368.
62. P. Bürgisser, C. Ikenmeyer, and J. Hüttenhain. Permanent versus determinant: not via saturations. arXiv:1501.05528, 2015.
63. S. R. Buss and R. Williams. Limits on alternation trading proofs for time-space lower bounds. *Computational Complexity*, 24(3):533–600, 2015. Earlier version in Proc. IEEE Complexity'2012, pp. 181–191.
64. J.-Y. Cai, X. Chen, and D. Li. Quadratic lower bound for permanent vs. determinant in any characteristic. 19(1):37–56, 2010. Earlier version in Proc. ACM STOC'2008, pp. 491–498.
65. A. Cobham. The intrinsic computational difficulty of functions. In *Proceedings of Logic, Methodology, and Philosophy of Science II*. North Holland, 1965.
66. S. Cook. The P versus NP problem, 2000. Clay Math Institute official problem description. At www.claymath.org/sites/default/files/pvsnp.pdf.
67. S. A. Cook. The complexity of theorem-proving procedures. In *Proc. ACM STOC*, pages 151–158, 1971.
68. S. A. Cook. A hierarchy for nondeterministic time complexity. In *Proc. ACM STOC*, pages 187–192, 1972.
69. D. Coppersmith. Rapid multiplication of rectangular matrices. *SIAM J. Comput.*, 11(3):467–471, 1982.
70. D. Coppersmith and S. Winograd. Matrix multiplication via arithmetic progressions. *Journal of Symbolic Computation*, 9(3):251–280, 1990. Earlier version in Proc. ACM STOC'1987.
71. C. Daskalakis, P. W. Goldberg, and C. H. Papadimitriou. The complexity of computing a Nash equilibrium. *Commun. of the ACM*, 52(2):89–97, 2009. Earlier version in Proc. ACM STOC'2006.
72. M. Davis, G. Logemann, and D. Loveland. A machine program for theorem proving. *Commun. of the ACM*, 5(7):394–397, 1962.
73. V. Deolalikar. $P \neq NP$. Archived version available at www.win.tue.nl/~gwoegi/P-versus-NP/Deolalikar.pdf, 2010.
74. J. Edmonds. Paths, trees, and flowers. *Canadian Journal of Mathematics*, 17(3):449–467, 1965.
75. M. R. Fellows. The Robertson-Seymour theorems: a survey of applications. *Contemporary Mathematics*, 89:1–18, 1989.
76. S. Fiorini, S. Massar, S. Pokutta, H. R. Tiwary, and R. de Wolf. Exponential lower bounds for polytopes in combinatorial optimization. *J. of the ACM*, 62(2):17, 2015. Earlier version in Proc. ACM STOC'2012, pp. 95–106.
77. M. A. Forbes and A. Shpilka. Explicit Noether normalization for simultaneous conjugation via polynomial identity testing. pages 527–542, 2013. arXiv:1303.0084.
78. L. R. Ford and D. R. Fulkerson. *Flows in Networks*. Princeton, 1962.
79. L. Fortnow. The status of the P versus NP problem. *Commun. of the ACM*, 52(9):78–86, 2009.
80. L. Fortnow. *The Golden Ticket: P, NP, and the Search for the Impossible*. Princeton University Press, 2009.
81. L. Fortnow and D. van Melkebeek. Time-space tradeoffs for nondeterministic computation. In *Proc. Conference on Computational Complexity*, pages 2–13, 2000.

82. L. Fortnow, A. Pavan, and S. Sengupta. Proving SAT does not have small circuits with an application to the two queries problem. *J. Comput. Sys. Sci.*, 74(3):358–363, 2008. Earlier version in Proc. IEEE Complexity'2003, pp. 347–350.
83. L. Fortnow and M. Sipser. Are there interactive protocols for co-NP languages? *Inform. Proc. Lett.*, 28:249–251, 1988.
84. H. M. Friedman. Mathematically natural concrete incompleteness. At u.osu.edu/friedman.8/files/2014/01/Putnam062115pdf-15ku867.pdf, 2015.
85. M. Furst, J. B. Saxe, and M. Sipser. Parity, circuits, and the polynomial time hierarchy. *Math. Systems Theory*, 17:13–27, 1984. Earlier version in Proc. IEEE FOCS'1981, pp. 260–270.
86. M. R. Garey and D. S. Johnson. *Computers and Intractability: A Guide to the Theory of NP-Completeness*. W. H. Freeman, 1979.
87. W. Gasarch. The P=?NP poll. *SIGACT News*, 33(2):34–47, June 2002.
88. O. Goldreich, S. Goldwasser, and S. Micali. How to construct random functions. *J. of the ACM*, 33(4):792–807, 1984. Earlier version in Proc. IEEE FOCS'1984, pp. 464–479.
89. O. Goldreich, S. Micali, and A. Wigderson. Proofs that yield nothing but their validity or all languages in NP have zero-knowledge proof systems. *J. of the ACM*, 38(1):691–729, 1991.
90. S. Goldwasser and M. Sipser. Private coins versus public coins in interactive proof systems. In *Randomness and Computation*, volume 5 of *Advances in Computing Research*. JAI Press, 1989.
91. R. Goodstein. On the restricted ordinal theorem. *J. Symbolic Logic*, 9:33–41, 1944.
92. B. Grenet. An upper bound for the permanent versus determinant problem. www.lirmm.fr/~grenet/publis/Gre11.pdf, 2011.
93. D. Grigoriev and M. Karpinski. An exponential lower bound for depth 3 arithmetic circuits. In *Proc. ACM STOC*, pages 577–582, 1998.
94. D. Grigoriev and A. A. Razborov. Exponential lower bounds for depth 3 arithmetic circuits in algebras of functions over finite fields. *Appl. Algebra Eng. Commun. Comput.*, 10(6):465–487, 2000. Earlier version in Proc. IEEE FOCS'1998, pp. 269–278.
95. J. A. Grochow. *Symmetry and equivalence relations in classical and geometric complexity theory*. PhD thesis, 2012.
96. J. A. Grochow. Unifying known lower bounds via Geometric Complexity Theory. *Computational Complexity*, 24(2):393–475, 2015. Earlier version in Proc. IEEE Complexity'2014, pp. 274–285.
97. L. K. Grover. A fast quantum mechanical algorithm for database search. In *Proc. ACM STOC*, pages 212–219, 1996. quant-ph/9605043.
98. A. Gupta, P. Kamath, N. Kayal, and R. Saptharishi. Arithmetic circuits: a chasm at depth three. In *Proc. IEEE FOCS*, pages 578–587, 2013.
99. A. Gupta, P. Kamath, N. Kayal, and R. Saptharishi. Approaching the chasm at depth four. *J. of the ACM*, 61(6):1–16, 2014. Earlier version in Proc. IEEE Complexity'2013, pp. 65–73.
100. L. Gurvits. On the complexity of mixed discriminants and related problems. In *Mathematical Foundations of Computer Science*, pages 447–458, 2005.
101. A. Haken. The intractability of resolution. *Theoretical Comput. Sci.*, 39:297–308, 1985.
102. J. Hartmanis and R. E. Stearns. On the computational complexity of algorithms. *Transactions of the American Mathematical Society*, 117:285–306, 1965.
103. J. Håstad. *Computational Limitations for Small Depth Circuits*. MIT Press, 1987.
104. J. Håstad. The shrinkage exponent of De Morgan formulas is 2. *SIAM J. Comput.*, 27(1):48–64, 1998. Earlier version in Proc. IEEE FOCS'1993, pp. 114–123.
105. J. Håstad. Some optimal inapproximability results. *J. of the ACM*, 48:798–859, 2001. Earlier version in Proc. ACM STOC'1997, pp. 1–10.
106. J. Håstad, R. Impagliazzo, L. A. Levin, and M. Luby. A pseudorandom generator from any one-way function. *SIAM J. Comput.*, 28(4):1364–1396, 1999.
107. M. Held and R. M. Karp. A dynamic programming approach to sequencing problems. *Journal of the Society for Industrial and Applied Mathematics*, 10(1):196–210, 1962.
108. C. Ikenmeyer, K. Mulmuley, and M. Walter. On vanishing of Kronecker coefficients. arXiv:1507.02955, 2015.

109. C. Ikenmeyer and G. Panova. Rectangular Kronecker coefficients and plethysms in geometric complexity theory. arXiv:1512.03798, 2015.
110. N. Immerman. Nondeterministic space is closed under complementation. *SIAM J. Comput.*, 17(5):935–938, 1988. Earlier version in Proc. IEEE Structure in Complexity Theory, 1988.
111. N. Immerman. *Descriptive Complexity*. Springer, 1998.
112. R. Impagliazzo, V. Kabanets, and A. Kolokolova. An axiomatic approach to algebrization. In *Proc. ACM STOC*, pages 695–704, 2009.
113. R. Impagliazzo, V. Kabanets, and A. Wigderson. In search of an easy witness: exponential time vs. probabilistic polynomial time. *J. Comput. Sys. Sci.*, 65(4):672–694, 2002. Earlier version in Proc. IEEE Complexity'2001, pp. 2–12.
114. R. Impagliazzo and N. Nisan. The effect of random restrictions on formula size. *Random Structures and Algorithms*, 4(2):121–134, 1993.
115. R. Impagliazzo and A. Wigderson. P=BPP unless E has subexponential circuits: derandomizing the XOR Lemma. In *Proc. ACM STOC*, pages 220–229, 1997.
116. M. Jerrum and M. Snir. Some exact complexity results for straight-line computations over semirings. *J. of the ACM*, 29(3):874–897, 1982.
117. V. Kabanets and J.-Y. Cai. Circuit minimization problem. In *Proc. ACM STOC*, pages 73–79, 2000. TR99-045.
118. V. Kabanets and R. Impagliazzo. Derandomizing polynomial identity testing means proving circuit lower bounds. *Computational Complexity*, 13(1-2):1–46, 2004. Earlier version in Proc. ACM STOC'2003. ECCC TR02-055.
119. D. M. Kane and R. Williams. Super-linear gate and super-quadratic wire lower bounds for depth-two and depth-three threshold circuits. arXiv:1511.07860, 2015.
120. R. Kannan. Circuit-size lower bounds and non-reducibility to sparse sets. *Information and Control*, 55:40–56, 1982. Earlier version in Proc. IEEE FOCS'1981, pp. 304–309.
121. M. Karchmer and A. Wigderson. Monotone circuits for connectivity require super-logarithmic depth. *SIAM J. Comput.*, 3:255–265, 1990. Earlier version in Proc. ACM STOC'1988, pp. 539–550.
122. R. M. Karp. Reducibility among combinatorial problems. In R. E. Miller and J. W. Thatcher, editors, *Complexity of Computer Computations*, pages 85–103. Plenum Press, 1972.
123. R. M. Karp and R. J. Lipton. Turing machines that take advice. *Enseign. Math.*, 28:191–201, 1982. Earlier version in Proc. ACM STOC'1980, pp. 302–309.
124. N. Kayal. Affine projections of polynomials: extended abstract. In *Proc. ACM STOC*, pages 643–662, 2012.
125. N. Kayal, N. Limaye, C. Saha, and S. Srinivasan. An exponential lower bound for homogeneous depth four arithmetic formulas. In *Proc. IEEE FOCS*, pages 61–70, 2014.
126. N. Kayal, C. Saha, and R. Saptharishi. A super-polynomial lower bound for regular arithmetic formulas. In *Proc. ACM STOC*, pages 146–153, 2014.
127. S. Khot. On the Unique Games Conjecture. In *Proc. Conference on Computational Complexity*, pages 99–121, 2010.
128. V. M. Khrapchenko. A method of determining lower bounds for the complexity of π schemes. *Matematischi Zametki*, 10:83–92, 1971. In Russian.
129. L. Kirby and J. Paris. Accessible independence results for Peano arithmetic. *Bulletin of the London Mathematical Society*, 14:285–293, 1982.
130. A. Klivans and D. van Melkebeek. Graph nonisomorphism has subexponential size proofs unless the polynomial-time hierarchy collapses. *SIAM J. Comput.*, 31:1501–1526, 2002. Earlier version in Proc. ACM STOC'1999.
131. D. E. Knuth and E. G. Daylight. *Algorithmic Barriers Falling: P=NP?* Lonely Scholar, 2014.
132. A. Knutson and T. Tao. The honeycomb model of $GL_n(\mathbb{C})$ tensor products I: proof of the saturation conjecture. *J. Amer. Math. Soc.*, 12(4):1055–1090, 1999.
133. P. Koiran. Arithmetic circuits: the chasm at depth four gets wider. *Theor. Comput. Sci.*, 448:56–65, 2012.
134. E. Kushilevitz and N. Nisan. *Communication Complexity*. Cambridge, 1997.

135. R. E. Ladner. On the structure of polynomial time reducibility. *J. of the ACM*, 22:155–171, 1975.
136. J. M. Landsberg. The border rank of the multiplication of two by two matrices is seven. *J. Amer. Math. Soc.*, 19(2):447–459, 2006. arXiv:math/0407224.
137. J. M. Landsberg. Geometric complexity theory: an introduction for geometers. *Annali dell'Universita di Ferrara*, 61(1):65–117, 2015. arXiv:1305.7387.
138. J. M. Landsberg and G. Ottaviani. New lower bounds for the border rank of matrix multiplication. *Theory of Computing*, 11:285–298, 2015. arXiv:1112.6007.
139. C. Lautemann. BPP and the polynomial hierarchy. *Inform. Proc. Lett.*, 17:215–217, 1983.
140. J. R. Lee, P. Raghavendra, and D. Steurer. Lower bounds on the size of semidefinite programming relaxations. In *Proc. ACM STOC*, pages 567–576, 2015.
141. L. A. Levin. Universal sequential search problems. *Problems of Information Transmission*, 9(3):115–116, 1973.
142. L. A. Levin. Laws of information conservation (non-growth) and aspects of the foundations of probability theory. *Problems of Information Transmission*, 10(3):206–210, 1974.
143. M. Li and P. M. B. Vitányi. Average case complexity under the universal distribution equals worst-case complexity. *Inform. Proc. Lett.*, 42(3):145–149, 1992.
144. N. Linial, Y. Mansour, and N. Nisan. Constant depth circuits, Fourier transform, and learnability. *J. of the ACM*, 40(3):607–620, 1993. Earlier version in Proc. IEEE FOCS'1989, pp. 574–579.
145. N. Linial and N. Nisan. Approximate inclusion-exclusion. *Combinatorica*, 10(4):349–365, 1990. Earlier version in Proc. ACM STOC'1990.
146. R. J. Lipton. New directions in testing. In *Distributed Computing and Cryptography*, pages 191–202. AMS, 1991.
147. R. J. Lipton. Galactic algorithms, 2010. rjlipton. wordpress.com/2010/10/23/galactic-algorithms/.
148. R. J. Lipton and K. W. Regan. Practically P=NP?, 2014. rjlipton.wordpress.com/2014/02/28/practically-pnp/.
149. R. J. Lipton and A. Viglas. On the complexity of SAT. In *Proc. IEEE FOCS*, pages 459–464, 1999.
150. D. Luna. Slices étales. *Mémoires de la Société Mathématique de France*, 33:81–105, 1973.
151. C. Lund, L. Fortnow, H. Karloff, and N. Nisan. Algebraic methods for interactive proof systems. *J. of the ACM*, 39:859–868, 1992. Earlier version in Proc. IEEE FOCS'1990, pp. 2–10.
152. D. van Melkebeek. A survey of lower bounds for satisfiability and related problems. *Foundations and Trends in Theoretical Computer Science*, 2:197–303, 2007. ECCC TR07-099.
153. T. Mignon and N. Ressayre. A quadratic bound for the determinant and permanent problem. *International Mathematics Research Notices*, (79):4241–4253, 2004.
154. G. L. Miller. Riemann's hypothesis and tests for primality. *J. Comput. Sys. Sci.*, 13:300–317, 1976. Earlier version in Proc. ACM STOC'1975.
155. C. Moore and S. Mertens. *The Nature of Computation*. Oxford University Press, 2011.
156. S. Moran. Some results on relativized deterministic and nondeterministic time hierarchies. *J. Comput. Sys. Sci.*, 22(1):1–8, 1981.
157. K. Mulmuley. GCT publications web page. ramakrishnadas.cs.uchicago.edu.
158. K. Mulmuley. Lower bounds in a parallel model without bit operations. *SIAM J. Comput.*, 28(4):1460–1509, 1999.
159. K. Mulmuley. Geometric complexity theory VII: nonstandard quantum group for the plethysm problem. Technical Report TR-2007-14, University of Chicago, 2007. arXiv:0709.0749.
160. K. Mulmuley. Geometric complexity theory VIII: on canonical bases for the non-standard quantum groups. Technical Report TR-2007-15, University of Chicago, 2007. arXiv:0709.0751.
161. K. Mulmuley. Explicit proofs and the flip. arXiv:1009.0246, 2010.

162. K. Mulmuley. Geometric complexity theory VI: the flip via positivity. Technical report, University of Chicago, 2011. arXiv:0704.0229.
163. K. Mulmuley. On P vs. NP and geometric complexity theory: dedicated to Sri Ramakrishna. *J. of the ACM*, 58(2):5, 2011.
164. K. Mulmuley. The GCT program toward the P vs. NP problem. *Commun. of the ACM*, 55(6):98–107, June 2012.
165. K. Mulmuley. Geometric complexity theory V: equivalence between blackbox derandomization of polynomial identity testing and derandomization of Noether's Normalization Lemma. In *Proc. IEEE FOCS*, pages 629–638, 2012. Full version available at arXiv:1209.5993.
166. K. Mulmuley, H. Narayanan, and M. Sohoni. Geometric complexity theory III: on deciding nonvanishing of a Littlewood-Richardson coefficient. *Journal of Algebraic Combinatorics*, 36(1):103–110, 2012.
167. K. Mulmuley and M. Sohoni. Geometric complexity theory I: An approach to the P vs. NP and related problems. *SIAM J. Comput.*, 31(2):496–526, 2001.
168. K. Mulmuley and M. Sohoni. Geometric complexity theory II: Towards explicit obstructions for embeddings among class varieties. *SIAM J. Comput.*, 38(3):1175–1206, 2008.
169. C. D. Murray and R. R. Williams. On the (non) NP-hardness of computing circuit complexity. In *Proc. Conference on Computational Complexity*, pages 365–380, 2015.
170. J. Naor and M. Naor. Number-theoretic constructions of efficient pseudo-random functions. *J. of the ACM*, 51(2):231–262, 2004. Earlier version in Proc. IEEE FOCS'1997.
171. J. Nash. Letter to the United States National Security Agency, 1950. Available at www.nsa.gov/public_info/_files/nash_letters/nash_letters1.pdf.
172. M. Nielsen and I. Chuang. *Quantum Computation and Quantum Information*. Cambridge University Press, 2000.
173. N. Nisan and A. Wigderson. Lower bounds on arithmetic circuits via partial derivatives. *Computational Complexity*, 6(3):217–234, 1997. Earlier version in Proc. IEEE FOCS'1995, pp. 16–25.
174. B. Aydınlıoğlu and E. Bach. Affine relativization: unifying the algebrization and relativization barriers. 2016.
175. C. H. Papadimitriou. *Computational Complexity*. Addison-Wesley, 1994.
176. M. Paterson and U. Zwick. Shrinkage of De Morgan formulae under restriction. *Random Structures and Algorithms*, 4(2):135–150, 1993.
177. W. J. Paul, N. Pippenger, E. Szemerédi, and W. T. Trotter. On determinism versus nondeterminism and related problems. In *Proc. IEEE FOCS*, pages 429–438, 1983.
178. C. Peikert and B. Waters. Lossy trapdoor functions and their applications. *SIAM J. Comput.*, 40(6):1803–1844, 2011. Earlier version in Proc. ACM STOC'2008.
179. G. Perelman. The entropy formula for the Ricci flow and its geometric applications. arXiv:math/0211159, 2002.
180. V. R. Pratt. Every prime has a succinct certificate. *SIAM J. Comput.*, 4(3):214–220, 1975.
181. M. O. Rabin. Probabilistic algorithm for testing primality. *J. Number Theory*, 12(1):128–138, 1980.
182. R. Raz. Multi-linear formulas for permanent and determinant are of super-polynomial size. *J. of the ACM*, 56(2):8, 2009. Earlier version in Proc. ACM STOC'2004, pp. 633–641. ECCC TR03-067.
183. R. Raz. Elusive functions and lower bounds for arithmetic circuits. *Theory of Computing*, 6:135–177, 2010. Earlier version in Proc. ACM STOC'2008.
184. R. Raz. Tensor-rank and lower bounds for arithmetic formulas. *J. of the ACM*, 60(6):40, 2013. Earlier version in Proc. ACM STOC'2010, pp. 659–666.
185. R. Raz and A. Yehudayoff. Lower bounds and separations for constant depth multilinear circuits. *Computational Complexity*, 18(2):171–207, 2009. Earlier version in Proc. IEEE Complexity'2008, pp. 128–139.
186. R. Raz and A. Yehudayoff. Multilinear formulas, maximal-partition discrepancy and mixed-sources extractors. *J. Comput. Sys. Sci.*, 77(1):167–190, 2011. Earlier version in Proc. IEEE FOCS'2008, pp. 273–282.

187. A. A. Razborov. Lower bounds for the monotone complexity of some Boolean functions. *Doklady Akademii Nauk SSSR*, 281(4):798–801, 1985. English translation in *Soviet Math. Doklady* 31:354–357, 1985.
188. A. A. Razborov. Lower bounds on monotone complexity of the logical permanent. *Mathematical Notes*, 37(6):485–493, 1985. Original Russian version in *Matematischi Zametki*.
189. A. A. Razborov. Lower bounds for the size of circuits of bounded depth with basis $\{\&, \oplus\}$. *Mathematicheskie Zametki*, 41(4):598–607, 1987. English translation in *Math. Notes. Acad. Sci. USSR* 41(4):333–338, 1987.
190. A. A. Razborov. Unprovability of lower bounds on circuit size in certain fragments of bounded arithmetic. *Izvestiya Math.*, 59(1):205–227, 1995.
191. A. A. Razborov and S. Rudich. Natural proofs. *J. Comput. Sys. Sci.*, 55(1):24–35, 1997. Earlier version in Proc. ACM STOC'1994, pp. 204–213.
192. K. W. Regan. Understanding the Mulmuley-Sohoni approach to P vs. NP. *Bulletin of the EATCS*, 78:86–99, 2002.
193. B. Rossman, R. A. Servedio, and L.-Y. Tan. An average-case depth hierarchy theorem for Boolean circuits. In *Proc. IEEE FOCS*, pages 1030–1048, 2015. ECCC TR15-065.
194. T. Rothvoß. The matching polytope has exponential extension complexity. In *Proc. ACM STOC*, pages 263–272, 2014.
195. R. Santhanam. Circuit lower bounds for Merlin-Arthur classes. In *Proc. ACM STOC*, pages 275–283, 2007.
196. S. Saraf. Recent progress on lower bounds for arithmetic circuits. In *Proc. Conference on Computational Complexity*, pages 155–160, 2014.
197. W. J. Savitch. Relationships between nondeterministic and deterministic tape complexities. *J. Comput. Sys. Sci.*, 4(2):177–192, 1970.
198. U. Schöning. A probabilistic algorithm for k-SAT and constraint satisfaction problems. In *Proc. IEEE FOCS*, pages 410–414, 1999.
199. U. Schöning and R. J. Pruim. *Gems of Theoretical Computer Science*. Springer, 1998.
200. A. Shamir. IP=PSPACE. *J. of the ACM*, 39(4):869–877, 1992. Earlier version in Proc. IEEE FOCS'1990, pp. 11–15.
201. C. Shannon. The synthesis of two-terminal switching circuits. *Bell System Technical Journal*, 28(1):59–98, 1949.
202. J. Shoenfield. The problem of predicativity. In Y. Bar-Hillel et al., editor, *Essays on the Foundations of Mathematics*, pages 132–142. Hebrew University Magnes Press, 1961.
203. P. W. Shor. Polynomial-time algorithms for prime factorization and discrete logarithms on a quantum computer. *SIAM J. Comput.*, 26(5):1484–1509, 1997. Earlier version in Proc. IEEE FOCS'1994. quant-ph/9508027.
204. A. Shpilka and A. Wigderson. Depth-3 arithmetic circuits over fields of characteristic zero. *Computational Complexity*, 10(1):1–27, 2001. Earlier version in Proc. IEEE Complexity'1999.
205. A. Shpilka and A. Yehudayoff. Arithmetic circuits: a survey of recent results and open questions. *Foundations and Trends in Theoretical Computer Science*, 5(3-4):207–388, 2010.
206. M. Sipser. A complexity theoretic approach to randomness. In *Proc. ACM STOC*, pages 330–335, 1983.
207. M. Sipser. The history and status of the P versus NP question. In *Proc. ACM STOC*, pages 603–618, 1992.
208. M. Sipser. *Introduction to the Theory of Computation (Second Edition)*. Course Technology, 2005.
209. R. Smolensky. Algebraic methods in the theory of lower bounds for Boolean circuit complexity. In *Proc. ACM STOC*, pages 77–82, 1987.
210. R. Solovay and V. Strassen. A fast Monte-Carlo test for primality. *SIAM J. Comput.*, 6(1):84–85, 1977.
211. A. Srinivasan. On the approximability of clique and related maximization problems. *J. Comput. Sys. Sci.*, 67(3):633–651, 2003. Earlier version in Proc. ACM STOC'2000, pp. 144–152.

212. L. J. Stockmeyer. The complexity of approximate counting. In *Proc. ACM STOC*, pages 118–126, 1983.
213. A. J. Stothers. *On the complexity of matrix multiplication*. PhD thesis, 2010.
214. V. Strassen. Gaussian elimination is not optimal. *Numerische Mathematik*, 14(13):354–356, 1969.
215. V. Strassen. Vermeidung von divisionen. *Journal für die Reine und Angewandte Mathematik*, 264:182–202, 1973.
216. B. A. Subbotovskaya. Realizations of linear functions by formulas using $+, \times, -$. *Doklady Akademii Nauk SSSR*, 136(3):553–555, 1961. In Russian.
217. E. R. Swart. P = NP. Technical report, University of Guelph, 1986. Revision in 1987.
218. R. Szelepcsényi. The method of forced enumeration for nondeterministic automata. *Acta Informatica*, 26(3):279–284, 1988.
219. A. Tal. Shrinkage of De Morgan formulae by spectral techniques. In *Proc. IEEE FOCS*, pages 551–560, 2014. ECCC TR14-048.
220. É. Tardos. The gap between monotone and non-monotone circuit complexity is exponential. *Combinatorica*, 8(1):141–142, 1988.
221. S. Tavenas. Improved bounds for reduction to depth 4 and depth 3. *Inf. Comput.*, 240:2–11, 2015. Earlier version in Proc. MFCS'2013, pp. 813–824.
222. S. Toda. PP is as hard as the polynomial-time hierarchy. *SIAM J. Comput.*, 20(5):865–877, 1991. Earlier version in Proc. IEEE FOCS'1989, pp. 514–519.
223. B. A. Trakhtenbrot. A survey of Russian approaches to perebor (brute-force search) algorithms. *Annals of the History of Computing*, 6(4):384–400, 1984.
224. L. G. Valiant. Graph-theoretic arguments in low-level complexity. In *Mathematical Foundations of Computer Science*, pages 162–176, 1977.
225. L. G. Valiant. Completeness classes in algebra. In *Proc. ACM STOC*, pages 249–261, 1979.
226. L. G. Valiant. The complexity of computing the permanent. *Theoretical Comput. Sci.*, 8(2):189–201, 1979.
227. L. G. Valiant. Accidental algorithms. In *Proc. IEEE FOCS*, pages 509–517, 2006.
228. L. G. Valiant, S. Skyum, S. Berkowitz, and C. Rackoff. Fast parallel computation of polynomials using few processors. *SIAM J. Comput.*, 12(4):641–644, 1983.
229. Various authors. Deolalikar P vs NP paper (wiki page). Last modified 30 September 2011. michaelnielsen.org/polymath1/index.php?title=Deolalikar_P_vs_NP_paper.
230. V. Vassilevska Williams. Multiplying matrices faster than Coppersmith-Winograd. In *Proc. ACM STOC*, pages 887–898, 2012.
231. N. V. Vinodchandran. A note on the circuit complexity of PP. *Theor. Comput. Sci.*, 347:415–418, 2005. ECCC TR04-056.
232. A. Wigderson. P, NP and mathematics - a computational complexity perspective. In *Proceedings of the International Congress of Mathematicians 2006 (Madrid)*, pages 665–712. EMS Publishing House, 2007. www.math.ias.edu/~avi/PUBLICATIONS/MYPAPERS/W06/w06.pdf.
233. A. Wiles. Modular elliptic curves and Fermat's Last Theorem. *Annals of Mathematics*, 141(3):443–551, 1995.
234. R. Williams. Better time-space lower bounds for SAT and related problems. In *Proc. Conference on Computational Complexity*, pages 40–49, 2005.
235. R. Williams. Time-space tradeoffs for counting NP solutions modulo integers. *Computational Complexity*, 17(2):179–219, 2008. Earlier version in Proc. IEEE Complexity'2007, pp. 70–82.
236. R. Williams. Guest column: a casual tour around a circuit complexity bound. *ACM SIGACT News*, 42(3):54–76, 2011.
237. R. Williams. Alternation-trading proofs, linear programming, and lower bounds. *ACM Trans. on Computation Theory*, 5(2):6, 2013. Earlier version in Proc. STACS'2010, pp. 669–680.
238. R. Williams. Improving exhaustive search implies superpolynomial lower bounds. *SIAM J. Comput.*, 42(3):1218–1244, 2013. Earlier version in Proc. ACM STOC'2010.

239. R. Williams. Natural proofs versus derandomization. In *Proc. ACM STOC*, pages 21–30, 2013.
240. R. Williams. Algorithms for circuits and circuits for algorithms: connecting the tractable and intractable. In *Proceedings of the International Congress of Mathematicians*, 2014.
241. R. Williams. New algorithms and lower bounds for circuits with linear threshold gates. In *Proc. ACM STOC*, pages 194–202, 2014.
242. R. Williams. Nonuniform ACC circuit lower bounds. *J. of the ACM*, 61(1):1–32, 2014. Earlier version in Proc. IEEE Complexity'2011.
243. C. B. Wilson. Relativized circuit complexity. *J. Comput. Sys. Sci.*, 31(2):169–181, 1985.
244. M. Yannakakis. Expressing combinatorial optimization problems by linear programs. *J. Comput. Sys. Sci.*, 43(3):441–466, 1991. Earlier version in Proc. ACM STOC'1988, pp. 223–228.
245. A. C-C. Yao. Separating the polynomial-time hierarchy by oracles (preliminary version). In *Proc. IEEE FOCS*, pages 1–10, 1985.
246. A. C-C. Yao. On ACC and threshold circuits. In *Proc. IEEE FOCS*, pages 619–627, 1990.

From Quantum Systems to L-Functions: Pair Correlation Statistics and Beyond

Owen Barrett, Frank W. K. Firk, Steven J. Miller, and Caroline Turnage-Butterbaugh

Abstract The discovery of connections between the distribution of energy levels of heavy nuclei and spacings between prime numbers has been one of the most surprising and fruitful observations in the twentieth century. The connection between the two areas was first observed through Montgomery's work on the pair correlation of zeros of the Riemann zeta function. As its generalizations and consequences have motivated much of the following work, and to this day remains one of the most important outstanding conjectures in the field, it occupies a central role in our discussion below. We describe some of the many techniques and results from the past sixty years, especially the important roles played by numerical and experimental investigations, that led to the discovery of the connections and progress towards understanding the behaviors. In our survey of these two areas, we describe the common mathematics that explains the remarkable universality. We conclude with some thoughts on what might lie ahead in the pair correlation of zeros of the zeta function, and other similar quantities.

1 Introduction

Montgomery's pair correlation conjecture posits that zeros of L-functions behave similarly to energy levels of heavy nuclei. The bridge between these fields is random matrix theory, a beautiful subject which has successfully modeled a large variety

O. Barrett
University of Chicago, Chicago, IL 60637, USA
e-mail: barrett@math.uchicago.edu; owen.barrett@aya.yale.edu

F.W.K. Firk
Yale University, New Haven, CT 06520, USA
e-mail: fwkfirk@aol.com

S.J. Miller (✉)
Williams College, Williamstown, MA 01267, USA
e-mail: Steven.Miller.MC.96@aya.yale.edu; sjm1@williams.edu

C. Turnage-Butterbaugh
Duke University, Durham, NC 27708, USA
e-mail: ctb@math.duke.edu

J.F. Nash, Jr., M.Th. Rassias (eds.), *Open Problems in Mathematics*,
DOI 10.1007/978-3-319-32162-2_2

of diverse phenomena (see [7, 87] for a great example of how varied the systems can be). It is impossible in a short chapter to cover all the topics and connections; fortunately there is no need as there is an extensive literature. Our goal is therefore to briefly describe the history of the subject and the correspondences, concentrating on some of the main objects of interest and past successes, ending with a brief tour through a *subset* of current work and a discussion of some of the open questions in mathematics. We are deliberately brief in areas that are well known or are extensively covered in the literature, and instead dwell at greater lengths on the inspiration from and interpretation through physics (see for example Sect. 2.6), as these parts of the story are not as well known but deserve to be (both for historical reasons as well as the guidance they can offer).

To this end, we begin with a short introduction to random matrix theory and a quick description of the main characters studied in this chapter. We then continue in Sect. 2 with a detailed exposition of the historical development of random matrix theory in nuclear physics in the 1950s and 1960s. We note the pivotal role played by the nuclear physics experimentalists in gathering data to support the theoretical conjectures; we will see analogues of these when we get to the work in the 1970s and 1980s on zeros of L-functions in Sect. 3.5. One of our main purposes is in fact to highlight the power of experimental data, be it data from a lab or a computer calculation, and show how attempts to explain such results influence the development and direction of subjects. We then shift emphasis to number theory in Sect. 3, and see how studies on the class number problem led Montgomery to his famous pair correlation conjecture for the zeros of the Riemann zeta function. This and related statistics are the focus of the rest of the chapter; we describe what they are, what progress has been made (theoretically and numerically), and then turn to some open questions. Most of these open questions involve how the arithmetic of L-functions influences the behavior; remarkably the main terms in a variety of problems are independent of the finer properties of L-functions, and it is only in lower order terms (or, equivalently, in the rates of convergence to the random matrix theory behavior) that the dependencies on these properties surface. We then conclude in Sect. 4 with current questions and some future trends.

1.1 The Early Days: Statistics and Biometrics

Though our main characters will be energy levels of nuclei and zeros of L-functions, the story of random matrix theory begins neither with physics nor with mathematics, but with statistics and biometrics. In 1928 John Wishart published an article titled *The Generalised Product Moment Distribution in Samples from a Normal Multivariate* [150] in Biometrika (see [149] for a history of the journal, which we briefly recap). The journal was founded at the start of the century by Francis Galton, Karl Pearson, and Walter Weldon for the study of statistics related to biometrics. In the editors' introduction in the first issue (see also [149]), they write:

> It is intended that Biometrika shall serve as a means not only of collecting or publishing under one title biological data of a kind not systematically collected or published elsewhere in any other periodical, but also of spreading a knowledge of such statistical theory as may be requisite for their scientific treatment.

The question of interest for Wishart was that of estimating covariance matrices. The paper begins with a review of work to date on samples from univariate and bivariate populations, and issues with the determination of correlation and regression coefficients. After summarizing some of the work and formulas from Fisher, Wishart writes:

> The distribution of the correlation coefficient was deduced by direct integration from this result. Further, K. Pearson and V. Romanovsky, starting from this fundamental formula, were able to deal with the regression coefficients. Pearson, in 1925, gave the mean value and standard deviation of the regression coefficient, while Romanovsky and Pearson, in the following year, published the actual distribution.

After talking about the new problems that arise when dealing with three or more variates, he continues:

> What is now asserted is that all such problems depend, in the first instance, on the determination of a fundamental frequency distribution, which will be a generalisation of Eq. (1.2). It will, in fact, be the simultaneous distribution in samples of the n variances (squared standard deviations) and the $\frac{n(n-1)}{2}$ product moment coefficients. It is the purpose of the present paper to give this generalised distribution, and to calculate its moments up to the fourth order. The case of three variates will first be considered in detail, and thereafter a proof for the general n-fold system will be given.

In his honor the distribution of the sample covariance matrix (arising from a sample from a multivariate normal distribution) is called the Wishart distribution. More specifically, if we have an $n \times p$ matrix X whose rows are independently drawn from a p-variate mean 0 normal distribution, the Wishart distribution is the density of the $p \times p$ matrices $X^T X$.

Several items are worth noting here. First, we have an ensemble (a collection) of matrices whose entries are drawn from a fixed distribution; in this case there are dependencies among the entries. Second, these matrices are used to model observable quantities of interest, in this case covariances. Finally, in his article he mentions an earlier work of his (published in the Memoirs of the Royal Meteorological Society, volume II, pages 29–37, 1928) which experimentally confirmed some of the results discussed, thus showing the connections between experiment and theory which play such a prominent role later in the story also played a key role in the founding.

It was not until almost thirty years later that random matrix theory, in the hands and mind of Wigner, bursts onto the physics scene, and then it will be almost another thirty years more before the connections with number theory emerge. Before describing these histories in detail, we end the introduction with a very quick tour of some of the quantities and objects we'll meet.

1.2 Cast of Characters: Nuclei and L-functions

The two main objects we study are energy levels of heavy nuclei on the physics side, and zeros of the Riemann zeta function (or more generally L-functions) on the number theory side, especially Montgomery's pair correlation conjecture and related statistics. We give a full statement of the pair correlation conjecture, and results towards its proof, in Sect. 3.2. Briefly, given an ordered sequence of events (such as zeros on the critical line, eigenvalues of Hermitian matrices, energy levels of heavy nuclei) one can look at how often a difference is observed. The remarkable conjecture is that these very different systems exhibit similar behavior.

We begin with a review of some facts about the these areas, from theories for their behavior to how experimental observations were obtained which shed light on the structures, and then finish the introduction with some hints at the similarities between these two very different systems. Parts of that section, as well as much of Sect. 2, are expanded with permission from the survey article [50] written by two of the authors of this chapter for the inaugural issue of the open access journal Symmetry. The goal of that article was similar to this chapter, though there the main quantity discussed was Wigner's semi-circle law and not pair correlation.

Many, if not all, of the other survey articles in the subject concentrate on the mathematics and ignore the experimental physics. When writing the survey [50] the authors deliberately sought a balance, with the intention of sharing and elaborating on that vantage again in a later work to give a wider audience a more complete description of the development of the subjects, as other approaches are already available in the literature. We especially recommend to the reader Goldston's excellent survey article *Notes on pair correlation of zeros and prime numbers* (see [64]) for an extended, detailed technical discussion; the purpose of this chapter is to complement this and other surveys by highlighting other aspects of the story, especially how Montgomery's work on the pair correlation of zeros of $\zeta(s)$ connects, through random matrix theory, a central object of study in number theory to our understanding of the physics of heavy nuclei.

1.2.1 Atomic Theory and Nuclei

Experiments and experimental data played a crucial role in our evolving understanding of the atom. For example, Ernest Rutherford's gold foil experiment (performed by Hans Geiger and Ernest Marsden) near the start of the twentieth century demonstrated that J. J. Thomson's plum pudding model of an atom with negatively charged electrons embedded in a positively charged region was false, and that the atom had a very small positively charged nucleus with the electrons far away. These experiments involved shooting alpha particles at thin gold foils. Alpha particles are helium atoms without the electrons and are thus positively charged. While this positive charge was responsible for disproving the plum pudding model, such particles could not deeply probe the positively charged nucleus due to the strong repulsion from like charges. To make further progress into the structure of the atom in general, and the nucleus in particular, another object was needed. A great candidate was the neutron (discovered by Chadwick in 1932); as it did not have a net charge, the electric force would play an immensely smaller role in its interaction with the nucleus than it did with the alpha particles.

The earliest studies of neutron induced reactions showed that the total neutron cross section[1] for the interaction of low-energy (electron-volt, eV) neutrons with a nucleus is frequently much greater than the geometrical area presented by the target nucleus to the incident neutron [44]. It was also found that the cross section varies rapidly as a function of the bombarding energy of the incident neutron. The appearance of these well-defined *resonances* in the neutron cross section is the most characteristic feature of low energy nuclear reactions.

In general, the low energy resonances were found to be closely spaced (spacing $\leq 10\,\text{eV}$ in heavy nuclei), and to be very narrow (widths $\leq 0.1\,\text{eV}$). These facts led Niels Bohr to introduce the *compound nucleus* model [15] that assumes the interaction between an incoming neutron and the target nucleus is so strong that the neutron rapidly shares its energy with many of the target nucleons. The nuclear state that results from the combination of incident neutron and target nucleus may therefore last until sufficient energy again resides in one of the nucleons for it to escape from the system. This is a statistical process, and a considerable time may elapse before it occurs. The long lifetime of the state (τ) (on a nuclear timescale) explains the narrow width (Γ) of the resonance.[2] Also, since many nucleons are involved in the formation of a compound state, the close spacing of the resonances is to be expected since there are clearly many ways of exciting many nucleons. The qualitative model outlined above has formed the basis of most theoretical descriptions of low-energy, resonant nuclear reactions [11].

If a resonant state can decay in a number of different ways (or channels), we can ascribe a probability per unit time for the decay into a channel, c, which can be expressed as a partial width $\Gamma_{\lambda c}$. The total width is the sum of the partial widths, i.e., $\Gamma_\lambda = \sum_c \Gamma_{\lambda c}$.

The appearance of well-defined resonances occurs in heavy nuclei (mass number $A \geq 100$, say) for incident neutron energies up to about 100 keV, and in light nuclei up to neutron energies of several MeV. As the neutron bombarding energies are increased above these energies, the total cross sections are observed to become smoother functions of neutron energy [81]. This is due to two effects: firstly,

---

[1] A total neutron cross section is defined as

$$\frac{\text{Number of events of all types per unit time per nucleus}}{\text{Number of incident neutrons per unit time per unit area}},$$

and has the dimensions of area (the standard unit is the *barn*, $10^{-24}\,\text{cm}^2$).

[2] The width, Γ, is related to the lifetime, τ, by the uncertainty relation $\Gamma = h/2\pi\tau$, where h is Planck's constant. The finite width (lack of energy definition) is due to the fact that a resonant state can decay by emitting a particle, or radiation, whereas a state of definite energy must be a stationary state.

the level density (i.e., the number of resonances per unit energy interval) increases rapidly as the excitation energy of the compound nucleus is increased, and secondly, the widths of the individual resonances tend to increase with increasing excitation energy so that, eventually, they overlap. The smoothed-out cross sections provide useful information on the average properties of resonances. One of the most significant features of these cross sections is the appearance of gross fluctuations that have been interpreted in terms of the single-particle nature of the neutron-nucleus interaction [92]. These *giant resonances* form one of the main sources of experimental evidence for introducing the successful *optical model* of nuclear reactions. This model represents the interaction between a neutron and a nucleus in terms of the neutron moving in a complex potential well [118] in which the imaginary part allows for the absorption of the incident neutron.

Experimental results show that, on increasing the bombarding energy above about 5 MeV, a different reaction mechanism may occur. For example, the energy spectra of emitted nucleons frequently contain too many high-energy nucleons compared with the predictions of the compound nucleus model. The mechanism no longer appears to be one in which the incident neutron shares its energy with many target nucleons but is one in which the neutron interacts with a single nucleon or, at most, a few nucleons. Such a mechanism is termed a *direct interaction*, which is defined as a nuclear reaction in which only a few of the available degrees of freedom of the system are involved [6].

The *optical model*, mentioned above, is an important example of a direct interaction that takes place even at low bombarding energies. The incident neutron is considered to move in the mean nuclear potential of all the nucleons in the target. This model also has been used to account for anomalies in the spectra of gamma-rays resulting from thermal neutron capture [89, 91].

At even higher bombarding energies, greater than 50 MeV, say, the mechanism becomes clearer in the sense that direct processes are the most important. The reactions then give information on the fundamental nucleon-nucleon interaction; these studies and their interpretation are, however, outside the scope of the present discussion.

When a low-energy neutron (energy $<$ 10 keV, say) interacts with a nucleus the excitation energy of the compound nucleus is greatly increased by the neutron binding energy that typically ranges from 5 to 10 MeV. In the late 1950s, experimental methods were developed for measuring low-energy neutrons with resolutions of a few electron-volts. This meant that, for the first time in any physical system, it became possible to study the fine structure of resonances at energies far above the ground state of the system. The relevant experimental methods are discussed in Sect. 2. Important information was thereby obtained concerning the properties that characterizes the resonances such as their peak cross sections, elastic scattering widths, and adjacent spacing. The results were used to test the predictions of various nuclear models used to describe the interactions. These models ranged from the Fermi Gas Model, a quantized version of classical Statistical Mechanics and Thermodynamics [10], to the sophisticated Nuclear Shell Model [11]. In the mid-1950s, all Statistical Mechanics Models predicted that the spacing distribution of nearest-neighbor resonances of the same spin and parity in a heavy nucleus (mass number $A \geq 100$, say) was an exponential distribution. By 1956, the experimental evidence on the spacing distribution of s-wave resonances in a number of heavy nuclei indicated a lack of very closely-spaced resonances, contradicting the predictions of an exponential distribution [71]. By 1960, two research groups [48, 124] showed, unequivocally, that the spacing distribution of resonances up to an energy of almost 2 keV followed the prediction of the random matrix model surmised by Wigner in 1956 [147]; in his model the probability of a zero spacing is zero! It is a model rooted in statistics, which interestingly is where our story on random matrix theory began!

1.2.2 *L*-Functions and Their Zeros

There are many excellent introductions, at a variety of levels, to number theory and *L*-functions. We assume the reader is familiar with the basics of the subject; for more details see among others [27, 39, 70, 82, 106, 129]. The discussion below is a quick review and is an abridgement (and slight expansion) of [50], which has additional details.

The primes are the building blocks of number theory: every integer can be written uniquely as a product of prime powers. Note that the role played by the primes mirrors that of atoms in building up molecules. One of the most important questions we can ask about primes is also one of the most basic: how many primes are there at most x? In other words, how many building blocks are there up to a given point?

Euclid proved over 2000 years ago that there are infinitely many primes; so, if we let $\pi(x)$ denote the number of primes at most x, we know $\lim_{x\to\infty} \pi(x) = \infty$. Though Euclid's proof is still used in courses around the world (and gives a growth rate on the order of $\log\log x$), one can obtain much better counts on $\pi(x)$.

The prime number theorem states that the number of primes at most x is $\mathrm{Li}(x) + o(\mathrm{Li}(x))$, where $\mathrm{Li}(x) = \int_2^x dt/\log t$ and for x large, $\mathrm{Li}(x)$ is approximately $x/\log x$, and $f(x) = o(g(x))$ means $\lim_{x\to\infty} f(x)/g(x) = 0$. While it is possible to prove the prime number theorem elementarily [43, 128], the most informative proofs use complex numbers and complex analysis, and lead to the fascinating connection between number theory and nuclear physics. One of the most fruitful approaches to understanding the primes is to understand properties of the Riemann zeta function, $\zeta(s)$, which is defined for $\mathrm{Re}(s) > 1$ by

$$\zeta(s) == \sum_{n=1}^{\infty} \frac{1}{n^s}; \tag{1.1}$$

the series converges for $\mathrm{Re}(s) > 1$ by the integral test. By unique factorization, we may also write $\zeta(s)$ as a product over primes. To see this, use the geometric series formula to expand $(1 - p^{-s})^{-1}$ as $\sum_{k=0}^{\infty} p^{-ks}$ and note that n^{-s} occurs exactly once on each side (and clearly every term from expanding the product is of the form n^{-s}

for some n). This is called the Euler product of $\zeta(s)$, and is one of its most important properties:

$$\zeta(s) = \sum_{n=1}^{\infty} \frac{1}{n^s} = \prod_{p \text{ prime}} \left(1 - \frac{1}{p^s}\right)^{-1}. \tag{1.2}$$

Initially defined only for $\mathrm{Re}(s) > 1$, using complex analysis the Riemann zeta function can be meromorphically continued to all of $\mathbb{C}$, having only a simple pole with residue 1 at $s = 1$. It satisfies the functional equation

$$\xi(s) = \frac{1}{2}s(s-1)\Gamma\left(\frac{s}{2}\right)\pi^{-\frac{s}{2}}\zeta(s) = \xi(1-s). \tag{1.3}$$

One proof is to use the Gamma function, $\Gamma(s) = \int_0^\infty e^{-t}t^{s-1}dt$. A simple change of variables gives

$$\int_0^\infty x^{\frac{1}{2}s-1}e^{-n^2\pi x}dx = \Gamma\left(\frac{s}{2}\right)/n^s\pi^{s/2}. \tag{1.4}$$

Summing over n represents a multiple of $\zeta(s)$ as an integral. After some algebra we find

$$\Gamma\left(\frac{s}{2}\right)\zeta(s) = \int_1^\infty x^{\frac{1}{2}s-1}\omega(x)dx + \int_1^\infty x^{-\frac{1}{2}s-1}\omega\left(\frac{1}{x}\right)dx, \tag{1.5}$$

with $\omega(x) = \sum_{n=1}^\infty e^{-n^2\pi x}$. Using Poisson summation, we see

$$\omega\left(\frac{1}{x}\right) = -\frac{1}{2} + \frac{1}{2}x^{\frac{1}{2}} + x^{\frac{1}{2}}\omega(x), \tag{1.6}$$

which yields

$$\pi^{-\frac{1}{2}s}\Gamma\left(\frac{s}{2}\right)\zeta(s) = \frac{1}{s(s-1)} + \int_1^\infty (x^{\frac{1}{2}s-1} + x^{-\frac{1}{2}s-\frac{1}{2}})\omega(x)dx, \tag{1.7}$$

from which the claimed functional equation follows.

The distribution of the primes is a difficult problem; however, the distribution of the positive integers is not and has been completely known for quite some time! The hope is that we can understand $\sum_n 1/n^s$ as this involves sums over the integers, and somehow pass this knowledge on to the primes through the Euler product.

Riemann [123] (see [19, 39] for an English translation) observed a fascinating connection between the zeros of $\zeta(s)$ and the error term in the prime number theorem. As this relation is the starting point for our story on the number theory side, we describe the details in some length. One of the most natural things to do to a complex function is to take contour integrals of its logarithmic derivative; this yields information about zeros and poles, and we will see later in (1.17) that we can get even more information if we weigh the integral with a test function. There are two expressions for $\zeta(s)$; however, for the logarithmic derivative it is clear that we should use the Euler product over the sum expansion, as the logarithm of a product is the sum of the logarithms. Let

$$\Lambda(n) = \begin{cases} \log p & \text{if } n = p^r \text{ for some integer } r \\ 0 & \text{otherwise.} \end{cases} \tag{1.8}$$

We find

$$\frac{\zeta'(s)}{\zeta(s)} = -\sum_p \frac{\log p \cdot p^{-s}}{1 - p^{-s}} = -\sum_{n=1}^{\infty} \frac{\Lambda(n)}{n^s} \tag{1.9}$$

(this is proved by using the geometric series formula to write $(1 - p^{-s})^{-1}$ as $\sum_{k=0}^{\infty} 1/p^s$, collecting terms and then using the definition of $\Lambda(n)$). Moving the negative sign over and multiplying by x^s/s, we find

$$\frac{1}{2\pi i} \int_{(c)} -\frac{\zeta'(s)}{\zeta(s)} \frac{x^s}{s}\, ds = \frac{1}{2\pi i} \int_{(c)} \sum_{n \le x} \Lambda(n) \left(\frac{x}{n}\right)^s \frac{ds}{s}, \tag{1.10}$$

where we are integrating over some line $\mathrm{Re}(s) = c > 1$. The integral on the right hand side is 1 if $n < x$ and 0 if $n > x$ (by choosing x non-integral, we do not need to worry about $x = n$), and thus gives $\sum_{n \le x} \Lambda(n)$. By shifting contours and keeping track of the poles and zeros of $\zeta(s)$, the residue theorem implies that the left hand side is

$$x - \sum_{\rho:\zeta(\rho)=0} \frac{x^\rho}{\rho}; \tag{1.11}$$

the x term comes from the pole of $\zeta(s)$ at $s = 1$ (remember we count poles with a minus sign), while the x^ρ/ρ term arises from zeros; in both cases we must multiply by the residue, which is x^ρ/ρ (it can be shown that $\zeta(s)$ has neither a zero nor a pole at $s = 0$). Some care is required with this sum, as $\sum 1/|\rho|$ diverges. The solution involves pairing the contribution from ρ with $\overline{\rho}$; see for example [27].

The Riemann zeta function vanishes whenever ρ is a negative even integer; we call these the *trivial* zeros. These terms contribute $\sum_{k=-1}^{\infty} x^{-2k}/(2k) = -\frac{1}{2}\log(1 - x^{-2})$. This leads to the following beautiful formula, known as the *explicit formula*:

$$x - \sum_{\substack{\rho:\mathrm{Re}(\rho)\in(0,1) \\ \zeta(\rho)=0}} \frac{x^\rho}{\rho} - \frac{1}{2}\log(1 - x^{-2}) = \sum_{n \le x} \Lambda(n) \tag{1.12}$$

If we write n as p^r, the contribution from all p^r pieces with $r \geq 2$ is bounded by $2x^{1/2}\log x$ for x large, thus we really have a formula for the sum of the primes at most x, with the prime p weighted by $\log p$. Through partial summation, knowing the weighted sum is equivalent to knowing the unweighted sum.

We can now see the connection between the zeros of the Riemann zeta function and counting primes at most x. The contribution from the trivial zeros is well-understood, and is just $-\frac{1}{2}\log(1-x^{-2})$. The remaining zeros, whose real parts are in $[0,1]$, are called the *non-trivial* or *critical* zeros. They are far more important and more mysterious. The smaller the real part of these zeros of $\zeta(s)$, the smaller the error. Due to the functional equation, however, if $\zeta(\rho) = 0$ for a critical zero ρ then $\zeta(1-\rho) = 0$ as well. Thus the 'smallest' the real part can be is 1/2. This is the celebrated *Riemann Hypothesis (RH)*, which is probably the most important mathematical aside ever in a paper. Riemann [19, 39, 123] wrote (translated into English; note when he talks about the roots being real, he's writing the roots as $1/2 + i\gamma$, and thus $\gamma \in \mathbb{R}$ is the Riemann Hypothesis):

> One now finds indeed approximately this number of real roots within these limits, and it is very probable that all roots are real. Certainly one would wish for a stricter proof here; I have meanwhile temporarily put aside the search for this after some fleeting futile attempts, as it appears unnecessary for the next objective of my investigation.

Though not mentioned in the paper, Riemann had developed a terrific formula for computing the zeros of $\zeta(s)$, and had checked (but never reported!) that the first few were on the critical line $\text{Re}(s) = 1/2$. His numerical computations were only discovered decades later when Siegel was looking through Riemann's papers.

RH has a plethora of applications throughout number theory and mathematics; counting primes is but one of many. The prime number theorem is in fact equivalent to the statement that $\text{Re}(\rho) < 1$ for any zero of $\zeta(s)$, and was first proved independently by Hadamard [69] and de la Vallée Poussin [28] in 1896. Each proof crucially used results from complex analysis, which is hardly surprising given that Riemann had shown $\pi(x)$ is related to the zeros of the meromorphic function $\zeta(s)$. It was not until almost 50 years later that Erdös [43] and Selberg [128] obtained elementary proofs of the prime number theorem (in other words, proofs that did not use complex analysis, which was quite surprising as the prime number theorem was known to be equivalent to a statement about zeros of a meromorphic function). See [62] for some commentary on the history of elementary proofs. It is clear, however, that the distribution of the zeros of the Riemann zeta function will be of primary (in both senses of the word!) importance.

The Riemann zeta function is the first of many similar functions that we can study. We assume the reader has seen L-functions before; in addition to the surveys mentioned earlier, see also the introductory remarks in [83, 127]. We can examine, for the real part of s sufficiently large,

$$L(s,f) := \sum_{n=1}^{\infty} \frac{a_f(n)}{n^s}; \tag{1.13}$$

of course, while we can create such a function for any sequence $\{a_f(n)\}$ of sufficient decay, only certain choices will lead to useful objects whose zeros encode the solution to questions of arithmetic interest. For example, if we chose a_f arising from Dirichlet characters we obtain information about primes in arithmetic progression, while taking $a_f(p)$ to count the number of solutions to an elliptic curve $y^2 = x^3 + Ax + B$ modulo p yields information about the rank of the group of rational solutions.

Our previous analysis, where many of our formulas are due to taking the logarithmic derivative and computing a contour integral, suggests that we insist that an Euler product hold:

$$L(s,f) \;=\; \sum_{n=1}^{\infty} \frac{a_f(n)}{n^s} \;=\; \prod_{p \text{ prime}} L_p(s,f). \tag{1.14}$$

Further, we want a functional equation relating the values of the completed L-function at s and $1-s$, which allows us to take the series expansion that originally converges only for real part of s large and obtain a function defined everywhere:

$$\Lambda(s,f) \;=\; L_\infty(s,f)L(s,f) \;=\; \epsilon_f \Lambda(1-s,f), \tag{1.15}$$

where ϵ_f, the sign of the functional equation, is of absolute value 1, and

$$\begin{aligned} L_p(s,f) &= \prod_{j=1}^{d} \left(1 - \alpha_{f;j}(p)p^{-s}\right)^{-1} \\ L_\infty(s,f) &= AQ^s \prod_{j=1}^{n} \Gamma\left(\frac{s}{2} + \alpha_{f;j}\right), \end{aligned} \tag{1.16}$$

with $A \neq 0$ a complex number, $Q > 0$, $\alpha_{f;j} \geq 0$ and $\sum_{j=1}^{n} \alpha_{f;j}(p)^\nu = a_f(p^\nu)$. For 'nice' L-functions, it is believed that the Generalized Riemann Hypothesis (GRH) holds: All non-trivial zeros real part equal to 1/2.

We end our introduction to our main number theoretic objects of interest by noting that (1.12) is capable of massive generalization, not just to other L-functions but we can multiply (1.9) by a nice test function $\phi(s)$ instead of the specific function x^s/s. The result of this choice is to have a formula that relates sums of ϕ at zeros of our L-function to sums of the Fourier transform of ϕ at the primes. For example (see Sect. 4 of [83]) one can show

$$\sum_{\rho} \phi\left(\frac{\gamma}{2\pi}\log R\right) \;=\; \frac{A}{\log R} - 2\sum_{p}\sum_{\nu=1}^{\infty} a_f(p^\nu)\widehat{\phi}\left(\frac{\log p^\nu}{\log R}\right)\frac{\log p}{p^{\nu/2}\log R}, \tag{1.17}$$

where R is a free scaling parameter chosen for the problem of interest, $A = 2\widehat{\phi}(0)\log Q + \sum_{j=1}^{n} A_j$ with

$$A_j = \int_{-\infty}^{\infty} \psi\left(\alpha_{f;j} + \frac{1}{4} + \frac{2\pi i x}{\log R}\right)\phi(x)dx \tag{1.18}$$

and the Fourier transform is defined by

$$\widehat{\phi}(y) := \int_{-\infty}^{\infty} \phi(x)e^{-2\pi i x y}dx. \tag{1.19}$$

1.2.3 From the Hilbert-Pólya Connection to Random Matrix Theory

As stated earlier, the Generalized Riemann Hypothesis asserts that the non-trivial zeros of the an L-function are of the form $\rho = 1/2 + i\gamma_\rho$ with γ_ρ real. Thus it makes sense to talk about the distribution between adjacent zeros. Around 1913, Pólya conjectured that the γ_ρ are the eigenvalues of a naturally occurring, unbounded, self-adjoint operator, and are therefore real.[3] Later, Hilbert contributed to the conjecture, and reportedly introduced the phrase 'spectrum' to describe the eigenvalues of an equivalent Hermitian operator, apparently by analogy with the optical spectra observed in atoms. This remarkable analogy pre-dated Heisenberg's Matrix Mechanics and the Hamiltonian formulation of Quantum Mechanics by more than a decade.

Not surprisingly, the Hilbert-Pólya conjecture was considered so intractable that it was not pursued for decades, and random matrix theory remained in a dormant state. To quote Diaconis [31]:

> Historically, random matrix theory was started by statisticians [150] studying the correlations between different features of population (height, weight, income...). This led to correlation matrices with (i,j) entry the correlation between the ith and jth features. If the data were based on a random sample from a larger population, these correlation matrices are random; the study of how the eigenvalues of such samples fluctuate was one of the first great accomplishments of random matrix theory.

Diaconis [32] has given an extensive review of random matrix theory from the perspective of a statistician. A strong argument can be made, however, that random matrix theory, as we know it today in the physical sciences, began in a formal mathematical sense with the Wigner surmise [147] concerning the spacing distribution of adjacent resonances (of the same spin and parity) in the interactions between low-energy neutrons and nuclei, which we describe in great detail in Sect. 2.

---

[3]If v is an eigenvector with eigenvalue λ of a Hermitian matrix A (so $A = A^*$ with A^* the complex conjugate transpose of A, then $v^*(Av) = v^*(A^*v) = (Av)^*v$); the first expression is $\lambda||v||^2$ while the last is $\overline{\lambda}||v||^2$, with $||v||^2 = v^*v = \sum |v_i|^2$ non-zero. Thus $\lambda = \overline{\lambda}$, and the eigenvalues are real. This is one of the most important properties of Hermitian matrices, as it allows us to order the eigenvalues.

2 The 'Birth' of Random Matrix Theory in Nuclear Physics

Below we discuss some of the history of investigations of the nucleus, concentrating on the parts that led to the introduction of random matrix theory to the subject. As mentioned earlier, this section is expanded with permission from [50]. Our goal is to provide the reader with both sides of the coin, highlighting the interplay between theory and experiment, and building the basis for applications to understanding zeros of L-functions; we have chosen to spend a good amount of space on these experiments and conjectures as these are less-well known to the general mathematician than the later parts of our story.

While other methods have since been developed, random matrix theory was the first to make truly accurate, testable predictions. The general idea is that the behavior of zeros of L-functions are well-modeled by the behavior of eigenvalues of certain matrices. This idea had previously been successfully used to model the distribution of energy levels of heavy nuclei (some of the fundamental papers and books on the subject, ranging from experiments to theory, include [18, 29, 37, 38, 48, 51, 52, 54, 57, 63, 71, 72, 78, 98–100, 106, 119, 136, 143–148]). We describe the development of random matrix theory in nuclear physics below, and then delve into more of the details of the connection between the two subjects.

2.1 Neutron Physics

The period from the mid-1930s to the late 1970s was the golden age of neutron physics; widespread interest in understanding the physics of the nucleus, coupled with the need for accurate data in the design of nuclear reactors, made the field of neutron physics of global importance in fundamental physics, technology, economics, and politics. In Sect. 1.2.1 we introduced some of the early models for nuclei, and discussed some of the original experiments. In this section we describe later work where better resolution was possible. Later we will show how a similar perspective and chain of progress holds in studies of zeros of the Riemann zeta function! Thus the material here, in addition to being of interest in its own right, will also provide a valuable vantage for study of arithmetic objects.

In the mid-1950s, a discovery was made that turned out to have far-reaching consequences beyond anything that those working in the field could have imagined. For the first time, it was possible to study the microstructure of the continuum in a strongly-coupled, many-body system, at very high excitation energies. This unique situation came about as the result of the following facts.

- Neutrons, with kinetic energies of a few electron-volts, excite states in compound nuclei at energies ranging from about five million electron-volts to almost ten million electron-volts—typical neutron binding energies. Schematically, see Fig. 1.

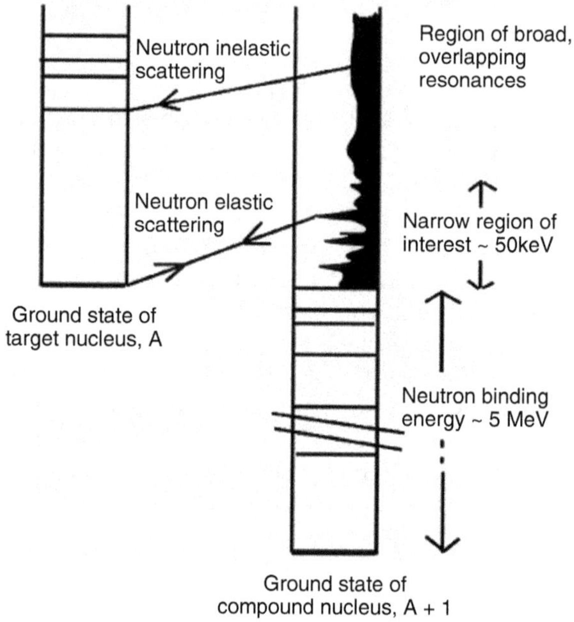

Fig. 1 An energy-level diagram showing the location of highly-excited resonances in the compound nucleus formed by the interaction of a neutron, n, with a nucleus of mass number A. Nature provides us with a narrow energy region in which the resonances are clearly separated, and are observable

- Low-energy resonant states in heavy nuclei (mass numbers greater than about 100) have lifetimes in the range 10^{-14}–10^{-15} s, and therefore they have widths of about 1 eV. The compound nucleus loses all memory of the way in which it is formed. It takes a relatively long time for sufficient energy to reside in a neutron before being emitted. This is a highly complex, statistical process. In heavy nuclei, the average spacing of adjacent resonances is typically in the range from a few eV to several hundred eV.
- Just above the neutron binding energy, the angular momentum barrier restricts the possible range of values of total spin of a resonance, $\mathbf{J}$ ($\mathbf{J} = \mathbf{I} + \mathbf{i} + \mathbf{l}$, where $\mathbf{I}$ is the spin of the target nucleus, $\mathbf{i}$ is the neutron spin, and $\mathbf{l}$ is the relative orbital angular momentum). This is an important technical point.
- The neutron time-of-flight method provides excellent energy resolution at energies up to several keV. (See Firk [47] for a review of time-of-flight spectrometers.)

The speed v_n of a neutron can be determined by measuring the time t_n that it takes to travel a measured distance ℓ in free space. Using the standard result of special relativity, the kinetic energy of the neutron can be deduced using the equation

$$\begin{aligned} E_n &= E_0[(1 - v_n^2/c^2)^{-1/2} - 1] \\ &= E_0[(1 - \ell^2/t_n^2c^2)^{-1/2} - 1], \end{aligned} \tag{2.1}$$

where $E_0 \approx 939.553\,\text{MeV}$ is the rest energy of the neutron and $c \approx 2.997925 \cdot 10^8$ m/s is the speed of light.

If the units of energy are MeV, and those of length and time are meters and nanoseconds, then

$$E_n = 939.553[(1 - 11.126496\ell^2/t_n^2)^{-1/2} - 1]\,\text{MeV}. \tag{2.2}$$

It is frequently useful to rearrange this equation to give the ratio t_n/ℓ for a given energy, E_n:

$$t_n/\ell = 3.3356404/\sqrt{1 - (939.553/(E_n + 939.553))^2}. \tag{2.3}$$

Typical values for this ratio are 72.355 ns/m for $E_n = 1\,\text{MeV}$ and 23.044 ns/m for $E_n = 10\,\text{MeV}$.

At energies below 1 MeV, the non-relativistic approximation to (2.3) is adequate:

$$(t_n/\ell)_{\text{NR}} = \sqrt{E_0/2E_n c^2} = 72.298/\sqrt{E_n}\,\mu\text{s/m}. \tag{2.4}$$

In the eV-region, it is usual to use units of μs/m: a 1 eV neutron travels 1 m in 72.3 μs. At non-relativistic energies, the energy resolution ΔE at an energy E is simply:

$$\Delta E \approx 2E\Delta t/t_E, \tag{2.5}$$

where Δt is the *total* timing uncertainty, and t_E is the flight time for a neutron of energy E.

In 1958, the two highest-resolution neutron spectrometers in the world had total timing uncertainties $\Delta t \approx 200\,\text{ns}$. For a flight-path length of 50 m the resolution was $\Delta E \approx 3\,\text{eV}$ at 1 keV.

In $^{238}\text{U} + \text{n}$, the excitation energy is about 5 MeV; the effective resolution for a 1 keV-neutron was therefore

$$\Delta E/E_{\text{effective}} \approx 6 \cdot 10^{-7} \tag{2.6}$$

(at 1 eV, the effective resolution was about 10^{-11}).

Two basic broadening effects limit the sensitivity of the method.

1. Doppler broadening of the resonance profile due to the thermal motion of the target nuclei; it is characterized by the quantity $\delta \approx 0.3\sqrt{E/A}$ (eV), where A is the mass number of the target. If $E = 1\,\text{keV}$ and $A = 200$, $\delta \approx 0.7\,\text{eV}$, a value that may be ten times greater than the natural width of the resonance.
2. Resolution broadening of the observed profile due to the finite resolving power of the spectrometer. For a review of the experimental methods used to measure neutron total cross sections see Firk and Melkonian [49]. Lynn [95] has given a detailed account of the theory of neutron resonance reactions.

In the early 1950s, the field of low-energy neutron resonance spectroscopy was dominated by research groups working at nuclear reactors. They were located at National Laboratories in the United States, the United Kingdom, Canada, and the former USSR. The energy spectrum of fission neutrons produced in a reactor is moderated in a hydrogenous material to generate an enhanced flux of low-energy neutrons. To carry out neutron time-of-flight spectroscopy, the continuous flux from the reactor is "chopped" using a massive steel rotor with fine slits through it. At the maximum attainable speed of rotation (about 20,000 rpm), and with slits a few thousandths-of-an-inch in width, it is possible to produce pulses each with a duration approximately 1 μs. The chopped beams have rather low fluxes, and therefore the flight paths are limited in length to less than 50 m. The resolution at 1 keV is then $\Delta E \approx 20$ eV, clearly not adequate for the study of resonance spacings about 10 eV.

In 1952, there were only four accelerator-based, low-energy neutron spectrometers operating in the world. They were at Columbia University in New York City, Brookhaven National Laboratory, the Atomic Energy Research Establishment, Harwell, England, and at Yale University. The performances of these early accelerator-based spectrometers were comparable with those achieved at the reactor-based facilities. It was clear that the basic limitations of the neutron-chopper spectrometers had been reached, and therefore future developments in the field would require improvements in accelerator-based systems.

In 1956, a new high-powered injector for the electron gun of the Harwell electron linear accelerator was installed to provide electron pulses with very short durations (typically less than 200 ns) [51]. The pulsed neutron flux (generated by the (γ, n) reaction) was sufficient to permit the use of a 56 m flight path; an energy resolution of 3 eV at 1 keV was achieved.

At the same time, Professors Havens and Rainwater (pioneers in the field of neutron time-of-flight spectroscopy) and their colleagues at Columbia University were building a new 385 MeV proton synchrocyclotron a few miles north of the campus (at the Nevis Laboratory). The accelerator was designed to carry out experiments in meson physics and low-energy neutron physics (neutrons generated by the (p, n) reaction). By 1958, they had produced a pulsed proton beam with duration of 25 ns, and had built a 37 m flight path [30, 124]. The hydrogenous neutron moderator generated an effective pulse width of about 200 ns for 1 keV-neutrons. In 1960, the length of the flight path was increased to 200 m, thereby setting a new standard in neutron time-of-flight spectroscopy [56].

2.2 *The Wigner Surmise*

At a conference on Neutron Physics by Time-of-Flight, held in Gatlinburg, Tennessee on November 1st and 2nd, 1956, Professor Eugene Wigner (Nobel Laureate in Physics, 1963) presented his surmise regarding the theoretical form of the spacing distribution of adjacent neutron resonances (of the same spin and parity) in heavy nuclei. At the time, the prevailing wisdom was that the spacing distribution had a

Poisson form (see, however, [68]). The limited experimental data then available was not sufficiently precise to fix the form of the distribution (see [78]). The following quotation, taken from Wigner's presentation at the conference, introduces the concept of random matrices in Physics, for the first time:

> Perhaps I am now too courageous when I try to guess the distribution of the distances between successive levels. I should re-emphasize that levels that have different J-values (total spin) are not connected with each other. They are entirely independent. So far, experimental data are available only on even-even elements. Theoretically, the situation is quite simple if one attacks the problem in a simple-minded fashion. The question is simply 'what are the distances of the characteristic values of a symmetric matrix with random coefficients?'
>
> We know that the chance that two such energy levels coincide is infinitely unlikely. We consider a two-dimensional matrix, $\begin{pmatrix} a_{11} & a_{12} \\ a_{21} & a_{22} \end{pmatrix}$, in which case the distance between two levels is $\sqrt{(a_{11}-a_{22})^2+4a_{12}^2}$. This distance can be zero only if $a_{11} = a_{22}$ and $a_{12} = 0$. The difference between the two energy levels is the distance of a point from the origin, the two coordinates of which are $(a_{11}-a_{22})$ and a_{12}. The probability that this distance is S is, for small values of S, always proportional to S itself because the volume element of the plane in polar coordinates contains the radius as a factor....
>
> The probability of finding the next level at a distance S now becomes proportional to SdS. Hence the simplest assumption will give the probability
>
> $$\frac{\pi}{2}\rho^2 \exp\left(-\frac{\pi}{4}\rho^2 S^2\right) SdS \tag{2.7}$$
>
> for a spacing between S and $S+dS$.
>
> If we put $x = \rho S = S/\langle S\rangle$, where $\langle S\rangle$ is the mean spacing, then the probability distribution takes the standard form
>
> $$p(x)dx = \frac{\pi}{2}\, x \exp\left(-\pi x^2/4\right) dx, \tag{2.8}$$
>
> where the coefficients are obtained by normalizing both the area and the mean to unity.

The form of the Wigner surmise had been previously discussed by Wigner [143], and by Landau and Smorodinsky [88], but not in the spirit of random matrix theory.

The Wigner form, in which the probability of zero spacing is zero, is strikingly different from the Poisson form

$$p(x)dx = \exp(-x)dx \tag{2.9}$$

in which the probability is a maximum for zero spacing. The form of the Wigner surmise had been previously discussed by Wigner himself [143], and by Landau and Smorodinsky [88], but not in the spirit of random matrix theory.

It is interesting to note that the Wigner distribution is a special case of a general statistical distribution, named after Professor E. H. Waloddi Weibull (1887–1979), a Swedish engineer and statistician [142]. For many years, the distribution has been in widespread use in statistical analyses in industries such as aerospace, automotive, electric power, nuclear power, communications, and life insurance.[4] The distribution gives the lifetimes of objects and is therefore invaluable in studies of the failure rates of objects under stress (including people!). The Weibull probability density function is

$$\mathrm{Wei}(x;k,\lambda) \;=\; \frac{k}{\lambda}\left(\frac{x}{\lambda}\right)^{k-1}\exp\left(-(x/\lambda)^k\right) \tag{2.10}$$

where $x \geq 0$, $k > 0$ is the *shape* parameter, and $\lambda > 0$ is the *scale* parameter. We see that $\mathrm{Wei}(x; 2, 2/\sqrt{\pi}) = p(x)$, the Wigner distribution. Other important Weibull distributions are given in the following list.

- $\mathrm{Wei}(x; 1, 1) = \exp(-x)$ the Poisson distribution;
- $\mathrm{Wei}(x; 2, \lambda) = \mathrm{Ray}(\lambda)$, the Rayleigh distribution;
- $\mathrm{Wei}(x; 3, \lambda)$ is approximately a normal distribution.[5]

For $\mathrm{Wei}(x; k, \lambda)$, the mean is $\lambda\Gamma\left(1 + (1/k)\right)$, the median is $\lambda \log(2)^{1/k}$, and the mode is $\lambda(k-1)^{1/k}/k^{1/k}$, if $k > 1$. As $k \to \infty$, the Weibull distribution has a sharp peak at λ. Historically, Frechet introduced this distribution in 1927, and Nuclear Physicists often refer to the Weibull distribution as the Brody distribution [18].

At the time of the Gatlinburg conference, no more than 20 s-wave neutron resonances had been clearly resolved in a single compound nucleus and therefore it was not possible to make a definitive test of the Wigner surmise. Immediately following the conference, Harvey and Hughes [71], and their collaborators, working at the fast-neutron-chopper-groups at the high flux reactor at the Brookhaven National Laboratory, and at the Oak Ridge National laboratory, gathered their own limited data, and all the data from neutron spectroscopy groups around the world, to obtain the first *global spacing distribution* of s-wave neutron resonances. Their combined results, published in 1958, showed a distinct lack of very closely spaced resonances, in agreement with the Wigner surmise.

By late 1959, the experimental situation had improved, greatly. At Columbia University, two students of Professors Havens and Rainwater completed their Ph.D. theses; one, Rosen [124], studied the first 55 resonances in $^{238}\mathrm{U} + \mathrm{n}$ up to 1 keV, and the other, Desjardins [30], studied resonances in two silver isotopes (of different spin) in the same energy region. These were the first results from the new high-resolution neutron facility at the Nevis cyclotron.

At Harwell, Firk et al. [48] completed their study of the first 100 resonances in $^{238}\mathrm{U} + \mathrm{n}$ at energies up to 1.8 keV; their measurement of the total neutron cross

---

[4]In fact, one of the authors has used Weibull distributions to model run production in major league baseball, giving a theoretical justification for Bill James' Pythagorean Won-Loss formula [103].

[5]Obviously this Weibull cannot be a normal distribution, as they have very different decay rates for large x, and this Weibull is a one-sided distribution! What we mean is that for $0 \leq x \leq 2$ this Weibull is well approximated by a normal distribution which shares its mean and variance, which are (respectively) $\Gamma(4/3) \approx 0.893$ and $\Gamma(5/3) - \Gamma(4/3)^2 \approx 0.105$.

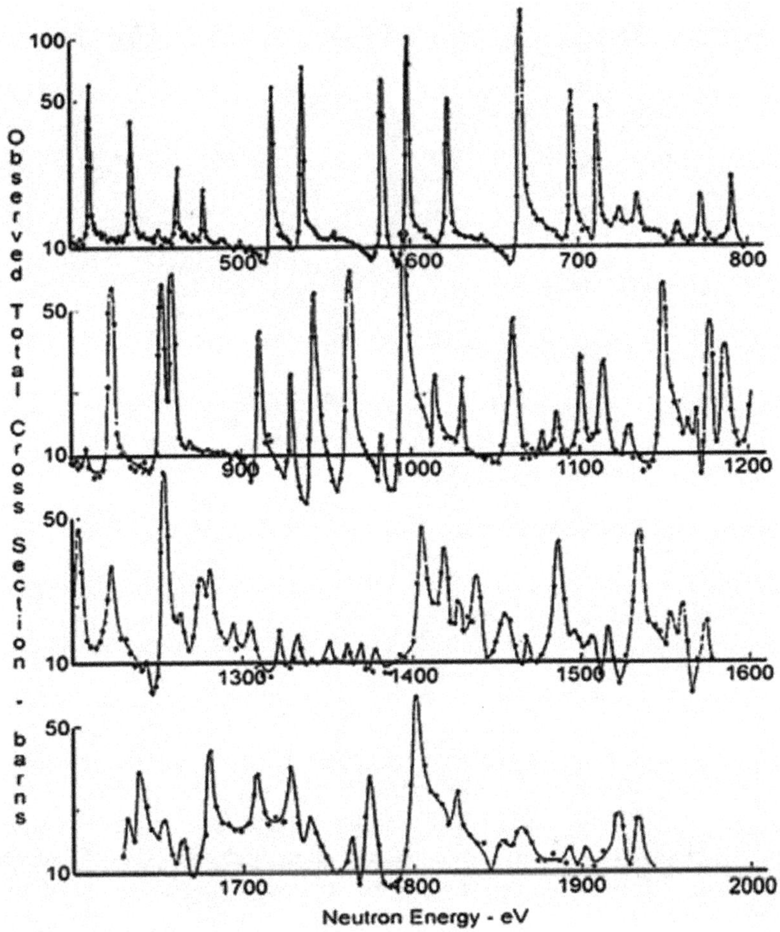

Fig. 2 High resolution studies of the total neutron cross section of ^{238}U, in the energy range 400–1800 eV. The *vertical scale* (in units of "barns") is a measure of the effective area of the target nucleus

section for the interaction $^{238}\mathrm{U} + \mathrm{n}$ in the energy range 400–1800 eV is shown in shown in Fig. 2.

When this experiment began in 1956, no resonances had been resolved at energies above 500 eV. The distribution of adjacent spacings of the first 100 resonances in the single compound nucleus, $^{238}\mathrm{U} + \mathrm{n}$, ruled out an exponential distribution and provided the best evidence (then available) in support of Wigner's proposed distribution.

Over the last half-century, numerous studies have not changed the basic findings. At the present time, almost 1000 s-wave neutron resonances in the compound nucleus ^{239}U have been observed in the energy range up to 20 keV. The latest results, with their greatly improved statistics, are shown in Fig. 3 [29].

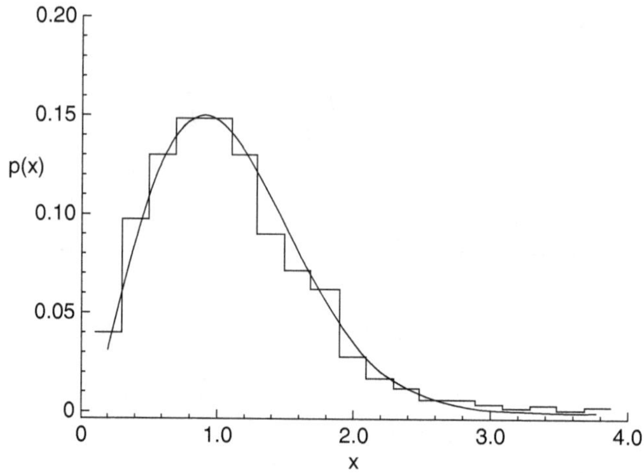

Fig. 3 A Wigner distribution fitted to the spacing distribution of 932 s-wave resonances in the interaction ^{238}U + n at energies up to 20 keV

2.3 *Some Nuclear Models*

It is interesting to note that, during the 1950s and 1960s, the study of the spacing distribution of neutron-induced resonances was far from the main stream of research in nuclear physics; almost all research was concerned with fundamental questions associated with nuclear structure and not with quantum statistical mechanics. The newly-discovered Shell Model [90, 97] of nuclei, and developments such as the Collective Model [14, 121] were popular, and quite rightly so, when the successes of these models in accounting for the observed energies, spins and parities, and magnetic moments of nuclear states, particularly in light nuclei (mass numbers < 20, say) were considered.

These models were not able to account for the spacing distributions in heavy nuclei (mass numbers $A > 150$); the complex nature of so many strongly interacting nucleons prevented any detailed analysis. However, the treatment of such complex problems had been considered in the mid-1930s, before the advent of the Shell-Model. The Fermi Gas Model and other approaches based upon quantum versions of classical statistical mechanics and thermodynamics, were introduced, particularly by Bethe [10]. The Fermi Gas Model treats the nucleons as non-interacting spin-$\frac{1}{2}$ particles in a confined volume of nuclear size. This, of course, seems at variance with the known strong interaction between pairs of nucleons. However, the argument is made that the nuclear gas is completely degenerate and therefore, because of the Pauli exclusion principle, the nucleons can be considered free! The model was the first to predict the energy-dependence of the density of states in the nuclear system.

The number of states that are available to a freely moving particle in a volume V (the nuclear volume) that has a linear momentum in the range p to $p + dp$ is

$$dn = (4\pi V/h^3)p^2 dp. \tag{2.11}$$

This leads to

$$n = (V/3\pi^2 h^3)p_{\max}^3, \tag{2.12}$$

where the result has been doubled because of the twofold spin degeneracy of the nucleons. The "Fermi energy" E_F corresponds to the maximum momentum:

$$E_F = p_{\max}^2/2m_{\text{Nucleon}}. \tag{2.13}$$

The level density $\rho(E^*)$ at an excitation energy E^* predicted by the model is

$$\rho(E^*) = \rho(0)\exp\left(2\sqrt{aE^*}\right), \tag{2.14}$$

where a is given by the equation

$$E^* = a(kT)^2 \tag{2.15}$$

in which k is Boltzmann's constant and T is the absolute temperature. The above expression for the level density is for states of all spins and parities.

In practical cases, E^* is about 6 MeV for low- energy neutron interactions; this value leads to the following ratio for the mean level spacing at $E^* = 6$ MeV and at $E^* = 0$ (the ground state):

$$\langle D(6\,\text{MeV})\rangle/\langle D(0)\rangle \approx 4 \cdot 10^{-8}. \tag{2.16}$$

For $\langle D(0)\rangle = 100$ keV (a practical value), the mean level spacing at $E^* = 6$ MeV is $\approx 4 \cdot 10^{-3}$ eV, which is more than three orders-of-magnitude smaller than typical values observed in heavy nuclei.

Many refinements of the model were introduced over the years; the models take into account spin, parity, and nucleon pairing effects. A frequently used refined form is

$$\rho(E^*, J) = \rho(E^*, 0)(2J + 1)\exp\left(-(J(J + 1))/2\sigma^2\right), \tag{2.17}$$

where σ is called the "spin-cut-off parameter"; the value of σ^2 is typically about 10. The predicted spacing distributions for two values of σ, and their comparison with a Wigner and an exponential distribution is shown in Fig. 4.

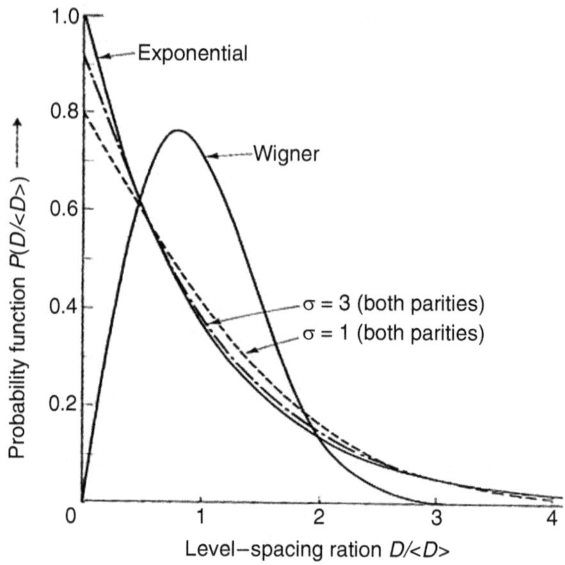

Fig. 4 The spacing distribution of adjacent levels of the same spin and parity follows a Wigner distribution. For a completely random distribution of levels (in both spin and parity) the distribution function is exponential. The distributions for random superpositions of several sequences (each of which is of a Wigner form with a characteristic spin and parity) are, for level densities given by (2.17) and $\sigma = 1$ and 3, found to approach the exponential distribution

2.4 *The Optical Model*

In 1936, Ostrofsky et. al. [118] introduced a model of nuclear reactions that employed a complex nuclear potential to account for absorption of the incoming nucleon. Later, Feshbach et al. [45] introduced an important development of the model that helped further our understanding of the average properties of parameters used to describe nuclear reactions at low energies.

The following discussion provides insight into the physical content of their model. Consider the plane-wave solutions of the Schrödinger equation:

$$\frac{d^2\phi}{dx^2} + (2m/h^2)[E + V_0 + iW]\phi = 0, \qquad \phi = \exp(\pm ikx), \tag{2.18}$$

where the + sign indicates outgoing waves and the − sign indicates incoming waves. The wave number, k is complex:

$$k = \sqrt{(2m/h^2)[(E + V0) + iW]}, \tag{2.19}$$

which can be written

$$k = k_{\mathrm{R}} + K_{\mathrm{IM}}. \tag{2.20}$$

For $W < (E + V_0)$ (a reasonable assumption) we have

$$\begin{aligned} k_{\mathrm{R}} &= 1/\lambda \approx \sqrt{(2m/h2)(E + V_0)} \\ K_{\mathrm{IM}} &= [W/(E + V_0)](k/2). \end{aligned} \tag{2.21}$$

Taking typical practical values $E = 10\,\text{MeV}$, $V_0 = 40\,\text{MeV}$ and $W = 10\,\text{MeV}$, the wave numbers are $k_\text{R} \approx 1.5\text{fm} - 1$ and $K_\text{IM} \approx k_\text{R}/10 \approx 0.15\text{fm}^{-1}$.

We see that the outgoing solution of the wave equation is

$$\phi = \exp\left(ik_\text{R}x\right)\exp\left(-K_\text{IM}x\right), \tag{2.22}$$

which represents an exponentially attenuated wave. The wave number K_IM is effectively an attenuation coefficient. The "decay length" associated with the probability function $|\phi|^2$ is the "mean free path":

$$\Lambda = 1/2K_\text{IM} = (E + V_0)/Wk_\text{R}. \tag{2.23}$$

Using the above values for the energies, we obtain $\Lambda \approx 3.2\text{fm}$. This value is of nuclear dimension, and supports the underlying hypothesis of the Compound Nucleus Model.

If the mean spacing of energy levels of a particle of mass m inside the compound nucleus is $\langle D\rangle$, and its wave number is K, then the particle covers a distance

$$d \approx (h/\langle D\rangle)((hK/2\pi m) = (h^2K)/(2\pi m\langle D\rangle) \tag{2.24}$$

inside the nucleus at an average speed $\langle v\rangle \approx hK/2\pi m$ before it is emitted (or before another indistinguishable particle is emitted). At an excitation energy of 10 MeV, a mean level spacing $\langle D\rangle \approx 40\,\text{eV}$, and a mean lifetime $h/\langle D\rangle \approx 10^{-16}\,\text{s}$ are predicted. These are reasonable values, considering the crudeness of the model.

The level density and level widths increase as the neutron bombarding energy increases; an energy region is therefore reached in which the levels completely overlap. Cross section measurements then provide information on the average properties of the levels and, in particular, on the *neutron strength function* [92] defined as

$$S = \langle\gamma_{\lambda n}\rangle^2/\langle D\rangle \tag{2.25}$$

in which = $\langle\gamma_{\lambda n}\rangle^2$ is the average reduced neutron width and $\langle D\rangle$ is the average spacing. For s-wave neutrons, $\gamma^2_{\lambda n} = 2ka\Gamma_{\lambda n}$, where k is the neutron wave number, a is the nuclear radius, and $\Gamma_{\lambda n}$ is the neutron width of the level λ.

The average absorption cross section $\langle\sigma_\text{abs}\rangle$ may be obtained by averaging over the collision function U [92]. The following expressions are then obtained:

$$\begin{aligned} 1 - |\langle U\rangle|^2 &= 2\pi\left(\langle\Gamma_{\lambda n}\rangle/\langle D\rangle\right) \\ \langle\sigma_\text{abs}\rangle &= (\pi/k^2)g\left(1 - |\langle U\rangle|^2\right), \end{aligned} \tag{2.26}$$

where g is a statistical "spin weighting factor".

The term $1 - |\langle U \rangle|^2$ is directly related to the cross section for the formation of a compound nucleus [45] which is, in turn, proportional to the strength function. *The importance of studying the spacing distribution of resonances, of a given spin and parity, originated in recognizing that the value of $\langle D \rangle$, the average spacing, appears as the denominator in the fundamental strength function.*

2.5 Further Developments

The first numerical investigation of the distribution of successive eigenvalues associated with random matrices was carried out by Porter and Rozenzweig in the late 1950s [120]. They diagonalized a large number of matrices where the elements are generated randomly but constrained by a probability distribution. The analytical theory developed in parallel with their work: Mehta [98], Mehta and Gaudin [100], and Gaudin [57]. At the time it was clear that the spacing distribution was not influenced significantly by the chosen form of the probability distribution. Remarkably, the $n \times n$ distributions had forms given *almost exactly* by the original Wigner 2×2 distribution.

The linear dependence of $p(x)$ on the normalized spacing x (for small x) is a direct consequence of the *symmetries* imposed on the Hamiltonian matrix, $H(h_{ij})$. Dyson [37] discussed the general mathematical properties associated with random matrices and made fundamental contributions to the theory by showing that different results are obtained when different *symmetries* are assumed for H. He introduced three basic distributions; in Physics, only two are important, they are:

- the Gaussian Othogonal Ensemble (GOE) for systems in which rotational symmetry and time-reversal invariance holds (the Wigner distribution): $p(x) = (\pi/2)\, x \exp\left(-(\pi/4)x^2\right)$;
- the Gaussian Unitary Ensemble (GUE) for systems in which time-reversal invariance does not hold (French et. al. [54]): $p(x) = (32/\pi^2)x^2 \exp(-(\pi/4)x^2)$.

The mathematical details associated with these distributions are given in [98].

The impact of these developments was not immediate in nuclear physics. At the time, the main research endeavors were concerned with the structure of nuclei–experiments and theories connected with Shell-, Collective-, and Unified models, and with the nucleon-nucleon interaction. The study of quantum statistical mechanics was far removed from the mainstream. Almost two decades went by before random matrix theory was introduced in other fields of physics (see, for example, Bohigas et al. [16] and Alhassid [1]).

2.6 Lessons from Nuclear Physics

We have discussed at great length the connections between nuclear physics and number theory, with random matrix theory describing the behavior in these two very different fields. Before we analyze in great detail the success it has had in modeling the zeros of L-functions, it's worth taking a few moments to create a dictionary comparing these two subjects.

In nuclear physics the main object of interest is the nucleus. It is a many-bodied system governed by complicated forces. We are interested in studying the internal energy levels. To do so, we shoot neutrons (which have no net charge) at the nucleus, and observe what happens. Ideally we would be able to send neutrons of any energy level; unfortunately in practice we can only handle neutrons whose energies are in a certain band. The more energies at our disposal, the more refined an analysis is possible. Finally, there is a remarkable universality from heavy nucleus to heavy nucleus, where the distribution of spacings between adjacent energy levels depends weakly on the quantum numbers.

Interestingly, there are analogues of all these quantities on the number theory side. The nucleus is replaced by an L-function, which is built up as an Euler product of many factors of arithmetic interest. We are interested in the zeros of this function. We can glean information about them by using the explicit formula, (1.17). We first choose an even Schwartz test function ϕ whose Fourier transform $\widehat{\phi}$ has compact support. The explicit formula relates sums of ϕ at the zeros of the L-function to weighted sums of $\widehat{\phi}$ at the primes. Thus the more functions $\widehat{\phi}$ where we can successfully execute the sums over the primes, the more information we can deduce about the zeros. Unfortunately, in practice we can only evaluate the prime sums for $\widehat{\phi}$ with small support (if we could do arbitrary $\widehat{\phi}$, we could take a sequence converging to the constant function 1, whose inverse Fourier transform would be a delta spike at the origin and thus tell us what is happening there). Similar to the weak dependence on the quantum numbers, the answers for many number theory statistics depend weakly on the Satake parameters (whose moments are the Fourier coefficients in the series expansion of the L-function). In particular, the spacing between adjacent zeros is independent of the distribution of these parameters, though other statistics (such as the distribution of the first zero or first few zeros above the central point) fall into several classes depending on their distribution.

We collect these correspondences in the table below. While the structures studied in the two fields are very different, we can unify the presentations. In both settings we study the spacings between objects. While there are exact rules that govern their behavior, these are complicated. We gain information through interactions of test objects with our system; as we can only analyze these interactions in certain windows, we gain only partial information on the items of interest.

We end by extracting some lessons from nuclear physics for number theory. The first is the importance of using the proper test function, or related to that the proper statistic. In the gold-foil experiments (1908 to 1913) positively charged alpha particles, which are helium nuclei, were used. Because they have a net positive

Item	Nuclear physics	Number theory
Object	Nucleus	L-function
Events	Energy levels	Zeros
Probe	Neutron (no net charge)	Test function ϕ (Schwartz)
Restriction	Neutron's energy	Supp($\widehat{\phi}$)
Individuality	Quantum numbers	Satake parameters

charge, they are repelled by the nucleus they are probing. With the discovery of the neutron in 1932, physicists had a significantly better tool for studying the nucleus. As the machinery improved, more and more neutron energy levels were available, which led to sharper resolutions of the internal structure. We see variants of these on the number theory side, from restrictions on the test function to the consequences of increasing support. For example, when Wigner made his bold conjectures the data was not sufficiently detailed to rule out Poissonian behavior; that was not done until later when better experiments were carried out. Similar situations arise in number theory, where some statistics are consistent with multiple models and only by increasing the support are we able to determine the true underlying behavior. Finally, while there is a remarkable universality in behavior of the zeros, as for statistics such as adjacent spacings or n-level correlations the exact form of the L-function coefficients do not matter, these distributions do affect the rate of convergence to the random matrix theory predictions, as well as govern other statistics.

3 From Class Numbers to Pair Correlation and Random Matrix Theory

The discovery that the pair correlation of the zeros of the Riemann zeta function (and other statistics of its zeros, and the zeros of other L-functions) are related to eigenvalues of random matrix ensembles has its beginnings with one of the most challenging problems in analytic number theory: the class number problem. Hugh Montgomery's investigation into the vertical distribution of the nontrivial zeros of $\zeta(s)$ arose during his work with Weinberger [108] on the class number problem. We give a short introduction to this problem to motivate Montgomery's subsequent work on the differences between zeros of $\zeta(s)$. We assume the reader is familiar with the basics of algebraic number theory and L-functions; an excellent introduction is Davenport's classic *Multiplicative Number Theory* [27]. For those wishing a more detailed and technical discussion of the class number problem and its history, see [61, 62]. We then continue with a discussion of Montgomery's work on pair correlation, followed by the work of Odlyzko and others on spacings between adjacent zeros. After introducing the number theory motivation and results, we reveal the connection to random matrix theory, and conclude with a discussion of the higher level correlations, other related statistics, and open problems.

As there are too many areas of current research to describe them all in detail in a short article, we have chosen to concentrate on two major areas: the main terms for the n-level correlations, and the lower order terms; thus we do not describe many other important areas of research, such as the determination of moments or value distribution. The main terms are believed to be described by random matrix theory; however, the lower order terms depend on subtle arithmetic of the L-functions, and there we can see different behavior. The situation is very similar to that of the Central Limit Theorem, and we will describe these connections and viewpoints in greater detail below.

3.1 The Class Number Problem

Let $K = \mathbb{Q}(\sqrt{-q})$ be the imaginary quadratic field associated to the negative fundamental discriminant $-q$. Here we have that $-q$ is congruent to 1 (mod4) and square-free or $-q = 4m$, where m is congruent to 2 or 3 (mod4) and square-free. The class number of K, denoted $h(-q)$, is the size of the group of ideal classes of K. When $h(-q) = 1$, the ring of integers of K, denoted $\mathcal{O}_K$, has unique factorization. Such an occurrence (the class number one problem, discussed below) is rare, and the class number $h(-q)$ may be thought of as a measure on the failure of unique factorization in $\mathcal{O}_K$.

One of the most difficult problems in analytic number theory is to estimate the size of $h(-q)$ effectively. Gauss [58] showed that $h(-q)$ is finite and further conjectured that h tends to infinity as $-q$ runs over the negative fundamental discriminants. This conjecture was proved by Heilbronn [75] in 1934. Thus, while it is settled that there are only finitely many imaginary quadratic fields with a given class number $h(-q)$, an obvious question remains: can we list all imaginary quadratic fields K with a given class number $h(-q)$? This is the class number problem.

One may easily deduce an upper bound on $h(-d)$ via Dirichlet's class number formula. For $\Re(s) > 1$, let $L(s, \chi_{-q})$ denote the Dirichlet L-function

$$L(s, \chi_{-q}) := \sum_{n=1}^{\infty} \frac{\chi_{-q}(n)}{n^s}, \tag{3.1}$$

where $\chi_{-q}(n)$ is the Kronecker symbol associated to the fundamental discriminant $-q$. In order to prove the equidistribution of primes in arithmetic progression, Dirichlet derived the class number formula,

$$h(-q) = \frac{w\sqrt{q}}{2\pi} L(1, \chi_{-q}), \tag{3.2}$$

where w denotes the number of roots of unity of $K = \mathbb{Q}(\sqrt{-q})$:

$$w = \begin{cases} 2 & \text{if} \quad q > 4 \\ 4 & \text{if} \quad q = 4 \\ 6 & \text{if} \quad q = 3. \end{cases} \tag{3.3}$$

Dirichlet needed to show $L(1, \chi_{-q}) \neq 0$, which is immediate from the class number formula as $h(-q) \geq 1$. This connection between class numbers and zeros of L-functions is almost 200 years old, and illustrates how knowledge of zeros of L-functions yields information on a variety of important problems.

Instead of using the class number formula to prove non-vanishing of L-functions, we can use results on the size of L-functions to obtain bounds on the class number. Combining (3.2) with that fact that $L(1, \chi_{-q}) \ll \log q$, it follows that $h(-q) \ll \sqrt{q} \log q$. On the other-hand, Siegel [132] proved that for every $\varepsilon > 0$ we have $L(1, \chi_{-q}) > c(\varepsilon) q^{-\varepsilon}$, where $c(\varepsilon)$ is a constant depending on ε that is not numerically computable for small ε. Upon inserting this lower bound in (3.2), it follows that $h(-q) \gg c(\varepsilon) q^{1/2-\varepsilon}$; however this does not help us solve the class number problem because the implied constant is ineffective.[6] Computing an effective lower bound on $h(-q)$ is very difficult task.

The class number one problem was eventually solved independently by Heegner [74], Stark [133] and Baker [8]. For $h(-d) = 2$, the class number problem was solved independently by Stark [134], Baker [9] and Montgomery and Weinberger [108]. In 1976, Goldfeld [59, 60] showed that if there exists an elliptic curve E whose Hasse-Weil L-function has a zero at the central point $s = 1$ of order at least three, then for any $\varepsilon > 0$, we have $h(-q) > c_{\varepsilon,E} \log(|-q|)^{1-\varepsilon}$, where the constant $c_{\varepsilon,E}$ is effectively computable. In other words, Goldfeld proved that if there exists an elliptic curve whose Hasse-Weil L-function has a triple zero at $s = 1$, then the class number problem is reduced to a finite amount of computations. In 1983, Gross and Zagier [66] showed the existence of such an elliptic curve. Combining this deep work of Gross-Zagier with a simplified version of Goldfield's argument to reduce the amount of necessary computations, Oesterlé [115] produced a complete list of imaginary quadratic fields with $h(-q) = 3$. To date, the class number problem is resolved for all $1 \leq h(-q) \leq 100$. (In addition to the previous references, see Arnon [4], Arnon et al. [5], Wanger [140] and Watkins [141].)

[6]In other words, while the above is enough to prove that the class number tends to infinity, we cannot use that argument to produce an explicit constant Q_n for each n so that we could assert that the class number is at least n if $q \geq Q_n$. One of the best illustrations of the importance of *effective* constants is the following joke: There is a constant T_0 such that if all the non-trivial zeros of $\zeta(s)$ in the critical strip up to height T_0 are on the critical line, then they all are and the Riemann Hypothesis is true; in other words, it suffices to check up to a finite height! To see this, if the Riemann Hypothesis is true we may take T_0 to be 0, while if it is false we take T_0 to be 1 more than the height of the first exemption. We have therefore shown a constant exists, but such information is completely useless!

Combining their work with results of Stark [134] and Lehmer et al. [93], Montgomery and Weinberger gave a complete proof for the class number two problem. Their proof is based on the curious Deuring-Heilbronn phenomenon, which implies that if $h(-d) < d^{1/4-\delta}$ then the low-lying nontrivial zeros of many quadratic Dirichlet L-functions are on the critical line, at least up to some height depending on d, δ, and the L-functions. For an overview of the Deuring-Heilbronn phenomenon, see the survey article by Stopple [135]. Montgomery and Weinberger also establish that if the class number is a bit smaller, then one can show that these nontrivial zeros on the critical line are very evenly spaced. Moreover, more precise information about the vertical distribution of these zeros would imply an effective lower bound on $h(-d)$. Montgomery and Weinberger write:

> Let $\rho = 1/2 + i\gamma$ and $\rho' = 1/2 + i\gamma'$ be consecutive zeros on the critical line of an L-function $L(s,\chi)$, where χ is a primitive character (mod$k$). Put
>
> $$\lambda(K) = \min \frac{1}{2\pi}|\gamma - \gamma'| \log K, \tag{3.4}$$
>
> where the minimum is over all $k \leq K$, all χ (mod$k$), and all $\rho = 1/2 + i\gamma$ of $L(s,\chi)$ with $|\gamma| \leq 1$. In this range the average of $|\gamma - \gamma'|$ is $2\pi/\log k$, so trivially $\limsup \lambda(K) \leq 1$. Presumably $\lambda(K)$ tends to 0 as K increases; if this could be shown effectively then the effective lower bound $h > d^{1/4-\varepsilon}$ would follow. In fact the weak inequality $\lambda(K) < 1/4-\delta$ for $K > K_0$ implies that $h > d^{(1/2)\delta-\varepsilon}$ for $d > C(K_0,\varepsilon)$; the function $C(K_0,\varepsilon)$ can be made explicit. Even $\lambda(K) < \frac{1}{2} - \delta$ has striking consequences.

3.2 Montgomery's Pair Correlation of the Zeros of $\zeta(s)$

We have seen that the class number problem is related to another very difficult question in analytic number theory: *What is the vertical distribution of the zeros of the Riemann zeta function (and general L-functions) on the critical line?*

Given an increasing sequence $\{\alpha_n\}_{n=1}^{\infty}$ and a box $B \subset \mathbb{R}^{n-1}$, the n-level correlation is defined by

$$\lim_{N\to\infty} \frac{\#\left\{\left(\alpha_{j_1} - \alpha_{j_2}, \ldots, \alpha_{j_{n-1}} - \alpha_{j_n}\right) \in B, j_i \neq j_k\right\}}{N}. \tag{3.5}$$

The pair correlation is the case $n = 2$, and through combinatorics knowing all the correlations yields the spacing between adjacent events (see for example [99]). In 1973, Montgomery [107] was able to partially determine the behavior for the pair correlation of zeros of the Riemann zeta function, $\zeta(s)$, which led to new results on the number of simple zeros of $\zeta(s)$ and the existence of gaps between zeros of $\zeta(s)$ that are closer together than the average. One of the most striking contributions in Montgomery's paper, however, is his now famous pair correlation conjecture. We first state his conjecture and then discuss related work on spacings between adjacent zeros in the next subsection; after these have been described in detail we then revisit these problems and describe the connections with random matrix theory in Sect. 3.5. See [25] for more on connections between spacings of zeros of $\zeta(s)$ and the class number.

Conjecture 1 (Montgomery's Pair Correlation Conjecture). Assume the Riemann hypothesis, and let γ, γ' denote the imaginary parts of nontrivial zeros of $\zeta(s)$. For fixed $0 < a < b < \infty$,

$$\lim_{T\to\infty} \frac{\#\{\gamma,\gamma' : 0 \le \gamma,\gamma' \le T, 2\pi a(\log T)^{-1} \le \gamma - \gamma' \le 2\pi b(\log T)^{-1}\}}{\frac{T}{2\pi}\log T}$$
$$= \int_a^b 1 - \left(\frac{\sin \pi u}{\pi u}\right)^2 du. \tag{3.6}$$

Thus Montgomery's pair correlation conjecture is the statement that the pair correlation of the zeros of $\zeta(s)$ is

$$1 - \left(\frac{\sin \pi u}{\pi u}\right)^2. \tag{3.7}$$

Notice that the factor $1 - (\sin \pi u/\pi u)^2$ suggests a 'repulsion' between the zeros of $\zeta(s)$. The notion that the zeros cannot be too close to one another was also revealed in the aforementioned work of Montgomery and Weinberger as a consequence of the Deuring-Heilbronn phenomenon.

To arrive at his conjecture, Montgomery introduced the function

$$F(x,T) = \sum_{0<\gamma,\gamma'\le T} x^{i(\gamma-\gamma')} w(\gamma-\gamma'), \tag{3.8}$$

where $w(u)$ is a weight function given by $w(u) = 4/(4+u^2)$. Let $F(\alpha)$ denote $F(x,T)$ with x set as $x = T^\alpha$; then

$$F(\alpha) = F(\alpha,T) = \left(\frac{T}{2\pi}\log T\right)^{-1} \sum_{0\le\gamma,\gamma'\le T} T^{i\alpha(\gamma-\gamma')} w(\gamma-\gamma'), \tag{3.9}$$

where α and $T \ge 2$ are real. $F(\alpha)$ is a real, even function. Let $r(u) \in L^1$, and define its Fourier transform by

$$\hat{r}(\alpha) = \int_{-\infty}^{\infty} r(u) e^{2\pi i \alpha u} du. \tag{3.10}$$

The function r is a test function that replaces the 'box' in the statement of the pair correlation Conjecture 1. One notable item about Montgomery's pair correlation conjecture is that there is no restriction on the length of the interval $[a, b]$; the difference $b - a$ is permitted to be arbitrarily small. In the language of smooth test functions, this translates to permitting arbitrarily large support on the Fourier transform $\hat{r}$.

If $\hat{r}(\alpha) \in L^1$, then upon multiplying (3.9) by $\hat{r}(\alpha)$ and integrating, we deduce

$$\sum_{0<\gamma,\gamma'\leq T} r\left(\frac{(\gamma'-\gamma)\log T}{2\pi} w(\gamma'-\gamma)\right) \sim \left(\frac{T}{2\pi}\log T\right)\int_{-\infty}^{\infty}\hat{r}(\alpha)F(\alpha)d\alpha \quad (3.11)$$

as T tends to infinity. If the Riemann hypothesis is true, the asymptotic (3.11) connects the pair correlation of $\zeta(s)$ to the function $F(\alpha)$ given in (3.9). Montgomery proceeded to prove an important special case of Conjecture 1 for a class of test functions with Fourier transform supported in $(-1, 1)$.

Theorem 1 (Montgomery's Theorem). *Assume the Riemann hypothesis. For real* α, $T \geq 2$, *let* $F(\alpha)$ *be defined by* (3.9). *Then* $F(\alpha)$ *is real, and* $F(\alpha) = F(-\alpha)$. *If* $T > T_0(\epsilon)$ *then* $F(\alpha) \geq -\epsilon$ *for all* α. *For fixed* α *satisfying* $0 \leq \alpha < 1$ *we have*

$$F(\alpha) = \alpha + o(1) + T^{-2\alpha}\log T(1 + o(1)) \quad (3.12)$$

uniformly for $0 \leq \alpha < 1$ *as* T *tends to infinity.*

Thus, for any function $r(u) \in L^1$ with Fourier transform $\hat{r}(\alpha)$ supported in $(-1, 1)$, one can use (3.12) to evaluate the sums appearing in (3.11). For $\alpha \geq 1$, Montgomery further conjectured, with heuristic arithmetic justification, that

$$F(\alpha) = 1 + o(1) \quad \text{uniformlyinboundedintervalsas} T \to \infty. \quad (3.13)$$

This conjecture, combined with (3.12) gives a complete picture of the function $F(\alpha)$, which led Montgomery to make his pair correlation conjecture.

3.3 *Proof of Montgomery's Pair Correlation Conjecture for Restricted* a, b

We now provide greater detail about Montgomery's original proof [107, Sect. 3, pp. 187–191] of his theorem (Theorem 1). The point of entry is an explicit formula due to him.

The role of explicit formulæ cannot be overstated when working with $\zeta(s)$ or L-functions, as these formulæ unlock the multiplicative structure implicit in the Euler product, usually via the argument principal applied to the logarithmic derivative. Assuming the Riemann hypothesis, and writing critical zeros of $\zeta(s)$ as $1/2 + i\gamma$ and γ real, with $1 < \sigma < 2$ and $x \geq 1$, Montgomery proved that

$$\begin{aligned}(2\sigma-1)\sum_{\gamma}&\frac{x^{i\gamma}}{\left(\sigma-\frac{1}{2}\right)^2+(t-\gamma)^2}\\&=-x^{-1/2}\left(\sum_{n\le x}\Lambda(n)\left(\frac{x}{n}\right)^{1-\sigma+it}+\sum_{n>x}\Lambda(n)\left(\frac{x}{n}\right)^{\sigma+it}\right)\\&\quad+x^{1/2-\sigma+it}(\log\tau+O_\sigma(1))+O_\sigma(x^{1/2}\tau^{-1}),\end{aligned}\tag{3.14}$$

where $\tau=|t|+2$ and the implied constants depend only on σ.

Proof (Proof of Montgomery's Theorem (Theorem 1); [107, Sect. 3, pp. 187–191]). Placing $\sigma=3/2$ in (3.14), and letting $L(x,t)$ and $R(x,t)$ denote the left and right sides, respectively, we now wish to evaluate the second moments of both sides; i.e. $\int_0^T|L(x,t)|^2\,dt$, $\int_0^T|R(x,t)|^2\,dt$. The reason to do this is that, as we will see, $F(\alpha)$ falls out of the second moment of the left side, and we end up with something tractable for the second moment of the right side. Thus the equation of the two moments gives us an identity for $F(\alpha)$.

By showing the contribution of those ordinates γ above height T is $O(\log^3 T)$, Montgomery obtained

$$\int_0^T|L(x,t)|^2\,dt=4\sum_{\substack{0<\gamma\le T\\0<\gamma'\le T}}x^{i(\gamma-\gamma')}\int_0^T\frac{dt}{(1+(t-y)^2)(1+(t-\gamma')^2)}+O(\log^3T).\tag{3.15}$$

Note that the range of integration may be extended to all of $\mathbb{R}$ at a penalty no greater in magnitude than $O(\log^2 T)$; we then have

$$\int_0^T|L(x,t)|^2\,dt=4\sum_{\substack{0<\gamma\le T\\0<\gamma'\le T}}x^{i(\gamma-\gamma')}\int_{-\infty}^{\infty}\frac{dt}{(1+(t-y)^2)(1+(t-\gamma')^2)}+O(\log^3T);\tag{3.16}$$

it then follows from the residue calculus that the definite integral evaluates to $w(\gamma-\gamma')\pi/2$ and

$$\int_0^T|L(x,t)|^2\,dt=2\pi\sum_{\substack{0<\gamma\le T\\0<\gamma'\le T}}x^{i(\gamma-\gamma')}w(\gamma-\gamma')+O(\log^3T).\tag{3.17}$$

Putting $x=T^\alpha$ yields

$$\int_0^T|L(x,t)|^2\,dt=F(\alpha)T\log T+O(\log^3T).\tag{3.18}$$

The non-negativity of the left side of (3.18) gives the statement in Theorem 1 of the positivity of $F(\alpha)$. (The evenness of $F(\alpha)$ follows from the fact that γ and γ' may be interchanged in the definition (3.9).) It then falls to evaluate $\int_0^T |R(x,t)|^2\,dt$. First,

$$\int_0^T \left|x^{-1+it}\log\tau\right|^2 dt = \frac{T}{x^2}(\log^2 T + O(\log T)) \tag{3.19}$$

for all $x \geq 1, T \geq 2$. Montgomery then applied a quantitative version of Parseval's identity for Dirichlet series to find

$$\int_0^T \left|\sum_n a_n n^{-it}\right|^2 dt = \sum_n |a_n|^2\,(T + O(n)). \tag{3.20}$$

Applying (3.20) to the explicit formula (3.14), we find

$$\begin{aligned}
&\frac{1}{x}\int_0^T \left|\sum_{n\leq x}\Lambda(n)\left(\frac{x}{n}\right)^{-1/2+it} + \sum_{n>x}\Lambda(n)\left(\frac{x}{n}\right)^{3/2+it}\right|^2 dt \\
&= \frac{1}{x}\sum_{n\leq x}\Lambda(n)^2\left(\frac{x}{n}\right)^{-1}(T+O(n) + \frac{1}{x}\sum_{n>x}\Lambda(n)^2\left(\frac{x}{n}\right)^{3}(T+O(n)) \\
&= T(\log x + O(1)) + O(x\log x),
\end{aligned} \tag{3.21}$$

where the last line follows from the prime number theorem with error term. It then follows from simple estimation of the error terms and a more delicate application of Cauchy-Schwarz that

$$\int_0^T |R(T^\alpha,t)|^2\,dt = ((1+o(1))T^{-2\alpha}\log T + \alpha + o(1))T\log T, \tag{3.22}$$

uniformly for $0 \leq \alpha \leq 1-\epsilon$. Combining (3.18) and (3.22) yields Montgomery's theorem. □

We end this section by describing the heuristic evidence that led Montgomery to conjecture (3.13) on the behavior of $F(\alpha)$ for $\alpha > 1$. The argument above for proving Montgomery's conjecture for $0 \leq \alpha < 1$ fails for $\alpha > 1$, since error terms such as in (3.21) and those arising from Cauchy-Schwarz and the last line of (3.14) are no longer dominated by the main term.

Examining the sum over primes from the explicit formula (3.14) with $\sigma = 3/2$,

$$\sum_{n\leq x}\Lambda(n)\left(\frac{x}{n}\right)^{-1/2+it} + \sum_{n>x}\Lambda(n)\left(\frac{x}{n}\right)^{3/2+it}, \tag{3.23}$$

the expected value is seen by the prime number theorem to be

$$\frac{2x^{1-it}}{\left(\frac{1}{2}+it\right)\left(\frac{3}{2}-it\right)}. \tag{3.24}$$

From the proof of Montgomery's theorem we have, with $F(x,T)$ as in (3.8), that

$$\begin{aligned} F(x,T) &= \frac{1}{2\pi x}\int_0^T \left| \sum_{n\le x}\Lambda(n)\left(\frac{x}{n}\right)^{-1/2+it} + \sum_{n>x}\Lambda(n)\left(\frac{x}{n}\right)^{3/2+it} \right. \\ &\qquad \left. - \frac{2x^{1-it}}{\left(\frac{1}{2}+it\right)\left(\frac{3}{2}-it\right)}\right|^2 dt + o(T\log T); \end{aligned} \tag{3.25}$$

it follows that we would like to know the size of

$$\int_0^T \left| \frac{1}{x}\sum_{n\le x}\Lambda(n)n^{1/2-it} + x\sum_{n>x}\Lambda(n)n^{-3/2-it} - \frac{2x^{1/2-it}}{\left(\frac{1}{2}+it\right)\left(\frac{3}{2}-it\right)}\right|^2 dt. \tag{3.26}$$

Montgomery proceeded to multiply out and integrate term-by-term, finding that the non-diagonal is non-neglectable. He collected terms in the form of sums of the sort

$$\sum_{n\le y}\Lambda(n)\Lambda(n+h); \tag{3.27}$$

invoking the Hardy-Littlewood k-tuple conjecture for 2-tuples with a strong error term, (3.27) should be $\asymp y$. This would give

$$F(x,T) \sim \frac{T}{2\pi}\log T \tag{3.28}$$

in $x \le T \le x^{2-\epsilon}$, and there is little reason to expect the behavior to change for bounded $\alpha \ge 2$. On this basis, Montgomery made his conjecture (3.13).

3.4 Spacings Between Adjacent Zeros

Motivated by Montgomery's pair correlation conjecture on the zeros of the Riemann zeta function, starting in the late 1970s Andrew Odlyzko began a large-scale computation of zeros of $\zeta(s)$ high in the critical strip. The average spacing between zeros of $\zeta(s)$ at height T in the critical strip is on the order of $1/\log T$; thus as we go higher and higher we have more and more zeros in regions of fixed size, and there is every reason to hope that, after an appropriate normalization, a limiting behavior exists.

The story of computing zeta zeros goes back to Riemann himself. As mentioned in Sect. 1.2.2, in his one paper on the zeta function [123], Riemann states the Riemann hypothesis (RH) in passing. He used a formula now known as the Riemann-Siegel formula to compute a few zeros of $\zeta(s)$ up to a height of probably no greater than 100 in the critical strip; though he did not mention these computations in the paper, the role of these computations was important in the development of mathematics and mirror the role played by the calculation of energy levels in nuclear physics in illuminating the internal structure of the nucleus. The formula was actually lost for almost 70 years, and did not enter the mathematics literature until Siegel was reading Riemann's works [131]. Siegel's role in understanding, collecting, and interpreting Riemann's notes should not be underestimated, since the expertise and insight needed to infer the ideas behind the notes was great.

The development of the Riemann-Siegel formula proceeds along the purely classical lines of complex analysis. Riemann had a formula for $\zeta(s)$ valid for all $s \in \mathbb{C}$; namely,

$$\zeta(s) \;=\; \frac{\Gamma(1-s)}{2\pi i}\int_{\mathscr{C}} \frac{(-x)^s}{e^x-1}\cdot\frac{dx}{x}, \tag{3.29}$$

where $\mathscr{C}$ is the contour that starts at $+\infty$, traverses the real axis towards the origin, circles the origin once with the positive orientation about 0, and then retraces its path along the real axis to $+\infty$.

By splitting off some finite sums from the contour integral above, Riemann arrived at the formula

$$\begin{aligned}\zeta(s) \;=\; &\sum_{n=1}^{N}\frac{1}{n^s} + \pi^{1/2-s}\frac{\Gamma\left(\frac{s}{2}\right)}{\Gamma\left(\frac{1}{2}(1-s)\right)}\sum_{n=1}^{M}\frac{1}{n^{1-s}}\\ &-\frac{\Gamma(1-s)}{2\pi i}\int_{\mathscr{C}_M}\frac{(-x)^s e^{-Nx}}{e^x-1}\cdot\frac{dx}{x},\end{aligned} \tag{3.30}$$

where here $s \in \mathbb{C}$, $N, M \in \mathbb{N}$ are arbitrary, and $\mathscr{C}_M$ is the contour that traces from $+\infty$ to $(2M+1)\pi$, circles the line $|s| = (2M+1)\pi$ once with positive orientation, and then returns to $+\infty$, thereby enclosing the poles $\pm 2\pi i M, \pm 2\pi i(M-1), \ldots, \pm 2\pi i$, and the singularity at 0. This formula for $\zeta(s)$ can be regarded as an approximate functional equation, where the remainder is expressed explicitly in terms of the contour integral over $\mathscr{C}_M$. The main task in developing the Riemann-Siegel formula then falls to estimating the contour integral over $\mathscr{C}_M$ using the saddle-point method.

Prior to Siegel's work, in 1903 Gram showed that the first 10 zeros of $\zeta(s)$ lie on the critical line, and showed that these 10 were the only zeros up to height 50. The development of the above, along with a cogent narrative of Riemann, Siegel, and Gram's contributions, may be found in Edwards [39].

In almost every decade in the last century, mathematicians have set new records for computations of critical zeros of $\zeta(s)$. Alan Turing brought the computer to bear on the problem of computing zeta zeros for the first time in 1950, when, as recounted by Hejhal and Odlyzko [77], Turing used the Manchester Mark 1 Electronic Computer, which had 25,600 bits of memory and punched its output on teleprint tape in base 32, to verify every zero up to height 1540 in the critical strip (he found there are 1104 such zeros). Turing also introduced a simplified algorithm to compute zeta zeros now known as Turing's method. Turing published on his computer computations and his new algorithm for the first time in 1953 [139].

Following Turing, the computation of zeros of $\zeta(s)$ took off thanks to the increasing power of the computer. At this time, the first 10^{13} nontrivial zeros of $\zeta(s)$, tens of billions of nontrivial zeros around the 10^{23} and 10^{24}, and hundreds of nontrivial zeros near zero number 10^{32} are known to lie on the critical line. Additionally, new algorithms by Schönhage and Odlyzko, and by Schönhage, Heath-Brown, and Hiary have sped up the verification of zeta zeros.

However, the aforementioned projects for numerically checking that zeros of $\zeta(s)$ lay on the critical line were not concerned with accurately recording the height along the critical line of the zeros computed; only with ensuring the zeros had real part exactly $1/2$. This changed in the late 1970s with a series of computations by Andrew Odlyzko, who was motivated not only by the Riemann Hypothesis but also by Montgomery's pair correlation conjecture.

Rather than verify consecutive zeros starting from the critical point, Odlyzko was interested in starting his search high up in the critical strip, in the hope that near zero number 10^{12}, the behavior of $\zeta(s)$ would be closer to its asymptotic behavior. For, as Montgomery's pair correlation conjecture is a statement about the limit as one's height in the critical strip passes to infinity, one would wish to know the ordinates of many consecutive zeta zeros in the regime where $\zeta(s)$ is behaving asymptotically if one wished to test the plausibility of the conjecture.

As he explains [110], his first computations [109] were in a window around zero number 10^{12}, and were done on a Cray supercomputer using the Riemann-Siegel formula. These computations motivated Odlyzko and Arnold Schönhage to develop a faster algorithm for computing zeros [111, 114], which was implemented in the late 1980s and was subsequently used to compute several hundred million zeros near zero number 10^{20} and some near number $2 \cdot 10^{20}$, as seen in [112, 113].

3.5 Number Theory and Random Matrix Theory Successes

After its introduction as a conjecture in the late 1950s to describe the energy levels of heavy nuclei, random matrix theory experienced successes on both the numerical and the experimental fronts. The theory was beautifully developed to handle a large number of statistics, and many of these predictions were supported as more and more data on heavy nuclei became available. While there was significant theoretical progress (see, among others, [37, 38, 57, 98, 100, 143–148]), there were some gaps that were not resolved until recently. For example, while the density of normalized eigenvalues in matrix ensembles (Wigner's semi-circle law) was known for all ensembles where the entries were chosen independently from nice distributions, the spacings between adjacent normalized eigenvalues resisted proof until this century (see, among others, [41, 42, 137, 138]).

The fact that random matrix theory also had a role to play in number theory only emerged roughly twenty years after Wigner's pioneering investigations. The cause of the connection was a chance encounter between Hugh Montgomery and Freeman Dyson at the Institute for Advanced Study at Princeton. As there are now many excellent summaries and readable surveys of their meeting, early years and statistics (see in particular [73] for a Hollywoodized version), and the story is now well known, we content ourselves with a quick summary. For more, see among others [20, 21, 31, 32, 82, 84–86, 106].

As described in Sect. 3.1, Montgomery was interested in the class number, which led him to study the pair correlation of zeros of the Riemann zeta function. Given an increasing sequence $\{\alpha_n\}_{n=1}^{\infty}$ and a box $B \subset \mathbb{R}^n$, the n-level correlation is defined by

$$\lim_{N\to\infty} \frac{\#\left\{\left(\alpha_{j_1}-\alpha_{j_2},\ldots,\alpha_{j_{n-1}}-\alpha_{j_n}\right)\in B, j_i\neq j_k\right\}}{N}; \tag{3.31}$$

the pair correlation is the case $n = 2$, and through combinatorics knowing all the correlations yields the spacing between adjacent events. Montgomery was partially able to determine the behavior for the pair correlation. When he told Dyson his result, Dyson recognized it as the pair correlation function of eigenvalues of random Hermitian matrices in a Gaussian Unitary Ensemble, GUE.

This observation was the beginning of a long and fruitful relationship between the two areas. At first it appeared that the GUE was the only family of matrices needed for number theory, as there was remarkable universality seen in statistics. This ranged from work by Hejhal [76] on the 3-level correlation of the zeros of $\zeta(s)$ and Rudnick and Sarnak [127] on the n-level correlation of general automorphic L-functions, to Odlyzko's [109, 110] striking experiments on spacings between adjacent normalized zeros. In all cases the behavior agreed with that of the GUE.

In particular, Odlyzko's computations of high zeta zeros showed that, high enough along the critical line, the empirical distribution of nearest-neighbor spacings for zeros of $\zeta(s)$ becomes more or less indistinguishable from that of eigenvalues of random matrices from the Gaussian Unitary Ensemble, or GUE. The agreement with the first million zeros is poor, but the agreement near zero number 10^{12} is close, near perfect near zero number 10^{16}, and even better near zero number 10^{20}. These results provide massive evidence for Montgomery's conjecture, and vindicate Odlyzko's choice of starting his search high along the critical line; see Fig. 5.

In all of these investigations, however, the statistics studied are insensitive to the behavior of finitely many zeros. This is a problem, as certain zeros of L-functions play an important role. The most important of these are those of elliptic curve L-functions. Numerical computations on the number of points on elliptic curves modulo p led to the Birch and Swinnerton-Dyer conjecture. Briefly, this states that the order of vanishing of the L-function at the central point equals the geometric

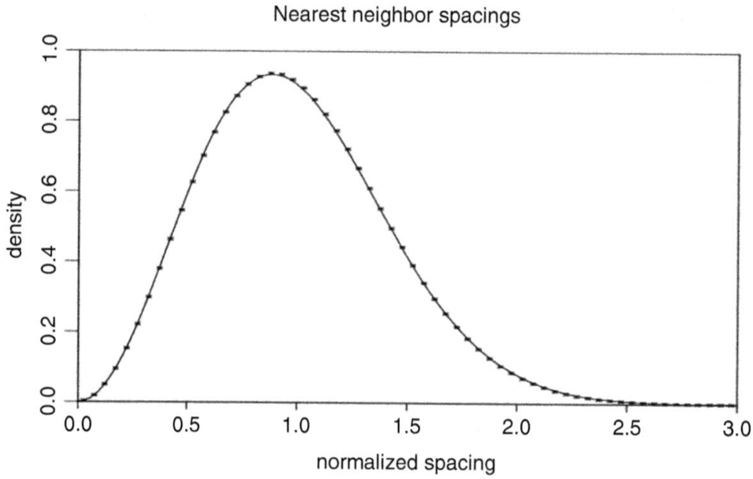

Fig. 5 Probability density of the normalized spacings δ_n. *Solid line*: GUE prediction. *Scatterplot*: empirical data based on Odlyzko's computation of a billion zeros near zero #1.3×10^{16}. (From Odlyzko [110, Fig. 1, p. 4])

rank of the Mordell-Weil group of rational solutions. The theorems on n-level correlations and spacings between adjacent zeros are all limiting statements; we may remove finitely many zeros without changing these limits. Thus these quantities cannot detect what is happening at the central point.

Unfortunately for those who were hoping to distinguish between different symmetry groups, Katz and Sarnak [84, 85] showed in the nineties that the n-level correlations of the scaling limits of the classical compact groups are all the same and equal that of the GUE. Thus when we were saying number theory agreed with GUE we could instead have said it agreed with unitary, symplectic or orthogonal matrices.

This led them to develop a new statistic that would be sensitive to finitely many zeros in general, and the important ones near the central point in particular. The resulting quantity is the n-level density. We assume the Generalized Riemann Hypothesis (GRH) for ease of exposition, so given an $L(s,f)$ all the zeros are of the form $1/2 + i\gamma_{j;f}$ with $\gamma_{j;f}$ real. The statistics are still well-defined if GRH fails, but we lose the interpretation of ordered zeros and connections with nuclear physics. For more detail on these statistics see the seminal work by Iwaniec et al. [83], who introduced them (or [2] for an expanded discussion).

Let ϕ_j even Schwartz functions such that the Fourier transforms

$$\widehat{\phi_j}(y) := \int_{-\infty}^{\infty} \phi_j(x) e^{-2\pi ixy} dx \tag{3.32}$$

are compactly supported, and set $\phi(x) = \prod_{j=1}^{n} \phi_j(x_j)$. The n-level density for f with test function ϕ is

$$D_n(f,\phi) \;=\; \sum_{\substack{j_1,\dots,j_n \\ j_\ell \neq j_m}} \phi_1\left(L_f \gamma_{j_1;f}\right)\cdots\phi_n\left(L_f \gamma_{j_n;f}\right), \tag{3.33}$$

where L_f is a scaling parameter which is frequently related to the conductor. Given a family $\mathcal{F} = \cup_N \mathcal{F}_N$ of L-functions with conductors tending to infinity, the n-level density $D_n(\mathcal{F},\phi,w)$ with test function ϕ and non-negative weight function w is defined by

$$D_n(\mathcal{F},\phi,w) \;:=\; \lim_{N\to\infty} \frac{\sum_{f\in\mathcal{F}_N} w(f) D_n(f,\phi)}{\sum_{f\in\mathcal{F}_N} w(f)}. \tag{3.34}$$

Katz and Sarnak [84, 85] conjecture that as the conductors tend to infinity, the n-level density of zeros near the central point in families of L-functions agree with the scaling limits of eigenvalues near 1 of classical compact groups. Determining *which* classical compact group governs the symmetry is one of the hardest problems in the subject, though in many cases through analogies with a function field analogue one has a natural candidate for the answer, arising from the monodromy group. Unlike the n-level correlations, the different classical compact groups all have different scaling limits. As the test functions are Schwartz and of rapid decay, this statistics *is* sensitive to the zeros at the central point. While it was possible to look at just one L-function when studying correlations, that is not the case for the n-level density. The reason is that while one L-function has infinitely many zeros, it only has a finite number within a small, bounded window of the central point (the size of the window is a function of the analytic conductor). We always need do perform some averaging; for the n-level correlations each L-function gives us enough zeros high up on the critical line for such averaging, while for the n-level density we must move horizontally and look at a *family* of L-functions. While the exact definition of family is still a work in progress, roughly it is a collection of L-functions coming from a common process. Examples include Dirichlet characters, elliptic curves, cuspidal newforms, symmetric powers of GL(2) L-functions, Maass forms on GL(3), and certain families of GL(4) and GL(6) L-functions; see for example [2, 3, 35, 36, 40, 46, 53, 55, 67, 79, 80, 83, 85, 94, 101, 105, 116, 117, 122, 125, 126, 151, 152]. This correspondence between zeros and eigenvalues allows us, at least conjecturally, to assign a definite symmetry type to each family of L-functions (see [36, 130] for more on identifying the symmetry type of a family).

There are many other quantities that can be studied in families. Instead of looking at zeros, one could look at values of L-functions at the central point, or moments along the critical line. There is an extensive literature here of conjectures and results, again with phenomenal agreement between the two areas. See for example [22].

4 Future Trends and Questions in Number Theory

The results above are just a small window of the great work that has been done with number theory and random matrix theory. Our goal above is not to write a treatise, but to quickly review the history and some of the main results, setting the stage for some of the problems we think will drive progress in the coming decades. As even that covers too large an area, we have chosen to focus on a few problems with a strong numeric component, where computational number theory is providing the same support and drive to the subject as experimental physics did years before. There are of course many other competing models for L-functions. One is the Ratios Conjectures of Conrey et al. [23, 24, 26]. Another excellent candidate is Gonek, Hughes and Keating's hybrid model [65], which combines random matrix theory with arithmetic by modeling the L-function as a partial Hadamard product over the zeros, which is modeled by random matrix theory, and a partial Euler product, which contains the arithmetic.

In all of the quantities studied, we have agreement (either theoretical or experimental) of the main terms with the main terms of random matrix theory in an appropriate limit. A natural question to ask is how this agreement is reached; in other words, what is the rate of convergence, and what affects this rate? In the interest of space we assume in parts of this section that the reader is familiar with the results and background material from [83, 127], though we describe the results in general enough form to be accessible to a wide audience.

4.1 Nearest Neighbor Spacings

We first look at spacing between adjacent zeros, where Odlyzko's work has shown phenomenal agreement for zeros of $\zeta(s)$ and eigenvalues of the GUE ensemble. We plot the *difference* between the empirical and 'theoretical,' or 'expected' GUE spacings in Fig. 6. In his paper [110], Odlyzko writes: *Clearly there is structure in this difference graph, and the challenge is to understand where it comes from.*

Recently, compelling work of Bogomolny et al. [12] provides a conjectural answer for the source of the additional structure in the form of lower-order terms in the pair correlation function for $\zeta(s)$. Though the main term is all that appears in the limit (where Montgomery's conjecture applies), the lower-order terms contribute to any computation outside the limit, and would therefore influence any numerical computations like those of Odlyzko. By comparing a conjectural formula for the two-point correlation function of critical zeros of $\zeta(s)$ of roughly height T due to Bogomolny and Keating in [13] with the known formula for the two-point correlation function for eigenvalues of unitary matrices of size N, Bogomolny et. al. deduce a recipe for picking a matrix size that will best model the lower-order terms in the two-point correlation function, and conjecture that it will be the best choice for all correlation functions, and therefore the nearest-neighbor spacing.

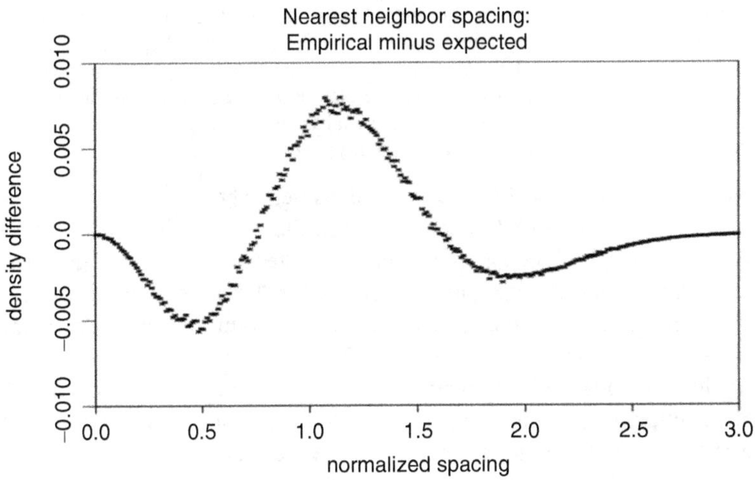

Fig. 6 Probability density of the normalized spacings δ_n. Difference between empirical distribution for a billion zeros near zero #1.3×10^{16}, as computed by Odlyzko, and the GUE prediction. (From Odlyzko [110, Fig. 2, p. 5])

More recently yet, Dueñez et al. [33, 34] have applied techniques of Bogomolny et. al. and others to studying lower-order terms in the behavior of the lowest zeros of L-functions attached to elliptic curves. Their results are currently being extended to other L-functions by the first and third named authors here and their colleagues.

4.2 n-Level Correlations and Densities

The results of the studies on spacings between zero suggest that, while the arithmetic of the L-function is not seen in the main term, it does arise in the lower order terms, which determine the *rate* of convergence to the random matrix theory predictions. Another great situation where this can be seen is through the n-level correlations and the work of Rudnick and Sarnak [127]. They proved that the n-level correlations of *all* cuspidal automorphic L-functions $L(s, \pi)$ have the same limit (at least in suitably restricted regions). Briefly, the source of the universality in the main term comes from the Satake parameters in the Euler product of the L-function, whose moments are the coefficients in the series expansion. In their Remark 3 they write (all references in the quote are to their paper):

> The universality (in π) of the distribution of zeros of $L(s, \pi)$ is somewhat surprising, the reason being that the distribution of the coefficients $a_\pi(p)$ in (1.6), as p runs over primes, is not universal. For example, for degree-two primitive L-functions, there are two conjectured possible limiting distributions for the $a_\pi(p)$'s: Sato-Tate or uniform distribution (with a Dirac mass term). As the degree increases, the number of possible limit distributions increases rapidly. However, it is a consequence of the theory of the Rankin-Selberg

> L-functions (developed by Jacquet, Piatetski-Shapiro, and Shalika for $m > 3$) that all these limiting distributions have the same second moment (at least under hypothesis (1.7)). It is the universality of the second moment that is eventually responsible for the universality in Theorems 1.1 and 1.2. For the case of pair correlation ($n = 2$), this is reasonably evident; for $n > 2$ it was (at least for us) unexpected, and it has its roots in a key feature of "diagonal pairings" that emerges as the main term in the asymptotics of $R_n(T, f, h)$.

Similar results are seen in the n-level densities. There we average the Satake parameters over a family of L-function, and in the limit as the conductors tend to infinity only the first and second moments contribute to the main term (at least under the assumption of the Ramanujan conjectures for the sizes of these parameters). The first moment controls the rank at the central point, and the second moment determines the symmetry type (see [36, 130]). For example, families of elliptic curves with very different arithmetic (complex multiplication or not, or different torsion structures) have the same limiting behavior *but* have different rates of convergence to that limiting behavior. This can be seen in terms of size one over the logarithm of the conductor; while these terms vanish as the conductors tend to infinity, they are present for finite values. See [102, 104] for several examples (as well as [96], where interesting biases are observed in lower order terms of the second moments in the families).

4.3 Conclusion

The number theory results above may be interpreted in a framework similar to that of the Central Limit Theorem. There, if we have 'nice' independent identically distributed random variables, their normalized sum (standardized to have mean zero and variance 1) converges to the standard normal distribution. The remarkable fact is the universality, and that the limiting distribution is independent of the shape of the distribution. We quickly review why this is the case and then interpret our number theory results in a similar vein.

Given a distribution with finite mean and variance, we can always perform a linear change of variables to study a related quantity where now the mean is zero and the variance one. Thus, the first moment where the *shape* of the distribution is noticeable is the third moment (or the fourth if the distribution is symmetric about the mean). In the proof of the Central Limit Theorem through moment generating functions or characteristic functions, the third and higher moments do not survive in the limit. Thus their effect is only on the rate of convergence to the limiting behavior (see the Berry-Esseen theorem), and not on the convergence itself.

The situation is similar in number theory. The higher moments of the Satake parameters (which control the coefficients of the L-functions) again surface only in terms which vanish in the limit, and their effect therefore is seen only in the rate of convergence.

This suggests several natural questions. We conclude with two below, which we feel will play a key role in studies in the years to come. These two questions provide a nice mix, with the first related to the main term and the second related to the rate of convergence.

- Is Montgomery's pair correlation true for all boxes (or test functions)? What about the n-level correlations, both for $\zeta(s)$ and cuspidal automorphic L-functions? Note agreement with random matrix theory for all these statistics implies the conjectures on spacings between adjacent zeros.
- For a given L-function (if we are studying n-level correlations) or a family of L-functions (if we are studying n-level densities), how does the arithmetic enter? Specifically, what are the possible lower order terms? How are these affected by properties of the L-functions? If we use Rankin-Selberg convolution to create new L-functions, how is the arithmetic of the lower order terms here a function of the arithmetic of the constituent pieces?

There are numerous resources and references for those wishing to pursue these questions further. For the n-level correlations, the starting point are the papers [76, 107, 127], while for the n-level densities it is [83–85].

Acknowledgements The third named author was partially supported by NSF grant DMS1265673. We thank our colleagues and collaborators over the years for many helpful discussions on these and related topics. One of us (Miller) was fortunate to be a graduate student at Princeton, and had numerous opportunities then to converse with John Nash on a variety of mathematical topics. It was always a joy sitting next to him at seminars. We are grateful for his kind invitation to contribute to this work, and his comments on an earlier draft.

References

1. Y. Alhassid, *The Statistical Theory of Quantum Dots*, Rev. Mod. Phys. **72** (2000), 895–968.
2. L. Alpoge, N. Amersi, G. Iyer, O. Lazarev, S. J. Miller and L. Zhang, *Maass waveforms and low-lying zeros*, in Analytic Number Theory: In Honor of Helmut Maier's 60th Birthday, Springer-Verlag, 2015.
3. L. Alpoge and S. J. Miller, *The low-lying zeros of level 1 Maass forms*, Int. Math. Res. Not. IMRN 2010, no. 13, 2367–2393.
4. S. Arno, *The imaginary quadratic fields of class number 4*, Acta. Arith. **60** (1992), no. 4, 321–334.
5. S. Arno, M. Robinson, and F. Wheeler, *Imaginary quadratic fields with small odd class number*, Acta. Arith. **83** (1998), no. 4, 296–330.
6. N. Austern, *Fast Neutron Physics* (Vol. 2), Interscience, 1963.
7. J. Baik, A. Borodin, P. Deiftn and T Suidan, *A Model for the Bus System in Cuernevaca (Mexico)*, J. Phys. A: Math. Gen. **39** (2006) 8965–8975. http://arxiv.org/abs/math/0510414.
8. A. Baker, *Linear forms in the logatirhms of algebraic numbers*, Mathematika **13** (1966) 204–216.
9. A. Baker, *Imaginary quadratic fields with class number two*, Ann. of Math. **2** (1971) 139–152.
10. H. A. Bethe, *Nuclear Physics, B. Nuclear Dynamics, Theoretical*, Rev. Mod. Phys. **9** (1937), 69–249.
11. J. M. Blatt and V. F. Weisskopf, *Theoretical Nuclear Physics*, Wiley, 1952.

12. E. Bogomolny, O. Bohigas, P. Leboeuf, and A.G. Monastra, *On the spacing distribution of the Riemann zeros: corrections to the asymptotic result*, J. Phys. A: Math. Gen. **39** (2006), 10743–10754.
13. E. Bogomolny and J.P. Keating, *Gutzwiller's trace formula and spectral statistics: beyond the diagonal approximation*, Phys. Rev. Lett. **77** (1996), 1472–1475.
14. A. Bohr and B. Mottelson, Dan. Nat. Fys. Medd. **27** (1953), no. 16.
15. N. Bohr, *Neutron Capture and Nuclear Constitution*, Nature **137** (1936), 344–348.
16. O. Bohigas, M. Giannoni and C. Schmit, *Characterization of chaotic quantum spectra and universality of level fluctuation laws*, Phys. Rev. Lett. **52** (1984), 1–4.
17. P. Borwein, S. Choi and B. Rooney, eds. *The Riemann Hypothesis: A Resource for the Afficionado and Virtuoso Alike*, CMS Books in Mathematics, Springer, New York, 2008.
18. T. Brody, J. Flores, J. French, P. Mello, A. Pandey, and S. Wong, *Random-matrix physics: spectrum and strength fluctuations*, Rev. Mod. Phys. **53** (1981), no. 3, 385–479.
19. Clay Mathematics Institute, webpage on the *Riemann Hypothesis*, http://www.claymath.org/millenium-problems/riemann-hypothesis.
20. J. B. Conrey, *L-Functions and random matrices*. Pages 331–352 in *Mathematics unlimited — 2001 and Beyond*, Springer-Verlag, Berlin, 2001.
21. J. B. Conrey, *The Riemann hypothesis*, Notices of the AMS, **50** (2003), no. 3, 341–353.
22. B. Conrey, D. Farmer, P. Keating, M. Rubinstein and N. Snaith, *Integral moments of L-functions*, Proc. London Math. Soc. (3) **91** (2005), no. 1, 33–104.
23. J. B. Conrey, D. W. Farmer and M. R. Zirnbauer, *Autocorrelation of ratios of L-functions*, Comm. Number Theor. Phys. **2** (2008), no. 3, 593–636
24. J. B. Conrey, D. W. Farmer and M. R. Zirnbauer, *Howe pairs, supersymmetry, and ratios of random characteristic polynomials for the classical compact groups*, preprint. http://arxiv.org/abs/math-ph/0511024.
25. J. B. Conrey and H. Iwaniec, *Spacing of Zeros of Hecke L-Functions and the Class Number Problem*, Acta Arith. **103** (2002) no. 3, 259–312.
26. J. B. Conrey and N. C. Snaith, *Applications of the L-functions Ratios Conjecture*, Proc. London Math. Soc. (3) **94** (2007), no. 3, 594–646.
27. H. Davenport, *Multiplicative Number Theory, 2nd edition*, Graduate Texts in Mathematics **74**, Springer-Verlag, New York, 1980, revised by H. Montgomery.
28. C. J. de la Vallée Poussin, *Recherches analytiques la théorie des nombres premiers*, Ann. Soc. scient. Bruxelles **20** (1896), 183–256. Reprinted in [17].
29. H. Derrien, L. Leal, and N. Larson, *Status of new evaluation of the neutron resonance parameters of* ^{238}U *at ORNL*, PHYSOR 2004, Available on CD-ROM. Amer. Nucl. Soc. LaGrange Park, IL.
30. J. S. Desjardins, J. Rosen, J. Rainwater and W. Havens Jr., *Slow neutron resonance spectroscopy II*, Phys. Rev. **120** (1960) 2214–2224.
31. Persi Diaconis, *"What is a random matrix?"*, Notices of the Amer. Math. Soc. **52** (2005) 1348–1349.
32. Persi Diaconis, *Patterns of Eigenvalues: the 70th Josiah Willard Gibbs Lecture*, Bull. Amer. Math. Soc. **40** (2003) 155–178.
33. E. Dueñez, D. K. Huynh, J. C. Keating, S. J. Miller and N. Snaith, *The lowest eigenvalue of Jacobi Random Matrix Ensembles and Painlevé VI*, Journal of Physics A: Mathematical and Theoretical **43** (2010) 405204 (27pp).
34. E. Dueñez, D. K. Huynh, J. C. Keating, S. J. Miller and N. Snaith, *Models for zeros at the central point in families of elliptic curves* (with Eduardo Dueñez, Duc Khiem Huynh, Jon Keating and Nina Snaith), J. Phys. A: Math. Theor. **45** (2012) 115207 (32pp).
35. E. Dueñez and S. J. Miller, *The low lying zeros of a* $GL(4)$ *and a* $GL(6)$ *family of L-functions*, Compositio Mathematica **142** (2006), no. 6, 1403–1425.
36. E. Dueñez and S. J. Miller, *The effect of convolving families of L-functions on the underlying group symmetries*, Proceedings of the London Mathematical Society, 2009; doi: 10.1112/plms/pdp018.

37. F. Dyson, *Statistical theory of the energy levels of complex systems: I, II, III*, J. Mathematical Phys. **3** (1962) 140–156, 157–165, 166–175.
38. F. Dyson, *The threefold way. Algebraic structure of symmetry groups and ensembles in quantum mechanics*, J. Mathematical Phys., **3** (1962) 1199–1215.
39. H. M. Edwards, *Riemann's Zeta Function*, Academic Press, New York, 1974.
40. A. Entin, E. Roditty-Gershon and Z. Rudnick, *Low-lying zeros of quadratic Dirichlet L-functions, hyper-elliptic curves and Random Matrix Theory*, Geometric and Functional Analysis **23** (2013), no. 4, 1230–1261.
41. L. Erdös, J. A. Ramirez, B. Schlein and H.-T. Yau, *Bulk Universality for Wigner Matrices*, Comm. Pure Appl. Math. **63** (2010), no. 7, 895–925
42. L. Erdös, B. Schlein and H.-T. Yau, *Wegner estimate and level repulsion for Wigner random matrices*, Int. Math. Res. Not. IMRN (2010), no. 3, 436–479
43. P. Erdös, *Démonstration élémentaire du théorème sur la distribution des nombres premiers*, Scriptum **1**, Centre Mathèmatique, Amsterdam, 1949.
44. E. Fermi and E. Amaldi, La Ricercio Scientifica **6** (1935), 544.
45. H. Feshbach, C. E. Porter, and V. F. Weisskopf, *Model for Nuclear Reactions with Neutrons*, Phys. Rev. **96** (1954), 448–464.
46. D. Fiorilli and S. J. Miller, *Surpassing the Ratios Conjecture in the 1-level density of Dirichlet L-functions*, Algebra & Number Theory **Vol. 9** (2015), No. 1, 13–52.
47. F. Firk, *Neutron Time-of-Flight Spectrometers* in Detectors in Nuclear Science (editor D. Allan Bromley, North-Holland, Amsterdam (1979)), special issue Nucl. Instr. and Methods, **162**(1979), 539–563.
48. F. Firk, J. E. Lynn and M. Moxon, *Parameters of neutron resonances in* U^{238} *below 1.8 keV*, Proc. Kingston Intern. Conf. on Nuclear Structure, University of Toronto Press, Toronto (1960), 757–759.
49. F. Firk and E. Melkonian, *Total Neutron Cross Section Measurements* in Experimental Neutron Resonance Spectroscopy (editor, J. A. Harvey), Academic Press, New York (1970), 101–154.
50. F. W. K. Firk and S. J. Miller, *Nuclei, Primes and the Random Matrix Connection*, Symmetry **1** (2009), 64–105; doi:10.3390/sym1010064. http://www.mdpi.com/2073-8994/1/1/64.
51. F. Firk, G. Reid and J. Gallagher, *High resolution neutron time- of-flight experiments using the Harwell 15 MeV linear electron accelerator*, Nucl. Instr. **3** (1958), 309–315.
52. P. Forrester, *Log-gases and random matrices*, London Mathematical Society Monographs **34**, Princeton University Press, Princeton, NJ 010.
53. E. Fouvry and H. Iwaniec, *Low-lying zeros of dihedral L-functions*, Duke Math. J. **116** (2003), no. 2, 189–217.
54. J. French, V. Kota, A. Pandey and S. Tomosovic, *Bounds on time-reversal non-invariance in the nuclear Hamiltonian*, Phys. Rev. Lett. **54** (1985), 2313–2316.
55. P. Gao, *N-level density of the low-lying zeros of quadratic Dirichlet L-functions*, Ph. D thesis, University of Michigan, 2005.
56. J. Garg, J. Rainwater, J. Peterson and W. Havens Jr., *Neutron resonance spectroscopy III.* Th^{232} *and* U^{238}, Phys. Rev. **134** (1964) B985–1009.
57. M. Gaudin, *Sur la loi limite de l'espacement des valeurs propres d'une matrice aléatoire*, Nucl. Phys. **25** (1961) 447–458.
58. C. Gauss, *Disquisitiones Arithmeticæ*, (1801)
59. D. Goldfeld, *The class number of quadratic fields and the conjectures of Birch and Swinnerton-Dyer*, Ann. Scuola Norm. Sup. Pisa Cl. Sci. 3, **4** (1976), 624–663.
60. D. Goldfeld, *The conjectures of Birch and Swinnerton-Dyer and the class number of quadratic fields*, Journées Arith. De Caen (Univ. Caen, Caen, 1976), Asteérisque nos. 41–42, Soc. Math. France, (1977) 219–227.
61. D. Goldfeld, *Gauss's class number problem for imaginary quadratic fields*, Bull. Amer. Math. Soc. (N.S.) **13** (1985), no. 1, 23–37.
62. D. Goldfeld, *The Elementary proof of the Prime Number Theorem, An Historical Perspective*. Pages 179–192 in *Number Theory, New York Seminar 2003*, eds. D. and G. Chudnovsky, M. Nathanson, Springer-Verlag, New York, 2004.

63. D. Goldfeld and A. Kontorovich, *On the* GL(3) *Kuznetsov formula with applications to symmetry types of families of L-functions*, In Automorphic Representations and *L*-Functions (ed. D. Prasad et al), Tata Institute (2013), 263–310.
64. D. A, Goldston, *Notes on pair correlation of zeros and prime numbers*, *Notes on pair correlation of zeros and prime numbers*, in Recent perspectives in random matrix theory and number theory, 79–110, London Math. Soc. Lecture Note Ser., 322, Cambridge Univ. Press, Cambridge, 2005. http://arxiv.org/pdf/math/0412313v1.
65. S. M. Gonek, C. Hughes and J. P. Keating, *A hybrid Euler-Hadamard product for the Riemann zeta function* , Duke Math. J. **136** (2007) 507–549.
66. B. Gross and D. Zagier, *Heegner points and derivatives of L-series*, Invent. Math. **84** (1986), no. 2, 225–320.
67. A. Güloğlu, *Low Lying Zeros of Symmetric Power L-Functions*, Int. Math. Res. Not. (2005), no. 9, 517–550.
68. I. I. Gurevich and M. I. Pevzner, *Repulsion of nuclear levels*, Physica **22** (1956), 1132.
69. J. Hadamard, *Sur la distribution des zéros de la fonction $\zeta(s)$ et ses conséquences arithmétiques*, Bull. Soc. math. France **24** (1896), 199–220. Reprinted in [17].
70. G. H. Hardy and E. Wright, *An Introduction to the Theory of Numbers*, 5th edition, Oxford Science Publications, Clarendon Press, Oxford, 1995.
71. J. Harvey and D. Hughes, *Spacings of nuclear energy levels*, Phys. Rev. **109** (1958), 471–479.
72. R. Haq, A. Pandey and O. Bohigas, *Fluctuation properties of nuclear energy levels: do theory and experiment agree?* Phys. Rev. Lett. **48** (1982), 1086–1089.
73. B. Hayes, *The spectrum of Riemannium*, American Scientist **91** (2003), no. 4, 296–300.
74. K. Heegner, *Diophantische Analysis und Modulfunktionen*, Math. Z. **56** (1952) 227–253.
75. H. Heilbronn, *On the class number in imaginary quadratic fields*, Quart. J. Math. Oxford Ser. 2 **5** (1934) 150–160.
76. D. Hejhal, *On the triple correlation of zeros of the zeta function*, Internat. Math. Res. Notices 1994, no. 7, 294–302.
77. D. A. Hejhal and A. M. Odlyzko, *Alan Turing and the Riemann zeta function*, Alan Turing – His Work and Impact (J. van Leeuwen and S.B. Cooper, eds.), Elsevier Science, 2012
78. D. Hughes, *Neutron Cross Sections*, Pergamon Press, New York (1957).
79. C. Hughes and S. J. Miller, *Low-lying zeros of L-functions with orthogonal symmetry*, Duke Math. J. **136** (2007), no. 1, 115–172.
80. C. Hughes and Z. Rudnick, *Linear statistics of low-lying zeros of L-functions*, Quart. J. Math. Oxford **54** (2003), 309–333.
81. D. J. Hughes and R. B. Schwartz, Brookhaven National Laboratory Report, No. BNL–325 (1958).
82. H. Iwaniec and E. Kowalski, *Analytic Number Theory*, AMS Colloquium Publications, Vol. 53, AMS, Providence, RI, 2004.
83. H. Iwaniec, W. Luo and P. Sarnak, *Low lying zeros of families of L-functions*, Inst. Hautes Études Sci. Publ. Math. **91** (2000), 55–131.
84. N. Katz and P. Sarnak, *Random Matrices, Frobenius Eigenvalues and Monodromy*, AMS Colloquium Publications **45**, AMS, Providence, 1999.
85. N. Katz and P. Sarnak, *Zeros of zeta functions and symmetries*, Bull. AMS **36** (1999), 1–26.
86. J. P. Keating and N. C. Snaith, *Random matrices and L-functions*, Random matrix theory, J. Phys. A **36** (2003), no. 12, 2859–2881.
87. M. Krbalek and P. Seba, *The statistical properties of the city transport in Cuernavaca (Mexico) and Random matrix ensembles*, J. Phys. A **33** (2000), 229–234.
88. L. Landau and Ya. Smorodinski, Lektsii po teori atomnogo yadra, Gos Izd. tex- teoreyicheskoi Lit. Moscow, (1955) 92–93.
89. A. M. Lane, *Theory of radiative capture reactions*, Nucl. Phys. **11** (1959), 625–645.
90. A. M. Lane and J. P. Elliott, *Handbuch der Physik* (Vol. 39), Springer-Verlag, 1957.
91. A. M. Lane and J. E. Lynn, *Analysis of experimental data on nucleon capture reactions*, Nucl. Phys. **11** (1959), 646–645.

92. A. M. Lane, R. G. Thomas and E. P. Wigner, *Giant Resonance Interpretation of the Nucleon-Nucleus Interaction*, Phys. Rev. **98** (1955), 693–701.
93. D. H. Lehmer, E. Lehmer, and D. Shanks, *Integer sequences having prescribed quadratic character*, Math. Comp. **24** (1970), 433–451
94. J. Levinson and S. J. Miller, *The n-level density of zeros of quadratic Dirichlet L-functions*, Acta Arithmetica **161** (2013), 145–182.
95. J. E. Lynn, *The Theory of Neutron Resonance Reactions*, The Clarendon Press, Oxford (1968).
96. B. Mackall, S. J. Miller, C. Rapti and K. Winsor, *Lower-Order Biases in Elliptic Curve Fourier Coefficients in Families*, to appear in the Conference Proceedings of the Workshop on Frobenius distributions of curves at CIRM in February 2014.
97. M. G. Mayer, *On Closed Shells in Nuclei. II*, Phys. Rev. **75** (1949), 1969–1970.
98. M. L. Mehta, *On the statistical properties of level spacings in nuclear spectra*, Nucl. Phys. **18** (1960), 395–419.
99. M. L. Mehta, *Random Matrices*, 3rd edition, Elsevier, San Diego, CA (2004)
100. M. L. Mehta and M. Gaudin, *On the density of the eigenvalues of a random matrix*, Nuclear Physics **18** (1960), 420–427.
101. S. J. Miller, 1- *and* 2-*level densities for families of elliptic curves: evidence for the underlying group symmetries*, Compositio Mathematica **140** (2004), 952–992.
102. S. J. Miller, *Variation in the number of points on elliptic curves and applications to excess rank*, C. R. Math. Rep. Acad. Sci. Canada **27** (2005), no. 4, 111–120.
103. S. J. Miller (with an appendix by E. Dueñez), *Investigations of zeros near the central point of elliptic curve L-functions*, Experimental Mathematics **15** (2006), no. 3, 257–279.
104. S. J. Miller, *Lower order terms in the* 1-*level density for families of holomorphic cuspidal newforms*, Acta Arithmetica **137** (2009), 51–98.
105. S. J. Miller and R. Peckner, *Low-lying zeros of number field L-functions*, Journal of Number Theory **132** (2012), 2866–2891.
106. S. J. Miller and R. Takloo-Bighash, *An Invitation to Modern Number Theory*, Princeton University Press, Princeton, NJ, 2006.
107. H. Montgomery, *The pair correlation of zeros of the zeta function*, Analytic Number Theory, Proc. Sympos. Pure Math. **24**, Amer. Math. Soc., Providence, 1973, 181–193.
108. H. Montgomery and P. Weinberger, *Notes on small class numbers*, Acta Arith. **24** (1973) 529–542.
109. A. Odlyzko, *On the distribution of spacings between zeros of the zeta function*, Math. Comp. **48** (1987), no. 177, 273–308.
110. A. Odlyzko, *The* 10^{22}*-nd zero of the Riemann zeta function*, Proc. Conference on Dynamical, Spectral and Arithmetic Zeta-Functions, M. van Frankenhuysen and M. L. Lapidus, eds., Amer. Math. Soc., Contemporary Math. series, 2001. http://www.research.att.com/~amo/doc/zeta.html.
111. A. M. Odlyzko, *New analytic algorithms in number theory*, Proc. Intern. Congress Math. 1986, Amer. Math. Soc., 1987, pp. 466–475.
112. A. M. Odlyzko, *The* 10^{20}*-th zero of the Riemann zeta function and 70 million of its neighbors* (1989), unpublished.
113. A. M. Odlyzko, *The* 10^{20}*-th zero of the Riemann zeta function and 175 million of its neighbors* (1992), unpublished.
114. A. M. Odlyzko and A. Schönhage, *Fast algorithms for multiple evalutions of the Riemann zeta function*, Trans. Amer. Math. Soc. **309** (1988), 797–809.
115. J. Oesterlé, *Nombre de classes des corps quadratiques imaginaires*, Séminaire Nicolas Bourbaki, Vol. 1983/84, Astérisque No. 121–122 (1985), 309–323.
116. A. E. Özlük and C. Snyder, *Small zeros of quadratic L-functions*, Bull. Austral. Math. Soc. **47** (1993), no. 2, 307–319.
117. A. E. Özlük and C. Snyder, *On the distribution of the nontrivial zeros of quadratic L-functions close to the real axis*, Acta Arith. **91** (1999), no. 3, 209–228.
118. M. Ostrofsky, G. Breit, and D. P. Johnson, *The Excitation Function of Lithium Under Proton Bombardment*, Phys. Rev. **49** (1936), 22–34.

119. C. Porter (editor), *Statistical Theories of Spectra: Fluctuations*, Academic Press, New York, 1965.
120. C. Porter and N. Rosenzweig, *Repulsion of energy levels in complex atomic spectra*, Phys. Rev. **120** (1960) 1698–1714.
121. L. J. Rainwater, *Nuclear Energy Level Argument for a Spheroidal Nuclear Model*, Phys. Rev. **79** (1950), 432–434.
122. G. Ricotta and E. Royer, *Statistics for low-lying zeros of symmetric power L-functions in the level aspect*, Forum Math. **23** (2011), no. 5, 969–1028.
123. G. F. B. Riemann, *Über die Anzahl der Primzahlen unter einer gegebenen Grösse*, Monatsber. Königl. Preuss. Akad. Wiss. Berlin, Nov. 1859, 671–680 (see [39] for an English translation).
124. J. Rosen, J. S. Desjardins, J. Rainwater and W. Havens Jr., *Slow neutron resonance spectroscopy I*, Phys. Rev. **118** (1960) 687–697.
125. E. Royer, *Petits zéros de fonctions L de formes modulaires*, Acta Arith. **99** (2001), 47–172.
126. M. Rubinstein, *Low-lying zeros of L-functions and Random Matrix Theory*, Duke Math. J. **109** (2001), no. 1, 147–181.
127. Z. Rudnick and P. Sarnak, *Zeros of principal L-functions and Random Matrix Theory*, Duke Math. J. **81** (1996), 269–322.
128. A. Selberg, *An Elementary Proof of the Prime Number Theorem*, Ann. Math. **50** (1949), 305–313.
129. J. P. Serre, *A Course in Arithmetic*, Springer-Verlag, New York, 1996.
130. S. W. Shin and N. Templier, *Sato-Tate theorem for families and low-lying zeros of automorphic L-functions*, Appendix A by Robert Kottwitz, and Appendix B by Raf Cluckers, Julia Gordon and Immanuel Halupczok. Invent. Math. **203** (2016), no. 1, 1–177.
131. C. L. Siegel, *Über Riemanns Nachlass zur analytischen Zahlentheorie*, Quellen und Studien zur Geschichte der Math. Astr. Phys., no. 2, 1932, pp. 45–80; reprinted in C.L. Siegel, *Gesammelte Abhandlungen*, Vol. 1, Springer, 1966.
132. C.L. Siegel, *Über die Klassenzahl quadratischer Zahlkörper*, Acta. Arith. **1** (1935) 83–86.
133. H. M. Stark, *A complete determination of the complex quadratic fields of class-number one*, Michigan Math. J. **14** (1969) 1–27.
134. H. M. Stark, *A transcendence theorem for class number problems*, Ann. of Math. **2** (1971) 153–173.
135. J. Stopple, *Notes on the Deuring-Heilbronn phenomenon*, Notices Amer. Math. Soc. **53** (2006), no. 8, 864–875
136. T. Tao, *Topics in Random Matrix Theory*, Graduate Studies in Mathematics, volume 132, American Mathematical Society, Providence, RI 2012.
137. T. Tao and V. Vu, *From the Littlewood-Offord problem to the Circular Law: universality of the spectral distribution of random matrices*, Bull. Amer. Math. Soc. **46** (2009), 377–396.
138. T. Tao and V. Vu, *Random matrices: universality of local eigenvalue statistics up to the edge*, Comm. Math. Phys. **298**, (2010), no. 2, 549–572
139. A. M. Turing, *Some calculations of the Riemann zeta-function*, Proc. London Math. Soc., ser. 3 **3** (1953), 99–117.
140. C. Wanger, *Class number 5, 6, and 7*, Math. Comp. **65** (1996), no. 214, 785–800.
141. M. Watkins, *Class numbers of imaginary quadratic fields*, Mathematics of Computation **73** (2004), 907–938.
142. W. Weibull , *A statistical distribution function of wide applicability*, J. Appl. Mech. Trans. ASME. **18** (1951) 293–297.
143. E. Wigner, *On the statistical distribution of the widths and spacings of nuclear resonance levels*, Proc. Cambridge Philo. Soc. **47** (1951), 790–798.
144. E. Wigner, *Characteristic vectors of bordered matrices with infinite dimensions*, Ann. of Math. **2** (1955), no. 62, 548–564.
145. E. Wigner, *Statistical Properties of real symmetric matrices*. Pages 174–184 in *Canadian Mathematical Congress Proceedings*, University of Toronto Press, Toronto, 1957.
146. E. Wigner, *Characteristic vectors of bordered matrices with infinite dimensions. II*, Ann. of Math. Ser. 2 **65** (1957), 203–207.

147. E. Wigner, *Results and theory of resonance absorption*, Gatlinburg Conference on Neutron Physics by Time-of-Flight, Oak Ridge National Lab. Report No. ORNL–2309 (1957), 59–70.
148. E. Wigner, *On the distribution of the roots of certain symmetric matrices*, Ann. of Math. Ser. **67** (1958), no. 2, 325–327.
149. Wikipedia Contributors, *Biometrika*, Wikipedia, The Free Encyclopedia, Date retrieved 29 March 2015 04:05 UTC, Page Version ID 609531691, http://en.wikipedia.org/w/index.php?title=Biometrika&oldid=609531691.
150. J. Wishart, *The generalized product moment distribution in samples from a normal multivariate population*, Biometrika **20 A** (1928), 32–52.
151. A. Yang, *Low-lying zeros of Dedekind zeta functions attached to cubic number fields*, preprint.
152. M. Young, *Low-lying zeros of families of elliptic curves*, J. Amer. Math. Soc. **19** (2006), no. 1, 205–250.

The Generalized Fermat Equation

Michael Bennett, Preda Mihăilescu, and Samir Siksek

Abstract We survey approaches to solving the generalized Fermat equation

$$x^p + y^q = z^r$$

in relatively prime integers x, y and z, and integers p, q and $r \geq 2$.

1 Le Roy est mort: Vive le Roy

Pythagoras' formula was purportedly kept secret within the closed circle of his initiates—but, as with any fact of nature, it became eventually widely known: the squares of the cathetes sum to the square of the hypotenuse. In the spirit of arithmetic, spread eight centuries later by Diophantus of Alexandria, one may instead phrase this statement in terms of integral solutions of the equation

$$x^2 + y^2 = z^2, \quad \text{withx, y, z} \in \mathbb{N} \text{ and gcd(x, y, z)} = 1, \tag{1}$$

or, equivalently, ask for the coordinates of all rational points on the unit circle. All these are variants of the problem appearing in the second book of Diophantus, in the chapter numbered VIII—often quoted accordingly as Diophantus II.VIII. We nowadays call the solutions to Eq. (1) *Pythagorean triples*. Already in Diophantus one finds parametrizations for these solutions, given by

M. Bennett
Department of Mathematics, University of British Columbia, Vancouver, BC, Canada V6T 1Z2
e-mail: bennett@math.ubc.ca

P. Mihăilescu (✉)
Mathematisches Institut der Universität Göttingen, Göttingen, Germany
e-mail: preda@uni-math.gwdg.de

S. Siksek
Mathematics Institute, University of Warwick, Coventry CV4 7AL, UK
e-mail: s.siksek@warwick.ac.uk

J.F. Nash, Jr., M.Th. Rassias (eds.), *Open Problems in Mathematics*,
DOI 10.1007/978-3-319-32162-2_3

$$x = 2uv; \quad y = u^2 - v^2; \quad z = u^2 + v^2, \tag{2}$$

where u and v are positive integers. This fact is easy to verify in one direction; the proof that all triples are parametrized in this form is one of the most popular of ancient mathematics and is widely taught to this day.

The work of Diophantus on Arithmetic, a collection of 12 volumes written in Greek, was lost for centuries. Only in the late sixteenth century were half of these books, namely I–III and VIII–X, rediscovered in the form of a later Byzantine transcription. These were translated into latin by Bombelli, and then subsequently published in Basel by Xylander. It was through the annotated translation of Gaspard Bachet from 1621 that the *Arithmetica* finally received a wide diffusion, capturing the interest of mathematicians of the epoch. Among them, Fermat, a lawyer from Toulouse, was particularly impressed by the beauty of Diophantus' solution to (1). It is on the margin of the text to Diophantus' Problem II.VIII that Fermat wrote in 1634 his historical note concerning a short proof of the fact that the equation

$$x^n + y^n = z^n, \quad \text{with } x, y, z \in \mathbb{N} \text{ and } \gcd(x, y, z) = 1, \tag{3}$$

has no solution for $n > 2$, a proof which the margin of Bachet's book was insufficiently large to contain.

The assertion, henceforth bearing the name Fermat's Last Theorem—hereafter denoted FLT for concision—remained an open problem for more than three centuries. Attempts to prove FLT led to some of the most significant developments in mathematics of the past three hundred years; it is fair to say that it is one of the problems that has generated the most mathematics in history. The first systematic approach, initiated by Kummer in the mid nineteenth century, was based on the theory of algebraic number fields and in particular *cyclotomic fields*. The conjecture was finally proved in 1994 by Wiles, with the help of Taylor, building on a series of ideas and results, due to Hellegouarch, Frey, Serre and Ribet, that connect the Fermat equation to elliptic curves, modular forms and Galois representations.

Even before Wiles announced his proof, various generalizations of Fermat's Last Theorem had already been considered, to equations of the shape

$$Ax^p + By^q = Cz^r,$$

for fixed integers A, B and C. In the case where $A = B = C = 1$, for reasons we discuss later, we focus our attention on the equation

$$x^p + y^q = z^r, \quad \text{with } x, y, z \in \mathbb{N}, \ \gcd(x, y, z) = 1 \text{ and } \frac{1}{p} + \frac{1}{q} + \frac{1}{r} < 1. \tag{4}$$

Perhaps the only solutions to this equation are those currently known; i.e. those with (x, y, z, p, q, r) coming from the solution to Catalan's equation $1^p + 2^3 = 3^2$, and from the following nine identities:

$$2^5 + 7^2 = 3^4, \quad 7^3 + 13^2 = 2^9, \quad 2^7 + 17^3 = 71^2, \quad 3^5 + 11^4 = 122^2,$$
$$17^7 + 76271^3 = 21063928^2, \; 1414^3 + 2213459^2 = 65^7, \; 9262^3 + 15312283^2 = 113^7,$$
$$43^8 + 96222^3 = 30042907^2 \text{ and } 33^8 + 1549034^2 = 15613^3.$$

In the mid 1990s, Andrew Beal, a graduate in mathematics with computational interests, in the course of carrying out calculations related to Fermat's equation and its variations, noted that the solutions listed here all have the property that $\min\{p, q, r\} = 2$ and made what is now termed the *Beal conjecture*, that there are no non-trivial solutions to (4), once we assume that $\min\{p, q, r\} \geq 3$. A prize for the solution to this problem now amounts to one million U.S. dollars. Other names attached to conjectures about Eq. (4) include the *Fermat–Catalan conjecture* and the *Tijdeman–Zagier conjecture*.

As we shall see, the few particular cases of these conjectures that have been investigated may well support the hope that this new generalization of Pythagoras' initial equation will also stimulate fascinating new mathematics.

Our main focus for the rest of the paper will be to describe two approaches to proving results about Eq. (4). The first of these is essentially a generalization of the cyclotomic methods that proved successful for Catalan's equation. The second is the adaptation of Wiles' proof of Fermat's Last Theorem to handle many special cases of Eq. (4).

2 Cyclotomic Approaches and Their Limitations

Let us start by noticing that in both Eqs. (3) and (4) one may assume all exponents to be prime: indeed, if there is a solution with non-prime exponents, by raising the variables to some power, one obtains a solution with prime exponents. It thus suffices to consider this case and show that it leads to no non-trivial solutions. By an elementary observation, sometimes attributed to Euler, we have that

$$G := \gcd\left(x \pm y, \; \frac{x^p \pm y^p}{x \pm y}\right)$$

is a divisor of p, provided that the integers x, y are coprime.[1] In this section, we will focus our attention on the special case of (4), given by the *Fermat-Catalan equation*

$$x^p + y^p = z^q, \quad \text{with x, y, z} \in \mathbb{N} \text{ and } \gcd(\text{x, y, z}) = 1, \tag{5}$$

where one may hope to apply cyclotomic theory in a way somewhat analogous to that of the Fermat equation. Equation (5) also serves as a generalization of the binary Catalan equation

$$x^p - y^q = 1 \tag{6}$$

[1]One verifies this by letting $t = x \pm y$ and $x = t \mp y$, as well as using the fact that $(t, y) = 1$.

which also resisted solution for more than a century (and was finally solved by the second author, eight years after Wiles' remarkable proof of Fermat's Last Theorem). Here *binary* refers to the presence of only two unknowns in this equation, a fact which facilitates an analytic approach to bounding the possible solutions.

Assuming that either (3) or (5) has a non-trivial solution for the prime p—or the prime pair (p, q)—one distinguishes the cases when $G = 1$ and $G = p$. The case where $p \nmid xyz$ has traditionally been termed the *First Case* of FLT, or FLT1; we retain this designation for the Fermat–Catalan equation. If $G = p$ and thus $p \mid xyz$, one may assume, in Eq. (3) at least, that $p \mid z$. It is further easy to show that, in this case,

$$p^2 \nmid \frac{x^p \pm y^p}{x \pm y}.$$

Since the Fermat equation is homogeneous, the assumption that x, y and z are pairwise coprime is straightforward—otherwise one could divide by the common divisor in a solution and obtain one in coprime integers. In the case of (5), however, with a little work, one can always construct infinitely many "trivial" solutions for which x, y and z fail to be coprime.

2.1 History of "Fermat's Last Theorem"

As is well known, Fermat left no published proof of his conjecture. He did, however, provide a beautiful argument in the biquadratic or quartic case. To be precise, he considered the more general equation

$$x^4 + y^4 = z^2 \tag{7}$$

and, using some astute manipulation of Pythagorean triples, proved that if (x, y, z) is a non-trivial solution of (7) in, say, positive integers, then one can construct a further positive solution (x', y', z') of the same equation, in which $z' < z$. By repeating the procedure one eventually finds a solution with $z' = 1$, which implies that $x' \cdot y' = 0$, a contradiction. This was the first instance of the method of *infinite descent* in number theory. Euler gave a proof of the cubic case using such an argument; Gauss gave later an alternative proof using congruences. Although elementary, both methods require quite intricate computations and are not easy to memorize. We provide here a short elementary proof, which uses some more recent ideas, that date back to Wieferich and Furtwängler[2]:

[2]The second author found this proof, confronted with the own incapacity to recall the details of the classical proofs, for a seminar. It is possible that the proof may have been known, but we found no reference to it in the literature

Lemma 1. *The equation* $x^3 + y^3 + z^3 = 0$ *has no non-trivial integer solutions.*

Proof. Assume that (x, y, z) is a non-trivial solution in coprime integers. Let

$$\rho = \frac{-1 + \sqrt{-3}}{2}$$

be a third root of unity and $\mathscr{E} = \mathbb{Z}[\rho]$ be the ring of Eisenstein integers, which is a Euclidean ring. We assume first that $3 \nmid xyz$, so that both $x + y$, and $x^2 - xy + y^2$ are cubes, say $x + y = s^3$, and

$$(x + \rho y) = (a + b\rho)^3$$

is the cube of a principal ideal. Exactly one of x, y, z must be even—we shall thus assume that $y = 2v$ is even. Since the units of $\mathscr{E}$ are the sixth roots of unity, there is some $\delta \in < -\rho >$ such that

$$\begin{aligned}\delta(x + y\rho) = (a + b\rho)^3 &= a^3 + b^3 + 3ab\rho(a + b\rho)\\ &= a^3 + b^3 - 3ab^2 + 3ab(a - b)\rho.\end{aligned}$$

If $\delta = \pm 1$, then comparing coefficients implies that $y \equiv 0 \bmod 3$, contradicting our assumption. We thus have that

$$x + y\rho = \pm\rho^c(a + b\rho)^3, \quad \text{with } c = \pm 1 .$$

We have chosen $y \equiv 0 \bmod 2$, whereby $x\rho^{-c} \equiv w^3 \bmod 2\mathscr{E}$ for some $w = \pm(a + b\rho) \in \mathscr{E}$. But $x \equiv x + y = s^3 \bmod 2$, whence we may conclude that

$$\rho^{-c} \equiv (w/s)^3 \bmod 2 .$$

The ideal $\mathfrak{p} := (2) \subset \mathscr{E}$ is prime; let $\pi : \mathscr{E} \to \mathbb{F}_{2^2}$ be the natural projection. Since $c \not\equiv 0 \bmod 3$, it follows that $\pi(w/s) \in \mathbb{F}_{2^2}$ is a primitive ninth root of unity, an impossibility. It remains, then, to consider the case where $3 \mid xyz$; we may suppose, without loss of generality, that $3 \mid z$. Appealing to (9), we find that there is a root of unity δ such that $(\alpha) = \delta\left(\frac{x+y\rho}{1-\rho}\right) = (a + b\rho)^3$ and $x + y = 9s^3$. Since $\frac{1-\rho}{1-\overline{\rho}} = -\rho$, we obtain after dividing the previous identity by its complex conjugate, that there is another root of unity, say δ', such that

$$\frac{x + y\rho}{x + y\overline{\rho}} = \frac{2x - y + y\sqrt{-3}}{2x - y - y\sqrt{-3}} = \delta' \cdot \left(\frac{a + b\rho}{a + b\overline{\rho}}\right)^3 \equiv \delta' \bmod 3\sqrt{-3} \cdot \mathscr{E}. \tag{8}$$

If $2x \not\equiv y \bmod 3$, letting $y' \equiv y/(2x-y) \bmod 3$, the last identities imply that

$$\frac{1+y'\sqrt{-3}}{1-y'\sqrt{-3}} \equiv \delta' \bmod 3\sqrt{-3}\cdot\mathscr{E}\ ,$$

whence

$$y' \equiv y \equiv 0 \bmod 3\ ,$$

contradicting the assumption that $3 \mid z$ and $\gcd(x, y, z) = 1$. Finally, suppose that $2x - y = -3d$. Inserting this value into (8) yields

$$\frac{d\sqrt{-3}+y}{d\sqrt{-3}-y} \equiv \delta' \bmod 3\sqrt{-3}\cdot\mathscr{E},$$

whereby $\delta' = -1$ and $d \equiv 0 \bmod 3$. Then $2x - y \equiv x + y \equiv 0 \bmod 9$. Summing these two congruences, we find $3x \equiv 0 \bmod 9$, and thus $3 \mid x$, a contradiction which completes the proof.

□

After Euler and Gauss, the quintic and septic cases of FLT were solved with contributions from Dirichlet, Lamé, Legendre and Cauchy. In his proof of the quintic case, Dirichlet distinguished the cases $p \nmid xyz$ and $p \mid xyz$, using descent for the proof of the second case. Lamé announced in 1841 a full proof of the general case of FLT. Unfortunately, his proof relied implicitly upon the (incorrect) assumption that the integers of the form

$$\alpha = \sum_{k=0}^{p-2} a_k \zeta^k\ ,$$

with $a_k \in \mathbb{Z}$ and ζ a pth root of unity, form a ring with unique factorization. Kummer demonstrated that this assumption is, in general, false and that the smallest prime for which it fails is $p = 23$. In order to circumvent this difficulty, Kummer proceeded with an investigation of divisibility in the rings of algebraic integers of the pth cyclotomic field, and introduced the notion of *ideal numbers*, a larger group in which unique factorization was recovered. This work, stemming from a desire to attack the Fermat problem, led, in the following decades, to the theory of ideals and the work of Dedekind, giving rise to the fundamental results underlying what we presently know as algebraic number theory.

If p is an arbitrary odd prime, we let $\mathbb{K} = \mathbb{Q}[\zeta]$ denote the pth cyclotomic extension. The algebraic integers of this field are $\mathscr{O}(\mathbb{K}) = \mathbb{Z}[\zeta]$ and the ideals of this ring factorize uniquely as products of prime ideals. If I is the semigroup of ideals and P the one of principal ideals, i.e. ideals generated by single elements, the quotient $\mathscr{C}(\mathbb{K}) = I/P$ is a finite abelian multiplicative group, the class group. The prime p is called *regular*, if p does not divide the size $h(\mathbb{K}) = |\mathscr{C}(\mathbb{K})|$ of the class group, and *irregular* otherwise. With respect to the Fermat equation, we have in $\mathbb{Z}[\zeta]$ one of the following factorizations:

$$x^p = (x+y) \cdot \left(\frac{x^p + y^p}{x+y} \right) = (x+y) \cdot \prod_{c=1}^{p-1} (x + y\zeta^c), \quad \text{or}$$

$$z^p = p(x+y) \cdot \left(\frac{x^p + y^p}{p(x+y)} \right) = p(x+y) \cdot \prod_{c=1}^{p-1} \left(\frac{x + y\zeta^c}{1-\zeta} \right),$$

in case of FLT1 or FLT2, respectively. In either situation, writing $\alpha = \frac{x+\zeta y}{(1-\zeta)^e}$, with $e = 0$ in the first case and $e = 1$ in the second, one finds that α is coprime to $p(x+y)$ and is, in fact, a pth power, albeit not one of an algebraic integer of $\mathbb{K}$, but rather of the ideal $\mathfrak{A} = (\alpha, z)$. We thus have

$$(\alpha) = \mathfrak{A}^p \quad \text{and} \quad N_{\mathbb{K}/\mathbb{Q}}(\alpha) = \frac{z^p}{p^e(x+y)}. \tag{9}$$

We note at this point that, in the case of the Fermat-Catalan equation, the same construction leads again to $(\alpha) = \mathfrak{A}^q$. Using the connection between class groups and factorization of ideals, Kummer proved his fundamental result on Fermat's Conjecture:

Theorem 1 (Kummer). *Equation (3) has no solutions for regular primes $p > 2$.*

For regular primes, it follows from (9) that there exists a $\rho \in \mathbb{Z}[\zeta]$ such that $(\alpha) = (\rho)^p$ is the pth power of a principal ideal. Starting from this, the proof of FLT1 is relatively simple. For the second case, however, Kummer appealed to a sophisticated variant of infinite descent in the real field $\mathbb{K}^+ \subset \mathbb{K}$—the method bears currently his name, Kummer descent. A modern, complete proof of this result can be found in the book of Washington [36], Chapter IX. One finds in Rassias' lovely introductory work for undergraduates [26], on page 147, more biographical details of Kummer's life.

Kummer's work was followed by a century of active research on the Fermat equation, which led to the establishment of a large number of conditions known to imply the truth of the Fermat Conjecture—see e.g. Ribenboim's famous survey [27]. However, before Wiles's breakthrough, there were only two known results known to hold for infinitely many exponents, namely the "elementary" proof [34] given by Terjanian in 1977 for the fact that (3) has no solution for *even* exponents $n > 2$ with $n \nmid xy$, as well as the deep analytic proof of Adelman, Fouvry and Heath-Brown, which showed that FLT1 holds for infinitely many primes.[3]

[3]One would expect, for various reasons, that regular primes occur more frequently than irregular ones. If we knew this, since it has been proved that there are infinitely many irregular primes, Kummer's result would already imply that there are infinitely many primes p for which FLT holds. However no proof of the fact that the set of regular primes is infinite is known, even now.

There are, however, a number of results of interest on FLT that were established via cyclotomic methods before Wiles' proof. We review here a few of the most important of them:

(i) Wieferich and then Mirimanoff and Furtwängler proved that if FLT1 has a non-trivial solution, then

$$a^{p-1} \equiv 1 \bmod p^2, \quad \text{for} a \in \{2, 3\}. \tag{10}$$

Variations on this theme were treated during the following decades by, among others, Morichima, Lehmer, Skula, Granville and Monagan, the last two of these eventually proving that if FLT1 had non-trivial solutions, then (10) holds for all primes $a \leq 89$. With this Granville and Monagan were able to prove that FLT1 has no solutions for

$$p < 714,591,416,091,389.$$

(ii) Eichler proved that if FLT1 has non-trivial solutions, then the p-rank of the p-part of the class group $A := \mathscr{C}(\mathbb{Q}[\zeta])_p$ of the pth cyclotomic field is necessarily *large*, namely $p\text{–rk}(A_p) \geq \sqrt{p} - 1$. He thus improved upon an earlier result of Krasner, who had proved that if FLT1 had solutions and $p > n_0 = (45!)^{88}$, then the Bernoulli numbers $B_{p-1-2i}, i = 1, 2, \ldots, \lfloor (\log p)^{1/3} \rfloor$ had numerators divisible by p; this implies in particular that $p\text{–rk}(A_p) \geq \lfloor (\log p)^{1/3} \rfloor$.

(iii) In the first half of the twentieth century, Harry Schulz Vandiver wrote extensively on Fermat's equation, partially fixing some gaps in earlier proofs of Kummer (and leaving a number of gaps himself, which were fixed only at the end of the century). We present below his main result as part of Theorem 2.

Bearing in mind the fact that the Fermat Conjecture has been proved, it is still of interest to analyze other approaches which may provide alternative proofs of this Theorem. There are currently two primes known, for which the Wieferich condition (10) is satisfied with $a = 2$; none are known with $a = 3$. If one admits the heuristic assumption that the vanishing of a Fermat quotient

$$\varphi_p(a) \equiv \frac{a^{p-1} - 1}{p} \bmod p$$

has probability $1/p$, one may expect on *average* $O(\log\log(X))$ primes $p < X$ for which the quotient $\varphi_p(a) = 0$ for some fixed $a < p$. However, the same heuristic argument suggests that one can find at most one prime for which two or more Fermat quotients vanish simultaneously. One may formulate the following:

Conjecture 1. There exists a constant $c \geq 2$ such that for every prime $p \in \mathbb{N}$ there are less than c values $a \in \{2, 3, \ldots, p-2\}$ with $\varphi_p(a) = 0$.

If $c < 87$, this conjecture implies FLT1.

Concerning the criteria in group (ii), Washington provides in [36] heuristic arguments suggesting that

$$p\text{–rk}(A_p) \ll O(\log(p))$$

for all primes. The Theorem of Eichler would imply FLT1 even if a rather weaker conjecture holds:

Conjecture 2. There is an integer $a > 2$ such that for all primes p, the p part of the class group of the pth cyclotomic field has rank $p\text{–rk}(A_p) < p^{1/a}$.

We have mentioned above that there are infinitely many irregular primes. The first of these were discovered by Kummer, the smallest one being $p = 37$. However, if one instead considers the largest *real* subfield contained in $\mathbb{K}$, which is $\mathbb{K}^+ = \mathbb{Q}[\zeta + \overline{\zeta}]$, the class number h^+ of this field is apparently much smaller and seems to never be divisible by p. Kummer was the first to suggest, in a letter to Kronecker from 1852 (see [23]), that this fact might hold for all p. The fact played an important role in many of the papers of Vandiver, who was seemingly unaware of Kummer's letter. The assumption that $p \nmid h_p^+ = |\mathscr{C}(\mathbb{K}^*)|_p$ is therefore called the *Kummer-Vandiver* Conjecture, or simply the Vandiver Conjecture. The conjecture has also deep implications in K-theory; it has been verified numerically for all primes $p < 2^{27}$. We should mention, however, that there are specialists who accept some heuristic arguments which suggest that the Conjecture might have counterexamples that are as scarce as the Wieferich primes. If this were true, there might be as many as $\log\log(X)$ primes $p < X$ for which the conjecture is false. Those who believe the Vandiver Conjecture are guided by the fact that there are numerous striking and rather improbable consequences to $p \mid h_p^+$, and therefore the heuristic assumption that the value of the residue $h_p^+ \mod p$ is uniformly distributed may be false.

In the context of Fermat's Last Theorem, two additional conditions of a rather specialized nature play a role; we formulate them also as assumptions: they have been computationally verified to hold within the same range as the Kummer–Vandiver Conjecture.

Assumption C. Assume that the exponent of A_p is p, whereby the p-part of the class group of the pth cyclotomic field is annihilated by p.

Assumption D. Assume that all the units $\delta \in \mathbb{Z}[\zeta_p + \overline{\zeta}_p]^\times$ for which there exists an algebraic integer $\rho \in \mathbb{Z}[\zeta_p + \overline{\zeta}_p]$ such that $\delta \equiv \rho^p \mod p^2\mathbb{Z}[\zeta_p + \overline{\zeta}_p]$ are global pth powers.

With these definitions, the following theorem holds:

Theorem 2. *Suppose that the Kummer–Vandiver Conjecture holds. If, additionally, Assumption C holds, then FLT1 has no solutions. If, instead, Assumption D holds, then FLT2 has no solutions.*

The first of these claims dates back to Vandiver, who, however supposed only that $p \nmid h_p^+$ and had not noticed the necessity of Assumption C. The correct statement was discovered by Sitaraman [32] in 1995. The second part of the theorem, together with its proof, are due to Kummer. We note that Theorem 2 provides the only known cyclotomic criterium which implies that there are no solutions to FLT2.

2.2 The Catalan Equation

Due to results of Victor Lebesgue (1853) and Chao Ko (1962), which eliminated the cases of even exponents, the Catalan Conjecture was reduced to proving that $x^p - y^q = 1$ has no solution with odd prime exponents (which are easily seen to be distinct). By considering cyclotomic factorizations, similar to the ones in (9), one obtains four cases. Cassels proved in 1962 that if (6) has a solution with odd exponents, then $p \mid y$ and $q \mid x$, while

$$\frac{x^p - 1}{p(x-1)} = v^q$$

for some rational integer v. One may define

$$\alpha = \frac{x - \zeta}{1 - \zeta} \quad \text{and} \quad \mathfrak{A} = (\alpha, y) ,$$

whence the analogue of Eq. (9) is $(\alpha) = \mathfrak{A}^q$.

Catalan's conjecture now follows from a combination of analytic methods with algebraic properties of cyclotomic fields. We present a brief exposition of some of the ideas that made this proof possible. Denoting by G the Galois group $\mathrm{Gal}(\mathbb{Q}[\zeta_p]/\mathbb{Q})$, one notices that the group ring $\mathbb{F}_q[G]$ acts on the class a of the ideal $\mathfrak{A}$; in other words, linear combinations of the type $\theta = \sum_{\sigma \in G} n_\sigma \cdot \sigma$, in which the integers n_σ are identified with their remainders modulo q, will act on the class a according to $a^\theta = \prod_{\sigma \in G} \sigma(a)^{n_\sigma}$. Since $a^q = 1$, we see that it suffices to consider $n_q \in \mathbb{F}_q$. We call an element $\theta \in \mathbb{F}_q[G]$ an *annihilator* of a if $a^\theta = 1$.

Suppose that we are able to find a non-trivial annihilator $(1 + j)\theta \in \mathbb{F}_q[G]$—here we denote the restricted action of complex conjugation to $\mathbb{K}$ by $j \in G$, so for instance $a^j = \overline{a}$. Then $\alpha^\theta = (\rho)^q$, for some $\rho \in \mathbb{K}^+$; the equality between principal ideals translates into an identity between algebraic numbers, involving an unknown unit ε: $\alpha^\theta = \varepsilon \cdot \rho^q$. Assume additionally, that there is a further $\theta' \in \mathbb{F}_q[G]$ such that $\varepsilon^{\theta'} \in (\mathbb{K}^\times)^q$. Given this, one is able to prove, using additional arguments about the structure of units in $\mathbb{K}$, that there is a number $\nu \in \mathbb{Z}[\zeta + \overline{\zeta}]$ such that $(x - \zeta)^{\theta \cdot \theta'} = \nu^q$: note that we eliminated the denominator of α! That these favourable assumptions situation can be shown to occur follows from an important theorem of Francisco Thaine [35] (which also leads to a cyclotomic proof of the Main Conjecture of one dimensional Iwasawa Theory).

Continuing, one now finds a multiple $\psi = \sum_{\sigma\in G} r_\sigma \sigma \in \mathbb{F}_q[G]$ of $\theta \cdot \theta'$ such that $\sum_{\sigma\in G} r_\sigma = hq$ for some $h \leq \frac{p-1}{2}$, leading to the following equation

$$\nu = x^h(1 - \zeta/x)^{\psi/q}.$$

The fact that $\nu \in \mathbb{R}$ has the important advantage that the rapidly converging binomial series expansion of the expression

$$(1 - \zeta/x)^{\psi/q} = \prod_{\sigma\in G} (\sigma(1 - \zeta/x))^{r_\sigma/q} \tag{11}$$

will in fact converge to ν/x^h (rather than to some number that differs from ν/x^h by a qth root of unity, as would be true generally). With this, we obtain $\nu = x^h g(1/x) + F(1/x)$, with $g \in \mathbb{K}[X]$ being a polynomial of degree h and $F(T) \in \mathbb{K}[[T]]$ a power series. Finally, appealing to some lower bounds on A which were obtained by Hyyrö, one can eventually show that, under the given arithmetic conditions, $F(1/x) = 0$. We thus have $\nu = x^h g(1/x)$, which leads to an arithmetic contradiction, completing the proof of Catalan's Conjecture.

2.3 *The Fermat: Catalan Equation*

As previously mentioned, in the case that (x, y, z) is a non-trivial solution to the Fermat-Catalan equation (5) with odd exponents p and q, if we let $\alpha = \frac{x+\zeta y}{(1-\zeta)^e}$ and $\mathfrak{A} = (\alpha, z)$—where $e = 1$ if $p \mid z$ and $e = 0$ otherwise, then we have $\mathfrak{A}^q = (\alpha)$, a situation which is reminiscent of both the Fermat and the Catalan equations, and a starting point for cyclotomic investigations of (5).

In this direction, the second author has tried to adapt the proof of Kummer's Theorem to the case of Eq. (5). It would take too long to explain here the main points in which this equation differs from (3), necessitating the introduction of additional methods. Let us only mention that it is possible to restrict our attention to six cases, depending on whether or not p or q divide any of the factors of $x \cdot y \cdot z \cdot (x \pm y)$. After proving a generalization of Kummer's descent to the $p \cdot q$th cyclotomic field, it was often possible to either discard cases, or reduce them to conditions about the vanishing of some Fermat quotient—e.g. $2^{q-1} \equiv 1 \bmod q^2$ or $p^{q-1} \bmod q^2$. This approach succeeds in five cases. Unfortunately, in the sixth case, all classical Kummerian methods apparently fail. As a consequence, the second author was unable to find conditions on p and q which imply that (5) has no solutions. By symmetry, a set of such conditions could however be derived for the *rational Catalan* equation, i.e. the Eq. (6) in which $x, y \in \mathbb{Q}$ is allowed. Note that, after clearing denominators, this equation is equivalent to

$$X^p + Y^{pq} + Z^q = 0\,,$$

which may be viewed as a "symmetrized Fermat—Catalan" equation.

This first attempt to apply cyclotomic methods thus appears to confirm the somewhat pessimistic expectation that they are not sufficient for solving Eq. (5) in any generality.

3 Fermat's Last Theorem

In the previous section, we discussed the cyclotomic approach to the Fermat equation and its potential limitations. We now sketch the approach of Hellegouarch, Frey, Serre and Ribet which culminated in Wiles' proof of Fermat's Last Theorem.

Theorem 3 (Wiles [37]). *The only integer solutions to the Fermat equation*

$$x^n + y^n = z^n$$

with $n \geq 3$ *satisfy* $xyz = 0$.

Recall that we call a solution *trivial* if $xyz = 0$, otherwise it is called *non-trivial.* Thus the theorem states that all solutions to the Fermat equation are trivial. As we have seen, the theorem is true for exponents $n = 3$ and 4. Thus it is sufficient to show, for primes $p \geq 5$, that all solutions to

$$x^p + y^p + z^p = 0 \tag{12}$$

are trivial. Of course, if (x, y, z) is a solution, we may by scaling suppose that $\gcd(x, y, z) = 1$; we call such a solution *primitive*. The purpose of this section is to sketch the proof of Fermat's Last Theorem and the ideas leading to it. The proof is based on three main pillars:

(i) Mazur's Theorem on irreducibility of Galois representations of elliptic curves;
(ii) The modularity theorem, due to Wiles, Breuil, Conrad, Diamond and Taylor;
(iii) Ribet's level lowering theorem.

Explaining these pillars will involve a detour into some of the most fascinating areas of modern number theory: elliptic curves, Galois representations, modular forms and modularity.

3.1 Elliptic Curves

There are many possible definitions of an elliptic curve. Let K be a field. An *elliptic curve* over K is a curve of genus 1 defined over K with a distinguished K-point. An alternative definition is: an *elliptic curve* over K is a 1-dimensional abelian variety over K. The simplest (though conceptually least enlightening) definition is:

an *elliptic curve* E over K is a smooth curve in $\mathbb{P}^2$ given by an equation of the form

$$E : y^2z + a_1xyz + a_3yz^2 = x^3 + a_2x^2z + a_4xz^2 + a_6z^3,$$

with a_1, a_2, a_3, a_4 and $a_6 \in K$. This in fact is a curve of genus 1, and the distinguished K-point is $(x : y : z) = (0 : 1 : 0)$. If the characteristic of K is not 2 or 3, then we can transform to a much simpler model given in $\mathbb{A}^2$ by

$$E : Y^2 = X^3 + aX + b, \tag{13}$$

where a and $b \in K$. We call this equation a *Weierstrass model*. Let

$$\Delta = -16(4a^3 + 27b^2)$$

which we call the *discriminant* of E (this is -16 times the discriminant of the polynomial on the right-hand side). The requirement that E is smooth is equivalent to the assumption that $\Delta \neq 0$. The distinguished K-point is now the 'point at infinity', which we denote by ∞ or $\mathcal{O}$. Given a field $L \supseteq K$, the set of L-points on E is given by

$$E(L) = \{(x, y) \in L^2 : y^2 = x^3 + ax + b\} \cup \{\mathcal{O}\}.$$

It turns out that the set $E(L)$ has the structure of an abelian group with $\mathcal{O}$ as the identity element. The group structure is easy to describe geometrically: three points P_1, P_2, $P_3 \in E(L)$ add up to the identity element if and only if there is a line ℓ defined over L meeting E in P_1, P_2, P_3 (with multiplicities counted appropriately). The fact that $E(L)$ is an abelian group (where the group operation has a geometric interpretation) ties in with the fact that E is an abelian variety.

Theorem 4 (The Mordell–Weil Theorem). *Let K be a number field and E an elliptic curve over K. Then $E(K)$ is a finitely generated abelian group.*

When K is a number field we refer to the group $E(K)$ as the *Mordell–Weil group of E over K*.

Example 1. As an example, consider the Fermat degree 3 equation over $\mathbb{Q}$:

$$x^3 + y^3 = z^3. \tag{14}$$

Viewed as a curve in $\mathbb{P}^2$, this is in fact a curve of genus 1. Let us choose the point $(1 : -1 : 0)$ to be the distinguished point. We now transform this into a Weierstrass model using the transformation

$$Y = \frac{36(x-y)}{x+y}, \qquad X = \frac{12z}{x+y}, \tag{15}$$

so that a solution to Eq. (14) corresponds to a rational point on the elliptic curve

$$E : Y^2 = X^3 - 432.$$

The solution $(1 : -1 : 0)$ to (14) corresponds to the point $\infty = \mathcal{O}$ on E. The model E is the elliptic curve denoted by 27A in Cremona's tables [8]. The group $E(\mathbb{Q})$ has rank zero and, in fact,

$$E(\mathbb{Q}) \simeq \mathbb{Z}/3\mathbb{Z}.$$

Indeed,

$$E(\mathbb{Q}) = \{\mathcal{O},\ (36, 12),\ (36, -12)\}$$

where in the group law on E we have

$$2 \cdot (36, 12) = (36, -12) \ \text{ and } \ 3 \cdot (36, 12) = \mathcal{O}\,.$$

Thus the degree 3 Fermat equation (14) has exactly three solutions, and we may obtain these by taking the three points belonging to $E(\mathbb{Q})$ and transferring them back to the model (14) using (15). Doing this, we find that the three solutions to (14) are $(1 : -1 : 0)$, $(1 : 0 : 1)$ and $(0 : 1 : 0)$—that is, just the trivial solutions.

Example 2. One can similarly transform the equation

$$y^2 = x^4 + z^4 \tag{16}$$

into the elliptic curve

$$E : Y^2 = X^3 - 4X$$

which has Mordell–Weil group

$$E(\mathbb{Q}) = \{\mathcal{O},\ (0,0),\ (2,0),\ (-2,0)\} \simeq \mathbb{Z}/2\mathbb{Z} \times \mathbb{Z}/2\mathbb{Z}$$

and use this information to deduce that the only solutions to (16) are the trivial ones.

It turns out that the proofs by Fermat and Euler of the degree 4 and degree 3 cases of Fermat's Last Theorem are simply special cases of what are now standard Mordell–Weil group computations.

The degree p Fermat equation (12), viewed as defining a curve in $\mathbb{P}^2$, has genus $(p-1)(p-2)/2$, and thus does not define an elliptic curve for $p \geq 5$. We do mention in passing the following celebrated theorem of Faltings.

Theorem 5 (Faltings [13]). *Let C be a curve of genus ≥ 2 over a number field K. Then $C(K)$ is finite.*

Faltings' theorem tells us that for each $p \geq 5$ the Fermat equation (12) has finitely many primitive solutions. Faltings' theorem is *ineffective*, in the sense that the proof does not yield an algorithm that is guaranteed to find all solutions.

3.2 Modular Forms

Let k and N be positive integers. We define

$$\Gamma_0(N) = \left\{ \begin{pmatrix} a & b \\ c & d \end{pmatrix} \in \mathrm{SL}_2(\mathbb{Z}) \ : \ c \equiv 0 \pmod{N} \right\}.$$

It is easy to see that $\Gamma_0(N)$ is a subgroup of $\mathrm{SL}_2(\mathbb{Z})$ of finite index. Let $\mathbb{H}$ be the complex upper-half plane

$$\mathbb{H} = \{z \in \mathbb{C} \ : \ \mathrm{Im}(z) > 0\}.$$

The group $\Gamma_0(N)$ acts on $\mathbb{H}$ via fractional linear transformations

$$\begin{pmatrix} a & b \\ c & d \end{pmatrix} : \mathbb{H} \to \mathbb{H}, \qquad z \mapsto \frac{az+b}{cz+d}.$$

The quotient $Y_0(N) = \Gamma_0(N)\backslash\mathbb{H}$ has the structure of a non-compact Riemann surface. This has a standard compactification denoted $X_0(N)$ and the difference $X_0(N) - Y_0(N)$ is a finite set of points called the *cusps*. In fact the Riemann surface $X_0(N)$ has the structure of an algebraic curve defined over $\mathbb{Q}$ and is an example of what is known as a *modular curve*.

A *modular form f of weight k and level N* is a function $f : \mathbb{H} \to \mathbb{C}$ that satisfies the following conditions

(i) f is holomorphic on $\mathbb{H}$;
(ii) f satisfies the property

$$f\left(\frac{az+b}{cz+d}\right) = (cz+d)^k f(z), \tag{17}$$

for all $z \in \mathbb{H}$ and $\left(\begin{smallmatrix} a & b \\ c & d \end{smallmatrix}\right) \in \Gamma_0(N)$;
(iii) f extends to a function that is holomorphic at the cusps.

Observe that $\left(\begin{smallmatrix} 1 & 1 \\ 0 & 1 \end{smallmatrix}\right) \in \Gamma_0(N)$. Thus by (ii), the function f satisfies $f(z+1) = f(z)$. Letting $q(z) = \exp(2\pi i z)$, we see, from the periodicity, that f must have a Fourier expansion

$$f(z) = \sum_{n \geq N_0} c_n q^n$$

for some integer N_0. In fact, one of the cusps is the cusp at $i\infty$ which we can think of as being arbitrarily high up on the imaginary axis. Note that $q(i\infty) = 0$. We see that for f to be holomorphic at the cusp $i\infty$ we require $c_n = 0$ for $n < 0$. Thus we may write

$$f(z) = \sum_{n \geq 0} c_n q^n. \tag{18}$$

It turns out that the set of modular forms of weight k and level N, denoted by $M_k(N)$, is a finite-dimensional vector space over $\mathbb{C}$.

A *cusp form* of weight k and level N is a modular form f of weight k and level N that vanishes at all the cusps. As $q(i\infty) = 0$ we see in particular that a cusp form must satisfy $c_0 = 0$. The cusp forms naturally form a subspace of $M_k(N)$ which we denote by $S_k(N)$. Of particular interest are the weight 2 cusp forms of level N: these can be interpreted as regular differentials on the modular curves $X_0(N)$. It follows that the dimension of $S_2(N)$ as a $\mathbb{C}$-vector space is equal to the genus of the modular curve $X_0(N)$.

There is a natural family of commuting operators $T_n : S_2(N) \to S_2(N)$ (with $n \geq 1$) called the *Hecke operators*. The *eigenforms* of level N are the weight 2 cusp forms that are simultaneous eigenvectors for all the Hecke operators. Such an eigenform is called normalized if $c_1 = 1$ and thus its Fourier expansion has the form

$$f = q + \sum_{n \geq 1} c_n q^n.$$

3.3 Modularity

Let E be an elliptic curve over $\mathbb{Q}$. Such an elliptic curve has a model of the form

$$E : y^2 + a_1 xy + a_3 y = x^3 + a_2 x^2 + a_4 x + a_6, \tag{19}$$

where the $a_i \in \mathbb{Z}$, and a (non-zero) discriminant Δ_E which is an integer given by a complicated polynomial expression in terms of the a_i. It is possible to change the model by carrying out a suitable linear substitution in x, y, and we generally work with a *minimal model*: that is one where the $a_i \in \mathbb{Z}$ with discriminant having the smallest possible absolute value. Associated to E is another, more subtle, invariant called the *conductor* N_E which we shall not define precisely, but we merely point that it is a positive integer sharing the same prime divisors as the discriminant; that it measures the 'bad behavior' of the elliptic curve E modulo primes; and that it can be computed easily through *Tate's algorithm* [31, Chap. IV].

Now let $p \nmid \Delta_E$ be a prime. Then we can reduce the Eq. (19) to obtain an elliptic curve $\tilde{E}$ over $\mathbb{F}_p$. The set $\tilde{E}(\mathbb{F}_p)$ is an abelian group as before, but now necessarily finite, and we denote its order by $\#\tilde{E}(\mathbb{F}_p)$. Let

$$a_p(E) = p + 1 - \#\tilde{E}(\mathbb{F}_p).$$

We are now ready to state a version of the modularity theorem due to Wiles, Breuil, Conrad, Diamond and Taylor [6, 37]. This remarkable theorem was previously known as the Taniyama–Shimura conjecture.

Theorem 6 (The Modularity Theorem). *Let E be an elliptic curve over $\mathbb{Q}$ with conductor N. There exists a normalized eigenform $f = q + \sum c_n q^n$ of weight 2 and level N such that $c_n \in \mathbb{Z}$ for all n, and if $p \nmid \Delta_E$ is prime then $c_p = a_p(E)$.*

In fact, for an eigenform f the Fourier coefficients are determined by the coefficients c_p with prime indices. Thus from the elliptic curve E we can construct the Fourier expansion of the corresponding eigenform f. What is astonishing is that f then satisfies the transformation properties (17).

Example 3. We consider the following elliptic curve E over $\mathbb{Q}$:

$$E \, : \, y^2 + y = x^3 - x^2 - 10x - 20.$$

This has conductor 11, the smallest possible conductor, and discriminant -11^5. The space $S_2(11)$ is 1-dimensional. Naturally every non-zero element of $S_2(11)$ is an eigenform (i.e. an eigenvector for the Hecke operators), and we take as our basis the unique normalized eigenform which has the following Fourier expansion:

$$f(z) = q - 2q^2 - q^3 + 2q^4 + q^5 + 2q^6 - 2q^7 - \cdots$$

According to the modularity theorem the eigenform f corresponds to the elliptic curve E and we may check a few of the coefficients to convince ourselves that this is the case. For example,

$$\tilde{E}(\mathbb{F}_2) = \{\mathcal{O}, \; (\bar{0}, \bar{0}), \; (\bar{0}, \bar{1}), \; (\bar{1}, \bar{0}), \; (\bar{1}, \bar{1})\}.$$

It follows that $a_2(E) = 2 + 1 - \#E(\mathbb{F}_2) = -2$ agrees with the coefficient $c_2 = -2$ for q^2 in the Fourier expansion for f. The reader can easily verify the relation $a_p(E) = c_p$ for the primes $p = 3, 5$ and 7.

3.4 Galois Representations

Let E be an elliptic curve over $\mathbb{C}$. The structure of the abelian group $E(\mathbb{C})$ is particularly easy to describe. There is a discrete lattice $\Lambda \subset \mathbb{C}$ of rank 2 (that is, as an abelian group $\Lambda \simeq \mathbb{Z}^2$) depending on E, and an isomorphism

$$E(\mathbb{C}) \simeq \mathbb{C}/\Lambda. \tag{20}$$

Let p be a prime. By the p-torsion of $E(\mathbb{C})$ we mean the subgroup

$$E[p] = \{Q \in E(\mathbb{C}) \,:\, pQ = 0\}.$$

It follows from (20) that

$$E[p] \simeq (\mathbb{Z}/p\mathbb{Z})^2. \tag{21}$$

This can be viewed as 2-dimensional $\mathbb{F}_p$-vector space.

Example 4. Let

$$E \,:\, y^2 = x^3 + x. \tag{22}$$

It turns out that the corresponding lattice is $\Lambda = \mathbb{Z} + \mathbb{Z}i$. The p-torsion subgroup of $\mathbb{C}/\Lambda$ is

$$\left\{ \frac{a+bi}{p} + \Lambda \,:\, a, b = 0, \ldots, p-1 \right\}.$$

The reader will see that this a 2-dimensional $\mathbb{F}_p$-vector space with basis $1/p + \Lambda$ and $i/p + \Lambda$.

Now let E be an elliptic curve over $\mathbb{Q}$. Then we may view E as an elliptic curve over $\mathbb{C}$, and with the above definitions obtain an isomorphism $E[p] \simeq (\mathbb{Z}/p\mathbb{Z})^2$. However, in this setting the points of $E[p]$ have algebraic coordinates, and are acted on by $G_{\mathbb{Q}} := \mathrm{Gal}(\overline{\mathbb{Q}}/\mathbb{Q})$, *the absolute Galois group of the rational numbers.* Via the isomorphism (21), the group $G_{\mathbb{Q}}$ acts on $(\mathbb{Z}/p\mathbb{Z})^2$. As noted, the latter is a 2-dimensional $\mathbb{F}_p$-vector space. We obtain a 2-dimensional representation that depends on the elliptic curve E and the prime p:

$$\overline{\rho}_{E,p} \,:\, G_{\mathbb{Q}} \to \mathrm{GL}_2(\mathbb{F}_p). \tag{23}$$

Example 5. We continue looking at the elliptic curve (22) but now regard it as an elliptic curve over $\mathbb{Q}$. The 2-torsion subgroup is

$$E[2] = \{\mathcal{O},\ (0,0),\ (i,0),\ (-i,0)\}.$$

Recall that $\mathcal{O}$ is the identity element. The three other elements of $E[2]$ are points of order 2. Moreover, they satisfy the additional relation

$$(0,0) + (i,0) + (-i,0) = \mathcal{O}.$$

We can see this from the geometric description of the group law: the three points on the left-hand side are the intersection of the line $y = 0$ with E. As $2 \cdot (-i,0) = \mathcal{O}$, we have that $(-i,0) = -(i,0)$ and thus

$$(-i, 0) = (0, 0) + (i, 0).$$

We now see that $E[2]$ is an $\mathbb{F}_2$-vector space with basis $(0,0)$ and $(i,0)$. Let us now use this to write down $\overline{\rho}_{E,2}$ explicitly. Let $\sigma \in G_{\mathbb{Q}}$. Then $\sigma(i) = i$ or $\sigma(i) = -i$. Suppose first that $\sigma(i) = i$. Then

$$\sigma(0,0) = (\sigma(0), \sigma(0)) = (0,0), \qquad \sigma(i,0) = (\sigma(i), \sigma(0)) = (i,0).$$

As σ leaves our chosen basis fixed, we have

$$\overline{\rho}_{E,2}(\sigma) = \begin{pmatrix} \overline{1} & \overline{0} \\ \overline{0} & \overline{1} \end{pmatrix} \in \mathrm{GL}_2(\mathbb{F}_2).$$

Suppose instead that $\sigma(i) = -i$. Then

$$\sigma(0,0) = (\sigma(0), \sigma(0)) = (0,0), \qquad \sigma(i,0) = (\sigma(i), \sigma(0)) = (-i,0) = (0,0)+(i,0).$$

Thus the action of σ with respect to our chosen basis is given by the matrix

$$\overline{\rho}_{E,2}(\sigma) = \begin{pmatrix} \overline{1} & \overline{1} \\ \overline{0} & \overline{1} \end{pmatrix} \in \mathrm{GL}_2(\mathbb{F}_2).$$

We record the image of the representation $\overline{\rho}_{E,2}$:

$$\overline{\rho}_{E,2}(G_{\mathbb{Q}}) = \left\{ \begin{pmatrix} \overline{1} & \overline{0} \\ \overline{0} & \overline{1} \end{pmatrix}, \begin{pmatrix} \overline{1} & \overline{1} \\ \overline{0} & \overline{1} \end{pmatrix} \right\}.$$

We note that the representation $\overline{\rho}_{E,2}$ is reducible, in the sense that all elements of the image share a common eigenvector $\begin{pmatrix} \overline{1} \\ \overline{0} \end{pmatrix}$.

We return to our general setting of an elliptic curve E over $\mathbb{Q}$ and a prime p. We say that the representation $\overline{\rho}_{E,p}$ is *reducible* if the matrices of the image $\overline{\rho}_{E,p}(G_{\mathbb{Q}})$ share some common eigenvector. Otherwise we say that $\overline{\rho}_{E,p}$ is irreducible.

We have now given enough definitions to be able to state Mazur's theorem; this is often considered as historically the first step in the proof of Fermat's Last Theorem.

Theorem 7 (Mazur [24]).

(i) Let E be an elliptic curve over $\mathbb{Q}$ and $p > 163$ be prime. Then $\overline{\rho}_{E,p}$ is irreducible.
(ii) Let E be an elliptic curve over $\mathbb{Q}$ with full 2-torsion (that is $E[2] \subseteq E(\mathbb{Q})$) and let $p \geq 5$ be prime. Then $\overline{\rho}_{E,p}$ is irreducible.

It turns out that an elliptic curve E over $\mathbb{Q}$ such that $\overline{\rho}_{E,p}$ is reducible corresponds to a rational point on the modular curve $X_0(p)$ that is not a cusp. Mazur proved his theorem by determining the rational points on this infinite family of curves.

In a sense, Mazur's theorem is not unlike Fermat's Last Theorem, which is also a statement about the rational points on an infinite family of curves.

We mention in passing the relationship between reducible mod p representations and isogenies. An *isogeny* of elliptic curves E, E' is a non-constant map $\phi : E \to E'$ defined by algebraic equations that take the point at infinity on E to the point at infinity on E'. A non-trivial consequence of the Riemann–Roch theorem is that isogenies respect the group law, and so are in a sense algebro-geometric homomorphisms. A p-isogeny is an isogeny $\phi : E \to E'$ such that the kernel of ϕ has order p. Let E be defined over $\mathbb{Q}$. Then the existence of an p-isogeny $\phi : E \to E'$ defined over $\mathbb{Q}$ is equivalent to the representation $\overline{\rho}_{E,p}$ being reducible. In fact, if Q is a non-zero element of the kernel of ϕ, then Q is a non-zero eigenvector for all the elements of the image of $\overline{\rho}_{E,p}$. We can restate (i) of Mazur's theorem as saying that an elliptic curve E defined over $\mathbb{Q}$ has no p-isogenies for $p > 163$.

3.5 Ribet's Level Lowering Theorem

Let E be an elliptic curve over $\mathbb{Q}$. We saw above that, for each prime p, the curve E gives rise to a mod p Galois representation $\overline{\rho}_{E,p} : G_{\mathbb{Q}} \to \mathrm{GL}_2(\mathbb{F}_p)$. Let f be an eigenform. Deligne and Serre showed that such an f gives rise, for each prime p, to a Galois representation $\overline{\rho}_{f,p} : G_{\mathbb{Q}} \to \mathrm{GL}_2(\mathbb{F}_{p^r})$, where $r \geq 1$ depends on f. If E corresponds to f via the Modularity theorem (Theorem 6) then, unsurprisingly, $\overline{\rho}_{E,p} \sim \overline{\rho}_{f,p}$ (the two representations are isomorphic). Thus the representation $\overline{\rho}_{E,p}$ is **modular** in the sense that it arises from a modular eigenform. Recall that if f corresponds to E via modularity, then the conductor of E is equal to the level of f. Sometimes it is possible to replace f by another eigenform of smaller level which has the same mod p representation. This process is called *level lowering*. We now state a special case of Ribet's level lowering theorem. Ribet's theorem is in fact part of Serre's modularity conjecture [29] that was proved by Khare and Wintenberger [20, 21].

Theorem 8 (Ribet's Level Lowering Theorem [28]). *Let E be an elliptic curve over $\mathbb{Q}$ with minimal discriminant Δ and conductor N. Let $p \geq 3$ be prime. Suppose*

(i) the curve E is modular;
(ii) the mod p representation $\overline{\rho}_{E,p}$ is irreducible.

Let

$$N_p = N \Big/ \prod_{\substack{\ell \| N, \\ p \mid \mathrm{ord}_\ell(\Delta)}} \ell. \tag{24}$$

Then $\overline{\rho}_{E,p} \sim \overline{\rho}_{g,p}$ for some eigenform g of weight 2 and level N_p.

Of course we now know, thanks to the Modularity theorem (Theorem 6) that all elliptic curves over $\mathbb{Q}$ are modular. Thus condition (i) in Ribet's theorem is automatically satisfied. But we still include it for historical interest.

We can make the relationship $\overline{\rho}_{E,p} \sim \overline{\rho}_{g,p}$ in Ribet's theorem more explicit. Write $g = q + \sum_{n\geq 1} d_n q^n$ for the Fourier expansion of g. It turns out that the d_n belong to the ring of integers $\mathfrak{O}_K$ of a number field K that depends on g. The relationship $\overline{\rho}_{E,p} \sim \overline{\rho}_{g,p}$ is equivalent to the existence of a prime ideal $\mathfrak{P}$ of $\mathfrak{O}_K$ that divides $p\mathfrak{O}_K$ such that $a_q(E) \equiv d_q \pmod{\mathfrak{P}}$ for all primes $q \nmid Np$.

Example 6. Consider the elliptic curve

$$E: \quad y^2 = x^3 - x^2 - 77x + 330$$

with Cremona reference `132B1`. Cremona's database [8] gives us the minimal discriminant and conductor

$$\Delta = 2^4 \times 3^{10} \times 11, \qquad N = 2^2 \times 3 \times 11. \tag{25}$$

The database also tells us that the only isogeny the curve E has is a 2-isogeny. Thus $\overline{\rho}_{E,p}$ is irreducible for $p \geq 3$. We apply Ribet's Theorem with $p = 5$. From the above recipe (24) for the level we find that $N_p = 44$. It is possible to check that $\overline{\rho}_{E,p} \sim \overline{\rho}_{g,p}$ where g is the following eigenform has weight 2 and level 44:

$$g = q + q^3 - 3q^5 + 2q^7 - 2q^9 - q^{11} + \cdots.$$

All the coefficients of g belong to $\mathbb{Z}$. We tabulate $a_q(E)$ and the coefficients d_q for primes $q < 50$. The reader will note that the relationship $a_q(E) \equiv d_q \pmod 5$ holds for all primes q in the range except for $q = 3$ which does divide N.

q	2	3	5	7	11	13	17	19	23	29	31	37	41	43	47
$a_q(E)$	0	−1	2	2	−1	6	−4	−2	−8	0	0	−6	0	10	0
d_q	0	1	−3	2	−1	−4	6	8	−3	0	5	−1	0	−10	0

3.6 *The Proof of Fermat's Last Theorem*

We are now able to sketch a proof of Fermat's Last Theorem. Suppose $p \geq 5$ is prime, and x, y and z are non-zero pairwise coprime integers such that $x^p + y^p + z^p = 0$. We may reorder (x, y, z) so that y is even and $x^p \equiv -1 \pmod 4$. We let E be the following elliptic curve which depends on the solution (x, y, z):

$$E : Y^2 = X \cdot (X - x^p) \cdot (X + y^p). \tag{26}$$

The curve E is called the *Frey–Hellegouarch curve*. The minimal discriminant and conductor of E are:

$$\Delta = \frac{x^{2p}y^{2p}z^{2p}}{2^8}, \qquad N = \prod_{\ell \mid \Delta} \ell.$$

The choice $2 \mid y$ and $x^p \equiv -1 \pmod 4$ ensures that $2 \parallel N$.

We now consider $\overline{\rho}_{E,p}$. The 2-torsion subgroup for E is

$$E[2] = \{\mathcal{O},\ (0,0),\ (x^p,0),\ (-y^p,0)\}.$$

Note that $E[2] \subseteq E(\mathbb{Q})$. As $p \geq 5$ we know by Mazur's theorem (Theorem 7) that $\overline{\rho}_{E,p}$ is irreducible. Moreover E is modular by the Modularity theorem. The hypotheses of Ribet's theorem are satisfied. We compute $N_p = 2$ using the recipe in (24). It follows that $\overline{\rho}_{E,p} \sim \overline{\rho}_{g,p}$ where g has weight 2 and level 2. A simple computation shows that there are no eigenforms of weight 2 and level 2. This contradiction completes the proof of Fermat's Last Theorem.

Some Historical Remarks In the early 1970s, Hellegouarch (e.g. [19]) had the idea of associating to a non-trivial solution of the Fermat equation the elliptic curve (26); he noted that the number field generated by its p-torsion subgroup $E[p]$ has surprisingly little ramification. In the early 1980s, Frey [18] observed that this elliptic curve enjoys certain remarkable properties that should rule out its modularity. Motivated by this, in 1985 Serre [29] made precise his modularity conjecture and showed that it implies Fermat's Last Theorem. Serre's remarkable paper also uses several variants of the Frey–Hellegouarch curve to link modularity to other Diophantine problems. Ribet announced his level-lowering theorem 1987, thereby proving that modularity of the Frey–Hellegouarch curve implies Fermat's Last Theorem. The proof of the Modularity theorem was completed around 1999 by Breuil, Conrad, Diamond and Taylor [6]. A *semistable* elliptic curve is one with squarefree conductor. We note from (25) that the Frey–Hellegouarch curve is semistable. In 1994 Wiles [37], with some help from Taylor [33], proved modularity of semistable elliptic curves over $\mathbb{Q}$, thereby proving Fermat's Last Theorem.

4 The (More) Generalized Fermat Equation

We now return to the *generalized Fermat equation*

$$x^p + y^q = z^r, \tag{27}$$

where x, y and z are integers, and the exponents p, q and r are (potentially distinct) positive integers. We restrict our attention to *primitive* solutions, i.e. those with $\gcd(x, y, z) = 1$, since, without such a restriction, it is easy to concoct uninteresting solutions in a fairly trivial fashion. Indeed, if we assume, say, that p, q and r are pairwise relatively prime, then we can choose integers u, v and w such that

$$uqr \equiv -1 \pmod{p}, \quad vpr \equiv -1 \pmod{q} \text{ and } wpq \equiv -1 \pmod{r}.$$

If we are given any integers a, b and c with $a + b = c$, multiplying this equation by $a^{uqr}b^{vpr}c^{wpq}$, we thus have that

$$\left(a^{(uqr+1)/p}b^{vr}c^{wq}\right)^p + \left(a^{ur}b^{(vpr+1)/q}c^{wp}\right)^q = \left(a^{uq}b^{vp}c^{(wpq+1)/r}\right)^r.$$

We call (p, q, r) the *signature* of Eq. (27). The behaviour of primitive solutions depends fundamentally upon the size of the quantity

$$\sigma(p, q, r) = \frac{1}{p} + \frac{1}{q} + \frac{1}{r},$$

in particular, whether $\sigma(p, q, r) > 1$, $\sigma(p, q, r) = 1$ or $\sigma(p, q, r) < 1$. If we set $\chi = \sigma(p, q, r) - 1$, then χ is the Euler characteristic of a certain stack associated to Eq. (27). It is for this reason that the cases $\sigma(p, q, r) > 1$, $\sigma(p, q, r) = 1$ and $\sigma(p, q, r) < 1$ are respectively termed *spherical*, *parabolic* and *hyperbolic*.

4.1 The Spherical Case $\sigma(p, q, r) > 1$

In this case, we may assume that (p, q, r) is one of $(2, 2, r)$, $(2, q, 2)$, $(2, 3, 3)$, $(2, 3, 4)$, $(2, 4, 3)$ or $(2, 3, 5)$. In each of these situations, the (infinitely many) relatively prime integer solutions to (27) come in finitely many two parameter families (the canonical model to bear in mind here is that of Pythagorean triples); in the (most complicated) $(2, 3, 5)$ case, there are precisely 27 such families, as proved by Johnny Edwards [11] in 2004 via an elegant application of classical invariant theory. In the case $(p, q, r) = (2, 4, 3)$, by way of example, we find that the relatively prime solutions x, y and z satisfy one of the following four parametrizations:

$$\begin{cases} x = 4ts(s^2 - 3t^2)(s^4 + 6t^2s^2 + 81t^4)(3s^4 + 2t^2s^2 + 3t^4), \\ y = \pm(s^2 + 3t^2)(s^4 - 18t^2s^2 + 9t^4), \\ z = (s^4 - 2t^2s^2 + 9t^4)(s^4 + 30t^2s^2 + 9t^4), \end{cases}$$

where $s \not\equiv t \pmod{2}$ and $3 \nmid s$,

$$\begin{cases} x = \pm(4s^4 + 3t^4)(16s^8 - 408t^4s^4 + 9t^8), \\ y = 6ts(4s^4 - 3t^4), \\ z = 16s^8 + 168t^4s^4 + 9t^8, \end{cases}$$

where t is odd and $3 \nmid s$,

$$\begin{cases} x = \pm(s^4 + 12t^4)(s^8 - 408t^4s^4 + 144t^8), \\ y = 6ts(s^4 - 12t^4), \\ z = s^8 + 168t^4s^4 + 144t^8, \end{cases}$$

where $s \equiv \pm 1 \pmod 6$, or

$$\begin{cases} x = 2(s^4 + 2ts^3 + 6t^2s^2 + 2t^3s + t^4)(23s^8 - 16ts^7 - 172t^2s^6 - 112t^3s^5 \\ \qquad -22t^4s^4 - 112t^5s^3 - 172t^6s^2 - 16t^7s + 23t^8), \\ y = 3(s-t)(s+t)(s^4 + 8ts^3 + 6t^2s^2 + 8t^3s + t^4), \\ z = 13s^8 + 16ts^7 + 28t^2s^6 + 112t^3s^5 + 238t^4s^4 \\ \qquad +112t^5s^3 + 28t^6s^2 + 16t^7s + 13t^8, \end{cases}$$

where $s \not\equiv t \pmod 2$ and $s \not\equiv t \pmod 3$. Here, s and t are relatively prime integers. Details on these parametrizations (and much more besides) can be found in Cohen's exhaustive work [7].

4.2 The Parabolic Case $\sigma(p,q,r) = 1$

If we have $\sigma(p,q,r) = 1$, then, up to reordering,

$$(p,q,r) = (2,6,3), (2,4,4), (4,4,2), (3,3,3) \text{ or } (2,3,6).$$

As in Examples 1 and 2, each equation now corresponds to an elliptic curve of rank 0 over $\mathbb{Q}$; the only primitive non-trivial solution comes from the signature $(p,q,r) = (2,3,6)$, corresponding to the Catalan solution $3^2 - 2^3 = 1$.

4.3 The Hyperbolic Case $\sigma(p,q,r) < 1$

It is the *hyperbolic case*, with $\sigma(p,q,r) < 1$, where most of our interest lies. Here, we are now once again considering the equation and hypotheses (4). As mentioned previously, it is expected that the only solutions to (4) are with (x,y,z,p,q,r) corresponding to the identity $1^p + 2^3 = 3^2$, for $p \geq 6$, or to

$$2^5 + 7^2 = 3^4, \quad 7^3 + 13^2 = 2^9, \quad 2^7 + 17^3 = 71^2, \quad 3^5 + 11^4 = 122^2,$$
$$17^7 + 76271^3 = 21063928^2, \; 1414^3 + 2213459^2 = 65^7, \; 9262^3 + 15312283^2 = 113^7,$$
$$43^8 + 96222^3 = 30042907^2 \quad \text{and} \quad 33^8 + 1549034^2 = 15613^3.$$

A less ambitious conjecture would be that (4) has at most finitely many solutions (where we agree to count those coming from $1^p + 2^3 = 3^2$ only once). In the rest of this section, we will discuss our current knowledge about this equation.

4.4 The Theorem of Darmon and Granville

What we know for sure in the hyperbolic case, is that, for a fixed signature (p, q, r), the number of solutions to Eq. (4) is at most finite:

Theorem 9 (Darmon and Granville [9]). *If A, B, C, p, q and r are fixed positive integers with*

$$\frac{1}{p} + \frac{1}{q} + \frac{1}{r} < 1,$$

then the equation

$$Ax^p + By^q = Cz^r$$

has at most finitely many solutions in coprime non-zero integers x, y and z.

Proof. The proof by Darmon and Granville is extremely elegant and we cannot resist giving a brief sketch. The hypothesis $1/p + 1/q + 1/r < 1$ is used to show the existence of a cover $\phi : D \to \mathbb{P}^1$ that is ramified only above 0, 1, ∞, where the curve D has genus ≥ 2. Moreover, this cover has the property that the ramification degrees above 0 are all divisors of p, above 1 are all divisors of q, and above ∞ are all divisors of r. Now let (x, y, z) be a non-trivial primitive solution to the equation $Ax^p + By^p = Cz^r$. The above properties of the cover ϕ imply that the points belonging to the fiber $\phi^{-1}(Ax^p/Cz^r)$ are defined over a number field K that is unramified away from the primes dividing $2ABCpqr$. It follows from a classical theorem of Hermite that there are only finitely many such number fields K. Moreover, by Faltings' theorem, for each possible K there are only finitely many K-points on D. It follows that the equation $Ax^p + By^p = Cz^r$ has only finitely many primitive solutions.

It is worth noting that the argument used in the proof is ineffective, due to its dependence upon Faltings' theorem; it is not currently known whether or not there exists an algorithm for finding all rational points on an arbitrary curve of genus ≥ 2.

4.5 A Brief Survey of What We Know

What we would really like to do goes rather further than what the theorem of Darmon and Granville tells us. Indeed, we would like to obtain finiteness results for Eq. (4) where we allow the exponent triples (p, q, r) to range over infinite families. In the following tables, we list all known (as of 2015) instances where Eq. (4) has been completely solved. For references to the original papers we recommend the exhaustive survey [4]. The first table collects all known infinite families treated to date:

(p, q, r)	Reference(s)
(n, n, n)	Wiles, Taylor-Wiles
$(n, n, k), k \in \{2, 3\}$	Darmon-Merel, Poonen
$(2n, 2n, 5)$	Bennett
$(2, 4, n)$	Ellenberg, Bennett-Ellenberg-Ng, Bruin
$(2, 6, n)$	Bennett-Chen, Bruin
$(2, n, 4)$	Bennett-Skinner, Bruin
$(2, n, 6)$	Bennett-Chen-Dahmen-Yazdani
$(3j, 3k, n),\ j, k \geq 2$	Immediate from Kraus
$(3, 3, 2n)$	Bennett-Chen-Dahmen-Yazdani
$(3, 6, n)$	Bennett-Chen-Dahmen-Yazdani
$(2, 2n, k),\ k \in \{9, 10, 15\}$	Bennett-Chen-Dahmen-Yazdani
$(4, 2n, 3)$	Bennett-Chen-Dahmen-Yazdani
$(2j, 2k, n),\ j, k \geq 5\text{prime}, n \in \{3, 5, 7, 11, 13\}$	Anni-Siksek

Our second table lists "sporadic" triples where the solutions to (4) have been determined, and infinite families of exponent triples where the (p, q, r) satisfy certain additional local conditions.

(p, q, r)	Reference(s)
$(3, 3, n)^*$	Chen-Siksek, Kraus, Bruin, Dahmen
$(2, 2n, 3)^*$	Chen, Dahmen, Siksek
$(2, 2n, 5)^*$	Chen
$(2m, 2n, 3)^*$	Bennett-Chen-Dahmen-Yazdani
$(2, 4n, 3)^*$	Bennett-Chen-Dahmen-Yazdani
$(3, 3n, 2)^*$	Bennett-Chen-Dahmen-Yazdani
$(2, 3, n),\ n \in \{6, 7, 8, 9, 10, 15\}$	Poonen-Schaefer-Stoll, Bruin, Zureick-Brown, Siksek, Siksek-Stoll
$(3, 4, 5)$	Siksek-Stoll
$(5, 5, 7),\ (7, 7, 5)$	Dahmen-Siksek

The asterisk here refers to conditional results. For instance, in case $(p, q, r) = (3, 3, n)$, we have no solutions if either $3 \leq n \leq 10^9$, or $n \equiv \pm 2$ modulo 5, or $n \equiv \pm 17$ modulo 78, or

$$n \equiv 51, 103, 105 \text{ (modulo 106)},$$

or for n (modulo 1296) one of

$$43, 49, 61, 79, 97, 151, 157, 169, 187, 205, 259, 265, 277, 295, 313, 367, 373, 385, 403, 421, 475, 481, 493, 511, 529, 583, 601, 619, 637, 691, 697, 709, 727, 745, 799, 805, 817, 835, 853, 907, 913, 925, 943, 961, 1015, 1021, 1033, 1051, 1069, 1123, 1129, 1141, 1159, 1177, 1231, 1237, 1249, 1267, 1285.$$

The results mentioned here have been proved by essentially two distinct methods. For a number of fixed triples, the problem has been reduced (via arguments similar to those of Darmon and Granville, or otherwise) to one of determining $\mathbb{Q}$-rational points on certain curves of genus 2 or higher. These points were subsequently found via Chabauty-type methods and appeal to a version of the Mordell-Weil sieve. In each case where Eq. (4) has been solved for an infinite family of triples (p, q, r), however, a different approach has been utilized, relying upon Frey–Hellegouarch curves and connections between them and modular forms.

5 The Modular Approach and the Generalized Fermat Equation

It is natural to ask if the proof of Fermat's Last Theorem can be adapted to resolve (4), at least for certain signatures (p, q, r). Roughly speaking a *Frey–Hellegouarch curve* is an elliptic curve E over $\mathbb{Q}$, attached to a solution of a Diophantine equation satisfying two conditions:

(i) the discriminant of E has the form $A \cdot B^p$ where A is a known (small) integer and p is a prime;
(ii) every prime $q \mid B$ divides the conductor exactly once.

Examining the recipe (24) in Ribet's theorem the reader will note that the level N_p depends only on the known quantity A. For example, in the proof of Fermat's Last Theorem, A is a power of 2 and the level $N_p = 2$.

Alas, only a few signatures have workable Frey–Hellegouarch curves. In the following table we record some of the known ones.

These and similar Frey–Hellegouarch curves have been used to prove many of the results surveyed in Sect. 4.5.

Equation	Frey–Hellegouarch Curve
$a^p + b^p = c^2$	$Y^2 = X^3 + 2cX^2 + a^pX$
$a^p + b^p = c^3$	$Y^2 = X^3 + 3cX^2 - 4b^p$
$a^3 + b^3 = c^p$	$Y^2 = X^3 + 3(a-b)X^2 + 3(a^2 - ab + b^2)X$
$a^2 + b^3 = c^p$	$Y^2 = X^3 + 3bX + 2a$

5.1 A Sample Signature: $(p, p, 2)$

To illustrate the approach we look specifically at the equation $x^p + y^p = z^2$ where $p \geq 7$ is prime. Here we follow the paper of Darmon and Merel [10] who showed that the only primitive solutions are $(\pm 1, \mp 1, 0)$, $(1, 0, \pm 1)$, $(0, 1, \pm 1)$. Let $(x, y, z) = (a, b, c)$ be a primitive solution satisfying $ab \neq 0$. As in the preceding table, we associate to this the Frey–Hellegouarch curve

$$E \ : \ Y^2 = X^3 + 2cX^2 + a^pX.$$

This is modular by Theorem 6. By a variant of Mazur's theorem (Theorem 7) the mod p representation $\overline{\rho}_{E,p}$ is irreducible. Now an application of the Ribet's theorem shows that $\overline{\rho}_{E,p} \sim \overline{\rho}_{g,p}$ where g is an eigenform of weight 2 and level 32. This is where we diverge from the proof of Fermat's Last Theorem: there is an eigenform of weight 2 and level 32. It turns however that this eigenform is rather special as it corresponds to an elliptic curve with complex multiplication. It follows from this fact that $\overline{\rho}_{g,p} \ : \ G_{\mathbb{Q}} \to \mathrm{GL}_2(\mathbb{F}_p)$ is not surjective. To complete the resolution of the equation $x^p + y^p = z^2$, Darmon and Merel needed to show that if $ab \neq -1$ then $\overline{\rho}_{E,p}$ is in fact surjective and hence cannot be isomorphic to $\overline{\rho}_{g,p}$. To do this, they showed that if $\overline{\rho}_{E,p}$ is not surjective then it gives rise to a rational point on one of a family of certain modular curves, and completed the proof by determining the rational points on this family. This last step is somewhat similar to the proof of Mazur's theorem.

6 Modularity over Number Fields

Even when we are interested in Diophantine equations in rational integer unknowns, factorization arguments often force us to consider Diophantine equations where the coefficients or unknowns lie in a number field. Consider for example the equation

$$a^4 + b^2 = c^p, \qquad \gcd(a, b, c) = 1 \tag{28}$$

where the exponent p is prime. This equation is not known to have a Frey–Hellegouarch curve defined over $\mathbb{Q}$. We can, however, factor the left-hand side as $(a^2 + bi)(a^2 - bi)$ where $i = \sqrt{-1}$. It is not hard to show using the arithmetic of the Gaussian ring $\mathbb{Z}[i]$ that

$$a^2 + bi = \alpha^p, \qquad a^2 - bi = \overline{\alpha}^p.$$

where $\alpha \in \mathbb{Z}[i]$ and $\overline{\alpha}$ is its conjugate. Eliminating b we obtain the equation

$$\alpha^p + \overline{\alpha}^p = 2a^2.$$

This is an $(p, p, 2)$ equation. However, unlike the equations we met in Sect. 5.1, some of the unknowns belong to $\mathbb{Z}[i]$. We ignore this uncomfortable fact for now and simply imitate the approach in the previous section to associate a Frey–Hellegouarch curve to this equation. The Frey–Hellegouarch curve is

$$E : y^2 = x^3 + 2ax^2 + 2\alpha^p x \tag{29}$$

which has discriminant $2^9(\alpha\overline{\alpha}^2)^p$. We note that the discriminant is close to being a perfect pth power. To solve our original generalized Fermat equation (28) with signature $(4, 2, p)$ and unknowns belonging to $\mathbb{Z}$, we need to consider an elliptic curve that is defined over $\mathbb{Q}(i)$. *It is natural to ask how much of modularity and level lowering carry over to the setting of number fields.* If we ask these questions for elliptic curves over general number fields then the answers are conjectural with almost no satisfactory theorems. However there are two situations where there are satisfactory theorems and these have been applied to certain generalized Fermat equations: $\mathbb{Q}$-curves and elliptic curves over totally real fields.

6.1 $\mathbb{Q}$-Curves

A $\mathbb{Q}$-*curve* is an elliptic curve E over a number field K that is isogenous to all its conjugates. The Frey curve (29) is an example of a $\mathbb{Q}$-curve: it is defined over the number field $\mathbb{Q}(i)$ and it happens to be isogenous to its conjugate $y^2 = x^3 + 2ax^2 + 2\overline{\alpha}^p x$ (the conjugate is obtained simply by conjugating the coefficients of the elliptic curve).

A consequence of the proof of Khare and Wintenberger of Serre's modularity conjecture is that $\overline{\mathbb{Q}}$-curves are modular. The modularity of the $\mathbb{Q}$-curve E given by (29) was used by Ellenberg [12] and by the first author, Ellenberg and Ng [5] to completely solve $a^4 + b^2 = c^p$ showing in fact that there are no non-trivial primitive solutions with $n \geq 4$ (here n does not have to be prime). The first author and Chen [3] have used modularity of Frey–Hellegouarch $\mathbb{Q}$-curves to show that the equation $a^2 + b^6 = c^n$ has no non-trivial solutions with $\gcd(a, b, c) = 1$ and $n \geq 3$.

6.2 Elliptic Curves over Totally Real Fields

A number field K of degree n has n embeddings into the complex numbers $\iota_j : K \hookrightarrow \mathbb{C}$ with $j = 1, \ldots, n$. For example if $K = \mathbb{Q}(\theta)$ is a number field of degree n, then θ is the root of an irreducible degree n polynomial with rational coefficients. Such a polynomial has n distinct complex roots $\theta_1, \ldots, \theta_n$, and the embedding ι_j satisfies $\iota_j(\theta) = \theta_j$. The embedding ι_j is real if $\iota_j(K) \subset \mathbb{R}$. Equivalently if $\theta_j \in \mathbb{R}$. If all the embeddings are real then we say that K is a totally real (number) field. An example of a totally real field is the cubic field $K = \mathbb{Q}(\theta)$ where $\theta = \zeta_7 + \zeta_7^{-1}$. Here θ is a root of the polynomial $x^3 + x^2 - 2x - 1$. The roots θ_j of the polynomial are $2\cos(2\pi j/7)$ with $j = 1, 2, 3$ which are all real.

Elliptic curves over totally real fields are conjecturally expected to be modular in the sense that they correspond to what are known as Hilbert modular forms (the classical modular forms of Sect. 3.2 are a special case of Hilbert modular forms). There has been substantial progress towards proving modularity for elliptic curves over totally real fields thanks to the work of Barnet-Lamb, Breuil, Diamond, Gee, Geraghty, Kisin, Skinner, Wiles and many others (many of these results in fact integrate level lowering with modularity). Building on this work, modularity of elliptic curves over real quadratic fields was recently proved by Freitas, Le Hung and Siksek [14]. To solve Diophantine problems via Frey curves that are defined over totally real fields one needs not only modularity and level lowering, but also irreducibility theorems for mod p representations of elliptic curves. Over the rationals, as we saw in Sect. 3.4, this is provided by Mazur's theorem. No such theorem is known over any number field other than $\mathbb{Q}$. However Frey curves are almost semistable and this fact can usually be used [17] together with the celebrated uniform boundedness theorem of Merel [25] to supply the required irreducibility result.

As an example we mention the equation

$$a^{2\ell} + b^{2m} = c^p, \qquad \gcd(a, b, c) = 1. \tag{30}$$

studied by Anni and Siksek [1]. Here $\ell, m \geq 5$ and $p \geq 3$ are primes. A complicated factorization argument is used to reduce this to a Fermat equation with signature (ℓ, ℓ, ℓ) with coefficients and unknowns belonging to the totally real field $\mathbb{Q}(\zeta_p + \zeta_p^{-1})$. The corresponding Frey curves over this field are shown to be modular using the above-mentioned works for $p = 3, 5, 7, 11$ and 13. This is then used to show that the only solutions to (30) are the trivial ones.

7 A Way Forward: Darmon's Program

The Frey–Hellegouarch approach used in the proof of Fermat's Last Theorem and in the resolution of many other equations (as sketched in previous sections) attaches an elliptic curve to a hypothetical solution of the equation in question and then uses modularity to make deductions about this solution. It is natural to ask:

(i) Are there geometric objects other than elliptic curves that are somehow modular?
(ii) If so, can these be used as an alternative, perhaps to add flexibility and tackle generalized Fermat equations for which no Frey–Hellegouarch elliptic curve is known?

An **abelian variety** is a connected and projective algebraic group. Roughly speaking this means that it is defined by algebraic equations in projective space and carries a group structure (that happens to be abelian). An abelian variety has a dimension $d \geq 1$, and abelian varieties of dimension 1 are simply elliptic curves. Are abelian varieties over $\mathbb{Q}$ modular? The answer should be 'yes', except that the precise meaning of word modular in this generality is not yet resolved.

An abelian variety of dimension d is said to be of GL_2-type if its endomorphism ring is an order in a number field of degree d. A Theorem of Khare and Wintenberger [20] states that abelian varieties of GL_2-type over $\mathbb{Q}$ are modular in a very precise sense (that is in fact very close to that of elliptic curves in Sect. 3.3). Abelian varieties of GL_2-type over totally real fields are expected to be modular in the sense that they correspond to Hilbert modular forms. Darmon exploits this idea to attack the equation $x^p + y^p = z^r$, where p and r are primes and $\gcd(x, y, z) = 1$ as usual. He attaches a hypothetical non-trivial solution to an abelian variety of GL_2-type over the totally real field $\mathbb{Q}(\zeta_r + \zeta_r^{-1})$. Using this he proves several beautiful theorems about possible solutions, though all are dependent on yet unproven conjectures. The biggest obstruction to Darmon's program is the absence of a Mazur-style irreducibility theorem for mod p representations of abelian varieties of GL_2-type.

The Darmon program holds the greatest promise for further substantial progress on the generalized Fermat equation. Just as Frey's original work was the spark that led to the formulation of Serre's modularity conjecture, and the proofs of Ribet's theorem and the Modularity theorem, so we hope that Darmon's program will supply the impetus for new theorems for abelian varieties of GL_2-type that in turn allow us to make deductions about the generalized Fermat equation.

References

1. S. Anni and S. Siksek, *On the generalized Fermat equation $x^{2\ell} + y^{2m} = z^p$ for $3 \leq p \leq 13$*, preprint, arXiv 1506.02860.
2. K. Belabas, F. Beukers, P. Gaudry, H. Lenstra, W. McCallum, B. Poonen, S. Siksek, M. Stoll, M. Watkins, *Explicit Methods in Number Theory: Rational Points and Diophantine Equations*, Panoramas et synthèses **36**, Société Mathématique de France, Paris, 2012.
3. M. A. Bennett and I. Chen, *Multi-Frey $\mathbb{Q}$-curves and the Diophantine equation $a^2 + b^6 = c^n$*, Algebra Number Theory **6** (2012), 707–730.
4. M. A. Bennett, I. Chen, S. R. Dahmen and S. Yazdani, *Generalized Fermat equations: a miscellany*, Int. J. Number Theory **11** (2015), 1–28.
5. M. A. Bennett, J. S. Ellenberg and N. Ng, *The Diophantine equation $A^4 + 2^\delta B^2 = C^n$*, Int. J. Number Theory **6** (2010), 311–338.

6. C. Breuil, B. Conrad, F. Diamond and R. Taylor, *On the modularity of elliptic curves over* $\mathbb{Q}$*: wild 3-adic exercises*, J. Amer. Math. Soc. **14** (2001), 843–939.
7. H. Cohen, *Number theory. Vol. II. Analytic and modern tools*, Graduate Texts in Mathematics, 240, Springer, New York, 2007.
8. J. Cremona, Elliptic Curve Data, September 2015.
9. H. Darmon and A. Granville, *On the equations* $z^m = F(x, y)$ *and* $Ax^p + By^q = Cz^r$, Bull. London Math. Soc. textbf27 (1995), 513–543.
10. H. Darmon and L. Merel, *Winding quotients and some variants of Fermat's last theorem*, J. Reine Angew. Math. **490** (1997), 81–100.
11. J. Edwards, *A complete solution to* $X^2 + Y^3 + Z^5 = 0$, J. Reine Angew. Math.**571** (2004), 213–236.
12. J. S. Ellenberg, *Galois representations attached to* $\mathbb{Q}$*-curves and the generalized Fermat equation* $A^4 + B^2 = C^p$, Amer. J. Math. **126** (2004), 763–787.
13. G. Faltings, *Endlichkeitssätze für abelsche Varietäten über Zahlkörpen*, Invent. Math. **73** (1983), 349–366.
14. N. Freitas, B. Le Hung and S. Siksek, *Elliptic curves over real quadratic fields are modular*, Invent. Math. **201** (2015), 159–206.
15. N. Freitas and S. Siksek, *The asymptotic Fermat's Last Theorem for five-sixths of real quadratic fields*, Compos. Math. **151** (2015), 1395–1415.
16. N. Freitas and S. Siksek, *Fermat's last theorem over some small real quadratic fields*, Algebra Number Theory **9** (2015), 875–895.
17. N. Freitas and S. Siksek, *Criteria for irreducibility of* mod *p representations of Frey curves*, J. Théor. Nombres Bordeaux **27**, 2015, 67–76.
18. G. Frey, *Links between stable elliptic curves and certain Diophantine equations*, Ann. Univ. Sarav. Ser. Math. **1** (1986), iv+40.
19. Y. Hellegouarch, *Points d'ordre* $2p^h$ *sur les courbes elliptiques*, Acta Arith. **26** (1974/75), 253–263.
20. C. Khare and J.-P. Wintenberger, *Serre's modularity conjecture. I*, Invent. Math. **178** (2009), 485–504.
21. C. Khare and J.-P. Wintenberger, *Serre's modularity conjecture. II*, Invent. Math. **178** (2009), 505–586.
22. A. Kraus, *Majorations effectives pour l'équation de Fermat généralisée*, Canad. J. Math. **49** (1997), 1139–1161.
23. S. Lang, *Cyclotomic fields I and II*, Graduate Texts in Mathematics, 121, Springer, New York, 1990.
24. B. Mazur, *Rational isogenies of prime degree*, Invent. Math. **44** (1978), 129–162.
25. L. Merel, *Bornes pour la torsion des courbes elliptiques sur les corps de nombres*, Invent. Math. **124** (1996), 437–449.
26. M. Th. Rassias, *Problem-solving and selected topics in number theory. In the spirit of the mathematical olympiads. With a foreword by Preda Mihilescu.*, Springer, NewYork, 2011.
27. P. Ribenboim, 13 *Lectures on Fermat's Last Theorem*, Springer, 1979.
28. K. Ribet, *On modular representations of* $\mathrm{Gal}(\overline{\mathbb{Q}}/\mathbb{Q})$ *arising from modular forms*, Invent. Math. **100** (1990), 431–476.
29. J.-P. Serre, *Sur les représentations modulaires de degré* 2 *de* $\mathrm{Gal}(\overline{\mathbf{Q}}/\mathbf{Q})$, Duke Math. J. **54** (1987), 179–230.
30. S. Siksek, *The modular approach to Diophantine equations*, pages 151–179 of [2].
31. J. H. Silverman, *Advanced topics in the arithmetic of elliptic curves*, Graduate Texts in Mathematics, 151, Springer-Verlag, New York, 1994.
32. S. Sitaraman, *Vandiver revisited*, J. Number Theory **57** (1996), 122–129.
33. R. Taylor and A. Wiles, *Ring-theoretic properties of certain Hecke algebras*, Ann. of Math. **141** (1995), 553–572.
34. G. Terjanian, *Sur l'équation* $x^{2p} + y^{2p} = z^{2p}$, Comptes Rendus Académie de Sciences Paris, **285** (1977), 973–975.

35. F. Thaine, *On the ideal class groups of real abelian number fields.*, Ann. of Math. **128** (1988), 1–18.
36. L. Washington, *Introduction to Cyclotomic Fields*, Graduate Texts in Mathematics, Springer, New York, 1996.
37. A. Wiles, *Modular elliptic curves and Fermat's Last Theorem*, Ann. Math. **141** (1995), 443–551.

The Conjecture of Birch and Swinnerton-Dyer

John Coates

Abstract The conjecture of Birch and Swinnerton-Dyer is one of the principal open problems of number theory today. Since it involves exact formulae rather than asymptotic questions, it has been tested numerically more extensively than any other conjecture in the history of number theory, and the numerical results obtained have always been in perfect accord with every aspect of the conjecture. The present article is aimed at the non-expert, and gives a brief account of the history of the conjecture, its precise formulation, and the partial results obtained so far in support of it.

1 History

The written history of the arithmetic of elliptic curves can be traced back at least to Arab manuscripts of over 1000 years ago, which were concerned with the problem of finding which positive integers are the areas of right-angled triangles, all of whose sides have rational length (traditionally, such positive integers are called *congruent numbers*). For example, 5 is a congruent number because it is the area of a right-angled triangle, whose sides have lengths 9/6, 40/6, 41/6. In fact no smaller congruent number was discovered by the ancients. It is easily seen that a positive integer N is a congruent number if and only if there is a point (x, y), with x and y rational numbers, and $y \neq 0$, on the curve

$$y^2 = x^3 - N^2x. \tag{1.1}$$

In the seventeenth century, Fermat gave the first proof that 1 is not a congruent number, by introducing his method of *infinite descent*, and carrying it out on the curve (1.1) with $N = 1$. Fermat also noted that an intermediate step in his proof showed that, when $n = 4$, the curve $x^n + y^n = z^n$ has no solution in integers x, y, z which are all non-zero, and presumably this is what led him to claim that the same assertion holds for all $n \geq 3$. More generally, by an elliptic curve over a field F, we

J. Coates (✉)
Emmanuel College, Cambridge CB2 3AP, UK
e-mail: jhc13@dpmms.cam.ac.uk

J.F. Nash, Jr., M.Th. Rassias (eds.), *Open Problems in Mathematics*,
DOI 10.1007/978-3-319-32162-2_4

mean an irreducible non-singular projective algebraic curve of genus 1 defined over F, which is endowed with a given F-rational point $\mathcal{O}$. Any such curve has a plane cubic model of the form

$$y^2 + a_1xy + a_3y = x^3 + a_2x^2 + a_4x + a_6 \quad (a_i \in F), \tag{1.2}$$

where $\mathcal{O}$ is now taken as the unique point at infinity (see, for example, [47]). Such an elliptic curve E is an abelian variety of dimension 1, meaning that the set of all points on such a curve with coordinates in some fixed extension field of F has a natural algebraic abelian group structure, with $\mathcal{O}$ as the zero element. In 1922, Mordell beautifully generalised Fermat's infinite descent argument and proved that the group of rational points on every elliptic curve defined over $\mathbb{Q}$ is always finitely generated as an abelian group. However, the big mystery left open by Mordell's proof was whether or not the procedure of infinite descent always terminated in a finite number of steps, thus enabling one to actually determine the group of rational points on the curve. In practice, this always seems to be the case, but, in fact, it has never been proven theoretically. The villain of the piece is a mysterious group, subsequently called the Tate-Shafarevich group of the elliptic curve, which is defined by

$$Ш(E/\mathbb{Q}) = Ker(H^1(\mathbb{Q}, E) \to \prod H^1(\mathbb{Q}_v, E)),$$

where v runs over all places of $\mathbb{Q}$, and $\mathbb{Q}_v$ is the completion of $\mathbb{Q}$ at v. This torsion abelian group is always conjectured to be finite, but today we can still only prove this under a very restrictive hypothesis discussed below.

The discoveries of Birch and Swinnerton-Dyer came as a great surprise to the mathematical world when they first became public around 1962. Starting in the autumn of 1958, they had carried out a series of brilliantly planned numerical experiments on the early EDSAC computers in Cambridge, whose aim was to uncover numerical evidence for the existence of some kind of analogue for elliptic curves of the mysterious exact analytic formulae proven by Dirichlet for the class numbers of binary quadratic forms, and powerfully extended to all quadratic forms by Siegel. Even though Siegel's work had been actively developed further for linear algebraic groups around this time by Kneser, Tamagawa, Weil, and others, it was Birch and Swinnerton-Dyer alone who first sought, and later found evidence for, an analogue for elliptic curves. It surely is one of the great mysteries of number theory, first uncovered by Birch and Swinnerton-Dyer, that purely arithmetic questions about the determination of $E(\mathbb{Q})$ and $Ш(E/\mathbb{Q})$ for an elliptic curve E over $\mathbb{Q}$ seem to be inextricably involved with the behaviour of the complex L-function of E.

In this survey article, we shall mainly discuss the conjecture of Birch and Swinnerton-Dyer in the most important and down to earth case of elliptic curves defined over $\mathbb{Q}$. However, the conjecture extends without difficulty to abelian varieties of arbitrary dimension defined over either a finite extension of $\mathbb{Q}$, or over a function field in one variable over a finite field (see [51]). To date, very little has been proven about the conjecture for general abelian varieties of dimension > 1

over number fields. However, for abelian varieties defined over a function field in one variable over a finite field, the remarkable work of Artin and Tate [51] makes great progress on the conjecture, apart from the mysterious question of the finiteness of the analogue of the Tate-Shafarevich group.

2 *L*-Functions

Let E be any elliptic curve defined over $\mathbb{Q}$. By a global minimal Weierstrass equation for E, we mean any equation for E of the form (1.2), whose coefficients a_i are all integers, and whose discriminant Δ is as small as possible in absolute value (for the definition of Δ, and other facts about the elementary geometry of elliptic curves see [47]). Such equations are not unique, but we fix any one of them for the discussion which follows. Like all the L-series of arithmetic geometry, the complex L-series of E is defined by an Euler product. For each prime number p, define N_p by letting $N_p - 1$ denote the number of solutions of the congruence

$$y^2 + a_1xy + a_3y \equiv x^3 + a_2x^2 + a_4x + a_6 \bmod p,$$

and then put

$$t_p = p + 1 - N_p.$$

If $(p, \Delta) = 1$, we have $|t_p| \leq 2\sqrt{p}$ by Hasse's theorem. If p divides Δ, then $t_p = 1$ if E has multiplicative reduction at p with tangents at the node defined over $\mathbb{F}_p$, $t_p = -1$ if E has multiplicative reduction at p with tangents at the node not defined over $\mathbb{F}_p$, and $t_p = 0$ when E had additive reduction at p. The complex L-series of E is then defined by the Euler product

$$L(E,s) = \prod_{p|\Delta} \left(1 - t_p p^{-s}\right)^{-1} \prod_{(p,\Delta)=1} \left(1 - t_p p^{-s} + p^{1-2s}\right)^{-1}. \tag{2.1}$$

This Euler product defines a Dirichlet series

$$L(E,s) = \sum_{n=1}^{\infty} c_n n^{-s},$$

where $c_p = t_p$ for every prime p, and which converges in the half plane $Re(s) > \frac{3}{2}$. When Birch and Swinnerton-Dyer first began their calculations, it was only known how to analytically continue this function to the entire complex plane when E has complex multiplication (i.e. the ring of endomorphisms of E, which are defined over $\mathbb{C}$, is strictly bigger than $\mathbb{Z}$), using ideas about Eisenstein series which go back to Eisenstein and Kronecker, and which were subsequently developed systematically by Deuring [17]. To prove the analytic continuation for all E, we need the following fundamental result, the essential idea behind the proof of which we owe to Wiles [57] (see also [5]). The conductor $C(E)$ of E is the positive integer defined by

$$C(E) = \prod_{p|\Delta} p^{f_p},$$

where $f_p = 1$ if E has multiplicative reduction at p, and $f_p = 2 + \delta_p$ for some integer $\delta_p \geq 0$ if E has additive reduction at p. Moreover, in this latter case, $\delta_p = 0$ when $p \geq 5$. Let $\Gamma_0(C(E))$ be the subgroup of $SL_2(\mathbb{Z})$ consisting of all matrices $\begin{pmatrix} a & b \\ c & d \end{pmatrix}$ with $c \equiv 0 \bmod C(E)$. Let $\mathcal{H}$ be the complex upper half plane, and put $q = e^{2\pi i\tau}$ with $\tau \in \mathcal{H}$. Define

$$f_E(\tau) = \sum_{n=1}^{\infty} c_n q^n.$$

Theorem 2.1. *The holomorphic function $f_E(\tau)$ is a primitive cusp form of weight 2 for $\Gamma_0(C(E))$.*

By a generalization of classical ideas of Hecke, this theorem not only proves that $L(E, s)$ can be extended to an entire holomorphic function of s, but it also establishes the following functional equation. Define

$$\Lambda(E, s) = C(E)^{s/2}(2\pi)^{-s}\Gamma(s)L(E, s).$$

Corollary 2.2. *The function $\Lambda(E, s)$ can be extended to an entire function of s, and satisfies the functional equation*

$$\Lambda(E, s) = w_E\Lambda(E, 2 - s), \tag{2.2}$$

where $w_E = \pm 1$.

The so called root number $w_E = \pm 1$ is important for us because we see immediately from (2.2) that $L(E, s)$ has a zero at $s = 1$ of even or odd multiplicity, according as $w_E = +1$ or $w_E = -1$. Moreover, the theory of L-functions shows that w_E can always be calculated as a product of purely local factors. For example, if E is taken to be the curve (1.1) with N a square free positive integer, then $w_E = +1$ when $N \equiv 1, 2, 3 \bmod 8$, and $w_E = -1$ when $N \equiv 5, 6, 7 \bmod 8$, whence, in particular, $L(E, s)$ always has a zero at $s = 1$ whenever $N \equiv 5, 6, 7 \bmod 8$.

We mention that one can, more generally, consider elliptic curves E which are defined over some finite extension F of $\mathbb{Q}$. Again the group of F-rational points on E is a finitely generated abelian group, and again no algorithm has ever been proven for infallibly determining this group, again thanks to our lack of knowledge of the finiteness of the Tate-Shafarevich group of such a curve. Of course, these elliptic curves also have a complex L-series, which we now denote by $L(E/F, s)$, which is defined in the region $Re(s) > 3/2$ by an entirely analogous Euler product to (2.1), but taken over all finite places of the field F. When E admits complex multiplication, the analytic continuation and functional equation of $L(E/F, s)$ follow immediately from Deuring's theorem [17], which identifies $L(E/F, s)$ with a product of Hecke L-series with Grossencharacters. However, when E does not have complex multiplication, our knowledge of the analytic continuation of $L(E/F, s)$ is still very limited, with the most striking results established so far being proofs of this assertion either when F is a real quadratic field [19], or when F is any finite extension of $\mathbb{Q}$ which is contained in the cyclotomic $\mathbb{Z}_p$-extension of $\mathbb{Q}$ for any prime number p [53].

3 The Birch-Swinnerton-Dyer Conjecture

We will now state the conjecture of Birch and Swinnerton-Dyer in both its weak and strong form, and discuss the evidence for it in subsequent sections. The conjecture, which was first published in [3], predicts a remarkable link between the arithmetic of an elliptic curve E defined over $\mathbb{Q}$, and the behaviour of its complex L-series $L(E, s)$ at the point $s = 1$. Let g_E denote the rank of $E(\mathbb{Q})$ (i.e. the $\mathbb{Q}$-dimension of $E(\mathbb{Q}) \otimes_{\mathbb{Z}} \mathbb{Q}$). We define

Definition 3.1. r_E is the order of the zero of $L(E, s)$ at $s = 1$.

Weak Birch-Swinnerton-Dyer Conjecture. *For all elliptic curves E over $\mathbb{Q}$, we have*

$$r_E = g_E. \tag{3.1}$$

The full Birch-Swinnerton-Dyer conjecture is the weak Birch-Swinnerton-Dyer conjecture, together with a purely arithmetic exact formula for the constant $\mathfrak{L}_E$ defined by

$$\mathfrak{L}_E = \lim_{s \to 1} L(E, s)/(s - 1)^{r_E}. \tag{3.2}$$

This formula involves the following arithmetic invariants. Firstly, there is a regulator term coming from the Neron-Tate height. If $\alpha = m/n$, with m and n relatively prime integers, is any non-zero rational number, we define its height $h(\alpha)$ by $h(\alpha) = log(max(|m|, |n|))$, and put $h(0) = 0$. Then Neron and Tate proved independently (see [47], Chap. 8) that there is a unique function

$$\hat{h} : E(\mathbb{Q}) \to \mathbb{R}$$

such that $\hat{h}(\mathcal{O}) = 0$, and, as P runs over the non-zero points in $E(\mathbb{Q})$, we have $\hat{h}(2P) = 4\hat{h}(P)$ and $|\hat{h}(P) - h(x(P))|$ is bounded independent of P, where $x(P)$ denotes the x-coordinate of P in any fixed generalised Weierstrass equation (1.2). Then the function on $E(\mathbb{Q}) \times E(\mathbb{Q})$ defined by

Definition 3.2. $\langle P, Q\rangle = \frac{1}{2}\left(\widehat{h}(P \oplus Q) - \widehat{h}(P) - \widehat{h}(Q)\right)$

is biadditive. Moreover, we have $\hat{h}(P) = 0$ if and only if P is a torsion point in $E(\mathbb{Q})$. If one uses, in addition, the fact that there are only finitely many points P in $E(\mathbb{Q})$ with $\hat{h}(P) \leq c$ for any constant $c > 0$, it follows, as was first remarked by Cassels, that $\hat{h}$ induces a positive definite quadratic form on $E(\mathbb{Q}) \otimes_{\mathbb{Z}} \mathbb{R}$. Hence, for any choice of g_E independent points $P_1, \ldots, P_{g_E}$ in $E(\mathbb{Q})$, we always have that the determinant $det\langle P_i, P_j\rangle$ is strictly positive. We then define $R_\infty(E) = 1/\#(E(\mathbb{Q})^2)$ if $g_E = 0$ and

Definition 3.3. $R_\infty(E) = det\langle P_i, P_j\rangle / [E(\mathbb{Q}) : \sum_{i=1}^{g_E} \mathbb{Z}P_i]^2$ if $g_E > 0$.

We assume that we have fixed any global minimal Weierstrass equation for E, and we define Ω_E to be the least positive real period of the Neron differential on E, which is given by

$$\omega = \frac{dx}{2y + a_1 x + a_3}.$$

The next subtle ingredient in the conjectural exact formula for $\mathfrak{L}_E$ are the so called *Tamagawa factors*, which are purely local terms occurring for the prime at infinity, and the finite primes q dividing the conductor $C(E)$ of E.

Definition 3.4. $c_\infty(E)$ is equal to 1 or 2, according as the group of points $E(\mathbb{R})$ on E with real coordinates has 1 or 2 connected components.

Next assume that q is any prime number dividing the conductor $C(E)$. Let $\mathbb{Q}_q$ be the completion of $\mathbb{Q}$ at q. Now, since q is a prime of bad reduction for E, the reduction of E modulo q will be a cubic curve with a singular point, and we define $E_0(\mathbb{Q}_q)$ to be the subgroup of $E(\mathbb{Q}_q)$ consisting of all points whose reduction modulo q is non-singular. Since we are working with a generalised Weierstrass equation which is minimal at q, the index of $E_0(\mathbb{Q}_q)$ in $E(\mathbb{Q}_q)$ will be independent of the choice of the Weierstrass equation.

Definition 3.5. For a prime q of bad reduction, $c_q(E) = [E(\mathbb{Q}_q) : E_0(\mathbb{Q}_q)]$.

In general, there is no simple formula for $c_q(E)$, but Tate [50] gave an explicit algorithm for computing $c_q(E)$ from any generalised Weierstrass equation for E which is minimal at q, and also proved:-

Lemma 3.6. *If E has split multiplicative reduction at q, then $c_q(E) = ord_q(\Delta)$. For all other primes q of bad reduction, $c_q(E) \leq 4$.*

We can now at last state the full Birch-Swinnerton-Dyer conjecture.

Full Birch-Swinnerton-Dyer Conjecture. *We have $r_E = g_E$. Moreover, $\text{Ш}(E/\mathbb{Q})$ is finite, and the following exact formula is valid*

$$\frac{\mathfrak{L}_E}{\Omega_E} = \#(\text{Ш}(E/\mathbb{Q}))R_\infty(E)c_\infty(E) \prod_{q|C(E)} c_q(E). \tag{3.3}$$

Note that elliptic curves which are $\mathbb{Q}$-isogenous have the same L-functions. However, it is not at all obvious that the exact formula (3.3) being valid for an elliptic curve implies that it is valid for any isogenous curve, but this was proven by Cassels [8] and Tate [51].

We shall spend most of the rest of this article discussing the fragmentary theoretical results proven so far, in the direction of both the weak and full Birch-Swinnerton-Dyer conjecture. However, to illustrate immediately the limits of our present knowledge, let us note three simple consequences of the conjecture which have never been established for a single elliptic curve E over $\mathbb{Q}$. Firstly, it has never been proven that there exists an elliptic curve E defined over $\mathbb{Q}$ with $r_E \geq 4$, even though there are many examples of such E with $g_E \geq 4$. Secondly, it has never been proven that $\text{Ш}(E/\mathbb{Q})$ is finite for a single elliptic curve E with $r_E \geq 2$. Thirdly, it has never been proven that $\mathfrak{L}_E/(\Omega_E R_\infty(E))$ is a rational number for a single E with $r_E \geq 2$.

The earliest numerical work in support of these conjectures is given in the papers [3, 49]. Today, the numerical evidence in support of both the weak and full Birch-Swinnerton-Dyer conjecture is overwhelming, and probably more extensive than for any other conjecture in the history of mathematics. Access to the vast amount of numerical data, which, to date, confirms experimentally every aspect of the conjecture, can be made at the website www.lmfdb.org/EllipticCurve/Q (see also the earlier book [15], which is available online on John Cremona's homepage at Warwick University). This website includes tables of all elliptic curves E over $\mathbb{Q}$ with conductor $C(E) < 360{,}000$. There are 2,247,187 elliptic curves in this table, lying in 1,569,126 $\mathbb{Q}$-isogeny classes. All such curves have $g_E \leq 4$, and in fact there is only one curve in the table with $g_E = 4$ (this curve has conductor 234,446). In addition, Miller, Stoll, and Creutz [14, 35, 36] have verified the full Birch-Swinnerton-Dyer conjecture for all E defined over $\mathbb{Q}$ with $C(E) < 5000$, which have $r_E \leq 1$. The analytic quantity $\mathfrak{L}_E$ can be computed numerically to great accuracy irrespective of the value of r_E, and the same is usually true for all quantities occurring in the exact formula (3.3), except for the order of the Tate-Shafarevich group of E. Even when $\text{Ш}(E/\mathbb{Q})$ is known to be finite, it is very difficult to actually compute its true order arithmetically. However, even granted this difficulty, there is one subtle sub-test of the order of $\text{Ш}(E/\mathbb{Q})$ as predicted by (3.3) being correct. As we shall explain in the next section, an important theorem of Cassels [9] proves that if $\text{Ш}(E/\mathbb{Q})$ is finite, then its order must be the square of an integer. Happily, in all of the vast number of numerical examples computed to date, the formula (3.3) has always produced a conjectural order for the Tate-Shafarevich group which is indeed the square of an integer.

4 Parity Questions

The only deep general results known about the conjecture of Birch and Swinnerton-Dyer, which do not involve in some fashion the hypothesis that the L-series of the curve has a zero at $s = 1$ of order at most 1, are parity theorems of two kinds. As always, E will be an elliptic curve defined over $\mathbb{Q}$, and $\text{Ш}(E/\mathbb{Q})$ will denote its Tate-Shafarevich group. If A is any abelian group and p a prime number, we write $A(p)$ for the p-primary subgroup of A, and $A[p]$ for the kernel of multiplication by p on A. Also, A_{div} will denote the maximal divisible subgroup of A. The following theorem is due to Cassels [9] and Tate [52].

Theorem 4.1. *There is a canonical non-degenerate, alternating, bilinear form on* $\text{Ш}(E/\mathbb{Q})/\text{Ш}(E/\mathbb{Q})_{div}$.

Corollary 4.2. *For every prime p, the $\mathbb{F}_p$-vector space given by the kernel of multiplication by p on $\text{Ш}(E/\mathbb{Q})/\text{Ш}(E/\mathbb{Q})_{div}$ has even dimension.*

Corollary 4.3. *If p is a prime such that $\text{Ш}(E/\mathbb{Q})(p)_{div} = 0$, then the order of $\text{Ш}(E/\mathbb{Q})(p)$ is a square.*

In particular, if $\text{Ш}(E/\mathbb{Q})$ is finite, its order must be a perfect square. We note that, for every prime p, classical Galois cohomology shows that, for some integer $t_{E,p} \geq 0$, one has

$$\text{Ш}(E/\mathbb{Q})(p) = (\mathbb{Q}_p/\mathbb{Z}_p)^{t_{E,p}} \oplus J_{E,p},$$

where $J_{E,p}$ is a finite group. Plainly $\text{Ш}(E/\mathbb{Q})(p)_{div} = (\mathbb{Q}_p/\mathbb{Z}_p)^{t_{E,p}}$. Of course, conjecturally $t_{E,p} = 0$ for every prime p, but the only far weaker general known result in this direction is the following parity theorem of the Dokchitser brothers [16]. Recall that g_E denotes the rank of $E(\mathbb{Q})$, and r_E denotes the order of zero of $L(E, s)$ at the point $s = 1$.

Theorem 4.4. *For every prime number p, we have $r_E \equiv g_E + t_{E,p} \bmod 2$. In particular, the parity of $t_{E,p}$ is independent of p.*

As a simple application of this theorem, we see that if there did exist a square free positive integer N with $N \equiv 5, 6, 7 \bmod 8$, which is not a congruent number, then the p-primary subgroup of the Tate-Shafarevich of the elliptic curve (1.1) would have to contain a copy of the divisible group $\mathbb{Q}_p/\mathbb{Z}_p$ for every prime p, and so a copy of $\mathbb{Q}/\mathbb{Z}$. We also note that the strong parity conjecture is the assertion that $r_E \equiv g_E \bmod 2$, but this has only been proven at present under the assumption that $r_E \leq 1$, when, as we shall see in the next section, we even have $r_E = g_E$, in accord with the weak Birch-Swinnerton-Dyer conjecture.

5 Main Results

As before, let E denote any elliptic curve defined over $\mathbb{Q}$. Define the modular curve $X_0(C(E))$ by

$$X_0(C(E)) = \Gamma_0(C(E)) \setminus (\mathcal{H} \cup \mathbb{P}^1(\mathbb{Q})),$$

where $\mathbb{P}^1(\mathbb{Q})$ denotes the projective line over $\mathbb{Q}$. Then $X_0(C(E))$ is the set of complex points of a curve defined over $\mathbb{Q}$, which we also denote by $X_0(C(E))$. Let $[\infty]$ denote the cusp ∞ of this curve. The modularity Theorem 2.1 of Wiles [5, 57], when combined with work of Shimura [46], shows that there exists an elliptic curve $E(f_E)$ defined over $\mathbb{Q}$, which is a factor up to isogeny of the Jacobian variety of $X_0(C(E))$, and has the same L-series as the elliptic curve E. Hence, by Faltings theorem [18], E and $E(f_E)$ must be isogenous over $\mathbb{Q}$, whence we obtain the following result.

Theorem 5.1. *There is a non-constant rational map defined over* $\mathbb{Q}$

$$\phi : X_0(C(E)) \to E \tag{5.1}$$

with $\phi([\infty]) = \mathcal{O}$.

The most important result to date in the direction of the conjecture of Birch and Swinnerton-Dyer is the following theorem of Kolyvagin and Gross-Zagier (see [23]). We again write r_E and g_E for the order of the zero of the complex L-series of E at $s = 1$, and for the rank of $E(\mathbb{Q})$.

Theorem 5.2. *If* $r_E \leq 1$, *then* $r_E = g_E$, *and* $\text{Ш}(E/\mathbb{Q})$ *is finite.*

The proof relies heavily on the earlier work of Gross-Zagier [24], relating the canonical height of Heegner points to derivatives of L-functions, as well as on Kolyvagin's highly original notion of an Euler system [27]. Heegner points were first discovered, in a special case, by Heegner in his celebrated paper [25], and it was Birch and Stephens who first conjectured that they should be related to the derivatives of L-series. We also note that the proof of the above theorem establishes the following rationality result.

Theorem 5.3. *If* $r_E \leq 1$, *then* $\mathfrak{L}_E/(\Omega_E R_\infty(E))$ *lies in* $\mathbb{Q}$.

When $r_E = 0$, we know by Theorem 5.2 that $g_E = 0$, whence $R_\infty(E) = 1/\#(E(\mathbb{Q}))^2$, and Theorem 5.3 in this case is just a consequence of the classical theory of modular symbols going back to the work of Hecke and others (see [15]). However, when $r_E = 1$, so that $g_E = 1$ by Theorem 5.2, the assertion of Theorem 5.3 can only be proven by using the Gross-Zagier theorem [24]. Moreover, we stress that, contrary to what is often stated in the literature, we still cannot prove the Birch-Swinnerton-Dyer conjectural exact formula for the order of $\text{Ш}(E/\mathbb{Q})$ under the hypothesis that $r_E \leq 1$. Finally, we note that some special cases of

Theorem 5.2 were proven earlier for elliptic curves with complex multiplication by rather different methods, which make use of elliptic units rather than Heegner points (see [12, 42–44]). In another direction, Zhang [59] has generalised the above results to the Jacobian varieties of Shimura curves defined over totally real number fields.

Theorem 5.2 has a surprising application to Gauss' class number problem for imaginary quadratic fields, as was discovered by Goldfeld [20] (see also [38]). If E is an elliptic curve with $g_E \geq 3$, Theorem 5.2 guarantees that necessarily $r_E > 1$, and so we must have that $r_E \geq 3$ when $w_E = -1$. It is usually easy to check numerically that $r_E \leq 3$ when $g_E = 3$, and thus one can prove that there exist elliptic curves E over $\mathbb{Q}$ with $r_E = 3$. For example, $r_E = 3$ for the curve

$$E : y^2 + y = x^3 - 7x + 3,$$

which has conductor $C(E) = 5077$, and $g_E = 3$. Goldfeld's work then enables one to give, for the first time, an explicit upper bound for the absolute value of the discriminants of all imaginary quadratic fields having any prescribed class number. The earlier work on this problem by Heilbronn and Siegel, while proving that the class number of an imaginary quadratic field tends to infinity with the absolute value of the discriminant of the field, was ineffective.

In view of Theorem 5.2, it is plainly an important problem to decide when the hypothesis $r_E \leq 1$ holds. In any particular numerical example, it is usually easy to settle this question, but our knowledge of theoretical results is still very limited. The most natural example of infinite families of elliptic curves defined over $\mathbb{Q}$ is given by the family of quadratic twists of a fixed elliptic curve E. If M is the discriminant of a quadratic field, $E^{(M)}$ will denote the quadratic twist of E by the extension $\mathbb{Q}(\sqrt{M})/\mathbb{Q}$ (in other words, $E^{(M)}$ is the unique elliptic curve defined over $\mathbb{Q}$, which is not isomorphic to E over $\mathbb{Q}$, but becomes isomorphic to E over $\mathbb{Q}(\sqrt{M})$). It is not difficult to see that, in the family of all quadratic twists $E^{(M)}$ of a given E defined over $\mathbb{Q}$, the root numbers $w_{E^{(M)}} = +1$ and $w_{E^{(M)}} = -1$ will each occur half the time. A folklore conjecture (see [21]) asserts that amongst those quadratic twists $E^{(M)}$ with $w_{E^{(M)}} = +1$ (respectively, with $w_{E^{(M)}} = -1$), we should have $r_{E^{(M)}} = 0$ (respectively, $r_{E^{(M)}} = 1$) outside a set of discriminants M of density zero, but this has never been proven for a single elliptic curve E. However, the papers [6] and [37] prove, by rather different methods, the important result that there always exist infinitely many discriminants M such that $r_{E^{(M)}} = 0$, and infinitely many discriminants M such that $r_{E^{(M)}} = 1$. From the point of view of diophantine equations, there is great interest in establishing conditions on the prime factors of M which guarantee that $r_{E^{(M)}} = 1$, since $E^{(M)}(\mathbb{Q})$ is infinite for such M by Theorem 5.2. The first result in this direction is due to Heegner [25], and subsequently Birch [4] generalised and reformulated it. We write $[0\ ]$ for the zero cusp on the modular curve $X_0(C(E))$.

Theorem 5.4. *Let $E/\mathbb{Q}$ be any elliptic such that $\phi([0\ ])$ is not contained in $2E(\mathbb{Q})$, and let p be any prime number such that $p \equiv 3 \bmod 4$, and every prime dividing $C(E)$ splits in the imaginary quadratic field $\mathbb{Q}(\sqrt{-p})$. Then $r_{E^{(-p)}} = 1$, and so, in particular, $E^{(-p)}(\mathbb{Q})$ is infinite.*

Recently, Tian discovered a method for generalising this result to quadratic twists by discriminants having an arbitrary prescribed number of prime factors, which he first applied to the classical congruent number problem [55]. More generally, his method yields the following theorem [13].

Theorem 5.5. *Let $E/\mathbb{Q}$ be any elliptic curve such that (i) $\phi([0])$ is not contained in $2E(\mathbb{Q})$, and (ii) there exists a good supersingular prime q with $q \equiv 1 \bmod 4$, and with $C(E)$ a square modulo q. Then, for each integer $k \geq 1$, there exist infinitely many square free discriminants M having exactly k prime factors, and with $(M, C(E)) = 1$, such that $r_{E^{(M)}} = 1$, whence, in particular, $E^{(M)}(\mathbb{Q})$ is infinite.*

Finally, we mention a recent result proven by Bertolini, Darmon, and Rotger [1], which is a key first step towards generalizing Theorem 5.2 to certain finite Galois extensions of $\mathbb{Q}$. Let ρ denote an odd, irreducible, 2-dimensional Artin representation of the absolute Galois group of $\mathbb{Q}$. As usual, we define $L(E, \rho, s)$ to be the Euler product attached to the tensor product of the Artin representation ρ with the l-adic Tate module of E. Since both E and ρ are known to be modular, it follows from the theory of modular forms that $L(E, \rho, s)$ is also entire.

Theorem 5.6. *If $L(E, \rho, 1) \neq 0$, then ρ does not occur in $E(F) \otimes_{\mathbb{Z}} \mathbb{C}$, where F is the fixed field of the kernel of ρ.*

Moreover, very recent work of Kings, Lei, Loeffler and Zerbes [29–31] constructs a new Euler system, and they use it both to give another proof of Theorem 5.6, and, in addition, to show that, when $L(E, \rho, 1) \neq 0$, the ρ-component of the p-primary subgroup of the Tate-Shafarevich group of E over F is finite for most primes p.

6 The Exact Formula Prime by Prime

In this section, we will discuss the partial results which are known about the conjectural exact Birch-Swinnerton-Dyer formula for the order of the Tate-Shafarevich group of an elliptic curve $E/\mathbb{Q}$ when we assume that $r_E \leq 1$. Theorem 5.2 assures us that, in this case, $E(\mathbb{Q})$ has rank r_E, and $\text{Ш}(E/\mathbb{Q})$ is finite. We note that the order of the torsion subgroup $E(\mathbb{Q})$ is easily determined, and has only one of 12 possibilities, thanks to the beautiful work of Mazur [32]. However, the classical theory of descent does not give a practical way to compute the order of the p-primary part of $\text{Ш}(E/\mathbb{Q})$ once $p > 5$, and the only techniques which work at present are p-adic methods related to Iwasawa theory. In view of this, it is convenient to break the conjecture up into a p-part for every prime number p. To simplify the notation, we define

$$Tam(E) = c_\infty(E) \prod_{q|C(E)} c_q(E).$$

It is also convenient to put

$$L^{(alg)}(E,1) = L(E,1)/\Omega_E,$$

which we know lies in $\mathbb{Q}$ by Theorem 5.3. Then, for every prime number p, the strong Birch-Swinnerton-Dyer conjecture predicts the following exact formula for the order of the p-primary subgroup $Ш(E/\mathbb{Q})(p)$ of $Ш(E/\mathbb{Q})$ when $r_E = 0$.

p-Part of the Birch-Swinnerton-Dyer Conjecture for Analytic Rank 0. *Assume* that $r_E = 0$. *Then, for each prime p,* we have

(6.1)
$$ord_p(\#(Ш(E/\mathbb{Q})(p))) = ord_p(L^{(alg)}(E,1)) + 2ord_p(\#(E(\mathbb{Q}))) - ord_p(Tam(E)).$$

The strongest general result known about this p-part of the Birch and Swinnerton-Dyer conjecture is for E with complex multiplication, and is proven using the Euler system of elliptic units, combined with arguments from Iwasawa theory. The result is due to Rubin [42], but also uses earlier work of Yager [58].

Theorem 6.1. *Assume that $r_E = 0$, and that E admits complex multiplication by an order in an imaginary quadratic field K. If p is any prime which does not divide the order of the group of roots of unity of K, then the p-part of the Birch-Swinnerton-Dyer conjecture is valid for E.*

As an example of this theorem, consider the modular curve $A = X_0(49)$, which is an elliptic curve with equation

$$A: y^2 + xy = x^3 - x^2 - 2x - 1.$$

Take $E = A^{(M)}$, with $M = q_1 \ldots q_r$, where the q_i are distinct primes, with $q_i \equiv 1 \bmod 4$ and $q_i \equiv 3, 5, 6 \bmod 7$, for $i = 1, \ldots, r$. It is shown in [13] that $ord_2(L^{(alg)}(E,1)) = r - 1$ for all $r \geq 0$. This proves that $L(E,1) \neq 0$, and it is easy to see that it establishes the 2-part of the Birch-Swinnerton-Dyer conjecture for E. Hence, applying Theorem 6.1, we conclude that $g_E = 0$, $Ш(E/\mathbb{Q})$ is finite, and the exact Birch-Swinnerton-Dyer formula is valid for the order of $Ш(E/\mathbb{Q})$.

For elliptic curves without complex multiplication, the only way of attacking the p-part of the conjecture of Birch and Swinnerton-Dyer when $r_E = 0$ is by considering the Iwasawa theory of E over the cyclotomic $\mathbb{Z}_p$-extension of $\mathbb{Q}$. We srecall that, for p any prime, the cyclotomic $\mathbb{Z}_p$-extension of $\mathbb{Q}$, which we denote by Φ_∞, is the unique subfield of the field generated over $\mathbb{Q}$ by all p-power roots of unity, whose Galois group Γ over $\mathbb{Q}$ is isomorphic to the additive group of $\mathbb{Z}_p$. Mazur and Swinnerton-Dyer [36] were the first to prove the existence of a p-adic L-function attached to E over Φ_∞ when p is a prime of good ordinary reduction for E, and to formulate a "main conjecture" relating this p-adic L-function to the Γ-module given by the p^∞-Selmer group of E over Φ_∞. As a special case of a more general result, Schneider [45] showed that, for p any odd prime number where E has good ordinary reduction, this "main conjecture" would indeed imply the p-part of the Birch-Swinnerton-Dyer conjecture for the order of $Ш(E/\mathbb{Q})$ when $r_E = 0$. The first major breakthrough in the direction of proving this main conjecture for odd good ordinary primes p was made by Kato [26]. He proved the existence of a remarkable new Euler system attached to E, and used it to prove a partial result in the direction of the "main conjecture" for sufficiently large good ordinary primes p. Subsequently, Skinner and Urban [48] have completed the proof of this "main conjecture" in many cases, by combining Kato's result with deep arguments from the theory of modular forms. This leads to the following specific theorem [48] about the p-part of the conjecture of Birch and Swinnerton-Dyer. Let E_p denote the Galois module of p-division points on E, and let

$$\nu_{E,p} : Gal(\overline{\mathbb{Q}}/\mathbb{Q}) \to Aut(E_p) = GL_2(\mathbb{F}_p)$$

be the associated Galois representation.

Theorem 6.2. *Assume that E does not admit complex multiplication, and that* $r_E = 0$*. Let p be any prime number such that (1)* $p \neq 2$*, (2) E has good ordinary reduction at p, (3)* $\nu_{E,p}$ *is surjective, and (4) there exists a prime q, where E has bad multiplicative reduction, such that the Galois module* E_p *is ramified at q. Then the p-part of the Birch-Swinnerton-Dyer conjecture is valid for E.*

In particular, if E is semistable (i.e. E has multiplicative reduction at all primes of bad reduction), and $r_E = 0$, then this theorem establishes the p-part of the Birch-Swinnerton-Dyer conjecture for all primes $p \geq 11$ of good ordinary reduction. For primes $p > 2$ where E has good supersingular reduction and $t_p = 0$, Wan [56] uses quite different methods in Iwasawa theory to give a proof of the p-part of the conjecture of Birch and Swinnerton-Dyer, assuming that $r_E = 0$ and E is semistable.

One can also formulate the p-part of the conjecture of Birch and Swinnerton-Dyer for every prime p when $r_E = 1$.

p-Part of the Birch-Swinnerton-Dyer Conjecture for Analytic Rank 1. *Assume that* $r_E = 1$*. Then, for each prime p, we have*

$$ord_p(\#(Ш(E/\mathbb{Q})(p))) = ord_p(\mathfrak{L}_E/(\Omega_E R_\infty(E))) - ord_p(Tam(E)). \tag{6.2}$$

When E admits complex multiplication and $r_E = 1$, the work of Kobayashi [28], Perrin-Riou [40], Pollak-Rubin [41], Rubin [42] establishes the following analogue of Theorem 6.1.

Theorem 6.3. *Assume that E admits complex multiplication and that* $r_E = 1$*. Let p be any odd prime where E has good reduction. Then the p-part of the Birch-Swinnerton-Dyer conjecture is valid for E.*

When E does not admit complex multiplication, we have the following recent theorem of Zhang [60].

Theorem 6.4. *Assume that E does not admit complex multiplication, and that $r_E = 1$. Let p be any prime number such that (1) $p \geq 5$, (2) E has good ordinary reduction at p, (3) $\nu_{E,p}$ is surjective, (4) there exist two primes $q_i (i = 1, 2)$ of bad multiplicative reduction for E such that the Galois module E_p is ramified at both q_1 and q_2, and (5) If q is any prime of bad multiplicative reduction for E with $q \equiv \pm 1 \mod p$, then E_p is ramified at q. Then the p-part of the Birch-Swinnerton-Dyer conjecture is valid for E.*

Finally, it is also interesting to note that Kato's work [27], combined with a theorem of Rohrlich [39], proves the following result, which was originally conjectured by Mazur [33]. Let μ_{p^∞} denote the group of all p-power roots of unity.

Theorem 6.5. *For all primes p, the abelian group $E(\mathbb{Q}(\mu_{p^\infty}))$ is finitely generated.*

7 A Numerical Example

Although this article is not directly concerned with the conjecture of Birch and Swinnerton-Dyer for elliptic curves over number fields, I want to end by briefly explaining a remarkable and naturally occurring numerical example, related to the elliptic curves of conductor 11. We recall that 11 is the smallest conductor for an elliptic curve defined over $\mathbb{Q}$, and there are three isogenous curves of conductor 11 defined over $\mathbb{Q}$. Two of these curves are given by

$$A_1 : y^2 + y = x^3 - x^2, \quad A_2 : y^2 + y = x^3 - x^2 - 7820x - 263580,$$

and they are linked by a $\mathbb{Q}$-isogeny $\psi : A_2 \to A_1$ of degree 25. It is well known that $A_1(\mathbb{Q}) = \mathbb{Z}/5\mathbb{Z}$, and $A_2(\mathbb{Q}) = 0$. As was pointed out to me by Fisher and Matsuno, the splitting field of the Galois representation given by the kernel of ψ, which we denote by J, is the field $\mathbb{Q}(\mu_5, r)$, where μ_5 is the group of fifth roots of unity, and r denotes any root of the abelian polynomial

$$x^5 - 65x^4 + 205x^3 + 140x^2 + 25x + 1.$$

Around the year 2000, Matsuno discovered that the complex L-series of either of these two curves, when viewed as curves over J, has a zero of order 4 at $s = 1$. It therefore became an interesting numerical challenge to show that the group of points of either of these two curves with coordinates in J also has rank 4, as predicted by the natural generalization of the weak Birch-Swinnerton-Dyer conjecture. Recently, S. Donnelly (private communication) finally found four linearly independent points, using the MAGMA system in Sydney University. I am very grateful to him for providing the following data. Define $E = A_1^{(5)}$ to be the quadratic twist of A_1 by $\mathbb{Q}(\sqrt{5})$, so that $C(E) = 275$. An equation for the curve E is given by

$$y^2 = x^3 - 10800x + 1026000.$$

Then Donnelly discovered that there is a point in $E(J)$ with x-coordinate

$$(1632096r^4 - 106533648r^3 + 363696696r^2 + 134074044r + 8312592)/41323 \qquad (7.1)$$

and the conjugates of this point under the Galois group of $\mathbb{Q}(r)/\mathbb{Q}$ span a subgroup of rank 4 in $E(\mathbb{Q}(r))$. Moreover, the torsion subgroup of $E(\mathbb{Q}(r))$ is trivial, and it is very probable that $E(\mathbb{Q}(r))$ is generated by any four of the conjugates of the point (7.1). Also, the Birch-Swinnerton-Dyer conjecture predicts that the Tate-Shafarevich group of E over $\mathbb{Q}(r)$ should be trivial. Note that $E(J) = A_1(J)$ because $\sqrt{5} \in J$. It also seems very likely that $A_1(J)$ is generated by any four of the conjugates of the point (7.1), together with the point $(0, 0)$ of order 5. Obviously, J is a subfield of the field F_∞ which is obtained by adjoining to $\mathbb{Q}$ the coordinates of all 5-power division points on any of the three curves of conductor 11. At present, these points found by Donnelly are the only known points of infinite order on the curves of conductor 11 with coordinates in F_∞ (see [11]).

References

1. M. Bertolini, H. Darmon, V. Rotger, *Beilinson-Flach elements and Euler systems II: The Birch-Swinnerton-Dyer conjecture for Hasse-Weil-Artin L-series*, J. Algebraic Geometry 24 (2015), 569–604.
2. B. Birch, P. Swinnerton-Dyer, *Notes on elliptic curves I*, Crelle 212 (1963), 7–25.
3. B. Birch, P. Swinnerton-Dyer, *Notes on elliptic curves II*, Crelle 218 (1965), 79–108.
4. B. Birch, *Elliptic curves and modular functions* in *Symposia Mathematica, Indam Rome 1968/1969*, Academic Press, 4 (1970), 27–32
5. C. Breuil, B. Conrad, F. Diamond, R. Taylor, *On the modularity of elliptic curves over* $\mathbb{Q}$*: wild 3-adic exercises*, J. Amer. Math. Soc. 14 (2001), 843–939.
6. D. Bump, S. Friedberg and J. Hoffstein, *Non-vanishing theorems for L-functions of modular forms and their derivatives*, Invent. Math. 102 (1990), 543–618.
7. L. Cai, J. Shu, Y. Tian, *Explicit Gross-Zagier and Waldpsurger formulae*, Algebra and Number Theory, 8 (2014), 2523–2572.
8. J. Cassels, *Arithmetic on curves of genus 1, VIII*, Crelle 217 (1965), 180–199.
9. J. Cassels, *Arithmetic on curves of genus 1, IV. Proof of the Hauptvermutung*, Crelle 211 (1962), 95–112
10. J. Coates, *Elliptic curves with complex multiplication and Iwasawa theory*, Bull. London Math. Soc. 23 (1991), 321–350.
11. J. Coates, *Elliptic curves - The crossroads of theory and computation* in ANTS 2002, Springer LNCS 2369 (2002), 9–19.
12. J. Coates, A. Wiles, *On the conjecture of Birch and Swinnerton-Dyer*, Invent. Math. 39 (1977), 233–251
13. J. Coates, Y. Li, Y. Tian, S. Zhai, Quadratic twists of elliptic curves, Proc. London Math. Soc. 110 (2015), 357–394.
14. B.Creutz, R. Miller, Second isogeny descents and the Birch-Swinnerton-Dyer conjectural formula, J. of Algebra 372 (2012), 673–701.

15. J. Cremona, *Algorithms for Modular Elliptic Curves*, second Edition, Cambridge University Press, 1997.
16. T. Dokchitser, V. Dokchitser, *On the Birch-Swinnerton-Dyer quotients modulo squares*, Ann. of Math. 172 (2010), 567–596.
17. M. Deuring, *Die Zetafunktionen einer algebraischen Kurve von Geschlechts Eins*, Nach. Akad. Wiss. Göttingen, (1953) 85–94, (1955) 13–42, (1956) 37–76, (1957) 55–80.
18. G. Faltings, *Endlichkeitssatze fur abelsche Varietten ber zahlkorpern*, Invent. Math. 73 (1983), 349–366.
19. N. Freitas, B. Le Hung, and S. Siksek, *Elliptic curves over real quadratic fields are modular*, Invent. Math., 201 (2015), 159–206.
20. D. Goldfeld *The conjectures of Birch and Swinnerton-Dyer and the class numbers of imaginary quadratic fields*, in *Journees arithmetiques de Caen*, Asterisque 41–42 (1977), 219–227.
21. D. Goldfeld *Conjectures on elliptic curves over quadratic fields*, in *Number Theory, Carbondale 1979*, Springer Lecture Notes 751 (1979), 108–118.
22. B. Gross, *Heegner Points on* $X_0(N)$, in *Modular Forms* (ed. R. A. Rankin). Ellis Horwood (1984).
23. B. Gross, *Kolyvagin's work on modular elliptic curves* in *L-functions and arithmetic (Durham 1989)*, London Math. Soc. Lecture Notes 153 (1991), 235–256.
24. B. Gross, D. Zagier, *Heegner points and derivatives of L-series*, Invent. Math. 84 (1986), 225–320.
25. K. Heegner, *Diophantische analysis und modulfunktionen*, Math. Z. 56 (1952), 227–253.
26. K. Kato, *p-adic Hodge theory and values of zeta functions and modular forms* in *Cohomologies p-adiques et applications arithmetiques III*, Asterisque 295 (2004), 117–290.
27. V. Kolyvagin, *Finiteness of* $E(\mathbb{Q})$ *and* $\text{Ш}(E/\mathbb{Q})$ *for a class of Weil curves*, Izv. Akad. Nauk SSSR 52 (1988), translation Math. USSR-Izv. 32 (1989), 523–541.
28. S. Kobayashi, *The p-adic Gross-Zagier formula for elliptic curves at supersingular primes*, Invent. Math. 191 (2013), 527–629.
29. G. Kings, D. Loeffler, S. Zerbes, *Rankin-Eisenstein classes and explicit reciprocity laws* arXiv.org/abs/1503.02888.
30. A. Lei, D. Loeffler, S. Zerbes *Euler systems for Rankin-Selberg convolutions of modular forms*, Ann. of Math., 180 (2014), 653–771.
31. D. Loeffler, S. Zerbes, *Rankin-Eisenstein classes in Coleman families*, arXiv.org/abs/1506.06703.
32. B. Mazur, *Modular curves and the Eisenstein ideal*, Publ. Math. IHES 47 (1977), 33–186.
33. B. Mazur, *Rational points of abelian varieties in towers of number fields*, Invent. Math. 18 (1972), 183–266.
34. B. Mazur, P. Swinnerton-Dyer, *Arithmetic of Weil curves*, Invent. Math. 25 (1974), 1–61.
35. R. Miller, *Proving the Birch-Swinnerton-Dyer conjecture for specific elliptic curves of analytic rank zero and one*, London Math. Soc. J. Comput. Math. 14(2011), 327–350.
36. R. Miller, M. Stoll, *Explicit isogeny descent on elliptic curves*, Math. Comp. 82 (2013), 513–529.
37. K. Murty and R. Murty, *Mean values of derivatives of modular L-series*, Ann. of Math., 133 (1991), 447–475.
38. J. Oesterle, *Nombres de classes de corps quadratiques imaginaires*, Seminaire N. Bourbaki, 1983–1984, 631, 309–323.
39. D. Rohrlich, *On L-functions of elliptic curves and cyclotomic towers*, Invent. Math. 75 (1984), 404–423
40. B. Perrin-Riou, *Fonctions L p-adiques, thorie d'Iwasawa, et points de Heegner*, Bull. Soc. Math. France, 115(1987), 399–456.
41. R. Pollack, K. Rubin *The main conjecture for CM elliptic curves at supersingular primes*, Ann. of Math. 159 (2004), 447–464.
42. K. Rubin, *The main conjectures of Iwasawa theory for imaginary quadratic fields*, Invent. Math. 103 (1991), 25–68.

43. K. Rubin, *Tate-Shafarevich groups and L-functions of elliptic curves with complex multiplication*, Invent. Math. 89 (1987), 527–560.
44. K. Rubin, *On the main conjecture of Iwasawa theory for imaginary quadratic fields*, Invent. Math. 93 (1988), 701–713.
45. P. Schneider, *p-adic height pairings II*, Invent. Math. 79 (1985), 329–374.
46. G. Shimura, *Introduction to the arithmetic theory of automorphic functions*, Publ. Math. Soc. Japan 11 (1971).
47. J. Silverman, *The arithmetic of elliptic curves*, Grad. Texts Math. 106, 1986, Springer.
48. C. Skinner, E. Urban, *The Iwasawa main conjecture for* GL_2, Invent. Math. 195 (2014), 1–277.
49. N. Stephens, *The Diophantine equation* $x^3 + y^3 = Dz^3$ *and the conjectures of Birch and Swinnerton-Dyer*, Crelle 231 (1968), 121–162.
50. J. Tate, *Algorithm for determining the type of singular fiber in an elliptic pencil*, Modular Functions of One Variable IV, Springer Lecture Notes 476 (1975), 33–52.
51. J. Tate, *On the conjectures of Birch and Swinnerton-Dyer and a geometric analog*, Seminaire N. Bourbaki, 1964–1966, 306, 415–440.
52. J. Tate, *Duality theorems in Galois cohomology over number fields*, Proc. Int. Cong. Math., Stockholm (1962), 288–295.
53. J. Thorne, *Elliptic curves over* $\mathbb{Q}_\infty$ *are modular*, arXiv:1505.04769
54. Y. Tian, *Congruent numbers with many prime factors*, Proc. Natl. Acad. Sci. USA 109 (2012), 21256–21258.
55. Y. Tian, *Congruent Numbers and Heegner Points*, Cambridge Journal of Mathematics, 2 (2014), 117–161.
56. X. Wan, *Iwasawa main conjectures for supersingular elliptic curves*, arXiv.org/abs/1411.6352
57. A. Wiles, *Modular elliptic curves and Fermat's Last Theorem*, Ann. of Math. 172 (2010), 567–596.
58. R. Yager, *On two variable p-adic L-functions*, Ann. of Math. 115 (1982), 411–449.
59. S. Zhang, *Heights of Heegner points on Shimura curves*, Ann. of Math. 153 (2001), 27–147.
60. W. Zhang, *Selmer group and the indivisibility of Heegner points*, Cambridge Journal of Mathematics 2 (2014), 191–253.

An Essay on the Riemann Hypothesis

Alain Connes

Abstract The Riemann hypothesis is, and will hopefully remain for a long time, a great motivation to uncover and explore new parts of the mathematical world. After reviewing its impact on the development of algebraic geometry we discuss three strategies, working concretely at the level of the explicit formulas. The first strategy is "analytic" and is based on Riemannian spaces and Selberg's work on the trace formula and its comparison with the explicit formulas. The second is based on algebraic geometry and the Riemann-Roch theorem. We establish a framework in which one can transpose many of the ingredients of the Weil proof as reformulated by Mattuck, Tate and Grothendieck. This framework is elaborate and involves noncommutative geometry, Grothendieck toposes and tropical geometry. We point out the remaining difficulties and show that RH gives a strong motivation to develop algebraic geometry in the emerging world of characteristic one. Finally we briefly discuss a third strategy based on the development of a suitable "Weil cohomology", the role of Segal's Γ-rings and of topological cyclic homology as a model for "absolute algebra" and as a cohomological tool.

1 Introduction

Let $\pi(x) := \#\{p \mid p \in \mathcal{P},\, p < x\}$ be the number of primes less than x with $\frac{1}{2}$ added when x is prime. Riemann [85] found for the counting function[1]

$$f(x) := \sum \frac{1}{n}\pi(x^{\frac{1}{n}}),$$

[1] Similar counting functions were already present in Chebyshev's work.

A. Connes (✉)
Collège de France, 3 Rue d'Ulm, Paris, France

IHÉS, 35 Route de Chartres, Bures-sur-Yvette, France

Ohio State University, Columbus, OH, USA
e-mail: alain@connes.org

J.F. Nash, Jr., M.Th. Rassias (eds.), *Open Problems in Mathematics*,
DOI 10.1007/978-3-319-32162-2_5

the following formula involving the integral logarithm function $\mathrm{Li}(x) = \int_0^x \frac{dt}{\log t}$,

$$f(x) = \mathrm{Li}(x) - \sum_\rho \mathrm{Li}(x^\rho) + \int_x^\infty \frac{1}{t^2-1}\frac{dt}{t\log t} - \log 2 \qquad (1)$$

in terms[2] of the non-trivial zeros ρ of the analytic continuation (shown as well as two proofs of the functional equation by Riemann at the beginning of his paper) of the Euler zeta function

$$\zeta(s) = \sum \frac{1}{n^s}$$

Reading Riemann's original paper is surely still the best initiation to the subject. In his lecture given in Seattle in August 1996, on the occasion of the 100th anniversary of the proof of the prime number theorem, Atle Selberg comments about Riemann's paper: [88]

> It is clearly a preliminary note and might not have been written if L. Kronecker had not urged him to write up something about this work (letter to Weierstrass, Oct. 26 1859). It is clear that there are holes that need to be filled in, but also clear that he had a lot more material than what is in the note.[3] What also seems clear : Riemann is not interested in an asymptotic formula, not in the prime number theorem, what he is after is an exact formula!

The Riemann hypothesis (RH) states that all the non-trivial zeros of ζ are on the line $\frac{1}{2} + i\mathbb{R}$. This hypothesis has become over the years and the many unsuccessful attempts at proving it, a kind of "Holy Grail" of mathematics. Its validity is indeed one of the deepest conjectures and besides its clear inference on the distribution of prime numbers, it admits relations with many parts of pure mathematics as well as of quantum physics.

It is, and will hopefully remain for a long time, a great motivation to uncover and explore new parts of the mathematical world. There are many excellent texts on RH, such as [10] which explain in great detail what is known about the problem, and the many implications of a positive answer to the conjecture. When asked by John Nash to write a text on RH,[4] I realized that writing one more encyclopedic text would just add another layer to the psychological barrier that surrounds RH. Thus I have chosen deliberately to adopt another point of view, which is to navigate between the many forms of the explicit formulas (of which (1) is the prime example) and possible strategies to attack the problem, stressing the value of the elaboration of new concepts rather than "problem solving".

---

[2]More precisely Riemann writes $\sum_{\Re(\alpha)>0}\left(\mathrm{Li}(x^{\frac{1}{2}+\alpha i}) + \mathrm{Li}(x^{\frac{1}{2}-\alpha i})\right)$ instead of $\sum_\rho \mathrm{Li}(x^\rho)$ using the symmetry $\rho \to 1-\rho$ provided by the functional equation, to perform the summation.

[3]See [44, Chap. VII] for detailed support to Selberg's comment.

[4]My warmest thanks to Michael Th. Rassias for the communication.

- *RH and algebraic geometry*
  We first explain the Riemann-Weil explicit formulas in the framework of adeles and global fields in Sect. 2.1. We then sketch in Sect. 2.3 the geometric proof of RH for function fields as done by Weil, Mattuck, Tate and Grothendieck. We then turn to the role of RH in generating new mathematics, its role in the evolution of algebraic geometry in the twentieth century through the Weil conjectures, proved by Deligne, and the elaboration by Grothendieck of the notions of scheme and of topos.
- *Riemannian Geometry, Spectra and trace formulas*
  Besides the proof of analogues of RH such as the results of Weil and of Deligne, there is another family of results that come pretty close. They give another natural approach of RH using analysis, based on the pioneering work of Selberg on trace formulas. These will be reviewed in Sect. 3 where the difficulty arising from the minus sign in front of the oscillatory terms will be addressed.
- *The Riemann-Roch strategy: A Geometric Framework*
  In Sect. 4, we shall describe a geometric framework, established in our joint work with C. Consani, allowing us to transpose several of the key ingredients of the geometric proof of RH for function fields recalled in Sect. 2.3. It is yet unclear if this is the right set-up for the final Riemann-Roch step, but it will illustrate the power of RH as an incentive to explore new parts of mathematics since it gives a clear motivation for developing algebraic geometry in characteristic 1 along the line of tropical geometry. This will take us from the world of characteristic p to the world of characteristic 1, and give us an opportunity to describe its relation with semi-classical and idempotent analysis, optimization and game theory,[5] through the Riemann-Roch theorem in tropical geometry [3, 48, 82].
- *Absolute Algebra and the sphere spectrum*
  The arithmetic and scaling sites which are the geometric spaces underlying the Riemann-Roch strategy of Sect. 4 are only the semiclassical shadows of a more mysterious structure underlying the compactification of Spec $\mathbb{Z}$ that should give a cohomological interpretation of the explicit formulas. We describe in this last section an essential tool coming from algebraic topology: Segal's Γ-rings and the sphere spectrum, over which all previous attempts at developing an absolute algebra organize themselves. Moreover, thanks to the results of Hesselholt and Madsen in particular, topological cyclic homology gives a cohomology theory suitable to treat in a unified manner the local factors of L-functions.

2 RH and Algebraic Geometry

I will briefly sketch here the way RH, once transposed in finite characteristic, has played a determining role in the upheaval of the very notion of geometric space in algebraic geometry culminating with the notions of scheme and topos

[5]One of the topics in which John Nash made fundamental contributions.

due to Grothendieck, with the notion of topos offering a frame of thoughts of incomparable generality and breadth. It is a quite remarkable testimony to the unity of mathematics that the origin of this discovery lies in the greatest problem of analysis and arithmetic.

2.1 The Riemann-Weil Explicit Formulas, Adeles and Global Fields

Riemann's formula (1) is a special case of the "explicit formulas" which establish a duality between the primes and the zeros of zeta. This formula has been extended by Weil in the context of global fields which provides a perfect framework for a generalization of RH since it has been solved, by Weil, for all global fields except number fields.

2.1.1 The Case of ζ

Let us start with the explicit formulas (cf. [9, 59, 83, 101, 103]). We start with a function $F(u)$ defined for $u \in [1, \infty)$, continuous and continuously differentiable except for finitely many points at which both $F(u)$ and $F'(u)$ have at most a discontinuity of the first kind,[6] and such that, for some $\epsilon > 0$, $F(u) = O(u^{-1/2-\epsilon})$. One then defines the Mellin transform of F as

$$\Phi(s) = \int_1^\infty F(u)\, u^{s-1} du \tag{2}$$

The explicit formula then takes the form

$$\Phi(\frac{1}{2}) + \Phi(-\frac{1}{2}) - \sum_{\rho \in \text{Zeros}} \Phi(\rho - \frac{1}{2}) = \sum_p \sum_{m=1}^{\infty} \log p \; p^{-m/2} F(p^m) + \tag{3}$$

$$+(\frac{\gamma}{2} + \frac{\log \pi}{2})F(1) + \int_1^\infty \frac{t^{3/2}F(t) - F(1)}{t(t^2-1)} dt$$

where $\gamma = -\Gamma'(1)$ is the Euler constant, and the zeros are counted with their multiplicities i.e. $\sum_{\rho \in \text{Zeros}} \Phi(\rho - \frac{1}{2})$ means $\sum_{\rho \in \text{Zeros}} \text{order}(\rho) \Phi(\rho - \frac{1}{2})$.

[6] And at which the value of $F(u)$ is defined as the average of the right and left limits there.

2.1.2 Adeles and Global Fields

By a result of Iwasawa [66] a field $\mathbb{K}$ is a algebraic number field, or an algebraic function field of one variable over a finite constant field, if and only if there exists a semi-simple (i.e. with trivial Jacobson radical [68]) commutative ring R containing $\mathbb{K}$ such that R is locally compact, but neither compact nor discrete and $\mathbb{K}$ is discrete and cocompact in R. This result gives a conceptual definition of what is a "global field" and indicates that the arithmetic of such fields is intimately related to analysis on the parent ring R which is called the ring of adeles of $\mathbb{K}$ [93, 102]. It is the opening door to a whole world which is that of automorphic forms and representations, starting in the case of GL_1 with Tate's thesis [93] and Weil's book [102]. Given a global field $\mathbb{K}$, the ring $\mathbb{A}_{\mathbb{K}}$ of adeles of $\mathbb{K}$ is the restricted product of the locally compact fields $\mathbb{K}_v$ obtained as completions of $\mathbb{K}$ for the different places v of $\mathbb{K}$. The equality $dax = |a|dx$ for the additive Haar measure defines the module $\mathrm{Mod} : \mathbb{K}_v \to \mathbb{R}_+$, $\mathrm{Mod}(a) := |a|$ on the local fields $\mathbb{K}_v$ and also as a group homomorphism $\mathrm{Mod} : C_{\mathbb{K}} \to \mathbb{R}_+^*$ where $C_{\mathbb{K}} = \mathrm{GL}_1(\mathbb{A}_{\mathbb{K}})/\mathbb{K}^\times$ is the idele class group. The kernel of the module is a compact subgroup $C_{\mathbb{K},1} \subset C_{\mathbb{K}}$ and the range of the module is a cocompact subgroup $\mathrm{Mod}(\mathbb{K}) \subset \mathbb{R}_+^*$. On any locally compact modulated group, such as $C_{\mathbb{K}}$ or the multiplicative groups $\mathbb{K}_v^*$, one normalizes the Haar measure d^*u uniquely so that the measure of $\{u \mid 1 \le |u| \le \Lambda\}$ is equivalent to $\log \Lambda$ when $\Lambda \to \infty$.

2.1.3 Weil's Explicit Formulas

As shown by Weil, in [103], adeles and global fields give the natural framework for the explicit formulas. For each character $\chi \in \widehat{C_{\mathbb{K},1}}$ one chooses an extension $\tilde{\chi}$ to $C_{\mathbb{K}}$ and one lets $Z_{\tilde{\chi}}$ be the set (with multiplicities and taken modulo the orthogonal of $\mathrm{Mod}(\mathbb{K})$, i.e. $\{s \in \mathbb{C} \mid q^s = 1, \forall q \in \mathrm{Mod}(\mathbb{K})\}$) of zeros of the L-function associated to $\tilde{\chi}$. Let then α be a nontrivial character of $\mathbb{A}_{\mathbb{K}}/\mathbb{K}$ and $\alpha = \prod \alpha_v$ its local factors. The explicit formulas take the following form, with $h \in \mathscr{S}(C_{\mathbb{K}})$ a Schwartz function with compact support:

$$\hat{h}(0) + \hat{h}(1) - \sum_{\chi \in \widehat{C_{\mathbb{K},1}}} \sum_{Z_{\tilde{\chi}}} \hat{h}(\tilde{\chi}, \rho) = \sum_v \int'_{\mathbb{K}_v^*} \frac{h(u^{-1})}{|1-u|} d^*u \tag{4}$$

where the principal value $\int'_{\mathbb{K}_v^*}$ is normalized by the additive character α_v (cf. [22, Chap. II, Sect. 8.5, Theorem 2.44] for the precise notations and normalizations) and for any character ω of $C_{\mathbb{K}}$ one lets

$$\hat{h}(\omega, z) := \int h(u)\, \omega(u)\, |u|^z \, d^*u, \quad \hat{h}(t) := \hat{h}(1, t) \tag{5}$$

For later use in Sect. 4.1 we compare (3) with the Weil way (4) of writing the explicit formulas. Let the function h be the function on $C_{\mathbb{Q}}$ given by $h(u) := |u|^{-\frac{1}{2}}F(|u|)$ (with $F(v) = 0$ for $v < 1$). Then $\hat{h}(\omega, z) = 0$ for characters with non-trivial restriction to $C_{\mathbb{Q},1} = \hat{\mathbb{Z}}^\times$, while $\hat{h}(1,z) = \Phi(z - \frac{1}{2})$. Moreover note that for the archimedean place v of $\mathbb{K} = \mathbb{Q}$ one has, disregarding the principal values for simplicity,

$$\int_{\mathbb{K}_v^*} \frac{h(u^{-1})}{|1-u|} d^*u = \int_{\mathbb{R}^*} \frac{h(u)}{|1-u^{-1}|} d^*u$$

$$= \frac{1}{2}\int_1^\infty h(t)\left(\frac{1}{|1-t^{-1}|} + \frac{1}{|1+t^{-1}|}\right)\frac{dt}{t} = \int_1^\infty \frac{t^{3/2}F(t)}{t(t^2-1)}dt$$

where the $\frac{1}{2}$ comes from the normalization of the multiplicative Haar measure of $\mathbb{R}^*$ viewed as a modulated group. In a similar way, the normalization of the multiplicative Haar measure on $\mathbb{Q}_p^*$ shows that for the finite place associated to the prime p one gets the term $\sum_{m=1}^\infty \log p\, p^{-m/2}F(p^m)$.

2.2 RH for Function Fields

When the module Mod($\mathbb{K}$) of a global field is a discrete subgroup of $\mathbb{R}_+^*$ it is of the form $\mathrm{Mod}(\mathbb{K}) = q^{\mathbb{Z}}$ where q is a prime power, and the field $\mathbb{K}$ is the function field of a smooth projective curve C over the finite field $\mathbb{F}_q$.

Already at the beginning of the twentieth century, Emil Artin and Friedrich Karl Schmidt have generalized RH to the case of function fields. We refer to the text of Cartier [16] where he explains how Weil's definition of the zeta function associated to a variety over a finite field slowly emerged, starting with the thesis of E. Artin where this zeta function was defined for quadratic extensions of $\mathbb{F}_q[T]$, explaining F. K. Schmidt's generalization to finite extensions of $\mathbb{F}_q[T]$ and the work of Hasse on the Riemann hypothesis for elliptic curves over finite fields.

When the global field $\mathbb{K}$ is a function field, geometry comes to the rescue. The problem becomes intimately related to the geometric one of estimating the number $N(q^r) := \#\,C(\mathbb{F}_{q^r})$ of points of C rational over a finite extension $\mathbb{F}_{q^r}$ of the field of definition of C. The analogue of the Riemann zeta function is a generating function: the Hasse-Weil zeta function

$$\zeta_C(s) := Z(C, q^{-s}), \quad Z(C,T) := \exp\left(\sum_{r\geq 1} N(q^r)\frac{T^r}{r}\right) \tag{6}$$

The analogue of RH for ζ_C was proved by André Weil in 1940. Pressed by the circumstances (he was detained in jail) he sent a Comptes-Rendus note to E. Cartan announcing his result. Friedrich Karl Schmidt and Helmut Hasse had previously been able to transpose the Riemann-Roch theorem in the framework of geometry over finite fields and shown its implications for the zeta function: it is a rational fraction (of the variable T) and it satisfies a functional equation. But it took André Weil several years to put on solid ground a general theory of algebraic geometry in finite characteristic that would justify his geometric arguments and allow him to transpose the Hodge index theorem in the form due to the Italian geometers Francesco Severi and Guido Castelnuovo at the beginning of the twentieth century.

2.3 The Proof Using Riemann-Roch on $\bar{C} \times \bar{C}$

Let C be a smooth projective curve over the finite field $\mathbb{F}_q$. The first step is to extend the scalars from $\mathbb{F}_q$ to an algebraic closure $\bar{\mathbb{F}}_q$. Thus one lets

$$\bar{C} := C \otimes_{\mathbb{F}_q} \bar{\mathbb{F}}_q \tag{7}$$

This operation of extension of scalars does not change the points over $\bar{\mathbb{F}}_q$, i.e. one has $\bar{C}(\bar{\mathbb{F}}_q) = C(\bar{\mathbb{F}}_q)$. The Galois action of the Frobenius automorphism of $\bar{\mathbb{F}}_q$ raises the coordinates of any point $x \in C(\bar{\mathbb{F}}_q)$ to the qth power and this transformation of $C(\bar{\mathbb{F}}_q)$ coincides with the *relative Frobenius* $\mathrm{Fr}_r := \mathrm{Fr}_C \times \mathrm{Id}$ of $\bar{C}$, where Fr_C is the *absolute Frobenius* of C (which is the identity on points of the scheme and the qth power map in the structure sheaf). The relative Frobenius Fr_r is $\bar{\mathbb{F}}_q$-linear by construction and one can consider its graph in the surface $X = \bar{C} \times_{\bar{\mathbb{F}}_q} \bar{C}$ which is the square of $\bar{C}$. This graph is the Frobenius correspondence Ψ. It is important to work over an algebraically closed field in order to have a good intersection theory. This allows one to express the right hand side of the explicit formula (4) for the zeta function ζ_C as an intersection number $D.\Delta$, where Δ is the diagonal in the square and $D = \sum a_k \Psi^k$ is the divisor given by a finite integral linear combination of powers of the Frobenius correspondence. The terms $\hat{h}(0)$, $\hat{h}(1)$ in the explicit formula are also given by intersection numbers $D.\xi_j$, where

$$\xi_0 = e_0 \times \bar{C}\,,\ \xi_1 = \bar{C} \times e_1 \tag{8}$$

where the e_j are points of $\bar{C}$. One then considers divisors on X up to the additive subgroup of principal divisors i.e. those corresponding to an element $f \in \mathscr{K}$ of the function field of X. The problem is then reduced to proving the negativity of $D.D$ (the self-intersection pairing) for divisors of degree zero. The Riemann-Roch theorem on the surface X gives the answer. To each divisor D on X corresponds an index problem and one has a finite dimensional vector space of solutions $H^0(X, \mathscr{O}(D))$ over $\bar{\mathbb{F}}_q$. Let

$$\ell(D) = \dim H^0(X, \mathscr{O}(D)) \tag{9}$$

The best way to think of the sheaf $\mathcal{O}(D)$ is in terms of Cartier divisors, i.e. a global section of the quotient sheaf $\mathcal{K}^\times/\mathcal{O}_X^\times$, where $\mathcal{K}$ is the constant sheaf corresponding to the function field of X and $\mathcal{O}_X$ is the structure sheaf. The sheaf $\mathcal{O}(D)$ associated to a Cartier divisor is obtained by taking the sub-sheaf of $\mathcal{K}$ whose sections on U_i form the sub $\mathcal{O}_X$-module generated by $f_i^{-1} \in \Gamma(U_i, \mathcal{K}^\times)$ where the f_i represent D locally. One has a "canonical" divisor K and Serre duality

$$\dim H^2(X, \mathcal{O}(D)) = \dim H^0(X, \mathcal{O}(K - D)) \tag{10}$$

Moreover the following Riemann-Roch formula holds

$$\sum_0^2 (-1)^j \dim H^j(X, \mathcal{O}(D)) = \frac{1}{2} D.(D - K) + \chi(X) \tag{11}$$

where $\chi(X)$ is the arithmetic genus. All this yields the Riemann-Roch inequality

$$\ell(D) + \ell(K - D) \geq \frac{1}{2} D.(D - K) + \chi(X) \tag{12}$$

One then applies Lemma 1 to the quadratic form $\mathfrak{s}(D, D') = D.D'$ using the ξ_j of (8). One needs three basic facts [54]

1. If $\ell(D) > 1$ then D is equivalent to a strictly positive divisor.
2. If D is a strictly positive divisor then

$$D.\xi_0 + D.\xi_1 > 0$$

3. One has $\xi_0.\xi_1 = 1$ and $\xi_j.\xi_j = 0$.

One then uses (12) to show (see [54]) that if $D.D > 0$ then after a suitable rescaling by $n > 0$ or $n < 0$ one gets $\ell(nD) > 1$ which shows that the hypothesis (2) of the following simple Lemma 1 is fulfilled, and hence that RH holds for ζ_C,

Lemma 1. *Let $\mathfrak{s}(x, y)$ be a symmetric bilinear form on a vector space E (over $\mathbb{Q}$ or $\mathbb{R}$). Let $\xi_j \in E$, $j \in \{0, 1\}$, be such that*

1. *$\mathfrak{s}(\xi_j, \xi_j) = 0$ and $\mathfrak{s}(\xi_0, \xi_1) = 1$.*
2. *For any $x \in E$ such that $\mathfrak{s}(x, x) > 0$ one has $\mathfrak{s}(x, \xi_0) \neq 0$ or $\mathfrak{s}(x, \xi_1) \neq 0$.*

Then one has the inequality

$$\mathfrak{s}(x, x) \leq 2\mathfrak{s}(x, \xi_0)\mathfrak{s}(x, \xi_1), \quad \forall x \in E \tag{13}$$

The proof takes one line but the meaning of this lemma is to reconcile the "naive positivity" of the right hand side of the explicit formula (4) (which is positive when $h \geq 0$ vanishes near $u = 1$) with the negativity of the left hand side needed to prove RH (cf. Sect. 3.1 (17) below).

At this point we see that it is highly desirable to find a geometric framework for the Riemann zeta function itself, in which the Hasse-Weil formula (6), the geometric interpretation of the explicit formulas, the Frobenius correspondences, the divisors, principal divisors, Riemann-Roch problem on the curve and the square of the curve all make sense.

Such a tentative framework will be explained in Sect. 4. It involves in particular the refinement of the notion of geometric space which was uncovered by Grothendieck and to which we now briefly turn.

2.4 *Grothendieck and the Notion of Topos*

The essential ingredients of the proof explained in Sect. 2.3 are the intersection theory for divisors on $\bar{C} \times \bar{C}$, sheaf cohomology and Serre duality, which give the formulation of the Riemann-Roch theorem. Both owe to the discovery of sheaf theory by J. Leray and the pioneering work of J. P. Serre on the use of sheaves for the Zariski topology in the algebraic context, with his fundamental theorem comparing the algebraic and analytic frameworks. The next revolution came from the elaboration by A. Grothendieck and M. Artin of etale ℓ-adic cohomology. It allows one to express the Weil zeta function of a smooth projective variety X defined over a finite field $\mathbb{F}_q$ i.e. the function $Z(X,t)$ given by (6) with $t = q^{-s}$ which continues to make sense in general, as an alternate product of the form

$$Z(X,t) = \prod_{j=0}^{2\dim X} \det(1 - tF^* \mid H^j(\bar{X}_{\text{et}}, \mathbb{Q}_\ell))^{(-1)^{j+1}} \tag{14}$$

where F^* corresponds to the action of the Frobenius on the ℓ-adic cohomology and ℓ is a prime which is prime to q. This equality follows from a Lefschetz formula for the number $N(q^r)$ of fixed points of the rth power of the Frobenius and when $X = C$ is a curve the explicit formulas reduce to the Lefschetz formula. The construction of the cohomology groups $H^j(\bar{X}_{\text{et}}, \mathbb{Q}_\ell)$ is indirect and they are defined as :

$$H^j(\bar{X}_{\text{et}}, \mathbb{Q}_\ell) = \varprojlim_n \left(H^j(\bar{X}_{\text{et}}, \mathbb{Z}/\ell^n\mathbb{Z})\right) \otimes_{\mathbb{Z}_\ell} \mathbb{Q}_\ell$$

where $\bar{X}_{\text{et}}$ is the etale site of $\bar{X}$. Recently the etale site of a scheme has been refined [4] to the *pro-etale* site whose objects no longer satisfy any finiteness condition. The cohomology groups $H^j(\bar{X}_{\text{proet}}, \bar{\mathbb{Q}}_\ell)$ are then directly obtained using the naive interpretation (without torsion coefficients). One needs to pay attention in (14) to the precise definition of F, it is either the relative Frobenius Fr_r or the *Geometric*

Frobenius Fr_g which is the inverse of the *Arithmetic Frobenius* Fr_a. The product $\mathrm{Fr}_a \circ \mathrm{Fr}_r = \mathrm{Fr}_r \circ \mathrm{Fr}_a$ is the absolute Frobenius Fr which acts trivially on the ℓ-adic cohomology. To understand the four different incarnations of "the Frobenius" it is best to make them explicit in the simplest example of the scheme $\mathrm{Spec}\, R$ where $R = \bar{\mathbb{F}}_q[T]$ is the ring of polynomials $P(T) = \sum a_j T^j$, $a_j \in \bar{\mathbb{F}}_q$

- Geometric Frobenius: $\sum a_j T^j \mapsto \sum a_j^{1/q} T^j$
- Relative Frobenius: $P(T) \mapsto P(T^q)$
- Absolute Frobenius: $P(T) \mapsto P(T)^q$
- Arithmetic Frobenius: $\sum a_j T^j \mapsto \sum a_j^q T^j$

The motivation of Grothendieck for developing etale cohomology came from the search of a Weil cohomology and the Weil conjectures which were solved by Deligne in 1973 [37].

In his quest Grothendieck uncovered several key concepts such as those of schemes and above all that of topos, see [2] and [78], in his own words:

> C'est le thème du topos, et non celui des schémas, qui est ce "lit", ou cette "rivière profonde", où viennent s'épouser la géométrie et l'algèbre, la topologie et l'arithmétique, la logique mathématique et la théorie des catégories, le monde du continu et celui des structures "discontinues" ou "discrètes". Si le thème des schémas est comme le *cœur* de la géométrie nouvelle, le thème du topos en est l'enveloppe, ou la *demeure*. Il est ce que j'ai conçu de plus vaste, pour saisir avec finesse, par un même langage riche en résonances géométriques, une "essence" commune à des situations des plus éloignées les unes des autres, provenant de telle région ou de telle autre du vaste univers des choses mathématiques.

3 Riemannian Geometry, Spectra and Trace Formulas

Riemannian Geometry gives a wealth of "spectra" of fundamental operators associated to a geometric space, such as the Laplacian and the Dirac operators.

3.1 The Selberg Trace Formula

In the case of compact Riemann surfaces X with constant negative curvature -1, the Selberg trace formula [86], takes the following form where the eigenvalues of the Laplacian are written in the form[7] $\lambda_n = -(\frac{1}{4} + r_n^2)$. Let $\delta > 0$, $h(r)$ be an analytic function in the strip $|\Im(r)| \leq \frac{1}{2} + \delta$ and such that $h(r) = h(-r)$ and with $(1 + r^2)^{1+\delta}|h(r)|$ being bounded. Then [60, 86, 87], with A the area of X,

[7] Where the argument of r_n is either 0 or $-\pi/2$.

$$\sum h(r_n) = \frac{A}{4\pi}\int_{-\infty}^{\infty} \tanh(\pi r)h(r)rdr + \sum_{\{T\}} \frac{\log N(T_0)}{N(T)^{\frac{1}{2}} - N(T)^{-\frac{1}{2}}} g(\log N(T)) \quad (15)$$

where g is the Fourier transform of h, i.e. more precisely $g(s) = \frac{1}{2\pi}\int_{-\infty}^{\infty} h(r)e^{-irs}dr$. The $\log N(T)$ are the lengths of the periodic orbits of the geodesic flow with $\log N(T_0)$ being the length of the primitive one. Already in 1950–51, Selberg saw the striking similarity of his formula with (3) which (cf. [60]) can be rewritten in the following form, with h and g as above and the non-trivial zeros of zeta expressed in the form $\rho = \frac{1}{2} + i\gamma$,

$$\sum_{\gamma} h(\gamma) = h(\frac{i}{2}) + h(-\frac{i}{2}) + \frac{1}{2\pi}\int_{-\infty}^{\infty} \omega(r)h(r)dr - 2\sum \Lambda(n)n^{-\frac{1}{2}}g(\log n) \quad (16)$$

where

$$\omega(r) = \frac{\Gamma'}{\Gamma}\left(\frac{1}{4} + i\frac{r}{2}\right) - \log\pi, \quad \frac{\Gamma'}{\Gamma}(s) = \int_0^1 \frac{1 - t^{s-1}}{1-t}dt - \gamma\,, \quad \forall s, \Re(s) > 0$$

and $\Lambda(n)$ is the von-Mangoldt function with value $\log p$ for powers p^ℓ of primes and zero otherwise. Moreover Selberg found that there is a zeta function which corresponds to (15) in the same way that $\zeta(s)$ corresponds to (16). The role of Hilbert space is crucial in the work of Selberg to ensure that the zeros of his zeta function satisfy the analogue of RH. This role of Hilbert space is implicit as well in RH which has been reformulated by Weil as the positivity of the functional $W(g)$ defined as both sides of (16). More precisely the equivalent formulation is that $W(g \star g^*) \geq 0$ on functions g which correspond to Fourier transforms of analytic functions h as above (i.e. even and analytic in a strip $|\Im z| \leq \frac{1}{2} + \delta$) where for even functions one has $g^*(s) := \overline{g(-s)} = \overline{g(s)}$. Moreover by [11, 15], it is enough, using Li's criterion (cf. [11, 73]), to check the positivity on a small class of explicit real valued functions with compact support. In fact for later purposes it is better to write this criterion as

$$RH \iff \mathfrak{s}(f,f) \leq 0\,,\ \forall f \mid \int f(u)d^*u = \int f(u)du = 0 \quad (17)$$

where for real compactly supported functions on $\mathbb{R}_+^*$, we let $\mathfrak{s}(f,g) := N(f \star \tilde{g})$ where $\star$ is the convolution product on $\mathbb{R}_+^*$, $\tilde{g}(u) := u^{-1}g(u^{-1})$, and

$$N(h) := \sum_{n=1}^{\infty} \Lambda(n)h(n) + \int_1^{\infty} \frac{u^2h(u) - h(1)}{u^2 - 1}d^*u + c\,h(1)\,,\ c = \frac{1}{2}(\log\pi + \gamma) \quad (18)$$

The Selberg trace formula has been considerably extended by J. Arthur and plays a key role in the Langland's program. We refer to [1] for an introduction to this vast topic.

3.2 The Minus Sign and Absorption Spectra

The Selberg trace formula [86, 87] for Riemann surfaces of finite area, acquires additional terms which make it look e.g. in the case of $X = H/PSL(2,\mathbb{Z})$ (where H is the upper half plane with the Poincaré metric) even more similar to the explicit formulas, since the parabolic terms now involve explicitly the sum

$$2\sum_{n=1}^{\infty} \frac{\Lambda(n)}{n} g(2\log n)$$

Besides the square root in the $\Lambda(n)$ terms in the explicit formulas (16)

$$-2\sum_{n=1}^{\infty} \frac{\Lambda(n)}{n^{\frac{1}{2}}} g(\log n)$$

there is however a striking difference which is that these terms occur with a positive sign instead of the negative sign in (16), as discussed in [60, Sect. 12]. This discussion of the minus sign was extended to the case of the semiclassical limit of Hamiltonian systems in physics in [5, 6]. In order to get some intuition of what this reveals, it is relevant to go back to the origin of spectra in physics, i.e. to the very beginning of spectroscopy. It occurred when Joseph Von Fraunhofer (1787–1826) could identify, using self-designed instruments, about 500 dark lines in the light coming from the sun, decomposed using the dispersive power of a spectroscope such as a prism (cf. Fig. 1). These dark lines constitute the "absorption spectrum" and it took about 45 years before Kirchhoff and Bunsen noticed that several of these Fraunhofer lines coincide (i.e. have the same wave length) with the bright lines of the "emission" spectrum of heated elements, and showed that they could be reobtained by letting white light traverse a cold gas. In his work on the trace formula in the finite covolume case, Selberg had to take care of a superposed continuous spectrum due to the presence of the non-compact cusps of the Riemann surface.

3.3 The Adele Class Space and the Explicit Formulas

I had the chance to be invited at the Seattle meeting in 1996 for the celebration of the proof of the prime number theorem. The reason was the paper [12] (inspired from [69]) in which the Riemann zeta function appeared naturally as the partition function of a quantum mechanical system (BC system) exhibiting phase transitions. The RH had been at the center of discussions in the meeting and I knew the analogy between the BC-system and the set-up that V. Guillemin proposed in [56] to explain the Selberg trace formula using the action of the geodesic flow on the horocycle foliation. To a foliation is associated a von Neumann algebra [18], and

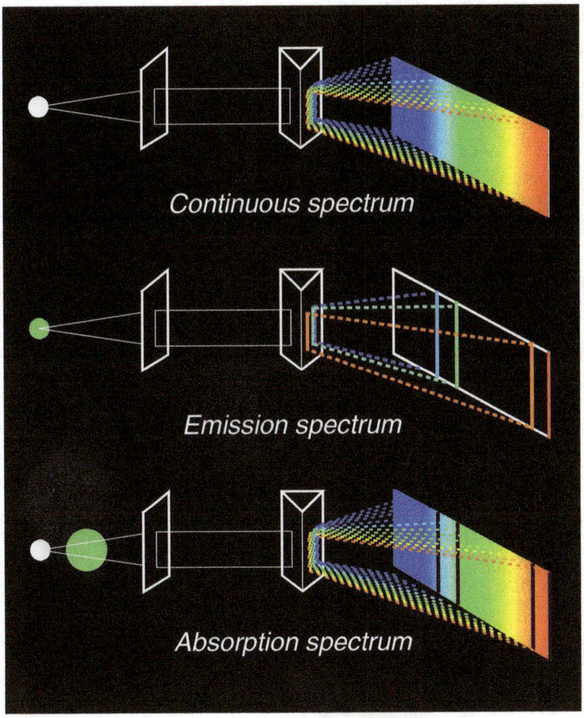

Fig. 1 The three kinds of spectra occurring in spectroscopy: (1) The *top one* is the "continuous spectrum" which occurs when white light is decomposed by passing through a prism. (2) The *middle one* is the "emission spectrum" which occurs when the light emitted by a heated gas is decomposed by passing through a prism and gives shining lines-a signature of the gas-over a dark background. (3) The *third one* is the "absorption spectrum" which occurs when white light traverses a cold gas and is then decomposed by passing through a prism. It appears as *dark lines* in a background continuous spectrum. The absorption lines occur at the same place as the emission lines

the horocycle foliation on the sphere bundle of a compact Riemann surface gives a factor of type II_∞ on which the geodesic flow acts by scaling the trace. An entirely similar situation comes canonically from the BC-system at critical temperature and after interpreting the dual system in terms of adeles, I was led by this analogy to consider the action of the idele class group of $\mathbb{Q}$ on the adele class space, i.e. the quotient $\mathbb{Q}^\times\backslash\mathbb{A}_\mathbb{Q}$ of the adeles $\mathbb{A}_\mathbb{Q}$ of $\mathbb{Q}$ by the action of $\mathbb{Q}^\times$. I knew from the BC-system that the action of $\mathbb{Q}^\times$, which preserves the additive Haar measure, is ergodic for this measure and gives the same factor of type II_∞ as the horocycle foliation. Moreover the dual action scales the trace in the same manner.

Let $\mathbb{K}$ be a global field and $C_\mathbb{K} = \mathrm{GL}_1(\mathbb{A}_\mathbb{K})/\mathbb{K}^\times$ the idele class group. The module $\mathrm{Mod} : C_\mathbb{K} \to \mathbb{R}_+^*$ being proper with cocompact range, one sees that the Haar measure on the Pontrjagin dual group of $C_\mathbb{K}$ is diffuse. Since a point is of measure 0 in a diffuse measure space there is no way one can see the absorption spectrum without introducing some smoothness on this dual which is done using a Sobolev space $L^2_\delta(C_{\mathbb{K}})$ of functions on $C_{\mathbb{K}}$ which (for fixed $\delta > 1$) is defined as

$$||\xi||^2 = \int_{C_{\mathbb{K}}} |\xi(x)|^2 \rho(x)\, d^*x, \quad \rho(x) := (1 + \log |x|^2)^{\delta/2} \tag{19}$$

Definition 1. Let $\mathbb{K}$ be a global field, the adele class space of $\mathbb{K}$ is the quotient $X_{\mathbb{K}} = \mathbb{A}_{\mathbb{K}}/\mathbb{K}^\times$ of the adeles of $\mathbb{K}$ by the action of $\mathbb{K}^\times$ by multiplication.

We then consider the codimension 2 subspace $\mathscr{S}(\mathbb{A}_{\mathbb{K}})_0$ of the Bruhat-Schwartz space $\mathscr{S}(\mathbb{A}_{\mathbb{K}})$ (cf. [13]) given by the conditions $f(0) = 0$, $\int f\, dx = 0$ The Sobolev space $L^2_\delta(X_{\mathbb{K}})_0$ is the separated completion of $\mathscr{S}(\mathbb{A}_{\mathbb{K}})_0$ for the norm with square

$$||f||^2 = \int_{C_{\mathbb{K}}} |\sum_{q \in \mathbb{K}^*} f(qx)|^2 \rho(x)\, |x| d^*x \tag{20}$$

Note that by construction all functions of the form $f(x) = g(x) - g(qx)$ for some $q \in \mathbb{K}^\times$ belong to the radical of the norm (20), which corresponds to the operation of quotient of Definition 1. In particular the representation of ideles on $\mathscr{S}(\mathbb{A}_{\mathbb{K}})$ given by

$$(\vartheta(\alpha)\xi)(x) = \xi(\alpha^{-1}x) \;\; \forall\, \alpha \in \mathrm{GL}_1(\mathbb{A}_{\mathbb{K}}),\; x \in \mathbb{A}_{\mathbb{K}} \tag{21}$$

induces a representation ϑ_a of $C_{\mathbb{K}}$ on $L^2_\delta(X_{\mathbb{K}})_0$. One has by construction a natural isometry $\mathfrak{E} : L^2_\delta(X_{\mathbb{K}})_0 \to L^2_\delta(C_{\mathbb{K}})$ which intertwines the representation ϑ_a with the regular representation of $C_{\mathbb{K}}$ in $L^2_\delta(C_{\mathbb{K}})$ multiplied by the square root of the module. This representation restricts to the cokernel of the map $\mathfrak{E}$, which splits as a direct sum of subspaces labeled by the characters of the compact group $C_{\mathbb{K},1} = \mathrm{Ker\,Mod}$ and its spectrum in each sector gives the zeros of L-functions with Grössencharakter. The shortcoming of this construction is in the artificial weight $\rho(x)$, which is needed to see this absorption spectrum but only sees the zeros which are on the critical line and where the value of δ artificially cuts the multiplicities of the zeros (cf. [19]).

This state of affairs is greatly improved if one gives up trying to prove RH but retreats to an interpretation of the explicit formulas as a trace formula. One simply replaces the above Hilbert space set-up by a softer one involving nuclear spaces [80]. The spectral side now involves all non-trivial zeros and, using the preliminary results of [14, 19, 20] one gets that the geometric side is given by:

$$\mathrm{Tr}_{\mathrm{distr}}\left(\int h(w)\vartheta(w) d^*w\right) = \sum_v \int_{\mathbb{K}_v^\times} \frac{h(w^{-1})}{|1-w|}\, d^*w \tag{22}$$

We refer to [19, 22, 80] for a detailed treatment. The subgroups $\mathbb{K}_v^\times \subset C_{\mathbb{K}} = \mathrm{GL}_1(\mathbb{A}_{\mathbb{K}})/\mathrm{GL}_1(\mathbb{K})$ arise as isotropy groups. One can understand why the terms

$\frac{h(w^{-1})}{|1-w|}$ occur in the trace formula by computing, formally as follows, the trace of the scaling operator $T = \vartheta_{w^{-1}}$ when working on the local field $\mathbb{K}_v$ completion of the global field $\mathbb{K}$ at the place v, one has

$$T\xi(x) = \xi(wx) = \int k(x,y)\xi(y)dy$$

so that T is given by the distribution kernel $k(x,y) = \delta(wx - y)$ and its trace is

$$\mathrm{Tr}_{\mathrm{distr}}(T) = \int k(x,x)\,dx = \int \delta(wx - x)\,dx = \frac{1}{|w-1|}\int \delta(z)\,dz = \frac{1}{|w-1|}$$

When working at the level of adeles one treats all places on the same footing and thus there is an overall minus sign in front of the spectral contribution. Thus the Riemann spectrum appears naturally as an absorption spectrum from the adele class space. As such, it is difficult to show that it is "real". While this solves the problem of giving a trace formula interpretation of the explicit formulas, there is of course still room for an interpretation as an emission spectrum. However from the adelic point of view it is unnatural to separate the contribution of the archimedean place.

4 The Riemann-Roch Strategy: A Geometric Framework

In this section we shall present a geometric framework which has emerged over the years in our joint work with C. Consani and seems suitable in order to transpose the geometric proof of Weil to the case of RH. The aim is to apply the Riemann-Roch strategy of Sect. 2.3. The geometry involved will be of elaborate nature inasmuch as it relies on the following three theories:

1. Noncommutative Geometry.
2. Grothendieck topoi.
3. Tropical Geometry.

4.1 The Limit $q \to 1$ and the Hasse-Weil Formula

In [91, cf. Sect. 6], C. Soulé, motivated by [79, cf. Sect. 1.5] and [38, 39, 70, 71, 92, 95], introduced the zeta function of a variety X over $\mathbb{F}_1$ using the *polynomial* counting function $N(x) \in \mathbb{Z}[x]$ associated to X. The definition of the zeta function is as follows

$$\zeta_X(s) := \lim_{q\to 1} Z(X, q^{-s})(q-1)^{N(1)}, \qquad s \in \mathbb{R} \tag{23}$$

where $Z(X, q^{-s})$ denotes the evaluation at $T = q^{-s}$ of the Hasse-Weil exponential series

$$Z(X,T) := \exp\left(\sum_{r\geq 1} N(q^r)\frac{T^r}{r}\right) \tag{24}$$

For instance, for a projective space $\mathbb{P}^n$ one has $N(q) = 1 + q + \ldots + q^n$ and

$$\zeta_{\mathbb{P}^n(\mathbb{F}_1)}(s) = \lim_{q\to 1}(q-1)^{n+1}\zeta_{\mathbb{P}^n(\mathbb{F}_q)}(s) = \frac{1}{\prod_0^n (s-k)}$$

It is natural to wonder on the existence of a "curve" C suitably defined over $\mathbb{F}_1$, whose zeta function $\zeta_C(s)$ is the complete Riemann zeta function $\zeta_{\mathbb{Q}}(s) = \pi^{-s/2}\Gamma(s/2)\zeta(s)$ (cf. also [79]). The first step is to find a counting function $N(q)$ defined for $q \in [1,\infty)$ and such that (23) gives $\zeta_{\mathbb{Q}}(s)$. But there is an obvious difficulty since as $N(1)$ represents the Euler characteristic one should expect that $N(1) = -\infty$ (since the dimension of H^1 is infinite). This precludes the use of (23) and also seems to contradict the expectation that $N(q) \geq 0$ for $q \in (1,\infty)$. As shown in [23, 24] there is a simple way to solve the first difficulty by passing to the logarithmic derivatives of both terms in Eq. (23) and observing that the Riemann sums of an integral appear from the right hand side. One then gets instead of (23) the equation:

$$\frac{\partial_s \zeta_N(s)}{\zeta_N(s)} = -\int_1^\infty N(u)\, u^{-s} d^*u \tag{25}$$

Thus the integral equation (25) produces a precise equation for the counting function $N_C(q) = N(q)$ associated to C:

$$\frac{\partial_s \zeta_{\mathbb{Q}}(s)}{\zeta_{\mathbb{Q}}(s)} = -\int_1^\infty N(u)\, u^{-s} d^*u \tag{26}$$

One finds that this equation admits a solution which is a *distribution* and is given with $\varphi(u) := \sum_{n<u} n\,\Lambda(n)$, by the equality

$$N(u) = \frac{d}{du}\varphi(u) + \kappa(u) \tag{27}$$

where $\kappa(u)$ is the distribution which appears in the explicit formula (3),

$$\int_1^\infty \kappa(u) f(u) d^*u = \int_1^\infty \frac{u^2 f(u) - f(1)}{u^2-1} d^*u + cf(1)\,, \qquad c = \frac{1}{2}(\log\pi + \gamma)$$

The conclusion is that the distribution $N(u)$ is positive on $(1,\infty)$ and is given by

$$N(u) = u - \frac{d}{du}\left(\sum_{\rho\in Z} \text{order}(\rho)\frac{u^{\rho+1}}{\rho+1}\right) + 1 \tag{28}$$

where the derivative is taken in the sense of distributions, and the value at $u = 1$ of the term $\omega(u) = \sum_{\rho\in Z} \text{order}(\rho)\frac{u^{\rho+1}}{\rho+1}$ is given by $\frac{1}{2} + \frac{\gamma}{2} + \frac{\log 4\pi}{2} - \frac{\zeta'(-1)}{\zeta(-1)}$.

The primitive $J(u) = \frac{u^2}{2} - \omega(u) + u$ of $N(u)$ is an increasing function on $(1,\infty)$, but tends to $-\infty$ when $u \to 1+$ while its value $J(1)$ is finite (Fig. 2). The tension between the positivity of the distribution $N(q)$ for $q > 1$ and the expectation that its value $N(1)$ should be $N(1) = -\infty$ is resolved by the theory of distributions: N is *finite* as a distribution, but when one looks at it as a function its value at $q = 1$ is formally given by

$$N(1) = 2 - \lim_{\epsilon\to 0}\frac{\omega(1+\epsilon)-\omega(1)}{\epsilon} \sim -\frac{1}{2}E\log E, \qquad E = \frac{1}{\epsilon}$$

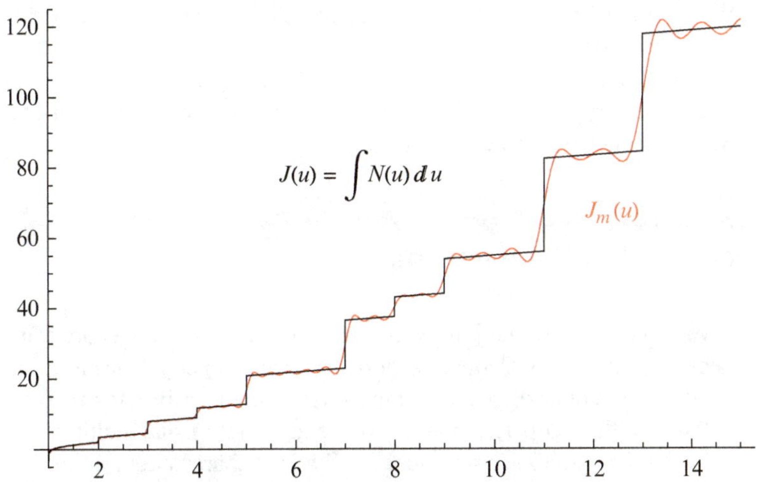

Fig. 2 This represents a function $J(u)$ which is a primitive of the counting distribution $N(u)$. This function is increasing and tends to $-\infty$ when $u \to 1$. The wiggly graph represents the approximation of $J(u)$ obtained using the symmetric set Z_m of the first $2m$ zeros, by

$$J_m(u) = \frac{u^2}{2} - \sum_{Z_m} \text{order}(\rho)\frac{u^{\rho+1}}{\rho+1} + u$$

Note that $J(u) \to -\infty$ when $u \to 1+$

which is $-\infty$ and in fact reflects, when $\epsilon \to 0$, the density of the zeros. Note that this holds independently of the choice of the principal value in the explicit formulas. This subtlety does not occur for function fields $\mathbb{K}$ since their module $\mathrm{Mod}(\mathbb{K})$ is discrete so that distributions and functions are the same thing. There is one more crucial nuance between the case $\mathbb{K} = \mathbb{Q}$ and the function fields: the distribution $\kappa(u)$ which is the archimedean contribution to $N(u)$ in (27), does not fulfill the natural inequality $N(q) \leq N(q^r)$ expected of a counting function. This is due to the terms $|1-u|^{-1}$ in the Weil explicit formula, which as explained in Sect. 2.1.3 contribute non-trivially at the archimedean place, and indicate that the counting needs to take into account an ambient larger space and transversality factors as in [56]. In fact, we have seen in Sect. 3.3 that the noncommutative space of adele classes of a global field provides a framework to interpret the explicit formulas of Riemann-Weil in number theory as a trace formula, and that the geometric contributions give the right answer. In [24], we showed that the quotient

$$X_{\mathbb{Q}} := \mathbb{Q}^\times \backslash \mathbb{A}_{\mathbb{Q}} / \hat{\mathbb{Z}}^\times \tag{29}$$

of the adele class space $\mathbb{Q}^\times \backslash \mathbb{A}_{\mathbb{Q}}$ of the rational numbers by the maximal compact subgroup $\hat{\mathbb{Z}}^\times$ of the idele class group, gives by considering the induced action of $\mathbb{R}_+^\times$, the above counting distribution $N(u)$, $u \in [1, \infty)$, which determines, using the Hasse-Weil formula in the limit $q \to 1$, the complete Riemann zeta function. The next step is to understand that the action of $\mathbb{R}_+^\times$ on the space $X_{\mathbb{Q}}$ is in fact the action of the Frobenius automorphisms Fr_λ on the points of the arithmetic site—an object of algebraic geometry—over $\mathbb{R}_+^{\max}$. To explain this we first need to take an excursion in the exotic world of "characteristic one".

4.2 *The World of Characteristic* 1

The key words here are: Newton polygons, Thermodynamics, Legendre transform, Game theory, Optimization, Dequantization, Tropical geometry. One alters the basic operation of addition of positive real numbers, replacing $x+y$ by $x \vee y := \max(x, y)$. When endowed with this operation as addition and with the usual multiplication, the positive real numbers become a semifield $\mathbb{R}_+^{\max}$. It is of characteristic 1, i.e. $1 \vee 1 = 1$ and contains the smallest semifield of characteristic 1, namely the Boolean semifield $\mathbb{B} = \{0, 1\}$. Moreover, $\mathbb{R}_+^{\max}$ admits non-trivial automorphisms and one has

$$\mathrm{Gal}_{\mathbb{B}}(\mathbb{R}_+^{\max}) := \mathrm{Aut}_{\mathbb{B}}(\mathbb{R}_+^{\max}) = \mathbb{R}_+^*, \;\; \mathrm{Fr}_\lambda(x) = x^\lambda\,, \;\; \forall x \in \mathbb{R}_+^{\max},\; \lambda \in \mathbb{R}_+^*$$

thus providing a first glimpse of an answer to Weil's query in [100] of an algebraic framework in which the connected component of the idele class group would appear as a Galois group. More generally, for any abelian ordered group H we let $H_{\max} = H \cup \{-\infty\}$ be the semifield obtained from H by the max-plus construction, i.e. the addition is given by the max, and the multiplication by $+$. In particular $\mathbb{R}_{\max}$ is isomorphic to $\mathbb{R}_+^{\max}$ by the exponential map (cf. [49]). Historically, and besides the uses of $\mathbb{R}_{\max}$ in idempotent analysis and tropical geometry which are discussed below, an early use of $\mathbb{R}_{\max}$ occurred in the late fifties in the work of R. Cuninghame-Green in Birmingham, who established the spectral theory of irreducible matrices with entries in $\mathbb{R}_{\max}$ (cf. [34]) and in the sixties, in Leningrad, where Vorobyev used the $\mathbb{R}_{\max}$ formalism in his work motivated by combinatorial optimization, and proved a fundamental covering theorem. A systematic use of the $\mathbb{R}_{\max}$ algebra was developed by the INRIA group at the beginning of the 80's in their work on the modelization of discrete event systems [17]. We refer to [49, 50] for a more detailed history of the subject, and for overwhelming evidence of its relevance in mathematics. We shall just give here a sample of this evidence starting by a really early occurrence in the work of C.G.J. Jacobi[8] and hoping to convince the reader that it would be a mistake to dismiss this algebraic formalism and the analogy with ordinary algebra as trivial.

4.2.1 Optimization, Jacobi

One of the early instances, around 1840, of the use of matrices over $\mathbb{R}_{\max}$ is the work of Jacobi [67] on optimal assignment problems, where he states

> Problema
>
> Disponantur nn quantitates $h_k^{(i)}$ quaecunque in schema Quadrati, ita ut habeantur n series horizontales et n series verticales, quarum quaeque est n terminorum. Ex illis quantitatibus eligantur n transversales, i.e. in seriebus horizontalibus simul atque verticalibus diversis positae, quod fieri potest n! modis; ex omnibus illis modis quaerendus est is, qui summam n numerorum electorum suppeditet maximam.

In other words, given a square matrix $m_{ik} = h_k^{(i)}$ he looks for the maximum over all permutations σ of the quantity $\sum m_{j\sigma(j)}$. Using the algebraic rules of $\mathbb{R}_{\max}$ one checks that he is in fact computing the analogue of the determinant for the matrix m_{ik}. In fact the perfect definition of the determinant is more subtle and was obtained in the work of Gondran-Minoux [53], instead of $\max \sum m_{j\sigma(j)}$ where σ runs over all permutations, one uses the signature of permutations and considers the pair

$$(\det_+(m_{ik}), \det_-(m_{ik})), \quad \det_\pm(m_{ik}) = \max \sum_{\mathrm{sign}(\sigma)=\pm} m_{j\sigma(j)}$$

The remarkable fact is that the Cayley-Hamilton theorem now holds, as the equality of two terms $P_+(m) = P_-(m)$ corresponding to the characteristic polynomial $P = (P_+, P_-)$. Each of the terms $P_\pm(m) \in M_n(\mathbb{R}_{\max})$ is computed from the original matrix $m \in M_n(\mathbb{R}_{\max})$ using the rules of matrices with entries in $\mathbb{R}_{\max}$ which turn $M_n(\mathbb{R}_{\max})$ into a semiring.

---

[8] I am grateful to S. Gaubert for pointing out this early occurrence.

4.2.2 Idempotent Analysis

The essence of the theory of semiclassical analysis in physics rests in the comparison of quantum systems with their semiclassical counterpart, [5, 46, 55, 57, 58]. In the eighties V. P. Maslov and his collaborators developed a satisfactory algebraic framework which encodes the semiclassical limit of quantum mechanics. They called it idempotent analysis. We refer to [72, 74] for a detailed account and just mention briefly some salient features here. The source of the variational formulations of mechanics in the classical limit is the behavior of sums of exponentials

$$\sum e^{-\frac{S_j}{\hbar}} \sim e^{-\frac{\inf S_j}{\hbar}}, \quad \text{when } \hbar \to 0$$

which are, when $\hbar \to 0$, dominated by the contribution of the minimum of S. The starting observation is that one can encode this fundamental principle by simply conjugating the addition of numbers by the power operation $x \mapsto x^\epsilon$ and passing to the limit when $\epsilon \to 0$. The new addition of positive real numbers is

$$\lim_{\epsilon \to 0} \left(x^{\frac{1}{\epsilon}} + y^{\frac{1}{\epsilon}} \right)^{\epsilon} = \max\{x, y\} = x \vee y$$

and one recovers $\mathbb{R}_+^{\max}$ as the natural home for semiclassical analysis. The superposition principle of quantum mechanics, i.e. addition of vectors in Hilbert space, now makes sense in the limit and moreover the "fixed point argument" proof of the Perron-Frobenius theorem works over $\mathbb{R}_+^{\max}$ and shows that irreducible compact operators have one and only one eigenvalue,[9] thus reconciling classical determinism with the quantum variability. But the most striking discovery of this school of Maslov, Kolokolstov and Litvinov [72, 74] is that the Legendre transform which plays a fundamental role in all of physics and in particular in thermodynamics in the nineteenth century, is simply the Fourier transform in the framework of idempotent analysis!

The contact between the INRIA school and the Maslov school was established in 92 when Maslov was invited in the Seminar of Jacques Louis Lions in College de France. At the BRIMS HP-Labs workshop on Idempotency in Bristol (1994) organized by J. Gunawardena, several of the early groups of researchers in the field were there, and an animated discussion took place on how the field should be named. The names max-plus, exotic, tropical, idempotent were considered, each one having its defaults.

[9] As mentioned above, this result was obtained already for matrices in 1962 by R. Cuninghame-Green.

4.2.3 Tropical Geometry, Riemann-Roch Theorems and the Chip Firing Game

The tropical semiring $\mathbb{N}_{\min} = \mathbb{N} \cup \{\infty\}$ with the operations min and $+$ was introduced by Imre Simon in [90] to solve a decidability problem in rational language theory. His work is at the origin of the term "tropical" coined by Marco schutzenberger and used in tropical geometry which is a vast subject, see e.g. [45, 51, 77, 81]. We refer to [96] for an excellent introduction starting from the sixteenth Hilbert problem. In its simplest form (cf. [48]) a tropical curve is given by a metric graph Γ (i.e. a graph with a usual line metric on its edges). The natural structure sheaf on Γ is the sheaf $\mathcal{O}$ of real valued functions which are continuous, convex, piecewise affine with integral slopes. The operations on such functions are given by the pointwise operations of $\mathbb{R}_{\max}$-valued functions, i.e. $(f \vee g)(x) = f(x) \vee g(x)$ for all $x \in \Gamma$ and similar for the product which is given by pointwise addition. One also adjoins the constant $-\infty$ which plays the role of the zero element in the semirings of sections. One proceeds as in the classical case with the construction of the sheaf $\mathcal{K}$ of semifields of quotients and finds the same type of functions as above but no longer convex. Cartier divisors make sense and one finds that the order of a section f of $\mathcal{K}$ at a point $x \in \Gamma$ is given by the sum of the (integer valued) outgoing slopes. The conceptual explanation of why the discontinuities of the derivative should be interpreted as zeros or poles is due to Viro, [97] who showed that it follows automatically if one understands that[10] the sum $x \vee x$ of two equal terms in $\mathbb{R}_{\max}$ should be viewed as ambiguous with all values in the interval $[-\infty, x]$ on equal footing. In their work Baker and Norine [3] proved in the discrete set-up of graphs (where g is the genus and K the canonical divisor) the Riemann-Roch equality in the form

$$r(D) - r(K - D) = \mathrm{Deg}(D) - g + 1 \tag{30}$$

where by definition $r(D) := \max\{k \mid H^0(D - \tau) \neq \{-\infty\}\,,\ \forall \tau \geq 0,\ \mathrm{Deg}(\tau) = k\}$ and $H^0(D)$ is the $\mathbb{R}_{\max}$-module of global sections f of the associated sheaf $\mathcal{O}_D$ i.e. sections of $\mathcal{K}$ such that $D + (f) \geq 0$. The essence of the proof of [3] is that the inequality $\mathrm{Deg}(D) \geq g$ for a divisor implies $H^0(D) \neq \{-\infty\}$. Once translated in the language of the chip firing game (*op.cit.*), this fact is equivalent to the existence of a winning strategy if one assumes that the total sum of dollars attributed to the vertices of the graph is $\geq g$ where g is the genus. We refer to [48, 82] for variants of the above Riemann-Roch theorem, and to [8, 40, 84] for early occurrences of these ideas in a different context (including sandpile models and parking functions!).

[10] As seen when using $\mathbb{R}_{\max}$ as the target of a valuation.

4.3 The Arithmetic and Scaling Sites

4.3.1 The Arithmetic Site and Frobenius Correspondences

The *arithmetic site* [30, 31] is an object of algebraic geometry involving two elaborate mathematical concepts: the notion of topos and of (structures of) characteristic 1 in algebra. A nice fact (cf. [52]) in characteristic 1 is that, provided the semiring R is multiplicatively cancellative (i.e. equivalently if it injects in its semifield of fractions) the map $x \mapsto x^n = \mathrm{Fr}_n(x)$ is, for any integer $n \in \mathbb{N}^\times$, an injective endomorphism Fr_n of R. One thus obtains a canonical action of the semigroup $\mathbb{N}^\times$ on any such R and it is thus natural to work in the topos $\widehat{\mathbb{N}^\times}$ of sets endowed with an action of $\mathbb{N}^\times$.

Definition 2. The arithmetic site $\mathscr{A} = (\widehat{\mathbb{N}^\times}, \mathbb{Z}_{\max})$ is the topos $\widehat{\mathbb{N}^\times}$ endowed with the *structure sheaf* $\mathscr{O} := \mathbb{Z}_{\max}$ viewed as a semiring in the topos using the action of $\mathbb{N}^\times$ by the Frobenius endomorphisms.

The topological space underlying the arithmetic site is the Grothendieck topos of sets endowed with an action of the multiplicative monoïd $\mathbb{N}^\times$ of non-zero positive integers. As we have seen above the semifield $\mathbb{R}_+^{\max}$ of tropical real numbers admits a one parameter group of Frobenius automorphisms Fr_λ, $\lambda \in \mathbb{R}_+^\times$, given by $\mathrm{Fr}_\lambda(x) = x^\lambda\ \forall x \in \mathbb{R}_+^{\max}$. Using a straightforward extension in the context of semi-ringed topos of the classical notion of algebraic geometry of a point over a ring, one then gets the following result which gives the bridge between the noncommutative geometry and topos points of view:

Theorem 1 ([30, 31]). *The set of points of the arithmetic site $\mathscr{A}$ over $\mathbb{R}_+^{\max}$ is canonically isomorphic with $X_\mathbb{Q} = \mathbb{Q}^\times \backslash \mathbb{A}_\mathbb{Q} / \hat{\mathbb{Z}}^\times$. The action of the Frobenius automorphisms Fr_λ of $\mathbb{R}_+^{\max}$ on these points corresponds to the action of the idele class group on $X_\mathbb{Q} = \mathbb{Q}^\times \backslash \mathbb{A}_\mathbb{Q} / \hat{\mathbb{Z}}^\times$.*

The square of the arithmetic site is the topos $\widehat{\mathbb{N}^{\times 2}}$ endowed with the structure sheaf defined globally by the multiplicatively cancellative semiring associated to the tensor square $\mathbb{Z}_{\min} \otimes_\mathbb{B} \mathbb{Z}_{\min}$ over the smallest Boolean semifield of characteristic one. In this way one obtains the semiring whose elements are Newton polygons and whose operations are given by the convex hull of the union and the sum. The points of the square of the arithmetic site over $\mathbb{R}_+^{\max}$ coincide with the product of the points of the arithmetic site over $\mathbb{R}_+^{\max}$. Then, we describe the Frobenius correspondences $\Psi(\lambda)$ as congruences on the square parametrized by positive real numbers $\lambda \in \mathbb{R}_+^\times$.

> The remarkable fact at this point is that while the arithmetic site is constructed as a combinatorial object of countable nature it possesses nonetheless a one parameter semigroup of "correspondences" which can be viewed as congruences in the square of the site.

In the context of semirings, the congruences i.e. the equivalence relations compatible with addition and product, play the role of the ideals in ring theory. The Frobenius correspondences $\Psi(\lambda)$, for a rational value of λ, are deduced from the diagonal of the square, which is described by the product structure of the semiring, by composition with the Frobenius endomorphisms. We interpret these correspondences geometrically, in terms of the congruence relation on Newton polygons corresponding to their belonging to the same half planes with rational slope λ. These congruences continue to make sense also for irrational values of λ and are described using the best rational approximations of λ, while different values of the parameter give rise to distinct congruences. The composition of the Frobenius correspondences is given for $\lambda, \lambda' \in \mathbb{R}_+^\times$ such that $\lambda\lambda' \notin \mathbb{Q}$ by the rule [30, 31]

$$\Psi(\lambda) \circ \Psi(\lambda') = \Psi(\lambda\lambda') \tag{31}$$

The same equality still holds if λ and λ' are rational numbers. When λ, λ' are irrational and $\lambda\lambda' \in \mathbb{Q}$ one has

$$\Psi(\lambda) \circ \Psi(\lambda') = \mathrm{Id}_\epsilon \circ \Psi(\lambda\lambda') \tag{32}$$

where Id_ϵ is the tangential deformation of the identity correspondence.

4.3.2 The Scaling Site and Riemann-Roch Theorems

The Scaling Site $\hat{\mathscr{A}}$, [33], is the algebraic geometric space obtained from the arithmetic site $\mathscr{A}$ of [30, 31] by extension of scalars from the Boolean semifield $\mathbb{B}$ to the tropical semifield $\mathbb{R}_+^{\max}$. The points of $\hat{\mathscr{A}}$ are the same as the points $\mathscr{A}(\mathbb{R}_+^{\max})$ of the arithmetic site over $\mathbb{R}_+^{\max}$. But $\hat{\mathscr{A}}$ inherits from its structural sheaf a natural structure of tropical curve, in a generalized sense, allowing one to define the sheaf of rational functions and to investigate an adequate version of the Riemann-Roch theorem in characteristic 1. In [33], we tested this structure by restricting it to the periodic orbits of the scaling flow, i.e. the points over the image of Spec $\mathbb{Z}$ under the canonical morphism of toposes $\Theta : \operatorname{Spec}\mathbb{Z} \to \mathscr{A}$ (cf. [31, Sect. 5.1]). We found that for each prime p the corresponding circle of length $\log p$ is endowed with a quasi-tropical structure which turns this orbit into the analogue $C_p = \mathbb{R}_+^*/p^{\mathbb{Z}}$ of a classical elliptic curve $\mathbb{C}^*/q^{\mathbb{Z}}$. In particular rational functions, divisors, etc all make sense. A new feature is that the degree of a divisor can now be any real number. The Jacobian of C_p (i.e. the quotient $J(C_p)$ of the group of divisors of degree 0 by principal divisors) is a cyclic group of order $p-1$. For each divisor D there is a corresponding Riemann-Roch problem with solution space $H^0(D)$ and the continuous dimension $\mathrm{Dim}_{\mathbb{R}}(H^0(D))$ of this $\mathbb{R}_{\max}$-module is defined as the limit

$$\mathrm{Dim}_{\mathbb{R}}(H^0(D)) := \lim_{n\to\infty} p^{-n}\dim_{\mathrm{top}}(H^0(D)^{p^n}) \tag{33}$$

where $H^0(D)^{p^n}$ is a natural filtration and $\dim_{top}(\mathscr{E})$ is the topological dimension of an $\mathbb{R}_{max}$-module $\mathscr{E}$. One has the following Riemann-Roch formula [33],

Theorem 2. *(i) Let $D \in \mathrm{Div}(C_p)$ be a divisor with $\deg(D) \geq 0$. Then the limit in (33) converges and one has $\mathrm{Dim}_{\mathbb{R}}(H^0(D)) = \deg(D)$.*
(ii) The following Riemann-Roch formula holds

$$\mathrm{Dim}_{\mathbb{R}}(H^0(D)) - \mathrm{Dim}_{\mathbb{R}}(H^0(-D)) = \deg(D)\,, \quad \forall D \in \mathrm{Div}(C_p)$$

The appearance of arbitrary positive real numbers as continuous dimensions in the Riemann-Roch formula is due to the density in $\mathbb{R}$ of the subgroup $H_p \subset \mathbb{Q}$ of fractions with denominators a power of p. This outcome is the analogue in characteristic 1 of what happens for modules over matroid C^*-algebras and the type II normalized dimensions as in [41].

At this point, what is missing is an intersection theory and a Riemann-Roch theorem on the square of the arithmetic site. One expects that the right hand side of the Riemann-Roch formula will be of the form $\frac{1}{2}D.D = \mathfrak{s}(f,f)$ when the divisor D is of the form

$$D(f) = \int \Psi(\lambda) f(\lambda) d^*\lambda$$

Here $f(\lambda)$ is a real valued function with compact support of the variable $\lambda \in \mathbb{R}_+^*$ and $\mathfrak{s}(f,f)$ is as in (17). More precisely $D.D$ should be obtained as the intersection number of $D \circ \tilde{D}$ (defined using composition of correspondences) with the diagonal Δ and hence as a suitably defined distributional trace as for the counting function $N(u)$ of Sect. 4.1 so that $\frac{1}{2}D(f).D(f) = \mathfrak{s}(f,f)$ with the notations of (17). So far the Riemann-Roch formula in tropical geometry is limited to curves and there is no Serre duality or good cohomological version of H^j for $j \neq 0$, but in the above context one can hope that a Riemann-Roch inequality of the type (12), i.e. of the form

$$\mathrm{Dim}_{\mathbb{R}}(H^0(D)) + \mathrm{Dim}_{\mathbb{R}}(H^0(-D)) \geq \frac{1}{2} D.D$$

would suffice to apply the strategy of Sect. 2.3 to prove the key inequality (17) (Table 1).

Table 1 Here are a few entries in the analogy

C curve over $\mathbb{F}_q$	Arithmetic site $\mathscr{A} = (\widehat{\mathbb{N}^\times}, \mathbb{Z}_{\max})$ over $\mathbb{B}$
Structure sheaf $\mathscr{O}_C$	Structure sheaf $\mathbb{Z}_{\max}$
$\bar{C} = C \otimes_{\mathbb{F}_q} \bar{\mathbb{F}}_q$	Scaling site $\hat{\mathscr{A}} = ([0,\infty) \rtimes \mathbb{N}^\times, \mathscr{O})$ over $\mathbb{R}_+^{\max}$
$C(\bar{\mathbb{F}}_q) = \bar{C}(\bar{\mathbb{F}}_q)$	$\mathscr{A}(\mathbb{R}_+^{\max}) = \hat{\mathscr{A}}(\mathbb{R}_+^{\max})$
Galois action on $C(\bar{\mathbb{F}}_q)$	Galois action on $\mathscr{A}(\mathbb{R}_+^{\max})$
Structure sheaf $\mathscr{O}_{\bar{C}}$ of $\bar{C} = C \otimes_{\mathbb{F}_q} \bar{\mathbb{F}}_q$	Structure sheaf $\mathscr{O} = \mathbb{Z}_{\max} \hat{\otimes}_{\mathbb{B}} \mathbb{R}_+^{\max}$ piecewise affine convex functions, integral slopes
Sheaf $\mathscr{K}$ of rational functions $\bar{C}$ on $\bar{C} = C \otimes_{\mathbb{F}_q} \bar{\mathbb{F}}_q$	Sheaf $\mathscr{K}$ of piecewise affine continuous functions with integral slopes
Cartier divisors = sections of $\mathscr{K}/\mathscr{O}^*$	Sections of $\mathscr{K}/\mathscr{O}^*$
$X = \bar{C} \times \bar{C}$	$\hat{\mathscr{A}} \times \hat{\mathscr{A}}$
$D = \sum a_k \Psi^k$ Frobenius correspondence Ψ	$D = \int \Psi(\lambda) f(\lambda) d^*\lambda$ correspondences $\Psi(\lambda)$

5 Absolute Algebra and the Sphere Spectrum

Even if the Riemann-Roch strategy of Sect. 4 happened to be successful, one should not view the arithmetic and scaling sites for more than what they are, namely a semiclassical shadow of a still mysterious structure dealing with compactifications of Spec $\mathbb{Z}$. An essential role in the unveiling of this structure should be played, for the reasons briefly explained below, by the discovery made by algebraic topologists in the 80's (see [42]) that in their world of "spectra" (in their sense) the sphere spectrum is a generalized ring $\mathbb{S}$ which is more fundamental than the ring $\mathbb{Z}$ of integers, while the latter becomes an $\mathbb{S}$-algebra. Over the years the technical complications of dealing with spaces "up to homotopy" have greatly been simplified, in particular for the smash product of spectra. For the purpose of arithmetic applications, Segal's Γ-rings provide a very simple algebraic framework which succeeds to unify several attempts pursued in recent times in order to define the meaning of "absolute algebra". In particular it contains the following three possible categories that had been considered previously to handle this unification: namely the category $\mathscr{M}$ of monoïds as in [24, 26, 35, 36], the category $\mathscr{H}$ of hyperrings of [25, 27, 28] and finally the category $\mathscr{S}$ of semirings as in [21, 30, 31, 33]. Thanks to the work of L. Hesselholt and I. Madsen briefly explained below in Sect. 5.2 one now has at disposal a candidate cohomology theory in the arithmetic context: topological cyclic homology.

5.1 Segal's Γ-Rings

Let Γ^{op} be the small, full subcategory of the category of finite pointed sets whose objects are the pointed finite sets[11] $k_+ := \{0, \ldots, k\}$, for $k \geq 0$. The object 0_+ is both initial and final so that Γ^{op} is a *pointed category*. The notion of a discrete Γ-space, i.e. of a Γ-set is as follows:

Definition 3. A Γ-set F is a functor $F : \Gamma^{\mathrm{op}} \longrightarrow \mathfrak{Sets}_*$ between pointed categories from Γ^{op} to the category of pointed sets.

The morphisms $\mathrm{Hom}_{\Gamma^{\mathrm{op}}}(M, N)$ between two Γ-sets are natural transformations of functors. The category $\Gamma\mathfrak{Sets}_*$ of Γ-sets is a symmetric closed monoidal category (cf. [42, Chap. II]). The monoidal structure is given by the smash product (denoted $X \wedge Y$) of Γ-sets which is a Day product. The closed structure property is shown in [76] (cf. also [42, Theorem 2.1.2.4]). The specialization of Definition 2.1.4.1. of [42] to the case of Γ-sets yields the following

Definition 4. A Γ-ring $\mathscr{A}$ is a Γ-set $\mathscr{A} : \Gamma^{\mathrm{op}} \longrightarrow \mathfrak{Sets}_*$ endowed with an associative multiplication $\mu : \mathscr{A} \wedge \mathscr{A} \to \mathscr{A}$ and a unit $1 : \mathbb{S} \to \mathscr{A}$, where $\mathbb{S} : \Gamma^{\mathrm{op}} \longrightarrow \mathfrak{Sets}_*$ is the inclusion functor.

Thus Γ-rings[12] make sense and the sphere spectrum corresponds to the simplest possible Γ-ring: $\mathbb{S}$. One can then easily identify the category $\Gamma\mathfrak{Sets}_*$ of Γ-sets with the category $\mathfrak{Mod}(\mathbb{S})$ of $\mathbb{S}$-modules. In [43], N. Durov developed a geometry over $\mathbb{F}_1$ intended for Arakelov theory applications by using monads as generalizations of classical rings. While in the context of [43] the tensor product $\mathbb{Z} \otimes_{\mathbb{F}_1} \mathbb{Z}$ produces an uninteresting output isomorphic to $\mathbb{Z}$, we showed in [32] that the same tensor square, re-understood in the theory of $\mathbb{S}$-algebras, provides a highly non-trivial object. The Arakelov compactification of $\mathrm{Spec}\,\mathbb{Z}$ is endowed naturally with a structure sheaf of $\mathbb{S}$-algebras and each Arakelov divisor provides a natural sheaf of modules over the structure sheaf. This new structure of $\overline{\mathrm{Spec}\,\mathbb{Z}}$ over $\mathbb{S}$ endorses a one parameter group of weakly invertible sheaves whose tensor product rules are the same as the composition rules (31), (32) of the Frobenius correspondences over the arithmetic site [30, 31]. The category $\mathfrak{Mod}(\mathbb{S})$ of $\mathbb{S}$-modules is not an abelian category and thus the tools of homological algebra need to be replaced along the line of the Dold-Kan correspondence, which for an abelian category $\mathscr{A}$ gives the correspondence between chain complexes in ≥ 0 degrees and simplicial objects i.e. objects of $\mathscr{A}^{\Delta^{\mathrm{op}}}$.

[11] Where 0 is the base point.

[12] Equivalently $\mathbb{S}$-algebras.

Table 2 Short dictionary homology–homotopy

$X \in Ch_{\geq 0}(\mathscr{A})$	$M \in \mathfrak{Mod}(\mathbb{S})^{\Delta^{op}}$
$H_q(X)$	$\pi_q(M)$
$H_q(f) : H_q(X) \simeq H_q(Y)$ quasi-isomorphism	$\pi_q(f) : \pi_q(M) \simeq \pi_q(N)$ weak equivalence
$f_n : X_n \overset{\subset}{\to} Y_n$ + projective cokernel	Cofibration (stable)
$f_n : X_n \to Y_n$ surjective if $n > 0$	Stable fibration

At this point one has the following simple but very important observation that Γ-spaces should be viewed as simplicial objects in $\Gamma\mathfrak{Sets}_* \equiv \mathfrak{Mod}(\mathbb{S})$, so that homotopy theory should be considered as the homological algebra corresponding to the "absolute algebra" taking place over $\mathbb{S}$.

We refer to Table 2 for a short dictionary. The category of Γ-spaces is the central tool of [42], while the relations between algebraic K-theory and topological cyclic homology is the main topic.

5.2 *Topological Cyclic Homology*

As shown in [32] the various attempts done in recent times to develop "absolute algebra" are all unified by means of the well established concept of $\mathbb{S}$-algebra, i.e. of Γ-rings. Moreover (cf. [42]) this latter notion is at the root of the theory of topological cyclic homology which can be understood as cyclic homology over the absolute base $\mathbb{S}$, provided one uses the appropriate Quillen model category. In particular, topological cyclic homology is now available to understand the new structure of $\overline{\mathrm{Spec}\,\mathbb{Z}}$ using its structure sheaf and modules. The use of cyclic homology in the arithmetic context is backed up by the following two results:

- At the archimedean places, and after the initial work of Deninger [38, 39] to recast the archimedean local factors of arithmetic varieties [89] as regularized determinants, we showed in [29] that cyclic homology in fact gives the correct infinite dimensional (co)homological theory for arithmetic varieties. The key operator Θ in this context is the generator of the λ-operations $\Lambda(k)$ [75, 98, 99] in cyclic theory. More precisely, the action u^Θ of the multiplicative group $\mathbb{R}_+^\times$ generated by Θ on cyclic homology, is uniquely determined by its restriction to the dense subgroup $\mathbb{Q}_+^\times \subset \mathbb{R}_+^\times$ where it is given by the formula

$$k^\Theta|_{HC_n} = \Lambda(k)\, k^{-n}\,, \quad \forall n \geq 0,\ k \in \mathbb{N}^\times \subset \mathbb{R}_+^\times \tag{34}$$

Let X be a smooth, projective variety of dimension d over an algebraic number field $\mathbb{K}$ and let $\nu|\infty$ be an archimedean place of $\mathbb{K}$. Then, the action of the operator Θ on the archimedean cyclic homology HC^{ar} (cf. [29]) of X_ν satisfies

$$\prod_{0\leq w\leq 2d} L_\nu(H^w(X),s)^{(-1)^w} = \frac{det_\infty(\frac{1}{2\pi}(s-\Theta)|_{HC^{ar}_{od}(X_\nu)})}{det_\infty(\frac{1}{2\pi}(s-\Theta)|_{HC^{ar}_{ev}(X_\nu)})} \tag{35}$$

The left-hand side of (35) is the product of Serre's archimedean local factors of the complex L-function of X (cf.[89]). On the right-hand side, det_∞ denotes the regularized determinant and one sets

$$HC^{ar}_{ev}(X_\nu) = \bigoplus_{n=2k\geq 0} HC^{ar}_n(X_\nu),\ HC^{ar}_{od}(X_\nu) = \bigoplus_{n=2k+1\geq 1} HC^{ar}_n(X_\nu)$$

- L. Hesselholt and I. Madsen have shown (cf. e.g. [61–63]) that the de Rham-Witt complex, an essential ingredient of crystalline cohomology (cf. [7, 65]), arises naturally when one studies the topological cyclic homology of smooth algebras over a perfect field of finite characteristic. One of the remarkable features in their work is that the arithmetic ingredients such as the Frobenius and restriction maps are naturally present in the framework of topological cyclic homology. Moreover L. Hesselholt has shown [64] how topological periodic cyclic homology with its inverse Frobenius operator may be used to give a cohomological interpretation of the Hasse-Weil zeta function of a scheme smooth and proper over a finite field in the form (cf. [64]):

$$\zeta(X,s) = \frac{det_\infty(\frac{1}{2\pi}(s-\Theta)|_{TP_{od}(X)})}{det_\infty(\frac{1}{2\pi}(s-\Theta)|_{TP_{ev}(X)})} \tag{36}$$

The similarity between (35) and (36) (applied to a place of good reduction) suggests the existence of a global formula for the L-functions of arithmetic varieties, involving cyclic homology of $\mathbb{S}$-algebras, and of a Lefschetz formula in which the local factors appear from the periodic orbits of the action of $\mathbb{R}_+^*$.

One of the stumbling blocks in order to reach a satisfactory cohomology theory is the problem of coefficients. Indeed, the natural coefficients at a prime p for crystalline cohomology are an extension of $\mathbb{Q}_p$ and it is traditional to relate them with complex numbers by an embedding of fields. Similarly, (36) uses an embedding of the Witt ring $W(\mathbb{F}_q) \to \mathbb{C}$. To an analyst it is clear that since such embeddings cannot be measurable[13] they will never be effectively constructed. This begs for a better construction, along the lines of Quillen's computation of the algebraic

[13] A measurable group homomorphism from $\mathbb{Z}_p^\times$ to $\mathbb{C}^\times$ cannot be injective.

K-theory of finite fields, which instead would only involve the ingredient of the Brauer lifting, i.e. a group injection of the multiplicative group of $\bar{\mathbb{F}}_p$ as roots of unity in $\mathbb{C}$.

5.3 Final Remarks

The Riemann hypothesis has been extended far beyond its original formulation to the question of localization of the zeros of L-functions. There are a number of constructions of L-functions coming from three different sources, Galois representations, automorphic forms and arithmetic varieties. André Weil liked to compare (cf. [10, Sect. 12] and also [104, vol. 1, p. 244–255 and vol. 2, p. 408–412]), the puzzle of these three different writings to the task of deciphering hieroglyphics with the help of the Rosetta Stone. In some sense the L-functions play a role in modern mathematics similar to the role of polynomials in ancient mathematics, while the explicit formulas play the role of the expression of the symmetric functions of the roots in terms of the coefficients of the polynomial. If one follows this line of thought, the RH should be seen only as a first step since in the case of polynomials there is no way one should feel to have understood the zeros once one proves that they are, say, real numbers. In fact Galois formulated precisely the problem as that of finding all numerical relations between the roots of an equation, with the trivial ones being given by the symmetric functions, while the others, when determined, will reveal a complete understanding of the zeros as obtained, in the case of polynomials, by Galois theory. In a fragment, page 103, of the complete works of Galois [47] concerning the memoir of February 1830, he delivers the essence of his theory:

> Remarquons que tout ce qu'une équation numérique peut avoir de particulier, doit provenir de certaines relations entre les racines. Ces relations seront rationnelles c'est-à-dire qu'elles ne contiendront d'irrationnelles que les coefficients de l'équation et les quantités adjointes. De plus ces relations ne devront pas être invariables par toute substitution opérée sur les racines, sans quoi on n'aurait rien de plus que dans les équations littérales. Ce qu'il importe donc de connaître, c'est par quelles substitutions peuvent être invariables des relations entre les racines, ou ce qui revient au même, des fonctions des racines dont la valeur numérique est déterminable rationnellement.[14]

Acknowledgements I am grateful to J. B. Bost for the reference [94], to J. B. Bost, P. Cartier, C. Consani, D. Goss, H. Moscovici, M. Th. Rassias, C. Skau and W. van Suijlekom for their detailed comments, to S. Gaubert for his help in Sect. 4.2 and to Lars Hesselholt for his comments and for allowing me to mention his forthcoming paper [64].

[14] In 2012 I had to give, in the French academy of Sciences, the talk devoted to the 200th anniversary of the birth of Evariste Galois. On that occasion I read for the $n+1$th time the book of his collected works and was struck by the pertinence of the above quote in the analogy with L-functions. In the case of function fields one is dealing with Weil numbers and one knows a lot on their Galois theory using results such as those of Honda and Tate cf. [94].

References

1. J. Arthur, *An introduction to the trace formula*. Harmonic analysis, the trace formula, and Shimura varieties, 1-263, Clay Math. Proc., 4, Amer. Math. Soc., Providence, RI, 2005.
2. M. Artin, A. Grothendieck, J-L. Verdier, eds. (1972), SGA_4, LNM 269-270-305, Berlin, New York, Springer-Verlag.
3. M. Baker, S. Norine, *Riemann-Roch and Abel-Jacobi theory on a finite graph*, Advances in Mathematics 215 (2007), 766–788.
4. B. Bhatt, P. Scholze, *The pro-etale topology for schemes* Preprint (2013), arXiv:1309.1198
5. M. Berry, *Riemann's zeta function: a model of quantum chaos*, Lecture Notes in Physics, Vol.263, Springer-Verlag, 1986.
6. M. Berry and J. Keating, $H = qp$ *and the Riemann zeros*, "Supersymmetry and Trace Formulae: Chaos and Disorder", edited by J.P. Keating, D.E. Khmelnitskii and I.V. Lerner, Plenum Press.
7. P. Berthelot, *Cohomologie cristalline des schémas de caractéristique* $p > 0$, Lecture Notes in Math., vol. 407, Springer-Verlag, New York, 1974.
8. A. Bjorner, L. Lovasz, P. W. Shor, *Chip-firing games on graphs*, European J. Combin., 12(4), (1991), 283–291.
9. E. Bombieri, *Problems of the Millenium: The Riemann Hypothesis*, Clay mathematical Institute (2000).
10. E. Bombieri, *The classical theory of Zeta and L-functions*, Milan J. Math. Vol. 78 (2010) 11–59.
11. E. Bombieri, J. Lagarias *Complements to Li's criterion for the Riemann hypothesis*. J. Number Theory 77 (1999), no. 2, 274–287.
12. J.B. Bost, A. Connes, *Hecke algebras, Type III factors and phase transitions with spontaneous symmetry breaking in number theory*, Selecta Math. (New Series) Vol.1 (1995) N.3, 411–457.
13. F. Bruhat, *Distributions sur un groupe localement compact et applications à l'étude des représentations des groupes p-adiques*. Bull. Soc. Math. France, 89 (1961), 43–75.
14. J.F. Burnol, *The explicit formula and the conductor operator*, math.NT/9902080.
15. J.F. Burnol, *Sur les formules explicites. I. Analyse invariante*. C. R. Acad. Sci. Paris Sér. I Math., 331 (2000), 423–428.
16. P. Cartier, *Des nombres premiers à la géométrie algébrique (une brève histoire de la fonction zêta)*. Analyse diophantienne et géométrie algébrique, 51–77, Cahiers Sém. Hist. Math. Sér. 2, 3, Univ. Paris VI, Paris, 1993.
17. G. Cohen, S. Gaubert, R. Nikoukhah, J.P. Quadrat, *Convex analysis and spectral analysis of timed event graphs*, Decision and Control, 1989, Proceedings of the 28th IEEE Conference.
18. A. Connes, *A survey of foliations and operator algebras*. In "Operator algebras and applications", Part I (Kingston, Ont., 1980), pp. 521–628, Proc. Sympos. Pure Math., 38, Amer. Math. Soc., Providence, R.I., 1982.
19. A. Connes, *Trace formula in noncommutative geometry and the zeros of the Riemann zeta function*. Selecta Math. (N.S.) 5 (1999), no. 1, 29–106.
20. A. Connes, *Formules explicites, formules de trace et réalisation spectrale des zéros de la fonction zéta*, Course at Collège de France, 1999.
21. A. Connes, *The Witt construction in characteristic one and Quantization*. Noncommutative geometry and global analysis, 83–113, Contemp. Math., 546, Amer. Math. Soc., Providence, RI, 2011.
22. A. Connes, M. Marcolli, *Noncommutative Geometry, Quantum Fields, and Motives*, Colloquium Publications, Vol.55, American Mathematical Society, 2008.
23. A. Connes, C. Consani, *Schemes over* $\mathbb{F}_1$ *and zeta functions*, Compositio Mathematica 146 (6), (2010) 1383–1415.
24. A. Connes, C. Consani, *From monoids to hyperstructures: in search of an absolute arithmetic*, in Casimir Force, Casimir Operators and the Riemann Hypothesis, de Gruyter (2010), 147–198.

25. A. Connes, C. Consani, *The hyperring of adèle classes*, Journal of Number Theory 131 (2011) 159–194.
26. A. Connes, C. Consani *On the arithmetic of the BC-system*, J. Noncommut. Geom. 8 (2014), no. 3, 873–945.
27. A. Connes, C. Consani, *Characteristic one, entropy and the absolute point*, "Noncommutative Geometry, Arithmetic, and Related Topics", the Twenty-First Meeting of the Japan-U.S. Mathematics Institute, Baltimore 2009, JHUP (2012), 75–139.
28. A. Connes, C. Consani, *The universal thickening of the field of real numbers*, Advances in the Theory of Numbers, Fields Institute Communications 77 (2015).
29. A. Connes, C. Consani, *Cyclic homology, Serre's local factors and the λ-operations*; J. K-Theory 14 (2014), no. 1, 1–45.
30. A. Connes, C. Consani, *The Arithmetic Site*, Comptes Rendus Mathématiques Ser. I 352 (2014), 971–975.
31. A. Connes, C. Consani, *Geometry of the Arithmetic Site*. Adv. Math. 291 (2016), 274–329. ArXiv: 1502.05580.
32. A. Connes, C. Consani, *Absolute algebra and Segal's Γ-rings*: au dessous de Spec(Z). J. Number Theory 162 (2016), 518–551.
33. A. Connes, C. Consani, *The Scaling Site*. C. R. Math. Acad. Sci. Paris 354 (2016), no. 1, 1–6.
34. R. Cuninghame-Green, *Minimax algebra*, Lecture Notes in Economics and Mathematical Systems, Volume 166, Springer, 1979.
35. A. Deitmar, *Schemes over $\mathbb{F}_1$*, in Number Fields and Function Fields Two Parallel Worlds. Ed. by G. van der Geer, B. Moonen, R. Schoof. Progr. in Math, vol. 239, 2005.
36. A. Deitmar, *$\mathbb{F}_1$-schemes and toric varieties*, Contributions to Algebra and Geometry Vol. 49, No. 2, pp. 517–525 (2008).
37. P. Deligne, *La conjecture de Weil. I.* Publ. Math. Inst. Hautes Études Sci. No. 43 (1974), 273–307.
38. C. Deninger, *On the Γ-factors attached to motives*, Invent. Math. 104 (1991) 245–261.
39. C. Deninger, *Motivic L-functions and regularized determinants*, in "Motives", Proceedings of Symposia in Pure Mathematics, Vol. 55 (1994) Part I, 707–743.
40. D. Dhar, *Self-organized critical state of sandpile automaton models*. Phys. Rev. Lett., 64(14):1613–1616, Apr 1990.
41. J. Dixmier, *On some C^*-algebras considered by Glimm*. J. Functional Analysis, 1, (1967), 182–203.
42. B. Dundas, T. Goodwillie, R. McCarthy, *The local structure of algebraic K-theory*. Algebra and Applications, 18. Springer-Verlag London, Ltd., London, 2013.
43. N. Durov, *New approach to Arakelov Geometry*. arXiv:0704.2030.
44. H.M. Edward, *Riemann's zeta function*, Dover, 2001.
45. M. Einsiedler, M. Kapranov, D. Lind, *Non-Archimedean amoebas and tropical varieties*. (English summary) J. Reine Angew. Math. 601 (2006), 139–157.
46. M. V. Fedoriuk, V. P. Maslov, *Semiclassical approximation in quantum mechanics*. Translated from the Russian by J. Niederle and J. Tolar. Mathematical Physics and Applied Mathematics, 7. Contemporary Mathematics, 5. D. Reidel Publishing Co., Dordrecht-Boston, Mass., 1981.
47. E. Galois, Oeuvres de Galois, Gauthier-Villars, Paris (1962).
48. A. Gathmann and M. Kerber, *A Riemann-Roch theorem in tropical geometry*. Math. Z., 259(1):217–230, 2008.
49. S. Gaubert, *Methods and applications of (max, +) linear algebra*, STACS 97 (Lubek), Lecture Notes in Comput. Sci., vol. 1200, Springer, Berlin, (1997), 261–282.
50. S. Gaubert, *Two lectures on the max-plus algebra*. Proceedings of the 26th Spring School of Theoretical Computer Science, (1998), 83–147.
51. I. Gelfand, M. Kapranov, A Zelevinsky, *Discriminants, resultants, and multidimensional determinants*. Mathematics: Theory and Applications. Birkhauser Boston, Inc., Boston, MA, 1994.
52. J. Golan, *Semi-rings and their applications*, Updated and expanded version of The theory of semi-rings, with applications to mathematics and theoretical computer science [Longman Sci. Tech., Harlow, 1992. Kluwer Academic Publishers, Dordrecht, 1999.

53. M. Gondran, M. Minoux, *L'indépendance linéaire dans les dioides*. (French) Bull. Direction Etudes Rech. Sér. C Math. Inform. 1978, no. 1, 67–90.
54. A. Grothendieck, *Sur une note de Mattuck-Tate* J. reine angew. Math. 200, 208–215 (1958).
55. V. Guillemin, S. Sternberg, *Geometric asymptotics*, Math. Surveys Vol. 14, American Mathematical Society, 1977.
56. V. Guillemin, *Lectures on spectral theory of elliptic operators*, Duke Math. J., Vol. 44, 3 (1977), 485–517.
57. M. Gutzwiller, *Classical Quantization of a Hamiltonian with Ergodic Behavior*, Physical Review Letters 45 (1980) 150–153.
58. M. Gutzwiller, *Chaos in classical and quantum mechanics*, Interdisciplinary Applied Mathematics, 1. Springer-Verlag, New York, 1990.
59. S. Haran, *Riesz potentials and explicit sums in arithmetic*, Invent. Math., 101 (1990), 697–703.
60. D. Hejhal, *The Selberg trace formula and the Riemann zeta function*. Duke Math. J. 43 (1976), no. 3, 441–482.
61. L. Hesselholt, *On the p-typical curves in Quillen's K-theory*. Acta Math. 177 (1996), no. 1, 1–53.
62. L. Hesselholt, *On the topological cyclic homology of the algebraic closure of a local field*, An Alpine Anthology of Homotopy Theory: Proceedings of the Second Arolla Conference on Algebraic Topology (Arolla, Switzerland, 2004), Contemp. Math., vol. 399, Amer. Math. Soc., Providence, RI, 2006, pp. 133–162.
63. L. Hesselholt, I. Madsen, *On the K-theory of finite algebras over Witt vectors of perfect fields*. Topology 36 (1997), no. 1, 29–102.
64. L. Hesselholt, *Periodic topological cyclic homology and the Hasse-Weil zeta function*
65. L. Illusie, *Complexe de Rham-Witt et cohomologie cristalline*, Ann. Scient. Ec. Norm. Sup. (4) 12 (1979), 501–661.
66. K. Iwasawa, *On the rings of valuation vectors* Ann. of Math. (2) 57, (1953). 331–356.
67. C.G.J. Jacobi, *De investigando ordine systematis aequationum differentialium vulgarium cujuscunque* C.G.J. Jacobi's gesammelte Werke, funfter Band, herausgegeben von K. Weierstrass, Berlin, Bruck und Verlag von Georg Reimer, 1890, p. 193–216.
68. N. Jacobson, *The radical and semi-simplicity for arbitrary rings*. Amer. J. Math. 67, (1945). 300–320.
69. B. Julia, Statistical theory of numbers, *Number Theory and Physics, Springer Proceedings in Physics,* **47** (1990).
70. M. Kapranov and A. Smirnov, *Cohomology determinants and reciprocity laws* Prepublication.
71. N. Kurokawa, *Multiple zeta functions: an example in Zeta functions in geometry* (Tokyo, 1990) Adv. Stud. Pure Math. Vol. 21 (1992), 219–226
72. V. Kolokoltsov, V. P. Maslov, *Idempotent analysis and its applications*. Mathematics and its Applications, 401. Kluwer Academic Publishers Group, Dordrecht, 1997.
73. X. J. Li, *The positivity of a sequence of numbers and the Riemann hypothesis*, J. Number Theory 65 (1997), 325–333.
74. G. Litvinov, *Tropical Mathematics, Idempotent Analysis, Classical Mechanics and Geometry*. Spectral theory and geometric analysis, 159–186, Contemp. Math., 535, Amer. Math. Soc., Providence, RI, 2011.
75. J.L. Loday, *Cyclic homology*. Grundlehren der Mathematischen Wissenschaften, 301. Springer-Verlag, Berlin, 1998.
76. M. Lydakis, *Smash products and Γ-spaces*, Math. Proc. Cambridge Philos. Soc. 126 (1999) 311–328.
77. D Maclagan, B. Sturmfels, *Introduction to tropical geometry*. Graduate Studies in Mathematics, 161. American Mathematical Society, Providence, RI, 2015.
78. S. Mac Lane, I Moerdijk, *Sheaves in geometry and logic. A first introduction to topos theory*. Corrected reprint of the 1992 edition. Universitext. Springer-Verlag, New York, 1994.
79. Yu.I. Manin, *Lectures on zeta functions and motives (according to Deninger and Kurokawa)*. Columbia University Number Theory Seminar (New York, 1992). Astérisque No. 228 (1995), 4, 121–163.

80. R. Meyer, *On a representation of the idele class group related to primes and zeros of L-functions*. Duke Math. J. Vol.127 (2005), N.3, 519–595.
81. G. Mikhalkin, *Enumerative tropical algebraic geometry in* $\mathbb{R}^2$. J. Amer. Math. Soc. 18 (2005), no. 2, 313–377.
82. G. Mikhalkin and I. Zharkov, *Tropical curves, their Jacobians and theta functions*. In Curves and abelian varieties, volume 465 of Contemp. Math., p 203–230. Amer. Math. Soc., Providence, RI, 2008.
83. S. Patterson, *An introduction to the theory of the Riemann Zeta-function*, Cambridge Univ. Press, 1988.
84. A. Postnikov and B. Shapiro, *Trees, parking functions, syzygies, and deformations of monomial ideals*. Trans. Amer. Math. Soc., 356(8):3109–3142 (electronic), 2004.
85. B. Riemann, *Über die Anzahl der Primzahlen unter einer gegebenen Grösse*, Monat der Königl. Preuss. Akad. der Wissen. zu Berlin aus der Jahre 1859 (1860) 671–680. (English translation in M.H.Edwards "Riemann's zeta function", Dover 2001.)
86. A. Selberg, *Harmonic analysis and discontinuous groups in weakly symmetric Riemannian spaces with applications to Dirichlet series*, Journal of the Indian Mathematical Society 20 (1956) 47–87.
87. A. Selberg, *Collected Papers*, Springer, 1989.
88. A. Selberg, *The history of the prime number theorem*. Lecture given in Seattle, Monday August 12, 1996, 5–6pm. Prime Number Theorem, A SYMPOSIUM on the Riemann Hypothesis, Seattle, Washington, August 12 15, 1996. In "publications.ias.edu/sites/default/files/seattle.pdf"
89. J. P. Serre, *Facteurs locaux des fonctions zêta des variétés algébriques (définitions et conjectures)*. Sém. Delange-Pisot-Poitou, exp. 19, 1969/70.
90. I. Simon, *Limited subsets of the free monoid*. Proc. of the 19-th Annual Symposium on computer Science (1978), 143–150.
91. C. Soulé, *Les variétés sur le corps à un élément*. Mosc. Math. J. 4 (2004), no. 1, 217–244.
92. R. Steinberg, *A geometric approach to the representations of the full linear group over a Galois field*, Transactions of the AMS, Vol. 71, No. 2 (1951), pp. 274–282.
93. J. Tate, *Fourier analysis in number fields and Hecke's zeta-function*, Ph.D. Thesis, Princeton, 1950. Reprinted in J.W.S. Cassels and A. Frölich (Eds.) "Algebraic Number Theory", Academic Press, 1967.
94. J. Tate, *Classes d'isogénie des variététs abéliennes sur un corps fini, d'après T. Honda*, Séminaire Bourbaki, 21-ème année, 352, 1968–1969.
95. J. Tits, *Sur les analogues algébriques des groupes semi-simples complexes*. Colloque d'algèbre supérieure, Bruxelles 19–22 décembre 1956, Centre Belge de Recherches Mathématiques Établissements Ceuterick, Louvain; Librairie Gauthier-Villars, Paris (1957), 261–289.
96. O. Viro, *From the sixteenth Hilbert problem to tropical geometry*. Jpn. J. Math. 3 (2008), no. 2, 185–214.
97. O. Viro, *On basic concepts of tropical geometry*. (Russian) Tr. Mat. Inst. Steklova 273 (2011), Sovremennye Problemy Matematiki, 271–303; translation in Proc. Steklov Inst. Math. 273 (2011), no. 1, 252–282
98. C. Weibel, *Cyclic Homology for schemes*. Proc. Amer. Math. Soc. 124 (1996), no. 6, 1655–1662.
99. C. Weibel, *The Hodge filtration and cyclic homology*. K-Theory 12 (1997), no. 2, 145–164.
100. A. Weil, *Sur la théorie du corps de classes* J. math. Soc. Japan, t. 3, 1951, p. 1–35.
101. A. Weil, *Sur les "formules explicites" de la théorie des nombres premiers*. (French) Comm. Sém. Math. Univ. Lund, (1952). Tome Supplémentaire, 252–265.
102. A. Weil, *Basic Number Theory*, Reprint of the second (1973) edition. Classics in Mathematics. Springer-Verlag, 1995.
103. A. Weil, *Sur les formules explicites de la théorie des nombres premiers*, Izv. Mat. Nauk., (Ser. Mat.) Vol.36 (1972) 3–18. (in Oeuvres compltes, Vol. 2, 48–62.)
104. A. Weil, *Oeuvres scientifiques/Collected papers*, I. II. III. Springer Collected Works in Mathematics. Springer, Heidelberg, 2014.

Navier Stokes Equations: A Quick Reminder and a Few Remarks

Peter Constantin

Abstract We describe briefly some mathematical problems related to the Navier-Stokes and Euler equations.

1 Introduction

The incompressible Navier-Stokes equations model viscous Newtonian fluids and can be viewed as an expression of Newton's second law (mass times acceleration equals force). In the simplest case, taking a fluid of unit density and considering only forces resulting from interior friction and incompressibility, the equations take the form

$$\partial_t u + u \cdot \nabla u - \nu \Delta u + \nabla p = 0. \tag{1}$$

Here $u = u(x,t)$ is the velocity of the fluid, $u(x,t) \in \mathbb{R}^d$ computed at the point $x \in \mathbb{R}^d$ and time $t \geq 0$, $\partial_t u$ is the time partial derivative and $u \cdot \nabla$ is the first order differential operator $u \cdot \nabla = \sum_{j=1}^d u_j(x,t)\partial_{x_j}$. The Laplacian of u, Δu, is multiplied by the kinematic viscosity ν, a positive constant. The gradient of the pressure $p(x,t)$, ∇p completes the equations. On components, the equations are thus

$$\partial_t u_i + u_j \partial_j u_i + \partial_i p - \nu \Delta u_i = 0, \quad i = 1, 2, \ldots d. \tag{2}$$

(We use the summation convention: repeated indices are summed.) The incompressibility is the constraint

$$\partial_i u_i = 0. \tag{3}$$

The constraint (3) results in an equation for the pressure. We will say more about this equation in the section about the pressure, but we mention already that the pressure responds instantaneously to far away disturbances: it is a non-local term.

P. Constantin (✉)
Department of Mathematics, Princeton University, Princeton, NJ 08544, USA
e-mail: const@math.princeton.edu

J.F. Nash, Jr., M.Th. Rassias (eds.), *Open Problems in Mathematics*,
DOI 10.1007/978-3-319-32162-2_6

The equation is usually studied in $d = 2$ or $d = 3$, and x may be restricted to belong to some domain $\Omega \subset \mathbb{R}^d$, in which case the most common boundary conditions are homogeneous Dirichlet

$$u_{|\partial\Omega} = 0. \tag{4}$$

In the case $\Omega = \mathbb{R}^d$ these are replaced by decay assumptions at spatial infinity, expressed by requiring u to belong to an appropriate class of functions. There is also a version in which $\mathbb{R}^d$ is replaced by the torus $\mathbb{T}^d$, that is the functions u and p are required to be spatially periodic with the same period.

When the kinematic viscosity is set equal to zero the equations are called the incompressible Euler equations

$$\partial_t u + u \cdot \nabla u + \nabla p = 0, \quad \nabla \cdot u = 0. \tag{5}$$

It is not possible to review even in passing the vast literature on Navier-Stokes and Euler equations. In this paper I touch on very few selected topics related to regularity of solutions and the inviscid limit. The relationship between the Euler equations [12] and the Navier-Stokes equations is at the core of both subjects.

2 Vorticity

A point of view on the subject centers on the vorticity, the antisymmetric part of the gradient of velocity, $(\nabla u) - (\nabla u)^*$. In three dimensions this can be written as the vector valued function

$$\omega = \nabla \times u, \tag{6}$$

and in two dimensions the vorticity can be viewed as a real scalar valued function,

$$\omega = \partial_1 u_2 - \partial_2 u_1. \tag{7}$$

If we think of two dimensional flow as depending only on the variables x_1, x_2 in three dimensional space with coordinates (x_1, x_2, x_3), and with zero velocity in the third direction, then the vorticity is a vector whose direction is perpendicular to the plane of $(x_1, x_2, 0)$ and whose signed magnitude is given by (7). The Navier-Stokes equations can be written in terms of the vorticity

$$\partial_t \omega + u \cdot \nabla \omega - \nu \Delta \omega = \omega \cdot \nabla u \tag{8}$$

and inverting (6) closes the system. The pressure disappeared. The nonlocality did not: the velocity depends on the vorticity in a non-local fashion. In two dimensions the vorticity equation is

$$\partial_t \omega + u \cdot \nabla \omega - \nu \Delta \omega = 0 \tag{9}$$

and the difference between two and three dimensions appears clearly, because (9) has important additional properties that are absent in (8). The right-hand side of (8) is called the "stretching term" and is absent in (9). Consequently, in two dimensions the vorticity has controlled magnitude: in the Euler equations, the distribution function of vorticity is conserved, and hence all L^p norms do not change in time; in the Navier-Stokes equations these norms are non-increasing functions of time. The question of regularity of the Navier-Stokes and Euler equations can be decided in terms of vorticity alone, and in two dimensions the decision is favorable. No singularities can be formed from smooth beginnings. This happens not because of some scaling subcriticality, but because of conservation laws in the Euler equations. The conservation of magnitude of vorticity on particle paths in the 2D Euler equations is a reflection of the coherence of the vorticity direction field. It is known since Leray [27] that the 3D Navier-Stokes equations have weak solutions whose gradients are square-summable in space-time. The Leray weak solutions retain the basic information that follows from the energy balance of the Navier-Stokes equations, namely that the kinetic energy is bounded in time,

$$u \in L^\infty(0, T; L^2(\mathbb{R}^3))$$

and that the rate of dissipation of energy is square integrable in time

$$u \in L^2(0, T; \dot{H}^1(\mathbb{R}^3)).$$

The latter is equivalent with the statement that

$$\int_0^T \|\omega\|_{L^2}^2 dt < \infty.$$

On the other hand, if a solution satisfies

$$\int_0^T \|\omega\|_{L^2}^4 dt < \infty$$

then no singularities can form from smooth initial data on the time interval $[0, T]$. This fact can be easily proved: first, the assumed bound implies directly from the evolution equation of vorticity a bound that gives

$$\omega \in L^\infty(0, T; L^2(\mathbb{R}^3)),$$

together with

$$\omega \in L^2(0,T;W^{1,2}(\mathbb{R}^3)).$$

These bounds signify that the solution is a "strong" solution. Once being a strong solution is established, then functional calculus inequalities can be used to control higher derivatives.
The Leray solutions permit the definition of basic turbulence quantities, the energy dissipation rate ϵ, and the Kolmogorov length scale ℓ_K. The Kolmogorov scale is the only quantity with units of length that can be formed with the energy dissipation and viscosity,

$$\ell_K = \nu^{\frac{3}{4}}\epsilon^{-\frac{1}{4}}$$

where

$$\epsilon = \nu\langle|\nabla u|^2\rangle$$

is the energy dissipation rate per unit mass. It is believed that this length is the typical length below which viscosity dominates. It is maybe appropriate here to discuss dimensions and scaling. The velocity has dimensions of $[u] = LT^{-1}$, the kinematic viscosity $[\nu] = L^2T^{-1}$. The energy per unit mass has then units L^2T^{-2} and its rate of dissipation (time derivative) has units $[\epsilon] = L^2T^{-3}$. Solving (as it were) for L knowing $[\epsilon]$ and $[\nu]$ yields ℓ_K. The scaling invariance of the equations means that if we change $x \mapsto y = \frac{x}{L}$, $t \mapsto s = \frac{t}{T}$, $\nu \mapsto \tilde{\nu} = \frac{L^2}{T}\nu$ then $u(x,t) = \frac{L}{T}v(\frac{x}{L}, \frac{t}{T})$ obeys the Navier-Stokes equations with viscosity $\tilde{\nu}$ if $v(y,s)$ obeyed Navier-Stokes with viscosity ν. If we fix the numerical value of ν we can rescale solutions of the same equation by keeping L^2T^{-1} constant. The Reynolds number $Re = \frac{UL}{\nu}$ where U is a typical velocity and L a typical length is a nondimensional number that helps organize solution classes. There is no deep mystery to this, but obviously, any correct statement should maintain its correctness after rescaling. Behavior under linear dilation is only one property of the equations. The nonlinear dynamical regularities are deeper and harder to exploit.

Solutions of Navier-Stokes equations at high Reynolds numbers are studied numerically and are believed to describe experimental situations. The phenomenologically observed fact is that regions of high vorticity organize themselves in coherent intense vortex tubes, separated by small distances proportional to the Kolmogorov scale [22]. The phenomenological observations (by which we mean both numerical and experimental) can be summarized as follows. During the time evolution of turbulent three dimensional flow, strong coherent vortices form, and as they stretch they become more intense narrow vortex tubes. These are distributed in space in a disorganized manner. When two such intense vortex tubes approach each other as to nearly collide, threatening to create thus a dynamic singularity in the direction field of vorticity, they reconnect, performing thus a topological change, prohibited in smooth Euler solutions, but permitted in the viscous evolution. The vorticity magnitude then (and there) locally decays. Several such events may occur. Topological change (vortex reconnection) is a nonlinear regularizing mechanism that might prevent singularity formation but it is difficult to formalize and to use in rigorous mathematical arguments. It is known that in 3D Navier-Stokes equations, if the direction of the vorticity

$$\xi = \frac{\omega}{|\omega|}$$

is locally well behaved (locally Lipschitz continuous in regions of high vorticity), then no singularities can arise from smooth initial data. The regularization is due to a geometric depletion of nonlinearity: if the direction of vorticity is coherent, then the stretching term is no longer of the order of ω^2 but rather of the order $\ell^{-1}u\omega$, where l is the coherence length. We use vaguely the term "coherence" but in this context it simply means "$|\nabla\xi|$ locally bounded", and the coherence length is $\ell \sim |\nabla\xi|^{-1}$. The theorem of [14] was generalized both in what concerns integrability conditions and geometric context [5, 23, 24]. The persistence for all time of a coherence length of the vorticity however is not known. What is known is an average bound for weak solutions:

$$\nu^2 \int_0^T \int_{\mathbb{R}^3} |\omega||\nabla\xi|^2 dxdt < E_0$$

where E_0 is proportional to the initial energy of the solution $\int_{\mathbb{R}^3} |u_0|^2 dx$ [9]. This bound says that, in regions of high vorticity, the direction of vorticity is coherent on average. If the direction of vorticity is coherent, then no singularities can form. There is a gap however between the sufficient condition on $\nabla\xi$ ensuring regularity and the known average bound above.

The fact that infinite vorticity is needed in order to produce a blow up in the Navier-Stokes equation is qualitatively similar to the situation in Euler equations, where singularities cannot form from smooth initial data on a time interval $[0, T]$ if [2]

$$\int_0^T \|\omega\|_{L^\infty} dt < \infty.$$

The amplification of vorticity by the inviscid vortex stretching mechanism can conceivably be a route to regularity in Navier-Stokes equations.

3 Velocity

Singularity formation in unforced Navier-Stokes equations in $\mathbb{R}^3$ requires infinite velocity. There is no such result for the Euler equations, without additional assumptions about the nature of the blow up. In fact, this represents a marked difference between the two equations. For Navier-Stokes equations, if

$$\int_0^T \|u\|^2_{L^\infty(\mathbb{R}^3)} dt < \infty$$

then no singularities can arise from smooth and localized data in the time interval $[0, T]$. Or, if

$$\sup_{t\in[0,T]} \|u\|_{L^3(\mathbb{R}^3)} < \infty$$

then no singularities can arise from smooth and localized data in the time interval $[0, T]$. The sufficient condition involving $\|u\|_{L^\infty(\mathbb{R}^3)}$ is easy to prove. In fact, one integration by parts in the stretching term of the vorticity equation and straightforward estimates show that this condition results in an a priori bound for $\omega \in L^\infty(0, T; L^2(\mathbb{R}^3)) \cap L^2(0, T; W^{1,2}(\mathbb{R}^3))$, and from then on we are in the case of strong solutions.

By contrast, the sufficient condition involving $\|u\|_{L^3(\mathbb{R}^3)}$ is hard to prove [19].

As is the case with the conditions involving vorticity, there are gaps between these sufficient conditions and generally known results on the corresponding quantities. For the L^∞ norm it is known that

$$\int_0^T \|u\|_{L^\infty(\mathbb{R}^3)} < \infty$$

and for the L^3 norm it is known that

$$\int_0^T \|u\|^4_{L^3(\mathbb{R}^3)} dt.$$

The first result was known for many years [21] (see also [10]). The second follows by interpolation directly from the known Leray bounds and Morrey's inequality. The celebrated result of [6] bounds the Hausdorff dimension of the singular set of suitable weak solutions of the Navier-Stokes equations. These solutions are not known to be unique. They are obtained as limits of good evolutionary approximations, they exist for arbitrary long time and they have numerous interesting properties. The singular set is the set in space-time where such a solution has infinite velocity. Its dimension is at most 1. The dimension of the singular set at the first blow up time ought to be smaller. (The first blow up time is the putative time T where $\int_0^t \|u\|^2_{L^\infty(\mathbb{R}^3)} ds < \infty$ for $t < T$ but $\int_0^T \|u\|^2_{L^\infty(\mathbb{R}^3)} ds = \infty$.) The singular set is then the set of points x such that $\limsup_{y\to x, s\to T, s<T} |u(y,s)| = \infty$.

4 Pressure

Singularity formation in unforced Navier-Stokes equations in $\mathbb{R}^3$ requires infinite pressure. Specifically, if

$$\int_0^T \|p\|^2_{L^3} dt < \infty$$

then no singularity can arise, from smooth and localized initial data [4] in the time interval $[0, T]$. Also if the pressure is bounded below

$$\inf_{0\le t\le T, x\in\mathbb{R}^3} p > -\infty$$

then no singularities can occur in weak solutions [29]. Of course, the pressure is actually defined only up to a constant, and these criteria require a convention determining it. The pressure obeys a Poisson equation

$$-\Delta p = \nabla\cdot(u\cdot\nabla u)$$

and in the whole space we may assume that the pressure decays at infinity. Then a natural choice for $(-\Delta)^{-1}$ is convolution with the Green's function for the whole space, and that gives an explicit formula for the pressure in terms of the velocity. Using the result that velocity bounded in $u \in L^\infty(0, T; L^3(\mathbb{R}^3))$ implies regularity, it is easy to recover the result that the pressure bounded in $p \in L^2(0, T; L^3(\mathbb{R}^3))$ implies regularity. Indeed, this follows from the inequality

$$\frac{d}{3dt}\int_{\mathbb{R}^3} |u|^3 dx + \nu\int_{\mathbb{R}^3} |\nabla u|^2|u| dx \le \left|\int_{\mathbb{R}^3} pu\cdot\nabla|u| dx\right|$$

by standard manipulations (Hölder inequality, Morrey inequality). Because the pressure controls the magnitude of velocity and because a Navier-Stokes blow up requires the velocity to become infinite, the study of pressure is important for the regularity problem. In fact, in the absence of pressure, or if the pressure would have been a local function of velocity magnitude, then no singularities could arise in the Navier-Stokes equations. The regularity criterion involving $\int_0^T \|u\|^2_{L^\infty(\mathbb{R}^3)} dt$ would suggest that a condition like $\int_0^T \|p\|_{L^\infty(\mathbb{R}^3)} < \infty$ would be a sufficient condition for regularity.

There exists an explicit, quasi-local decomposition of the pressure [11] using the spherical averages

$$\overline{p}_r(x) = \frac{1}{4\pi r^2}\int_{|x-y|=r} p(y)dS(y),$$

at arbitrary $r > 0$. The pressure is expressed as

$$p(x) = \beta_r(x) + \pi_r(x)$$

where β_r is a local average of the pressure,

$$\beta_r(x) = \frac{1}{r}\int_r^{2r} \overline{p}_\rho(x)d\rho.$$

The expression for $\pi_r(x)$ is computed from integrals in the ball of radius $2r$ of velocity increments

$$\delta_z u_i(x) = u_i(x+z) - u_i(x)$$

squared:

$$\pi_r(x) = \int w\left(\frac{|z|}{r}\right) k_{ij}(z)(\delta_z u_i(x))(\delta_z u_j(x))dz$$

where k_{ij} is symmetric in i, j, homogeneous of order -3 in the ball of radius one around the origin, smooth away from the origin, and it is explicit. The weight w is supported in $[0, 2]$, equals one for $0 \le \lambda \le 1$, and $2 - \lambda$ on $[1, 2]$. The function π_r vanishes for the harmonic part of p, and is quadratically small in r:

$$\|\pi_r\|_{L^q(\mathbb{R}^3)} \le Cr^2\|\nabla u\|^2_{L^{2q}(\mathbb{R}^3)}$$

for $1 \le q \le \infty$. On the other hand, the function β_r obeys very strong bounds for positive r:

$$\|\beta_r\|_{L^\infty(\mathbb{R}^3)} \le Cr^{-3}\|u\|^2_{L^2(\mathbb{R}^3)},$$
$$\|\nabla\beta_r\|_{L^\infty(\mathbb{R}^3)} \le Cr^{-4}\|u\|^2_{L^2(\mathbb{R}^3)}.$$

These bounds follow from the equation for the pressure, using a kind of monotonicity equation for a modified object

$$b_r(x) = \overline{p}_r(x) + \frac{1}{4\pi r^2}\int_{|x-y|=r}\left(\frac{y-x}{|y-x|}\cdot u(y)\right)^2 dS(y).$$

Bounds of the form

$$\|\beta_r\|_{L^q(\mathbb{R}^3)} \leq C_q \|u\|^2_{L^{2q}(\mathbb{R}^3)}$$

for $1 < q < \infty$ are true for $r = 0$ as well, and follow from the Calderon-Zygmund bounds on p. The bounds for β_r are obviously sufficient for regularity. The bounds on π_r are not. The smallness of π_r is remarkable though. For instance, it follows from the well-known bound [11, 21]

$$\int_0^T \|\Delta u\|^{\frac{2}{3}}_{L^2(\mathbb{R}^3)} dt < \infty$$

that

$$\|\pi_r\|_{L^3(\mathbb{R}^3)} \leq C(t) r^2$$

holds almost everywhere in $[0, T]$ with $\int_0^T C(t)^{\frac{1}{3}} dt$ depending only on the initial energy $\|u_0\|^2_{L^2(\mathbb{R}^3)}$, T and ν. A stronger form of smallness of π_r is all that is needed for regularity. Indeed, if there would exist $r > 0$ (independent of time, although this requirement could be relaxed) and a constant C such that

$$u \cdot \nabla \pi_r + \nu |\nabla u|^2 + C \geq 0$$

pointwise, then we would deduce regularity. This condition is a form of the requirement of the existence of positive r such that the local Reynolds number at scale r is small. If this condition is assumed, then regularity follows from the decomposition of the pressure at scale r, $p = \beta_r + \pi_r$ because

$$\int_0^T \|u \cdot \nabla \beta_r\|_{L^\infty(\mathbb{R}^3)} dt$$

is a priori finite (for fixed positive r) and because the inequality

$$(\partial_t + u \cdot \nabla - \nu\Delta) |u|^2 \leq F(t)$$

with $\int_0^T F(t) dt < \infty$ results in

$$\|u(t)\|^2_{L^\infty(\mathbb{R}^3)} \leq F(t)$$

by the maximum principle applied to $|u(x, t)|^2 - \int_0^t F(s) ds$.

5 Zero Viscosity

Many of the interesting mathematical problems about Navier-Stokes and Euler equations originate in the attempt to understand questions related to turbulence. One of the basic tenets of turbulence is the non-vanishing of energy dissipation ϵ in the limit of high Reynolds numbers. The average energy dissipation of energy per unit mass is an important parameter in turbulence theory, where the Kolmogorov energy spectrum is given by

$$E(k) = C_K \epsilon^{\frac{2}{3}} k^{-\frac{5}{3}}$$

in the inertial range. The power law is deduced on dimensional grounds. If the answer depends only on the wave number k (which has dimensions of L^{-1}) and ϵ (which has dimensions $[\epsilon] = L^2T^{-3}$), and if the answer is a product of powers, $E = \epsilon^a k^b$, then because the energy spectrum which has units of energy per wave number, i.e. $[E] = L^3T^{-2}$, equating powers of L and T we deduce the Kolmogorov spectrum. The trouble with this argument is not that it is simplistic, but rather that it gives the "right" result, for an astounding range of turbulent experiments and numerical simulations. The constant C_K is not dependent on Reynolds number. Because $\epsilon = \nu\langle|\nabla u|^2\rangle$, no matter what interpretation we give to the average, somehow the average gradient of velocity should saturate the nonzero bound, in the limit of zero viscosity. This is termed by physicists "anomalous dissipation".

There are two distinct approaches to the question of anomalous dissipation. In the first, the limit of zero viscosity is taken on solutions of the initial value problem with fixed initial data. Under appropriate conditions this leads to a solution of the corresponding initial value problem of the Euler equation. This equation conserves energy if solutions are smooth, but might dissipate energy if solutions are not sufficiently smooth. This circle of ideas, and specifically the precise degree of smoothness needed, goes by the name of "Onsager conjecture" [20]. This approach is therefore about the initial value problem for the limit equations and it requires lack of smoothness of solutions. The blow up problem is open for 3D incompressible Euler equations, and this allows to envision the possibility of existence of dissipative solutions arising from smooth initial data. Anomalous dissipation of energy can be proven for incompressible 2D Euler equations as well, for very non-smooth solutions, although in 2D non-smooth solutions cannot arise spontaneously from smooth ones. The Onsager conjecture states roughly that energy is conserved on solutions of Euler equations if they are smoother than $C^{\frac{1}{3}}$, and that there are solutions of the Euler equations with exactly $C^{\frac{1}{3}}$ smoothness which dissipate energy. The first part of the statement is pretty much proved [7, 13]. Much progress has been made on the second part of the statement [3, 18].

The second way of looking at the anomalous dissipation issue is to take long time averages first, in order to achieve a "permanent regime" of the viscous equations, and only then send the viscosity to zero. This procedure has the advantage of being more appropriate to a turbulence setting which requires a statistical description.

Turbulence is generated at boundaries or through other forcing. Anomalous dissipation is much harder to establish in this case. The difficulty is conceptual, because it is necessary to produce long lived solutions that achieve an equilibrium that is not obtained by the balance of forcing and viscosity, rather by the balance of nonlinearity and forcing. In two dimensions if nonvanishing linear damping is imposed then a dissipation anomaly cannot exist [16, 17].

The finite time relationship between the Navier-Stokes equations and Euler equations is understood only for as long as the solutions of the Euler equations are smooth, and only if no boundaries are present (i.e. $\mathbb{R}^d$ or $\mathbb{T}^d$, $d = 2$, or $d = 3$ under the assumption of smoothness.) In these cases, if the Euler solution starting from a smooth initial datum remains smooth on a time interval $[0, T]$, then there exists a viscosity ν_0, depending on the Euler solution and on T, such that for all $\nu \leq \nu_0$, the solution of the Navier-Stokes equation with the same initial datum exists on $[0, T]$, is smooth, and ν-close in strong norms to the Euler solution [8, 25, 28].

In the case of boundaries, the mathematical problem is wide open, despite more than a century of effort. The difficulty stems from the presence of boundary layers, vanishingly small regions near the boundary where high gradients of the solutions are concentrated. In many studies, a reference smooth Euler flow is supposed to be known and used to set units of time and length. In other words, it is assumed that a smooth Euler solution is given and has velocity of order one and derivatives of order one. The classical Prandtl boundary layer length scale associated to the Eulerian solution is of order $\sqrt{\nu}$ and an asymptotic description based on this given Euler flow is attempted. An asymptotic description of the Navier-Stokes equation, based on a given "nearby" smooth Euler flow is difficult because the connection between the imposed Euler flow and the Navier-Stokes equation is illusory near the boundary.

It was shown by Kato [26] that if the rate of dissipation vanishes on a much smaller scale,

$$\lim_{\nu \to 0} \int_0^T \int_{dist(x \partial\Omega) < c\nu} \nu |\nabla u^{(\nu)}|^2 dx dt = 0$$

for Navier-Stokes solutions with viscosity ν, then inviscid limits solve the Euler equation. In fact, in this case any weak limit of Navier-Stokes solutions is a weak dissipative up to boundary [1] solution of the Euler equations. There are variants of this sufficient condition ensuring convergence to an Eulerian solution which allow for more singular behavior, if there are no turning points in the flow [15]. The possibility of the inviscid limit to be a solution of the Euler equations exists, but is very limited. In general, the behavior of the inviscid limit might fail to be purely Eulerian. How does the laminar Eulerian picture break down (if indeed, as I suspect, it does) is still a mystery. Much remains to be done.

References

1. C. Bardos, E.S. Titi, Mathematics and turbulence: where do we stand? Journal of Turbulence **14** (2013), 42–76.
2. J. T. Beale, T. Kato and A. J. Majda, Remarks on the breakdown of smooth solutions for the 3-D Euler equations, Comm. Math. Phys., **94**(1984), 61–66.
3. T. Buckmaster, C. De Lellis, P. Isett and L. Székelyhidi Jr., Anomalous dissipation for 1/5-Hölder Euler flows, Annals of Mathematics, 2015
4. L. Berselli and G. Galdi, Regularity criteria involving the pressure for the weak solutions to the Navier-Stokes equations, Proceedings of the AMS, **130** (12), (2002), 3585–3595.
5. H. Beirao da Veiga, Direction of Vorticity and Regularity up to the Boundary: On the Lipschitz-Continuous Case, Journal of Mathematical Fluid Mechanics **15** (2013), 55–63.
6. L. Caffarelli, R. Kohn and L. Nirenberg, Partial regularity of suitable weak solutions of the Navier-Stokes equations, Commun. Pure Appl. Math. **35** (1982), 771–831.
7. A. Cheskidov, P. Constantin, S. Friedlander, and R. Shvydkoy, Energy conservation and Onsager's conjecture for the Euler equations, Nonlinearity, **21** (2008) 1233–1252.
8. P. Constantin, Note on loss of regularity for solutions of the 3-D incompressible Euler and related equations, Commun. Math. Phys. **104** (1986), 311–326.
9. P. Constantin, Navier-Stokes equations and area of interfaces, Commun.Math. Phys. **129** (1990), 241–266.
10. P. Constantin, An Eulerian-Lagrangian approach to the Navier-Stokes equations, Commun. Math. Phys., **216** (2001), 663–686.
11. P. Constantin, Local formulas for the hydrodynamic pressure and applications, Russian Mathematical Surveys, **69** (2014), 395–418.
12. P. Constantin, On the Euler equations of incompressible fluids, Bulletin of the AMS **44**, (2007), 603–621.
13. P. Constantin, W. E, and E.S. Titi, Onsager's conjecture on the energy conservation for solutions of Euler's equation, Comm. Math. Phys., **165** (1994) 207–209.
14. P. Constantin, C. Fefferman, Direction of vorticity and the problem of global regularity for the Navier-Stokes equations, Indiana Univ. Math. Journal, **42** (1993), 775–787.
15. P. Constantin, I. Kukavica, V. Vicol, On the inviscid limit of the Navier-Stokes equations, Proc. Amer. Math. Soc. 143 (2015), 3075–3090.
16. P. Constantin and F. Ramos, Inviscid limit for damped and driven incompressible Navier-Stokes equations in $\mathbb{R}^2$, Comm. Math. Phys., **275** (2007) 529–551.
17. P. Constantin, A. Tarfulea, V. Vicol, Absence of anomalous dissipation of energy in forced two dimensional fluid equations, ARMA **212** (2014), 875–903.
18. C. De Lellis and L. Székelyhidi, Jr., The *h*-principle and the equations of fluid dynamics, Bull. Amer. Math. Soc. (N.S.), **49**(2012) 347–375.
19. L. Escauriaza, G. Seregin, V. Sverak, $L^{3,\infty}$-solutions to the Navier-Stokes Equations and Backward Uniqueness. Uspekhi Mat. Nauk **58**, 2 (2003) 3–44. English Translation: Russ. Math. Surv. **58**, 2 (2003) 211–250.
20. G.L. Eyink and K.R. Sreenivasan, Onsager and the theory of hydrodynamic turbulence, Rev. Modern Phys., **78** (2006) 87–135.
21. C. Foias, C. Guillopé, R. Temam, New apriori estimates for the Navier-Stokes equations in dimension 3, Comm. PDE **6** (3), (1981) 329–359.
22. U. Frisch, *Turbulence, the legacy of A.N. Kolmogorov*, C.U.P., Cambridge (1995).
23. Y. Giga, P.-Y. Hsu and Y. Maekawa, A Liouville theorem for the planar Navier-Stokes equations with the no-slip boundary condition and its application to a geometric regularity criterion, Comm. in Partial Differential Equations, **39** (2014), 1906–1935.
24. Z. Grujic, Localization and geometric depletion of vortex-stretching in the 3D NSE, Comm. Math. Phys. **290** (2009), 861–870.
25. T. Kato, Nonstationary flows of viscous and ideal fluids in $\mathbb{R}^3$, J. Funct. Anal. **9** (1972) 296–305.

26. T. Kato, Remarks on zero viscosity limit for nonstationary Navier-Stokes flows with boundary, Seminar on Nonlinear PDE Berkeley, California, Math. Sci. Res. Inst. Publ. **2** (1984), 85–98.
27. J. Leray, Sur le mouvement d'un liquide visqeux emplissant l'espace, Acta Math. **63** (1934) 183–248.
28. N. Masmoudi, Remarks about the inviscid limit of the Navier-Stokes system, Commun. Math. Phys, **270** (2007), 777–788.
29. G. Seregin, V. Sverak, Navier-Stokes equations with lower bounds on the pressure, ARMA **163** (1) (2002), 65–86.

Plateau's Problem

Jenny Harrison and Harrison Pugh

Abstract Plateau's problem is not a single conjecture or theorem, but rather an abstract framework, encompassing a number of different problems in several related areas of mathematics. In its most general form, Plateau's problem is to find an element of a given collection $\mathscr{C}$ of "surfaces" specified by some boundary constraint, which minimizes, or is a critical point of, a given "area" function $F : \mathscr{C} \to \mathbb{R}$. In addition, one should also show that any such element satisfies some sort of regularity, that it be a sufficiently smooth manifold away from a well-behaved singular set. The choices apparent in making this question precise lead to a great many different versions of the problem. Plateau's problem has generated a large number of papers, inspired new fields of mathematics, and given rise to techniques which have proved useful in applications further afield. In this review we discuss a few highlights from the past hundred years, with special attention to papers of Federer, Fleming, Reifenberg and Almgren from the 1960s, and works by several groups, including ourselves, who have made significant progress on different aspects of the problem in recent years. A number of open problems are presented.

1 Introduction

Plateau's problem has intrigued mathematicians and scientists alike for over 200 years. It remains one of the most accessible problems in mathematics, yet retains a subtle difficulty in its formulation. Many different versions of Plateau's problem have been solved, but even today there are still important questions left unanswered, and deep mysteries about the problem still remain.

Plateau's problem was first posed by Lagrange, who in 1760 derived the minimal surface equation and asked if one could find a surface of minimal area with a prescribed boundary. The problem was later named after Plateau [74] who

J. Harrison (✉)
Department of Mathematics, University of California, Berkeley, CA 94705-3860, USA
e-mail: harrison@math.berkeley.edu

H. Pugh
Department of Mathematics, Stony Brook University, Stony Brook, NY 11794-3651, USA
e-mail: hpugh@math.stonybrook.edu

J.F. Nash, Jr., M.Th. Rassias (eds.), *Open Problems in Mathematics*,
DOI 10.1007/978-3-319-32162-2_7

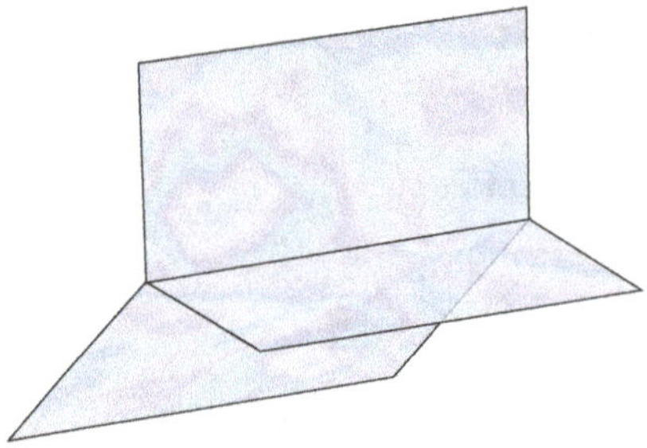

Fig. 1 Three surfaces meeting at a triple junction singularity

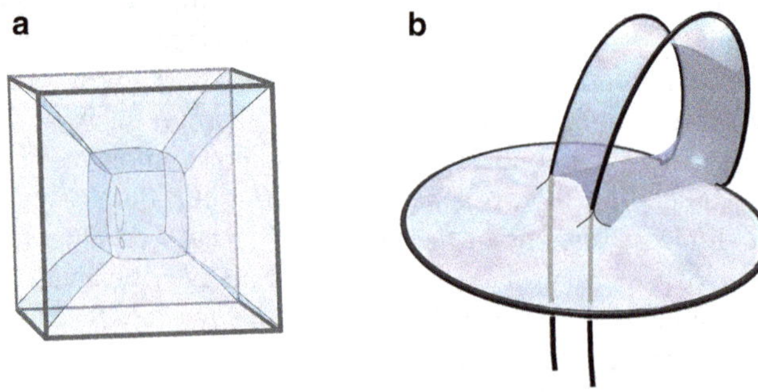

Fig. 2 Non-standard boundaries. (**a**) A wireframe as a boundary. (**b**) A non-closed boundary

undertook a physical study of soap films and characterized their properties, most notably their singularities. It was actually Lebesgue who coined the term "Plateau's problem," the crux of which was to describe these soap films in mathematical terms. These objects do not behave like classical surfaces such as embedded, or even immersed manifolds with branch points. Three sheets can come together along a line and form what is known as a triple junctions (Fig. 1). Soap films can span wires that are not cycles (Fig. 2). Some films are local area minimizers, yet can retract onto their boundaries (Fig. 3).

The long journey towards a full understanding of these phenomena began with a much simpler question about the existence of a function with prescribed values on the boundary of a domain $\Omega \subset \mathbb{R}^2$, such that the graph of the function on the interior of Ω is a minimal surface. This problem was studied by Weierstass and Riemann and evolved into the classical theory of minimal surfaces. The next step up in generality came with the study of surfaces defined as images of disks. Jesse Douglas won the first Fields Medal for his solution, which proved the existence of a minimally immersed disk in $\mathbb{R}^3$ with a prescribed contour boundary. Many others continued on with striking results of existence and regularity along the way, and the Douglas-Plateau problem for surfaces with higher (non-infinite) genus in arbitrary dimension and codimension was finally solved by Jost in 1985.

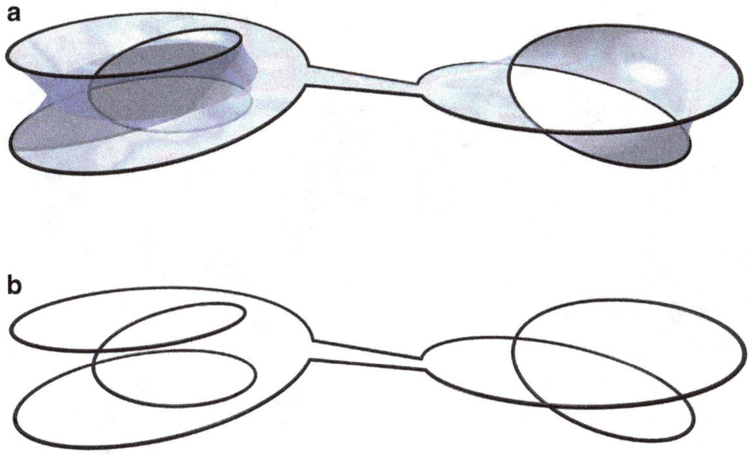

Fig. 3 The Adams surface (**a**) retracts onto its boundary (**b**). The left portion of (**a**) is a triple Möbius band (a "Y" cross an interval, glued along the ends with a $1/3$ twist, see also Fig. 9b) and the right portion is a classical Möbius band. They are joined by a bridge, so that the boundary is a single Jordan curve

However, Fleming demonstrated the existence of a contour boundary which bounds a minimal surface of infinite genus (Fig. 4). In 1960, Federer and Fleming introduced objects known as integral currents which could model these somewhat pathological surfaces. Their novel approach won them the Steele prize and helped launch the modern study of geometric measure theory. They proved the existence of an integral current with a given boundary which minimizes mass, a quantity which can be thought of as area weighted by an integer multiplicity. It later became known that in low enough dimension, their minimizing current corresponded to an embedded minimal submanifold.

At the same time, Reifenberg thought of a completely different approach. Building on results of Besicovitch and aided by Adams, he defined what it meant for a surface to span a bounding set A using Čech homology. Using his theory, he proved the existence of a surface with minimal area amongst those surfaces which can be written as a nested union of manifolds whose boundaries converge to the contour. These surfaces include non-orientable surfaces, as well as the example of Fleming, amongst others. His work is considered to be a masterpiece and deeply influenced several mathematicians, including Morrey, Almgren, Fomenko, as well as ourselves.

Almgren proposed three approaches to Plateau's problem. The first used varifolds [4], which Young [93, 94] had discovered but called "generalized surfaces." Integral varifolds have a compactness theorem [2] which can be used to prove the existence of a stationary varifold with smallest area. Integral varifolds model just about any imaginable minimal surface, including the example of Adams (Fig. 3). However,

a

b

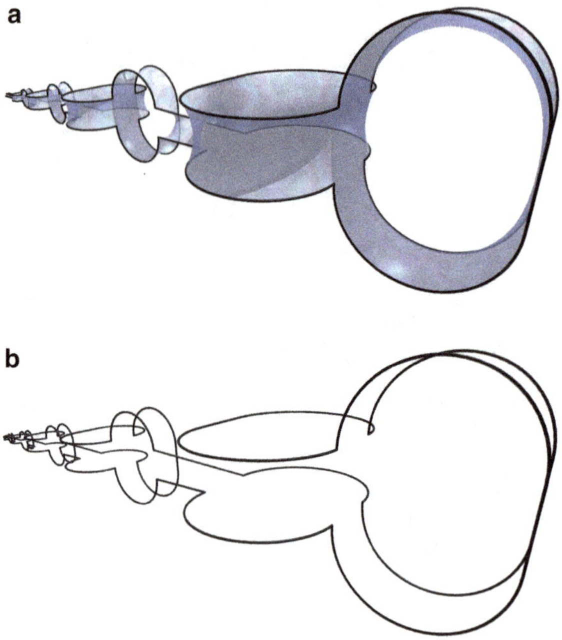

Fig. 4 Fleming's example of a minimal surface with infinite genus. (**a**) A minimal surface with infinite genus. (**b**) The boundary wire can be made smooth except at a single point

Almgren did not prove that this smallest stationary varifold was the smallest among a class of surfaces which also included non-stationary varifolds.

Almgren's second approach [7] was an attempt to generalize Reifenberg's results to elliptic integrands. To read his paper requires expertise in methods of geometric measure theory, varifolds, integral currents, and flat chains, for it blends them all. It has some gaps, one of which seems serious (see [57] for more details.)

Almgren's third and final approach [9] was to define a new class of surfaces which later became known as quasiminimal sets, which, roughly speaking, have a controllable increase of area under small deformations. Although he was unable to prove an existence theorem of an area minimizer in this category, his regularity results are of major importance.

The authors have recently announced the first solution to the full elliptic Plateau problem [57]. Our proof of existence of minimizers builds upon classical measure theory, and techniques of Reifenberg [77], Federer and Fleming [45]. Our proof of regularity relies upon Almgren [9], although Reifenberg [78, 79] is closely related. Spanning sets can be defined using homology, cohomology or homotopy. An axiomatic approach without requiring any definition of a spanning set is also provided. Our results carry over to ambient spaces of Lipschitz neighborhood retracts, including manifolds with boundaries and manifolds with singularities.

In the last few years there has been a flurry of other activity in Plateau's problem and in related fields, some of which we shall discuss below. For example, papers

Fig. 5 A piece of the helicoid, which extends to infinity in all directions

have recently appeared on sliding boundaries, where the soap film's interface within a larger boundary is permitted to move freely, on flexible boundaries, where the boundary itself is permitted to move subject to forces created by the spanning soap film, and on axiomatic theory, ellipticity, spanning conditions, and classical minimal surface theory. Indeed, very recently, a beautiful and central theorem in classical minimal surfaces was proved by Meeks and Rosenberg, namely that every simply connected, properly embedded minimal surface in $\mathbb{R}^3$ must be either a plane or a helicoid (Fig. 5.)

There are still many major questions in the world of Plateau's problem, indeed the list seems to be growing, not shrinking. We have enumerated several of our favorites at the end of this paper, some of which are newly posed.

Disclaimer

Before diving in, the reader should be aware that the following exposition is far from complete and will often be imprecise. We hope that it will be useful to give non-specialists some idea of the history of Plateau's problem, a few lines of current theoretical development, and some open problems, both enduring and emerging. We have endeavored to give a broad overview of many different aspects of the problem, and in doing so, have, by necessity, left out many important contributions by numerous mathematicians. If we have neglected to mention your favorite result, it was not due to malice, but rather due to the constraints of writing this article. If we have incorrectly stated your favorite result, know that we are experts in only a small portion of the Plateau problem, and would welcome any corrections you might provide. Lastly, the authors would like to thank Emanuele Paolini,[1] Ken Brakke,[2] and Claire-Audrey Bayan[3] for the use of their soap-film figures.

2 Classical Minimal Surfaces

2.1 *The Minimal Surface Equation*

Let Σ be a surface in $\mathbb{R}^3$. We say that Σ is a ***minimal surface*** if every point in Σ has an ϵ-neighborhood U which has least area among all surfaces $S \subset \mathbb{R}^3$ with boundary ∂U. This condition is equivalent to Σ having vanishing mean curvature, and to the condition that $\frac{\partial}{\partial t}\text{Area}(\Sigma_t)\lfloor_{t=0}= 0$ for all compactly supported variations Σ_t of Σ. These are the two (and higher) dimensional analogs of geodesics, but one must be careful in this comparison: even in $\mathbb{R}^n$, a minimal surface with a given boundary may not have smallest area amongst all surfaces with that boundary. For example, consider two horizontal disks in $\mathbb{R}^3$ separated vertically by a small amount. Their union forms a minimal surface, but the cylinder has smaller area, and the catenoid, smaller still (Fig. 6).

A special kind of minimal surface Σ is one which occurs as a graph:

Suppose Ω is a bounded open set in $\mathbb{R}^2$ with locally Lipschitz boundary $\partial\Omega$, and suppose $g : \partial\Omega \to \mathbb{R}$ is continuous. Let F denote the set of continuous extensions of g to Ω which are continuously differentiable in Ω and whose derivative is integrable. Let $A(f)$ denote the surface area of the graph of such a function f and suppose $f \in F$ solves Plateau's problem for this setup. That is, f satisfies

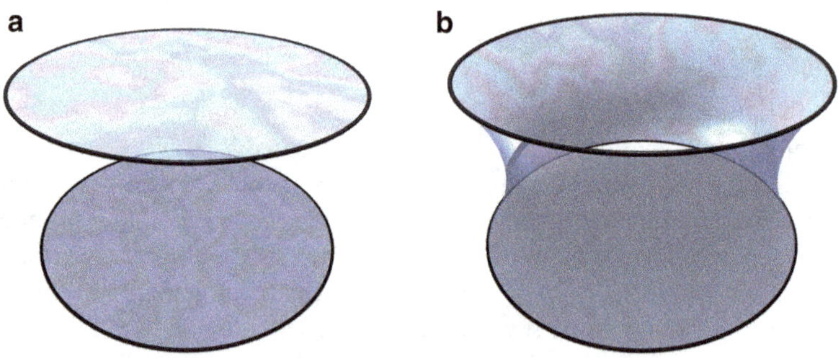

Fig. 6 Minimal surfaces with the same boundary. (**a**) Two disks. (**b**) Catenoid

[1]Figures 2b, 3–7, 9.

[2]Figure 11.

[3]Figure 8.

$$A(f) \leq A(h) \qquad \text{for all } h \in F. \tag{1}$$

Then,

$$div\left(\frac{\nabla(f)}{(1+|\nabla f|^2)^{1/2}}\right) = 0. \tag{2}$$

This differential equation is called the ***minimal surface equation***. It is the Euler-Lagrange equation for the area functional.

Theorem 1. *A function* $f \in F$ *satisfies* (1) *if and only if* f *satisfies the minimal surface equation* (2).

Such a function f is unique, and is in fact analytic on Ω [60].

Suppose now the domain Ω is a disk D of radius r. This produces a unique minimal surface for each continuous function g on S^1, and thus there are uncountably many minimal surfaces that are graphs over disks of radius $r < \infty$. However, if $r = \infty$, the situation simplifies dramatically:

Theorem 2 (Bernstein [13]). *Any solution of the minimal surface equation* (2) *which is defined on all* $\mathbb{R}^2$ *must be linear.*

Bernstein conjectured that this was also the case in higher dimension. Indeed this is the case up to dimension seven:

Theorem 3 (de Giorgi [27], Almgren [6]). *The Bernstein conjecture holds for minimal graphs* $\Gamma = \{(x, f(x)) \in \mathbb{R}^{n+1} : x \in \mathbb{R}^n\}$ *for* $n \leq 4$.

Theorem 4 (Simons [81]). *The Bernstein Conjecture holds for minimal graphs when* $n \leq 7$.

2.2 Recent Developments

In the classical theory, non-compact minimal surfaces might not be graphs and might not have a boundary. An example in $\mathbb{R}^3$ is the helicoid (Fig. 5). A long outstanding question posed by Osserman was the following generalization of Bernstein's Conjecture: The plane and the helicoid are the only properly embedded, simply-connected, minimal surfaces in $\mathbb{R}^3$.

Osserman's conjecture has recently been solved after many years of effort by Meeks and Rosenberg [70] who built on the work of Colding and Minicozzi [21, 22] as well as a number of other mathematicians (see [68] for a detailed discussion and a more complete list of citations.)

Combining this with work of Collin [24], López and Ros [63], Meeks et al. [69] give the following classification theorem:

Theorem 5. *Up to scaling and rigid motion, any connected, properly embedded, minimal surface in* $\mathbb{R}^3$ *is a plane, a helicoid, a catenoid or one of the Riemann minimal examples. In particular, for every such surface there exists a foliation of* $\mathbb{R}^3$ *by parallel planes, each of which intersects the surface transversely in a connected curve which is a circle or a line.*

This beautiful theorem is only just the beginning of what looks to be a new era in classical minimal surface theory. A few particularly intriguing open problems are listed in Sect. 7 and can be found as part of a larger list in [68].

3 The Douglas-Plateau Problem: Immersions of Disks and Surfaces of Higher Genus

The Plateau problem, in its original formulation, was to find a minimal immersed disk whose boundary was a given Jordan curve in $\mathbb{R}^n$. For any immersed disk, coordinates on the disk D can be chosen so that the immersion is conformal, in which case minimality is equivalent to the immersion being harmonic.

Independently, Douglas [39] and Radó [75] proved the existence of such a surface, with (possibly) isolated singularities:

Theorem 6. *If* C *is a Jordan curve in* $\mathbb{R}^n$*, there exists a continuous map* $\iota : D \to \mathbb{R}^n$ *which is conformal and harmonic away from a set of isolated singularities (i.e. branch points), such that* $\iota\lfloor_{\partial D}$ *parameterizes* C.

The solutions produced by Douglas and Radó also had minimal area in the class of branched immersions. However, Douglas could prove slightly more than Radó, principally his theorem allowed for certain pathological boundaries C which could only be spanned by disks of infinite area.[4] In addition, Douglas's methods signaled a significant and promising departure from the classical techniques, and for these reasons he was awarded the first Fields medal in 1936.

Osserman proved [72] that if $n = 3$, then branch points did not exist in such minimal disks. Thus,

Theorem 7. *If* C *is a smooth Jordan curve in* $\mathbb{R}^3$*, there exists a (conformal, harmonic) immersion* $\iota : D \to \mathbb{R}^3$ *such that* $\iota\lfloor_{\partial D}$ *parameterizes* C*, whose area is minimal among all immersed disks whose boundaries parameterize* C.

[4]Douglas was well known for his displeasure at having to share credit with Radó for Theorem 6. When teaching subsequent geometry courses, he eschewed those books which, in covering the theorem, contained the attribution "Douglas-Radó." Unfortunately, as the years went on, he was forced to use increasingly antiquated texts, since virtually no book published after the 1930s failed to give this (correct) attribution (See [85]. The second author has also heard a similar story from Martin Bendersky, who was a student in one of Douglas's courses at City College.)

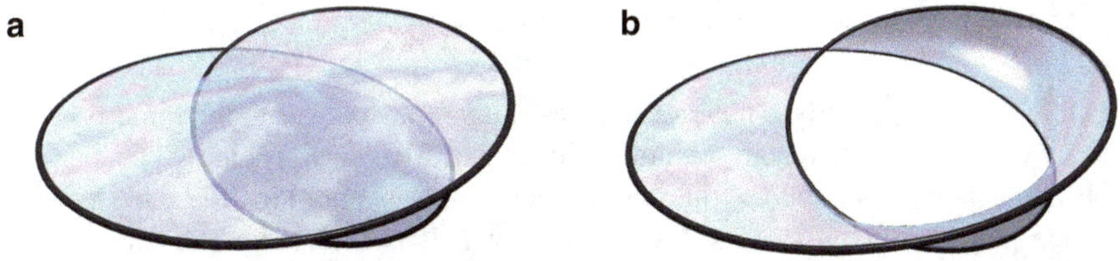

Fig. 7 A non-orientable surface with smaller area than any immersed disk with the same boundary. (**a**) An immersed disk spanning the boundary of the Möbius band. (**b**) A Möbius band

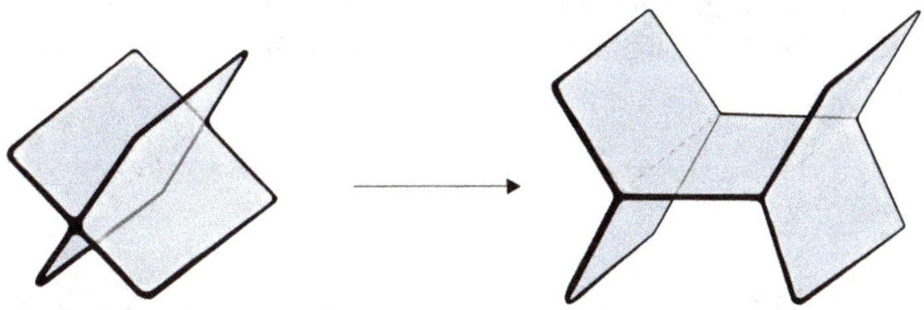

Fig. 8 The transverse intersection of two sheets can be replaced with a pair of triple junctions with smaller area. These singularities do not, however, show up in the mass minimization problem (see Sect. 4.2), as the *horizontal portion of the right hand figure* would require higher mass to cancel out any contribution of the triple junction to the boundary

However, there still could be immersed surfaces with boundary C whose area is strictly less than those of the solutions produced above. Consider a thin Möbius band. Its area is less than that of any immersed disk spanning the boundary curve (Fig. 7). Higher genus surfaces could also have less area. Douglas had made attempts to generalize his techniques to account for possibly non-orientable surfaces and those of higher genus, but the consensus seems to be that his arguments were incomplete [52]. It took until the 1980s for a complete solution to the higher genus problem to appear in a paper by Jost [61] (see [14] for the non-orientable case), who built upon ideas of Schoen-Yau [84] and Sacks-Uhlenbeck [88]. Tomi-Tromba [87] soon after offered a different solution to the higher genus problem based on a development of Teichmüller theory from the viewpoint of differential geometry.

Although not a mathematical shortcoming, self-intersections in which two sheets intersect transversally can easily show up in Douglas's and others' immersed solutions. Such solutions are not physically realistic as soap-films, since transverse intersections resolve into pairs of triple junctions (Fig. 8) with smaller total area when one is allowed to consider surfaces more general than immersed manifolds. These generalized surfaces with triple junctions became important in the 1960s in the work of Reifenberg [77], Almgren [7] and Taylor [86], who studied soap-film regularity and classified the singularities for these general size-minimizing surfaces.

4 Orientable Generalized Surfaces

Fleming's example (Fig. 4) shows that one must consider all topological types to find a true minimizer for an orientable version of Plateau's problem. The curve in the figure is an unknotted simple closed curve and is smoothly embedded except at one point. It bounds the shaded surface which is orientable and clearly area minimizing. This shows that one should not insist that competitors have finite topological type when looking for absolute area minimizers. The proof of the existence of an orientable surface minimizing the area among all possible surfaces without restriction on their topological type requires other techniques, namely the results of Reifenberg [77] and the integral currents of Federer and Fleming [45].

4.1 Hausdorff Measure

Readers will recall that for any $E \subset \mathbb{R}^n$, and any non-negative real number k, the k-dimensional Hausdorff measure of E is

$$\mathscr{H}^k(E) = \lim_{\delta \to 0} \mathscr{H}^k_\delta(E)$$

where

$$\mathscr{H}^k_\delta(E) = \alpha_k \inf \left\{ \sum_{i \in I} \operatorname{diam}(U_i)^k : E \subset \cup_{i \in I} U_i \right\},$$

α_k is a normalizing constant, and the infimum is taken over all coverings of E by a collection $\{U_i\}_{i \in I}$ of sets with $\operatorname{diam}(U_i) < \delta$. The normalizing constant α_k is chosen so that when k is an integer, the k-dimensional Hausdorff measure of the unit cube in $\mathbb{R}^k$ is one.

Note that $\mathscr{H}^k_\delta(E)$ is monotone decreasing in δ, so the limit $\lim_{\delta \to 0} \mathscr{H}^k_\delta(E)$ exists but may be infinite. $\mathscr{H}^k$ is a Borel regular outer measure and coincides with Lebesgue measure when E is a k-dimensional submanifold.

The ***Hausdorff dimension*** of E is the infimum over all $k \geq 0$ such that $\mathscr{H}^k(E) = 0$.

4.2 Integral Currents and Mass Minimization

A ***k-dimensional current*** T on an n-dimensional smooth manifold M is a linear functional on the space of compactly supported smooth k-forms $\mathscr{D}^k(M)$, continuous in the following sense:

If ω_i is a sequence of k-forms supported in a single compact set K contained in a coordinate neighborhood and $\partial^r \omega_i \to 0$ uniformly for all $0 \leq |r| < \infty$, then $T(\omega_i) \to 0$. Here $r = (r_1, \ldots, r_n)$ is a n-tuple of non-negative integers, $|r| = \sum r_i$, and ∂^r is shorthand for the coordinate-wise differentiation operator

$$\frac{\partial^{|r|}}{\partial x_1^{r_1} \ldots \partial x_n^{r_n}}.$$

It is important to note that the topology on $\mathscr{D}^k(M)$ is strictly finer than the subspace topology induced by the inclusion of $\mathscr{D}^k(M)$ into the space $\mathscr{E}^k(M)$ of C^∞ differential k-forms on M, in which a sequence of forms η_i converges to zero whenever $\partial^r \eta_i \lfloor_K \to 0$ uniformly for all $0 \leq r < i$ and all compact sets K contained in a coordinate neighborhood. The difference is subtle, but has extremely important consequences (the full description of the space $\mathscr{D}^0$, its topology, and continuous dual was the major component of L. Schwartz's Fields medal.) For example, in $\mathscr{E}^0(\mathbb{R})$, any sequence of bump functions f_i equal to 1 on the interval $[-i, i]$ converges to the function 1, which is no longer compactly supported. Such a sequence is not convergent in $\mathscr{D}^0(\mathbb{R})$. As a matter of fact, $\mathscr{D}^k(\mathbb{R}^n)$ is complete, so it is not even Cauchy.

4.2.1 Examples

- If $S \subset M$ is an oriented k-dimensional submanifold, then a k-current $[[S]]$ is defined, setting $[[S]](\omega) \equiv \int_S \omega$.
- If M is equipped with a volume form dV, then a k-vector field X on M defines a current $[[X]]$, whereby $[[X]](\omega) \equiv \int_M \omega(X) dV$.
- A generalized Dirac delta is a current: if $p \in M$ and $\alpha \in \Lambda^k(T_p M)$, then a current $[[(p, \alpha)]]$ is defined, where $[[(p, \alpha)]](\omega) \equiv \omega_p(\alpha)$.
- An $(n-k)$-form η in $\mathscr{E}^{n-k}(M)$ defines a k-current $[[\eta]]$, where $[[\eta]](\omega) = \int_M \eta \wedge \omega$. Such a current is called a ***smooth current***. Through convolution, it is possible to construct a smoothing operator which approximates any current by a smooth current (See [76, Sect. 15]).

Denote the space of k-dimensional currents by $\mathscr{D}_k(M)$. The operator dual to exterior differentiation on forms, denoted ∂, turns $\mathscr{D}_\bullet(M)$ into a chain complex. The image and kernel of ∂ are closed. When given the opposite grading (i.e. give $\mathscr{D}_k(M)$ degree $n - k$), the resulting cochain complex $(\mathscr{D}_{n-\bullet}, \partial)$ is quasi-isomorphic, via application of the aforementioned smoothing operator, to the cochain complex $(\mathscr{E}^\bullet(M), d)$. Thus, Poincaré duality holds in this setting: the homology of currents in degree k (which is dual to the compactly supported de Rham cohomology in degree k) is isomorphic to de Rham cohomology in degree $n - k$.

Before we can describe a sub-complex of $(\mathscr{D}_\bullet, \partial)$ which computes the integral homology of M, it will be necessary to define the mass of a current. If M is equipped with a Riemannian metric and $W \subset M$, define $\|T\|(W) \equiv \sup\{T(\omega) : supp(\omega) \subset$

W, $\|\omega\|_0 \leq 1\}$, where $\|\omega\|_0$ is the supremum of $\omega_p \alpha$, where $p \in M$ and α is a unit simple k-vectors in $\Lambda^k T_p M$. The (possibly infinite) ***mass*** of T, denoted $\mathbf{M}(T)$, is the quantity $\|T\|(\mathbb{R}^n)$. If $\|T\|(W)$ is finite for every $W \subset\subset M$, we say T has ***locally finite mass***.

A k-dimensional current T is called ***(integer) rectifiable*** if it has locally finite mass and there exists a sequence S_i of C^1 oriented k-dimensional submanifolds of M, a sequence of pairwise disjoint closed subsets $K_i \subset S_i$ and a sequence of positive integers k_i such that

$$T(\omega) = \sum_i k_i \int_{K_i} \omega$$

for all $\omega \in \mathscr{D}^k(M)$. If T and ∂T are rectifiable, we say that T is an ***integral current***. One can show, e.g., using sheaf theory, that the homology of the chain complex $(\mathscr{I}_\bullet(M), \partial)$ of integral currents computes the homology of M with $\mathbb{Z}$ coefficients.

One can also show that the mass $\mathbf{M}(T)$ of an integral current T is the same as the quantity $\sum_i k_i \mathscr{H}^k(K_i)$.

Central to the utility of integral currents is the following compactness theorem:

Theorem 8 (Federer-Fleming). *If $\{T_i\} \subset \mathscr{I}_k(M)$ is a sequence of integral currents such that*

$$\sup_i \|T_i\|(W) + \|\partial T_i\|(W) < \infty$$

for all $W \subset\subset M$, then there exists an integral current T and a subsequence of $\{T_i\}$ which converges weakly to T.

Since mass is weakly lower-semicontinuous, Federer-Fleming produced the following corollary:

Corollary 1. *If $T \in \mathscr{I}_k(M)$, then there exists $T_0 \in \mathscr{I}_k(M)$ with $T - T_0 = \partial R_0$ for some $R_0 \in \mathscr{I}_{k+1}$ such that*

$$\mathbf{M}(T_0) = \inf_{R \in \mathscr{I}_{k+1}} \mathbf{M}(T_0 + \partial R).$$

As a special case, setting $Q = \partial T$:

Corollary 2. *If $Q \in \mathscr{I}_{k-1}(\mathbb{R}^n)$, there exists $T_0 \in \mathscr{I}_k(\mathbb{R}^n)$ with $\partial T_0 = Q$ such that*

$$\mathbf{M}(T_0) = \inf_{T \in \mathscr{I}_k, \partial T = Q} \mathbf{M}(T).$$

Another special case occurs when T is a cycle:

Corollary 3. *Each class in $H_k(\mathscr{I}_\bullet(M), \partial)$ contains a representative of least mass.*

4.3 Regularity

The ***support*** $supp(T)$ of a current $T \in \mathscr{D}_k$ is the complement of the largest open set U for which $supp(\omega) \subset U \Rightarrow T(\omega) = 0$. We say $p \in supp(T) \setminus supp(\partial T)$ is an ***interior regular point*** if there exists $\epsilon > 0$, a positive integer κ and an oriented k-dimensional smooth submanifold S such that $T(\omega) = \kappa[[S]](\omega)$ for all forms ω supported in the ball of radius ϵ about p. The remaining points in $supp(T)\setminus supp(\partial T)$ are called ***interior singular points***, the set of which will be denoted $\mathscr{P}(T)$.

Theorem 9 (Complete Interior Regularity). *If $T_0 \in \mathscr{I}_n(\mathbb{R}^{n+1})$, where $2 \leq n \leq 6$, and the mass of T_0 is minimal among all integral currents with the same boundary, then $supp(T_0) \setminus supp(\partial T_0)$ is an embedded minimal hypersurface in $\mathbb{R}^n \setminus supp(\partial T_0)$, and $\mathscr{P}(T_0)$ is empty.*

In 1962, Fleming proved the result for $n = 2$, so other than regularity at the boundary which was to take another 17 years [59], this result completed the solution of the oriented Plateau Problem in $\mathbb{R}^3$ for surfaces of all topological types.

Almgren [6] extended Fleming's theorem to $n = 3$, and Simons extended it up to $n = 6$ in [81].

Also in [81] Simons constructed an example which showed that singularities could in fact occur in dimension 7 and higher. The "Simons cone" is the cone over $S^3 \times S^3 \subset S^7 \subset \mathbb{R}^8$. He showed it was locally mass minimizing, yet has an isolated interior singularity. Immediately after Simons published his example, Bombieri, de Giorgi, and Giusti [12] showed in a marathon three-day session[5] that S is in fact globally mass minimizing. As a corollary, they also showed that, for any $n \geq 8$, there exist functions which satisfy the minimal surface equation and are not affine, finally settling the Bernstein problem in all dimensions.

Not long after, Federer [44] put a bound on the size of the singular set $\mathscr{P}(T)$:

Theorem 10. *The singular set $\mathscr{P}(T_0)$ has Hausdorff dimension at most $n - 7$. Singularities are isolated points if $n = 7$.*

Bombieri, de Giorgi, and Giusti [12] showed that this bound is sharp: there exist mass minimizers T_0 in every dimension $n \geq 7$ such that $\mathscr{H}^{n-7}(\mathscr{P}(T_0)) > 0$. In the 90's, Simon [83] proved this singular set is well-behaved:

Theorem 11. *Except for a set of $\mathscr{H}^{m-7}$ measure zero, the singularity set of a codimension one mass minimizer T_0 is covered by a countable collection of C^1 submanifolds of dimension $m - 7$.*

A new and simpler proof of Simon's theorem has been recently found by Naber and Valtorta [71]. To the best of our knowledge, however, it is still an open question whether or not the remainder of the singularity set stratifies as lower-dimensional submanifolds.

Surprisingly, codimension one mass minimizers do not have boundary singularities, as Hardt and Simon [59] established:

[5] This story was recently communicated by Simons to the second author.

Theorem 12. *If $T_0 \in \mathscr{I}_n(\mathbb{R}^{n+1})$, $\partial T_0 = [[S]]$ for some oriented embedded C^2 submanifold $S \subset \mathbb{R}^n$, and the mass of T_0 is minimal among all integral currents with the same boundary, then there exists an open neighborhood V of S such that $V \cap supp(T_0)$ is an embedded $C^{1,\alpha}$ hypersurface with boundary for all $0 < \alpha < 1$.*

The story is more complicated and incomplete in higher codimension. Almgren in his 1700 page "big regularity paper" [10] proved the following theorem:

Theorem 13. *The singular set $\mathscr{P}(T_0)$ of an m-dimensional mass-minimizing integral current T_0 in $\mathbb{R}^n$ has Hausdorff dimension at most $m-2$.*

Again, this bound is sharp in codimension ≥ 2 [42]. Chang [20] built upon Almgren's work to show that if $m = 2$, then the singularity set consists of isolated branch points. More recently, in a series of papers [30–34], De Lellis and Spadaro took on the monumental task of modernizing and simplifying Almgren's work. For an excellent overview of their approach, see [28].

5 Non-orientable Generalized Surfaces

Much of the above story can be repeated using chains with coefficients in a finite group, and in particular in $\mathbb{Z}/2\mathbb{Z}$ to account for non-orientable surfaces. Fleming [47] has a beautiful theory of flat chains with coefficients (see also [95]), the homology of which recovers the mod-p homology of the ambient space.

However, soap films that occur in nature are not only non-orientable, but possess singularities such as triple junctions which are not amenable to mass-minimization. To ensure that the triple junction not be part of the algebraic boundary, one must assign one of the three surfaces a higher multiplicity. This in turn increases the total mass of the surface, and as a result triple junctions do not show up in solutions to the mass minimization problem.

To get around this issue, there is a different approach one can take, and that is to ignore multiplicity when measuring area. Instead of minimizing mass, one can instead minimize ***size***, which for an integral current $\sum_i k_i \int_{K_i}$ is the quantity $\sum_i \mathscr{H}^k(K_i)$. The k-dimensional size of an arbitrary subset $E \subset M$ is just $\mathscr{H}^k(E)$. Note that the size of an integral current may be smaller than the size of its support. The primary difficulty with working with size is that unlike mass, it is not weakly lower semicontinuous. Extreme care must be taken with the minimizing sequence to account for this. The payoff is that size is better suited to the study of soap films than mass.

5.1 Reifenberg's 1960 Paper

The same year that Federer and Fleming's seminal paper appeared [45], Reifenberg published a work [77] which dealt with the Plateau problem for non-orientable manifolds of arbitrary genus. This paper is also famous for a result that later became known as "Reifenberg's disk theorem," which placed sufficient conditions on the approximate tangencies of a surface to guarantee that it was a topological disk. A set satisfying these conditions is now known as ***Reifenberg flat.*** Reifenberg was well known for his prowess with tricky if not quirky estimates, and indeed his disk theorem did not disappoint: a condition involved in the statement required $\epsilon \leq 2^{-2000n^2}$. Reifenberg's paper has sparked a number of subsequent results: Almgren provided a generalization to so-called "elliptic integrands" in [7]. Morrey generalized Reifenberg's result to ambient manifolds in [65] (see also [66]).

Reifenberg's approach to proving his main theorem was built on work by Besicovitch and was purely set-theoretic, not involving any fancy machinery such as currents or varifolds. His main result was the following:

Consider a finite collection A of pairwise disjoint Jordan curves in $\mathbb{R}^3$. A compact subset X of $\mathbb{R}^3$ is said to be a ***surface spanning*** A if X can be written as an increasing union of manifolds X_i with boundary, such that for each i, there exists a manifold Y_i with boundary $A \cup \partial X_i$ such that $Y_i \to A$ in the Hausdorff distance. Reifenberg then proved:

Theorem 14. *There is a surface spanning A of least area.*

His class of surfaces included those with infinite genus such as in Fleming's example (Fig. 4), and non-orientable surfaces as well. However, it did not include soap-film type surfaces with triple junctions (Fig. 9b). Reifenberg proved a secondary result which did minimize amongst this larger category, in the case that $A \subset \mathbb{R}^n$ is homeomorphic to the $(m-1)$-sphere:

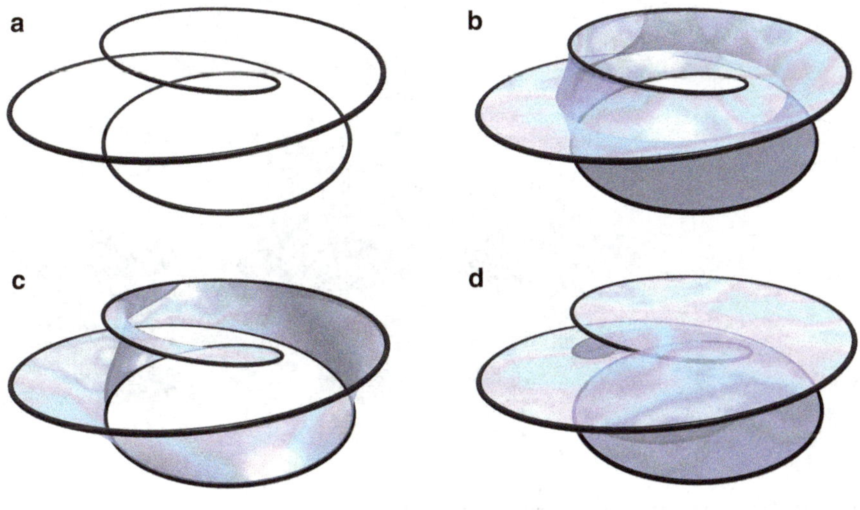

Fig. 9 Depending on the configuration of the boundary wire, any of (**b**), (**c**) or (**d**) can have smaller area. (**a**) The boundary of a triple Möbius band. (**b**) A Triple Möbius band. (**c**) A non-orientable embedded surface. (**d**) An immersed disk

Theorem 15. *If A is a topological $(m-1)$-sphere in $\mathbb{R}^n$, $2 \leq m \leq n$ and $\mathcal{G}^*$ is the collections of all compact sets $X \supset A$ which do not retract onto A, then there exists a set $X \in \mathcal{G}^*$ with least m-dimensional Hausdorff spherical measure. Any such minimizer is locally Euclidian almost everywhere.*

Both of these theorems were special cases of a general result involving "surfaces with algebraic boundary," which were defined by Reifenberg and developed by Adams in an appendix of [77].

Let G be a compact[6] abelian group and suppose A is a compact subset of $\mathbb{R}^n$ with $\mathcal{H}^{m-1}(A) < \infty$. Suppose L is a subgroup of the $(m-1)$-dimensional Čech homology $\check{H}_{m-1}(A;G)$ of A with coefficients in G. We say that a compact set $X \supset A$ is a ***surface with (algebraic) boundary*** $\supset L$ if L is in the kernel of the inclusion homomorphism $\iota_* : \check{H}_{m-1}(A;G) \to \check{H}_{m-1}(X;G)$. Reifenberg proved existence of a surface with algebraic boundary $\supset L$ with least m-dimensional Hausdorff spherical measure. The case that $G = \mathbb{Z}/2\mathbb{Z}$ implies Reifenberg's first theorem, and the case that $G = S^1$ implies, via a theorem of Hopf, the second.

A shortcoming of Reifenberg's theory is that for boundaries more general than a sphere, he did not defined a single, unifying collection surfaces with soap-film singularities. For example, consider the disjoint union of a disk and a circle in $\mathbb{R}^3$. There is no retraction to the pair of circles, yet we might not want to consider this as an admissible spanning set. As another example, consider the surfaces X_i, $i = 1, 2, 3$, in Fig. 10. Any one could be a surface with minimal area, depending on the distance between the circles, but a simple computation shows there is no non-trivial collection of Reifenberg surfaces which contains all three simultaneously. Thus, one would have to find an appropriate subgroup L which would produce the correct minimizer, and this task would change depending on the configuration of the circles in the ambient space.

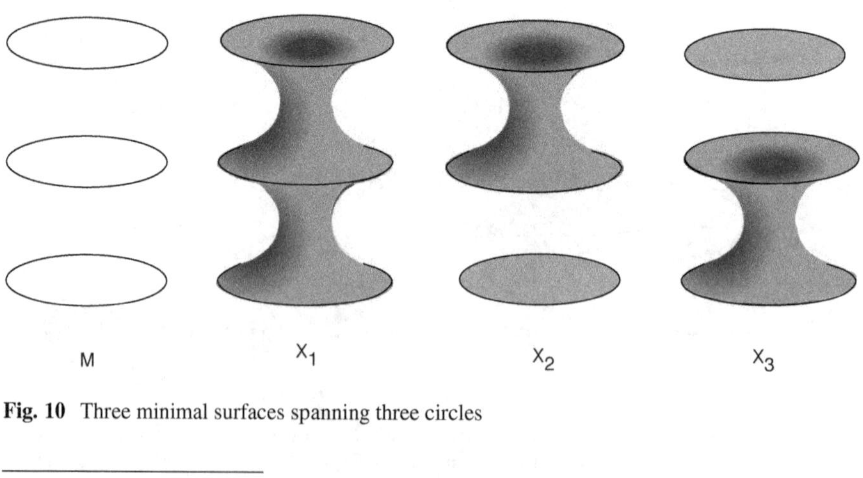

Fig. 10 Three minimal surfaces spanning three circles

[6]The exactness axiom is used in the proof and Čech homology only satisfies exactness when the coefficients are compact, so we will need this assumption.

In a recent paper [55, 56], the authors found a way around this problem using linking number to define spanning sets, and later in [58], using Čech cohomology in higher codimension. We also generalized Reifenberg's result so as to minimize a Lipschitz density functional.

5.1.1 Spanning Sets via Linking Numbers

Definition 1. Suppose A is an $(n-2)$-dimensional compact orientable submanifold of $\mathbb{R}^n$, $n \geq 2$. We say that a circle S embedded in $\mathbb{R}^n \setminus A$ is a ***simple link of*** A if the absolute value of the linking number $L(S, A_i)$ of S with one of the connected components A_i of A is equal to one, and $L(S, A_j) = 0$ for the other connected components A_j of A, $j \neq i$. We say that a compact subset $X \subset \mathbb{R}^n$ ***spans*** A if every simple link of A intersects X (see Fig. 11).

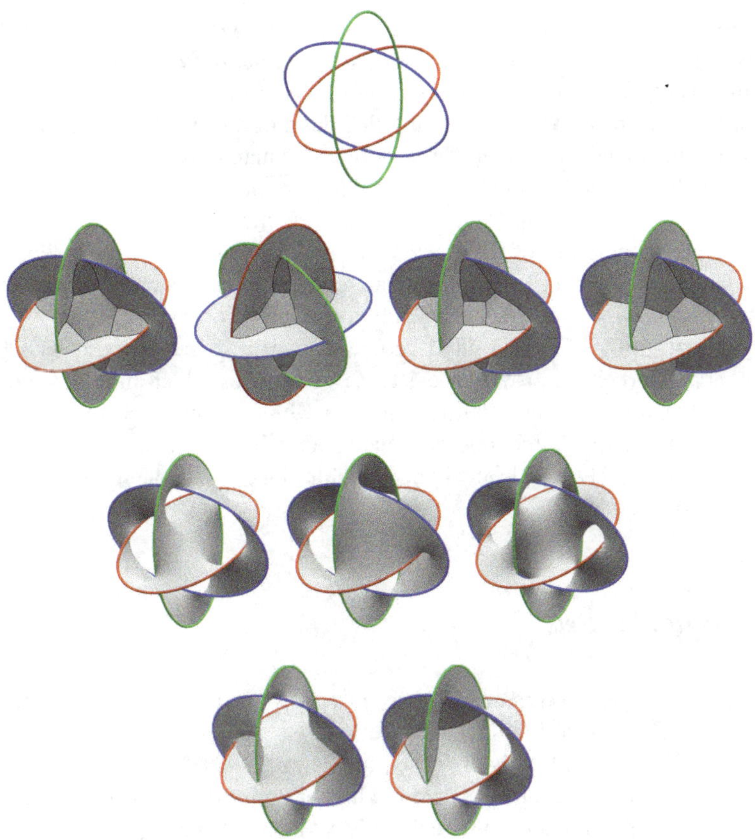

Fig. 11 Each row of surfaces depicts a distinct type of minimal surface spanning the Borromean rings. A surface in the *first row* spans the Borromean rings using any linking test. That is, every simple link of any number of curves must meet the surface. In the *second row*, every simple link of one curve or all three curves must meet the surface. The third row has mixed types

If A is a topological $(n-2)$-sphere then the set of spanning surfaces is the same as the collection $\mathscr{G}^*$ above. Any orientable $(n-1)$-manifold with boundary A spans A. The set A can be a frame such as the $(n-2)$-skeleton of an n-cube, in which case one can specify $(n-2)$-cycles which the simple links need to link. This procedure generalizes to higher dimension using linking spheres, or alternatively, via Alexander duality, to Čech cohomology.

This idea was first proposed for connected smooth boundaries in the existence paper of [54] which was followed by the more substantial [55, 56] which established lower semicontinuity of Hausdorff measure for a minimizing sequence $X_k \to X_0$ in codimension one and applied to any number of boundary components. One of us (HP) realized that linking number tests could be naturally viewed a cohomological spanning condition in higher codimension,[7] and while we were writing this generalization [58], two papers appeared [35] and [37] which built upon our linking number test for spanning sets. See [57] for a new homotopy spanning condition, building on [35] and [37].

The cohomological spanning condition is stated as follows: if L is a *subset* of the $(m-1)$-st (reduced) Čech cohomology group $\check{H}^{m-1}(A;G)$ (G need not be compact), we say that $X \supset A$ is a ***surface with (algebraic) coboundary*** $\supset L$ if L is disjoint from the image of $\iota^* : \check{H}^{m-1}(X;G) \to \check{H}^{m-1}(A;G)$.

One of the primary benefits of using this definition over the covariant "surface with algebraic boundary" is that if A is an oriented manifold, then there is a natural choice for the subset L, namely the collection $L^{\mathbb{Z}}$ of those cocycles on A which evaluate to 1 on the fundamental cycle of a particular component of A, and zero on the rest. By naturality of the Alexander duality isomorphism, the collection of surfaces with coboundary $\supset L^{\mathbb{Z}}$ is equivalent in codimension one to the collection of compact sets which span A in the sense of linking number.

Eight years after [77] was published, Almgren proposed an extension [7] of Reifenberg's theorem to prove the existence of surfaces which minimize not only area, but area weighted by a density function which is permitted to vary in both spacial and tangential directions, subject to an ellipticity condition. To discuss Almgren's papers [7] and [9] we will need to introduce rectifiable sets and varifolds, which we now define.

5.2 Rectifiable Sets

A subset E of $\mathbb{R}^n$ is m-***rectifiable*** if there exist a countable collection of Lipschitz maps $\{f_i : \mathbb{R}^m \to \mathbb{R}^n\}$ such that the m-dimensional Hausdorff measure of $E \setminus \cup_{i=0}^{\infty} f_i(\mathbb{R}^m)$ is zero. If E is $\mathscr{H}^m$ measurable and $\mathscr{H}^m(E) < \infty$, then the maps f_i can be taken to be C^1. Such sets are the higher dimensional analog of rectifiable curves. The defining property of a rectifiable set is that it is equipped with a unique

[7] Similar "surfaces with coboundary" were discovered independently by Fomenko [49].

"approximate tangent m-plane" almost everywhere, a consequence of Rademacher's theorem. If these approximate tangent spaces are equipped with an orientation, it becomes possible to integrate m-forms: An integer rectifiable current is just an integer weighted rectifiable E together with an orientation, i.e. an $\mathcal{H}^m$ measurable field of m-vectors on E such that for almost every $p \in E$, the m-vector at p is unit simple in the direction of the approximate tangent space at p.

A set F is ***purely m-unrectifiable*** if $\mathcal{H}^m(F \cap E) = 0$ for every m-rectifiable set E. Every subset of $\mathbb{R}^n$ with finite $\mathcal{H}^m$ measure can be written, uniquely up to $\mathcal{H}^m$ measure zero sets, as the disjoint union of a m-rectifiable set an a purely m-unrectifiable set. The beautiful Besicovitch-Federer structure theorem says that if F is purely m-unrectifiable, then for almost every m-plane V in the Grassmannian $\mathrm{Gr}(m, n)$, the orthogonal projection of F onto V has $\mathcal{H}^m$ measure zero.

Morally, every subset of Euclidian space can be decomposed almost uniquely into a countable collection of C^1 submanifolds, and a remainder which casts no shadows.

5.3 *Integral and Stationary Varifolds*

Varifolds were first introduced by Young [93, 94] as "generalized surfaces" and developed by Young and Fleming [46, 51]. Fleming, who had been Young's student, in turn, taught Almgren what he knew [48] when Almgren was a student at Brown 1958–62. Almgren took an interest in generalized surfaces and changed the name to "varifolds," a mnemonic for manifolds in the calculus of variations. He produced a set of mimeographed notes [4] on varifolds that were circulated amongst his students but never published. Allard, who had been Fleming's student, produced the definitive reference on varifolds [2] in which he proved the compactness theorem for integral varifolds.

Definition 2. Let M be a smooth n-dimensional Riemannian manifold. ***A k-varifold V in M*** is a Radon measure on the total space of the Grassmannian bundle π : $\mathrm{Gr} \to M$, whose fiber above a point $p \in M$ is the Grassmannian of un-oriented linear k-planes in T_pM. The pushforward of V by π is denoted $\|V\|$. The ***mass of*** V is the quantity $\|V\|(M)$. The ***support of*** V is the support of the measure $\|V\|$.

For example, an embedded k-dimensional submanifold $S \subset M$, together with a $\mathcal{H}^k$-measurable function $\theta : S \to \mathbb{R}^+$ determine a varifold V as follows:

$$V(A) := \int_{S \cap \{p:(p,T_pS) \in A\}} \theta(p) d\mathcal{H}^k(p).$$

More generally, S can be replaced by a k-rectifiable set. In this case V is called a rectifiable k-varifold. If θ takes integer values, the varifold is called an integral varifold. Integral varifolds are the non-orientable analogs of integral currents. There is no notion of integration of differential forms on a varifold, and unlike currents the space of varifolds does not possess a boundary operator. However, integral k-varifolds can be pushed forward by a Lipschitz map.

Almgren saw these features as an advantage, for he wanted to model non-orientable surfaces and those with triple junctions. In [5] Almgren credits Federer-Fleming for proving the Plateau problem for mass minimization of oriented surfaces, and Reifenberg for size minimization of non-oriented surfaces "subject to certain topological restraints." Almgren sought at this time to prove a mass minimization result for non-oriented surfaces. He did not specify what it meant for a varifold to span a given contour in [4] or [5], but his focus at the time was on stationary varifolds where the definition seemed self-evident as we shall next see.

The ***first variation*** δV of a compactly supported varifold V is a function which assigns to a smooth compactly supported vector field Y on M the rate of change of the mass of the pushforward of V by the time-t map of the flow of Y at $t = 0$. A varifold is ***stationary*** if $\delta V = 0$.

In [4], Almgren proved the following theorem:

Theorem 16. *Let M be a smooth compact n-dimensional Riemannian manifold. For each $0 < k < n$ there exists a stationary integral k-varifold in M.*

Allard in [2] proved a beautiful regularity result for such varifolds:

Theorem 17. *If V is a stationary integral k-varifold in a smooth compact n-dimensional Riemannian manifold M, $0 < k < n$, then there is an open dense subset of the support of V which is a smooth k-dimensional minimal submanifold of M.*

Even many years later, this theorem remains state-of-the-art in terms of what is known about the singularity set of stationary varifolds. For example, it is not known if the singularity set has zero Hausdorff measure in dimension k. Indeed, regularity theory for stationary varifolds is still at an early stage, even compared to what is known about mass minimizing integral currents in higher codimension. See [17] where Bombieri mentions this problem, as well as [29] for an accounting of progress.

5.4 Elliptic variational problems

In [7] Almgren initiated the study of elliptic variational problems for non-orientable surfaces by providing the first definition of an elliptic integrand and a proof of regularity, depending on the degree of smoothness of the integrand. His definitions and main regularity result follow:

Let A be a compact $(m-1)$-rectifiable subset of $\mathbb{R}^n$ with $\mathscr{H}^{m-1}(A) < \infty$, G a finitely generated abelian group, and $\sigma \in \check{H}_m(\mathbb{R}^n, A; G)$. We say a compact m-rectifiable set $X \supset A$ is a ***surface which spans*** σ if σ is in the kernel of the homomorphism on homology induced by the inclusion $(\mathbb{R}^n, A) \hookrightarrow (\mathbb{R}^n, X)$.

A C^k ***(resp. real analytic) integrand*** is a C^k (resp. real analytic) function $f : \mathbb{R}^n \times \mathrm{Gr}(m, n) \to [a, b]$, where $0 < a < b < \infty$. We say f is ***elliptic with respect to*** G

if there exists a continuous function $c : \mathbb{R}^n \to \mathbb{R} \cap \{t : t > 0\}$ such that if $D \subset \mathbb{R}^n$ is an m-disk, $\tau \in \check{H}_m(\mathbb{R}^n, \partial D; G) \setminus \{0\}$, and $\tilde{D}$ is any surface which spans τ, then

$$\int_{\tilde{D}} f(x, T_y\tilde{D})\, d\mathcal{H}^m(y) - \int_D f(x, T_yD)\, d\mathcal{H}^m(y) \geq c(x)\left(\mathcal{H}^m(\tilde{D}) - \mathcal{H}^m(D)\right)$$

for all $x \in \mathbb{R}^n$.

Theorem 18. *Let $3 \leq k \leq \imath$ and $G \in \mathbf{G}$. If f is a C^k (resp. real analytic) integrand, elliptic with respect to G, and S is a surface which spans σ such that*

$$\int_S f(x, T_xS)\, d\mathcal{H}^m(x) \leq \int_T f(x, T_xT)\, d\mathcal{H}^m(x)$$

for all surfaces T which span σ, then S is $\mathcal{H}^m$ almost everywhere a C^{k-1} (resp. real analytic) submanifold of $\mathbb{R}^n$.

The authors proved there exists such a surface S. These results marked a significant advance[8] over Reifenberg's paper which only dealt with the functional $f = 1$, and bring Plateau's problem, which had grown well beyond the classical theory of minimal surfaces and the minimal surface equation, squarely back into the realm of PDE's.

5.4.1 (f, ϵ, δ)-minimal sets

In his memoir [9] (see also [8]), Almgren defined new classes of surfaces to model soap bubbles as well as many types of soap films. His regularity theory for minimizers in such a class is often cited.

Fix $A \subset \mathbb{R}^n$. If $\phi : \mathbb{R}^n \to \mathbb{R}^n$ is Lipschitz, let $W_\phi = \{x : \phi(X) \neq x\}$. If $W_\phi \cup \phi(W_\phi)$ is disjoint from A and contained in a ball of radius δ for some $0 < \delta < \imath$, we say that ϕ is a ***δ-deformation fixing*** A.

Let $1 \leq \gamma < \infty$. A compact set $X \subset \mathbb{R}^n$ with $\mathcal{H}^m(X) < \infty$ is ***(γ, δ)-restricted with respect to*** A if

$$\mathcal{H}^m(X \cap W_\phi) < \gamma \mathcal{H}^m(\phi(X \cap W_\phi))$$

for all δ-deformations ϕ fixing A.

[8]Readers should be warned that the existence portion of [7] contains a serious gap. Briefly, a minimizing convergent sequence for a bounded elliptic integrand does not automatically yield a uniformly quasiminimal subsequence, but [7] assumes that it does. This is a critical part of the argument for existence of a minimizer (see [57] for a more detailed discussion.) In [9] spanning surfaces are chosen to be a priori uniformly quasiminimal so that the problem disappears. However, he was not able to prove a general existence theorem (see [67] for further details.).

If $\epsilon : [0, \infty) \to [0, \infty)$ with $\epsilon(0) = 0$ is a continuous non-decreasing function, the set X is called ***(f, ϵ, δ)-minimal*** if in addition for every r-deformation ϕ fixing A, $0 < r < \delta$,

$$\int_X f(x, T_x X)\, d\mathscr{H}^m(x) \leq (1 + \epsilon(r)) \int_{\phi(X)} f(x, T_x\phi(X))\, d\mathscr{H}^m(x).$$

An important fact about (γ, δ)-restricted sets is that they are m-rectifiable [43, 3.2.14(4)] and have both upper and lower bounds on density ratios. Almgren proved regularity results a.e for (f, ϵ, δ) minimal sets in [9]:

Theorem 19. *Suppose f is elliptic and C^3 and X is (f, ϵ, δ)-minimal with respect to A where ϵ satisfies*

$$\int_0^1 t^{-(1+\alpha)} \epsilon(t)^{1/2} dt < \infty$$

for some $0 \leq \alpha < 1$. Then there exists an open set $U \subset \mathbb{R}^n$ such that $\mathscr{H}^m(X \backslash U) = 0$ and $X \cap U$ is a C^1 m-dimensional submanifold of $\mathbb{R}^n$.

Almgren states however that "These hypotheses and conclusions, incidentally, do not imply that $S \cap U$ locally can be represented as the graph of a function which satisfies any of the various Euler equations associated with f." Indeed, any C^2 m-dimensional submanifold S with boundary is[9] (M, ϵ, δ) minimal with respect to ∂S if δ is sufficiently small and ϵ is a linear map with large slope.

For the three-dimensional case, Taylor [86] relied upon Theorem 19 to prove a beautiful soap film regularity result for (M, ϵ, δ)-minimal sets. However, Morgan [67] points out that the class of $(M, 0, \delta)$-minimal sets, taken over all $\delta > 0$, is not compact. It remains an open problem of whether a smoothly embedded closed curve in $\mathbb{R}^3$ bounds a film with minimal area in the class of all $(M, 0, \delta)$-minimal sets.

Another open problem motivated by [35, 37] and [55, 56] is to prove the same regularity theorems as above in the case that ϕ is also required to be uniformly close to a diffeomorphism.

6 Variable Boundaries

6.1 Sliding Boundaries

The notion of a sliding boundary has had a long history in the study of elasticity in mechanical engineering (see Sect. 24 of [73], for example). David brought the attention of this problem to those in geometric measure theory [26]. We shall mention a formulation of the problem found in [35] and [37] which was influenced

[9] Here and in the literature, M denotes the constant function 1.

by [26]. These works assume that the bounding set A has zero $\mathscr{H}^m$ measure and often $(m-1)$-rectifiable, but others do not (see, e.g., [23, 50]) as applications often require a large, and even rough, bounding set.

Definition 3. Let $A \subset \mathbb{R}^n$ be compact and $S_* \subset \mathbb{R}^n \setminus A$ be relatively compact. Let Lip(A) denote the collection of Lipschitz maps $\phi : \mathbb{R}^n \to \mathbb{R}^n$ such that there exists a continuous map $\Phi : [0,1] \times \mathbb{R}^n \to \mathbb{R}^n$ with $\Theta(1,\cdot) = \phi$, $\Theta(0,\cdot) = Id$ and $\Theta(t,A) \subset A$ for each $t \in [0,1]$. Define

$$\mathscr{C}(A,S_*) = \{S : S = \phi(S_*) \text{ for some } \phi \in \text{Lip}(A)\}$$

and call S_* a ***sliding minimizer*** if $\mathscr{H}^m(S_*) = \inf\{\mathscr{H}^m(S) : S \in \mathscr{C}(A,S_*)\}$.

Note that $\mathscr{C}(A,S_*)$ does not form an equivalence class. It is not known if $\mathscr{C}(A,S_*)$ is compact. However it can be shown with some assumptions on A (see [35] for codimension one, [37] for higher codimension) that if $\{S_k\} \subset \mathscr{C}(A,S_*)$ is a minimizing sequence, then the measures $\mathscr{H}^m \lfloor_{S_k}$ converge weakly to a measure $g\mathscr{H}^m \lfloor_{S_0}$ where S_0 is m-rectifiable and $g \geq 1$. In particular, $\mathscr{H}^m(S_0) \leq \liminf \mathscr{H}^m(S_k)$. It is not known if $S_0 \in \mathscr{C}(A,S_*)$, but [35] and [37] proved nonetheless that S_0 is a sliding minimizer. We address a slightly different sliding boundary problem in [57].

It is an open question if this result extends to Lipschitz or elliptic integrands. It is similarly open to prove the result if the aforementioned assumptions on A (e.g., $\mathscr{H}^m(A) < \infty$) are removed.

6.2 *Euler-Plateau Problem*

Mahadevan and Giomi [53] proposed a type of Plateau problem in which a rigid boundary is replaced by a soft boundary such as a flexible wire. Specifically, in the language of Kirchhoff's theory of rods [38], permissible boundaries are circular rods which resist bending yet are inextensible, unshearable, without intrinsic curvature, and without resistance to twisting about their centerlines. Mahadevan and Giomi formulated an energy functional which measured not only the area of a spanning surface, but also the energy of the boundary. The resulting Euler-Lagrange equations are equivalent, in the zero surface tension case, to those derived by Langer and Singer [64] (see [41]). This minimization problem is called the ***Euler-Plateau problem*** after Euler's study of column buckling, but might more appropriately be called the ***Kirchhoff-Plateau problem***.[10]

Chen and Fried [41] rigorously derived the equilibrium conditions for the minimization problem, and provided geometric and physical interpretations of these conditions. Briefly, the surface on the interior must have zero mean curvature, and the boundary is required to bend elastically in response to a force exerted by the

[10]The authors would like to thank Eliot Fried for helpful remarks.

spanning film. The class of competitors for minimization are those surfaces which occur as images of the disk. However, since the boundary is permitted to vary, the maps cannot, in contrast to Douglas, Radó and Courant, be assumed to be conformal. See also [16, 50].

These papers are closely related to earlier work by Bernatzki [15] and Bernatzki-Ye [19].

7 Open Problems

Though we have labeled the following problems as "open," some may have been solved without our knowledge. If you have solved one of these, please accept our apologies (and our congratulations!)

7.1 *Classical Minimal Surfaces*

The following problems are part of a longer list in [68]. Let C be the space of connected, complete, embedded minimal surfaces and let $P \subset C$ be the subspace of properly embedded surfaces.

- Isolated Singularity Conjecture (Lawson and Gulliver): The closure of a properly embedded minimal surface in the punctured closed unit ball is a compact embedded minimal surface.
- Convex Curve Conjecture (Meeks): Two convex Jordan curves in parallel planes cannot bound a compact minimal surface of positive genus.
- 4π-Conjecture (Meeks, Yau, Nitsche): If Γ is a simple closed curve in $\mathbb{R}^3$ with total curvature at most 4π, then Γ bounds a unique compact, orientable, branched minimal surface and this unique minimal surface is an embedded disk.
- Liouville Conjecture (Meeks): If $M \in P$ and $h : M \to \mathbb{R}$ is a positive harmonic function, then h is constant.
- Finite Genus Properness Conjecture (Meeks, Pérez, Ros): If $M \in C$ and M has finite genus, then $M \in P$.

7.2 *Integral Currents*

These problems are adapted from a longer list in [11].

- Establish the uniqueness of tangent cones to an mass-minimizing current. Uniqueness for 2-dimensional currents was proved in [90], and partial results in the general case in [1] and [82].

- Does the singular set of a mass-minimizing current have locally finite $\mathcal{H}^{m-2}$ measure? Chang [20] proved that it does if $m = 2$.
- Is the singular set of an mass minimizing current rectifiable? Does it have other geometric structure such as a stratification? See e.g., Theorem 11.

7.3 *Reifenberg Problems*

Reifenberg posed ten open problems in [77]. Three of particular interest are these:

- Let M be a manifold with boundary and D_k be discs with boundary ∂M. Let μ be the infimum of the areas of discs with boundary ∂M. Suppose $D_k \to M$ and $\mathcal{H}^2(D_k) \to \mathcal{H}^2(M) = \mu$. Prove that M is a disc. Prove the same for m-dimensional disks.
- Generalize Theorem 2 of [77] to the case where the boundary is any manifold. For example, let A be the 2-torus and X the solid torus with a small interior ball removed. Then A is not a retract of X, as one can deformation retract X onto the union of A and a transverse disk. If the torus is made to be narrower in some region, the solution to the generalized Theorem 2 in this case would be a transverse disk at the narrowest location.
- Find a class of surfaces which includes those such as the Adams example which retract onto their boundary, and also includes some class of deformations thereof; then prove a compactness theorem for such surfaces. Do those sets which do not admit a deformation retraction onto the boundary forms such a class?

7.4 *Elliptic integrands*

- Show by example that interesting non-smooth solutions can arise which represent observed phenomena in nature if an elliptic integrand is not smooth.
- Prove a version of the main result in [57] for mass, instead of size, weighted by an elliptic density functional.
- What restrictions on the competing class of surfaces can be made that carry over to minimizing solutions? E.g., one can restrict the problem to graphs, disks, continuous embeddings, bordisms, topological type, etc. Each problem presents its own existence and regularity questions.
- Axiomatic approach: Let $\mathcal{S}$ be a collection of surfaces such that if $S \in \mathcal{S}$ and ϕ is a Lipschitz map fixing A which is C^0 close to a diffeomorphism, then $\phi(S) \in \mathcal{S}$. What are minimal conditions needed on $\mathcal{S}$ to guarantee existence of a minimizer for an elliptic integrand in S? See [35–37, 57].

7.5 Non-closed Curves

- Find models for surfaces spanning non-closed curves and prove a compactness theorem (Fig. 2b). In [58] we proposed using relative (co)homology as follows: If one replaces the boundary set A with a pair (A, B), the definition of a surface with coboundary can be repeated: A pair $(X, Y) \supset (A, B)$ is a surface with coboundary $\supset L$ if L is disjoint from the image of $\iota^* : \tilde{H}^{m-1}(X, Y) \to \tilde{H}^{m-1}(A, B)$. To what extent can this be adapted if B is permitted to vary in some restricted fashion? See also [40] and [67, 11.3].

7.6 Varifolds

- Does a smoothly embedded closed curve in $\mathbb{R}^3$ bound a film with minimal area in the class of all $(M, 0, \delta)$-minimal sets?
- Does the singular set of a stationary varifold have measure zero?

7.7 Dynamics and Deformations

We close with three increasingly open-ended problems

- Euler-Plateau for sliding boundaries: State and solve the Euler-Plateau problem for sliding boundaries. Not only is the bounding set B allowed to be flexible, but frontiers of solutions can slide around within B as it flexes.
- To what extent can mean curvature flow detect soap-film solutions including triple junctions and non-orientable surfaces, starting with a given spanning set? See [18, 91, 92].
- The problem of lightning (Harrison and Pugh): Formulate a dynamic version of Plateau's problem which models the formation and evolution of branched solutions. Applications would be numerous: lightning, formation of capillaries, branches, fractures, etc. One should permit boundaries with higher Hausdorff dimension and solve the elliptic integrand problem, as solutions would be branched minimizers of the corresponding action principle, and thus should be relevant to physics.

References

1. William K. Allard and Frederick Almgren, *On the radial behavior of minimal surfaces and the uniqueness of their tangent cones*, Annals of Mathematics **113** (1981), no. 2, 215–265.
2. William K. Allard, *On the first variation of a varifold*, Annals of Mathematics **95** (1972), no. 3, 417–491.
3. ———, *On the first variation of a varifold: Boundary behavior*, Annals of Mathematics **101** (1975), no. 3, 418–446.

4. Frederick J. Almgren, *Theory of varifolds*, mimeographed notes, 1965.
5. ———, *Plateau's Problem: An Invitation to Varifold Geometry*, Benjamin, 1966.
6. ———, *Some interior regularity theorems for minimal surfaces and an extension of Bernstein's theorem*, Annals of Mathematics **84** (1966), 277–292.
7. ———, *Existence and regularity almost everywhere of solutions to elliptic variational problems among surfaces of varying topological type and singularity structure*, Annals of Mathematics **87** (1968), no. 2, 321–391.
8. ———, *Existence and regularity almost everywhere of solutions to elliptic variational problems with constraints*, Bulletin of the American Mathematical Society **81** (1975), no. 1, 151–154.
9. ———, *Existence and regularity almost everywhere of solutions to elliptic variational problems with constraints*, vol. 4, Mem. Amer. Math. Soc., 1976.
10. ———, *Q-valued functions minimizing Dirichlet's integral and the regularity of area minimizing rectifiable currents up to codimension two.*, Bulletin of the American Mathematical Society **8** (1983), 327–328.
11. Luigi Ambrosio, *Regularity theory for mass-minimizing currents (after Almgren-De Lellis-Spadaro)*, Calculus of Variations and Geometric Measure Theory (2015), 1–23.
12. Enrico Bombieri, Ennio de Giorgi, and Enrico Giusti, *Minimal cones and the Bernstein problem*, Inventiones mathematicae **7** (1969), 243–268.
13. Serge Bernstein, *Sur une théorème de géometrie et ses applications aux équations dérivées partielles du type elliptique*, Comm. Soc. Math. Kharkov **15** (1915–1917), 38–45.
14. Felicia Bernatzki, *The Plateau problem for nonorientable minimal surfaces*, Manuscripta Mathematica **79** (1993), 73–80.
15. ———, *On the existence and regularity of mass-minimizing currents with an elastic boundary*, Annals of Global Analysis and Geometry **15** (1997), 379–399.
16. Aisa Biria and Eliot Fried, *Buckling of a soap film spanning a flexible loop resistant to bending and twisting*, Proceedings of the Royal Society of London. Series A, Mathematical and Physical Sciences **470** (2014), 1–18.
17. Enrico Bombieri, *Recent progress in the theory of minimal surfaces*, L'Enseignement Mathématique **25** (1979), 1–9.
18. Kenneth Brakke, *Soap films and covering space*, Journal of Geometric Analysis **5** (1995), no. 4, 445–514.
19. Felicia Bernatzki and Rugang Ye, *Minimal surfaces with an elastic boundary*, Annals of Global Analysis and Geometry **19** (2001), 1–9.
20. Sheldon Chang, *Two-dimensional area minimizing integral currents are classical minimal surfaces*, J. Amer. Math. Soc. **4** (1988), no. 1, 699–788.
21. Tobias Colding and William Philip Minicozzi II, *The space of embedded minimal surfaces of fixed genus in a 3-manifold IV; locally simply-connected*, Annals of Mathematics **160** (2004), 573–615.
22. ———, *The Calabi-Yau conjectures for embedded surfaces*, Annals of Mathematics **167** (2008), 211–243.
23. Simon Cox and Sian Jones, *Instability of stretched and twisted soap films in a cylinder*, J Eng Math (2013), 1–7.
24. Pascal Collin, *Topologie et courbure des surfaces minimales de R3*, Annals of Mathematics **145** (1997), no. 1, 1–31.
25. Guy David, *Quasiminimal sets for Hausdorff measure*, Recent Developments in Nonlinear Partial Differential Equations (Donatella Danielli, ed.), Contemporary Mathematics series, vol. 439, Purdue University, AMS, 2007, p. 133.
26. ———, *Local regularity properties of almost- and quasiminimal sets with a sliding boundary condition*, arXiv eprints, January 2014.
27. Ennio de Giorgi, *Una estensione del teorema di Bernstein*, Ann. Scuola Norm. Sup. Pisa **19** (1965), no. 3, 79–85.
28. Camillo De Lellis, *The size of the singular set of area-minimizing currents.*
29. ———, *Allard's interior regularity theorem: An invitation to stationary varifolds*, 2012.

30. Camillo De Lellis and Emanuele Spadaro, *Q-valued functions revisited*, Memoirs of the AMS **211** (2011), no. 991, vi+79.
31. ———, *Multiple valued functions and integral currents*, preprint, 2013.
32. ———, *Regularity of area-minimizing currents I: gradient L^p estimates*, preprint, 2013.
33. ———, *Regularity of area-minimizing currents II: center manifold*, preprint, 2013.
34. ———, *Regularity of area-minimizing currents III: blow-up*, preprint, 2013.
35. Camillo De Lellis, C., Francesco Ghiraldin, and Francesco Maggi, *A direct approach to Plateau's problem*, Journal of the European Mathematical Society (2015), 1–17.
36. Camillo De Lellis, Antonio De Rosa, and Francesco Ghiraldin, *A direct approach to the anisotropic Plateau's problem*, arxiv http://arxiv.org/abs/1602.08757.
37. Guido De Philippis, Antonio De Rosa, and Francesco Ghiraldin, *A direct approach to Plateau's problem in any codimension*, arXiv eprints, January 2015.
38. Ellis Dill, *Kirchoff's theory of rods*, Archive for History of Exact Sciences **44** (1992), 1–23.
39. Jesse Douglas, *Solutions of the problem of Plateau*, Transactions of the American Mathematical Society **33** (1931), 263–321.
40. Jordan Drachman and Brian White, *Soap films bounded by non-closed curves*, Journal of Geometric Analysis **8** (1998), no. 2, 239–250.
41. Eliot Fried and Yi-chao Chen, *Stability and bifurcation of a soap film spanning a flexible loop*, 2013.
42. Herbert Federer, *Some theorems on integral currents*, Transactions of the American Mathematical Society **117** (1965), pp. 43–67.
43. ———, *Geometric Measure Theory*, Springer, Berlin, 1969.
44. ———, *The singular sets of area minimizing rectifiable currents with codimension one and of area minimizing flat chains modulo two with arbitrary codimension*, Bulletin of the American Mathematical Society **76** (1970), 767–771.
45. Herbert Federer and Wendell H. Fleming, *Normal and integral currents*, The Annals of Mathematics **72** (1960), no. 3, 458–520.
46. Wendell H. Fleming, *Irreducible generalized surfaces*, Riv. Mat. Univ. Parma **8** (1957), 251–281.
47. ———, *Flat chains over a finite coefficient group*, Transactions of the American Mathematical Society **121** (1966), no. 1, 160–186.
48. ———, *Geometric measure theory at Brown in the 1960s*, 2015.
49. Anatoly T. Fomenko, *The Plateau Problem*, Studies in the Development of Modern Mathematics, Gordon and Breach, 1990.
50. Eliot Fried and Brian Seguin, *Stable and unstable helices: Soap films in cylindrical tubes*, Calculus of Variations and PDE;s **54** (2015), no. 1, 969–988.
51. Wendell H. Fleming and Laurence Chisholm Young, *A generalized notion of boundary*, Transactions of the American Mathematical Society **76** (1954), 457–484.
52. Jeremy Gray and Mario Micallef, *The work of Jesse Douglas on minimal surface*, Bulletin of the American Mathematical Society **45** (2008), no. 2, 293–302.
53. Luca Giomi and Lakshminarayanan Mahadevan, *Minimal surfaces bounded by elastic lines*, Proceedings of the Royal Society of London. Series A, Mathematical and Physical Sciences **471** (2015), 21–81.
54. Jenny Harrison, *Soap film solutions of Plateau's problem*, Journal of Geometric Analysis **24** (2014), 271–297.
55. Jenny Harrison and Harrison Pugh, *Existence and soap film regularity of solutions to Plateau's problem*, arXiv eprints, October 2013.
56. ———, *Existence and soap film regularity of solutions to Plateau's problem*, Advances in Calculus of Variations (2015).
57. ———, *General methods of elliptic minimization*, submitted, March 2016, http://arxiv.org/abs/1603.04492.
58. ———, *Solutions to the Reifenberg Plateau problem with cohomological spanning conditions*, accepted by Calculus of Variations and Partial Differential Equations, May 2016.

59. Robert Hardt and Leon Simon, *Boundary regularity and embedded solutions for the oriented Plateau problem*, Annals of Mathematics **110** (1979), 439–486.
60. Eberhard Hopf, *Ober den funktionalen, insbesondere den analytischen cha rakter der lösungen elliptischer differentialgleichungen zweiter ordnung*, Math. Zeit. **34** (1932), 194–233.
61. Jürgen Jost, *Conformal mappings and the Plateau-Douglas problem in riemannian manifolds*, Journal fur die reine und angewandte Mathematik **359** (1985), 37–54.
62. Xiangyu Liang, *On the topological minimality of unions of planes of arbitrary dimension*, International Mathematics Research Notices (2015).
63. Francisco J. López and Antonio Ros, *On embedded complete minimal surfaces of genus zero*, Journal of Differential Geometry **33** (1991), no. 1, 293–300.
64. Joel Langer and David Singer, *Knotted elastic curves in R3*, Journal London Math. Soc. **30** (1984), 512–520.
65. Charles Morrey, *The higher-dimensional Plateau problem on a riemannian manifold*, Proceedings of the National Academy of Sciences **54** (1965), no. 4, 1029–1035.
66. ______, *Multiple integrals in the calculus of variations*, Springer, 1966.
67. Frank Morgan, *Geometric Measure Theory: A Beginners Guide*, Academic Press, London, 1988.
68. Williams Meeks and Jose Pérez, *The classical theory of minimal surfaces*, Bulletin of the American Mathematical Society **48** (2011), 325–407.
69. William Meeks, Jose Pérez, and Antonio Ros, *Properly embedded minimal planar domains*, Annals of Mathematics **181** (2015), 1–74.
70. William Meeks and Harold Rosenberg, *The theory of minimal surfaces in M x R*, Comment. Math. Helv. **80** (2005), 811–858.
71. Aaron Naber and Daniele Valtorta, *The singular structure and regularity of stationary and minimizing varifolds*, preprint, 2015.
72. Robert Osserman, *A proof of the regularity everywhere of the classical solution to Plateau's problem*, Annals of Mathematics **91** (1970), 550–569.
73. Paolo Podio-Guidugli, *A primer in elasticity*, Journal of Elasticity **58** (2000), 1–104.
74. Joseph Plateau, *Experimental and theoretical statics of liquids subject to molecular forces only*, Gauthier-Villars, 1873.
75. Tibor Radó, *On Plateau's problem*, Annals of Mathematics **Vol 31** (1930), no. 3, 457–469.
76. Georges de Rham, *Variétés Différentiables: Formes, Courants, Formes Harmoniques*, Hermann, Paris, 1973.
77. Ernst Robert Reifenberg, *Solution of the Plateau problem for m-dimensional surfaces of varying topological type*, Acta Mathematica **80** (1960), no. 2, 1–14.
78. ______, *An epiperimetric inequality related to the analyticity of minimal surfaces*, Annals of Mathematics **80** (1964), 1–14.
79. ______, *On the analyticity of minimal surfaces*, Annals of Mathematics **80** (1964), 15–21.
80. Reinger Schatzle, *Quadratic tilt-excess decay and strong maximum principle for varifolds*, Ann. Scuola Norm. Sup. Pisa **5** (2004), no. 3, 171–231.
81. James Simons, *Minimal varieties in riemannian manifolds*, Annals of Mathematics **88** (1968), 62–105.
82. Leon Simon, *Asymptotics for a class of nonlinear evolution equations, with applications to geometric problems*, Annals of Mathematics **118** (1983), no. 3, 525–571.
83. ______, *Rectifiability of the singular sets of multiplicity 1 minimal surfaces and energy minimizing maps*, Surveys in differential geometry, vol. II, International Press, 1995, pp. 246–305.
84. Richard Schoen and Shing-Tung Yau, *Existence of incompressible minimal surfaces and the topol- ogy of three-dimensional manifolds with nonnegative scalar curvature,*, Annals of Mathematics **110** (1979), no. 2, 127–142.
85. Fritz Steinhardt, *Jesse Douglas as teacher and colleague*, The Problem of Plateau - A Tribute to Jesse Douglas and Tibor Radó (Themistocles M. Rassias, ed.), World Scientific, 1992, pp. 37–40.
86. Jean Taylor, *The structure of singularities in soap-bubble-like and soap-film-like minimal surfaces*, Annals of Mathematics **103** (1976), no. 2, 489–539.

87. Friedrich Tomi and Anthony Tromba, *Existence theorems for minimal surfaces of non-zero genus spanning a contour*, Memoirs of the AMS **71** (1988), no. 382, 1–83.
88. Karen Uhlenbeck and Jonathan Sacks, *The existence of minimal immersions of 2-spheres*, Annals of Mathematics **113** (1981), no. 2, 1–24.
89. Hassler Whitney, *Geometric Integration Theory*, Princeton University Press, Princeton, NJ, 1957.
90. Brian White, *Tangent cones to two-dimensional area-minimizing integral currents are unique*, Duke Math. J. **50** (1983), 143–160.
91. ———, *A local regularity theorem for mean curvature flow*, Annals of Mathematics **161** (2005), 1487–1519.
92. ———, *Currents and flat chains associated to varifolds, with an application to mean curvature flow*, Duke Math. J. **148** (2009), no. 1, 41–62.
93. Laurence Chisholm Young, *Generalized surfaces in the calculus of variations*, Annals of Mathematics **43** (1942), 84–103.
94. ———, *Generalized surfaces in the calculus of variations II*, Annals of Mathematics **38** (1942), 530–544.
95. William P. Ziemer, *Integral currents mod 2*, Transactions of the American Mathematical Society **105** (1962), 496–524.

The Unknotting Problem

Louis H. Kauffman

Abstract This paper tells the story of knots and the search to detect their knottedness.

1 Introduction

We say that a closed loop embedded in three dimensional space is *knotted* if there is no continuous deformation of the loop, through embeddings, that changes it to a circle embedded in a plane. The fundamental problem in knot theory is to determine whether a closed loop embedded in three dimensional space is knotted. The loop in Fig. 1 is a trefoil knot, the simplest knot. View Fig. 2. Can you tell whether this loop is knotted or not? It requires a special intuition for topology to just look at a loop and know if it is really knotted.

In order to analyze the knottedness of a knot, a mathematical representation is required. In this paper we shall use the method of knot and link diagrams, and the equivalence relation generated by the Reidemeister moves (see Fig. 3 for an illustration of these moves). Knot diagrams are graphs with extra structure that encode the embedding type of the knot. Each diagram is a pictorial representation of the knot, and so appeals to the intuition of the viewer. The Reidemeister moves are a set of simple combinatorial moves, proved in the 1920s to capture the notion of topological equivalence of knots and links in three dimensional space. Single applications of these moves can leave the diagram with the same number of crossings (places where a weaving of two segments occurs), or increase or decrease the number of crossings. Some unknottings can be accomplished without increasing the number of crossings in the diagram. We call such unknot diagrams *easy* since the fact that they are unknotted can be determined by a finite search for simplifying moves. However, there are culprit diagrams [28, 56, 58, 60] that require moves that *increase* the number of crossings before the diagram can be simplified

L.H. Kauffman (✉)
Department of Mathematics, Statistics, and Computer Science,
University of Illinois at Chicago, Chicago, IL 60607-7045, USA
e-mail: kauffman@uic.edu

J.F. Nash, Jr., M.Th. Rassias (eds.), *Open Problems in Mathematics*,
DOI 10.1007/978-3-319-32162-2_8

Fig. 1 The trefoil knot

Fig. 2 Knotted?

to an unknotted circle with no crossings. It is the structure of such culprits that is the subject matter of Sect. 3 of this paper.

This paper is an exploration of the theme of detecting knottedness. This exposition is an outgrowth of a series of lectures [30] that I gave in Tokyo at the *Knots 96* conference in the summer of 1996. The present exposition goes considerably farther than those lectures, as in the intervening time there has been much progress on this problem. In particular there is a beautiful combinatorial solution to the problem of detecting whether a knot diagram is knotted, due to Dynnikov [10] that we sketch in Sect. 3 of this paper.

The problem of detecting the unknot has been investigated by many people. The three dimensional work of Haken and Hemion [33] solves these problems in

Fig. 3 Reidemeister moves

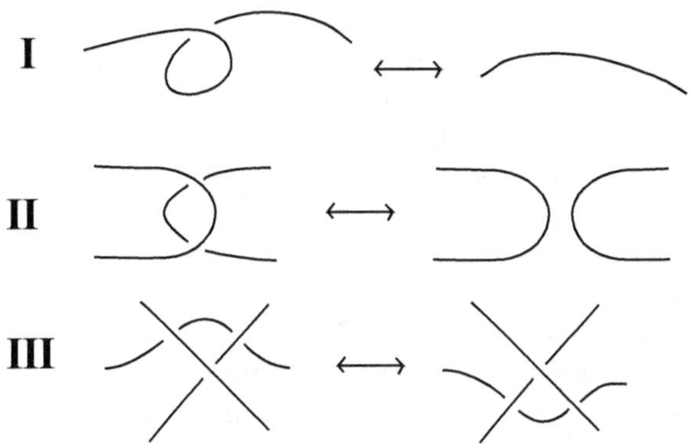

principle, giving definite algorithms to tell whether two knots are equivalent, or whether a given knot is unknotted. The algorithms are given in terms of the structure of a triangulation of the complement of the knot or link, and these algorithms are unwieldy. Nevertheless, the major problem that we have discussed is solved by the Haken-Hemion method. This result did not end the field of study of knot invariants. In fact, it spurred it on and the methods that we have discussed in this paper are just the tip of the iceberg of the revolutions that have engulfed the theory of knots and links since the early 1980s. For example, the papers by Birman and Hirsch [54] and Birman and Moody [55] study the unknotting problem. More recently it has been shown that both Khovanov Homology [17] (a generalization of the Jones polynomial) and Heegard Floer Homology (a generalization of the Alexander polynomial) detect the unknot. Heegaard Floer Homology not only detects the unknot, but can be used to calculate the least genus of an orientable spanning surface for any knot. This is an outstanding result. The reader can examine the paper by Manolescu et al. [62] for more information. In that work, the Heegard Floer homology is expressed via a chain complex that is associated to a rectangular diagram of just the type that Dynnikov uses.

This paper is organized as follows. Section 2 discusses the Reidemeister moves and the combinatorial approach to the theory of knots and links. We sketch the proof of Reidemeister's Theorem that expresses equivalence of knots and links in terms of diagram moves (the Reidemeister moves). We give examples of knot diagrams that are unknotted but must be made more complicated in order to be undone by a sequence of Reidemeister moves. If it were not for this phenomenon, the problem of detecting an unknotted diagram would be over at once. One would only need to try to simplify the diagram by Reidemeister moves (that is, to try to reduce the number of crossings in the diagram). This is a finite search and that would be an algorithm to determine the knottedness of the diagram. Section 2 explains the solution by Dynnikov to this conundrum of unknotting by moves. We reformulate knot theory in terms of *arc-diagrams* and explain how Dynnikov constructs moves on these diagrams that can simplify an unknotted knot diagram. In this way, the problem of detecting the unknot is solved. There are many questions still remaining in the combinatorial domain, and the end of Sect. 3 discusses them. We give upper bounds on the number of crossings needed to unknot an unknotted diagram (work with Henrich [28] based on Dynnikov's work) and upper bounds on the number of Reidemeister moves needed for unknotting. Section 4 turns to the question of detection of knotting by the Jones polynomial and gives examples of links that cannot be detected by the Jones polynomial. It is still an unsolved problem whether the Jones polynomial detects the unknot. Section 5 is an introduction to Vassiliev invariants. We include this section in order to give the reader a flavor of the wider context the includes the Jones polynomial with invariants that are related to physics and to Lie algebras. We include a description of the relationship of Vassiliev invariants with Witten's approach to knot invariants via quantum field theory. We end this chapter with a statement of the remarkable problem to distinguish knots from their reverses. There are knots that are not isotopic to the curve obtained by reversing the orientation of the loop. At this writing it is not known if Vassiliev invariants can detect reversibility.

There are many more problems about the detection of knottedness. We have not touched on the question of distinguishing distinct knots in this paper. It is possible that if two knots are not isotopic, then there are Vassiliev invariants that distinguish them. But that is another country.

2 Reidemeister Moves

Reidemeister [39] discovered a simple set of moves on link diagrams that captures the concept of ambient isotopy of knots in three-dimensional space. There are three basic Reidemeister moves. Reidemeister's theorem states that two diagrams represent ambient isotopic knots (or links) if and only if there is a sequence of Reidemeister moves taking one diagram to the other. The Reidemeister moves are illustrated in Fig. 3.

Reidemeister's three moves are interpreted as performed on a larger diagram in which the small diagram shown is a literal part. Each move is performed without disturbing the rest of the diagram. Note that this means that each move occurs, up to topological deformation, just as it is shown in the diagrams in Fig. 3. There are no extra lines in the local diagrams. For example, the equivalence (A) in Fig. 4 is **not** an instance of a single first Reidemeister move. Taken literally, it factors into a move *II* followed by a move *I*.

Diagrams are always subject to topological deformations in the plane that preserve the structure of the crossings. These deformations could be designated as "Move Zero". See Fig. 4.

A few exercises with the Reidemeister moves are in order. First of all, view the diagram in Fig. 5. It is unknotted and you can have a good time finding a sequence of Reidemeister moves that will do the trick. Diagrams of this type are produced by tracing a curve and always producing an undercrossing at each return crossing.

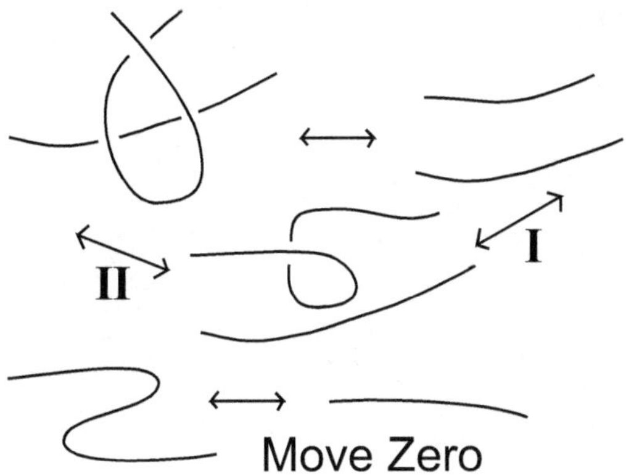

Fig. 4 Factorable move, move zero

Fig. 5 Standard unknot

This type of knot is called a *standard unknot*. Of course we see clearly that a standard unknot is unknotted by just *pulling* on it, since it has the same structure as a coil of rope that is wound down onto a flat surface.

Can one recognize unknots by simply looking for sequences of Reidemeister moves that undo them? This would be easy if it were not for the case that there are examples of unknots that require some moves that increase the number of crossings before they can be subsequently decreased. Such an culprit is illustrated in Fig. 6.

It is generally not so easy to recognise unknots. However, here is a tip: Look for *macro moves* of the type shown in Fig. 7. In a macro move, we identify an arc that passes entirely under some piece of the diagram (or entirely over) and shift this part of the arc, keeping it under (or over) during the shift. In Fig. 7, we illustrate a macro move on an arc that passes under a piece of the diagram that is indicated by arcs going into a circular region. A more general macro move is possible where the moving arc moves underneath one layer of diagram, and at the same time, over another layer of diagram. Macro moves often allow a reduction in the number of crossings even though the number of crossings will increase during a sequence of Reidemeister moves that generates the macro move.

Fig. 6 A culprit

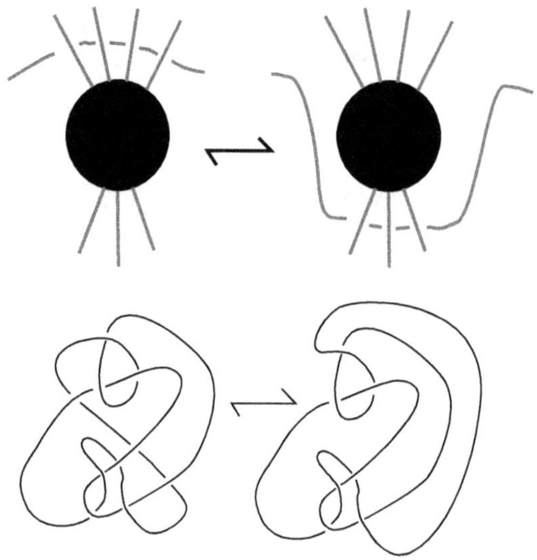

Fig. 7 Macro move

As shown in Fig. 7, the macro-move includes as a special case both the second and the third Reidemeister moves, and it is not hard to verify that a macro move can be generated by a sequence of type II and type III Reidemeister moves. It is easy to see that the type I moves can be left to the end of any deformation. The demon of Fig. 6 is easily demolished by macro moves, and from the point of view of macro moves the diagram never gets more complicated.

Let's say that a knot can be *reduced* by a set of moves if it can be transformed by these moves to the unknotted circle diagram through diagrams that never have more crossings than the original diagram. Then we have shown that there are diagrams representing the unknot that cannot be reduced by the Reidemeister moves. On the other hand, I do not know whether unknotted diagrams can always be reduced by (appropriately generalized) macro moves in conjunction with the first Reidemeister move. If this were true it would give a combinatorial way to recognise the unknot.

Remark. In fact, there is a combinatorial way to recognise the unknot based on a diagrams and moves. In [10] I. A. Dynnikov finds just such a result, using piecewise linear knot diagrams with all ninety degree angles in the diagrams, and all arcs in the diagram either horizontal or vertical. We shall discuss Dynnikov's work in Sect. 3 of this paper.

2.1 Reidemeister's Theorem

We now indicate how Reidemeister proved his Theorem.

An embedding of a knot or link in three-dimensional space is said to be *piecewise linear* if it consists in a collection of straight line segments joined end to end. Reidemeister started with a *single* move in three-dimensional space for piecewise linear knots and links. Consider a point in the complement of the link, and an edge in the link such that the surface of the triangle formed by the end points of that edge and the new point is not pierced by any other edge in the link. Then one can replace the given edge on the link by the other two edges of the triangle, obtaining a new link that is ambient isotopic to the original link. Conversely, one can remove two consecutive edges in the link and replace them by a new edge that goes directly from initial to final points, whenever the triangle spanned by the two consecutive edges is not pierced by any other edge of the link. This triangle replacement constitutes Reidemeister's three-dimensional move. See Fig. 8. It can be shown that two piecewise linear knots or links are ambient isotopic in three-dimensional space if and only if there is a sequence of Reidemeister triangle moves from one to the other. This will not be proved here. At the time when Reidemeister wrote his book, equivalence via three-dimensional triangle moves was taken as the definition of topological equivalence of links.

It can also be shown that tame knots and links have piecewise linear representatives in their ambient isotopy class. It is sufficient for our purposes to work with piecewise linear knots and links. Reidemeister's planar moves then follow from an analysis of the shadows projected into the plane by Reidemeister triangle moves in space. Figure 9 gives a hint of this analysis. The result is a reformulation of the three-dimensional problems of knot theory to a combinatorial game in the plane.

To go beyond the hint in Fig. 9 to a complete proof that Reidemeister's planar moves suffice involves preliminary remarks about subdivision. The simplest subdivision that one wants to be able to perform on a piecewise linear link is the

Fig. 8 Triangle move

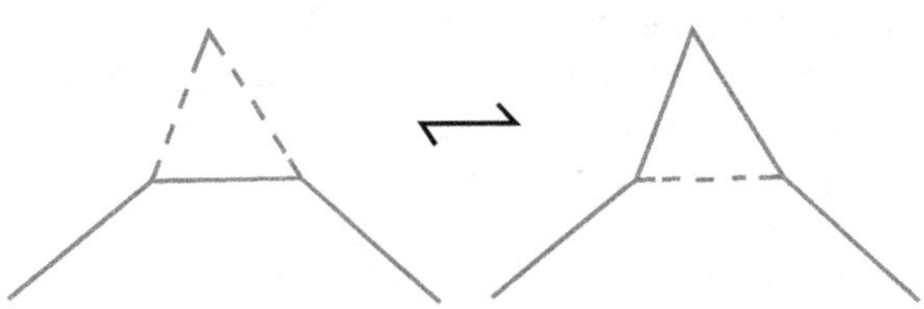

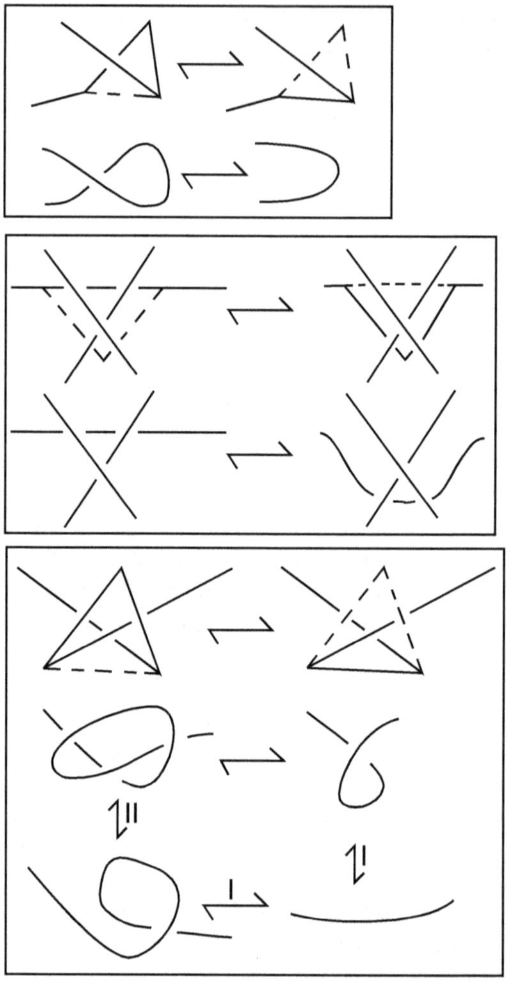

Fig. 9 Shadows

placement of a new vertex at an interior point of an edge—so that edge becomes two edges in the subdivided link. Figure 10 shows how to accomplish this subdivision via triangle moves.

Any triangle move can be factored into a sequence of smaller triangle moves corresponding to a simplicial subdivision of that triangle. This is obvious, since the triangles in the subdivision of the large triangle that is unpierced by the link are themselves unpierced by the link.

To understand how the Reidemeister triangle move behaves on diagrams it is sufficient to consider a projection of the link in which the triangle is projected to a non-singular triangle in the plane. Of course, there may be many arcs of the link also projected upon the interior of the projected triangle. However, by using subdivision, we can assume that the cases of the extra arcs are as shown in Fig. 11. In Fig. 11

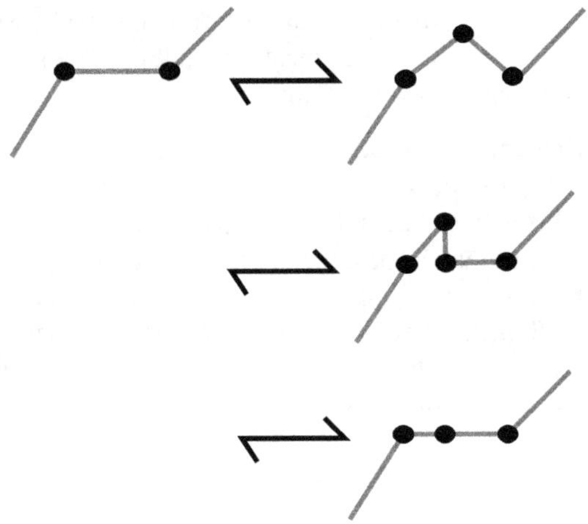

Fig. 10 Subdivision of an edge

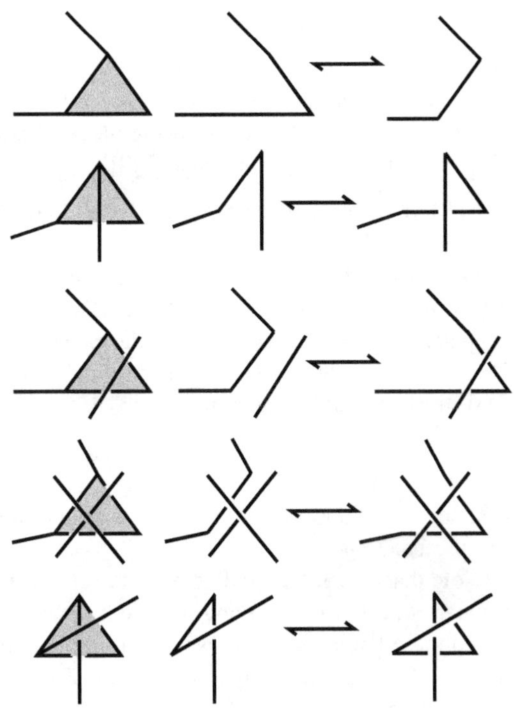

Fig. 11 Projections of triangle moves

we have also shown how each of these cases can be accomplished by (combinations of) the three Reidemeister moves. This proves that a projection of a single triangle move can be accomplished by a sequence of Reidemeister diagram moves.

A piecewise linear isotopy consists in a finite sequence of triangle moves. There exists a direction in three-dimensional space that makes a non-zero angle with each of theses triangles and is in general position with the link diagram. Projecting to the plane along this direction makes it possible to perform the entire ambient isotopy in the language of projected triangle moves. Now apply the results of the previous paragraph and we conclude

Reidemeister's Theorem If two links are piecewise linearly equivalent (ambient isotopic), then there is a sequence of Reidemeister diagram moves taking a projection of one link to a projection of the other.

Note that the proof tells us that the two diagrams can be obtained from one spatial projection direction for the entire spatial isotopy. It is obvious that diagrams related by Reidemeister moves represent ambient isotopic links. Reidemeister's Theorem gives a complete combinatorial description of the topology of knots and links in three-dimensional space.

3 Dynnikov's Solution of the Problem of Knot Detection

We now discuss a powerful result proven by Dynnikov in [10]. Dynnikov uses, a diagram called an *arc-presentation* for the knot. We define this below, and show how one can detect the unknot using moves that preserve this type of presentation. This section is based on our paper [28].

Definition 1. An *arc–presentation* of a knot is a knot diagram comprised of horizontal and vertical line segments such that at each crossing in the diagram, the horizontal arc passes under the vertical arc. Furthermore, we require that no two edges in an arc–diagram are colinear. Two arc–presentations are *combinatorially equivalent* if they are isotopic in the plane via an ambient isotopy of the form $h(x, y) = (f(x), g(y))$. The *complexity* $c(L)$ of an arc–presentation is the number of vertical arcs in the diagram.

We say more generally that a link diagram is *rectangular* if it has only vertical and horizontal edges. In Fig. 12 we give an example of a rectangular diagram that is an arc–presentation and another example of a rectangular diagram that is not an arc–presentation.

Note that a rectangular diagram can naturally be drawn on a rectangular grid. If we start with such a grid and represent rectangular diagrams on the grid we have called these knots *mosaic knots* and used them to define a notion of *quantum knot.* See [61] for more about quantum knots. For now, we focus our attention on arc–presentations.

Proposition 1 (Dynnikov). *Every knot has an arc–presentation. Any two arc–presentations of the same knot can be related to each other by a finite sequence of* elementary moves, *pictured in Figs. 13 and 14.*

The proof of this proposition is elementary, based on the Reidemeister moves. One shows that each Reidemeister move can be represented by (a sequence of)

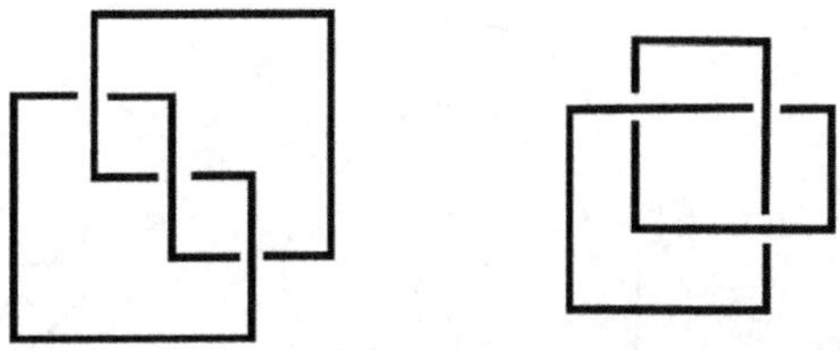

Fig. 12 The picture on the *left* is an example of an arc–presentation of a trefoil. The picture on the *right* is an example that is *not* an arc–presentation (since not all horizontal arcs pass under vertical arcs)

Fig. 13 Elementary (de)stabilization moves. Stabilization moves increase the complexity of the arc–presentation while destabilization moves decrease the complexity

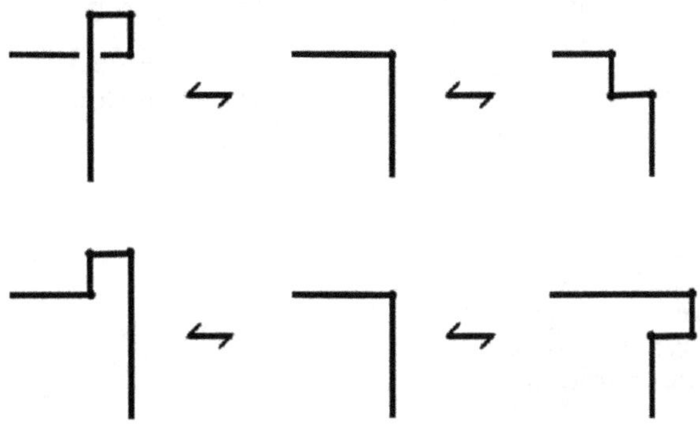

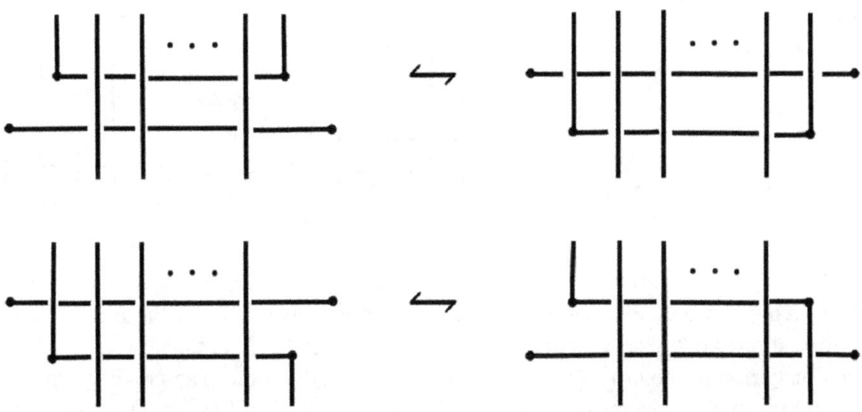

Fig. 14 Some examples of exchange moves. Other allowed exchange moves include switching the heights of two horizontal arcs that lie in distinct halves of the diagram

elementary moves. See [10]. We will show how to convert a usual knot diagram to an arc–presentation in the next few paragraphs, making use of the concept of Morse diagrams of knots.

Definition. A knot diagram is in *Morse form* if it has

1. no horizontal lines,
2. no inflection points,
3. a single singularity at each height, and
4. each crossing is oriented to create a 45 degree angle with the vertical axis.

Fig. 15 A Morse diagram of a knot and a corresponding rectangular diagram

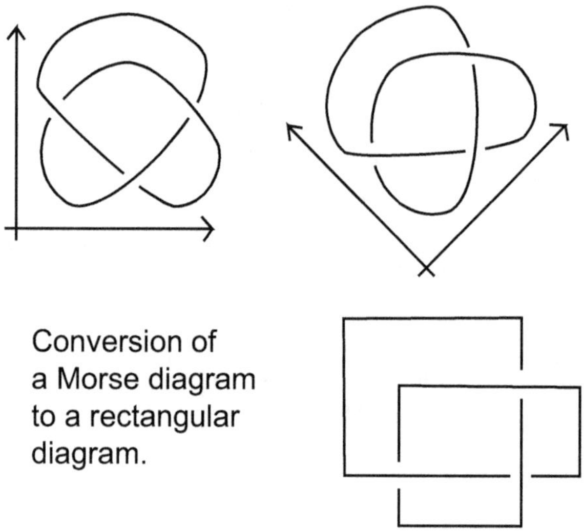

Fig. 16 Rotating a crossing to convert a rectangular diagram into an arc–presentation

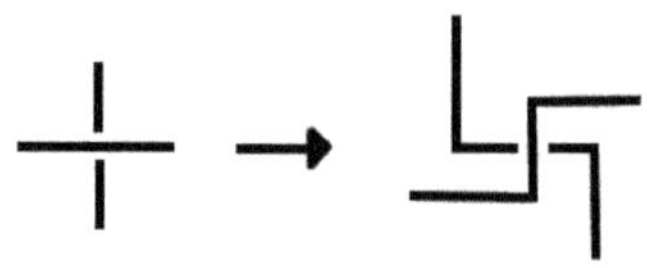

Fig. 17 Converting a rectangular diagram into an arc–presentation by rotating a crossing

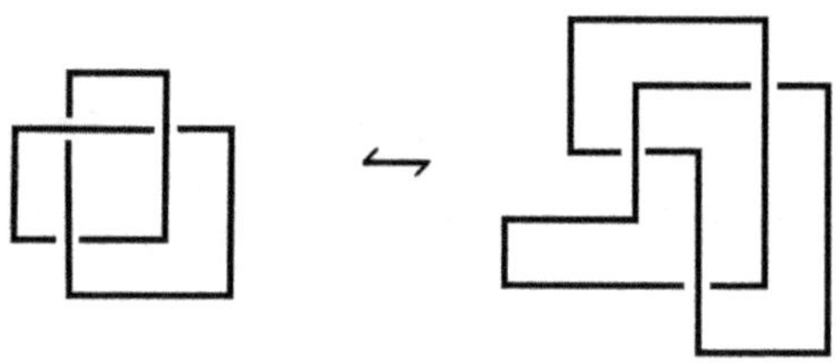

We note that converting an arbitrary knot diagram into a diagram in Morse form requires no Reidemeister moves, only ambient isotopies of the plane. More information about Morse diagrams can be found in [59]. See Figs. 15, 16 and 17 for an illustration of the process of conversion of a knot diagram to an arc diagram.

In Fig. 18 we show the example found by Goeritz [56] in 1934 of a knot diagram that is unknotted, but requires Reidemeister moves that create more crossings before it can be simplified. In Fig. 19 we show another example of this same type. We shall refer to the latter as the "culprit" and analyse it below.

Here is Dynnikov's solution to the problem of recognizing the unknot.

Theorem 1 (Dynnikov). *If L is an arc–presentation of the unknot, then there exists a finite sequence of exchange and destabilization moves*

$$L \to L_1 \to L_2 \to \cdots \to L_m$$

such than L_m is trivial.

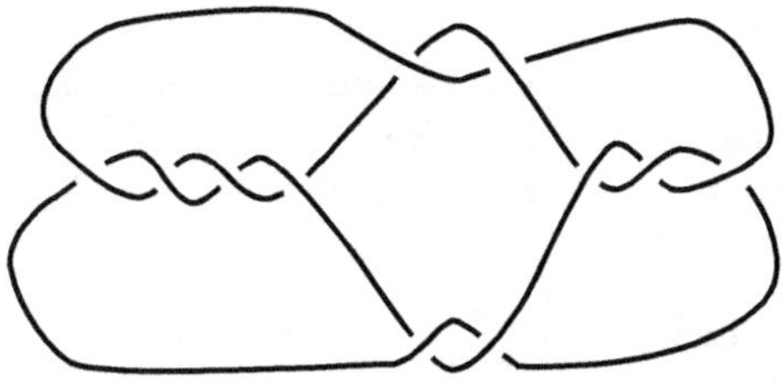

Fig. 18 Goeritz unknot

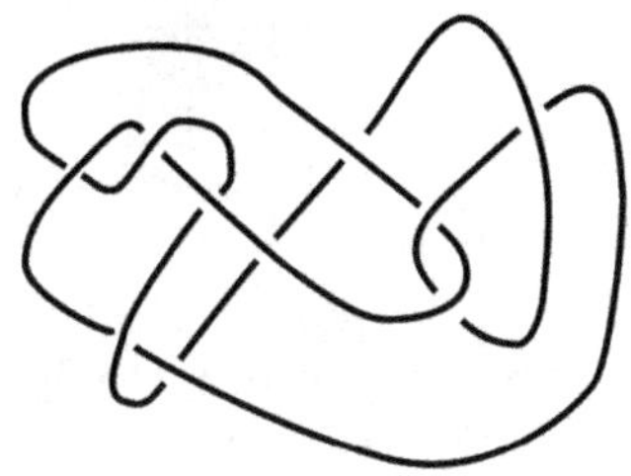

Fig. 19 The culprit

What is particularly interesting about this result is that the unknot can be simplified **without increasing the complexity of the arc–presentation**, that is, without the use of stabilization moves. This means that a finite search will reveal a diagram to be unknotted if that is the case. Furthermore, if we apply Dynnikov's method to a knotted knot, it will stop on a diagram that is not a planar circle. Thus Dynnikov's diagrammatic method can detect the unknot.

We can go further and ask how large a diagram is needed to unknot a knot by Reidemeister moves, and how many Reidemeister moves are needed.

Theorem 2 ([28]). *Suppose K is a diagram (in Morse form) of the unknot with crossing number $cr(K)$ and number of maxima $b(K)$. Then, for every i, the crossing number $cr(K_i)$ is no more than $(M-2)^2$ where $M = 2b(K) + cr(K)$ and $K = K_0, K_1, K_2, \ldots, K_N$ is a sequence of knot diagrams such that K_{i+1} is obtained from K_i by a single Reidemeister move and K_N is a trivial diagram of the unknot.*

To find our upper bound on the number of Reidemeister moves, we must first specify an upper bound on the number m of exchange and destabilization moves required to trivialize an arc–presentation. This bound will depend on the complexity $c(L) = n$ of the arc–diagram L. We also must provide an upper bound on the number of Reidemeister moves required for a destabilization or exchange move.

In [10], Dynnikov provides the following bounds on the number of combinatorially distinct arc–presentations of complexity n.

Proposition 2. *Let $N(n)$ denote the number of combinatorially distinct arc–presentations of complexity n. Then the following inequality holds.*

$$N(n) \leq \frac{1}{2}n[(n-1)!]^2$$

Using this count on the number of distinct arc–presentations of a given size, we can find a bound (albeit a large one) on the number of arc–presentation moves we need. This is simply by virtue of the fact that any reasonable sequence of moves will contain mutually distinct arc–presentations that don't exceed the complexity of the original, and there are a limited number of such diagrams. With this we obtain

Theorem 3 ([28]). *Suppose K is a diagram (in Morse form) of the unknot with crossing number $cr(K)$ and number of maxima $b(K)$. Let $M = 2b(K) + cr(K)$. Then the number of Reidemeister moves required to unknot K is less than or equal to*

$$\sum_{i=2}^{M} \frac{1}{2} i[(i-1)!]^2 (M-2).$$

We have provided several upper bounds regarding the complexity of the Reidemeister sequence required to simplify an unknot. The bound that Dynnikov's work helps us obtain for the number of Reidemeister moves required to unknot an unknot is superexponential. Using a different technique, Hass and Lagarias were able to find a bound that is exponential in the crossing number of the diagram [57]. They prove the following result.

Theorem 4 (Hass and Lagarias). *There is a positive constant c_1, such that for each $n \geq 1$, any unknotted knot diagram $\mathcal{D}$ with n crossings can be transformed to the trivial knot diagram using at most $2^{c_1 n}$ Reidemeister moves. In fact one can take $c_1 = 10^{11}$.*

Hass and Lagarias use the same technique to find an exponential bound for the number of crossings required for unknotting. For bounds of this second sort, the one presented here is a comparatively sharper estimate. Our bound on the number of Reidemeister moves required to unknot an unknot eventually becomes larger than the previously known bound from the above Theorem, but it does remain significantly smaller for knots with up to $10^{10^{10}}$ crossings. We can see this by assuming that the number of maxima in a Morse diagram of a knot is approximately the same size as the crossing number (in practice the number of maxima is significantly smaller than the crossing number). We can estimate the size of our bound by computing the quantity

$$\frac{M-2}{2}(M!)^2.$$

This larger estimate remains smaller than the bound proposed by Hass and Lagarias until the knots have $10^{10^{10}}$ crossings. It is possible that these methods may be improved to find smaller unknotting bounds for unknot diagrams. In fact, polynomial bounds are now shown by Lackenby [34].

Let us return to the Culprit. Recall that hard unknots are difficult to unknot by virtue of the fact that no simplifying type I or type II Reidemeister moves and no type III moves are available. In Fig. 20, we picture a Morse diagram of the Culprit, a

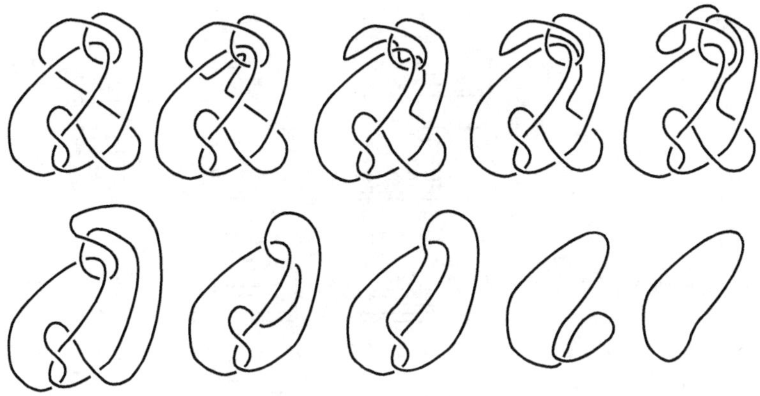

Fig. 20 Undoing the culprit by Reidemeister moves

corresponding arc-presentation and a sequence of Dynnikov moves that simplify it to a standard unknot. Bounds for the size of diagrams that are needed using Reidemeister's moves can be deduced using Dynnikov's work. See [10, 28]. In fact, we [28] derive a quadratic upper bound on the crossing number of diagrams in an unknotting sequence. A similar result can be found in [10]. We saw that the Culprit may be unknotted with ten Reidemeister moves in Fig. 21 (see also [60]). The maximum crossing number of all diagrams in the given Reidemeister sequence is 12, two more than the number of crossings in the Culprit. On the other hand, we can compute our upper bound on the number of crossings required for unknotting as follows. Since the crossing number $cr(K) = 10$ and the number of maxima in the diagram is $b(K) = 5$, we see that $M = cr(K) + 2b(K) = 20$. Thus, our bound is $(M-2)^2 = 18^2 = 324$. There is room for improvement!

We can also use M to find our bound for the number of Reidemeister moves required to unknot the Culprit.

$$\sum_{i=2}^{M} \frac{1}{2} i[(i-1)!]^2 (M-2) = 9 \sum_{i=2}^{20} i[(i-1)!]^2.$$

The largest term in this expression is roughly 10^{35}, unfortunately quite a bit larger than ten.

In this section we have sketched results given in more detail in [10, 28], showing Dynnikov's remarkable solution to the unknotting problem. Remarkable as this solution is, we are not happy since we believe that better bounds on the number of Reidemeister moves needed to unknot a knot are surely possible, and better bounds on the complexity of diagrams is also possible. Could there be a simple algorithm in the form of a calculation from a diagram that would tell if a knot was knotted? This is the subject of the next section where we discuss the Jones polynomial.

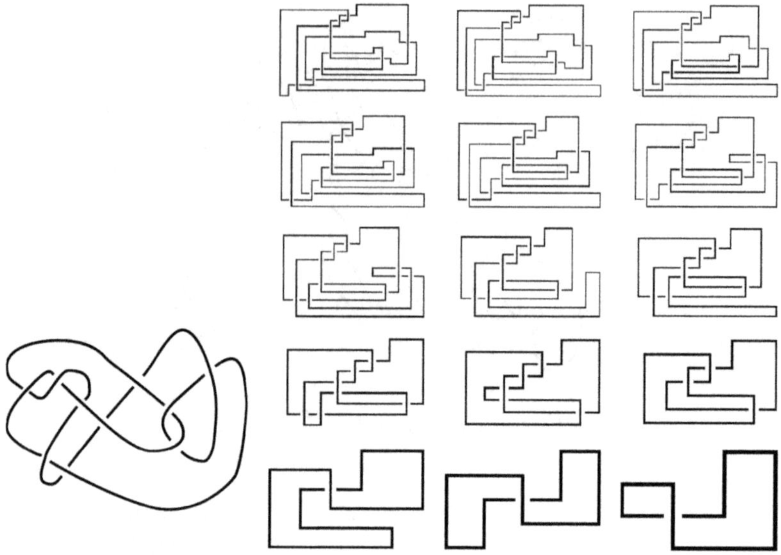

Fig. 21 Undoing the culprit by Dynnikov moves

4 The Bracket Polynomial and the Jones Polynomial

Now that we have exhibited a solution to problem of the detection of the unknot, we turn to some unsolved problems related to other methods of detection. It is an open problem whether there exist classical knots (single component loops) that are knotted and yet have unit Jones polynomial [14]. In other words, it is an open problem whether the Jones polynomial can detect all knots. There do exist families of links whose linkedness is undetectable by the Jones polynomial [47, 48]. It is the purpose of this section of the paper to give a summary of some of the information that is known in this arena. We begin with a sketch of ways to calculate the bracket polynomial model of the Jones polynomial, and then discuss how to construct classical links that are undetectable by the Jones polynomial.

The bracket polynomial [20–23, 25] model for the Jones polynomial [14–16, 52] is described by the expansion

$$\langle \times \rangle = A \langle \asymp \rangle + A^{-1} \langle)(\rangle \tag{1}$$

and we have

$$\langle K \bigcirc \rangle = (-A^2 - A^{-2}) \langle K \rangle \tag{2}$$

$$\langle \gamma \rangle = (-A^3) \langle \smile \rangle \tag{3}$$

$$\langle \gamma \rangle = (-A^{-3}) \langle \smile \rangle \tag{4}$$

A state S of a link diagram K is obtained by choosing a smoothing for each crossing in the diagram and labelling that smoothing with either A or A^{-1} according to the convention indicated in the bracket expansion above. Then, given a state S, one has the evaluation $< K|S >$ equal to the product of the labels at the smoothings, and one has the evaluation $||S||$ equal to the number of loops in the state. One then has the formula

$$< K >= \Sigma_S < K|S > d^{||S||-1}$$

where the summation runs over the states S of the diagram K, and $d = -A^2 - A^{-2}$. This state summation is invariant under all classical and virtual moves except the first Reidemeister move. The bracket polynomial is normalized to an invariant $f_K(A)$ of all the moves by the formula $f_K(A) = (-A^3)^{-w(K)} < K >$ where $w(K)$ is the writhe of the (now) oriented diagram K. The writhe is the sum of the orientation signs (± 1) of the crossings of the diagram. The Jones polynomial, $V_K(t)$ is given in terms of this model by the formula

$$V_K(t) = f_K(t^{-1/4}).$$

The state sum is part of a wider approach to invariants of knots and links that we do not concentrate upon in this paper. First of all, the Alexander polynomial [53] was the first polynomial invariant of knots and links. It was not until 1981 that the Alexander polynomial was reformulated as a state summation [18–20]. In 1983 Jones discovered his polynomial and showed that it satisfied a skein relation similar to that for the Alexander-Conway polynomial [9]. Along with this there arose relations with statistical mechanics [5] in the work of Jones. The bracket model was the first direct relationship of the Jones polynomial with statistical mechanics. In the wake of the discovery of the Jones polynomial came more skein polynomials, particularly the Homflypt polynomial [20, 35] named after the people who discovered it—Hoste, Ocneanu, Freyd, Lickorish, Yetter, Przytycki and Trawczk, and the Kauffman Polynomial [24]. These invariants are more powerful than the Jones polynomial, but it is still conjectural that they detect the unknot. After the skein polynomials, came more algebraic state sums based on the work of Yang and Baxter in statistical mechanics, and then arose relationships with Lie algebras and gauge theoretic physics. We shall sketch some of these developments in the sections to follow. It remains to be seen how powerful all these new invariants are as detectors of knottedness, but it is known that certain families of these invariants do detect knottedness and those results are found in the relationships of the knot theory with physics

It remains on open problem whether the Jones polynomial can detect the unknot. We can make the conjecture as follows:

Knot Detection Conjecture If K is a knot diagram of one component and $V_K(t) = 1$, then K is equivalent by Reidemeister moves to the unknot.

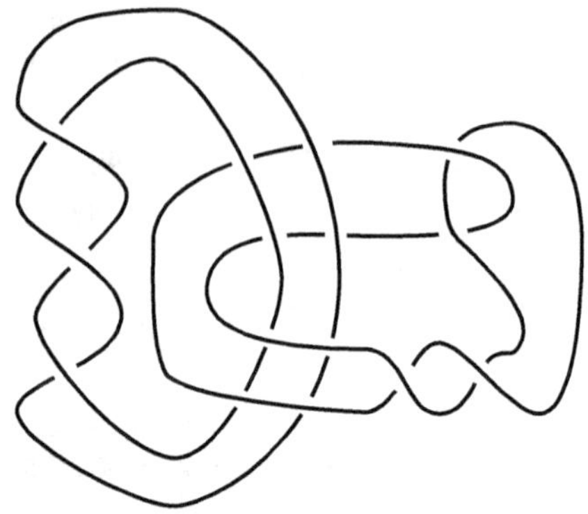

Fig. 22 Thistethwaite's link

This knot detection conjecture is false for links. View Fig. 22. Here we have a version of a link L discovered by Thistlethwaite [47] in December 2000. One can verify that this link is indeed non-trivial, but it has the same Jones polynomial as the unlink of two circles. In [48] we produce infinite families of distinct links that appear to be unlinked to the Jones polynomial.

4.1 Present Status of Links Not Detectable by the Jones Polynomial

In this section we give a quick review of the status of our work

A tangle (2-tangle) consists in an embedding of two arcs in a three-ball (and possibly some circles embedded in the interior of the three-ball) such that the endpoints of the arcs are on the boundary of the three-ball. One usually depicts the arcs as crossing the boundary transversely so that the tangle is seen as the embedding in the three-ball augmented by four segments emanating from the ball, each from the intersection of the arcs with the boundary. These four segments are the *exterior edges* of the tangle, and are used for operations that form new tangles and new knots and links from given tangles. Two tangles in a given three-ball are said to be *topologically equivalent* if there is an ambient isotopy from one to the other in the given three-ball, fixing the intersections of the tangles with the boundary.

It is customary to illustrate tangles with a diagram that consists in a box (within which are the arcs of the tangle) and with the exterior edges emanating from the box in the NorthWest (NW), NorthEast (NE), SouthWest (SW) and SouthEast (SE) directions. Given tangles T and S, one defines the *sum*, denoted $T + S$ by placing the diagram for S to the right of the diagram for T and attaching the NE edge of T to the NW edge of S, and the SE edge of T to the SW edge of S. The resulting tangle $T + S$ has exterior edges corresponding to the NW and SW edges of T and the NE and SE edges of S. There are two ways to create links associated to a tangle T. The *numerator* T^N is obtained, by attaching the (top) NW and NE edges of T

together and attaching the (bottom) SW and SE edges together. The denominator T^D is obtained, by attaching the (left side) NW and SW edges together and attaching the (right side) NE and SE edges together. We denote by [0] the tangle with only unknotted arcs (no embedded circles) with one arc connecting, within the three-ball, the (top points) NW intersection point with the NE intersection point, and the other arc connecting the (bottom points) SW intersection point with the SE intersection point. A ninety degree turn of the tangle [0] produces the tangle $[\infty]$ with connections between NW and SW and between NE and SE. One then can prove the basic formula for any tangle T

$$< T >= \alpha_T < [0] > + \beta_T < [\infty] >$$

where α_T and β_T are well-defined polynomial invariants (of regular isotopy) of the tangle T. From this formula one can deduce that

$$< T^N >= \alpha_T d + \beta_T$$

and

$$< T^D >= \alpha_T + \beta_T d.$$

We define the *bracket vector* of T to be the ordered pair (α_T, β_T) and denote it by $br(T)$, viewing it as a column vector so that $br(T)^t = (\alpha_T, \beta_T)$ where v^t denotes the transpose of the vector v. With this notation the two formulas above for the evaluation for numerator and denominator of a tangle become the single matrix equation

$$\begin{bmatrix} < T^N > \\ < T^D > \end{bmatrix} = \begin{bmatrix} d & 1 \\ 1 & d \end{bmatrix} br(T).$$

We then use this formalism to express the bracket polynomial for our examples. The class of examples that we considered are each denoted by $H(T, U)$ where T and U are each tangles and $H(T, U)$ is a satellite of the Hopf link that conforms to the pattern shown in Fig. 23, formed by clasping together the numerators of the tangles T and U. Our method is based on a transformation $H(T, U) \longrightarrow H(T, U)^\omega$, whereby the tangles T and U are cut out and reglued by certain specific homeomorphisms of the tangle boundaries. This transformation can be specified by a modification described by a specific rational tangle and its mirror image. Like mutation, the transformation ω preserves the bracket polynomial. However, it is more effective than mutation in generating examples, as a trivial link can be transformed to a prime link, and repeated application yields an infinite sequence of inequivalent links.

Specifically, the transformation $H(T, U)^\omega$ is given by the formula

$$H(T, U)^\omega = H(T^\omega, U^{\bar{\omega}})$$

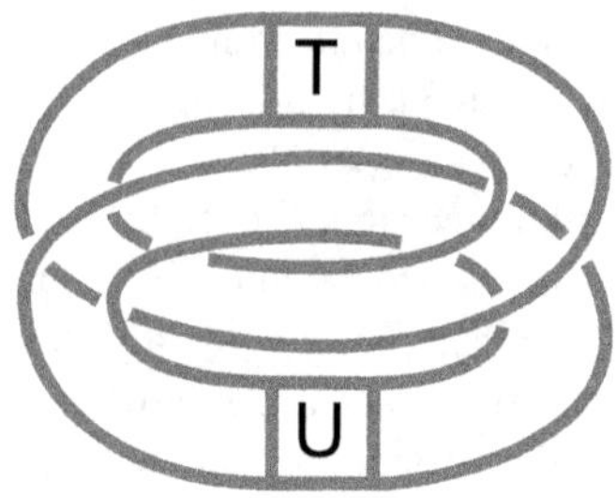

Fig. 23 Hopf link satellite $H(T, U)$

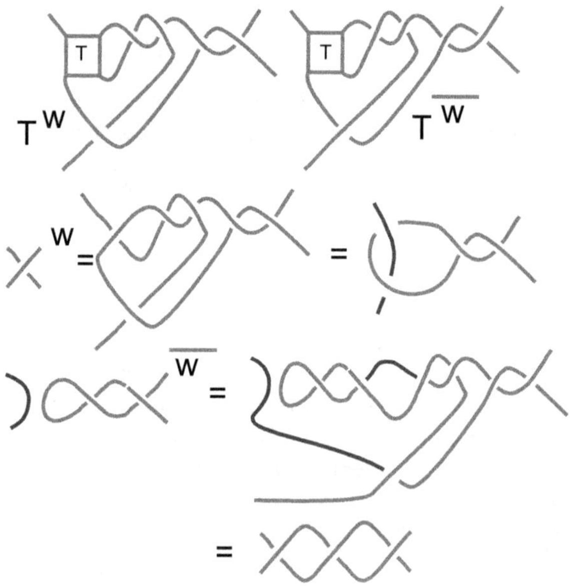

Fig. 24 The omega operations

where the tangle operations T^{ω} and $U^{\bar{\omega}})$ are as shown in Fig. 24. By direct calculation, there is a matrix M such that

$$< H(T, U) >= br(T)^{t} M br(U)$$

and there is a matrix Ω such that

$$br(T^{\omega}) = \Omega br(T)$$

and

$$br(T^{\bar{\omega}}) = \Omega^{-1} br(T).$$

One verifies the identity

$$\Omega^{t} M \Omega^{-1} = M$$

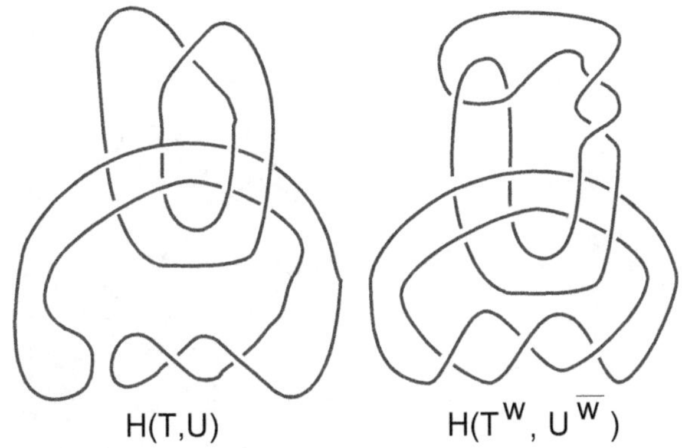

Fig. 25 Applying omega operations to an unlink

from which it follows that $< H(T,U) >^{\omega} = < H(T,U) >$. This completes the sketch of our method for obtaining links that whose linking cannot be seen by the Jones polynomial. Note that the link constructed as $H(T^{\omega}, U^{\bar{\omega}})$ in Fig. 25 has the same Jones polynomial as an unlink of two components. This shows how the first example found by Thistlethwaite fits into our construction.

5 Vassiliev Invariants and Invariants of Rigid Vertex Graphs

If $V(K)$ is a (Laurent polynomial valued, or more generally—commutative ring valued) invariant of knots, then it can be naturally extended to an invariant of rigid vertex graphs by defining the invariant of graphs in terms of the knot invariant via an "unfolding" of the vertex. That is, we can regard the vertex as a "black box" and replace it by any tangle of our choice. Rigid vertex motions of the graph preserve the contents of the black box, and hence entail ambient isotopies of the link obtained by replacing the black box by its contents. Invariants of knots and links that are evaluated on these replacements are then automatically rigid vertex invariants of the corresponding graphs. If we set up a collection of multiple replacements at the vertices with standard conventions for the insertions of the tangles, then a summation over all possible replacements can lead to a graph invariant with new coefficients corresponding to the different replacements. In this way each invariant of knots and links implicates a large collection of graph invariants. See [25, 26].

The simplest tangle replacements for a 4-valent vertex are the two crossings, positive and negative, and the oriented smoothing. Let $V(K)$ be any invariant of knots and links. Extend V to the category of rigid vertex embeddings of 4-valent graphs by the formula (see Fig. 26)

$$V(K_*) = aV(K_+) + bV(K_-) + cV(K_0)$$

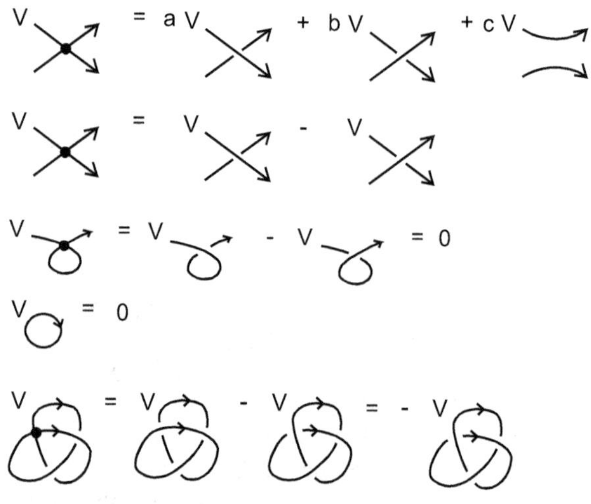

Fig. 26 Graphical vertex formulas

Here K_* indicates an embedding with a transversal 4-valent vertex. This formula means that we define $V(G)$ for an embedded 4-valent graph G by taking the sum

$$V(G) = \sum_S a^{i_+(S)} b^{i_-(S)} c^{i_0(S)} V(S)$$

with the summation over all knots and links S obtained from G by replacing a node of G with either a crossing of positive or negative type, or with a smoothing (denoted 0). Here $i_+(S)$ denotes the number of positive crossings in the replacement, $i_-(S)$ the number of negative crossings in the replacement, and $i_0(S)$ the number of smoothings in the replacement. It is not hard to see that if $V(K)$ is an ambient isotopy invariant of knots, then, this extension is a rigid vertex isotopy invariant of graphs. In rigid vertex isotopy the cyclic order at the vertex is preserved, so that the vertex behaves like a rigid disk with flexible strings attached to it at specific points. See the previous section.

There is a rich class of graph invariants that can be studied in this manner. The Vassiliev Invariants [4, 6, 50] constitute the important special case of these graph invariants where $a = +1$, $b = -1$ and $c = 0$. Thus $V(G)$ is a Vassiliev invariant if

$$V(K_*) = V(K_+) - V(K_-).$$

Call this formula the *exchange identity* for the Vassiliev invariant V. V is said to be of finite type k if $V(G) = 0$ whenever $|G| > k$ where $|G|$ denotes the number of 4-valent nodes in the graph G. The notion of finite type is of paramount significance in studying these invariants. One reason for this is the following basic Lemma.

Lemma. *If a graph G has exactly k nodes, then the value of a Vassiliev invariant v_k of type k on G, $v_k(G)$, is independent of the embedding of G.*

Proof. The different embeddings of G can be represented by link diagrams with some of the 4-valent vertices in the diagram corresponding to the nodes of G. It suffices to show that the value of $v_k(G)$ is unchanged under switching of a crossing. However, the exchange identity for v_k shows that this difference is equal to the evaluation of v_k on a graph with $k+1$ nodes and hence is equal to zero. This completes the proof.

The upshot of this Lemma is that Vassiliev invariants of type k are intimately involved with certain abstract evaluations of graphs with k nodes. In fact, there are restrictions (the four-term relations) on these evaluations demanded by the topology (we shall articulate these restrictions shortly) and it follows from results of Kontsevich [4] that such abstract evaluations actually determine the invariants. The invariants derived from classical Lie algebras are all built from Vassiliev invariants of finite type. All this is directly related to Witten's functional integral [52].

Definition. Let v_k be a Vassiliev invariant of type k. The *top row* of v_k is the set of values that v_k assigns to the set of (abstract) 4-valent graphs with k nodes. If we concentrate on Vassiliev invariants of knots, then these graphs are all obtained by marking $2k$ points on a circle, and choosing a pairing of the $2k$ points. The pairing can be indicated by drawing a circle and connecting the paired points with arcs. Such a diagram is called a *chord diagram*. Some examples are indicated in Fig. 27.

Note that a top row diagram cannot contain any isolated pairings since this would correspond to a difference of local curls on the corresponding knot diagram (and these curls, being isotopic, yield the same Vassiliev invariants.

The Four-Term Relation (Compare [45].) Consider a single embedded graphical node in relation to another embedded arc, as illustrated in Fig. 28. The arc underlies the lines incident to the node at four points and can be slid out and isotoped over the top so that it overlies the four nodes. One can also switch the crossings one-by-one

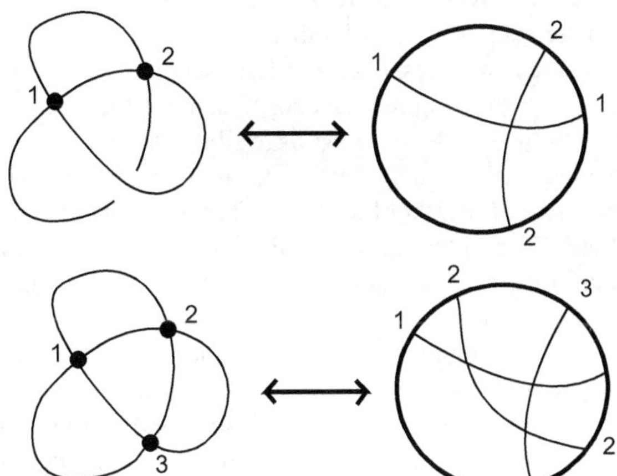

Fig. 27 Chord diagrams

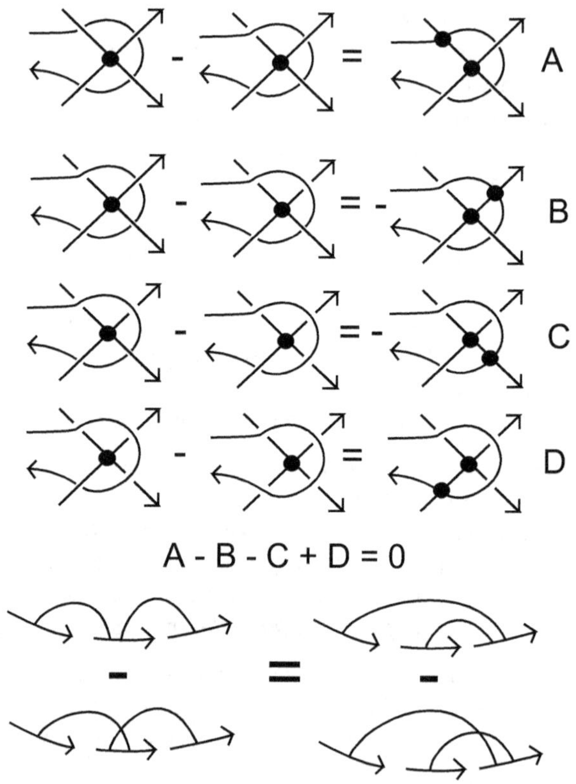

Fig. 28 The four term relation

to exchange the arc until it overlies the node. Each of these four switchings gives rise to an equation, and the left-hand sides of these equations will add up to zero, producing a relation corresponding to the right-hand sides. Each term in the right-hand side refers to the value of the Vassiliev invariant on a graph with two nodes that are neighbors to each other.

There is a corresponding 4-term relation for chord diagrams. This is the 4-term relation for the top row. In chord diagrams the relation takes the form shown at the bottom of Fig. 28. Here we have illustrated only those parts of the chord diagram that are relevant to the two nodes in question (indicated by two pairs of points on the circle of the chord diagram). The form of the relation shows the points on the chord diagram that are immediate neighbors. These are actually neighbors on any chord diagram that realizes this form. Otherwise there can be many other pairings present in the situation.

As an example, consider the possible chord diagrams for a Vassiliev invariant of type 3. There are two possible diagrams as shown in Fig. 29. One of these has the projected pattern of the trefoil knot and we shall call it the *trefoil graph*. These diagrams satisfy the 4-term relation. This shows that one diagram must have twice the evaluation of the other. Hence it suffices to know the evaluation of one of these

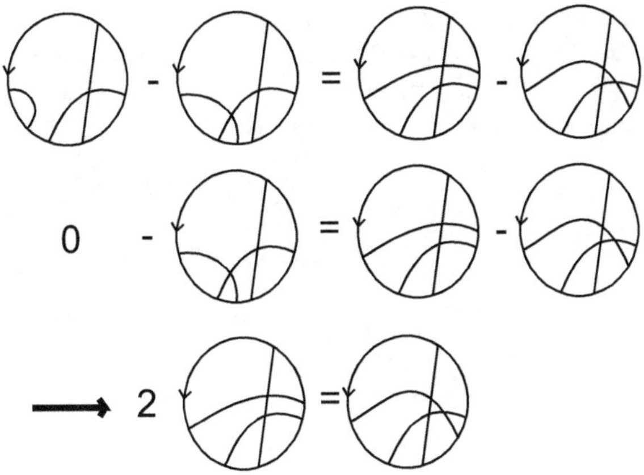

Fig. 29 The four term relation for a type three invariant

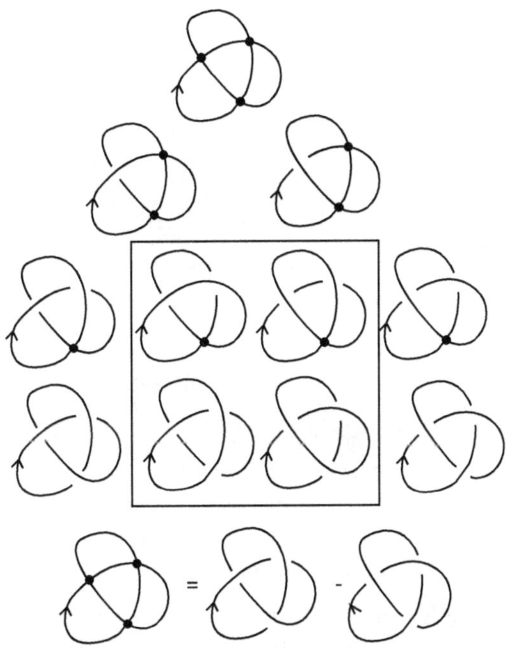

Fig. 30 Trefoil graph

two diagrams to know the top row of a Vassiliev invariant of type 3. We can take this generator to be the trefoil graph

Now one more exercise: Consider any Vassiliev invariant v and let's determine its value on the trefoil graph as in Fig. 30.

The value of this invariant on the trefoil graph is equal to the difference between its values on the trefoil knot and its mirror image. Therefore any Vassiliev invariant that assigns a non-zero value to the trefoil graph can tell the difference between the trefoil knot and its mirror image.

Example. This example shows how the original Jones polynomial is composed of Vassiliev invariants of finite type. Let $V_K(t)$ denote the original Jones polynomial [14]. Recall the oriented state expansion for the Jones polynomial [27] with the basic formulas (δ is the loop value.)

$$V_{K_+} = -t^{1/2}V_{K_0} - tV_{K_\infty}$$
$$V_{K_-} = -t^{-1/2}V_{K_0} - t^{-1}V_{K_\infty}.$$
$$\delta = -(t^{1/2} + t^{-1/2}).$$

Let $t = e^x$. Then

$$V_{K_+} = -e^{x/2}V_{K_0} - e^{x}V_{K_\infty}$$
$$V_{K_-} = -e^{-x/2}V_{K_0} - e^{-x}V_{K_\infty}.$$
$$\delta = -(e^{x/2} + e^{-x/2}).$$

Thus

$$V_{K_*} = V_{K_+} - V_{K_-} = -2sinh(x/2)V_{K_0} - 2sinh(x)V_{K_\infty}.$$

Thus x divides V_{K_*}, and therefore x^k divides V_G whenever G is a graph with at least k nodes. Letting

$$V_G(e^x) = \sum_{k=0}^{\infty} v_k(G)x^k,$$

we see that this condition implies that $v_k(G)$ vanishes whenever G has more than k nodes. Hence *the coefficients of the powers of x in the expansion of $V_K(e^x)$ are Vassiliev invariants of finite type!* This result was first observed by Birman and Lin [6] by a different argument.

Let's look a little deeper and see the structure of the top row for the Vassiliev invariants related to the Jones polynomial. By our previous remarks the top row evaluations correspond to the leading terms in the power series expansion. Since

$$\delta = -(e^{x/2} + e^{-x/2}) = -2 + [higher],$$
$$-e^{x/2} + e^{-x/2} = -x + [higher],$$
$$-e^{x} + e^{-x} = -2x + [higher],$$

it follows that the top rows for the Jones polynomial are computed by the recursion formulas

$$v(K_*) = -v(K_0) - 2V(K_\infty)$$
$$v([loop]) = -2.$$

The reader can easily check that this recursion formula for the top rows of the Jones polynomial implies that v_3 takes the value 24 on the trefoil graph and hence it is the Vassiliev invariant of type 3 in the Jones polynomial that first detects the difference between the trefoil knot and its mirror image.

This example gives a good picture of the general phenomenon of how the Vassiliev invariants become building blocks for other invariants. In the case of the Jones polynomial, we already know how to construct the invariant and so it is possible to get a lot of information about these particular Vassiliev invariants by looking directly at the Jones polynomial. This, in turn, gives insight into the structure of the Jones polynomial itself.

5.1 Lie Algebra Weights

Consider the diagrammatic relation shown in Fig. 31. Call it (after Bar-Natan [4]) the *STU* relation.

Lemma. *STU implies the 4-term relation.*

Proof. View Fig. 32.

STU is the smile of the Cheshire cat. That smile generalizes the idea of a Lie algebra. Take a (matrix) Lie algebra with generators T^a. Then

$$T^aT^b - T^bT^a = if_{abc}T^c$$

expresses the closure of the Lie algebra under commutators. Translate this equation into diagrams as shown in Fig. 33, and see that this translation is *STU* with Lie algebraic clothing!

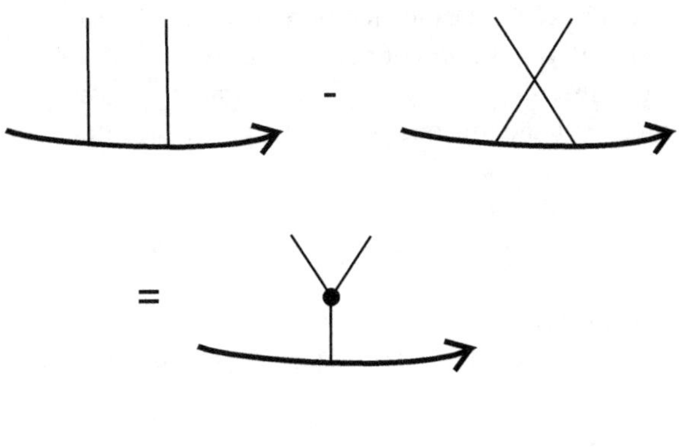

Fig. 31 The *STU* relation

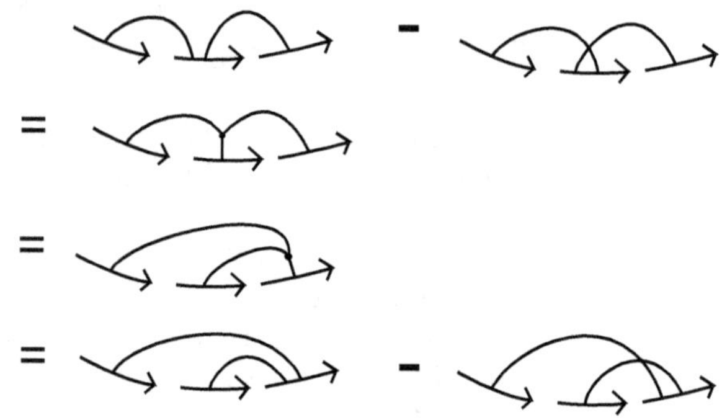

Fig. 32 A diagrammatic proof

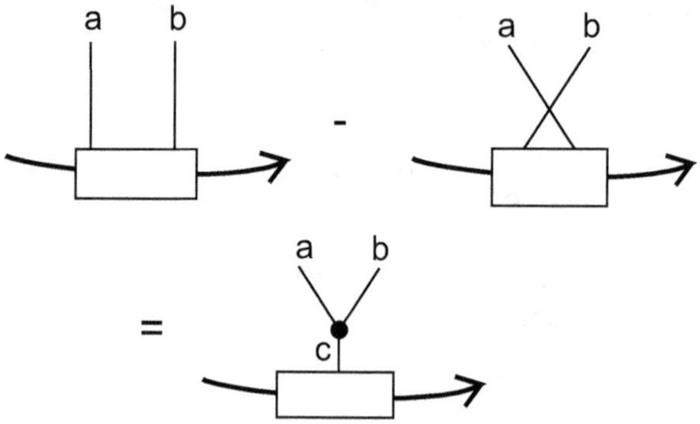

Fig. 33 Algebraic clothing

Here the structure tensor of the Lie algebra has been assumed (for simplicity) to be invariant under cyclic permutation of the indices. This invariance means that our last Lemma applies to this Lie algebraic interpretation of *STU*. The upshot is that we can manufacture weight systems for graphs that satisfy the 4-term relation by replacing paired points on the chord diagram by an insertion of T^a in one point of the pair and a corresponding insertion of T^a at the other point in the pair and summing over all a. The result of all such insertions on a given chord diagram is a big sum of specific matrix products along the circle of the diagram, each of which (being a circular product) is interpreted as a trace.

Let's say this last matter more precisely: Regard a graph with k nodes as obtained by identifying *k pairs* of points on a circle. Thus a code such as 1212 taken in cyclic order specifies such a graph by regarding the points $1, 2, 1, 2$ as arrayed along a circle with the first and second 1's and 2's identified to form the graph. Define,for a code $a_1a_2 \ldots a_m$

$$wt(a_1a_2 \ldots a_m) = trace(T^{a_1}T^{a_2}T^{a_1} \ldots T^{a_m})$$

where the Einstein summation convention is in place for the double appearances of indices on the right-hand side. This gives the weight system.

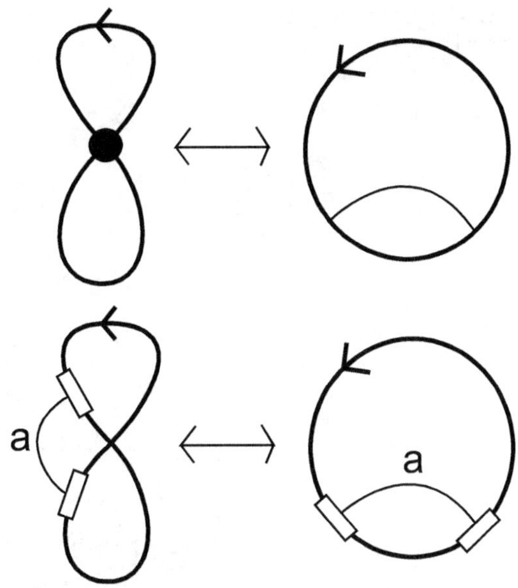

Fig. 34 Weight system and Casimir insertion

The weight system described by the above procedure satisfies the 4-term relation, but does not necessarily satisfy the vanishing condition for isolated pairings. This is because the framing compensation for converting an invariant of regular isotopy to ambient isotopy has not yet been introduced. We will show how to do this in the course of the discussion in the next paragraph. The main point to make here is that by starting with the idea of extending an invariant of knots to a Vassiliev invariant of embedded graphs and searching out the conditions on graph evaluation demanded by the topology, we have inevitably entered the domain of relations between Lie algebras and link invariants. Since the *STU* relation does not demand Lie algebras for its satisfaction we see that the landscape is wider than the Lie algebra context, but it is not yet understood how big is the class of link invariants derived from Lie algebras.

In fact, we can line up this weight system with the formalism related to the knot diagram by writing the Lie algebra insertions back on the 4-valent graph. We then get a Casimir insertion at the node. See Fig. 34.

To get the framing compensation, note that an isolated pairing corresponds to the trace of the Casimir. Let γ denote this trace. See Fig. 34.

$$\gamma = tr(\sum_a T^a T^a)$$

Let D be the trace of the identity. Then it is easy to see that we must compensate the given weight system by subtracting (γ/D) multiplied by the result of dropping the identification of the two given points. We can diagram this by drawing two crossed arcs without a node drawn to bind them. Then the modified recursion formula becomes as shown in Fig. 35.

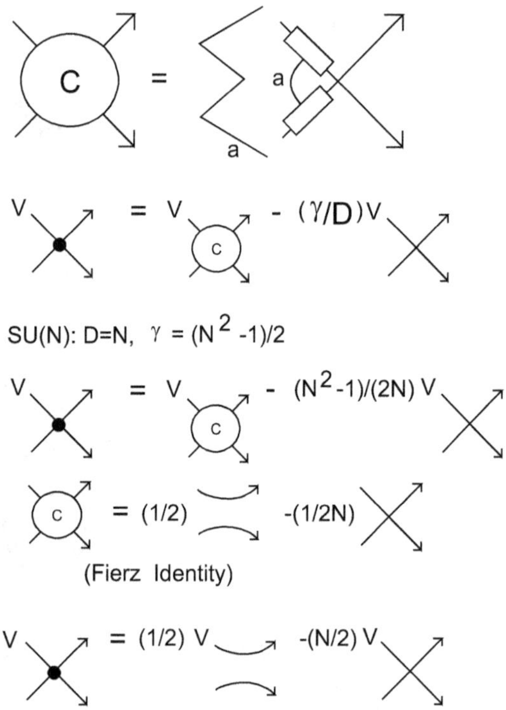

Fig. 35 Modified recursion formula

For example, in the case of $SU(N)$ we have $D = N$, $\gamma = (N^2 - 1)/2$ so that we get the transformation shown in Fig. 35, including the use of the Fierz identity.

For $N = 2$ the final formula of Fig. 35 is,up to a multiple, exactly the top row formula that we deduced for the Jones polynomial from its combinatorial structure.

5.2 Witten's Functional Integral and Vassiliev Invariants

In [52] Edward Witten proposed a formulation of a class of 3-manifold invariants as generalized Feynman integrals taking the form $Z(M)$ where

$$Z(M) = \int dAexp[(ik/4\pi)S(M,A)].$$

Here M denotes a 3-manifold without boundary and A is a gauge field (also called a gauge potential or gauge connection) defined on M. The gauge field is a one-form on a trivial G-bundle over M with values in a representation of the Lie algebra of G. The group G corresponding to this Lie algebra is said to be the gauge group. In this integral the "action" $S(M,A)$ is taken to be the integral over M of the trace of the Chern-Simons three-form $CS = AdA + (2/3)AAA$. (The product is the wedge product of differential forms.)

$Z(M)$ integrates over all gauge fields modulo gauge equivalence (see [2] for a discussion of the definition and meaning of gauge equivalence.)

The formalism and internal logic of Witten's integral supports the existence of a large class of topological invariants of 3-manifolds and associated invariants of knots and links in these manifolds.

The invariants associated with this integral have been given rigorous combinatorial descriptions [31, 35, 41, 49, 51], but questions and conjectures arising from the integral formulation are still outstanding (see for example [3, 11–13, 40]). Specific conjectures about this integral take the form of just how it involves invariants of links and 3-manifolds, and how these invariants behave in certain limits of the coupling constant k in the integral. Many conjectures of this sort can be verified through the combinatorial models. On the other hand, the really outstanding conjecture about the integral is that it exists! At the present time there is no measure theory or generalization of measure theory that supports it. It is a fascinating exercise to take the speculation seriously, suppose that it does really work like an integral and explore the formal consequences. Here is a formal structure of great beauty. It is also a structure whose consequences can be verified by a remarkable variety of alternative means. Perhaps in the course of the exploration there will appear a hint of the true nature of this form of integration.

We now look at the formalism of the Witten integral in more detail and see how it involves invariants of knots and links corresponding to each classical Lie algebra. In order to accomplish this task, we need to introduce the Wilson loop. The Wilson loop is an exponentiated version of integrating the gauge field along a loop K in three-space that we take to be an embedding (knot) or a curve with transversal self-intersections. For this discussion, the Wilson loop will be denoted by the notation $W_K(A) =< K|A >$ to denote the dependence on the loop K and the field A. It is usually indicated by the symbolism $tr(Pexp(\int_K A))$. Thus

$$W_K(A) =< K|A >= tr(Pexp(\int_K A)).$$

Here the P denotes path ordered integration—we are integrating and exponentiating matrix valued functions, and so must keep track of the order of the operations. The symbol tr denotes the trace of the resulting matrix.

With the help of the Wilson loop functional on knots and links, Witten writes down a functional integral for link invariants in a 3-manifold M:

$$\begin{aligned} Z(M, K) &= \int dAexp[(ik/4\pi)S(M, A)]tr(Pexp(\int_K A)) \\ &= \int dAexp[(ik/4\pi)S] < K|A > . \end{aligned}$$

Here $S(M, A)$ is the Chern-Simons Lagrangian, as in the previous discussion.

We abbreviate $S(M,A)$ as S and write $< K|A >$ for the Wilson loop. Unless otherwise mentioned, the manifold M will be the three-dimensional sphere S^3

An analysis of the formalism of this functional integral reveals quite a bit about its role in knot theory. This analysis depends upon key facts relating the curvature of the gauge field to both the Wilson loop and the Chern-Simons Lagrangian. The idea for using the curvature in this way is due to Smolin [43, 44] (see also [38]). To this end, let us recall the local coordinate structure of the gauge field $A(x)$, where x is a point in three-space. We can write $A(x) = A^k_a(x)T^a dx_k$ where the index a ranges from 1 to m with the Lie algebra basis $\{T^1, T^2, T^3, \ldots, T^m\}$. The index k goes from 1 to 3. For each choice of a and k, $A^k_a(x)$ is a smooth function defined on three-space. In $A(x)$ we sum over the values of repeated indices. The Lie algebra generators T^a are matrices corresponding to a given representation of the Lie algebra of the gauge group G. We assume some properties of these matrices as follows:

1. $[T^a, T^b] = if_{abc}T^c$ where $[x,y] = xy - yx$, and f_{abc} (the matrix of structure constants) is totally antisymmetric. There is summation over repeated indices.
2. $tr(T^aT^b) = \delta^{ab}/2$ where δ^{ab} is the Kronecker delta ($\delta^{ab} = 1$ if $a = b$ and zero otherwise).

We also assume some facts about curvature. (The reader may enjoy comparing with the exposition in [27]. But note the difference of conventions on the use of i in the Wilson loops and curvature definitions.) The first fact is the relation of Wilson loops and curvature for small loops:

Fact 1. The result of evaluating a Wilson loop about a very small planar circle around a point x is proportional to the area enclosed by this circle times the corresponding value of the curvature tensor of the gauge field evaluated at x. The curvature tensor is written

$$F^{rs}_a(x)T^a dx_r dy_s.$$

It is the local coordinate expression of $AdA + AA$.

Application of Fact 1 Consider a given Wilson line $< K|S >$. Ask how its value will change if it is deformed infinitesimally in the neighborhood of a point x on the line. Approximate the change according to Fact 1, and regard the point x as the place of curvature evaluation. Let $\delta < K|A >$ denote the change in the value of the line. $\delta < K|A >$ is given by the formula

$$\delta < K|A >= dx_r dx_s F^{rs}_a(x)T^a < K|A > .$$

This is the first order approximation to the change in the Wilson line.

In this formula it is understood that the Lie algebra matrices T^a are to be inserted into the Wilson line at the point x, and that we are summing over repeated indices. This means that each $T^a < K|A >$ is a new Wilson line obtained from the original line $< K|A >$ by leaving the form of the loop unchanged, but inserting the matrix T^a into that loop at the point x. A Lie algebra generator is diagrammed by a little

Fig. 36 Wilson loop insertion

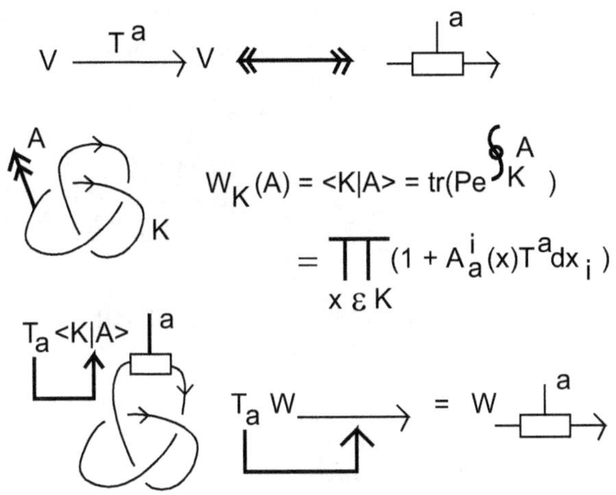

box with a single index line and two input/output lines which correspond to its role as a matrix (hence as mappings of a vector space to itself). See Fig. 36.

Remark. In thinking about the Wilson line $< K|A >= tr(Pexp(\int_K A))$, it is helpful to recall Euler's formula for the exponential:

$$e^x = lim_{n \to \infty}(1 + x/n)^n.$$

The Wilson line is the limit, over partitions of the loop K, of products of the matrices $(1 + A(x))$ where x runs over the partition. Thus we can write symbolically,

$$< K|A >= \prod_{x \in K}(1 + A(x)) = \prod_{x \in K}(1 + A^k_a(x)T^a dx_k).$$

It is understood that a product of matrices around a closed loop connotes the trace of the product. The ordering is forced by the one-dimensional nature of the loop. Insertion of a given matrix into this product at a point on the loop is then a well-defined concept. If T is a given matrix then it is understood that $T < K|A >$ denotes the insertion of T into some point of the loop. In the case above, it is understood from context in the formula

$$dx_r dx_s F^{rs}_a(x)T^a < K|A >$$

that the insertion is to be performed at the point x indicated in the argument of the curvature.

Remark. The previous remark implies the following formula for the variation of the Wilson loop with respect to the gauge field:

$$\delta < K|A > /\delta(A^k_a(x)) = dx_k T^a < K|A > .$$

Varying the Wilson loop with respect to the gauge field results in the insertion of an infinitesimal Lie algebra element into the loop.

Proof.

$$\delta < K|A > /\delta(A_a^k(x))$$

$$= \delta \prod_{y \in K}(1 + A_a^k(y)T^a dy_k)/\delta(A_a^k(x))$$

$$= \prod_{y<x \in K} (1 + A_a^k(y)T^a dy_k)[T^a dx_k] \prod_{y>x \in K} (1 + A_a^k(y)T^a dy_k)$$

$$= dx_k T^a < K|A > .$$

Fact 2. The variation of the Chern-Simons Lagrangian S with respect to the gauge potential at a given point in three-space is related to the values of the curvature tensor at that point by the following formula:

$$F_a^{rs}(x) = \epsilon_{rst}\delta S/\delta(A_a^t(x)).$$

Here ϵ_{abc} is the epsilon symbol for three indices, i.e. it is $+1$ for positive permutations of 123 and -1 for negative permutations of 123 and zero if any two indices are repeated.

With these facts at hand we are prepared to determine how the Witten integral behaves under a small deformation of the loop K.

In accord with the theme of this paper, we shall use a system of abstract tensor diagrams to look at the differential algebra related to the functional integral. The translation to diagrams is accomplished with the aid of Figs. 37 and 38. In Fig. 37 we give diagrammatic equivalents for the component parts of our machinery. Tensors become labelled boxes. Indices become lines emanating from the boxes. Repeated indices that we intend to sum over become lines from one box to another. (The eye can immediately apprehend the repeated indices and the tensors where they are repeated.) Note that we use a capital D with lines extending from the top and the bottom for the partial derivative with respect to the gauge field, a capital W with a link diagrammatic subscript for the Wilson loop, a cubic vertex for the three-index epsilon, little triangles with emanating arcs for the differentials of the space variables.

The Lie algebra generators are little boxes with single index lines and two input/output lines which correspond to their roles as matrices (hence as mappings of a vector space to itself). The Lie algebra generators are, in all cases of our calculation, inserted into the Wilson line either through the curvature tensor or through insertions related to differentiating the Wilson line.

In Fig. 38 we give the diagrammatic calculation of the change of the functional integral corresponding to a tiny change in the Wilson loop. The result is a double insertion of Lie Algebra generators into the line, coupled with the presence of a

Fig. 37 Notation

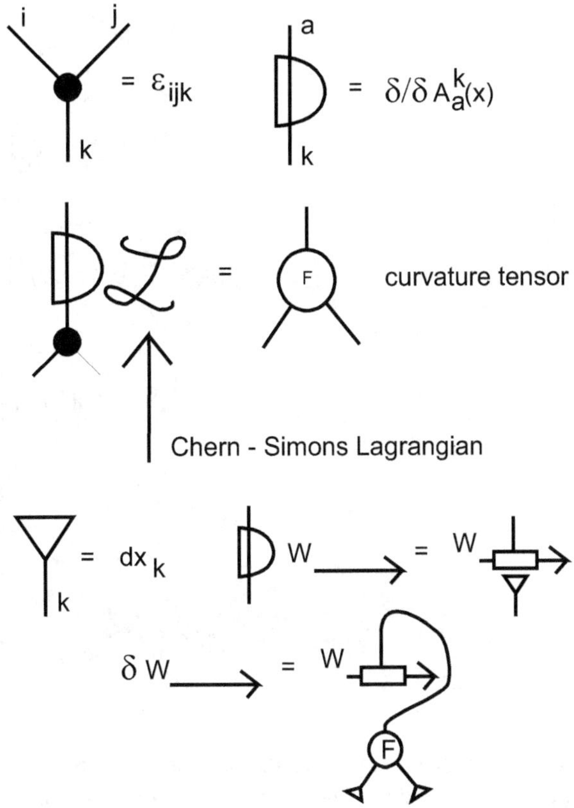

volume form that will vanish if the deformation does not twist in three independent directions. This shows that the functional integral is formally invariant under regular isotopy since the regular isotopy moves are changes in the Wilson line that happen entirely in a plane. One does not expect the integral to be invariant under a Reidemeister move of type one, and it is not. This framing compensation can be determined by the methods that we are discussing [32], but we will not go into the details of those calculations here.

In Fig. 39 we show the application of the calculation in Fig. 38 to the case of switching a crossing. The same formula applies, with a different interpretation, to the case where x is a double point of transversal self-intersection of a loop K, and the deformation consists in shifting one of the crossing segments perpendicularly to the plane of intersection so that the self-intersection point disappears. In this case, one T^a is inserted into each of the transversal crossing segments so that $T^a T^a < K|A >$ denotes a Wilson loop with a self-intersection at x and insertions of T^a at $x + \epsilon_1$ and $x + \epsilon_2$ where ϵ_1 and ϵ_2 denote small displacements along the two arcs of K that intersect at x. In this case, the volume form is nonzero, with two directions coming from the plane of movement of one arc, and the perpendicular

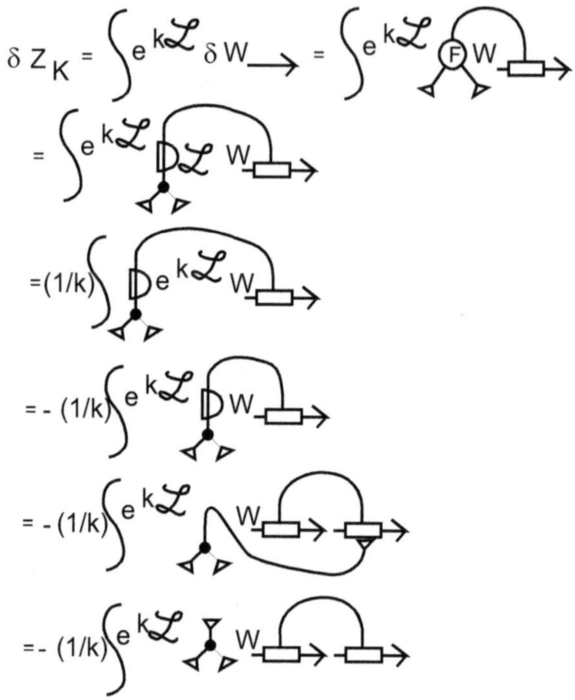

Fig. 38 Derivation

direction is the direction of the other arc. The reason for the insertion into the two lines is a direct consequence of the calculational form of Fig. 38: The first insertion is in the moving line, due to curvature. The second insertion is the consequence of differentiating the self-touching Wilson line. Since this line can be regarded as a product, the differentiation occurs twice at the point of intersection, and it is the second direction that produces the non-vanishing volume form.

Up to the choice of our conventions for constants, the switching formula is, as shown in Fig. 39,

$$Z(K_+) - Z(K_-) = (4\pi i/k) \int dA exp[(ik/4\pi)S] T^a T^a < K_{**}|A >$$

$$= (4\pi i/k) Z(T^a T^a K_{**}).$$

The key point is to notice that the Lie algebra insertion for this difference is exactly what we did to make the weight systems for Vassiliev invariants (without the framing compensation). Thus the formalism of the Witten functional integral takes us directly to these weight systems in the case of the classical Lie algebras. The functional integral is central to the structure of the Vassiliev invariants.

Fig. 39 Crossing switch

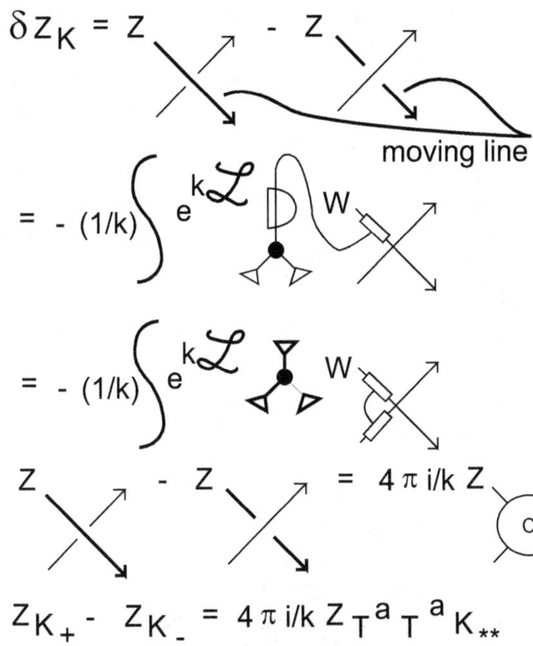

5.3 *Combinatorial Constructions for Vassiliev Invariants*

Perhaps the most remarkable thing about this story of the structure of the Vassiliev invariants is the way that Lie algebras are so naturally involved in the structure of the weight systems. This shows the remarkably close nature of the combinatorial structure of Lie algebras and the combinatorics of knots and links via the Reidemeister moves. A really complete story about the Vassiliev invariants at this combinatorial level would produce their existence on the basis of the weight systems with entirely elementary arguments.

As we have already mentioned, one can prove that a given set of weights for the top row, satisfying the abstract four-term relation does imply that there exists a Vassiliev invariant of finite type n realizing these weights for graphs with n nodes. Proofs of this result either use analysis [1, 4] or non-trivial algebra [4, 7]. There is no known elementary combinatorial proof of the existence of Vassiliev invariants for given top rows.

Of course quantum link invariants (see Sect. 4 of these lectures.) do give combinatorial constructions for large classes of link invariants. These constructions rest on solutions to the Yang-Baxter equations, and it is not known how to describe the subset of finite type Vassiliev invariants that are so produced.

It is certainly helpful to look at the structure of Vassiliev invariants that arise from already-defined knot invariants. If $V(K)$ is an already defined invariant of knots (and possibly links), then its extension to a Vassiliev invariant is calculated on embedded graphs G by expanding each graphical vertex into a difference by resolving the vertex into a positive crossing and a negative crossing. If we know that $V(K)$ is of finite type n and G has n nodes then we can take any embedding of G that is convenient, and calculate $V(G)$ in terms of all the knots that arise in resolving the nodes of this chosen embedding. This is a finite collection of knots. Since there is a finite collection of 4-valent graphs with n nodes, it follows that the top row evaluation for the invariant $V(K)$ is determined by the values of $V(K)$ on a finite collection of knots. Instead of asking for the values of the Vassiliev invariant on a top row, we can ask for this set of knots and the values of the invariant on this set of knots. A minimal set of knots that can be used to generate a given Vassiliev invariant will be called a *knots basis* for the invariant. Thus we have shown that the set consisting of the unknot, the right-handed trefoil and the left handed-trefoil is a knots basis for a Vassiliev invariant of type 3. See [36] for more information about this point of view.

A tantalizing combinatorial approach to Vassiliev invariants is due to Polyak and Viro [37]. They give explicit formulas for the second, third and fourth Vassiliev invariants and conjecture that their method will work for Vassiliev invariants of all orders. The method is as follows.

First one makes a new representation for oriented knots by taking *Gauss diagrams*. A Gauss diagram is a diagrammatic representation of the classical *Gauss code* of the knot. The Gauss code is obtained from the oriented knot diagram by first labelling each crossing with a naming label (such as 1,2,...) and also indicating the crossing type ($+1$ or -1). Then choose a basepoint on the knot diagram and begin walking along the diagram, recording the name of the crossings encountered, their sign and whether the walk takes you over or under that crossing. For example, if you go under crossing 1 whose sign is $+$ then you will record $o+1$. Thus the Gauss code of the positive trefoil diagram is

$$(o1+)(u2+)(o3+)(u1+)(o2+)(u3+).$$

For prime knots the Gauss code is sufficient information to reconstruct the knot diagram. See [29] for a sketch of the proof of this result and for other references.

To form a Gauss diagram from a Gauss code, take an oriented circle with a basepoint chosen on the circle. Walk along the circle marking it with the labels for the crossings in the order of the Gauss code. Now draw chords between the points on the circle that have the same label. Orient each chord from overcrossing site to undercrossing site. Mark each chord with $+1$ or -1 according to the sign of the corresponding crossing in the Gauss code. The resulting labelled and basepointed graph is the Gauss diagram for the knot. See Fig. 40 for examples.

The Gauss diagram is deliberately formulated to have the structure of a chord diagram (as we have discussed for the weight systems for Vassiliev invariants). If $G(K)$ is the Gauss diagram for a knot K, and D is an oriented (i.e. the chords as well as the circle in the diagram are oriented) chord diagram, let $|G(K)|$ denote the number of chords in $G(K)$ and $|D|$ denote the number of chords in D. If $|D| \leq |G(K)|$ then we may consider oriented embeddings of D in $G(K)$. For a given embedding $i : D \longrightarrow G(K)$ define

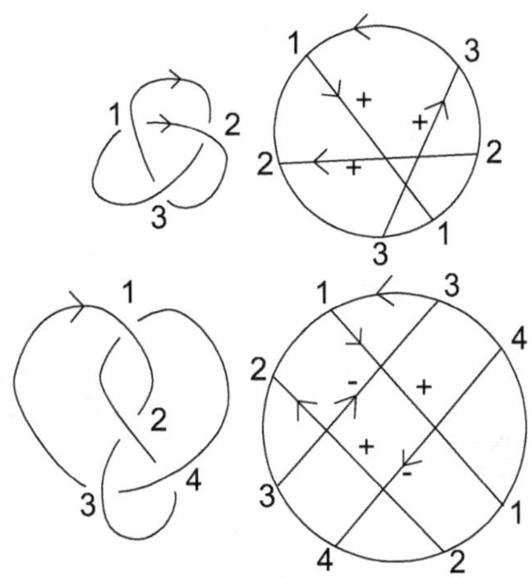

Fig. 40 Gauss diagrams

$$< i(D)|G(K) >= sign(i)$$

where $sign(i)$ denotes the product of the signs of the chords in $G(K) \cap i(D)$. Now suppose that C is a collection of *oriented* chord diagrams, each with n chords, and that

$$eval : C \longrightarrow R$$

is an evaluation mapping on these diagrams that satisfies the four-term relation at level n. Then we can define

$$< D|K >= \sum_{i:D \longrightarrow G(K)} < i(D)|G(K) >$$

and

$$v(K) = \sum_{D \in C} < D|K > eval(D).$$

For appropriate oriented chord subsets this definition can produce Vassiliev invariants $v(K)$ of type n. For example, in the case of the Vassiliev invariant of type three taking value 0 on the unknot and value 1 on the right-handed trefoil, -1 on the left-handed trefoil, Polyak and Viro give the specific formula

$$v_3(K) =< A|K > +(1/2) < B|K >$$

where A denotes the trefoil chord diagram as we described it in Sect. 3 and B denotes the three-chord diagram consisting of two parallel chords pierced by a third chord. In Fig. 41 we show the specific orientations for the chord diagrams A and B. The key to this construction is in the choice of orientations for the chord diagrams in $C = \{A, B\}$. It is a nice exercise in translation of the Reidemeister moves to Gauss diagrams to see that $v_3(K)$ is indeed a knot invariant.

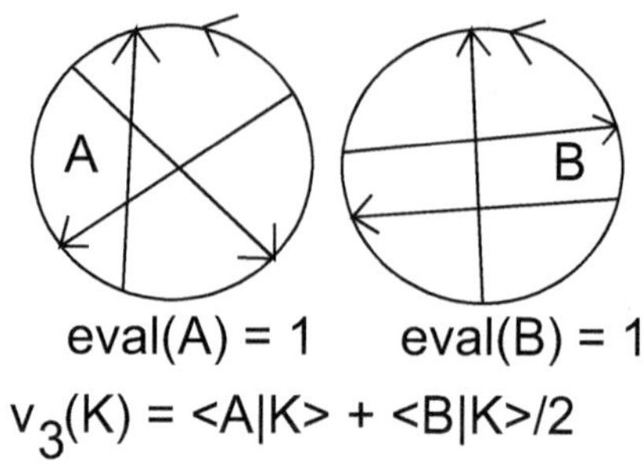

Fig. 41 Oriented chord diagrams for $\mathbf{v_3}$

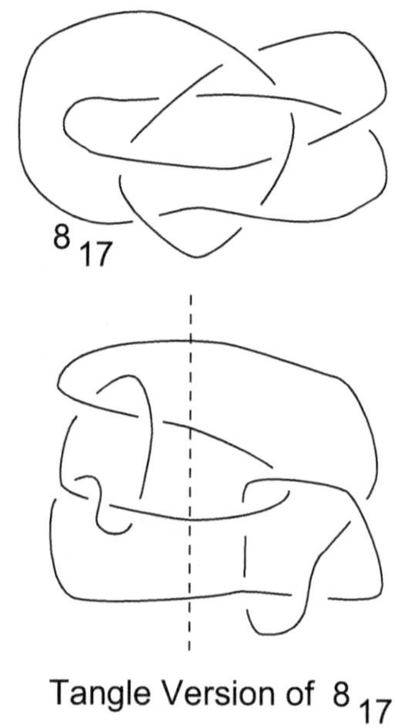

Fig. 42 Tangle decomposition of 8_{17}

It is possible that all Vassiliev invariants can be constructed by a method similar to the formula $v(K) = \sum_{D \in C} < D|K > eval(D)$. This remains to be seen.

5.4 *Invertibility and the knot* 8_{17}

It is an open problem whether there are Vassiliev invariants that can detect the difference between a knot and its reverse (The reverse of an oriented knot is obtained by flipping the orientation.). The smallest instance of a non-invertible knot is the knot 8_{17} depicted in Fig. 42. Hale Trotter [46] was the first person to give proofs that some knots are non-invertible. We have not discussed his methods in this paper.

Thus, at the time of this writing there is no known Vassiliev invariant that can detect the non-invertibility of 8_{17}. On the other hand, the tangle decomposition shown in Fig. 42 can be used in conjunction with the results of Siebenmann and Bonahon [42] and the formulations of John Conway [8] to show this non-invertibility. These tangle decomposition methods use higher level information about the diagrams than is easy to encode in Vassiliev invariants. The purpose of this section is to underline this discrepancy between different levels in the combinatorial topology.

References

1. D. Altshuler and L. Friedel, Vassiliev knot invariants and Chern-Simons perturbation theory to all orders. *Comm. Math. Phys.* 187 (1997), no. 2, 261–287.
2. M.F. Atiyah, *Geometry of Yang-Mills Fields*, Accademia Nazionale dei Lincei Scuola Superiore Lezioni Fermiare, Pisa ,1979.
3. M.F. Atiyah, *The Geometry and Physics of Knots*, Cambridge University Press, 1990.
4. D. Bar-Natan, On the Vassiliev knot invariants, *Topology* 34 (1995), no. 2, 423–472.
5. R.J. Baxter, *Exactly Solved Models in Statistical Mechanics*, Acad. Press, 1982.
6. J. Birman and X.S. Lin, Knot polynomials and Vassiliev's invariants. *Invent. Math.* 111 (1993), no. 2, 225–270.
7. P. Cartier, Construction combinatoire des invariants de Vassiliev - Kontsevich des noeuds, *C. R. Acad. Sci. Paris* **316**, Série I, (1993), pp. 1205–1210.
8. J.H. Conway, (private conversation)
9. J.H. Conway, An enumeration of knots and links and some of their algebraic properties, in *Computational Problems in Abstract Algebra*, Pergammon Press, N.Y., (1970), pp. 329–358.
10. I.A. Dynnikov, Arc presentations of links - Monotonic simplification, Fund. Math. 190 (2006), 29–76. *www.arxiv.org/math.GT/0208153 v2 8 Sept 2003.*
11. S. Garoufalidis, Applications of TQFT to invariants in low dimensional topology, (preprint 1993).
12. D.S. Freed and R.E. Gompf, Computer calculation of Witten's three-manifold invariant, *Commun. Math. Phys.*, No. 141, (1991), pp. 79–117.
13. L.C. Jeffrey, Chern-Simons-Witten invariants of lens spaces and torus bundles, and the semi-classical approximation, *Commun. Math. Phys.*, No. 147, (1992), pp. 563–604.
14. V.F.R. Jones, A polynomial invariant of links via von Neumann algebras, *Bull. Amer. Math. Soc.*, 1985, No. 129, pp. 103–112.
15. V.F.R.Jones, Hecke algebra representations of braid groups and link polynomials, *Ann. of Math.*,Vol.126, 1987, pp. 335–338.
16. V.F.R.Jones, On knot invariants related to some statistical mechanics models, *Pacific J. Math.*, Vol. 137, no. 2,1989, pp. 311–334.
17. Kronheimer, P. B.; Mrowka, T. S. Khovanov homology is an unknot-detector. Publ. Math. Inst. Hautes tudes Sci. No. 113 (2011), 97–208. math.GT.arXiv:1005.4346.
18. L.H. Kauffman, The Conway polynomial, *Topology*, **20** (1980), pp. 101–108.
19. L.H. Kauffman, *Formal Knot Theory*, Princeton University Press, Lecture Notes Series 30 (1983).
20. L.H. Kauffman, *On Knots*, Annals Study No. 115, *Princeton University Press* (1987)
21. L.H. Kauffman, State Models and the Jones Polynomial, *Topology*,Vol. 26, 1987,pp. 395–407.
22. L.H. Kauffman, Statistical mechanics and the Jones polynomial, *AMS Contemp. Math. Series*, Vol. 78,1989, pp. 263–297.
23. L.H. Kauffman, Map coloring, q-deformed spin networks, and Turaev-Viro invariants for 3-manifolds, *Int. J. of Modern Phys. B*, Vol. 6, Nos. 11, 12 (1992), pp. 1765–1794.
24. L.H. Kauffman, An invariant of regular isotopy, *Trans. Amer. Math. Soc.*, Vol. 318. No. 2 ,1990, pp 417–471.

25. L.H. Kauffman, New invariants in the theory of knots, *Amer. Math. Monthly*, Vol. 95, No.3, March 1988. pp 195–242.
26. L.H. Kauffman and P. Vogel, Link polynomials and a graphical calculus, *Journal of Knot Theory and Its Ramifications*, Vol. 1, No. 1, March 1992.
27. L.H. Kauffman, "Knots and Physics", World Scientific, Singapore/New Jersey/London/Hong Kong, 1991, 1994, 2001.
28. A. Henrich and L. H. Kauffman, Unknotting Unknots, American Mathematical Monthly, Vol. 121, May 2014, pp. 379–390.
29. L.H. Kauffman, Gauss Codes, quantum groups and ribbon Hopf algebras, *Reviews in Mathematical Physics* **5** (1993), 735–773. (Reprinted in [27], 551–596.
30. L.H. Kauffman, Knots and Diagrams, in "Lectures at Knots 96", ed. by Shin'ichi Suzuki (1997), World Scientific Pub. Co. pp. 123–194.
31. L.H. Kauffman and S. L. Lins, *Temperley-Lieb Recoupling Theory and Invariants of 3-Manifolds*, Annals of Mathematics Study 114, Princeton Univ. Press,1994.
32. L.H. Kauffman, Functional Integration and the theory of knots, J. Math. Physics, Vol. 36 (5), May 1995, pp. 2402–2429.
33. G. Hemion, "The Classification of Knots and 3-Dimensional Spaces", Oxford University Press, 1992.
34. M. Lackenby, A polynomial upper bound on Reidemeister moves. Ann. of Math. (2) 182 (2015), no. 2, 491–564.
35. W.B.R. Lickorish, The Temperley-Lieb Algebra and 3-manifold invariants, *Journal of Knot Theory and Its Ramifications*, Vol. 2,1993, pp. 171–194.
36. J. Mathias, *Ph.D. Thesis,* University of Illinois at Chicago, 1996.
37. Michael Polyak and Oleg Viro, Gauss diagram formulas for Vassiliev invariants, *Intl. Math. Res. Notices*, No. 11, (1994) pp. 445–453.
38. P. Cotta-Ramusino, E. Guadagnini, M. Martellini, M. Mintchev, Quantum field theory and link invariants, (preprint 1990)
39. K. Reidemeister, *Knotentheorie*, Chelsea Pub. Co., New York, 1948, Copyright 1932, Julius Springer, Berlin.
40. L. Rozansky, Witten's invariant of 3-dimensional manifolds: loop expansion and surgery calculus, In *Knots and Applications*, edited by L. Kauffman, (1995), World Scientific Pub. Co.
41. N.Y. Reshetikhin and V. Turaev, Invariants of Three-Manifolds via link polynomials and quantum groups, *Invent. Math.*,Vol.103,1991, pp. 547–597.
42. L. Siebenmann and F. Bonahon, (Unpublished Manuscript)
43. L. Smolin, Link polynomials and critical points of the Chern-Simons path integrals, *Mod. Phys. Lett. A*, Vol. 4,No. 12, 1989, pp. 1091–1112.
44. L. Smolin,The physics of spin networks. The geometric universe (Oxford, 1996), 291–304, Oxford Univ. Press,n Oxford, 1998.
45. T. Stanford, Finite-type invariants of knots, links and graphs, (preprint 1992).
46. H. Trotter, Non-invertible knots exist, *Topology*,Vol.2,1964,pp. 275–280.
47. M.B. Thistlethwaite, Links with trivial Jones polynomial, JKTR, Vol. 10, No. 4 (2001), 641–643.
48. S. Eliahou, L. Kauffman and M.B. Thistlethwaite, Infinite families of links with trivial Jones polynomial, *Topology*, **42**, pp. 155–169.
49. V.G. Turaev and H. Wenzl, Quantum invariants of 3-manifolds associated with classical simple Lie algebras, *International J. of Math.*, Vol. 4, No. 2,1993, pp. 323–358.
50. V. Vassiliev, Cohomology of knot spaces, In *Theory of Singularities and Its Applications*, V.I. Arnold, ed., Amer. Math. Soc.,1990, pp. 23–69.
51. K. Walker, On Witten's 3-Manifold Invariants, (preprint 1991).
52. Edward Witten, Quantum field Theory and the Jones Polynomial, *Commun. Math. Phys.*,vol. 121, 1989, pp. 351–399.
53. J. W. Aexander. Topological invariants of knots and links. *Trans. Amer. Math. Soc.*, 20:275–306, 1923.

54. J. Birman and M. Hirsch. A new algorithm for recognizing the unknot. *Geom. Top.*, 2:175–220, 1998.
55. J. Birman and J. Moody. Obstructions to trivializing a knot. *Israel J. Math.*, 142:125–162, 2004.
56. L. Goeritz. Bemerkungen zur knotentheorie. *Abh. Math. Sem. Univ. Hamburg*, 18:201–210, 1997.
57. J. Hass and J. Lagarias. The number of Reidemeister moves needed for unknotting. *J. Amer. Math. Soc*, 14:399–428, 2001.
58. L. Kauffman. Knots and physics. In *Series on Knots and Everything*, volume 1, page 18. World Scientific Pub. Co., 1991, 1993, 2001.
59. L. Kauffman. Knot diagrammatics. In W. Menasco and M. Thistlethwaite, editors, *The Handbook of Knot Theory*, chapter 6, pages 233–318. Elsevier, 2005.
60. L. Kauffman and S. Lambropoulou. Hard unknots and collapsing tangles. In L. Kauffman, S. Lambropoulou, S. Jablan, and J. H. Przytycki, editors, *Introductory Lectures on Knot Theory – Selected Lectures presented at the Advanced School and Conference on Knot Theory and its Applications to Physics and Biology, ICTP, Trieste, Italy, 11–29 May 2009*. World Scientific Pub. Co., 2011.
61. L. Kaufman and S. Lomonaco. Quantum knots and mosaics. *J. Quantum Info. Processing*, 7:85–115, 2008.
62. C. Manolescu, P. Ozsvath, Z. Szabo, and D. Thurston. Combinatorial link floer homology. *Geom.Top.*, 11:2339–2412, 2007.

How Can Cooperative Game Theory Be Made More Relevant to Economics? : An Open Problem

Eric Maskin

Abstract Game Theory pioneers J. von Neumann and O. Morgenstern gave most of their attention to the cooperative side of the subject. But cooperative game theory has had relatively little effect on economics. In this essay, I suggest why that might be and what is needed for cooperative theory to become more relevant to economics.

Cooperative game theory is the part of game theory that pertains when players can sign binding contracts determining their actions and payoffs. J. von Neumann and O. Morgenstern devoted most of their seminal book [6] to cooperative theory, with subsequent major contributions by Nash [4] and Shapley [5].

But despite its auspicious beginnings, cooperative game theory has been used far less than noncooperative theory as a predictive tool in economics. Indeed, inspection of the current leading game theory textbooks used in graduate economics programs reveals that the ratio of cooperative to noncooperative theory is remarkably low (in one such text, [1], the ratio is 0). And all Nobel Memorial Prizes awarded for game theory to date have recognized work exclusively on the noncooperative side.

This imbalance may seem strange. Cooperative theory seems to offer the important advantage of giving insight into how coalitions behave, i.e., how subsets of players bargain over which actions are played. Such bargaining seems basic to many aspects of economic and political life from the European Union, to the Paris climate change agreement, to the OPEC cartel. Moreover, on the face of it, cooperative theory appears to be far less dependent on particular details about strategies—and, therefore, more robust and general—than noncooperative theory.

To understand the sense in which noncooperative is more detail dependent, let us briefly go over the basic noncooperative and cooperative models. In a noncooperative game, each player $i, i = 1, \ldots, n$, chooses a strategy s_i from a strategy set S_i, and the payoffs of the game are given by the mapping

E. Maskin (✉)
Department of Economics, Harvard University, Littauer Center, Room 312, 1805 Cambridge Street, Cambridge, MA, USA

Higher School of Economics, Moscow, Russia
e-mail: emaskin@fas.harvard.edu

J.F. Nash, Jr., M.Th. Rassias (eds.), *Open Problems in Mathematics*,
DOI 10.1007/978-3-319-32162-2_9

$$g : S_1 \times \cdots \times S_n \to \mathbb{R}^n ,$$

where $g_i(s_1, \ldots, s_n)$ is player i's payoff if strategies $(s_1, \ldots, s_n)$ are played. The standard prediction for what will happen in game g is that players will choose *Nash equilibrium* strategies [3]. Strategies $(s_1^*, \ldots, s_n^*)$ constitute a Nash equilibrium if

$$g_i(s_1^*, \ldots, s_n^*) \geq g_i(s_1^*, \ldots, s_{i-1}^*, s_i, s_{i+1}^*, \ldots, s_n^*), \text{ for all } i \text{ and all } s_i \in S_i . \quad (1)$$

As formula (1) makes clear, Nash equilibrium depends crucially on what strategies are or are not in each player's strategy set; for example, adding a single strategy s_i' to S_i can destroy $(s_1^*, \ldots, s_n^*)$ as an equilibrium and change the predicted outcome of the game discontinuously, even if s_i' generates payoffs quite similar to those of s_i^*.

By contrast, cooperative games are typically described by a characteristic function v. Given a coalition of players $S \subseteq \{1, \ldots, n\}$, $v(S)$ is the sum of payoffs that the members of S can get on their own. An often-used predictive concept in cooperative game theory is the *Shapley value* (unlike Nash equilibrium in noncooperative theory, the Shapley value has serious competition as a predictive concept; there are some other leading notions, such as the core and the bargaining set). Given characteristic function v, player i's Shapley value payoff is

$$\sum_{S \subseteq \{1,\ldots,i-1,i+1,\ldots,n\}} \frac{|S|!(n-|S|-1)!}{n!} (v(S \cup \{i\}) - v(S)) ,$$

i.e., player i gets his expected marginal contribution to coalitions, where the expectation is taken over all possible coalitions that he might join.

Notice that in the cooperative-game setting, players' strategies no longer are modeled explicitly—only the resulting payoffs matter. Thus, many different noncooperative games can be associated with the same characteristic function. In that sense, the characteristic function approach is more general than the noncooperative model. Moreover, in cooperative games, the discontinuities that arise in noncooperative games no longer occur: the characteristic function and Shapley value vary continuously with the payoff possibilities. In that sense, cooperative games are more robust than noncooperative games.

So why, despite these advantages, is cooperative game theory currently dominated by noncooperative theory as applied to economics? Perhaps one answer is that the characteristic function, by assumption, rules out *externalities*—situations in which a coalition's payoff depends on what other coalitions are doing. Yet, interactions between coalitions are at the very heart of economics, e.g., bargaining between unions and management, competition between companies, and trade between nations. Moreover, even in the (relatively small) cooperative literature that *does* accommodate externalities (the partition-function approach; see [2]), extensions of the Shapley value and of other leading cooperative concepts do not predict competition between coalitions; instead, they assume as a matter of definition that the grand coalition—the coalition of all players—always forms. Of course this flies in the face of *reality*, where, in most settings, we don't typically see just a single big coalition, but rather several smaller coalitions. Furthermore, there is a good *theoretical* reason why, in a model with externalities, we should *not* expect the grand coalition to form.

To illustrate this point, let us consider the following three-player game, in which coalitions can produce public goods. The coalition of players 1 and 2—$\{1,2\}$—can produce a total payoff of 12 for itself, $\{1,3\}$ can produce 13, and $\{2,3\}$ can produce 14. The grand coalition $\{1,2,3\}$ can produce 24. A player can produce nothing on his own. However, if the other two players form a coalition, he can free-ride on the public good they produce and enjoy a payoff of 9 (which is the externality that the coalition confers on him).

I claim that, we should not expect the grand coalition to form in this game. To see why not, imagine that all bargaining is conducted at a particular site and player 1 arrives there first, followed by 2, and finally by 3. When player 2 arrives, player 1 can make him offer to join 1 in a coalition. Let us explore what 2 must be offered to be willing to join. Notice that if he does not join with 1, he will be in competition with 1 for signing up 3. In this competition, 1 will be willing to bid 13 (the gross value of the coalition with 3) minus 9 (which he would get as a free-rider if 3 signed up with 2), i.e., 4. Similarly, 2 will be willing to bid $14-9 = 5$. Hence, 2 will win the bidding war for 3 and will pay 4 (notice that because, in this thought experiment, 1 and 2 don't form a coalition, 3 has no possibility of free-riding and so will be willing to accept 4). Hence 2's payoff if he refuses to join with 1 is $14 - 4 = 10$. Thus, player 1 must offer him 10 in order to sign him up.

Assuming 2 is signed up, 1 must then offer 3 a payoff of 9 to attract him to coalition $\{1,2\}$ (because 3 has the option to free-ride on $\{1,2\}$ and get 9 that way). Hence, altogether player 1 must pay $10 + 9 = 19$ in order to form the grand coalition. But this leaves only $24 - 19 = 5$ for himself. Clearly, he would be better off refraining from signing up 2—in which case, as analyzed above, 2 will form a coalition with 3. And 1 obtains a free-riding payoff of 9.

I conclude that with arrival order 1, 2, 3, two separate coalitions will form: $\{2,3\}$ and $\{1\}$. A similar conclusion follows for the five other possible arrival orders.

Unfortunately, cooperative game theory in its current state does not allow for such a two-coalition outcome. In my view, it remains an open problem—perhaps the most important open problem in cooperative theory—to develop an approach that properly accommodates the formation of multiple coalitions. Only by solving this problem can we make cooperative game theory relevant to economics.

Acknowledgements I acknowledge research support from the Rilin Fund at Harvard University.

References

1. Fudenberg, D. and Tirole, J. (1991), *Game Theory*, Cambridge, MA: MIT Press.
2. Myerson, R. (1977), "Values for Games in Partition Function Form", *International Journal of Game Theory*, 6, pp. 23–31.

3. Nash, J. (1950), "Equilibrium Points in *n*-person Games", *Proceedings of the National Academy of Sciences USA*, 36(1), pp. 48–49.
4. Nash, J. (1953), "Two Person Cooperative Games", *Econometrica*, 21, pp.128–140.
5. Shapley L. (1953), "A Value for *n*-person Games", in H. W. Kuhn and A. W. Tucker (eds), Contributions to the Theory of Game, II, *Annals of Mathematical Studies*, Vol 28. Princeton University Press pp. 307–317.
6. von Neumann, J. and O. Morgenstern (1944), *Theory of Games and Economic Behavior*, Princeton, NJ: Princeton University Press.

The Erdős-Szekeres Problem

Walter Morris and Valeriu Soltan

Abstract Erdős and Szekeres proved in their 1935 paper that for every integer $n \geq 3$ there exists a smallest positive integer $N(n)$ such that any set of at least $N(n)$ points in general position in the plane contains n points which are the vertices of a convex n-gon. They also posed the problem to determine the value of $N(n)$ and conjectured that $N(n) = 2^{n-2} + 1$ for all $n \geq 3$. Despite the efforts of many mathematicians, the Erdős-Szekeres problem is still far from being solved. This chapter describes recent achievements towards the solution of this problem and some of its close relatives.

1 Introduction

In 1932, the young Hungarian mathematician Esther Klein observed that any set of five points in general position in the plane (that is, no three of the points belong to a line) contains the vertices of a convex quadrilateral.

Indeed, there are three distinct types of placement of five points in general position in the plane, as shown in Fig. 1. In any of these cases, we can pick out at least one convex quadrilateral determined by the points. Furthermore, the chosen quadrilateral may contain no other point of the set in its interior (see Sect. 3 for a related problem).

Esther Klein realized that this elementary fact could be a particular case of the following interesting problem, which she posed to her friends Paul Erdős and George Szekeres: given an integer $n \geq 3$, does there exist a positive integer $N(n)$ such that from any set containing at least $N(n)$ points in general position in the plane, it is possible to choose n points which form a convex n-gon?

In their enormously influential paper [16] from 1935, Erdős and Szekeres proved the existence of the number $N(n)$ by two different methods. The first method uses Ramsey's theorem (independently discovered by Szekeres for the purpose of solving Klein's problem) and gives the estimate $N(n) \leq R_4(5, n)$, where $R_4(5, n)$ is a Ramsey number (see Sect. 2.4 for details). The second method, based on

W. Morris (✉) • V. Soltan
George Mason University, 4400 University Drive, Fairfax, VA 22030, USA
e-mail: wmorris@gmu.edu; vsoltan@gmu.edu

J.F. Nash, Jr., M.Th. Rassias (eds.), *Open Problems in Mathematics*,
DOI 10.1007/978-3-319-32162-2_10

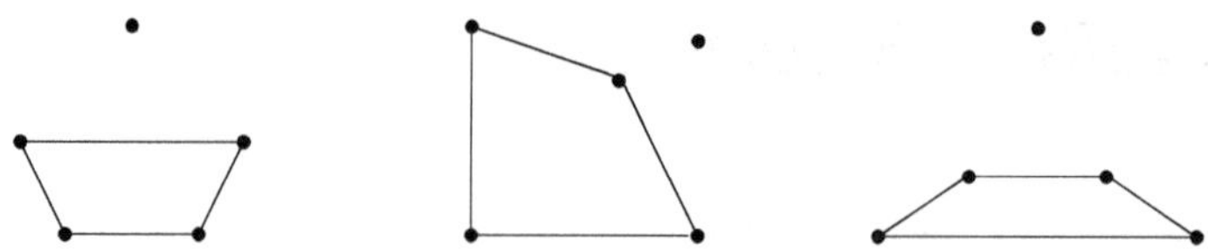

Fig. 1 Any five points in general position in the plane determine a convex quadrilateral

consideration of convex and concave sequences of points, produces a better upper bound, $N(n) \leq \binom{2n-4}{n-2} + 1$. In the same paper, Erdős and Szekeres posed their famous conjecture, often formulated as a problem.

Problem 1. Is it true that $N(n) = 2^{n-2} + 1$ for all $n \geq 3$?

The conjecture that $N(n) = 2^{n-2} + 1$ for all $n \geq 3$ is attributed to Klein by Szekeres and Peters in [43]. In 1961, Erdős and Szekeres [17] returned to their problem; they gave an example of 2^{n-2} points in general position in the plane containing no vertex set of a convex n-gon. In other words, they established the inequality $N(n) \geq 2^{n-2} + 1$.

Their work on Problem 1 preceded and perhaps led to the marriage of Klein and Szekeres, prompting Erdős to call Problem 1 the "Happy End Problem." Their passing (see [13]) was shortly before the Szekeres-Peters paper [43] appeared and established that $N(6) = 17$. Thus Problem 1 framed their marriage of nearly seventy years. Additional historic and bibliographic comments related to this topic can be found in [36].

Problem 1 is steadily gaining interest, generating new original questions and directions of research. An essential part of the existing results were summarized around 2000 in the surveys of Bárány and Károlyi [2] and Morris and Soltan [36]. This topic also is considered in separate chapters of the books of Matoušek [35] and Brass et al. [8]. At present we witness a further development of the field, with research to find the broadest generalizations and the deepest essence of Erdős and Szekeres' original results.

This chapter is based on the authors' survey [36]. It gives an updated account of results immediately related to the Erdős-Szekeres problem, leaving more distant topics for further analysis. The content of the chapter is indicated by section headings as follows.

1. Introduction.
2. The Erdős-Szekeres problem on convex polygons.
3. The Erdős problem on empty convex polygons.
4. Higher dimensional extensions.

Some words about notation: $|X|$ stands for the cardinality of a finite set X; if X is a subset of the Euclidean space E^d, then aff X and conv X mean, respectively, the affine and convex hull of X. A set $X \subset E^d$ is said to be *convexly independent* if no point of X lies in the convex hull of the remaining points. Furthermore, $\langle p, q \rangle$ denotes the line through distinct points p and q, and $[p, q\rangle$ stands for the closed halfline through

q with endpoint p. Given functions $f(x)$ and $g(x)$ on the halfline $[0,\infty)$, we write $f = O(g)$ and $g = \Omega(f)$ provided there are positive constants C and k such that $|f(x)| \leq C|g(x)|$ for all $x \geq k$. Finally, $[m]$ means the set $\{1,2,\dots,m\}$, and $\binom{[m]}{k}$ denotes the family of all subsets of $[m]$ of cardinality k.

2 The Erdős-Szekeres Problem on Convex Polygons

This section describes the advances in solving the Erdős-Szekeres problem in the plane. We start with the "caps and cups" technique from [16], which gives upper bounds on the number $N(n)$, that are $O(4^n)$.

2.1 *Caps and Cups*

Let $X = \{p_1, p_2, \dots, p_m\}$ be a set of points in general position in the coordinate plane. If necessary, we can rotate X by a small angle to ensure that no two points of X have the same x-coordinate. Renumbering these points, we also may assume that p_j has a greater x-coordinate than p_i whenever $i < j$. For a subset $I = \{i_1, i_2, \dots, i_n\}$ of the set $[m] \equiv \{1,2,\dots,m\}$, we say that the set $\{p_i : i \in I\}$ is an *n-cap* if $p_{i_{j+1}}$ lies above the line through p_{i_j} and $p_{i_{j+2}}$ for all $j = 1,2,\dots,n-2$. Similarly, the set $\{p_i : i \in I\}$ is an *n-cup* if $p_{i_{j+1}}$ lies below the line through p_{i_j} and $p_{i_{j+2}}$ for all $j = 1,2,\dots,n-2$ (see examples in Fig. 2). We will call p_{i_1} the left endpoint and p_{i_n} the right endpoint of the n-cap (or an n-cup) $\{p_{i_1}, p_{i_2}, \dots, p_{i_n}\}$.

Given integers $k, \ell \geq 3$, define $f(k,\ell)$ as the smallest positive integer for which a set X in general position in the plane contains a k-cup or an ℓ-cap whenever X has at least $f(k,\ell)$ points. The following lemma gives a key argument in known proofs for upper bounds on $N(n)$.

Lemma 1. *Suppose that a set X in general position in the plane contains at least $f(k-1,\ell)+f(k,\ell-1)-1$ points, where $k,\ell \geq 3$. If the left endpoint p_{i_1} of a $(k-1)$-cup $\{p_{i_1},\dots,p_{i_{k-1}}\}$ is the right endpoint $p_{j_{\ell-1}}$ of an $(\ell-1)$-cap $\{p_{j_1},p_{j_2},\dots,p_{j_{\ell-1}}\}$, then X contains a k-cup $\{p_{j_{\ell-2}},p_{i_1},p_{i_2},\dots,p_{i_{k-1}}\}$ or an ℓ-cap $\{p_{j_1},p_{j_2},\dots,p_{j_{\ell-1}}, p_{i_2}\}$, depending on whether p_{i_1} is below or above the line through $p_{j_{\ell-2}}$ and p_{i_2}. The same conclusion is reached if the left endpoint of an $(\ell-1)$-cap is the right endpoint of a $(k-1)$-cup (Fig. 3).* □

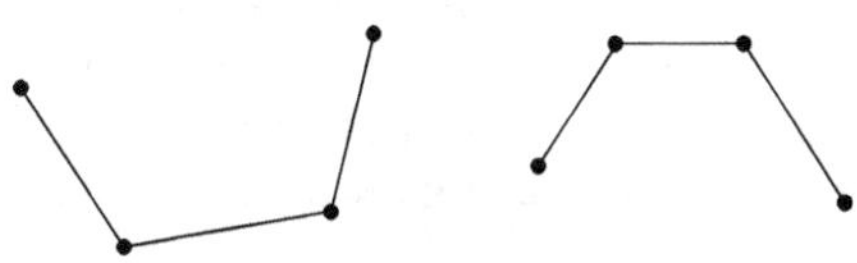

Fig. 2 A cup and a cap

Fig. 3 A cap followed by a cup

Theorem 1. *$f(k,\ell) \le \binom{k+\ell-4}{k-2} + 1$ for all $k,\ell \ge 3$.*

Proof. The inequality follows from the boundary conditions

$$f(k,3) = f(3,k) = k = \binom{k-1}{k-2} + 1$$

and the recurrence

$$f(k,\ell) \le f(k-1,\ell) + f(k,\ell-1) - 1.$$

Here is a proof of the recurrence. Choose a set X in general position in the plane with at least $f(k-1,\ell)+f(k,\ell-1)-1$ points. Let Y be the set of right endpoints of $(\ell-1)$-caps of X. If $X\setminus Y$ contains $f(k,\ell-1)$ points, then it contains a k-cup, because $X\setminus Y$ contains no $(\ell-1)$-cap. Otherwise Y contains $f(k-1,\ell)$ points. Suppose that Y contains a $(k-1)$-cup $\{p_{i_1}, p_{i_2}, \dots, p_{i_{k-1}}\}$. Let $\{p_{j_1}, p_{j_2}, \dots, p_{j_{\ell-1}}\}$ be an $(\ell-1)$-cap with $j_{\ell-1} = i_1$. By Lemma 1, X contains a k-cup $\{p_{j_{\ell-2}}, p_{i_1}, p_{i_2}, \dots, p_{i_{k-1}}\}$ or an ℓ-cap $\{p_{j_1}, p_{j_2}, \dots, p_{j_{\ell-1}}, p_{i_2}\}$. Finally,

$$f(k-1,\ell) + f(k,\ell-1) - 1 \le \binom{k+\ell-5}{k-3} + 1 + \binom{k+\ell-5}{k-2} + 1 - 1$$
$$= \binom{k+\ell-4}{k-2} + 1. \qquad \square$$

The inequality $N(n) \le f(n,n)$ immediately implies the following upper bound on $N(n)$, obtained by Erdős and Szekeres [16] in 1935.

Corollary 1 ([16]). *$N(n) \le \binom{2n-4}{n-2} + 1$ for all $n \ge 3$.* $\square$

In this regard we observe that the argument, which show the existence of a convex n-gon, actually proves the stronger statement that either an n-cap or an n-cup is contained in the set. It is obvious, but perhaps worth mentioning, that if $|X| \ge f(n,n)$, then every set obtained by rotating X will contain an n-cup or an n-cap; thus the set X contains several sets that are rotations of n-cups or n-caps.

The idea for the lower bound construction comes from Erdős and Szekeres' 1961 paper [17]. A correction to an error in that paper was made by Kalbfleisch and Stanton [27]. The following proof is taken from Lovasz [34, Sect. 14].

Theorem 2. $f(k,\ell) = \binom{k+\ell-4}{k-2} + 1$ *for all* $k,\ell \geq 3$.

Proof. By Theorem 1, it suffices to prove the inequality $f(k,\ell) \geq \binom{k+\ell-4}{k-2} + 1$.

We argue by double induction on k and ℓ. As previously observed, this inequality is true for $k = 3$ and $\ell = 3$. Suppose that we have a set A of $\binom{k+\ell-5}{k-3}$ points in general position, with no $(k-1)$-cup and no ℓ-cap, and a set B of $\binom{k+\ell-5}{k-2}$ points in general position, with no k-cup and no $(\ell-1)$-cap. Translate these sets so that every point of B has greater x-coordinate than the x-coordinates of points of A, and the slope of any line connecting a point of A to a point of B is greater than the slope of any line connecting two points of B. Let $X = A \cup B$ be the resulting set. Any cup in X that contains elements of both A and B may have only one element of B. Thus X contains no k-cup. We similarly see that X contains no ℓ-cap. Consequently,

$$f(k,\ell) \geq |A| + |B| + 1 = \binom{k+\ell-4}{k-2} + 1. \qquad \square$$

The next theorem gives the lower bound on $N(n)$ obtained by Erdős and Szekeres [17].

Theorem 3 ([17]). $N(n) \geq 2^{n-2} + 1$ *for all* $n \geq 3$.

Proof. For $i = 0, 1, \ldots, n-2$, let T_i be a set of $\binom{n-2}{i}$ points in general position in the plane, containing no $(i+2)$-cap and no $(n-i)$-cup and having the property that no two points in the set are connected by a line having slope of absolute value greater than 1. (The existence of T_i follows from Theorem 2, and the condition on slopes can be satisfied by "compressing" T_i vertically.) Place a small copy of T_i in a neighborhood of the point on the unit circle making an angle of $\frac{\pi}{4} - \frac{i\pi}{2(n-2)}$ with the positive x-axis. Let X be the union of $T_1, T_2, \ldots, T_{n-2}$. Clearly, $|X| = 2^{n-2}$.

Suppose that Y is a convexly independent subset of X. Let k and ℓ be the smallest and the largest values of i so that $Y \cap T_i \neq 0$. If $k = \ell$, then Y contains no $(k+2)$-cap and no $(n-k)$-cup. The construction guarantees that:

1. $Y \cap T_k$ is a cap of at most $k+1$ points,
2. $Y \cap T_\ell$ is a cup of at most $n - \ell - 1$ points,
3. $|Y \cap T_i| \leq 1$ for all $i = k+1, k+2, \ldots, \ell - 1$.

Summing up, $|Y| \leq (k+1) + (l-k-1) + (n-\ell-1) = n-1$. $\square$

A 16-point set with no convex hexagon and with an appealing circular symmetry is shown in the book of Bokowski [6, p. 283]. It would be interesting to see this example generalized to larger sets with 2^{n-2} points and no convex n-gon.

2.2 Improvements of the Upper Bounds on $N(n)$

The first improvement of the inequality $N(n) \leq \binom{2n-4}{n-2} + 1$ was made in 1998 by Chung and Graham [11], who reduced the number $\binom{2n-4}{n-2} + 1$ by one. This was followed shortly afterward by Kleitman and Pachter's [30] improvement to $N(n) \leq \binom{2n-4}{n-2} - 2n + 7$. A key new idea of a projectively transformed set led Tóth and Valtr [44] to lower the bound further to $N(n) \leq \binom{2n-5}{n-2} + 2$. In 2005, the same authors proved the following result.

Theorem 4 ([45]). $N(n) \leq \binom{2n-5}{n-2} + 1$ *for all* $n \geq 5$.

Proof. Let X be a set of $\binom{2n-5}{n-2} + 1$ points in general position in the plane. Choose an extreme point p of $\operatorname{conv} X$, and let q be a point outside $\operatorname{conv} X$ such that no line determined by a pair of points in $X \setminus \{p\}$ meets the segment $[p, q]$. Also, choose a line L through q disjoint from $\operatorname{conv} X$.

Let T be a projective transformation that maps L to the line at infinity and also maps the segment $[p, q]$ to the vertical halfline emanating downward from $T(p)$. Clearly, the set $T(X)$ has the following properties:

1. a subset Y of X is convexly independent if and only if $T(Y)$ is,
2. a subset Z of X containing p is convexly independent if and only if $T(Z \setminus \{p\})$ is a cap.

For the rest of the proof, assume that $T(X \setminus \{p\})$ does not contain any $(n-1)$-cap or n-cup. Let Y be the set of points in $T(X \setminus \{p\})$ that are right endpoints of $(n-2)$-caps. If $|Y| > \binom{2n-6}{n-3}$ then Y contains an $(n-1)$-cap or an $(n-1)$-cup. Since the first case contradicts the assumption, Y contains an $(n-1)$-cup. The left endpoint of this cup is the right endpoint of an $(n-2)$-cap. By Lemma 1, $T(X \setminus \{p\})$ would have an $(n-1)$-cap or an n-cup, and we are assuming that neither of these may occur. If $|T(X \setminus \{p\}) \setminus Y| > \binom{2n-6}{n-2}$, then $T(X \setminus \{p\}) \setminus Y$ contains an $(n-2)$-cap or an n-cup. The first of these cannot happen because Y contains all the right endpoints of $(n-2)$-caps.

Summing up, we have established that

$$|Y| = \binom{2n-6}{n-3} \quad \text{and} \quad |T(X \setminus \{p\}) \setminus Y| = \binom{2n-6}{n-2}.$$

If $t \in T(X \setminus \{p\}) \setminus Y$, then $Y \cup \{t\}$ contains $\binom{2n-6}{n-3} + 1$ points; so it contains an $(n-1)$-cup. If the left endpoint of this cup is in Y, then we have an $(n-1)$-cap or n-cup in $T(X \setminus \{p\})$ by Lemma 1. Therefore, every element of $T(X \setminus \{p\}) \setminus Y$ is the left endpoint of an $(n-1)$-cup with right endpoint in Y. This implies, by the same line of reasoning, that every element of Y is the right endpoint of an $(n-2)$-cap with left endpoint in $T(X \setminus \{p\}) \setminus Y$.

Let S be the set of all segments $[v, v']$, where $v \in T(X \setminus \{p\}) \setminus Y$ and $v' \in Y$, and there is an $(n-2)$-cap or $(n-1)$-cup with left endpoint v and right endpoint v'.

Assume that a segment $[v, v']$ from S has the largest slope. Suppose $[v, v']$ represents an $(n-1)$-cup with elements $v = v_1, v_2, \ldots, v_{n-1} = v'$ listed in order from left to right. There is an $(n-2)$-cap with right endpoint v' and left endpoint in $T(X \setminus \{p\}) \setminus Y$. Let $u_1, u_2, \ldots, u_{n-2}$ be its points listed from left to right. If u_{n-3} lies above the line through v_1, v_2, then $\{u_{n-3}, v_1, v_2, \ldots, v_{n-1}\}$ is a convexly independent set. Otherwise, $\{u_1, u_2, \ldots, u_{n-3}, v_1, v_2\}$ is an $(n-1)$-cap in $T(X \setminus \{p\})$. The argument for the case where $[v, v']$ represents an $(n-2)$-cap is similar to the previous case. □

Strunk [42] proves some improved bounds for point sets satisfying some conditions. For example, he proves that for $n \geq 4$, any set of $\binom{2n-5}{n-2} - 2n + 10$ points in general position in the plane with an $(n-1)$-gon as its convex hull contains a convex n-gon.

2.3 The Values of $N(n)$ for Small n

Esther Klein's observation (see Sect. 1) shows that $N(4) = 5$. The next value, $N(5) = 9$, was first published in 1970 by Kalbfleisch et al. [26]. The proof below is due to Bonnice [7].

We will say that a set of nine points X in general position in the plane is of type (k_1, k_2, k_3) if the set X_1 of vertices of conv X has k_1 elements, the set X_2 of vertices of conv $(X \setminus X_1)$ has k_2 elements, and $|X \setminus (X_1 \cup X_2)| = k_3$ (see Fig. 4).

Suppose a set of points $Q = \{x, y, z, w\}$ is convexly independent and the points are encountered in the order x, y, z, w as one traverses the convex hull of Q. The convex polygonal region bounded by the halflines $[y, x\rangle$, $[z, w\rangle$ and the segment $[y, z]$ is called the *beam* $yz : xw$ (see Fig. 5). Note that the beam $yz : xw$ is bounded if the halflines $[y, x\rangle$ and $[z, w\rangle$ meet. Similarly, for non-collinear points x, y, w, the convex cone bounded by the halflines $[y, x\rangle$ and $[y, w\rangle$ is called a beam and is denoted $y : xw$.

Suppose that X is of type $(3, 3, 2)$, and let $X_3 = \{z_1, z_2\}$. There are two points of $X_2 = \{v_1, v_2, v_3\}$, say v_1 and v_2, on the same side of the line containing X_3. We

Fig. 4 A set of type (4,3,2)

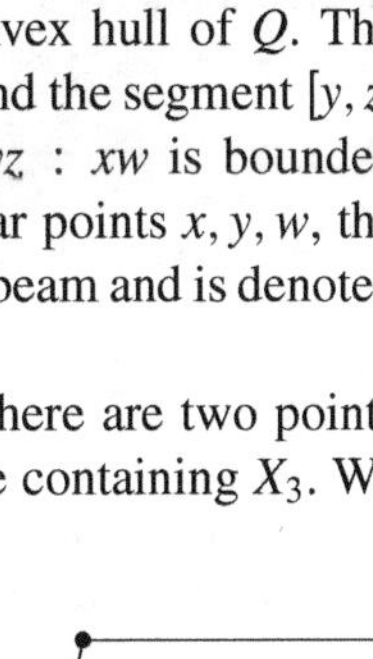

Fig. 5 A beam $yz : xw$

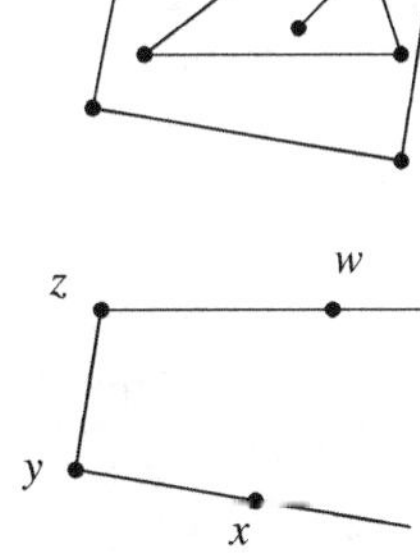

can assume that the halfline $[z_1, z_2\rangle$ meets the segment $[v_1, v_3]$. If there is a point of $X_1 = \{y_1, y_2, y_3\}$ in the beam $z_1z_2 : v_1v_2$, then this point, together with z_1, z_2, v_1, v_2, forms a convex pentagon. Otherwise, we can assume that there are two points of X_1, say y_1 and y_2, that are both in one of the beams $z_1 : v_2v_3$ and $z_2 : v_1v_3$. In either case, y_1 and y_2, together with the three points defining the beam, form a convexly independent set of five points.

Suppose that X is of type $(4, 3, 1)$. Let $X_3 = \{z\}$. Then one of the three beams formed by z and two points of X_2 must contain two points of X_1. As in the previous case, these two points, together with the three points defining the beam, form a convex pentagon.

Finally, suppose that X is of type $(3, 4, 2)$. Put $X_3 = \{z_1, z_2\}$, and let the points of X_2 be v_1, v_2, v_3, v_4, appearing in the order v_1, v_2, v_3, v_4 as one traverses the boundary of $\operatorname{conv} X_2$. If the line through z_1 and z_2 has one point of X_2, say v_1, on one side and the other three points of X_2 on the other side, then the points z_1, z_2, v_2, v_3, v_4 form a convex pentagon. Thus we may assume that v_1 and v_2 are on one side of the line $\langle z_1, z_2\rangle$, while v_3 and v_4 are on the other side. We let the halfline $[z_1, z_2\rangle$ meet the segment $[v_2, v_3]$. If the beam $z_1z_2 : v_1v_2$ or the beam $z_1z_2 : v_3v_4$ contains a point of X_1, then this point, together with the four defining the beam, forms a convex pentagon. Otherwise, there are two points of X_1 in one of the beams $z_1 : v_1v_4$ and $z_2 : v_2v_3$. In both cases, the two points of X_1 together with the three points defining the beam yield a convex pentagon.

Theorem 5. $N(5) = 9$.

Proof. Let X be a set of 9 points in general position in the plane. If $\operatorname{conv} X$ has less than 5 points, it must be of one of the types

$$(4, 4, 1),\ \ (4, 3, 2),\ \ (3, 4, 2),\ \ (3, 3, 3).$$

Clearly, each set of the first two types contains a subset of type $(4, 3, 1)$. The third type, $(3, 4, 2)$, has been considered above, and a set of type $(3, 3, 3)$ contains a subset of type $(3, 3, 2)$. In every case, X contains a subset of 5 convexly independent points. □

The main virtue of this proof is its brevity. A similar approach was used by Dehnhardt et al. [14] to show, in seventeen pages, that if X is a set of 17 points in general position in the plane and $\operatorname{conv} X$ has 5 vertices, then X contains 6 convexly independent points. The rapid increase in the number of cases seems to limit the method of classifying point sets by the sequence of sizes of the nested convex hulls. This method, however, was surprisingly effective in the solution of the empty hexagon problem discussed in Sect. 3.

The next value of $N(n)$, namely $N(6) = 17$, follows from a much stronger result of Szekeres and Peters [43], described below.

Suppose that $X = \{p_1, p_2, \ldots, p_m\}$ is a set in general position in the plane such that the x-coordinate of p_i is less than the x-coordinate of p_j whenever $i < j$. Define a function $\sigma_X : \binom{[m]}{3} \to \{-1, 1\}$ by $\sigma_X(\{i, j, k\}) = 1$ if $i < j < k$ and p_j is above the line through p_i and p_j, and $\sigma_X(\{i,j,k\}) = -1$ otherwise. A function $\sigma : \binom{[m]}{3} \to \{-1,1\}$ will be called *realizable* if there is a planar set X in general position such that $\sigma = \sigma_X$.

The geometric meaning of this approach can be explained as follows: if a function $\sigma : \binom{[m]}{3} \to \{-1,1\}$ is realizable, then for a set

$$I = \{i_1, i_2, \dots, i_n\} \subseteq [m] \equiv \{1,2,\dots,m\}$$

and sets $\{i_1, i_n\} \subseteq A \subseteq I$ and $B = (I \setminus A) \cup \{i_1, i_n\}$, expressed as

$$A = \{i_1 = a_1 < a_2 < \cdots < a_k = i_n\}, \quad B = \{i_1 = b_1 < b_2 < \cdots < b_{n-k+2} = i_n\},$$

the equality

$$\sum_{i=1}^{k-2} \sigma(\{a_i, a_{i+1}, a_{i+2}\}) - \sum_{j=1}^{n-k} \sigma(\{b_i, b_{i+1}, b_{i+2}\}) = n-2 \qquad e(I,A)$$

implies the existence of a convex n-gon made up of an $|A|$-cap indexed by elements of A and a $|B|$-cup indexed by elements of B.

Denote by $\tilde{N}$ the smallest integer m such that there is no function $\sigma : \binom{[m]}{3} \to \{-1,1\}$ satisfying the following condition: the inequality

$$\sum_{i=1}^{k-2} \sigma(\{a_i, a_{i+1}, a_{i+2}\}) - \sum_{j=1}^{n-k} \sigma(\{b_i, b_{i+1}, b_{i+2}\}) < n-2 \qquad i(I,A)$$

holds for all $I = \{i_1, i_2, \dots, i_n\} \subseteq [m]$ and $\{i_1, i_n\} \subseteq A \subseteq I$. We will denote by $i(I)$ the conjunction of all inequalities $i(I,A)$ for a given set I.

Theorem 6 ([43]). $\tilde{N}(4) = 5$.

Proof. This tiny argument from [43] can serve as a model for later computer searches. Suppose that a function $\sigma : \binom{[5]}{3} \to \{-1,1\}$ satisfies the inequality $i(I,A)$ for all

$$I = \{i_1, i_2, i_3, i_4\} \subseteq [5] \quad \text{and} \quad \{i_1, i_4\} \subseteq A \subseteq I.$$

Assume, without loss of generality, that $\sigma(\{1,2,3\}) = 1$. From the obvious inequality $i(\{1,2,3,5\},\{1,2,3,5\})$ it follows that $\sigma(\{2,3,5\} = -1$. Then

$$i(\{2,3,4,5\},\{2,4,5\}) \Rightarrow \sigma(\{2,4,5\}) = -1,$$

$$i(\{1,2,4,5\},\{1,5\}) \Rightarrow \sigma(\{1,2,4\}) = 1,$$

$$i(\{1,2,3,4\},\{1,2,4\}) \Rightarrow \sigma(\{1,3,4\}) = 1,$$

$$i(\{1,3,4,5\},\{1,3,4,5\}) \Rightarrow \sigma(\{3,4,5\}) = -1,$$
$$i(\{2,3,4,5\},\{2,5\}) \Rightarrow \sigma(\{2,3,4\}) = 1.$$

Now $i(\{1,2,3,4\},\{1,2,3,4\})$ implies $\sigma(\{1,2,3\}) = -1$, contradicting our original assumption. □

By a computer search, Szekeres and Peters established the following statement.

Theorem 7 ([43]). $\tilde{N}(5) = 9$.

It should be emphasized that Theorem 7 is much stronger than Theorem 5, because it holds for the class of all functions $\sigma : \binom{[9]}{3} \to \{-1,1\}$, which is much larger than the class of realizable functions $\sigma_X : \binom{[9]}{3} \to \{-1,1\}$.

Szekeres and Peters raise the fascinating possibility that $\tilde{N}(n) = 2^{n-2}+1$ might hold for all n. This would not contradict the lower bound on the Ramsey number $R_3(n,n)$, as discussed in Sect. 2.4. Indeed, the equality $e(I,A)$ for some $A \subseteq I$ determines the values of $\sigma(\{i,j,k\})$ for all $\{i,j,k\} \subseteq I$ when the function σ is realizable, while this is not the case for non-realizable functions σ.

Szekeres and Peters were not able to determine whether $\tilde{N}(6) = 17$, because they could not utilize the constraints $i(I,A)$ efficiently enough to force a contradiction within the time they had allotted for the computation. This led them to the introduction of the 4-realizability constraints, which hold for realizable functions σ.

For a given $\sigma : \binom{[m]}{3} \to \{-1,1\}$ and a 4-element set $J = \{i_1 < i_2 < i_3 < i_4\} \subseteq [m]$, a realizable σ must satisfy either

$$\sigma(\{i_1,i_2,i_3\}) = \sigma(\{i_1,i_2,i_4\}) \quad \text{and} \quad \sigma(\{i_1,i_3,i_4\}) = \sigma(\{i_2,i_3,i_4\})$$

if the points indexed by i_1, i_2, i_3, i_4 are convexly independent, or

$$\sigma(\{i_1,i_2,i_3\}) = -\sigma(\{i_2,i_3,i_4\}) \quad \text{and} \quad \sigma(\{i_1,i_2,i_4\}) = \sigma(\{i_1,i_3,i_4\})$$

if the point indexed by i_2 or i_3 is in the convex hull of the other three. The function satisfies one or the other of these if and only if the following system $i_4(J)$ of inequalities holds:

$$\begin{aligned} -3 &< \sigma(\{i_1,i_2,i_4\}) - \sigma(\{i_1,i_3,i_4\}) + \sigma(\{i_2,i_3,i_4\}) < 3, \\ -3 &< \sigma(\{i_1,i_2,i_3\}) - \sigma(\{i_1,i_2,i_4\}) + \sigma(\{i_2,i_3,i_4\}) < 3. \end{aligned} \qquad i_4(J)$$

If σ satisfies the condition $i_4(J)$ for all $J \subseteq [m]$, we say that σ is 4-*realizable*.

We define $N^*(n)$ as the smallest integer m such that there is no 4-realizable function $\sigma : \binom{[m]}{3} \to \{-1,1\}$ so that all inequalities $i(I,A)$ hold. These additional restrictions were enough to allow Szekeres and Peters to establish the following result.

Theorem 8. $N^*(6) = 17$.

Theorem 8 is equivalent to the statement that the generalization of the conjecture $N(6) = 17$ to rank 3 uniform oriented matroids, put forward by Goodman and Pollack [21], is true.

We will say a few words about the strategy for the computer proof. A function $\sigma : \binom{[m]}{3} \to \{-1, 1, *\}$ will be called a *partial function*. We say that a partial function σ is *consistent* if for each pair (I, A) the inequality $i(I, A)$ holds whenever the restriction of σ to the variables involved in the inequality has range contained in $\{-1, 1\}$. A partial function σ' *agrees with* σ if $\sigma'(\tau) = \sigma(\tau)$ whenever $\sigma(\tau) \in \{-1, 1\}$. We say that a partial function σ is *feasible* if there is a consistent function σ' with range $\{-1, 1\}$ that agrees with σ. The constant function σ_0 that sends every triple to $*$ is consistent, and the goal is to show that it is not feasible.

In the proof of Theorem 6 above, we started with the constant partial function σ_0, and picked a function σ' that agrees with it, by changing $\sigma_0(\{1, 2, 3\})$ to 1. The subsequent derivation established that σ' was infeasible. If we went on to show that the function σ'' obtained from σ_0 by changing $\sigma_0(\{1, 2, 3\})$ to -1 is infeasible, then we would have established that σ_0 is infeasible. This last step is not necessary to verify, by symmetry.

Szekeres and Peters determined that there are only 892 possible restrictions of a function $\sigma : \binom{[m]}{3} \to \{-1, 1\}$ to the set of triples in a 6-element subset I of $[m]$ that are 4-realizable and satisfy the system $i(I)$. (Compare 892 to the cardinality $2^{\binom{6}{3}}$ of the family of all $\sigma : \binom{I}{3} \to \{-1, 1\}$). Their computer search alternates between finding, for a given consistent partial function σ, a partial function σ' that agrees with both σ and the restriction of a 4-realizable function satisfying $i(I)$ to a contiguous subset $I = \{i_1, i_1 + 1, \ldots, i_1 + 5\}$ of [17], and determining if the σ' obtained is consistent. Three independent implementations of this strategy, each differing slightly in the details, were made, one each by Szekeres, Peters, and by Brendan McKay. Each arrived at the conclusion that there is no feasible function $\sigma : \binom{[17]}{3} \to \{-1, 1\}$. Subsequent work by Koshelev [32] (see Sect. 3) independently verified the result.

The work of Szekeres and Peters leaves us with the challenges:

1. Find a noncomputer proof that $\tilde{N}(5) = 9$ holds,
2. Find a proof (computer or otherwise) that $\tilde{N}(6) = 17$ holds.

Lemma 1 can be applied directly to show that $\tilde{N}(n) \leq \binom{2n-4}{n-2} + 1$, as was pointed out in [18]. It is not obvious that the improved arguments from [44] can be adapted to show that $\tilde{N}(n)$ is bounded by any smaller function of n.

2.4 Ramsey Theory

The first proof of the existence of $N(n)$ involved rediscovering Ramsey's theorem. The bounds given by Ramsey's Theorem tend to be much larger than the ones given in previous sections.

Theorem 9 ([39]). *For any positive integers $k_1, \ell_1, \ell_2, \ldots, \ell_r$, there exists a smallest positive integer m_0 satisfying the following condition. For any integer $m \geq m_0$, if the k-element subsets of $[m]$ are colored with colors $1, 2, \ldots, r$, then there exists an i, $1 \leq i \leq r$, and an ℓ_i-element subset $T \subset [m]$ so that each of the k-element subsets of T is i-colored.*

The smallest number m_0 for which the conclusion of Ramsey's theorem holds is usually denoted by $R_k(\ell_1, \ell_2, \ldots, \ell_r)$ or, if all of the ℓ_i are equal to ℓ, $R_k(r; \ell)$.

Theorem 10. *For any positive integer $n \geq 3$, the number $N(n)$ exists and*

$$N(n) \leq \min\{R_4(n, 5), R_3(n, n)\}.$$

Proof. Suppose X is a set of $R_4(n, 5)$ points in general position in the plane. Color a subset of four points from X with color 1 if the four points are convexly independent, and with color 2 otherwise. Because $N(4) = 5$, there is no set of 5 points with all 4-element subsets colored with color 2. Therefore, there is a set of n points of X with every 4-element subset convexly independent. It is easy to see (using a planar version of Carathéodory's theorem [10]) that this n-element set must itself be convexly independent. Summing up, $N(n) \leq R_4(n, 5)$.

For the second inequality, let $X = \{p_1, p_2, \ldots, p_m\}$ be a set of points in general position in the plane, with $m \geq R_3(n, n)$. Color a 3-element subset $\{p_i, p_j, p_k\} \subset X$, $i < j < k$, with color 1 if one encounters the points in the order (p_i, p_j, p_k) by passing clockwise around their convex hull. Color the subset with color 2 otherwise. An n-element subset of X is convexly independent if all of its 3-element subsets are colored with the same color. Hence $N(n) \leq R_3(n, n)$. □

The first statement of Theorem 10 gives Szekeres' original proof of the existence of $N(n)$ for all $n \geq 3$, while the second statement is due to Tarsy (see [33]). It seems that Tarsy's proof is most relevant for the evolution of the theory. Another way to see that there is no set of five points with all 4-element subsets colored with color 2 in Szekeres' proof is that the no two of the six line segments formed by pairs of vertices in a 4-element set of color 2 cross. A set of five points with all 4-element subsets of color 2 would therefore yield a planar drawing of the complete graph K_5. This proof that $N(4) = 5$ was shown to us by H. Tverberg (private communication).

The particular set of n convexly independent points guaranteed by the monochromatic set of triples depends on the ordering of the elements of X. If the points of X are listed in order of increasing x coordinate, then the 3-element subsets of an n-element subset Y of X are all of color 1 (color 2) if and only if the points of Y form an n-cap (n-cup). Thus $|X| \geq R_3(n, n)$ implies the existence of an n-cup or an n-cap, but also many other convexly independent sets.

The best known bounds on $R_3(n, n)$ are $2^{bn^2} \leq R_3(n, n) \leq 2^{c^n}$ for some constants b and c (see [12]).

One reason why the number $R_3(n, n)$ is so much larger than $N(n)$ is the requirement of a monochromatic clique in the hypergraph. For a realizable set, there are sparse monochromatic subhypergraphs which, together with the 4-realizability conditions, guarantee the existence of a monochromatic clique in the 3-regular hypergraph. Szekeres and Peters [43] looked for a solution to an equality $e(I, A)$, as shown in the previous section. The special cases $A = I$ and $A = \{i_1, i_n\}$ correspond to minimal evidence for existence of an n-cup and an n-cap. The other values of A involve triples for which σ is both 1 and -1, so they do not fit so neatly into the Ramsey theory framework in which monochromatic subsets of triples are sought.

Fox et al. [18] pursued this direction further, going beyond the planar case and allowing more than 2 colors. For a sequence of positive integers $j_1 < j_2 < \cdots < j_n$, the k-tuples

$$(j_i, j_{i+1}, \ldots, j_{i+k-1}), \quad i = 1, 2, \ldots, n-k+1,$$

are said to form a *monotone path* of length n. Note that the case $k = 3$ gives the collection of triples appearing in $e(I, I)$ and $e(I, \{j_1, j_n\})$ for $I = \{j_1, j_2, \ldots, j_n\}$. Let $N_k(q, n)$ be the smallest integer N such that for every coloring of the edges of the complete k-regular hypergraph on N vertices there is a monochromatic monotone path of length n. It is proved in [18] that

$$2^{(n/q)^{q-1}} \leq N_3(q, n) \leq 2^{n^{q-1} \log n}.$$

One should also mention that paths in 3-regular hypergraphs, without the monotonicity requirement, have also been studied. The paper of Haxell et al. [23] shows that only a linear number in n of points is needed to guarantee the existence of a monochromatic sequence $(\{j_i, j_{i+1}, j_{i+2}\})$, $i = 1, 2, \ldots, n-2$. However, the absence of the condition $j_1 < j_2 < \cdots < j_n$ means that this type of sequence is not sufficient to guarantee a convex n-gon in the realizable case.

Johnson's [25] Ramsey-theoretic proof colors a 3-element subset S of a planar point set X in general position with color 1 if there is an even number of points of X in the interior of conv S, and with color 2 otherwise. Again it is easy to show that a complete monochromatic hypergraph with n vertices corresponds to a convexly independent n-element subset of X.

The paper of Alon et al. [1] considered colorings of ordered triples rather than unordered triples. It defines a *happy-end space* to be a set S along with a function $f : S \times S \times S \to \{1, -1\}$ such that $f(x, y, z) = f(y, x, z)$ for all x, y, z. A subset C of S is called convex if and only if for every subset B of C such that $|B| > 2$ and every point x of B there is a $y \neq x$ in B so that $f(x, y, z)$ is constant on $B \setminus \{x, y\}$. Every set S of points in general position in the plane where no two have the same first coordinate defines a happy-end space by $f(x, y, z) = 1$ if point z lies above the line xy and $f(x, y, z) = -1$ if point z lies below the line xy. Here a set is convex if and only if it consists of vertices of a convex polygon. Alon et al. show that if S is a happy-end space with at least $R_3(n; 8)$ points, then it contains a convex n-set. They also show that for every positive integer n there is a happy-end space on $2^{\Omega(n^2)}$ points that contains no convex set of n points.

3 The Erdős Problem on Empty Convex Polygons

In 1978 Erdős [15] posed a new problem on convex polygons.

Problem 2. For any positive integer $n \geq 3$, determine the smallest positive integer $H(n)$, if it exists, such that any finite set X of at least $H(n)$ points in general position in the plane contains n points which are the vertices of an empty convex polygon, i. e., a polygon whose interior does not contain any point of X.

Bukh and Matoušek [9] express the suspicion that Problem 2 is a substantially harder kind of problem than Problem 1. We will see that known values of $N(n)$ serve as tools for bounding corresponding values $H(n)$.

Trivially, $H(3) = 3$, and the argument of Esther Klein (see Fig. 1) shows that $H(4) = 5$. The next value of $H(n)$ was determined by Harborth [22].

Theorem 11 ([22]). $H(5) = 10$.

The argument of Harborth is as follows. The inequality $H(5) \geq 10$ immediately follows from Fig. 6, where a set of nine points in general position determines no empty convex pentagon (there are still two convex pentagons, neither being empty).

For the opposite inequality, choose a finite set X of n (≥ 10) points in general position in the plane, then, using Theorem 5, one can find in X_n a vertex set $F = \{p_1, p_2, p_3, p_4, p_5\}$ of a convex pentagon $P = \operatorname{conv} F$. Among all such pentagons, there is at least one whose interior contains at most one point from X. If no such point exists in the interior of P, then F is the desired set. Assuming that the interior of P contains exactly one point from X, all other points of X should be outside P. A carefully study of possible placements of the latter points reveals the existence of five points from X which are the vertices of an empty convex pentagon, which results in the inequality $H(5) \leq 10$.

In 1983 Horton [24] showed that $H(n)$ does not exist for all $n \geq 7$. Valtr [46] gave a simple inductive construction of point sets, called *Horton sets*, that do not contain empty 7-gons. The empty set and a one-point set are Horton sets. A Horton set H is in general position in the plane, with distinct x-coordinates. Furthermore, H can be partitioned into two sets A and B such that

1. Each of A and B is a Horton set.

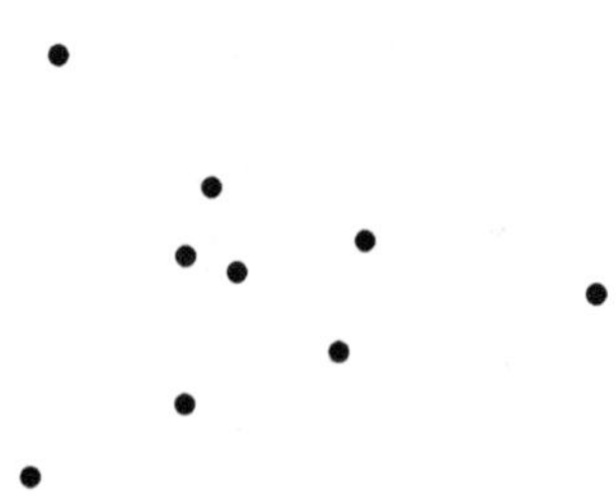

Fig. 6 A set of nine points with no empty convex pentagon

Fig. 7 A Horton set

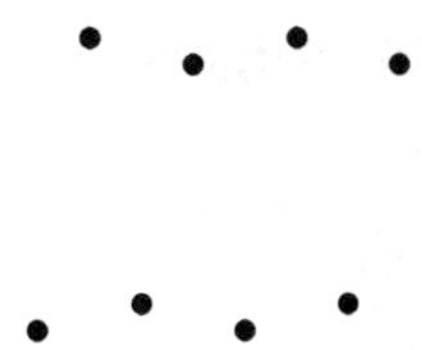

2. The set A is below any line connecting two points of B, and the set B lies above any line connecting two points of A.
3. The x-coordinates of the points of A and B alternate.

It follows by induction on n that H contains no empty 7-gon. If a 7-gon P in H contained points from both A and B, then one of $|P \cap A|$ and $|P \cap B|$ would be at least four. One could then find a point of H in the interior of P (Fig. 7).

The existence of $H(6)$ was a major open problem for many years. The lower bound, $H(6) \geq 30$, was established by Overmars [38], whose computer program produced a set of 29 points in the plane with no empty convex 6-gon. The finiteness of $H(6)$ was established independently in 2005 by Gerken [19] and Nicolás [37]. The proof of [37] shows that $H(6) \leq N(25)$, while that of [19] gives a better inequality $H(6) \leq N(9)$.

A more easily readable manuscript of Valtr [48] simplified Gerken's proof and provided a key lemma to describe point sets with no empty 6-gon. A convex h-gon H with vertices in X is called *minimal* if there is no convex h-gon Y, with vertices in X, contained in H and distinct from H. Let H be a minimal h-gon in X, and let I be the intersection of X with the interior of $\operatorname{conv} H$. Denote by J the intersection of X with the interior of $\operatorname{conv} I$. Valtr's lemma then states that if $h \geq 7$ and X contains no empty 6-gon, then there are no points of X in the interior of $\operatorname{conv} J$. Combined with the bound on $N(8)$ obtained by Tóth and Valtr [45], this argument gives the inequality $H(6) \leq 463$.

This result was followed by the improvement of Koshelev [31], who showed that $H(6) \leq \max\{N(8), 400\}$. The proof of Koshelev investigates each of the possible triples (h, i, j) of cardinalities of the boundaries of the above sets H, I, J with $h = 8$, and in each case shows that either X has an empty 6-gon or $|X| \leq 400$.

In this regard, we observe that the approach of Szekeres and Peters [43] to prove that $N(6) = 17$ was followed by Koshelev [32]. He proves that every set X of 17 points contains a convex 6-gon with at most 2 points of X in its interior. He also gives an example of a 17-point set X which does not have a convex 6-gon with at most one point of X in its interior, and proves that every 18-point set X contains a convex 6-gon with at most one point of X in its interior. One can speculate on the existence of a number $H(n, k)$ for which every point set X in general position in the plane, with at least $H(n, k)$ points, contains the vertex set of a convex n-gon with at most k points of X in its interior. Some results for $n = 7$ were obtained by Sendov [40].

4 Higher Dimensional Extensions

This section deals with a generalization of the Erdős-Szekeres problem (as well as the problem on empty convex polygons) for higher dimensions.

4.1 Convex Polytopes

An observation that Problem 1 can be generalized for higher dimensions was already mentioned by its authors (see [16]) and later rediscovered by Grünbaum [20, pp. 22–23]. We recall that a set X of points in the Euclidean space E^d is in *general position* if no $d + 1$ points of X lie in a hyperplane. Furthermore, X is said to be *convexly independent* if no point of X lies in the convex hull of the remaining points. In other words, a set X in general position is convexly independent provided it is the vertex set of a convex d-polytope in E^d.

Following Grünbaum [20], we define $N_d(n)$, $n \geq d + 1$, as the smallest positive integer such that any set of $N_d(n)$ points in general position in E^d contains a subset of n convexly independent points. Similarly to the planar case, one can pose the following two questions.

1. Do the numbers $N_d(n)$ exist for all $n \geq d + 1$?
2. If yes, what are the values of $N_d(n)$?

The theorem below shows the existence of $N_d(n)$, establishing an upper bound in terms of Ramsey numbers. It is a direct generalization of the proof of [16] for the planar case.

Theorem 12. *For any positive integer* $n \geq d + 1$, *the number* $N_d(n)$ *exists and*

$$N_d(n) \leq R_{d+2}(n, d + 3).$$

Proof. We observe first that any set $Z \subset E^d$ of $d + 3$ points in general position contains a subset of $d+2$ convexly independent points. Indeed, consider the convex hull P of Z and denote by V the vertex set of P. Clearly, $V \subset Z$. Since the case $|V| \geq d + 2$ is obvious, we may assume that $|V| = d + 1$, that is, P is a d-simplex. Let u, v be the points from $Z \setminus V$. The line l through u, v meets at most two facets of P. If Q is a facet of P disjoint from l, then the union of $\{u, v\}$ and the d vertices of Q form a desired convexly independent subset of Z of cardinality $d + 2$.

Next, we state that a finite set $F \subset E^d$ in general position is convexly independent if and only if each of its subsets of $d + 2$ points is convexly independent. The "if part" is trivial, so it suffices to prove the "only if" part. Indeed, if F is not convexly independent, then $p \in \operatorname{conv}(F \setminus \{p\})$ for a certain point $p \in F$. According to Carathéodory's theorem [10], $F \setminus \{p\}$ contains a subset G of $d + 1$ points such that $p \in \operatorname{conv} G$. Consequently, $\{p\} \cup G$ is a convexly dependent subset of F of cardinality $d + 2$.

Finally, let $X \subset \mathrm{E}^d$ be a set in general position consisting of $R_{d+2}(n, d+3)$ points. Color the $(d+2)$-element subsets of X red if the points are convexly independent, and color them blue otherwise. The above argument shows that it is impossible for all of the $(d+2)$-element subsets of a $(d+3)$-element subset of X to be blue. Hence it must be true that X contains an n-element subset Y for which all $(d+2)$-element subsets are red. Therefore Y is convexly independent. Hence $N_d(n) \le R_{d+2}(n, d+3)$. □

Tarsy's Ramsey-theoretic proof from the planar case was generalized in Theorem 9.4.7 of [5]. If a point set $X \subset \mathrm{E}^d$ in general position contains $R_{d+1}(n, n)$ points, then X contains the n vertices of a *cyclic* d-polytope. Thus we know that X not only contains a convexly independent set of n points, but the combinatorial type of the convex hull of this point set is prescribed. Thus one can define a function $N_d^{cyc}(n)$ which is the smallest positive integer such that any set of $N_d^{cyc}(n)$ points in general position in E^d contains the vertex set of a cyclic polytope with n vertices. Clearly, $N_d^{cyc}(n) \ge N_d(n)$ and $N_2^{cyc}(n) = N_2(n)$. The theorem of [5] also states that if a combinatorial type of polytopes has the property that a large enough point set in general position must contain it, then it is a cyclic polytope.

Johnson [25] also pointed out that his Ramsey-theoretic argument for the planar case can be generalized to higher dimensions.

Valtr [47] gave another approach for establishing the existence of $N_d(n)$. He considers any set X of at least $N_2(n)$ points in general position in E^d and its projection Y onto a two-dimensional plane $L \subset \mathrm{E}^d$ such that Y is in general position in L. Since $|Y| \ge N_2(n)$, one can choose in Y a subset of n convexly independent points. It is easily seen that the prototypes of these points in X are convexly independent. This argument implies the inequality $N_d(n) \le N_2(n)$, $d \ge 2$.

A similar consideration is true for the case of projections of X on m-dimensional planes in E^d, $2 < m < d$. Using Theorem 4 and Valtr's argument, one obtains the estimates

$$N_d(n) \le N_{d-1}(n) \le \cdots \le N_2(n) \le \binom{2n-5}{n-3} + 1.$$

Another upper bound on $N_d(n)$ was obtained by Károlyi [28], as shown in the theorem below.

Theorem 13 ([28]). *If $n > d \ge 3$, then $N_d(n) \le N_{d-1}(n-1) + 1$. Consequently,*

$$N_d(n) \le N_2(n-d+2) + d - 2.$$

Proof. Let $X \subset \mathrm{E}^d$ be a set of $N_{d-1}(n-1) + 1$ points in general position. Choose a vertex p of $\operatorname{conv} X$, and denote by H a hyperplane strictly separating p from $X \setminus \{p\}$. Let $Y = \{[p, u] \cap H : u \in X \setminus \{p\}\}$. It is easy to see that the set Y is in general position in H, and the mapping $\varphi : X \setminus \{p\} \to Y$ defined by $\varphi(u) = [p, u] \cap H$ is a bijection. Since H can be identified with E^{d-1}, and since $|Y| \ge N_{d-1}(n-1)$, there is a convexly independent subset Z of Y of size $n-1$. Clearly, the set $\{p\} \cup \varphi^{-1}(Z)$ is a convexly independent subset of X of size n. □

A combination of Theorems 4 and 13 results in the following inequality.

Corollary 2. *If* $n > d \geq 2$, *then* $N_d(n) \leq \binom{2n-2d-1}{n-d} + d - 1$.

In contrast to what is known about the planar case, it is not known if the function $N_d(n)$ is exponential for any fixed $d > 2$. The only known general lower bound for $N_d(n)$ is due to Károlyi and Valtr [29], given in Theorem 14 below.

Theorem 14 ([29]). *For all integers* $N > d \geq 2$, *there exists a configuration of* N *points in general position in* E^d *which contains at most* $c'_d(\log N)^{d-1}$ *convexly independent points, where* c'_d *is a constant depending on d. Equivalently, there is a constant* $c_d > 1$ *such that*

$$N_d(n) = \Omega\left(c_d^{n^{1/(d-1)}}\right).$$

Sketch of the proof of Theorem 14. A set $X \subset \mathrm{E}^d$ is said to be in *strongly general position* if it is in general position and, for every $f = 1, 2, \ldots, e-1$, any $f+1$ points of X determine an f-dimensional affine subspace which is not parallel to the $(e-f)$-dimensional subspace of E^e spanned by its last $e-f$ coordinate axes, $e \leq d$. Denote by $\mathrm{mc}(X)$ the maximum size of a convexly independent subset of X.

Next, we say that two finite sets in general position and of equal size, *have the same order type* if there is a one-to-one correspondence between them which preserves the orientation of each $(e+1)$-tuple. It is clear that small perturbations of the sets do not affect the order type. More precisely, it is possible to show that for every finite set $X = \{p_1, p_2, \ldots, p_t\}$ in general position in E^e, $e \leq d$, there is a largest scalar $\delta = \delta_d(X) > 0$ such that whenever $Y = \{q_1, q_2, \ldots, q_t\} \subset \mathrm{E}^e$ satisfies $|p_i^j - q_i^j| < \delta$ for all $1 \leq i \leq t$ and $1 \leq j \leq e$, then Y also is in general position and has the same order type as X. (Here $p_i = (p_i^1, p_i^2, \ldots, p_i^n)$ and $q_i = (q_i^1, q_i^2, \ldots, q_i^n)$.) In particular, X is convexly independent if and only if Y is.

For a set $X = \{p_1, \ldots, p_t\}$, let

$$\epsilon_e(X) = \min\{\delta_f(\pi_f(X)) : 1 \leq f \leq e\},$$

where π_f is the orthogonal projection of E^d onto E^f, and $\delta_f(\pi_f(X))$ is defined as above. Given a scalar $0 < \epsilon < \epsilon_e(X)$, chose for every point $p \in X$ a vector $v(p) = (v^1(p), \ldots, v^e(p))$ such that

$$0 < v^1(p) < \cdots < v^e(p) \text{ and } v^f(p) < \epsilon v^{f+1}(p) \text{ for all } 1 \leq f < e.$$

These vectors $v(p_1), \ldots, v(p_t)$ can be chosen in such a way that the set $X' = \{p \pm v(p) : p \in X\}$ of size $2|X|$ is in strongly general position; in the latter case X' is called an ϵ-*double* of X. Important properties of ϵ-doubles:

1. If X' is an ϵ-double of X, then $\pi_f(X')$ is an ϵ-double of $\pi_f(X)$ for all $1 \le f \le e$,
2. If X is in strongly general position and $0 < \epsilon \le \epsilon_e(X)$ is small enough, then

$$\mathrm{mc}(X') \le \mathrm{mc}(X) + \mathrm{mc}(\pi_{e-1}(X)).$$

Finally, a desired set is constructed inductively, starting with a singleton $X_0 \subset \mathrm{E}^d$. If for some $i \ge 0$, a set X_i of points in strongly general position is selected, then a very small $\epsilon_i > 0$ and an ϵ_i-double X_i' of X_i can be chosen so that $\pi_e(X_i')$ is an ϵ_i-double of $\pi_e(X_i)$ for all $1 \le e \le d$ and

$$\mathrm{mc}(\pi_e(X')) \le \mathrm{mc}(\pi_e(X_i)) + \mathrm{mc}(\pi_{e-1}(X_i)), \quad 2 \le e \le d.$$

Based on the latter inequality, a double induction on e and i gives

$$\mathrm{mc}(\pi_e(X_i)) \le 2i^{e-1} \quad \text{for all} \quad 1 \le e \le d \quad \text{and} \quad i \ge 1. \quad \square$$

As shown in [28], a more careful calculation yields that

$$\mathrm{mc}(\pi_e(X_i)) \le (2/(e-1)!)i^{e-1} + O(i^{e-2}).$$

Thus, for large n and N, Theorem 14 is valid with $c_d \approx 2^{0.37d}$ and $c_d' \approx 2/(d-1)!$.

There are quite few cases when the exact values of $N_d(n)$ is known. For small values of $n \ge d+1$, we can mention the following equalities, derived by Morris and Soltan [36] from a stronger result of Bisztriczky and Soltan [4] (see Theorem 16 below):

$$N_d(n) = 2n - d - 1 \quad \text{for all} \quad 2 \le d \quad \text{and} \quad d+2 \le n \le \lfloor 3d/2 \rfloor + 1.$$

These equalities show that $N_d(n)$ behaves as a linear function with respect to n for small values of n.

For $n > \lfloor 3d/2 \rfloor + 1$ and $d \ge 3$, there is only one known value: $N_3(6) = 9$, due to Bisztriczky and Soltan [4] (see Theorem 18 below).

Grünbaum [20, pp. 22–23] discussed a variant of the Erdős-Szekeres problem in higher dimensions. Namely, he established the existence of a minimum number $B_d(n)$, $d \ge 2$ and $n \ge d+1$, such that any set $X \subset \mathrm{E}^d$ of at least $B_d(n)$ points in general position contains a subset of n points lying on the boundary of a convex body. His proof is based on Ramsey's theorem and the following assertion: a finite set $Y \subset \mathrm{E}^d$ lies on the boundary of a convex body in E^d if and only if each of its subsets of at most $2d+1$ points lies on the boundary of a convex body in E^d. The last statement is a direct consequence of the Steinitz theorem (see [41, Sect. 10]): a point $p \in \mathrm{E}^d$ belongs to the interior of the convex hull of a set $S \subset \mathrm{E}^d$ if and only if p belongs to the interior of the convex hull of at most $2d$ points of S. Later Bisztriczky and Soltan [4] showed that in the definition of $B_d(n)$ the set $X \subset \mathrm{E}^d$ can be arbitrary (not necessarily in general position); they also proved the equality $B_d(n) = N_d(n)$ for all $d \geq 2$ and $n \geq d + 1$, based on the simple argument that any finite set in E^d can be approximated by a set in general position.

4.2 Empty Convex Polytopes

Generalizing the Erdős problem on empty convex polygons (see Sect. 3), Bisztriczky and Soltan [4] defined $H_d(n)$ as the smallest positive integer, if it exists, such that any set X of at least $H_d(n)$ points in general position in E^d contains a subset of n points that are the vertices of an empty convex polytope, that is, of a polytope whose interior does not contain any point of X. We observe that in this definition one may consider only sets X of precisely $H_d(n)$ points. Indeed, if X contains more than $H_d(n)$ points, we can partition X into subset Y and Z by a suitable hyperplane such that $|Y| = H_d(n)$. Then, if some n points from Y form an empty convex polytope with respect to Y, then this polytope also is empty with respect to X.

The following theorem of Valtr [47] (formulated in slightly distinct terms) shows the existence of numbers $H_d(n)$ for all $d + 1 \leq n \leq 2d + 1$.

Theorem 15. *$H_d(2d + 1) \leq N_d(4d + 1)$ for all $d \geq 2$.*

Proof. Let $X \subset E^d$ be a set of $N_d(4d + 1)$ points in general position. Choose a subset $Y \subset X$ of $4d + 1$ convexly independent points such that the interior of the convex polytope $P = \operatorname{conv} Y$ contains the smallest number of points from X. Let $Z = X \cap \operatorname{int} P$ and $k = |Z|$.

If $k \leq d$, then we consider a hyperplane H containing Z. Since the set $X \setminus Z$ contains at least $4d - k + 1$ points and lies in the union of two open halfspaces determined by H, at least one of these halfspaces contains a subset V of $X \setminus Z$ consisting of at least $\lceil (4d - k + 1)/2 \rceil$ points. It is easy to see that the set $V \cup Z$ is convexly independent, contains at least $\lceil (4d - k + 1)/2 \rceil + k \geq 2d + 1$ points, and is the vertex set of an empty polytope.

Suppose that $k \geq d + 1$ and let $Q = \operatorname{conv} Z$. Choose a facet F of Q and consider the set Y_0 of all points of Y which belong to the open halfspace C determined by the hyperplane H containing F but not Q. We state that $|Y_0| \geq d + 1$. Indeed, assume for a moment that $|Y_0| = m \leq d$. If G is a subset of $F \cap Z$ of m points, then the set $Y' = G \cup (Y \cap C')$, where C' is the second open halfspace determined by H, consists of $4d + 1$ convexly independent points such that the interior of the polytope $\operatorname{conv} Y'$ contains less then k points of X. Since the latter contradicts the choice of Y, we conclude that $|Y_0| \geq d + 1$. This argument shows that $Y_0 \cup (F \cap Z)$ is a convexly independent set of at least $2d + 1$ points which form an empty convex polytope. □

Using an n-dimensional version of Horton sets (see [24]), Valtr [47] gave an example of an arbitrary large set $X \subset E^d$ in general position containing no convexly independent subset of size $2^{d-1}(r_2 r_3 \cdots r_d + 1)$ which determines an empty convex polytope with respect to X (here $r_2 = 2, r_3 = 3, r_4 = 5, \ldots$ is the sequence of prime numbers). Consequently, the numbers $H_d(n)$ do not exist for all $n \geq 2^{d-1}(r_2 r_3 \cdots r_d + 1)$. If $d = 3$, then the latter statement can be improved: $H_3(n)$ do not exist for all $n \geq 23$ (instead of $n \geq 2^2(2 \cdot 3 + 1) = 28$ due to the above argument).

For small values of n, Bisztriczky and Soltan [4] determined a sharp upper bound on $H_d(n)$, formulated here in slightly distinct terms.

Theorem 16 ([4]). *If* $d \geq 2$ *and* $d+1 \leq n \leq \lfloor 3d/2 \rfloor + 1$, *then* $H_d(n) \leq 2n-d-1$.

Proof. Let $m = 2n - d - 1$ and $X \subset \mathrm{E}^d$ be a set of m points in general position. Set $P = \operatorname{conv} X$. Denote by v the number of vertices of the polytope P. Since the case $v = m$ is obvious, we assume that $d + 1 \leq v < m$. Then $v = d + t$, where $1 \leq t < 2n - 2d - 1$. Consequently, $\operatorname{int} P$ contains $m - v = 2n - 2d - t - 1$ points from X. Choose any $3d - 2n + t + 1$ vertices of P (which is possible due to $n \leq \lfloor 3d/2 \rfloor + 1$), and let H be the hyperplane through the selected

$$(2n - 2d - t - 1) + (3d - 2n + t + 1) = d$$

points from X. Denote by Y the set of these d vertices. Clearly, H contains no other point of X. Consequently, $Y = H \cap X$ and the remaining

$$v - (3d - 2n + t + 1) = 2n - 2d - 1$$

vertices of P lie in $\mathrm{E}^d \setminus H$. Hence there is a subset Z of at least $n - d$ vertices of P which belong to an open halfplane determined by H. It is easy to see that the n-element set $Y \cup Z$ is the vertex set of an empty convex polytope. □

Later Bisztriczky and Harborth [3] proved the opposite inequality.

Theorem 17 ([3]). *If* $H_d(n)$ *exists, then* $H_d(n) \geq 2n - d - 1$, *where* $d \geq 2$ *and* $n \geq d + 1$.

Proof. Since $H_d(d + 1) = d + 1$ and $H_d(d + 2) = d + 3$, we may assume that $n \geq d + 3$. Let $X \subset \mathrm{E}^d$ be a set in general position, and $1 \leq t < |X|$ be an integer. We will say that X is *t-fat* provided

$$\cap \big(\operatorname{int}(\operatorname{conv} Y) : Y \subset X,\ |Y| = |X| - t\big) \neq \emptyset.$$

Let $L_d(t)$ denote the smallest integer l such that there exists a t-fat set $X \subset \mathrm{E}^d$ of cardinality l. As shown in [4],

$$L_d(t) = d + 2t + 1 \ \text{ whenever } \ d \geq 2 \ \text{ and } \ t \geq 1.$$

Put $t = n - d - 2$ and choose a t-fat set $X \subset \mathrm{E}^n$ of cardinality $d + 2t + 1$. Then there is a point

$$p \in U \equiv \cap \big(\operatorname{int}(\operatorname{conv} Y) : Y \subset X,\ |Y| = d + t + 1\big).$$

Since U is an open set, we may choose p so that the set $X^* = X \cup \{p\}$ is in general position. Then $|X^*| = d + 2t + 2$ and p belongs to the interior of every convex polytope $\operatorname{conv} Y$, where $Y \subset X^*$ and $|Y| = d + t + 2 = n$. Hence no subset of n points from X^* forms an empty convex polytope. Consequently,

$$H_d(n) \geq |X^*| + 1 = d + 2t + 3 = 2n - d - 1. \quad \square$$

Combining Theorems 16 and 17, we obtain the following equalities:

$$H_d(n) = 2n - d - 1 \quad \text{for all} \quad d \geq 2 \quad \text{and} \quad d + 1 \leq n \leq \lfloor 3d/2 \rfloor + 1.$$

For $d > 2$ and $n > \lfloor 3d/2 \rfloor + 1$, only one value of $H_d(n)$ is known: $H_3(6) = 9$, as shown in the theorem below.

Theorem 18 ([4]). $N_3(6) = H_3(6) = 9$.

Proof. Due to the obvious inequality $N_3(6) \leq H_3(6)$, it suffices to show that $9 \leq N_3(6)$ and $H_3(6) \leq 9$.

Case 1. $9 \leq N_3(6)$. For this, let

$$p_1 = (-64, 0, 0), \ p_2 = (0, 64, 0), \ p_3 = 64, 0, 0), \ p_4 = (0, 0, 128),$$

$$p_5 = \left(-\tfrac{32}{10}, 8 + \tfrac{8}{30}, 24 - \tfrac{8}{30}\right), p_6 = \left(-\tfrac{32}{5}, 8 + \tfrac{8}{15}, 24 - \tfrac{8}{15}\right), p_7 = \left(16, \tfrac{128}{6}, \tfrac{64}{6}\right),$$

and $p_8 = (0, 32(1 - \epsilon), 64(1 - \epsilon))$, where ϵ is an arbitrary small real number. It is a matter of computation to verify that the set $X = \{p_1, p_2, \dots, p_8\}$ is in general position and has the following properties:

1. $\operatorname{conv} X = \operatorname{conv} \{p_1, p_2, p_3, p_4\}$,
2. $\{p_5, p_6\} \subset \operatorname{conv} \{p_1, p_3, p_4, p_7\} \cap \operatorname{conv} \{p_1, p_3, p_4, p_8\}$,
3. $p_5 \in \operatorname{conv} \{p_2, p_3, p_4, p_6\} \cap \operatorname{conv} \{p_3, p_4, p_6, p_7\}$,
4. $p_6 \in \operatorname{conv} \{p_1, p_2, p_5, p_8\} \cap \operatorname{conv} \{p_1, p_3, p_5, p_8\} \cap \operatorname{conv} \{p_1, p_4, p_5, p_7\}$,
5. $p_7 \in \operatorname{conv} \{p_1, p_2, p_3, p_8\} \cap \operatorname{conv} \{p_2, p_3, p_4, p_6\} \cap \operatorname{conv} \{p_2, p_3, p_5, p_6\}$,
6. $p_8 \in \operatorname{conv} \{p_2, p_4, p_5, p_7\} \cap \operatorname{conv} \{p_2, p_4, p_6, p_7\}$.

Consequently, no six-element subset of X is the vertex set of an empty convex polytope.

Case 2. $H_3(6) \leq 9$. Let $X = \{p_1, p_2, \dots, p_9\}$ be any set of nine points in general position in E^3. Put $P = \operatorname{conv} X$, and denote by Y the vertex set of P. Without loss of generality, we assume that $Y = \{p_1, p_2, \dots, p_r\}$, where $4 \leq r \leq 9$. Since the case $r = 9$ is obvious (any six points from Y are convexly independent), we let $r \leq 8$. Put $Z = X \setminus Y$ and $T = \operatorname{conv} Z$.

a) If $r \geq 6$, then we argue as in the proof of Theorem 16 and obtain the existence of an empty polytope with six vertices from X.
b) Let $r = 5$. Then $|Z| = 4$ and T is a 3-simplex. Obviously, $T \subset \operatorname{int} P$.

Suppose first that the line $l = \langle p_i, p_j \rangle$ meets $\operatorname{int} T \neq \emptyset$ for some $p_i \in X$ and $p_j \in Y$. Let $l = \langle p_5, p_6 \rangle$. Then l meets the relative interior of conv $\{p_7, p_8, p_9\}$. Denote by L_i the closed halfplane bounded by l and containing p_i, $i = 7, 8, 9$. It is clear that $E^3 \setminus (L_7 \cup L_8 \cup L_9)$ is the union of three open convex regions, A_1, A_2, A_3, one of them containing at least two points from $\{p_1, p_2, p_3, p_4\}$. Assume that namely A_1 is bounded by L_7, L_8 and contains two points, say p_1, p_2, from $\{p_1, p_2, p_3, p_4\}$. Then, as easy to see, the set $\{p_1, p_2, p_5, p_6, p_7, p_8\}$ generates an empty polytope with six vertices.
Suppose now that

$$\langle p_i, p_j \rangle \cap \operatorname{int} T = \emptyset \text{ for all } p_i \in X \text{ and } p_j \in Y. \tag{1}$$

Translating X on a suitable vector, we may assume that p_6 is the origin of E^3. Since X is in general position, $\{p_7, p_8, p_9\}$ is a basis for E^3. Choosing a new coordinate system, we let $p_7 = (1, 0, 0)$, $p_8 = (0, 1, 0)$, and $p_9 = (0, 0, 1)$. The three coordinate planes decompose E^3 into eight pairwise disjoint open octants $Q(\alpha_1, \alpha_2, \alpha_3)$, where $\alpha_i = \pm 1$ is the sign of the ith coordinate of a point in $Q(\alpha_1, \alpha_2, \alpha_3)$.
Next, let H denote the plane through $\{p_7, p_8, p_9\}$; also, H_1 and H_2 be the opposite open halfspace determined by H such that $p_6 \in H^2$. It is a direct consequence of (1) that

$$X \cap H^1 \subset Q(+,+,-) \cup Q(+,-,+) \cup Q(-,+,+),$$
$$X \cap H^2 \subset Q(-,-,+) \cup Q(-,+,-) \cup Q(+,-,-).$$

Put

$$Q^k(\alpha_1, \alpha_2, \alpha_3) = H^k \cap Q(\alpha_1, \alpha_2, \alpha_3), \;\; k = 1, 2.$$

Then Y lies in the union of six regions,

$$Q^1(+,+,-), \;\; Q^1(+,-,+), \;\; Q^1(-,+,+),$$
$$Q^2(-,-,+), \;\; Q^2(-,+,-), \;\; Q^2(+,-,-).$$

If one of these six regions contains two points, say p_1 and p_2, from $\{p_1, p_2, p_3\}$, then conv$(\{p_1, p_2\} \cup Z)$ is an empty convex polytope with six vertices. Assume this is not the case, and let, for instance,

$$p_1 \in Q^1(+,+,-), \;\; p_2 \in Q^2(-,+,-), \;\; p_3 \in Q^2(+,-,-).$$

In this case, the plane H strictly separates $\{p_9\}$ and $\{p_1, p_2, p_3\}$, which shows that the convex polytope conv $\{p_1, p_2, p_3, p_6, p_7, p_8\}$ is empty.

c) Let $r = 4$. Then $|Z| = 5$. If the polytope $T = \operatorname{conv} Z$ has five vertices, then it is the union of two non-overlapping simplices

$$\operatorname{conv}\{p_5, p_7, p_8, p_9\} \quad \text{and} \quad \operatorname{conv}\{p_6, p_7, p_8, p_9\}.$$

Then the line $l = \langle p_5, p_6 \rangle$ meets the relative interior of the triangle $\operatorname{conv}\{p_7, p_8, p_9\}$.

If T has four vertices, then it is a simplex. In this case, we may assume $p_5 \in \operatorname{int} T$ and whence the line $l = \langle p_5, p_6 \rangle$ still has the above property. In either case, we define the halfplanes L_i, $i = 7, 8, 9$, and argue as in part b). □

Since the number $H_3(6)$ can be viewed as a particular case of $H_d(\lfloor 3d/2 \rfloor + 2)$, we pose the question on the values of $H_d(\lfloor 3d/2 \rfloor + 2)$ for all $d \geq 4$.

References

1. Alon, N., Chiniforooshan, E., Chvatal, V., Genest, F.: Another abstraction of the Erdős-Szekeres happy end theorem. Electron. J. Combin. **17**, no. 1, Note 11, 6 pp. (2010)
2. Bárány, I., Károlyi, G.: Problems and results around the Erdős-Szekeres convex polygon theorem. Lecture Notes in Computer Science, 2098, pp. 91–105, Springer-Verlag, Berlin (2001)
3. Bisztriczky, T., Harborth, H.: On empty convex polytopes. J. Geom. **52**, 25–29 (1995)
4. Bisztriczky, T., Soltan, V.: Some Erdős-Szekeres type results about points in space. Monatsh. Math. **118**, 33–40 (1994)
5. Björner, A., Las Vergnas, M., Sturmfels, B., White, N., Ziegler, G.: Oriented Matroids. Encyclopedia of Mathematics and its Applications **46**, Cambridge University Press, Cambridge (1993)
6. Bokowski, J. G.: Computational Oriented Matroids. Cambridge University Press, Cambridge (2006)
7. Bonnice, W. E.: On convex polygons determined by a finite planar set. Amer. Math. Monthly **81**, 749–752 (1974)
8. Brass, P., Moser, W., Pach, J.: Research Problems in Discrete Geometry. Springer, New York (2005)
9. Bukh, B, Matoušek, J.: Erdős-Szekeres-type statements: Ramsey function and decidability in dimension 1. Duke Math. J. **163**, 2243–2270 (2014)
10. Carathéodory, C.: Über den Variabilitätsbereich der Fourierschen Konstanten von positiven harmonischen Funktionen. Rend. Circ. Mat. Palermo **32**, 193–217 (1911)
11. Chung, F. R. L., Graham, R. L.: Forced convex n-gons in the plane. Discrete Comput. Geom. **19**, 367–371 (1998)
12. Conlon, D., Fox, J., Sudakov, B.: Hypergraph Ramsey numbers. J. Amer. Math. Soc. **23**, 247–266 (2010)
13. Cowling, M.: Obituary: George Szekeres, 29/5/1911–28/8/2005; Esther Szekeres-Klein, 20/2/1910–28/8/2005. Austral. Math. Soc. Gaz. **32**, 221–224 (2005)
14. Dehnhardt, K., Harborth, H., Langi, Z.: A partial proof of the Erdős-Szekeres conjecture for hexagons. J. Pure Appl. Math. Adv. Appl. **2**, 69–86 (2009)
15. Erdős, P.: Some more problems on elementary geometry. Austral. Math. Soc. Gaz. **5**, 52–54 (1978)
16. Erdős, P., Szekeres, G.: A combinatorial problem in geometry. Compositio Math. **2**, 463–470 (1935)
17. Erdős, P., Szekeres, G.: On some extremum problems in elementary geometry. Ann. Univ. Sci. Budapest. Eötvös Sect. Math. **3–4**, 53–62 (1961)

18. Fox, J., Pach, J., Sudakov, B., Suk, A.: Erdős-Szekeres-type theorems for monotone paths and convex bodies. Proc. Lond. Math. Soc. **105**, 953–982 (2012)
19. Gerken, T.: Empty convex hexagons in planar point sets. Discrete Comput. Geom. **39**, 239–272 (2008)
20. Grünbaum, B.: Convex Polytopes. Interscience Publ., New York (1967)
21. Goodman, J. E. Pollack, R.: A combinatorial perspective on some problems in geometry. Congr. Numer. **32**, 383–394 (1981)
22. Harborth, H.: Konvexe Fünfecke in ebenen Punktmengen. Elem. Math. **33**, 116–118 (1978)
23. Haxell, P., Luczak, T., Peng, Y., Rödl, V., Ruciński, A., Skokan, J.: The Ramsey number for 3-uniform tight hypergraph cycles. Combin. Probab. Comput. **18**, 165–203 (2009)
24. Horton, J. D.: Sets with no empty convex 7-gon. Canad. Math. Bull. **26**, 482–484 (1983)
25. Johnson, S.: A new proof of the Erdős-Szekeres convex k-gon result. J. Combin. Theory Ser. A **42**, 318–319 (1986)
26. Kalbfleisch, J. D., Kalbfleisch J. G., Stanton R. G.: A combinatorial problem on convex regions. Congr. Numer. **1**, 180–188 (1970)
27. Kalbfleisch, J. G., Stanton, R. G.: On the maximum number of coplanar points containing no convex n-gons. Utilitas Math. **47**, 235–245 (1995)
28. Károlyi, G.: Ramsey-remainder for convex sets and the Erdős-Szekeres theorem. Discrete Appl. Math. **109**, 163–175 (2001)
29. Károlyi, G., Valtr, P.: Point configurations in d-space without large subsets in convex position. Discrete Comput. Geom. **30**, 277–286 (2003)
30. Kleitman, D., Pachter, L.: Finding convex sets among points in the plane. Discrete Comput. Geom. **19**, 405–410 (1998)
31. Koshelev, V. A.: The Erdős-Szekeres problem. Dokl. Math. **76**, 603–605 (2007)
32. Koshelev, V. A.: Computer solution of the almost empty hexagon problem. Math. Notes **89**, 455–458 (2011)
33. Lewin, M.: A new proof of a theorem of Erdős and Szekeres. Math. Gaz. **60**, 136–138 (1976)
34. Lovász, L.: Combinatorial Problems and Exercises. North-Holland, Amsterdam (1979)
35. Matoušek, J.: Lectures on Discrete Geometry. Springer-Verlag, New York (2002)
36. Morris, W. D., Soltan, V.: The Erdős-Szekeres problem on points in convex position–a survey. Bull. Amer. Math. Soc. **37**, 437–458 (2000)
37. Nicolás, C.: The empty hexagon theorem. Discrete Comput. Geom. **38**, 389–397 (2007)
38. Overmars, M.: Finding sets of points without empty convex 6-gons. Discrete Comput. Geom. **29**, 153–158 (2002)
39. Ramsey, F. P.: On a problem of formal logic. Proc. London Math. Soc. **30**, 264–286 (1930)
40. Sendov, B.: Compulsory configurations in the plane. Fundam. and Prikl. Math. **1**, 491–516 (1995)
41. Steinitz, E.: Bedingt konvergente Reihen und konvexe Systeme. J. Reine Angew. Math. **143**, 128–175 (1913)
42. Strunk, F.: Two upper bounds for the Erdős–Szekeres number with conditions. Discrete Comput. Geom. **49**, 183–188 (2013)
43. Szekeres, G., Peters, L.: Computer solution to the 17-point Erdős-Szekeres problem. ANZIAM J. **48**, 151–164 (2006)
44. Tóth, G., Valtr, P.: Note on the Erdős-Szekeres theorem. Discrete Comput. Geom. **19**, 457–459 (1998)
45. Tóth, G., Valtr, P.: The Erdős-Szekeres theorem: upper bounds and related results. In: Goodman, J. E., Pach, J., Welzl E. (eds.) Combinatorial and Computational Geometry, pp. 557–568, Math. Sci. Res. Inst. Publ., 52, Cambridge Univ. Press, Cambridge (2005)
46. Valtr, P.: Convex independent sets and 7-holes in restricted planar point sets. Discrete Comput. Geom. **7**, 135–152 (1992)
47. Valtr, P.: Sets in R^d with no large empty convex subsets. Discrete Math. **108**, 115–124 (1992)
48. Valtr, P.: On empty hexagons. In: Goodman, J. E., Pach, J., Pollack, R. (eds.) Surveys on Discrete and Computational Geometry, pp. 433–441, Contemp. Math., 453, Amer. Math. Soc., Providence, RI (2008)

Novikov's Conjecture

Jonathan Rosenberg

Abstract We describe Novikov's "higher signature conjecture," which dates back to the late 1960s, as well as many alternative formulations and related problems. The Novikov Conjecture is perhaps the most important unsolved problem in high-dimensional manifold topology, but more importantly, variants and analogues permeate many other areas of mathematics, from geometry to operator algebras to representation theory.

1 Origins of the Original Conjecture

The Novikov Conjecture is perhaps the most important unsolved problem in the topology of high-dimensional manifolds. It was first stated by Sergei Novikov, in various forms, in his lectures at the International Congresses of Mathematicians in Moscow in 1966 and in Nice in 1970, and in a few other papers [85–88]. For an annotated version of the original formulation, in both Russian and English, we refer the reader to [37]. Here we will try instead to put the problem in context and explain why it might be of interest to the average mathematician. For a nice book-length exposition of this subject, we recommend [66]. Many treatments of various aspects of the problem can also be found in the many papers in the collections [38, 39].

For the typical mathematician, the most important topological spaces are smooth manifolds, which were introduced by Riemann in the 1850s. However, it took about 100 years for the tools for classifying manifolds (except in dimension 1, which is trivial, and dimension 2, which is relatively easy) to be developed. The problem is that manifolds have no local invariants (except for the dimension); all manifolds of the same dimension look the same *locally*. Certainly many different manifolds were known, but how can one tell whether or not the known examples are "typical"? How can one distinguish one manifold from another?

J. Rosenberg (✉)
Department of Mathematics, University of Maryland, College Park, MD 20742-4015, USA
e-mail: jmr@math.umd.edu

J.F. Nash, Jr., M.Th. Rassias (eds.), *Open Problems in Mathematics*,
DOI 10.1007/978-3-319-32162-2_11

With big leaps forward in topology in the 1950s, it finally became possible to answer these questions, at least in part. Here were a few critical ingredients:

1. the development of the theory of Reidemeister and Whitehead torsion and the related notion of "simple homotopy equivalence" (see [77] for a good survey of all of this);
2. the theory of characteristic classes of vector bundles, developed by Chern, Weil, Pontrjagin, and others;
3. the notion of cobordism, introduced by Thom [112], who also provided a method for computing it;
4. the Hirzebruch signature theorem $\operatorname{sign}(M) = \langle \mathscr{L}(M), [M] \rangle$ [54], giving a formula for the signature of an oriented closed manifold M^{4k} (this is the algebraic signature of the nondegenerate symmetric bilinear form $(x, y) \mapsto \langle x \cup y, [M] \rangle$ on H^{2k} coming from Poincaré duality), in terms of a certain polynomial $\mathscr{L}(M)$ in the rational Pontrjagin classes of the tangent bundle.

Using just these ingredients, Milnor [74] was able to show that there are at least 7 different diffeomorphism classes of 7-manifolds homotopy equivalent to S^7. (Actually there are 28 diffeomorphism classes of such manifolds, as Milnor and Kervaire [65] showed a bit later.) This and the major role played by items 2 and 4 on the above list[1] came as a big surprise, and showed that the classification of manifolds, even within a "standard" homotopy type, has to be a hard problem.

The final two ingredients came just a bit later. One was Smale's famous *h-cobordism theorem*, which was the main ingredient in his proof [109] of the high-dimensional Poincaré conjecture in the topological category. (In other words, if M^n is a smooth compact n-manifold, $n \geq 5$, homotopy equivalent to S^n, then M is homeomorphic to S^n, even though it may not be diffeomorphic to it.) But from the point of view of the general manifold classification program, Smale's important contribution was a criterion for telling when two manifolds really are diffeomorphic to one another. An h-cobordism between compact manifolds M and M' is a compact manifold with boundary W, such that $\partial W = M \sqcup M'$ and such that W has deformation retractions down to both M and M'. The h-cobordism theorem [76] says that if $\dim M = \dim M' \geq 5$ and if M, M', and W are simply connected, then W is diffeomorphic to $M \times [0, 1]$, and in particular, M and M' are diffeomorphic. The advantage of this is that diffeomorphisms between different manifolds are usually very hard to construct directly; it is much easier to construct an h-cobordism.

If one dispenses with simple connectivity, then an h-cobordism between M and M' need not be diffeomorphic to a product $M \times [0, 1]$. However, the s-cobordism theorem, due to Barden, Mazur, and Stallings, with simplifications due to Kervaire

[1] Spheres have stably trivial tangent bundle and no interesting cohomology, so one's first guess might be that the theory of vector bundles and the signature theorem might be irrelevant to studying homotopy spheres. Milnor, however, showed that one can construct lots of manifolds with the homotopy type of a 7-sphere as unit sphere bundles in rank-4 vector bundles over S^4. He also showed that the signature of an 8-manifold bounded by such a manifold yields lots of information about the homotopy sphere.

[64], says that the h-cobordisms themselves are classifiable by the Whitehead torsion $\tau(W,M)$, which takes values in the Whitehead group $\mathrm{Wh}(\pi)$, where $\pi = \pi_1(M)$, and all values in $\mathrm{Wh}(\pi)$ can be realized by h-cobordisms. (The Whitehead group is the quotient of the algebraic K-group $K_1(\mathbb{Z}\pi)$ by its "obvious" subgroup $\{\pm 1\} \times \pi_{\mathrm{ab}}$.) Thus an h-cobordism is a product if $\mathrm{Wh}(\pi) = 0$, which is the case for π free abelian, and in fact is conjectured to be the case if π is torsion-free. But for π finite, for example, $\mathrm{Wh}(\pi)$ is a finitely generated group of rank $r - q$, where r is the number of irreducible real representations of π, and q is the number of irreducible rational representations of π [77, Theorem 6.2]. This number $r - q$ is usually positive (for example, when π is finite cyclic, it vanishes only if $|\pi| = 1, 2, 3, 4$, or 6). Bass and Murthy have even shown [8] that there are finitely generated abelian groups π for which $\mathrm{Wh}(\pi)$ is not finitely generated.

The last major ingredient for the classification of manifolds is the method of *surgery*. Surgery on an n-manifold M^n means cutting out a neighborhood $S^k \times D^{n-k}$ of a k-sphere $S^k \hookrightarrow M$ (with trivial normal bundle) and replacing it by $D^{k+1} \times S^{n-k-1}$, which has the same boundary. This can be used to modify a manifold without changing its bordism class, and was first introduced by Milnor [75] and Wallace [117].

With the help of all of these techniques, Browder [20, 21] and Novikov [81, 82] finally introduced a general methodology for classifying manifolds in high dimensions. The method gave complete results for simply connected manifolds in dimensions ≥ 5, and only partial information in dimensions 3 and 4, which have their own peculiarities we won't discuss here. With the help of additional contributions by Sullivan [111], Novikov [86], and above all, Wall [115], this method grew into what we know today as *surgery theory*, codified by Wall in his book [116], which originally appeared in 1970. There are now fairly good expositions of the theory, for example in Ranicki's books [94, 95], in the book by Kreck and Lück [66], in the first half of Weinberger's book [119], and in Browder's colloquium lectures from 1977 [22], so we won't attempt to compete by going into details, which anyway would take far too many pages. Instead we will just outline enough of the ideas to set the stage for Novikov's conjecture.

As we indicated before, surgery theory addresses the uniqueness question for manifolds: given (closed and connected, say) manifolds M and M' of the same dimension n, when are they diffeomorphic (or homeomorphic)? It also addresses an existence question: given a connected topological space X (say a finite CW complex), when is it homotopy equivalent to a (closed) manifold?

A few necessary conditions are evident from a first course in topology. If M and M' are diffeomorphic, then certainly they are homotopy equivalent, and so they have the same fundamental group π. Furthermore, if a finite connected CW complex X has the homotopy type of a closed manifold, then it has to satisfy Poincaré duality, even in the strong sense of (possibly twisted) Poincaré duality of the universal cover with coefficients in $\mathbb{Z}\pi$. Homotopy equivalences preserve homology and cohomology groups and cup products, so an orientation-preserving homotopy equivalence also preserves the signature (in dimensions divisible by 4 when the signature is defined). However, these conditions are not nearly enough.

For one thing, for a homotopy equivalence to be homotopic to a diffeomorphism (or even a homeomorphism), it has to be *simple*, i.e., to have vanishing torsion in $\mathrm{Wh}(\pi)$. Depending on the fundamental group π, this may or may not be a serious restriction.

But the most serious conditions involve characteristic classes of the tangent bundle. Via a very ingenious argument using surgery theory and the Hirzebruch signature theorem, Novikov [83, 84] showed that the rational Pontrjagin classes of the tangent bundle of a manifold are preserved under homeomorphisms.[2] (Incidentally, Gromov [45, Sect. 7] has given a totally different short argument for this.) The rational Pontrjagin classes do *not* have to be preserved under homotopy equivalences. So if $\varphi\colon M \to M'$ is a homotopy equivalence not preserving rational Pontrjagin classes, it cannot be homotopic to a homeomorphism.

In the simply connected case, this is (modulo finite ambiguity) just about all: if $M' \to M$ is an orientation-preserving homotopy equivalence of closed simply connected oriented manifolds, the rational Pontrjagin classes of M' have to satisfy the constraint $\langle \mathscr{L}(M'), [M'] \rangle = \mathrm{sign}(M') = \mathrm{sign}(M)$ imposed by the Hirzebruch signature theorem, but otherwise they are effectively unconstrained (assuming the dimension of the manifold is at least 5).[3] And if the map does preserve rational Pontrjagin classes, then there are only finitely many possibilities for M' up to diffeomorphism.

When M is not simply connected, the situation is appreciably more complicated. Suppose one wants to check if two n-manifolds M and M' are diffeomorphic. As we indicated before, that means we need to have a simple homotopy equivalence $\varphi\colon M' \to M$. If φ were homotopic to a diffeomorphism, it would preserve the classes of the tangent bundles, so it's convenient to assume that φ has been promoted to a *normal map* $\varphi\colon (M', \nu') \to (M, \nu)$. Here ν and ν' are the stable normal bundles defined via the Whitney embedding theorem: if k is large enough ($n+1$ suffices), then M and M' have embeddings into Euclidean space $\mathbb{R}^{n+k}$, and any two such embeddings are isotopic, so the isomorphism class of the normal bundle ν or ν' for such an embedding is well defined. (Because of the Thom-Pontrjagin construction, it's better to work with the normal bundle than with the tangent bundle, but they contain the same information.) Being a normal map means that φ has been extended to a bundle map from ν' to ν, which we can assume is an isomorphism on fibers. The idea of trying to show that M and M' are diffeomorphic is to start with a *normal bordism* from φ to id_M, i.e., a manifold W^{n+1} with boundary $M \sqcup M'$ and a map $\Phi\colon W \to M \times [0,1]$ restricting to φ and to id_M on the two boundary components, and with a compatible map of bundles, and then to try to modify (W, Φ) by surgery

[2]The same does not hold for the torsion part of the Pontrjagin classes, as one can see from calculations with lens spaces [87, Sect. 3].

[3]A precise statement to this effect may be found in [31, Theorem 6.5]. It says for example that if M is a closed simply connected manifold and $\dim M$ is not divisible by 4, then for *any* set of elements $x_j \in H^{4j}(M, \mathbb{Q})$, $1 \le j \le \left\lfloor \frac{\dim M}{4} \right\rfloor$, there is a positive integer R such that for any integer m, there is a homotopy equivalence of manifolds $\varphi_m\colon M'_m \to M$ such that $p_j(M'_m) = \varphi_m^*\big(p_j(M) + m R x_j\big)$.

to make it into an s-cobordism. Once this is accomplished, then M and M' are diffeomorphic by the s-cobordism theorem. It turns out that doing the surgery is not difficult until one gets up to the middle dimension (if $n + 1$ is even) or the "almost middle" dimension $\lfloor \frac{n+1}{2} \rfloor$ (if $n + 1$ is odd). At this point a *surgery obstruction* appears, taking its value in a group $L_{n+1}(\mathbb{Z}\pi)$ constructed purely algebraically out of quadratic forms on $\mathbb{Z}\pi$. (Roughly speaking, the L-groups are groups of stable equivalence classes of forms on finitely generated projective or free $\mathbb{Z}\pi$-modules, and the type of the form—symmetric, skew-symmetric, etc.—depends only on the value of n mod 4. The original construction may be found in [116].) The existence problem (telling if one can find a manifold homotopy equivalent to a given finite complex with Poincaré duality) works in a very similar way, just down in dimension by 1, and the surgery obstruction in that case takes its values in $L_n(\mathbb{Z}\pi)$.

Ultimately, the result of this surgery process is to prove that there is a *surgery exact sequence* for computation of the *structure set* $\mathscr{S}(M)$, the set of (simple) homotopy equivalences $\varphi\colon M' \to M$, where M' is a smooth compact manifold, modulo equivalence. We say that two such maps $\varphi\colon M' \to M$ and $\varphi'\colon M'' \to M$ are equivalent if there is a commuting diagram

$$\begin{array}{ccc} M' & \xrightarrow{\varphi} & M \\ & {\scriptstyle\cong}\searrow \quad \nearrow{\scriptstyle\varphi'} & \\ & M'' & \end{array} .$$

The surgery exact sequence then takes the form

$$\cdots \xrightarrow{\alpha} L_{n+1}(\mathbb{Z}\pi) \dashrightarrow \mathscr{S}(M) \xrightarrow{\eta} \mathscr{N}(M) \xrightarrow{\alpha} L_n(\mathbb{Z}\pi)\,. \qquad (1)$$

Here $\mathscr{N}(M)$ is the set of *normal invariants*, the normal bordism classes of all normal maps $\varphi\colon (M', \nu') \to (M, \nu)$ (not necessarily homotopy equivalences as before) modulo linear automorphisms of ν. This can also be identified with homotopy classes of maps from M into a classifying space called G/O. If one works instead in the PL or the topological category, the same sequence (1) is valid, but G/O is replaced by G/PL or G/Top, which are easier to deal with,[4] and in fact look a lot like BO, the classifying space for real K-theory. The natural maps $G/O \to G/PL \to G/Top$ are rational homotopy equivalences. The map $\eta\colon \mathscr{S}(M) \to \mathscr{N}(M)$ sends a homotopy equivalence $\varphi\colon M' \to M$ to the associated normal data.

The groups $L_\bullet(\mathbb{Z}\pi)$ are 4-periodic, and only depend on the fundamental group and some "decorations" which we are suppressing here, which only affect the torsion. The map $\alpha\colon \mathscr{N}(M) \to L_n(\mathbb{Z}\pi)$ takes the bordism class of a normal map $\varphi\colon (M', \nu') \to (M, \nu)$ to its associated *surgery obstruction*. When this vanishes, exactness of (1) says we can lift φ to an element of $\mathscr{S}(M)$, or in other words, we

[4] Once the dimension is bigger than 4!

can do surgery to convert it to a homotopy equivalence. The dotted arrow from $L_{n+1}(\mathbb{Z}\pi)$ to $\mathcal{S}(M)$ signifies that the surgery group operates on $\mathcal{S}(M)$ (which is just a pointed set, not a group) and that two elements of the structure set have the same normal invariant if and only if they lie in the same orbit for the action of $L_{n+1}(\mathbb{Z}\pi)$.

The exact sequence (1) is closely related to an *algebraic surgery exact sequence*

$$\cdots \to L_{n+1}(\mathbb{Z}\pi) \to \mathcal{S}_n(M) \to H_n(M, \mathbb{L}(\mathbb{Z})) \xrightarrow{A} L_n(\mathbb{Z}\pi) \tag{2}$$

constructed in [93, 95], where the map A, called the *assembly map*, corresponds to local-to-global passage. We will come back to this later.

For most groups π, the L-groups $L_\bullet(\mathbb{Z}\pi)$ are not easy to calculate, so a lot of the literature on surgery theory emphasizes things related to the exact sequence (1) which don't rely on explicit calculation of all the groups. For example, sometimes one can compare two related surgery problems, or rely on other invariants, such as η- and ρ-invariants for finite groups. These (as well as direct calculation from (1)) show that there are infinitely many manifolds with the homotopy type of $\mathbb{RP}^{4k+3}$, $k \geq 1$. In fact, it's shown in [27] that in dimension $4k+3$, $k \geq 1$, any closed manifold M with torsion in its fundamental group has infinitely many distinct manifolds simple homotopy-equivalent to it.

Now we are ready to explain Novikov's conjecture. We can rewrite the Hirzebruch signature theorem as saying that for a closed connected oriented manifold M, the 0-degree component of $\mathcal{L}(M) \cap [M]$ in $H_0(M, \mathbb{Q}) \cong \mathbb{Q}$ coincides with $\operatorname{sign} M$, which is preserved by orientation-preserving homotopy equivalences. The components of $\mathcal{L}(M) \cap [M]$ in other degrees have no such invariance property, and knowing them is equivalent to knowing the rational Pontrjagin classes. However, Novikov discovered in [83] (see [31, Theorem 2.1 and its proof] for a simplified version of his argument) that if $\pi_1(M) \cong \mathbb{Z}$, then the degree-1 component of $\mathcal{L}(M) \cap [M]$ is also an oriented homotopy invariant. This theorem is the simplest special case of Novikov's conjecture.

Definition 1.1. Let M be a closed connected oriented manifold, and let π be a countable discrete group (usually taken to be the fundamental group of M). Let $B\pi$ be a classifying space for π, a CW complex with contractible universal cover and fundamental group π, and let $f\colon M \to B\pi$ be a continuous map. (Up to homotopy, it's determined by the induced homomorphism $\pi_1(M) \to \pi$.) The associated *higher signature* of M is $f_*(\mathcal{L}(M) \cap [M]) \in H_\bullet(B\pi, \mathbb{Q})$.

Conjecture 1.2 (Novikov's Conjecture). *Any higher signature $f_*(\mathcal{L}(M) \cap [M]) \in H_\bullet(B\pi, \mathbb{Q})$ is always an oriented homotopy invariant. In other words, if M and M' are closed connected oriented manifolds and if $\varphi\colon M' \to M$ is*

(continued)

Conjecture 1.2 (continued)
an orientation-preserving homotopy equivalence and $f: M \to B\pi$, *then*

$$f_*(\mathscr{L}(M) \cap [M]) = (f \circ \varphi)_*(\mathscr{L}(M') \cap [M']) \in H_\bullet(B\pi, \mathbb{Q}).$$

The utility of the conjecture can be illustrated by an example.

Problem 1.3. Classify smooth compact 5-manifolds homotopy equivalent to $\mathbb{CP}^2 \times S^1$. (Note: the diffeomorphism classification of smooth 4-manifolds homotopy equivalent to $\mathbb{CP}^2$ is not known, since surgery breaks down in the smooth category in dimension 4. It is known by work of Freedman [41] that up to *homeomorphism*, there are exactly two closed topological 4-manifolds homotopy equivalent to $\mathbb{CP}^2$, but for the "exotic" one, the product with S^1 does not have a smooth structure.)

Proof. Suppose M is a smooth closed manifold of the homotopy type of $\mathbb{CP}^2 \times S^1$. There is a smooth map $f: M \to S^1$ inducing an isomorphism on π_1, and we can take this to be the map $f: M \to B\pi$, $\pi = \mathbb{Z}$, for the case of the conjecture proven by Novikov himself. So the conjecture implies that if $K = f^{-1}(\text{pt})$, the inverse image of a regular value of f, then K has signature 1. This fixes the first Pontrjagin class of M. Furthermore, K being a smooth 4-manifold with signature 1, it is in the same oriented bordism class as $\mathbb{CP}^2$. From this we can get a normal bordism W^6 between M (with its stable normal bundle ν) and $\mathbb{CP}^2 \times S^1$ (with its stable normal bundle ξ). We plug into the surgery machine and try to do surgery to convert this to an h-cobordism (and thus automatically an s-cobordism, since $\text{Wh}(\mathbb{Z}) = 0$). The surgery obstruction lives in $L_6(\mathbb{Z}[\mathbb{Z}])$. This group turns out to be $\mathbb{Z}/2$ (coming from the image of the Arf invariant in $L_6(\mathbb{Z}) \cong \mathbb{Z}/2$). So there are not a lot of possibilities. In fact one can show by studying the continuation of the sequence (1) to the left that M is diffeomorphic to $\mathbb{CP}^2 \times S^1$. But note that the key ingredient in the whole argument is the Novikov Conjecture, which pins down the first Pontrjagin class. □

2 Methods of Proof

Work on the Novikov Conjecture began almost as soon as the conjecture was formulated. Roughly speaking, methods fall into three different categories: topological, analytic, and algebraic. The *topological* approach began with Novikov's own work on the free abelian case of the conjecture, which we already mentioned in the case $\pi = \mathbb{Z}$, and which only uses transversality and basic homology theory. This method was generalized in work of Kasparov, Farrell-Hsiang, and Cappell [23, 33, 58], who used codimension-one splitting methods to deal with free abelian and poly-$\mathbb{Z}$ groups, and certain kinds of amalgamated free products.

Subsequent topological approaches to the conjecture have been based on *controlled topology* (if you like, a blend of analysis and topology since it amounts to topology with δ-ε estimates) or on various methods in stable homotopy theory. There is a lot more in this area than we can possibly summarize here, but it is discussed in detail in [37], which includes a long bibliography.

The *analytic* approach began with the important contribution of Lusztig [72]. The key idea here is to realize the higher signature of Definition 1.1 as the index of a family of elliptic operators, just as Atiyah and Singer [2, Sect. 6] had reproven Hirzebruch's signature theorem by realizing the signature as the index of a certain elliptic operator, now universally called the signature operator. (This is just the operator $d + d^*$ operating on differential forms, but with a grading on the forms coming from the Hodge $*$-operator.) A major step forward from the work of Lusztig came with the work of Mishchenko [78, 79] and Kasparov [57, 61, 62], who realized that one could generalize this construction by using "noncommutative" families of elliptic operators, based on a C^*-algebra completion $C^*(\pi)$ of the algebraic group ring $\mathbb{C}\pi$. Underlying this method was the idea [79, 99] that because of the inclusions $\mathbb{Z}\pi \hookrightarrow \mathbb{C}\pi \hookrightarrow C^*(\pi)$, there is a natural map $L_n(\mathbb{Z}\pi) \to L_n(C^*(\pi))$, and that because the spectral theorem enables one to diagonalize quadratic forms over a C^*-algebra, the L-groups and topological K-groups of a C^*-algebra essentially coincide. As we will see in the next section, the analytic approach to the Novikov conjecture is the one that has attracted the most recent attention, though there is still plenty of work being done on topological and algebraic methods.

Algebraic approaches to proving the Novikov conjecture depend on a finer understanding of the surgery exact sequence (1) and the L-groups. For a homotopy equivalence of manifolds $\varphi\colon M' \to M$, the difference $\varphi_*(\mathscr{L}(M') \cap [M']) - (\mathscr{L}(M) \cap [M]) \in H_\bullet(M, \mathbb{Q})$ is basically $\eta([M' \to M]) \otimes_{\mathbb{Z}} \mathbb{Q}$ in (1). The Novikov conjecture says that this should vanish when we apply $f_*, f\colon M \to B\pi$. Since we could also apply (1) with M replaced by $B\pi$ (at least if $B\pi$ can be chosen to be a manifold—but there is a way of getting around this), exactness in (1) shows that the Novikov Conjecture is equivalent to rational injectivity of the map α in (1), when we replace M by $B\pi$.

More precisely, we need to make use of an idea of Quinn [91], that the L-groups are the homotopy groups of a spectrum:

$$L_n(\mathbb{Z}\pi) = \pi_n(\mathbb{L}_\bullet(\mathbb{Z}\pi))$$

and that the map α in the surgery exact sequence (1) comes from an *assembly map* which is the induced map on homotopy groups of a map of spectra

$$A_M\colon M_+ \wedge \mathbb{L}_\bullet(\mathbb{Z}) \to \mathbb{L}_\bullet(\mathbb{Z}\pi).$$

This map factors (via $f\colon M \to B\pi$) through a similar map

$$A_\pi\colon B\pi_+ \wedge \mathbb{L}_\bullet(\mathbb{Z}) \to \mathbb{L}_\bullet(\mathbb{Z}\pi). \tag{3}$$

If A_π in (3) induces a rational injection on homotopy groups, then the Novikov Conjecture follows from exactness of (1). On the other hand, if A_π is not rationally injective, then one can construct an M and a higher signature for it that is not homotopy invariant. So the Novikov Conjecture is reduced to a statement which at least in principle is purely algebraic, as Ranicki in [93, 95] gives a purely algebraic construction of the surgery spectra and of the map A_π, leading to the exact sequence (2).[5]

3 Variations on a Theme

One of the most interesting features of the Novikov Conjecture is that it is closely related to a number of other useful conjectures. Some of these are known to be true, some are known to be false, and most are also unsolved. But even the ones that are false are false for somewhat subtle reasons, and still carry some "element of truth." Here we mention a number of these related conjectures and something about their status.

Conjecture 3.1 (Borel's Conjecture). *Any two closed aspherical (i.e., having contractible universal covers) manifolds M and M' with the same fundamental group are homeomorphic. In fact, any homotopy equivalence $\varphi\colon M' \to M$ of such manifolds is homotopic to a homeomorphism.*

This conjecture is known to have been posed informally by Armand Borel, before the formulation of Novikov's Conjecture, and was motivated by the Mostow Rigidity Theorem. It amounts to a kind of topological rigidity for aspherical manifolds. Note that if M is aspherical with fundamental group π and $n = \dim M \geq 5$, then we can take $M = B\pi$, and Borel's conjecture amounts to saying that in the surgery sequence (1) in the topological category, $\mathscr{S}(M)$ is just a single point, or by exactness, the assembly map A_π is an equivalence. This implies the Novikov Conjecture for π, but is stronger.

Incidentally, it is known now that the analogue of Borel's Conjecture, but with homeomorphism replaced by diffeomorphism, is false. The simplest

[5]It turns out that (2) coincides with the analogue of (1) in the topological, rather than smooth, category, but the difference between these is rather small since all homotopy groups of *Top*/*O* are torsion.

counterexample is with $M = T^7$, the 7-torus. Since a torus is parallelizable, Wall pointed out in [116, Sect. 15A] that the set of smooth structures on T^n compatible with the standard PL structure is parameterized by $[T^n, PL/O]$ (for $n \geq 5$). It is known that the classifying space PL/O is 6-connected and that (for $j \geq 7$) its jth homotopy group can be identified with the group Θ_j of smooth homotopy j-spheres.[6] Since $\Theta_7 \cong \mathbb{Z}/28$ by [65, 74], the differentiable structures on T^7 are parameterized by $[T^7, PL/O] \cong [T^7, K(\Theta_7, 7)] \cong H^7(T^7, \Theta_7) \cong \mathbb{Z}/28$ and there are 28 different differentiable structures on T^7. A series of counterexamples with negative curvature to the smooth Borel conjecture was constructed in [34, 35].

The fundamental group π of an aspherical manifold M (even if noncompact) has to be torsion-free, since if $g \in \pi$ has finite order $k > 1$, it would act freely on the universal cover $\widetilde{M}$, and $\widetilde{M}/\langle g \rangle$ would be a finite-dimensional model for $B\mathbb{Z}/k$, contradicting the fact that $\mathbb{Z}/k$ has homology in all positive odd dimensions. So Conjecture 3.1 can't apply to groups with torsion. In fact, the result of [27] shows that for groups with torsion, A_π in (3) is never an equivalence. We will come back to this shortly.

However, we have already mentioned the role of the Whitehead group, which comes from the algebraic K-theory of $\mathbb{Z}\pi$, in studying manifolds with fundamental group π. An important conjecture which we have already mentioned is:

Conjecture 3.2 (Vanishing of Whitehead Groups). *If* π *is torsion-free, then* $\mathrm{Wh}(\pi) = 0$.

Note that if Conjecture 3.2 fails and π is the fundamental group of a closed manifold M, then by the s-cobordism theorem, there is an h-cobordism W with $\partial W = M \sqcup (-M')$ which is not a product, and we have a homotopy equivalence $M' \to M$ which is not simple, hence Borel's Conjecture, Conjecture 3.1, fails for M.

More generally, one can ask what one can say about the algebraic K-theory of $\mathbb{Z}\pi$ in all degrees. Loday [69] constructed an assembly map $B\pi_+ \wedge \mathbb{K}(\mathbb{Z}) \to \mathbb{K}(\mathbb{Z}\pi)$, and this being an equivalence would say that all of the algebraic K-theory of $\mathbb{Z}\pi$ comes in some sense from homology of π and K-theory of $\mathbb{Z}$. This is known in some cases—for π free abelian, it follows from the "Fundamental Theorem of K-theory." The assembly map being an equivalence in degrees ≤ 1 for torsion-free groups π and $R = \mathbb{Z}$ implies Conjecture 3.2. The analogue of Novikov's Conjecture for K-theory is

[6]The group operation is the connected sum; inversion comes from reversing the orientation.

Conjecture 3.3 (Novikov Conjecture for K-Theory). *Let $R = \mathbb{Z}, \mathbb{Q}, \mathbb{R}$, or $\mathbb{C}$ and let π be a discrete group. Then the assembly map $B\pi_+ \wedge \mathbb{K}(R) \to \mathbb{K}(R\pi)$ induces an injection of rational homotopy groups.*

Conjecture 3.3 was proved (with $R = \mathbb{Z}$, the most important case) for groups π with finitely generated homology in [16]. It was also proved (without rationalizing) in [25], when π is a discrete, cocompact, torsion-free discrete subgroup of a connected Lie group. Subsequently, Carlsson and Pedersen [26] proved it (without rationalizing) for any group π for which there is a finite model for $B\pi$, such that the universal cover $E\pi$ of $B\pi$ admits a contractible metrizable π-equivariant compactification X such that compact subsets of $E\pi$ become small near the "boundary" $X \setminus E\pi$. This was recently improved [92] to the case where there is a finite model for $B\pi$ and π has finite decomposition complexity, which is a tameness condition on π viewed as a metric space with the word length metric (for some finite generating set).

As we have already mentioned, for groups with torsion, the assembly map A_π of (3) is never an equivalence. For similar reasons, one also can't expect the K-theory assembly map to be an equivalence for groups with torsion. The correct replacement seems to be the following.[7]

Conjecture 3.4 (Farrell-Jones Conjecture). *Let π be a discrete group and let $\mathcal{F}$ be its family of virtually cyclic subgroups (subgroups that contain a cyclic subgroup of finite index). Such subgroups are either finite or else admit a surjection with finite kernel onto either $\mathbb{Z}$ or the infinite dihedral group $(\mathbb{Z}/2) * (\mathbb{Z}/2)$. Let $E_{\mathcal{F}}(\pi)$ denote the universal π-space with isotropy in $\mathcal{F}$. This is a contractible π-CW-complex X with all isotropy groups in $\mathcal{F}$ (for the π-action) and with X^H contractible for each $H \in \mathcal{F}$. It is known to be uniquely defined up to π-homotopy equivalence. Then the assembly maps*

$$H^\pi_\bullet(E_{\mathcal{F}}(\pi); \mathbb{L}(\mathbb{Z})) \to \mathbb{L}(\mathbb{Z}\pi) \quad \text{and} \quad H^\pi_\bullet(E_{\mathcal{F}}(\pi); \mathbb{K}(R)) \to \mathbb{K}(R\pi) \tag{4}$$

are isomorphisms for $R = \mathbb{Z}, \mathbb{Q}, \mathbb{R}$, or $\mathbb{C}$.

[7] Just for the experts: one needs to use the $-\infty$ decoration on the L-spectra here.

When π is torsion-free, (4) is just the assembly map (3) or its K-theory version, and the conjecture says that the assembly map is an equivalence. Conjecture 3.4 implies Conjectures 3.1, 1.2, and 3.3, even for groups with torsion, as well as Conjecture 3.2. More details on Conjecture 3.4 may be found in [70], in [66, Chaps. 19–24], or in [71]. The K-theory version of the conjecture has been proven in [7] for fundamental groups of manifolds of negative curvature and in [6] for hyperbolic groups, and both the K-theory and L-theory versions have been proven for certain groups acting on trees in [5, 107] and for cocompact lattice subgroups of Lie groups in [4]. Split injectivity of (4) has been proved for groups with finite quotient finite decomposition complexity (a condition weaker than that of [92]) in [63]. Rational injectivity of (4) holds under much weaker conditions; see for example [30].

Another variation on the Novikov Conjecture is to consider the situation where a finite group G acts on a manifold, and one wants to study G-equivariant invariants of M. Under suitable circumstances, one finds that the fundamental group of M leads to a certain extra amount of equivariant topological rigidity. To formulate the analogue of Conjecture 1.2, one needs a substitute for the homology L-class $\mathscr{L}(M) \cap [M]$. The easiest way to formulate this is in K-homology, since Kasparov [59, 60], following ideas of Atiyah and Singer, showed that an elliptic differential operator D on M naturally leads to a K-homology class $[D] \in K_\bullet(M)$ (see also [50] for an exposition), and when D is G-invariant, the class naturally lives in $K^G_\bullet(M)$. The image of $[D]$ in $K^G_\bullet(\mathrm{pt}) = R(G)$ under the map induced by $M \to \mathrm{pt}$ is the equivariant index $\mathrm{ind}_G D \in R(G)$ in the sense of Atiyah and Singer. When D is the signature operator, $\mathscr{L}(M) \cap [M]$ is basically (except for some powers of 2, not important here) the Chern character of $[D] \in K_\bullet(M)$, and so if $f\colon M \to B\pi$, the higher signature of Definition 1.1, is basically the Chern character of $f_*([D])$. That motivates the following.

Conjecture 3.5 (Equivariant Novikov Conjecture [105]). *Let M be a closed oriented manifold admitting an action of a finite group G, and suppose $f\colon M \to X$ is a G-equivariant smooth map to a finite G-CW complex which is G-equivariantly aspherical (i.e., X^H is aspherical for all subgroups H of G). Let $\varphi\colon M' \to M$ be a G-equivariant map of closed G-manifolds which, non-equivariantly, is a homotopy equivalence. Then if $[D_M]$ and $[D_{M'}]$ denote the equivariant K-homology classes of the signature operators on M and M', respectively,*

$$f_*([D_M]) = (f \circ \varphi)_*([D_{M'}]) \in K^G_\bullet(X).$$

Various generalizations and applications to rigidity theorems are possible (see for example [36, 104]), but we won't go into details here. Conjecture 3.4 was proven in [105] for X a closed manifold of nonpositive curvature and in [43] for X a Euclidean building, in both cases with G acting by isometries.

4 New Directions

The conjectures we discussed in Sect. 3 are fairly directly linked to the original Novikov Conjecture, and it is easy to see how they are connected with topological rigidity of highly connected manifolds. But in this section, we will discuss a number of other conjectures which grew out of work on Novikov's Conjecture but which go somewhat further afield, to the point where the connection with the original conjecture may not be immediately obvious. However, we will try to explain the relationships as we go along.

We have already mentioned the assembly map and the Farrell-Jones Conjecture (Conjecture 3.4), which gives a conjectural calculation of the L-groups $L_\bullet(\mathbb{Z}\pi)$ for a discrete group π. However, work on Novikov's Conjecture by analytic techniques (see Sect. 2) already required passing from the integral group ring to the complex group ring (this only affects 2-torsion in the L-groups) and then completing $\mathbb{C}\pi$ to a C^*-algebra. For C^*-algebras, L-theory is basically the same as topological K-theory, and even for real C^*-algebras, they agree after inverting 2 [99, Theorem 1.11]. So it's natural to ask if assembly can be used to compute the topological K-theory of $C^*(\pi)$. For the full group C^*-algebra this seems to be impossible, but for the *reduced* group C^*-algebra $C^*_r(\pi)$,[8] the completion of $\mathbb{C}\pi$ for its action on $L^2(\pi)$, there is a good guess for a purely topological calculation of $K_\bullet(C^*_r(\pi))$. (Here $K_\bullet$ denotes *topological* K-theory for Banach algebras, which satisfies Bott periodicity. This is much more closely related to L-theory, which is 4-periodic, than is algebraic K-theory in the sense of Quillen.) This guess is given by the *Baum-Connes Conjecture*, originally formulated in [9, 10] and further refined in [11] (see also [47] for a nice quick survey). The conjecture applies to far more than just discrete groups; it applies to locally compact groups, to such groups "with coefficients" (i.e., acting on a C^*-algebra), and even to groupoids [113]. In its greatest generality the conjecture is known to be false [48], though a patch which might repair it has been proposed [13]. However, the original version of the conjecture is still open, though the literature on the conjecture has grown to more than 300 items. To avoid having to talk about Kasparov's KK-theory, we will omit discussion of the conjecture with coefficients, and will just stick to the original conjecture for groups.

[8] It is known that the natural map $C^*(\pi) \twoheadrightarrow C^*_r(\pi)$ is an isomorphism if and only if π is amenable.

Conjecture 4.1 (Baum-Connes Conjecture). *Let G be a second countable locally compact group, and let $C_r^*(G)$ denote the completion of $L^1(G)$ for its action by left convolution on $L^2(G)$. Then there is a natural assembly map*

$$\mu\colon K_\bullet^G(\mathscr{E}G) \to K_\bullet(C_r^*(G)),$$

where $\mathscr{E}G$ is the universal proper G-space (a contractible space on which G acts properly), and this map is an isomorphism. If G has no nontrivial compact subgroups, then the assembly map simplifies to

$$\mu\colon K_\bullet(BG) \to K_\bullet(C_r^*(G)).$$

Proposition 4.2. *Conjecture 4.1 implies Conjecture 1.2.*

Proof. For this we take $G = \pi$ to be discrete and countable. For simplicity, we also work with the periodic L-theory spectra instead of the connective ones. (The difference only affects the bottom of the surgery sequence (1).) If π is torsion-free, the domain of μ is $K_\bullet(B\pi) = H_\bullet(B\pi;\mathbb{K}^{\text{top}})$. But after inverting 2, $\mathbb{K}^{\text{top}}$ is just a direct sum of two copies of $\mathbb{L}(\mathbb{Z})$, one of them shifted in degree by 2. So if Conjecture 4.1 holds for π and π is torsion-free, we have the commuting diagram

$$\begin{array}{ccc}
H_\bullet(B\pi;\mathbb{L}(\mathbb{Z}))\otimes\mathbb{Q} & \xrightarrow{A_\pi} & L_\bullet(\mathbb{Z}\pi)\otimes\mathbb{Q} \\
\big\downarrow & & \downarrow \\
 & & L_\bullet(C_r^*(\pi))\otimes\mathbb{Q} \\
 & & \downarrow\cong \\
H_\bullet(B\pi;\mathbb{K}^{\text{top}})\otimes\mathbb{Q} & \underset{\cong}{\xrightarrow{\mu}} & K_\bullet(C_r^*(\pi))\otimes\mathbb{Q}.
\end{array} \tag{5}$$

Diagram (5) immediately implies that the rational L-theory assembly map A_π [the same map as the map induced on rational homotopy groups by (3)] is injective.

If π is not torsion-free, then $\mathscr{E}\pi$ and $E\pi$ are not the same,[9] but there is always a π-equivariant map $E\pi \to \mathscr{E}\pi$. Thus we need only replace (5) by the diagram

[9]In the extreme case where π is a torsion group, $\mathscr{E}\pi = \text{pt}$, while if π is nontrivial, $E\pi$ is necessarily infinite dimensional.

$$
\begin{array}{ccccc}
 & & \xrightarrow{\qquad\qquad A_\pi \qquad\qquad} & & \\
H_\bullet(B\pi;\mathbb{L}(\mathbb{Z}))\otimes\mathbb{Q} & \longrightarrow & H_\bullet^\pi(\mathcal{E}\pi;\mathbb{L}(\mathbb{Z}))\otimes\mathbb{Q} & \longrightarrow & L_\bullet(\mathbb{Z}\pi)\otimes\mathbb{Q} \\
 & & & & \downarrow \\
\downarrow & & \downarrow & & L_\bullet(C_r^*(\pi))\otimes\mathbb{Q} \\
 & & & & \downarrow\cong \\
H_\bullet(B\pi;\mathbb{K}^{\mathrm{top}})\otimes\mathbb{Q} & \xrightarrow{\alpha} & H_\bullet^\pi(\mathcal{E}\pi;\mathbb{K}^{\mathrm{top}})\otimes\mathbb{Q} & \xrightarrow[\cong]{\mu} & K_\bullet(C_r^*(\pi))\otimes\mathbb{Q}.
\end{array}
\tag{6}
$$

Since points in $\mathcal{E}\pi$ have finite isotropy, and since the π-map $\pi \twoheadrightarrow \pi/\sigma$, σ a finite subgroup of π, induces the map $\mathbb{Z} \hookrightarrow R(\sigma)$ on equivariant K-homology, a spectral sequence argument shows that the bottom left map α in (6) is injective, and so by a diagram chase, A_π is injective. □

Thus Conjecture 4.1 (for the case of discrete groups) implies Conjecture 1.2. However, Conjecture 4.1 for *non-discrete* groups is also quite interesting and important. There are two main reasons for this:

1. There are "change of group methods" that enable one to pass from results for a group to results for a closed subgroup. Many of the significant early results on Novikov's Conjecture were proved by considering discrete groups π that embed in a Lie group (or p-adic Lie group) and then using these change of group methods to pass from the Lie group to the discrete subgroup.
2. The Baum-Connes Conjecture for connected Lie groups (also known as the Connes-Kasparov Conjecture) and the same conjecture for p-adic groups are both quite interesting in their own right, and say a lot about representation theory. For an introduction to this topic, see [11, 47]. For some of the more significant results, see [14, 67, 110, 118]. For recent applications to harmonic analysis on reductive groups, see [3, 73, 90, 102].

Another direction arising out of both the controlled topology and the analytic approaches to Novikov's Conjecture leads to the so-called *coarse Baum-Connes Conjecture* [49, 96, 120]. This conjecture deals with the large-scale geometry of metric spaces X of bounded geometry (think of complete Riemannian manifolds with curvature bounds, or of finitely generated groups with a word-length metric). Roughly speaking, the coarse Novikov Conjecture says that indices of generalized elliptic operators capture all of the coarse (i.e., "large-scale") rational homology of such a space X.

Conjecture 4.3 (Coarse Baum-Connes and Novikov). *Let X be a uniformly contractible locally compact complete metric space of bounded*

(continued)

Conjecture 4.3 (continued)
geometry, in which all metric balls are compact. Let $KX_\bullet(X)$ be the coarse K-homology of X (the direct limit of the K-homologies of successively coarser Rips complexes) and let $C^*(X)$ be the C^*-algebra of locally compact, finite propagation operators on X. Then Roe defined a natural assembly map

$$\mu\colon\ KX_\bullet(X) \to K_*(C^*(X)). \tag{7}$$

The coarse Baum-Connes Conjecture is that μ is an isomorphism; the coarse Novikov Conjecture is that μ is rationally injective.

Positive results on Conjecture 4.3 may be found in [28, 29, 40, 42, 49, 96, 114, 120].

However, it is known that the conjecture fails in various situations [32, 48, 121], especially if one drops the bounded geometry assumption.

The coarse Baum-Connes conjecture implies the Novikov conjecture under mild conditions. To see this, suppose for example that there is a compact metrizable model Y for $B\pi$, and let $X = E\pi$ be its universal covering. Then there is a commutative diagram

$$\begin{array}{ccc} K_*(B\pi) & \xrightarrow{\ \alpha\ } & K_*(C_r^*(\pi)) \\ {\scriptstyle\cong}\downarrow{\scriptstyle \mathrm{tr}} & & \downarrow{\scriptstyle \mathrm{tr}} \\ \pi_*(\mathbb{K}X_*(X)^{h\pi}) & \xrightarrow{\ \mu^{h\pi}\ } & \pi_*(\mathbb{K}_*(C^*(X))^{h\pi}), \end{array}$$

where α is usual Baum-Connes assembly, μ is as in Conjecture 4.3, $h\pi$ denotes homotopy fixed points, and tr is a transfer map. Then μ being an isomorphism implies that $\mu^{h\pi}$ is an isomorphism, and so we get a splitting for α. Refinements of this argument, as well as generalizations of the coarse Baum-Connes conjecture, may be found in [80].

Thinking of $C_r^*(\pi)$ as being (up to Morita equivalence) the same thing as the fixed points of π on $C^*(X)$ also gives rise to a nice way of relating the surgery exact sequence (2) to the Baum-Connes assembly map. This was accomplished in the series of papers [51–53, 89], which set up a natural transformation from the surgery sequence to a long exact sequence where the C^*-algebraic assembly map corresponds to the L-theory assembly map in the original sequence. This gives an even more direct connection between coarse Baum-Connes and surgery theory.

Other "new directions" from Novikov's Conjecture arise from replacing the higher signature of Definition 1.1 with other sorts of "higher indices." For example, an important case is obtained by replacing $\mathscr{L}(M)$ with $\widehat{\mathscr{A}}(M)$, the total $\widehat{A}$ class.

This is again a certain polynomial in the rational Pontrjagin class, and has the property that when M is a spin manifold, $\widehat{\mathscr{A}}(M) \cap [M]$ is the Chern character of the class $[D]$ defined by the Dirac operator on M. (Here the reader doesn't need to know much about the Dirac operator D except for the fact that it's an elliptic first-order differential operator canonically defined on a Riemannian manifold with a spin structure.) It was pointed out by Lichnerowicz [68] that when M is closed and has positive scalar curvature, then the spectrum of D must be bounded away from 0, and thus $\operatorname{ind}(D) = \langle \widehat{\mathscr{A}}(M), [M] \rangle$ has to vanish. When M is not simply connected, a major strengthening of this is possible:

Conjecture 4.4 (Gromov-Lawson Conjecture [46]). *Let M be a connected closed spin Riemannian manifold of positive scalar curvature, let π be a discrete group, and let $f: M \to B\pi$ be a continuous map (determined up to homotopy by a homomorphism $\pi_1(M) \to \pi$). Then the higher $\widehat{A}$-genus $f_*(\widehat{\mathscr{A}}(M) \cap [M]) \in H_\bullet(B\pi, \mathbb{Q})$ vanishes.*

This conjecture is still open in general, but it is known to be closely related to Novikov's Conjecture. For example, it was shown in [97] that Conjecture 4.4 is true whenever the K-theory assembly map $K_\bullet(B\pi) \to K_\bullet(C^*_r(\pi))$ is rationally injective, and thus *a fortiori* whenever Conjecture 4.1 holds. It also can be deduced from certain cases of Conjecture 4.3, by a descent argument similar to the one above. The Lichnerowicz argument also applies to complete noncompact spin manifolds M of *uniformly* positive scalar curvature, and when Conjecture 4.3 holds, one gets obstructions to existence of such metrics living in $K_\bullet(C^*(X))$ whenever there is a coarse map $M \to X$.

Conjecture 4.4 can be refined to conjectures about necessary and sufficient conditions for positive scalar curvature. Here we just mention a few of several possible versions. For these it's necessary to go beyond ordinary homology and to consider KO-homology, the homology theory dual to the (topological) K-theory of real vector bundles. This theory is 8-periodic and has coefficient groups $KO_j = \mathbb{Z}$ when j is divisible by 4 (this part is detected by the Chern character to ordinary homology), $\mathbb{Z}/2$ when $j \equiv 1, 2 \pmod 8$, 0 otherwise. The class $[D]$ of the Dirac operator on a spin manifold M lives in $KO_n(M)$, $n = \dim(M)$. While the actual operator D depends on a choice of a Riemannian metric, the class $[D] \in KO_n(M)$ does not, so that the following conjecture makes sense.

Conjecture 4.5 (Gromov-Lawson-Rosenberg Conjecture). *Let M be a connected closed spin manifold with fundamental group π and Dirac operator D_M, and let $f\colon M \to B\pi$ be the classifying map for the universal cover. Let $A\colon KO_\bullet(B\pi) \to KO_\bullet(C_r^*(\pi))$ be the assembly map in real K-theory. Then M admits a Riemannian metric of positive scalar curvature if and only if $A \circ f_*([D_M]) = 0$ in $KO_n(C_r^*(\pi))$, $n = \dim M \geq 5$.*

The restriction to $n \geq 5$ is needed only to use surgery methods to construct a metric of positive scalar curvature when the obstruction vanishes; it is not needed to show that there is a genuine obstruction to positive scalar curvature when $A \circ f_*([D_M]) \neq 0$, which was proven in [98]. For the next conjecture, we need to introduce a choice of *Bott manifold*, a geometric representative for Bott periodicity in *KO*-homology. This is a simply connected closed spin manifold Bt^8 of dimension 8 with $\langle \widehat{\mathscr{A}}(\mathrm{Bt}^8), [\mathrm{Bt}^8]\rangle = 1$. It may be chosen to be Ricci flat. Since scalar curvature is additive on Riemannian products, Bt^8 being Ricci flat implies that taking a product with the Bott manifold does not change the scalar curvature.

Conjecture 4.6 (Stable Gromov-Lawson-Rosenberg Conjecture). *Let M be a connected closed spin manifold with fundamental group π and Dirac operator D_M, and let $f\colon M \to B\pi$ be the classifying map for the universal cover. Let Bt^8 be a Bott manifold as above. Then M* **stably** *admits a Riemannian metric of positive scalar curvature, in the sense that $M \times \overbrace{\mathrm{Bt}^8 \times \cdots \times \mathrm{Bt}^8}^{k}$ admits such a metric for some k, if and only if $A \circ f_*([D_M]) = 0$ in $KO_n(C_r^*(\pi))$, $n = \dim M$.*

There are simple implications

$$\text{Conj. 4.5} \Rightarrow \text{Conj. 4.6}, \quad \text{Conj. 4.6} + \text{injectivity of } A \Rightarrow \text{Conj. 4.4}.$$

The (very strong) Conjecture 4.5 is known to hold for especially nice groups, such as free abelian groups [98], hyperbolic groups of low dimension [55], and finite groups with periodic cohomology [18], but it fails in general [55, 108]. Conjecture 4.6 is weaker, and holds for all the known counterexamples to Conjecture 4.5. It was formulated and proven for finite groups in [103]. Subsequently, Stolz [unpublished] showed that it follows from the Baum-Connes Conjecture, Conjecture 4.1. For a survey on this entire field, see [100].

The last "new direction" we would like to discuss here comes from replacing the higher signature in Novikov's Conjecture by the higher Todd genus or the higher elliptic genus. This seems to be quite relevant for understanding the interaction between topological invariants and algebraic geometry invariants for algebraic varieties defined over $\mathbb{C}$.

The Todd class $\mathscr{T}(M)$ is still another polynomial in characteristic classes, this time the rational Chern classes of a complex (or almost complex) manifold. Suppose for simplicity that M is a smooth projective variety over $\mathbb{C}$, viewed as a complex manifold via an embedding into some complex projective space. The Hirzebruch Riemann-Roch Theorem then says that

$$\langle \mathscr{T}(M), [M]\rangle = \chi(M, \mathscr{O}_M) = \sum_{j=0}^{n}(-1)^j \dim H^j(M, \mathscr{O}_M), \tag{8}$$

where $\mathscr{O}_M$ is the structure sheaf of M, the sheaf of germs of holomorphic functions, and n is the complex dimension of M. The right-hand side of (8) is called the *arithmetic genus*. (The original definition of the latter by algebraic geometers like Severi turned out to be $(-1)^n(\chi(M, \mathscr{O}_M) - 1)$, but the normalization here is a bit more convenient.) The left-hand side of (8) is called the *Todd genus*, and is known to be a birational invariant.[10] Once again, if one has a map $f\colon M \to B\pi$, then we can define the associated *higher Todd genus* as $f_*(\mathscr{T}(M) \cap [M]) \in H_\bullet(B\pi, \mathbb{Q})$.

Conjecture 4.7 (Algebraic Geometry Novikov Conjecture [101]). *Let M be a smooth complex projective variety, and let $f\colon M \to B\pi$ be a continuous map (for the topology of M as a complex manifold). Let $M' \overset{\varphi}{\dashrightarrow} M$ be a birational map. Then the corresponding higher Todd genera agree, i.e.,*

$$f_*(\mathscr{T}(M) \cap [M]) = (f \circ \varphi)_*(\mathscr{T}(M') \cap [M']) \in H_\bullet(B\pi, \mathbb{Q}).$$

Note the obvious similarity with Conjecture 1.2. However, unlike Novikov's original conjecture, this statement is actually a *theorem* [15, 19]. That follows from

[10]Recall that two varieties are said to be birationally equivalent if there are rational maps between them which are inverses of each. Since rational maps do not have to be everywhere defined (this is why we denote rational maps below by dotted lines), two varieties are birationally equivalent if and only if they have Zariski-open subsets which are isomorphic as varieties.

the fact that if $M' \overset{\varphi}{\dashrightarrow} M$ is a birational map, then $\varphi_*([D_{M'}]) = [D_M] \in K_0(M)$, where $[D_M]$ denotes the K-homology class of the Dolbeault operator, whose Chern character is $\mathcal{T}(M) \cap [M]$.[11] The corresponding statement for the signature operator is *not* true; a homotopy equivalence does not have to preserve the class of the signature operator. (However, the mod 8 reduction of this class *is* preserved [106].)

However, there is another similarity with Novikov's Conjecture which is pointed out in [101]. By [112, Théorème IV.17], $\Omega_\bullet$, the graded ring of cobordism classes of oriented manifolds, is, after tensoring with $\mathbb{Q}$, a polynomial ring in the classes of the complex projective spaces $\mathbb{CP}^{2k}$, $k \in \mathbb{N}$. Then if $I_\bullet$ is the ideal in $\Omega_\bullet$ generated by all $[M] - [M']$ with M and M' homotopy equivalent (in a way preserving orientation), Kahn [56] proved that $\Omega_\bullet/I_\bullet \cong \mathbb{Q}$, with the quotient map identified with the Hirzebruch signature. Similarly, $\Omega^U_\bullet$, the graded ring of cobordism classes of almost complex manifolds, is, after tensoring with $\mathbb{Q}$, a polynomial ring in the classes of all complex projective spaces, and the quotient of $\Omega^U_\bullet$ by the ideal generated by all $[M]-[M']$ with M and M' birationally equivalent smooth projective varieties is again $\mathbb{Q}$, this time with the quotient map identifiable with the Todd genus.

These results effectively say that, up to multiples, the signature is the only homotopy-invariant genus on oriented manifolds, and the arithmetic genus is the only birationally invariant genus on smooth projective varieties. But if one considers manifolds with large fundamental group, the situation changes. By [101, Theorem 4.1], a linear functional on $\Omega_\bullet(B\pi) \otimes \mathbb{Q}$ that is an oriented homotopy invariant must come from the higher signature, and by [101, Theorem 4.3], a linear functional on $\Omega^U_\bullet(B\pi) \otimes \mathbb{Q}$ that is a birational invariant must (under a certain technical condition satisfied in many cases) come from the higher Todd genus.

Finally, the papers [17, 24, 44] consider still more analogues of higher genera with the Todd genus replaced by the elliptic genus. The result of [17] is particularly nice; it is the exact analogue of Conjecture 4.7, but with the Todd genus replaced by the elliptic genus and with birational equivalence replaced by K-equivalence (a birational equivalence preserving canonical bundles).

Acknowledgements Work on this paper was partially supported by the United States National Science Foundation, grant number 1206159. I would like to thank Greg Friedman, Daniel Kasprowski, Andrew Ranicki, and Shmuel Weinberger for useful feedback on an earlier draft of this paper.

[11] It takes a bit of work to make sense of φ_* here, since φ may not be everywhere defined, but this can be done. The point is that by the factorization theorem for birational maps [1], we can factor φ into a sequence of blow-ups and blow-downs, and φ_* is clearly defined for a blow-down (since it is a continuous map) and is an isomorphism in this case by the Baum-Fulton-MacPherson variant of Grothendieck-Riemann-Roch [12]. In the case of a blow-up, let φ_* be given by the inverse of the map induced by the reverse blow-down.

References

1. Abramovich, D., Karu, K., Matsuki, K., Włodarczyk, J.: Torification and factorization of birational maps. J. Amer. Math. Soc. **15**(3), 531–572 (electronic) (2002). DOI 10.1090/S0894-0347-02-00396-X
2. Atiyah, M.F., Singer, I.M.: The index of elliptic operators. III. Ann. of Math. (2) **87**, 546–604 (1968)
3. Aubert, A.M., Baum, P., Plymen, R., Solleveld, M.: On the local Langlands correspondence for non-tempered representations. Münster J. Math. **7**, 27–50 (2014)
4. Bartels, A., Farrell, F.T., Lück, W.: The Farrell-Jones conjecture for cocompact lattices in virtually connected Lie groups. J. Amer. Math. Soc. **27**(2), 339–388 (2014). DOI 10.1090/S0894-0347-2014-00782-7
5. Bartels, A., Lück, W.: Isomorphism conjecture for homotopy K-theory and groups acting on trees. J. Pure Appl. Algebra **205**(3), 660–696 (2006). DOI 10.1016/j.jpaa.2005.07.020
6. Bartels, A., Lück, W., Reich, H.: The K-theoretic Farrell-Jones conjecture for hyperbolic groups. Invent. Math. **172**(1), 29–70 (2008). DOI 10.1007/s00222-007-0093-7
7. Bartels, A., Reich, H.: On the Farrell-Jones conjecture for higher algebraic K-theory. J. Amer. Math. Soc. **18**(3), 501–545 (2005). DOI 10.1090/S0894-0347-05-00482-0
8. Bass, H., Murthy, M.P.: Grothendieck groups and Picard groups of abelian group rings. Ann. of Math. (2) **86**, 16–73 (1967)
9. Baum, P., Connes, A.: Chern character for discrete groups. In: A fête of topology, pp. 163–232. Academic Press, Boston, MA (1988). DOI 10.1016/B978-0-12-480440-1.50015-0
10. Baum, P., Connes, A.: Geometric K-theory for Lie groups and foliations. Enseign. Math. (2) **46**(1–2), 3–42 (2000)
11. Baum, P., Connes, A., Higson, N.: Classifying space for proper actions and K-theory of group C^*-algebras. In: C^*-algebras: 1943–1993 (San Antonio, TX, 1993), *Contemp. Math.*, vol. 167, pp. 240–291. Amer. Math. Soc., Providence, RI (1994). DOI 10.1090/conm/167/1292018
12. Baum, P., Fulton, W., MacPherson, R.: Riemann-Roch and topological K theory for singular varieties. Acta Math. **143**(3–4), 155–192 (1979). DOI 10.1007/BF02392091
13. Baum, P., Guentner, E., Willett, R.: Expanders, exact crossed products, and the Baum-Connes conjecture. Ann. of K-Theory, **1**(2), 155–208 (2016). URL http://arxiv.org/abs/1311.2343.
14. Baum, P., Higson, N., Plymen, R.: A proof of the Baum-Connes conjecture for p-adic $\mathrm{GL}(n)$. C. R. Acad. Sci. Paris Sér. I Math. **325**(2), 171–176 (1997). DOI 10.1016/S0764-4442(97)84594-6
15. Block, J., Weinberger, S.: Higher Todd classes and holomorphic group actions. Pure Appl. Math. Q. **2**(4, Special Issue: In honor of Robert D. MacPherson. Part 2), 1237–1253 (2006). DOI 10.4310/PAMQ.2006.v2.n4.a13
16. Bökstedt, M., Hsiang, W.C., Madsen, I.: The cyclotomic trace and algebraic K-theory of spaces. Invent. Math. **111**(3), 465–539 (1993). DOI 10.1007/BF01231296
17. Borisov, L., Libgober, A.: Higher elliptic genera. Math. Res. Lett. **15**(3), 511–520 (2008). DOI 10.4310/MRL.2008.v15.n3.a10
18. Botvinnik, B., Gilkey, P., Stolz, S.: The Gromov-Lawson-Rosenberg conjecture for groups with periodic cohomology. J. Differential Geom. **46**(3), 374–405 (1997). URL http://projecteuclid.org/euclid.jdg/1214459973
19. Brasselet, J.P., Schürmann, J., Yokura, S.: Hirzebruch classes and motivic Chern classes for singular spaces. J. Topol. Anal. **2**(1), 1–55 (2010). DOI 10.1142/S1793525310000239
20. Browder, W.: Poincaré spaces, their normal fibrations and surgery. Invent. Math. **17**, 191–202 (1972)
21. Browder, W.: Surgery on simply-connected manifolds. Springer-Verlag, New York-Heidelberg (1972). Ergebnisse der Mathematik und ihrer Grenzgebiete, Band 65

22. Browder, W.: Differential topology of higher-dimensional manifolds. In: Surveys on surgery theory, Vol. 1, *Ann. of Math. Stud.*, vol. 145, pp. 41–71. Princeton Univ. Press, Princeton, NJ (2000)
23. Cappell, S.E.: On homotopy invariance of higher signatures. Invent. Math. **33**(2), 171–179 (1976)
24. Cappell, S.E., Libgober, A., Maxim, L.G., Shaneson, J.L.: Hodge genera of algebraic varieties. II. Math. Ann. **345**(4), 925–972 (2009). DOI 10.1007/s00208-009-0389-6
25. Carlsson, G.: Bounded K-theory and the assembly map in algebraic K-theory. In: Novikov conjectures, index theorems and rigidity, Vol. 2 (Oberwolfach, 1993), *London Math. Soc. Lecture Note Ser.*, vol. 227, pp. 5–127. Cambridge Univ. Press, Cambridge (1995). DOI 10.1017/CBO9780511629365.004
26. Carlsson, G., Pedersen, E.K.: Controlled algebra and the Novikov conjectures for K- and L-theory. Topology **34**(3), 731–758 (1995). DOI 10.1016/0040-9383(94)00033-H
27. Chang, S., Weinberger, S.: On invariants of Hirzebruch and Cheeger-Gromov. Geom. Topol. **7**, 311–319 (electronic) (2003). DOI 10.2140/gt.2003.7.311
28. Chen, X., Wang, Q., Yu, G.: The maximal coarse Baum-Connes conjecture for spaces which admit a fibred coarse embedding into Hilbert space. Adv. Math. **249**, 88–130 (2013). DOI 10.1016/j.aim.2013.09.003
29. Chen, X., Wang, Q., Yu, G.: The coarse Novikov conjecture and Banach spaces with Property (H). J. Funct. Anal. **268**(9), 2754–2786 (2015). DOI 10.1016/j.jfa.2015.02.001
30. Cortiñas, G., Tartaglia, G.: Trace class operators, regulators, and assembly maps in K-theory. Doc. Math. **19**, 439–456 (2014). URL https://www.math.uni-bielefeld.de/documenta/vol-19/14.pdf
31. Davis, J.F.: Manifold aspects of the Novikov conjecture. In: Surveys on surgery theory, Vol. 1, *Ann. of Math. Stud.*, vol. 145, pp. 195–224. Princeton Univ. Press, Princeton, NJ (2000)
32. Dranishnikov, A.N., Ferry, S.C., Weinberger, S.: Large Riemannian manifolds which are flexible. Ann. of Math. (2) **157**(3), 919–938 (2003). DOI 10.4007/annals.2003.157.919
33. Farrell, F.T., Hsiang, W.C.: Manifolds with $\pi_1 = G \times_\alpha T$. Amer. J. Math. **95**, 813–848 (1973)
34. Farrell, F.T., Jones, L.E.: Negatively curved manifolds with exotic smooth structures. J. Amer. Math. Soc. **2**(4), 899–908 (1989). DOI 10.2307/1990898
35. Farrell, F.T., Jones, L.E.: Exotic smoothings of hyperbolic manifolds which do not support pinched negative curvature. Proc. Amer. Math. Soc. **121**(2), 627–630 (1994). DOI 10.2307/2160446
36. Ferry, S., Rosenberg, J., Weinberger, S.: Phénomènes de rigidité topologique équivariante. C. R. Acad. Sci. Paris Sér. I Math. **306**(19), 777–782 (1988)
37. Ferry, S.C., Ranicki, A., Rosenberg, J.: A history and survey of the Novikov conjecture. In: Novikov conjectures, index theorems and rigidity, Vol. 1 (Oberwolfach, 1993), *London Math. Soc. Lecture Note Ser.*, vol. 226, pp. 7–66. Cambridge Univ. Press, Cambridge (1995). DOI 10.1017/CBO9780511662676.003
38. Ferry, S.C., Ranicki, A., Rosenberg, J. (eds.): Novikov conjectures, index theorems and rigidity. Vol. 1, *London Mathematical Society Lecture Note Series*, vol. 226. Cambridge University Press, Cambridge (1995). Including papers from the conference held at the Mathematisches Forschungsinstitut Oberwolfach, Oberwolfach, September 6–10, 1993
39. Ferry, S.C., Ranicki, A., Rosenberg, J. (eds.): Novikov conjectures, index theorems and rigidity. Vol. 2, *London Mathematical Society Lecture Note Series*, vol. 227. Cambridge University Press, Cambridge (1995). Including papers from the conference held at the Mathematisches Forschungsinstitut Oberwolfach, Oberwolfach, September 6–10, 1993
40. Finn-Sell, M.: Fibred coarse embeddings, a-T-menability and the coarse analogue of the Novikov conjecture. J. Funct. Anal. **267**(10), 3758–3782 (2014). DOI 10.1016/j.jfa.2014.09.012
41. Freedman, M.H.: The topology of four-dimensional manifolds. J. Differential Geom. **17**(3), 357–453 (1982). URL http://projecteuclid.org/euclid.jdg/1214437136
42. Fukaya, T., Oguni, S.i.: Coronae of product spaces and the coarse Baum–Connes conjecture. Adv. Math. **279**, 201–233 (2015). DOI 10.1016/j.aim.2015.01.022

43. Gong, D.: Equivariant Novikov conjecture for groups acting on Euclidean buildings. Trans. Amer. Math. Soc. **350**(6), 2141–2183 (1998). DOI 10.1090/S0002-9947-98-01990-4
44. Gong, D., Liu, K.: Rigidity of higher elliptic genera. Ann. Global Anal. Geom. **14**(3), 219–236 (1996). DOI 10.1007/BF00054471
45. Gromov, M.: Positive curvature, macroscopic dimension, spectral gaps and higher signatures. In: Functional analysis on the eve of the 21st century, Vol. II (New Brunswick, NJ, 1993), *Progr. Math.*, vol. 132, pp. 1–213. Birkhäuser Boston, Boston, MA (1996). DOI 10.1007/s10107-010-0354-x
46. Gromov, M., Lawson Jr., H.B.: Positive scalar curvature and the Dirac operator on complete Riemannian manifolds. Inst. Hautes Études Sci. Publ. Math. **58**, 83–196 (1984) (1983). URL http://www.numdam.org/item?id=PMIHES_1983__58__83_0
47. Higson, N.: The Baum-Connes conjecture. In: Proceedings of the International Congress of Mathematicians, Vol. II (Berlin, 1998), Extra Vol. II, pp. 637–646 (electronic) (1998). URL https://www.math.uni-bielefeld.de/documenta/xvol-icm/08/08.html
48. Higson, N., Lafforgue, V., Skandalis, G.: Counterexamples to the Baum-Connes conjecture. Geom. Funct. Anal. **12**(2), 330–354 (2002). DOI 10.1007/s00039-002-8249-5
49. Higson, N., Roe, J.: On the coarse Baum-Connes conjecture. In: Novikov conjectures, index theorems and rigidity, Vol. 2 (Oberwolfach, 1993), *London Math. Soc. Lecture Note Ser.*, vol. 227, pp. 227–254. Cambridge Univ. Press, Cambridge (1995). DOI 10.1017/CBO9780511629365.008
50. Higson, N., Roe, J.: Analytic K-homology. Oxford Mathematical Monographs. Oxford University Press, Oxford (2000). Oxford Science Publications
51. Higson, N., Roe, J.: Mapping surgery to analysis. I. Analytic signatures. K-Theory **33**(4), 277–299 (2005). DOI 10.1007/s10977-005-1561-8
52. Higson, N., Roe, J.: Mapping surgery to analysis. II. Geometric signatures. K-Theory **33**(4), 301–324 (2005). DOI 10.1007/s10977-005-1559-2
53. Higson, N., Roe, J.: Mapping surgery to analysis. III. Exact sequences. K-Theory **33**(4), 325–346 (2005). DOI 10.1007/s10977-005-1554-7
54. Hirzebruch, F.: The signature theorem: reminiscences and recreation. In: Prospects in mathematics (Proc. Sympos., Princeton Univ., Princeton, N.J., 1970), pp. 3–31. Ann. of Math. Studies, No. 70. Princeton Univ. Press, Princeton, N.J. (1971)
55. Joachim, M., Schick, T.: Positive and negative results concerning the Gromov-Lawson-Rosenberg conjecture. In: Geometry and topology: Aarhus (1998), *Contemp. Math.*, vol. 258, pp. 213–226. Amer. Math. Soc., Providence, RI (2000). DOI 10.1090/conm/258/04066
56. Kahn, P.J.: Characteristic numbers and oriented homotopy type. Topology **3**, 81–95 (1965)
57. Kasparov, G.: Novikov's conjecture on higher signatures: the operator K-theory approach. In: Representation theory of groups and algebras, *Contemp. Math.*, vol. 145, pp. 79–99. Amer. Math. Soc., Providence, RI (1993). DOI 10.1090/conm/145/1216182
58. Kasparov, G.G.: The homotopy invariance of rational Pontrjagin numbers. Dokl. Akad. Nauk SSSR **190**, 1022–1025 (1970)
59. Kasparov, G.G.: Topological invariants of elliptic operators. I. K-homology. Izv. Akad. Nauk SSSR Ser. Mat. **39**(4), 796–838 (1975)
60. Kasparov, G.G.: The operator K-functor and extensions of C^*-algebras. Izv. Akad. Nauk SSSR Ser. Mat. **44**(3), 571–636, 719 (1980)
61. Kasparov, G.G.: Equivariant KK-theory and the Novikov conjecture. Invent. Math. **91**(1), 147–201 (1988). DOI 10.1007/BF01404917
62. Kasparov, G.G.: K-theory, group C^*-algebras, and higher signatures (conspectus). In: Novikov conjectures, index theorems and rigidity, Vol. 1 (Oberwolfach, 1993), *London Math. Soc. Lecture Note Ser.*, vol. 226, pp. 101–146. Cambridge Univ. Press, Cambridge (1995). DOI 10.1017/CBO9780511662676.007
63. Kasprowski, D.: On the K-theory of groups with finite decomposition complexity. Proc. Lond. Math. Soc. (3) **110**(3), 565–592 (2015). DOI 10.1112/plms/pdu062
64. Kervaire, M.A.: Le théorème de Barden-Mazur-Stallings. Comment. Math. Helv. **40**, 31–42 (1965)

65. Kervaire, M.A., Milnor, J.W.: Groups of homotopy spheres. I. Ann. of Math. (2) **77**, 504–537 (1963)
66. Kreck, M., Lück, W.: The Novikov conjecture, Geometry and algebra, *Oberwolfach Seminars*, vol. 33. Birkhäuser Verlag, Basel (2005).
67. Lafforgue, V.: K-théorie bivariante pour les algèbres de Banach et conjecture de Baum-Connes. Invent. Math. **149**(1), 1–95 (2002). DOI 10.1007/s002220200213
68. Lichnerowicz, A.: Spineurs harmoniques. C. R. Acad. Sci. Paris **257**, 7–9 (1963)
69. Loday, J.L.: K-théorie algébrique et représentations de groupes. Ann. Sci. École Norm. Sup. (4) **9**(3), 309–377 (1976)
70. Lück, W.: K- and L-theory of group rings. In: Proceedings of the International Congress of Mathematicians. Volume II, pp. 1071–1098. Hindustan Book Agency, New Delhi (2010)
71. Lück, W., Reich, H.: The Baum-Connes and the Farrell-Jones conjectures in K- and L-theory. In: Handbook of K-theory. Vol. 1, 2, pp. 703–842. Springer, Berlin (2005). DOI 10.1007/978-3-540-27855-9_15. URL http://k-theory.org/handbook/2-0703-0842.pdf
72. Lusztig, G.: Novikov's higher signature and families of elliptic operators. J. Differential Geometry **7**, 229–256 (1972)
73. Mendes, S., Plymen, R.: Base change and K-theory for $GL(n)$. J. Noncommut. Geom. **1**(3), 311–331 (2007). DOI 10.4171/JNCG/9
74. Milnor, J.: On manifolds homeomorphic to the 7-sphere. Ann. of Math. (2) **64**, 399–405 (1956)
75. Milnor, J.: A procedure for killing homotopy groups of differentiable manifolds. In: Proc. Sympos. Pure Math., Vol. III, pp. 39–55. American Mathematical Society, Providence, R.I (1961)
76. Milnor, J.: Lectures on the h-cobordism theorem. Notes by L. Siebenmann and J. Sondow. Princeton University Press, Princeton, N.J. (1965)
77. Milnor, J.: Whitehead torsion. Bull. Amer. Math. Soc. **72**, 358–426 (1966)
78. Mishchenko, A.S.: Infinite-dimensional representations of discrete groups, and higher signatures. Izv. Akad. Nauk SSSR Ser. Mat. **38**, 81–106 (1974)
79. Mishchenko, A.S.: C^*-algebras and K-theory. In: Algebraic topology, Aarhus 1978 (Proc. Sympos., Univ. Aarhus, Aarhus, 1978), *Lecture Notes in Math.*, vol. 763, pp. 262–274. Springer, Berlin (1979)
80. Mitchener, P.D.: The general notion of descent in coarse geometry. Algebr. Geom. Topol. **10**(4), 2419–2450 (2010). DOI 10.2140/agt.2010.10.2419. URL http://msp.org/agt/2010/10-4/p18.xhtml
81. Novikov, S.P.: A diffeomorphism of simply connected manifolds. Dokl. Akad. Nauk SSSR **143**, 1046–1049 (1962)
82. Novikov, S.P.: Homotopically equivalent smooth manifolds. I. Izv. Akad. Nauk SSSR Ser. Mat. **28**, 365–474 (1964)
83. Novikov, S.P.: Rational Pontrjagin classes. Homeomorphism and homotopy type of closed manifolds. I. Izv. Akad. Nauk SSSR Ser. Mat. **29**, 1373–1388 (1965)
84. Novikov, S.P.: Topological invariance of rational classes of Pontrjagin. Dokl. Akad. Nauk SSSR **163**, 298–300 (1965)
85. Novikov, S.P.: Pontrjagin classes, the fundamental group and some problems of stable algebra. In: Amer. Math. Soc. Translations Ser. 2, Vol. 70: 31 Invited Addresses (8 in Abstract) at the Internat. Congr. Math. (Moscow, 1966), pp. 172–179. Amer. Math. Soc., Providence, R.I. (1968)
86. Novikov, S.P.: Algebraic construction and properties of Hermitian analogs of K-theory over rings with involution from the viewpoint of Hamiltonian formalism. Applications to differential topology and the theory of characteristic classes. I. II. Izv. Akad. Nauk SSSR Ser. Mat. **34**, 253–288; ibid. 34 (1970), 475–500 (1970)
87. Novikov, S.P.: Pontrjagin classes, the fundamental group and some problems of stable algebra. In: Essays on Topology and Related Topics (Mémoires dédiés à Georges de Rham), pp. 147–155. Springer, New York (1970)

88. Novikov, S.P.: Analogues hermitiens de la K-théorie. In: Actes du Congrès International des Mathématiciens (Nice, 1970), Tome 2, pp. 39–45. Gauthier-Villars, Paris (1971)
89. Piazza, P., Schick, T.: The surgery exact sequence, K-theory and the signature operator. Ann. of K-Theory, **1**(2), 109–154 (2016). URL http://arxiv.org/abs/1309.4370, DOI 10.2140/akt.2016.1.109.
90. Plymen, R.J.: Reduced C^*-algebra of the p-adic group $\mathrm{GL}(n)$. II. J. Funct. Anal. **196**(1), 119–134 (2002). DOI 10.1006/jfan.2002.3980
91. Quinn, F.: A geometric formulation of surgery. In: Topology of Manifolds (Proc. Inst., Univ. of Georgia, Athens, Ga., 1969), pp. 500–511. Markham, Chicago, Ill. (1970)
92. Ramras, D.A., Tessera, R., Yu, G.: Finite decomposition complexity and the integral Novikov conjecture for higher algebraic K-theory. J. Reine Angew. Math. **694**, 129–178 (2014). DOI 10.1515/crelle-2012-0112
93. Ranicki, A.: The total surgery obstruction. In: Algebraic topology, Aarhus 1978 (Proc. Sympos., Univ. Aarhus, Aarhus, 1978), *Lecture Notes in Math.*, vol. 763, pp. 275–316. Springer, Berlin (1979)
94. Ranicki, A.: Algebraic and geometric surgery. Oxford Mathematical Monographs. The Clarendon Press, Oxford University Press, Oxford (2002). DOI 10.1093/acprof:oso/9780198509240.001.0001. Oxford Science Publications
95. Ranicki, A.A.: Algebraic L-theory and topological manifolds, *Cambridge Tracts in Mathematics*, vol. 102. Cambridge University Press, Cambridge (1992)
96. Roe, J.: Coarse cohomology and index theory on complete Riemannian manifolds. Mem. Amer. Math. Soc. **104**(497), x+90 (1993). DOI 10.1090/memo/0497
97. Rosenberg, J.: C^*-algebras, positive scalar curvature, and the Novikov conjecture. Inst. Hautes Études Sci. Publ. Math. **58**, 197–212 (1984) (1983). URL http://www.numdam.org/item?id=PMIHES_1983__58__197_0
98. Rosenberg, J.: C^*-algebras, positive scalar curvature, and the Novikov conjecture. III. Topology **25**(3), 319–336 (1986). DOI 10.1016/0040-9383(86)90047-9
99. Rosenberg, J.: Analytic Novikov for topologists. In: Novikov conjectures, index theorems and rigidity, Vol. 1 (Oberwolfach, 1993), *London Math. Soc. Lecture Note Ser.*, vol. 226, pp. 338–372. Cambridge Univ. Press, Cambridge (1995). DOI 10.1017/CBO9780511662676.013
100. Rosenberg, J.: Manifolds of positive scalar curvature: a progress report. In: Surveys in differential geometry. Vol. XI, *Surv. Differ. Geom.*, vol. 11, pp. 259–294. Int. Press, Somerville, MA (2007). DOI 10.4310/SDG.2006.v11.n1.a9
101. Rosenberg, J.: An analogue of the Novikov conjecture in complex algebraic geometry. Trans. Amer. Math. Soc. **360**(1), 383–394 (2008). DOI 10.1090/S0002-9947-07-04320-6
102. Rosenberg, J.: Structure and applications of real C^*-algebras. In: R.S. Doran, E. Park (eds.) Operator Algebras and Their Applications: A Tribute to Richard V. Kadison. Contemporary Math., Amer. Math. Soc., 2016. URL http://arxiv.org/abs/1505.04091
103. Rosenberg, J., Stolz, S.: A "stable" version of the Gromov-Lawson conjecture. In: The Čech centennial (Boston, MA, 1993), *Contemp. Math.*, vol. 181, pp. 405–418. Amer. Math. Soc., Providence, RI (1995). DOI 10.1090/conm/181/02046
104. Rosenberg, J., Weinberger, S.: Higher G-indices and applications. Ann. Sci. École Norm. Sup. (4) **21**(4), 479–495 (1988). URL http://www.numdam.org/item?id=ASENS_1988_4_21_4_479_0
105. Rosenberg, J., Weinberger, S.: An equivariant Novikov conjecture. K-Theory **4**(1), 29–53 (1990). DOI 10.1007/BF00534192. With an appendix by J. P. May
106. Rosenberg, J., Weinberger, S.: The signature operator at 2. Topology **45**(1), 47–63 (2006). DOI 10.1016/j.top.2005.06.001
107. Roushon, S.K.: The isomorphism conjecture in L-theory: graphs of groups. Homology Homotopy Appl. **14**(1), 1–17 (2012). DOI 10.4310/HHA.2012.v14.n1.a1
108. Schick, T.: A counterexample to the (unstable) Gromov-Lawson-Rosenberg conjecture. Topology **37**(6), 1165–1168 (1998). DOI 10.1016/S0040-9383(97)00082-7

109. Smale, S.: Generalized Poincaré's conjecture in dimensions greater than four. Ann. of Math. (2) **74**, 391–406 (1961)
110. Solleveld, M.: Periodic cyclic homology of reductive p-adic groups. J. Noncommut. Geom. **3**(4), 501–558 (2009). DOI 10.4171/JNCG/45
111. Sullivan, D.P.: Triangulating and smoothing homotopy equivalences and homeomorphisms. Geometric Topology Seminar Notes. In: The Hauptvermutung book, *K-Monogr. Math.*, vol. 1, pp. 69–103. Kluwer Acad. Publ., Dordrecht (1996). DOI 10.1007/978-94-017-3343-4_3
112. Thom, R.: Quelques propriétés globales des variétés différentiables. Comment. Math. Helv. **28**, 17–86 (1954). URL http://retro.seals.ch/digbib/view?rid=comahe-002:1954:28::8
113. Tu, J.L.: The Baum-Connes conjecture for groupoids. In: C^*-algebras (Münster, 1999), pp. 227–242. Springer, Berlin (2000) www.e-periodica.ch/digbib/view?pid=com-001:1954:28::8
114. Tu, J.L.: The coarse Baum-Connes conjecture and groupoids. II. New York J. Math. **18**, 1–27 (2012). URL http://nyjm.albany.edu:8000/j/2012/18_1.html
115. Wall, C.T.C.: Surgery of non-simply-connected manifolds. Ann. of Math. (2) **84**, 217–276 (1966)
116. Wall, C.T.C.: Surgery on compact manifolds, *Mathematical Surveys and Monographs*, vol. 69, second edn. American Mathematical Society, Providence, RI (1999). DOI 10.1090/surv/069. Edited and with a foreword by A. A. Ranicki
117. Wallace, A.H.: Modifications and cobounding manifolds. Canad. J. Math. **12**, 503–528 (1960)
118. Wassermann, A.: Une démonstration de la conjecture de Connes-Kasparov pour les groupes de Lie linéaires connexes réductifs. C. R. Acad. Sci. Paris Sér. I Math. **304**(18), 559–562 (1987)
119. Weinberger, S.: The topological classification of stratified spaces. Chicago Lectures in Mathematics. University of Chicago Press, Chicago, IL (1994)
120. Yu, G.: Coarse Baum-Connes conjecture. K-Theory **9**(3), 199–221 (1995). DOI 10.1007/BF00961664
121. Yu, G.: The coarse Baum-Connes conjecture for spaces which admit a uniform embedding into Hilbert space. Invent. Math. **139**(1), 201–240 (2000). DOI 10.1007/s002229900032

The Discrete Logarithm Problem

René Schoof

Abstract For large prime numbers p, computing discrete logarithms of elements of the multiplicative group $(\mathbf{Z}/p\mathbf{Z})^*$ is at present a very difficult problem. The security of certain cryptosystems is based on the difficulty of this computation. In this expository paper we discuss several generalizations of the discrete logarithm problem and we describe various algorithms to compute discrete logarithms.

1 Introduction

For a prime number p, the multiplicative group $(\mathbf{Z}/p\mathbf{Z})^*$ is cyclic of order $p-1$. Generators of $(\mathbf{Z}/p\mathbf{Z})^*$ are called *primitive roots mod p*. Let p be a prime and let g denote a primitive root modulo p. Then for every $x \in (\mathbf{Z}/p\mathbf{Z})^*$ we have

$$x = g^a,$$

for some integer a. This integer is called the *discrete logarithm of* x and is denoted by $\log x$. Since it depends on the primitive root g, one often writes $\log_g x$ rather than $\log x$. Since the discrete logarithm is only unique modulo $p-1$, we view it as an element of the additive group $\mathbf{Z}/(p-1)\mathbf{Z}$. Just as for the usual logarithm, we have for $x, y \in (\mathbf{Z}/p\mathbf{Z})^*$ that

$$\log xy = \log x + \log y.$$

Here is an explicit example. Let $p = 10000000259$. Since $(p-1)/2$ is prime, it is easy to see that $g = 2$ is a primitive root modulo p. We have

$$\log 3 = 9635867242,$$
$$\log 5 = 227891530,$$

R. Schoof (✉)
Dipartimento di Matematica, Università di Roma "Tor Vergata", Via della Ricerca Scientifica, I-00133 Roma, Italy
e-mail: schoof@mat.uniroma2.it

J.F. Nash, Jr., M.Th. Rassias (eds.), *Open Problems in Mathematics*,
DOI 10.1007/978-3-319-32162-2_12

$$\log 7 = 1803320787,$$
$$\vdots$$

illustrating the fact that there is no simple minded formula for $\log x$ in terms of x. Indeed, the discrete logarithms of the first few primes appear like random numbers in the interval from 0 to $p - 1$.

Given a prime number p, a primitive root $g \in (\mathbf{Z}/p\mathbf{Z})^*$ and an exponent a in $\mathbf{Z}/(p-1)\mathbf{Z}$, the element $x = g^a$ can be computed efficiently by repeated squarings and multiplications. On the other hand, given a large prime number p and a primitive root $g \in (\mathbf{Z}/p\mathbf{Z})^*$, there are at present no good methods to compute the discrete logarithm of a given element $x \in (\mathbf{Z}/p\mathbf{Z})^*$. In other words, computing the exponent $a \in \mathbf{Z}/(p-1)\mathbf{Z}$ for which $x = g^a$, is a very difficult problem. In particular, there is no polynomial time algorithm known to perform this calculation. There do, however, exist subexponential algorithms.

Designing good algorithms to compute discrete logarithms is a problem that is of interest in itself. It is also relevant for applications in cryptography. The security of the Diffie-Hellmann key exchange protocol [6] relies on the assumption that computing discrete logarithms is very hard. More precisely, the working hypothesis of this protocol is that given a prime number p, a primitive root g modulo p and elements x and y in $(\mathbf{Z}/p\mathbf{Z})^*$, but not their discrete logarithms a and b, it is very hard to compute the element g^{ab}.

In this expository paper we describe some methods for computing discrete logarithms. We do not pretend to present the best known ones, but merely try to give the main ideas behind the most commonly used algorithms. In Sect. 2 we describe a natural generalization of the discrete logarithm problem and we discuss the baby-step-giant-step method due to Dan Shanks and a probabilistic method due to John Pollard. Section 3 is dedicated to the subexponential index calculus algorithm. In Sect. 4 we discuss the discrete logarithm problem for multiplicative groups of finite fields of relatively small characteristic. In particular, we describe recent work of Antoine Joux and others. Finally in Sect. 5, we discuss the discrete logarithm problem for groups of points of elliptic curves over finite fields. This is relevant for applications in cryptography.

I thank Hendrik Lenstra for several useful comments on an earlier version of this paper.

2 Exponential Algorithms

Following ideas in [11, Sect. 7], we consider the following general problem. It is a natural extension of the discrete logarithm problem.

Problem. Given a finite set S, a finite abelian group A and a group homomorphism

$$f : \mathbf{Z}^S \longrightarrow A,$$

determine the kernel of f.

In applications, the group A is the multiplicative group of a finite field or the multiplicative group $(\mathbf{Z}/n\mathbf{Z})^*$ of the finite ring $\mathbf{Z}/n\mathbf{Z}$. It can also be an ideal class group of a number field, or the group of points of an elliptic curve or of an abelian variety over a finite field. In general, we assume that A is given to us in such a way that we can efficiently compose elements, calculate inverses and test for equality. Usually we even require that every element in A has a unique, easily computable "reduced" representative. But in general we do not suppose that we know the structure or even the cardinality of A. In most applications we know an upper bound for $\#A$. Since $\mathrm{Hom}(\mathbf{Z}^S, A)$ is naturally isomorphic to A^S, the map f can be specified by giving $\#S$ elements in A indexed by $s \in S$. The kernel of f is a free group of the same rank as $\mathbf{Z}^S$. It can be described by giving $\#S$ generators.

If S consists of one element, then f is determined by $f(1) = x \in A$. Determining the kernel of f is the same problem as determining the order of the element $x \in A$. An algorithm that solves this problem for the group $A = (\mathbf{Z}/n\mathbf{Z})^*$ can be used to factor the integer n. See [13, Lemma 5]. If $\#S = 2$ and f is therefore given by two elements $x, y \in A$, determining the kernel of f is the same as finding all relations between x and y that are of the form $x^k y^m = 1$ with $(k, m) \in \mathbf{Z}^2$. In particular, if A is a cyclic group of order m generated by x, so that $y = x^a$ for some $a \in \mathbf{Z}$, then the kernel of f is the subgroup generated by $(m, 0)$ and $(a, -1) \in \mathbf{Z}^2$. Determining $\ker f$ is therefore the same problem as computing the 'discrete logarithm' a of y.

In general, the difficulty in computing the kernel of a group homomorphism $f : \mathbf{Z}^S \longrightarrow A$ does not depend on the group structure of the finite abelian group A, but rather on the way it is presented. For instance, if A is the additive group $\mathbf{Z}/n\mathbf{Z}$ of integers modulo n, presented in the usual way, the problem is very easy. It can be solved using at most $O(\#S)$ gcd computations in the ring $\mathbf{Z}$, which can be efficiently calculated by means of the Euclidean algorithm. Indeed, when $\#S = 1$ the kernel of $f : \mathbf{Z} \longrightarrow \mathbf{Z}/n\mathbf{Z}$ is generated by n/d, where $d = \gcd(f(1), n)$. For larger S one proceeds inductively, dealing in a similar way with one element of S at the time. On the other hand, if A is the cyclic multplicative group $\mathbf{F}_p^*$ or the group $E(\mathbf{F}_p)$ of an elliptic curve over $\mathbf{F}_p$ for some large p, there are no methods known to compute the kernel of a homomorphism $f : \mathbf{Z}^S \longrightarrow A$ that are of a comparable efficiency.

Let $f : \mathbf{Z}^S \longrightarrow A$ be a homomorphism. Since both A and S are finite sets, the problem of determining the kernel of f is a finite issue. The straightforward naive algorithm to solve it, runs as follows. First assume that $\#S = 1$. In other words, we are given an element in $x \in A$ and we wish to compute its order. This can be done by computing the powers $1, x, x^2, \ldots$ of x until x^i is equal to the neutral element $1 \in A$. The exponent i is then the order of x and generates $\ker f$. When S is larger and f is determined by elements $x, y, z, \ldots \in A$, we first list the elements of the subgroup H generated by x as above. Next we list all elements in the cosets $y^i H$ for $i = 0, 1, 2, \ldots$ The smallest exponent i for which $y^i H = H$ gives rise to a relation of the form $y^i = x^j$ and hence to an element in the kernel of f. Next one puts $H = \langle x, y \rangle$ and lists all elements of the cosets $z^i H \dots$ etc. This method uses $O(\#A\#S)$ operations in A. Since eventually all elements of A may be listed, the amount of memory required is $O(\#A)$.

If we can compute a proper non-trivial subgroup B of A, then the problem of computing the kernel K of $f : \mathbf{Z}^S \longrightarrow A$ can be reduced to *two* similar problems involving the subgroup B and the quotient group $B' = A/B$. More precisely, writing π for the canonical map $A \longrightarrow B'$ and K' for the kernel of the composite map $\pi \cdot f : \mathbf{Z}^S \longrightarrow A \longrightarrow B$, we have the following commutative diagram with exact rows and columns

$$
\begin{array}{ccccccccc}
 & & 0 & & 0 & & & & \\
 & & \downarrow & & \downarrow & & & & \\
 & & \ker f & \stackrel{\cong}{\longrightarrow} & K & & & & \\
 & & \downarrow & & \downarrow & & & & \\
0 & \longrightarrow & K' & \longrightarrow & \mathbf{Z}^S & \stackrel{\pi\cdot f}{\longrightarrow} & B' & & \\
 & & \downarrow{\scriptstyle f} & & \downarrow{\scriptstyle f} & & \| & & \\
0 & \longrightarrow & B & \longrightarrow & A & \stackrel{\pi}{\longrightarrow} & B' & \longrightarrow & 0.
\end{array}
$$

The group K' is isomorphic to $\mathbf{Z}^{S'}$ for some finite set S' having the same cardinality as S. The map f maps K' to B. Since the kernel of the homomorphism $f : K' \longrightarrow B$, is isomorphic to K, we can compute K by first computing the kernel K' of $\pi \cdot f : \mathbf{Z}^S \longrightarrow B'$ and then the kernel of $f : K' \longrightarrow B$.

The groups B and B' are smaller than A. Since B is contained in A, one can efficiently compose elements, calculate inverses and test for equality in B. Therefore, if one is also able to do this in the group $B' = A/B$, then it is usually a good idea to make this reduction. This observation is due to Pohlig-Hellmann [16]. It applies for instance, if one knows a section $j : B' \longrightarrow A$ of π so that $A \cong B \times B'$. In this case equality tests in B' can be performed in A. Another example is the case when A is cyclic and we know a proper divisor d of the order of $n = \#A$. Then we can take $B = dA$ and compute in $B' = A/dA$ exploiting the isomorphism $B' \cong (n/d)A$ given by multiplication by n/d. It follows that computing the kernel of $f : \mathbf{Z}^S \longrightarrow A$ is relatively easy when all prime divisors of $\#A$ are small. Therefore, in this paper one should keep in mind groups A, for which $\#A$ is divisible by at least one large prime number.

Next we describe a more efficient algorithm to compute the kernel of a homomorphism $f : \mathbf{Z}^S \longrightarrow A$. It is the baby-step-giant-step algorithm due to Dan Shanks [20]. It is deterministic and uses $O(\#S\sqrt{\#A})$ operations and equality tests in the group A. It also requires the storage of $O(\sqrt{\#A})$ elements of A. We explain the algorithm in the case $\#S = 1$. For larger S, the idea remains the same, but, as in our description above of the naive algorithm, the details are more cumbersome to write down [3]. Any homomorphism $f : \mathbf{Z} \longrightarrow A$ is determined by the element $x = f(1)$ of A. Let a be the integer part of $\sqrt{\#A} + 1$. We first make baby-steps. This means that we make a list of the elements x^i for $0 \leq i < a$. If for some i in this range, x^i is the neutral element of A, we are done: the smallest such i is the order of x and generates the kernel of f. If this is not the case, we make giant steps: we put $y = x^a$ and compute y^j for $1 \leq j \leq a$. Each time we check whether y^j is in the list that we made. In order to be able to do this efficiently, we assume that the elements in A are presented in some unique "reduced" way and that the list of the elements x^i for $0 \leq i \leq a$ is sorted with respect to this presentation. Since the order of x is at most $\#A < a^2$, it is bound to happen that for some j, the element y^j is in the list. If it does, we have $x^i = y^j = x^{aj}$ for some $i = 0, 1, \ldots, a$. The first value of j for which this happens has the property that $aj - i$ is the order of x and hence generates the kernel of f.

There are also probabilistic algorithms to compute the kernel of $f : \mathbf{Z}^S \longrightarrow A$. They have the same running time as the baby-step-giant-step algorithm. The advantage of the probabilistic algorithms is, that they do not require making lists of size $\sqrt{\#A}$. We describe the so-called ϱ-algorithm, due to John Pollard [17]. Once again we explain the algorithm only in the case $\#S = 1$. In this case, we put $x = f(1)$ and make a random, or rather pseudorandom, walk by evaluating elements of the form x^{n_i} for $i = 0, 1, 2, \ldots$ with $1 = n_0 < n_1 < n_2 \ldots$. This means that for each i, the next element $x^{n_{i+1}}$ is computed in a pseudorandom fashion from the element $x^{n_i} \in A$. By the birthday paradox, one expects that $x^{n_i} = x^{n_j}$ for two distinct values of i, j that are $O(\sqrt{\#A})$. Moreover, this can be detected efficiently using the cycle detection algorithm, attributed to R.W. Floyd by Knuth [9, p. 7]. The order of x divides $n_i - n_j$. In practice the quotient is small, so that the order of x can be computed easily.

3 Index Calculus

Index calculus is a method to compute discrete logarithms and, more generally, to determine kernels of homomorphisms $f : \mathbf{Z}^S \longrightarrow A$, that applies when A is the multiplicative group of a finite field. In this section we assume that $A = \mathbf{F}_p^*$ for some large prime p. In the next section we consider the case where A is the multiplicative group of a finite field of small characteristic.

Before considering the map f, we do a precomputation and use the index calculus algorithm to compute the kernel of

$$h : \mathbf{Z}^T \longrightarrow \mathbf{F}_p^*,$$

where T is the set of the primes $l \leq X$ for some bound $X < p$ and h is the homomorphism that, for any prime $l \leq X$, maps the lth basis vector of $\mathbf{Z}^T$ to $l \pmod{p}$. The kernel of h consists of the vectors $(x_l)_{l \in T} \in \mathbf{Z}^T$ that satisfy

$$\prod_{l \in T} l^{x_l} = 1 \text{ in } \mathbf{Z}_p^*$$

and hence

$$\sum_{l \in T} x_l \log l \equiv 0 \pmod{p-1},$$

where $\log l$ denotes the discrete logarithm of l with respect to any fixed primitive root in $\mathbf{F}_p^*$. The algorithm to determine $\ker h$ runs as follows. Pick random exponents $e(l) \geq 0$ with $\sum_{l \in T} e(l)$ bounded by some power of $\log p$, that is sufficiently large in the sense that the products $\prod_{l \in T} l^{e(l)}$ exceed p. Then check whether the remainder modulo p of $\prod_{l \in T} l^{e(l)}$ is "X-smooth". In other words, check whether its factorization in the ring $\mathbf{Z}$ is of the form $\prod_{l \in T} l^{f(l)}$. If it is, we obtain the relation

$$\prod_{l \in T} l^{e(l)} = \prod_{l \in T} l^{f(l)}, \qquad \text{in } \mathbf{F}_p^*.$$

It follows that the vector $(e(l) - f(l))_{l \in S}$ is in the kernel of h:

$$\sum_{l \in T} (e(l) - f(l)) \log l = 0, \qquad \text{in } \mathbf{Z}/(p-1)\mathbf{Z}.$$

Repeating this procedure, we occasionally find that $\prod_{l \in T} l^{e(l)}$ is X-smooth, hence obtain a non-trivial relation and thus a non-zero vector in the kernel of h. Once we have obtained a bit more than $\#T$ vectors, it is reasonable to expect that the vectors that we found, generate the kernel.

It remains to choose the value of X. If X is very small with respect to p, there are very few X-smooth numbers in the set $\{1, 2, \ldots, p-1\}$. Since the remainders of the products $\prod_{l \in T} l^{e(l)}$ appear to be distributed randomly in this set, it is difficult to obtain relations and the algorithm may be time consuming. On the other hand, if X is very large, it is much easier to find X-smooth numbers and vectors in $\ker h$. However, since we need more than $\#T$ relations, we need to find many more of them and the algorithm may also be time consuming.

The optimal value of X is somewhere in the middle. It depends on the probability that a random natural number less than p is X-smooth. Writing $X = p^{1/u}$ for some $u > 1$, this probability is roughly u^{-u}. See [4]. A back of an envelope computation shows that the optimal value for u is approximately $u = 2\sqrt{\log p / \log\log p}$. With this choice of u, computing the kernel of f involves

$$\exp(2\sqrt{\log p \log\log p})$$

elementary operations with numbers that have $O(\log p)$ digits. Therefore this is a subexponential algorithm.

With this choice of u, the set of primes $l \leq X$ almost certainly generates $\mathbf{F}_p^*$, so that f is surjective. Therefore the induced map

$$\overline{h} : (\mathbf{Z}/(p-1)\mathbf{Z})^T \longrightarrow \mathbf{F}_p^*$$

is also surjective and hence split. This means that the kernel of $\overline{h}$ is the zero set of a single linear equation $\sum_{l\in T} a_l X_l \equiv 0 \pmod{p-1}$, with coefficients a_l equal to $\log l$ with respect to the primitive root g that is given by $g = \prod_{l\in T} l^{y_l}$. Here $(y_l)_{l\in T}$ is any vector for which one has $\sum_{l\in T} a_l y_l = 1$ in $\mathbf{Z}/(p-1)/\mathbf{Z}$. The equation can be computed efficiently using linear algebra over the ring $\mathbf{Z}/(p-1)\mathbf{Z}$. This completes the description of the precomputation.

In order to explain how to determine the kernel of the given homomorphism

$$f : \mathbf{Z}^S \longrightarrow \mathbf{F}_p^*,$$

we consider first the case that $\#S = 1$. In this case f is determined by the element $x = f(1) \in \mathbf{F}_p^*$ and the kernel of f is generated by the order of x in the group $\mathbf{F}_p^*$. To compute the kernel, pick random products $x\prod_{l\in T} l^{e(l)}$ with T as above and check whether the factorization in $\mathbf{Z}$ of the remainder modulo p is of the form $\prod_{l\in T} l^{f(l)}$. If this happens, we obtain the relation

$$x\prod_{l\in T} l^{e(l)} = \prod_{l\in T} l^{f(l)}, \qquad \text{in } \mathbf{F}_p^*.$$

This implies that

$$\log x = \sum_{l\in T} (f(l) - e(l)) \log l, \quad \text{in } \mathbf{Z}/(p-1)\mathbf{Z}.$$

Since we already have computed $\log l$ for every $l \in T$, we can now evaluate $\log x$. The order of x in the group $\mathbf{F}_p^*$ is equal to the order of $\log x$ in the additive group $\mathbf{Z}/(p-1)\mathbf{Z}$. It is therefore equal to $p-1$ divided by $\gcd(p-1, \log x)$.

The method for $\#S > 1$ is based on this. One computes the discrete logarithm of $f(s)$ for each element $s \in S$. Composing $f : \mathbf{Z}^S \longrightarrow \mathbf{F}_p^*$ with the discrete logarithm gives a homomorphism from $\mathbf{Z}^S$ to the additive group $\mathbf{Z}/(p-1)\mathbf{Z}$. As remarked above, determining the kernel of such a homomorphism is easy and can be done by means of linear algebra over $\mathbf{Z}$.

4 Finite Fields

Recently there has been great progress in solving the discrete logarithm problem for finite fields of small characteristic. Indeed, in [1, 5, 7, 8] algorithms are described that almost run in polynomial time. As in the previous section, the algorithms proceed by first computing the logarithms of a set of elements—*the factor base*—that are, in some sense, small. Next one uses this to solve the problem of computing the discrete logarithm of an individual element that is not in the factor base. Here we describe the first phase following Antoine Joux and his collaborators [1]. For the second phase we refer to the papers mentioned above for more details.

As a typical example of the method, we discuss the case of a finite field of $Q = q^{2k}$ elements, where q is a prime power and k is of the same order of magnitude as q. See [1] for a precise description of the range of finite fields for which the algorithm is effective. Let $\mathbf{F}_{q^2}$ denote the subfield of q^2 elements of $\mathbf{F}_Q$. Note that if k and q are approximately equal, q^2 is small with respect to $Q = q^{2k}$. Therefore, making a list of *all* elements of $\mathbf{F}_{q^2}$ can be done in time polynomial in $\log Q$. As a consequence, computing discrete logarithms of elements in $\mathbf{F}_{q^2}^*$ can be done in time polynomial in $\log Q$ as well. Therefore, the Pohlig-Hellmann argument of Sect. 2 reduces the problem of computing discrete logarithms in the group $\mathbf{F}_Q^*$ to the problem of computing discrete logarithms in the group $A = \mathbf{F}_Q^*/\mathbf{F}_{q^2}^*$.

We assume that the field with $Q = q^{2k}$ elements is represented as $\mathbf{F}_{q^2}[X]/(\phi(X))$, where $\phi(X)$ is an irreducible degree k polynomial in $\mathbf{F}_{q^2}[X]$. In order to have an efficient algorithm, the polynomial $\phi(X)$ in $\mathbf{F}_Q = \mathbf{F}_{q^2}[X]/(\phi(X))$ is supposed to have a special shape: we want that

$$X^q \equiv r(X) \bmod \phi(X),$$

for some rational function $r(X) \in \mathbf{F}_{q^2}(X)$ whose numerator and denominator have very small degrees. Examples are provided by the polynomials $\phi(X) = X^{q-1} - g$ or $X^{q+1} - g$, where g is a generator of the cyclic group $\mathbf{F}_{q^2}^*$. In the first case we have $r(X) = gX$ and in the second $r(X) = g/X$. See [23]. Numerical experiments suggest [1] that when k is close to q, one can find a rational function $r(X)$ with denominator and numerator of degree at most 2, for which $X^q - r(X)$ is divisible by an irreducible degree k polynomial $\phi(X)$. Since an algorithm of Lenstra [10] allows one to compute an isomorphism between any presentation of the finite field $\mathbf{F}_Q$ and $\mathbf{F}_{q^2}[X]/(\phi(X))$ in polynomial time, requiring $\phi(X)$ to have this special shape, is not a serious restriction.

In the first phase of the algorithm we compute the discrete logarithms of the elements in a factor base, which in this case consists of the images of all monic linear polynomials in $\mathbf{F}_{q^2}[X]$ in the group $A = \mathbf{F}_{q^2}[X]/(\phi(X))^*/\mathbf{F}_{q^2}^*$. Putting $T = \mathbf{F}_{q^2}$, this means that we compute the kernel of the homomorphism

$$h : \mathbf{Z}^T \longrightarrow A,$$

that maps the basisvector e_u corresponding to $u \in T$ to the image of $X - u$ in the group $A = (\mathbf{F}_{q^2}[X]/(\phi(X)))^*/\mathbf{F}_{q^2}^*$.

The identity

$$X^q - X = \prod_{u \in \mathbf{F}_q} (X - u)$$

implies that

$$r(X) - X = \prod_{u \in \mathbf{F}_q} (X - u), \qquad \text{in } \mathbf{F}_{q^2}[X]/(\phi(X)).$$

The denominator of the rational function $r(X) - X$ is equal to the one of $r(X)$. For the sake of exposition, we suppose that it factors into a product of linear polynomials in $\mathbf{F}_{q^2}[X]$. Its numerator has degree at most 3 and may or may not factor into a product of linear polynomials in $\mathbf{F}_{q^2}[X]$. If it does, we obtain a multiplicative relation in the group A between the elements of our factor base. The relation gives then rise to an element in the kernel of the homomorphism $h : \mathbf{Z}^T \longrightarrow A$.

In order to get more relations, we apply automorphisms of the fraction field $\mathbf{F}_{q^2}(X)$ of $\mathbf{F}_{q^2}[X]$. The group $\mathrm{PGL}_2(\mathbf{F}_{q^2})$ acts on the right on $\mathbf{F}_{q^2}(X)$ as follows:

$$f^{\sigma}(X) = f(\frac{aX + b}{cX + d}), \qquad \text{for any } \sigma = \begin{pmatrix} a & b \\ c & d \end{pmatrix} \in \mathrm{PGL}_2(\mathbf{F}_{q^2}).$$

Applying $\sigma \in \mathrm{PGL}_2(\mathbf{F}_{q^2})$ to the identity above, we obtain the equality

$$(X^{\sigma})^q - X^{\sigma} = (X^q - X)^{\sigma} = \prod_{u \in \mathbf{F}_q} (X^{\sigma} - u), \qquad \text{in } \mathbf{F}_{q^2}(X).$$

The group $\mathrm{PGL}_2(\mathbf{F}_{q^2})$ acts via linear fractional transformations on the left on the projective line $\mathbf{P}_1$ and preserves the set of $\mathbf{F}_{q^2}$-points $\mathbf{P}_1(\mathbf{F}_{q^2})$. We view $\mathbf{F}_{q^2}$ as a subset of $\mathbf{P}_1(\mathbf{F}_{q^2})$. So we have $\mathbf{P}_1(\mathbf{F}_{q^2}) = \mathbf{F}_{q^2} \cup \{\infty\}$. For a function $f \in \mathbf{F}_{q^2}(X)$, a point $u \in \mathbf{P}_1(\mathbf{F}_{q^2})$ and an automorphism $\sigma \in \mathrm{PGL}_2(\mathbf{F}_{q^2})$, we have $f^{\sigma}(u) = f(\sigma(u))$.

The above identity for $\sigma = \begin{pmatrix} a & b \\ c & d \end{pmatrix} \in \mathrm{PGL}_2(\mathbf{F}_{q^2})$ then becomes

$$(cX + d)(aX + b)^q - (aX + b)(cX + d)^q = \prod_{u \in \sigma^{-1}(\mathbf{P}_1(\mathbf{F}_q)) - \{\infty\}} (X - u).$$

It holds in the function field $\mathbf{F}_{q^2}(X)$ up to multiplication by some $\lambda \in \mathbf{F}_{q^2}^*$. Note that the set $\sigma^{-1}(\mathbf{P}_1(\mathbf{F}_q))$ consists of $q + 1$ points and may or may not contain the point ∞. Since $X^q \equiv r(X)$ modulo $\phi(X)$, we have

$$(aX + b)^q \equiv \overline{a}r(X) + \overline{b} \quad \text{and} \quad (cX + d)^q \equiv \overline{c}r(X) + \overline{d}$$

in $\mathbf{F}_{q^2}[X]/(\phi(X))$. Here we put $\overline{t} = t^q$ for $t \in \mathbf{F}_{q^2}$. This leads to the following relation

$$(cX+d)(\overline{a}r(X)+\overline{b})-(aX+b)(\overline{c}r(X)+\overline{d}) = \prod_{u\in\sigma^{-1}(\mathbf{P}_1(\mathbf{F}_q))-\{\infty\}} (X-u). \quad (*)$$

It holds in the ring $\mathbf{F}_{q^2}[X]/(\phi(X))$ up to multiplication by some $\lambda \in \mathbf{F}_{q^2}^*$. The denominator of the left hand side of this equation is equal to the one of $r(X)$. The numerator has degree at most 3 and it seems reasonable to expect that it varies randomly when we vary σ. Numerical experiments have confirmed this [1]. Under this assumption a positive proportion factors into a product of linear polynomials in $\mathbf{F}_{q^2}[X]$. Indeed, this proportion is approximately $1/6$. For other choices of $r(X)$, see [15]. A positive proportion of the relations $(*)$ are therefore multiplicative relations between the elements $X-u$ of the factor base in the group $A = (\mathbf{F}_{q^2}[X]/(\phi(X)))^*/\mathbf{F}_{q^2}^*$. They give rise to elements in the kernel of the homomorphism $h : \mathbf{Z}^T \longrightarrow A$.

The question is how many independent multiplicative relations between the elements $X-u$ of the factor base we obtain in this way. The discrete logarithms of the elements $X-u$ of the factor base are a solution of the system of linear equations that we obtain from the relations $(*)$. There is at present no proof that we obtain sufficiently many relations for the linear system to have a unique solution over $\mathbf{Z}/M\mathbf{Z}$, where $M = \#A$. However, there are some heuristic arguments in this direction that seem to be confirmed by experiments [1].

The subgroup of $\mathrm{PGL}_2(\mathbf{F}_{q^2})$ that preserves the subset $\mathbf{P}_1(\mathbf{F}_q)$ of $\mathbf{P}_1(\mathbf{F}_{q^2})$, is equal to $\mathrm{PGL}_2(\mathbf{F}_q)$. Therefore, the set $\sigma^{-1}(\mathbf{P}_1(\mathbf{F}_q))$ and hence the right hand side of the relation $(*)$, depends only on the left coset $\sigma^{-1}\mathrm{PGL}_2(\mathbf{F}_q)$ rather than on σ^{-1} itself. The number of cosets is equal to $\#\mathrm{PGL}_2(\mathbf{F}_{q^2})/\#\mathrm{PGL}_2(\mathbf{F}_q) = q^3+q$. It follows that there are q^3+q different subsets $\sigma^{-1}(\mathbf{P}_1(\mathbf{F}_q))$ and hence q^3+q possibilities for the right hand sides of the relations $(*)$. For a positive proportion the left hand sides of $(*)$ factor into products of linear polynomials in $\mathbf{F}_{q^2}[X]$. Therefore one expects many more than q^2 relations between the elements $X-u$ of the factor base, at least when q is not very small. As a consequence we obtain many more than q^2 elements in the kernel of the homomorphism $h : \mathbf{Z}^T \longrightarrow A$.

The subsets $\sigma^{-1}(\mathbf{P}_1(\mathbf{F}_q))$ and therefore the right hand sides of the relations, are very different from one another. Indeed, it easy to see that any two distinct subsets $\sigma^{-1}(\mathbf{P}_1(\mathbf{F}_q))$ intersect in at most two points. Moreover, when σ runs over representatives of the cosets of $\mathrm{PGL}_2(\mathbf{F}_q)$ in $\mathrm{PGL}_2(\mathbf{F}_{q^2})$ and P runs over the points of $\mathbf{P}_1(\mathbf{F}_{q^2})$, the q^3+q by q^2+1 matrix $m_{\sigma,P}$ given by

$$m_{\sigma,P} = \begin{cases} 1, \text{ when } P \in \sigma^{-1}(\mathbf{P}_1(\mathbf{F}_q)); \\ 0, \text{ otherwise.} \end{cases}$$

has maximal rank q^2+1. See [1]. In fact, it is not difficult to show that its rows span a subgroup of index $q+1$ in $\mathbf{Z}^{q^2+1}$. The matrix of the homogeneous linear system we want to solve consists of the subset of rows of the matrix $m_{\sigma,P}$ for which the left hand side of $(*)$ factors completely, somewhat perturbed by the few non-zero coefficients that come from the left hand side of $(*)$. Since the matrix $m_{\sigma,P}$ has maximal rank, it is perhaps not unreasonable to expect that our linear system has a unique solution over $\mathbf{Z}/M\mathbf{Z}$, where $M = \#A$.

It was pointed out by Wan et al. [5] that there is a problem with linear polynomials $X - u$ that divide $(X^q - r(X))/\phi(X)$. Indeed, the relations that we find, not only hold modulo $\phi(X)$, but also modulo $X - u$. Typically the multiplicity of $X - u$ is the same on both sides of the relations $(*)$. This cancellation implies that the logarithm of $X - u$ does not appear in the linear system. Therefore it cannot be computed this way. For instance, in the case $\phi(X) = X^{q-1} - g$, we have $r(X) = gX$ and $X^q - gX = X\phi(X)$. In this case, the logarithm of X may not at all appear in the linear system. In this special case however, computing the logarithm of X is easy since its order in the group A is $q - 1$, which is very small. See the original papers [1, 8] for ways to get around this problem in general.

5 Elliptic Curves

Elliptic curve cryptography is based on the difficulty of solving the discrete logarithm problem in the finite group $E(\mathbf{F}_q)$ of points of an elliptic curve E over a finite field $\mathbf{F}_q$. More generally, determining the kernel of a homomorphism

$$f : \mathbf{Z}^S \longrightarrow E(\mathbf{F}_q),$$

is a difficult problem. Apart from some exceptional situations that we describe below, the only methods that are available at present, are the baby-step-giant-step method and the Pollard ϱ-method that were discussed in Sect. 2. Since $\#E(\mathbf{F}_q) \approx q$, both methods require $O(\sqrt{q})$ operations in the group $E(\mathbf{F}_q)$. This is much more than the number of operations required by the subexponential index calculus algorithm that was described in Sect. 3. Therefore, in cryptographical systems based on elliptic curves, key sizes can be made smaller, so that encryption and decryption algorithms are faster.

Suppose that E is an elliptic curve over a finite field $\mathbf{F}_q$ given by a Weierstrass equation

$$Y^2 + a_1XY + a_3 = X^3 + a_2X^2 + a_4X + a_6,$$

with coefficients $a_i \in \mathbf{F}_q$. See [21] for the basic properties of elliptic curves. We let $E(\overline{\mathbf{F}}_q)$ denote the group of points on E with coordinates in an algebraic closure $\overline{\mathbf{F}}_q$ of $\mathbf{F}_q$. It is an infinite torsion group. The set $E(\mathbf{F}_q)$ of points on E with coordinates in $\mathbf{F}_q$ is a finite subgroup. For every natural number n, we let $E[n]$ denote the group of points on E that are annihilated by n. In other words, we have

$$E[n] = \{P \in E(\overline{\mathbf{F}}_q) : nP = 0\}.$$

If n is not divisible by the characteristic p of $\mathbf{F}_q$, the group $E[n]$ is isomorphic to $\mathbf{Z}/n\mathbf{Z} \times \mathbf{Z}/n\mathbf{Z}$. The *Weil pairing* is a bilinear, antisymmetric and non-degenerate pairing

$$e_n : E[n] \times E[n] \longrightarrow \mu_n.$$

Here μ_n denotes the subgroup of nth roots of unity of $\overline{\mathbf{F}}_q^*$. The pairing e_n is Galois equivariant.

Suppose now that the group $E(\mathbf{F}_q)$ is cyclic of order n, coprime to p. Let $Q \in E[n]$ be a point of order n with the property that the subgroup it generates has trivial intersection with $E(\mathbf{F}_q)$. Then the map

$$g : E(\mathbf{F}_q) \longrightarrow \mu_n$$

given by $g(P) = e_n(P, Q)$ is an injective group homomorphism. It can be efficiently computed by means of an algorithm invented by Victor Miller [14]. In this way the kernel of $f : \mathbf{Z}^S \longrightarrow E(\mathbf{F}_q)$ can be calculated by computing the kernel of the composite homomorphism

$$\mathbf{Z}^S \xrightarrow{f} E(\mathbf{F}_q) \xrightarrow{g} \mu_n \hookrightarrow \mathbf{F}_{q^d}^*.$$

Here d is the order of q modulo n. This approach is due to Menezes et al. [12]. It reduces the problem of computing the kernel of a homomorphism $\mathbf{Z}^S \longrightarrow E(\mathbf{F}_q)$ to a similar problem involving the multiplicative group $\mathbf{F}_{q^d}^*$ rather than $E(\mathbf{F}_q)$.

Since the Weil pairing is Galois equivariant, the field of definition of the point Q contains $\mathbf{F}_{q^d}$. Since d is typically very large, computing in the group $E(\mathbf{F}_{q^d})$ is very costly and this approach is usually not very successful. However, in certain special cases it can be very effective. An important example is provided by *supersingular* elliptic curves over prime fields $\mathbf{F}_p$. When $p \equiv 1 \pmod 4$, the group $E(\mathbf{F}_p)$ of a supersingular curve E is cyclic of order $p + 1$. In this case μ_{p+1} is contained in an extension of $\mathbf{F}_p$ that has only degree $d = 2$. Therefore this method is very efficient. With small modifications it also works when $p \equiv 3 \pmod 4$ and more generally when the order of q modulo $n = \#E(\mathbf{F}_q)$ is small. In this situation, this use of the Weil pairing is an efficient way to compute discrete logarithms or, more generally, to compute the kernels of homomorphisms $f : \mathbf{Z}^S \longrightarrow E(\mathbf{F}_q)$.

An even faster algorithm was invented by Semaev [19] for elliptic curves E over prime fields $\mathbf{F}_p$, for which the group $E(\mathbf{F}_p)$ has order p. In this case an isomorphism

$$g : E[p] \longrightarrow \mathbf{Z}/p\mathbf{Z}$$

is constructed as follows. We fix a non-zero point Q in $E(\mathbf{F}_p)$. For $P \in E(\mathbf{F}_p)$ we put

$$g(P) = \frac{f_P'}{f_P}(Q).$$

Here f_P is a function on E whose divisor is equal to $p(Q - \infty)$. We let f'_P denote the function for which we have the following equality of Kähler differentials: $df_P = f'_P dX$. The map g can be efficiently computed by means of Miller's algorithm. In this way we can compute the kernel of $f : \mathbf{Z}^S \longrightarrow E(\mathbf{F}_p)$ by computing the kernel of

$$\mathbf{Z}^S \xrightarrow{f} E(\mathbf{F}_p) \xrightarrow{g} \mathbf{Z}/p\mathbf{Z}.$$

Similar related algorithms have been proposed by Satoh and Araki [18] and by Smart [22]. See Belding's paper [2] for a relation between the algorithms described in this section.

References

1. Barbulescu, R., Gaudry, P., Joux, A. and Thomé, E.: A quasi-polynomial algorithm for discrete logarithm in finite fields of small characteristic, In Nguyen, P., Oswald, E. (Eds) Eurocrypto 2014, LNCS **8441**, 1–16, Springer 2014.
2. Belding, J.V.: A Weil pairing on the p-torsion of ordinary elliptic curves over $K[\varepsilon]$, *J. of Number Theory*, **128** (2008), 1874–1888.
3. Buchmann, J., Jacobson, M. and Teske, E.: On some computational problems in finite abelian groups, *Math. Comp.* **66** (1997), 1663–1687.
4. Canfield, E.R., Pomerance C. and Erdős, P.: On a problem of Oppenheim concerning 'Factorisation Numerorum', *J. Number Theory* **17** (1981), 1–28.
5. Cheng, Q., Wan, D. and Zhuang, J: Traps to the BGJT-algorithm for discrete logarithms, *LMS Journal of Computation and Mathematics* **17** (2014), 218–229.
6. Diffie, W. and Hellman, M.: New directions in cryptography. *IEEE Transactions on Information Theory* **22** (1976), 587–594.
7. Göloğlu, F., Granger, R., McGuire, G. and Zumbrägel, J.: On the function field sieve and the impact of higher splitting probabilities. In Canetti, R. and Garay, J. editors, *Advances in Cryptology—CRYPTO 2013*, LNCS **8043**, 109–128. Springer 2013.
8. Granger, R., Kleinjung, T. and Zumbrägel, J.: On the discrete logarithm problem in finite fields of fixed characteristic, Cryptology ePrint Archive: Report 2015/685.
9. Knuth, Donald E.: *The Art of Computer Programming*, vol. II: Seminumerical Algorithms, Addison-Wesley 1969.
10. Lenstra, H.W.: Finding isomorphisms between finite fields, *Math. Comp.* **56** (1991), 329–347.
11. Lenstra, H.W. and Silverberg, A.: Roots of unity in orders, *Foundations of Computational Mathematics*, to appear (2016).
12. Menezes, A., Okamoto, T., Vanstone, S. A.: Reducing elliptic curve logarithms to logarithms in a finite field, *IEEE Transactions on Information Theory* **39** (1993), 1639–1646.
13. Miller, G.: Riemann's Hypothesis and tests for primality *J. of Computer and System science* **13** (1976), 300–317.
14. Miller, V.: The Weil pairing, and its efficient calculation, *J. Cryptology* **17** (2004), 235–261.
15. Panario, D., Gourdon, X. and Flajolet, P.: An analytic approach to smooth polynomials over finite fields. In J. Buhler, editor, *Algorithmic Number Theory*, Proceedings of the ANTS-III conference, **1423**, 226–236. Springer 1998.
16. Pohlig, S. and Hellman, M.: An improved algorithm for computing logarithms over GF(p) and its cryptographic significance. *IEEE Trans. Inform. Theory* IT-24 (1978), 106–110.
17. Pollard, J.: Monte Carlo methods for index computation mod p, *Mathematics of Computation*, **32** (1978), 918–924.

18. Satoh T. and Araki, K.: Fermat quotients and the polynomial time discrete log algorithm for anomalous elliptic curves, *Comment. Math. Univ. St. Paul.* (1998), 81–92.
19. Semaev, I.A.: Evaluation of discrete logarithms in a group of p-torsion points of an elliptic curve in characteristic p, *Math. Comp.* **67** (1998), 353–356.
20. Shanks, D.: Class number, a theory of factorization and genera. In *Proc. Symp. Pure Math.* **20** (1971), 415–440. AMS, Providence, R.I.
21. Silverman, J.H.: *The arithmetic of elliptic curves*, Graduate Texts in Mathematics **106**, 2nd Ed. Springer–Verlag, 2009.
22. Smart, N.: The discrete logarithm problem on elliptic curves of trace one, *J. Cryptology* **12** (1999), 193–196.
23. Xiao, D., Zhuang, J. and Cheng, Q.: Factor base discrete logarithms in Kummer Extensions, Cryptology ePrint Archive: Report 2015/859.

Hadwiger's Conjecture

Paul Seymour

Abstract This is a survey of Hadwiger's conjecture from 1943, that for all $t \geq 0$, every graph either can be t-coloured, or has a subgraph that can be contracted to the complete graph on $t + 1$ vertices. This is a tremendous strengthening of the four-colour theorem, and is probably the most famous open problem in graph theory.

1 Introduction

The four-colour conjecture (or theorem as it became in 1976), that every planar graph is 4-colourable, was the central open problem in graph theory for a hundred years; and its proof is still not satisfying, requiring as it does the extensive use of a computer. (Let us call it the 4CT.) We would very much like to know the "real" reason the 4CT is true; what exactly is it about planarity that implies that four colours suffice? Its statement is so simple and appealing that the massive case analysis of the computer proof surely cannot be the book proof.

So there have been attempts to pare down its hypotheses to a minimum core, in the hope of hitting the essentials; to throw away planarity, and impose some weaker condition that still works, and perhaps works with greater transparency so we can comprehend it. This programme has not yet been successful, but it has given rise to some beautiful problems.

Of these, the most far-reaching is Hadwiger's conjecture. (One notable other attempt is Tutte's 1966 conjecture [79] that every 2-edge-connected graph containing no subdivision of the Petersen graph admits a "nowhere-zero 4-flow", but that is beyond the scope of this survey.) Before we state it, we need a few definitions. All graphs in this paper have no loops or parallel edges, and are finite unless we say otherwise. If G is a graph, any graph that can be obtained by moving to a subgraph of G and then contracting edges is called a *minor* of G. The complete graph on t vertices is denoted by K_t, and the complete bipartite graph with sides of cardinalities a, b is denoted by $K_{a,b}$.

P. Seymour (✉)
Princeton University, Fine Hall, Washington Road, Princeton, NJ 08544-1000, USA
e-mail: pds@math.princeton.edu

J.F. Nash, Jr., M.Th. Rassias (eds.), *Open Problems in Mathematics*,
DOI 10.1007/978-3-319-32162-2_13

By the Kuratowski-Wagner theorem [56, 83], planar graphs are precisely the graphs that do not contain K_5 or $K_{3,3}$ as a minor; so the 4CT says that every graph with no K_5 or $K_{3,3}$ minor is 4-colourable. If we are searching for the "real" reason for the four-colour theorem, then it is natural to exclude K_5 here, because it is not four-colourable; but why are we excluding $K_{3,3}$? What if we just exclude K_5, are all graphs with no K_5 minor four-colourable? And does the analogous statement hold if we change K_5 to K_{t+1} and four-colouring to t-colouring? That conjecture was posed by Hadwiger in 1943 [37] and is still open:

Hadwiger's Conjecture 1.1. *For every integer $t \geq 0$, every graph with no K_{t+1} minor is t-colourable.*

Let HC(t) denote the statement "every graph with no K_{t+1} minor is t-colourable". Hadwiger proved HC(t) for $t \leq 3$ in 1943 when he introduced his conjecture. Wagner [83] had already shown that HC(4) is equivalent to the 4CT in 1937; and so HC(4) was finally proved when the 4CT was proved by Appel and Haken [4, 5] in 1976. Then in 1993, Robertson, Thomas and I proved HC(5) [71]; one step further than the 4CT! And the proof did not use a computer (although it did assume the 4CT itself). HC(6) remains open.

There have been numerous weakenings and variations proved, of various types, and strengthenings proposed, some of which still survive; and this is an attempt to survey them. Incidentally, there is an excellent 1996 survey on Hadwiger's conjecture by Toft [78], which is particularly informative on the early history of the problem.

2 The Proved Special Cases

Let us first go through the results just mentioned more carefully. HC(0) and HC(1) are trivial. Graphs with no K_3 minor are forests, which are 2-colourable, so HC(2) holds. The first case that is not quite obvious is HC(3). How do we show that graphs with no K_4 minor are 3-colourable? Hadwiger [37] showed that every non-null graph with no K_4 minor has a vertex of degree at most two, which implies that all such graphs are 3-colourable; and there are later theorems of Dirac [22] and Duffin [24] on the same topic. This assembly of results can be expressed in several different ways, but here is one that is convenient for us. Take two graphs G_1, G_2, and for $i = 1, 2$ let C_i be a clique (that is, a subset of vertices, all pairwise adjacent) of G_i, where $|C_1| = |C_2|$. Choose some bijection between the cliques, and identify each vertex of C_1 with the corresponding vertex of C_2. We obtain a graph H say, with two subgraphs isomorphic to G_1, G_2 respectively, overlapping on a clique. Now let G be obtained from H by deleting some edges (or none) of the clique; we say that G is a *clique-sum* of G_1, G_2, and if the clique has size k, we also call it a *k-sum*.

It is easy to see that if G is a clique-sum of G_1, G_2, and both G_1, G_2 are t-colourable, then so is G. So if G can be built by repeated clique-sums starting from some basic class of graphs that are all t-colourable, then so is G. This gives us a slick proof of HC(t) for $t \leq 3$, because of the following:

Theorem 2.1. *For $0 \leq t \leq 3$, the graphs with no K_{t+1} minor are precisely the graphs that can be built by repeated clique-sums, starting from graphs with at most t vertices.*

HC(4) implies the 4CT, so we should not expect Theorem 2.1 to extend to $t = 4$. And it doesn't; large grids have no K_5 minor and yet cannot be built from 4-vertex graphs by clique-sums. (Indeed, let us say G has *tree-width* k if k is minimum such that G can be built by clique-sums from pieces with at most $k + 1$ vertices; then the $n \times n$ grid has tree-width n.) Nevertheless, we can describe all the graphs with no K_5 minor in this language. Let V_8 be the graph obtained from a cycle of length 8 by adding four edges joining the four opposite pairs of vertices of the cycle. Wagner [83] essentially proved the following in 1937.

Theorem 2.2. *The graphs with no K_5 minor are precisely the graphs that can be built by repeated 0-, 1-, 2-, and 3-sums, starting from planar graphs and copies of V_8.*

Consequently the 4CT implies HC(4), as Wagner points out in his 1937 paper [83]. (Of course, this does not yet provide the profound insight into the four-colour theorem we hope for, because not only does the *proof* of HC(4) use the 4CT, but the graphs it concerns are themselves basically planar.)

What about HC(5)? One might imagine that since the curve of difficulty versus t has recently had such a steep slope, HC(5) would be impossible (or false); but that is not so. Suppose it is false, and let G be a smallest counterexample. Robertson et al. [71] showed, without using a computer and without assuming the four-colour theorem, that G must be an *apex* graph, that is, there is a vertex whose deletion makes it planar. If so, then since the 4CT implies that the planar part of G is 4-colourable, we still have a colour left for the vertex we deleted, so G is 5-colourable after all.

The proof that G is apex is (very roughly) as follows. One can show that G is 6-connected, and in particular all vertices have degree at least six; and vertices of degree six belong to K_4 subgraphs, and it follows that there are not many of them (in fact at most two), or else we could piece together all these K_4's to make a K_6 minor. On the other hand, a theorem of Mader says that the average degree of G is less than eight, and we cannot make the average degree bigger then eight even if we cleverly contract edges. That implies that there are edges that are in several triangles or squares. If, say, there is an edge uv in four triangles, then there is no K_4 minor of $G \setminus \{u, v\}$ on the four surviving vertices of the triangles (since G has no K_6 minor), and graphs with this property are well-understood; basically they have to be planar with the four special vertices on the infinite region. So $G \setminus \{u, v\}$ is planar, and now a little more thought shows that one of $G \setminus u, G \setminus v$ is planar, and hence G is apex.

Proving that graphs with no K_7-minor are 6-colourable is thus the first case of Hadwiger's conjecture that is still open. Albar and Gonçalves[2] proved:

Theorem 2.3. *Every graph with no K_7 minor is 8-colourable, and every graph with no K_8 minor is 10-colourable.*

3 Average Degree

Since we are stuck trying to prove Hadwiger's conjecture itself, let us see what we *can* show about the chromatic number of graphs with no K_{t+1} minor. As Wagner [82] proved in 1964, all graphs with no K_{t+1} minor are 2^t-colourable. The proof is as follows: we may assume G is connected; fix some vertex z, and for each i let L_i be the set of vertices at distance i from z; since G has no K_{t+1} minor, the subgraph induced on L_i has no K_t minor (because the union of all the earlier levels would provide one more vertex in the minor); inductively each level L_i induces a subgraph that is 2^{t-1}-colourable; and now alternate colours in even and odd levels to get a 2^t-colouring of G.

Wagner's result has been considerably improved, but most of these improvements depend on "degeneracy", so let us first discuss that. We say G is *k-degenerate* if every non-null subgraph has a vertex of degree at most k. For instance, forests are 1-degenerate, series-parallel graphs (the graphs with no K_4 minor) are 2-degenerate, and planar graphs are 5-degenerate. By deleting a vertex of degree at most k and applying an inductive hypothesis, we have:

Theorem 3.1. *If G is k-degenerate then its chromatic number is at most $k + 1$.*

So, if we can bound the degeneracy of the graphs with no K_{t+1} minor, we also bound their chromatic number. (This gives us another proof of HC(t) for $t \leq 3$, because for $t \leq 3$ every graph with no K_{t+1} minor is $(t - 1)$-degenerate.)

The simplest way to bound the degeneracy is to bound the average degree. How many edges an n-vertex graph with no K_t minor can have is a much-studied question. Mader [59, 60] showed in 1967 that:

Theorem 3.2. *For every graph H there exists c such that $|E(G)| \leq c|V(G)|$ for every graph G with no H minor.*

But when $H = K_t$ for small values of t, we know the answer exactly:

- for $n \geq 1$, n-vertex graphs with no K_3 minor (forests) have at most $n - 1$ edges;
- for $n \geq 2$, graphs with no K_4 minor have at most $2n - 3$ edges;
- for $n \geq 3$, graphs with no K_5 minor have at most $3n - 6$ edges.

Here is an example: for $n \geq t - 2$, take the complete bipartite graph $K_{t-2,n-t+2}$, and add edges joining all pairs of vertices on the side of cardinality $t - 2$. This has no K_t minor, and has n vertices and $(t-2)n - (t-1)(t-2)/2$ edges. Thus for $t \leq 5$, this graph has the maximum number of edges possible, and if this were so for all t, it would prove Hadwiger's conjecture within a factor of 2. Mader [60] showed that the same holds for $t = 6, 7$:

Theorem 3.3. *For $t \leq 7$ and all $n \geq t-2$, every n-vertex graph G with no K_t minor satisfies*

$$|E(G)| \leq (t - 2)n - (t - 1)(t - 2)/2.$$

But for $t \geq 8$ the pattern fails. If $n_1, \ldots, n_t > 0$, we denote by $K_{n_1,\ldots,n_t}$ the complete t-partite graph with parts of cardinality $n_1, \ldots, n_t$. Mader pointed out that $K_{2,2,2,2,2}$ has no K_8 minor, and does not satisfy the formula of Theorem 3.3.

On the other hand, for $t = 8$ we understand all counterexamples to the formula. In the definition of a k-sum we are permitted to delete edges from the clique involved; if we do not delete any such edges let us call it a *pure* k-sum. Jørgensen [39] proved:

Theorem 3.4. *Let G be an n-vertex graph with no K_8 minor, with $n \geq 6$ and $|E(G)| > 6n - 21$; then $|E(G)| = 6n - 20$, and G can be built by pure 5-sums from copies of $K_{2,2,2,2,2}$.*

The same holds for K_9; Song and Thomas [72] proved:

Theorem 3.5. *Let G be an n-vertex graph with no K_9 minor, with $n \geq 7$ and $|E(G)| > 7n - 28$; then $|E(G)| = 7n - 27$, and either $G = K_{2,2,2,3,3}$, or G can be built by pure 6-sums from copies of $K_{1,2,2,2,2,2}$.*

But as t grows, the formula of Theorem 3.3 becomes completely wrong. For a graph H, let $\phi(H)$ be the infimum of all d such that every graph G with no H minor has average degree at most d, that is, satisfies $|E(G)| \leq d|V(G)|/2$. (We are particularly concerned here with the case when H is a complete graph K_t, but $\phi(H)$ is of interest for non-complete graphs too.) Kostochka [49, 51] and Fernandez de la Vega [28] proved that $\phi(K_t)$ is at least of order $t(\log t)^{1/2}$, and Kostochka [49, 51] and Thomason [73] proved the same was an upper bound; and in particular Kostochka [51] showed (logarithms are to base e):

Theorem 3.6. *For every integer $t \geq 4$, $\phi = \phi(K_t)$ satisfies:*

$$\frac{0.032\,\phi}{(\log(\phi/2))^{1/2}} \leq t.$$

Later Thomason [74] found the limit exactly: he proved (again with logarithms to base e):

Theorem 3.7. *Let $\lambda < 1$ be the solution of the equation $1 - \lambda + 2\lambda \log \lambda = 0$ and let*

$$\alpha = (1 - \lambda) \log(1/\lambda)^{-1/2} \simeq 0.63817.$$

Then as $t \to \infty$, $\phi(K_t) = (\alpha + o(1))t(\log t)^{1/2}$.

This was extended to non-complete graphs by Myers and Thomason [63], who proved the following ($\mathbb{R}^+$ denotes the set of nonnegative real numbers, and α is as before):

Theorem 3.8. *Let H be a graph with t vertices, and let $\gamma(H)$ be the minimum of $\frac{1}{t}\sum_{u\in V(H)} w(u)$ over all functions $w : V(H) \to \mathbb{R}^+$ such that*

$$\sum_{uv\in E(H)} t^{-w(u)w(v)} \leq t.$$

Then as $t \to \infty$, $\phi(H) = (\alpha\gamma(H) + o(1))t(\log t)^{1/2}$.

For classes of graphs H with $\gamma(H)$ bounded away from zero (such as regular graphs with degree ct^{ϵ} where $c, \epsilon > 0$), this determines $\phi(H)$ asymptotically; but for some classes of graphs (such as those with a linear number of edges) it does not. This gap is addressed by two theorems of Reed and Wood [67]:

Theorem 3.9. *There is a constant d_0 such that $\phi(H) \leq 3.895(\log d)^{1/2}t$ for every graph H with t vertices and average degree $d \geq d_0$.*

Theorem 3.10. *For every graph H, $\phi(H) \leq |V(H)| + 6.291|E(H)|$.*

The Myers-Thomason theorem implies that $\phi(H)$ is not linear in t for graphs with t vertices and with a quadratic number of edges; but the second Reed-Wood theorem implies that if $|E(H)|$ is linear in t then so is $\phi(H)$.

For some graphs H we can determine exactly the maximum number of edges in graphs with no H minor, but those theorems are thinner on the ground. We already mentioned the cases when $H = K_t$; and the same can be done for many graphs H with at most six vertices, such as $K_{3,3}$; and there are two theorems doing it for larger graphs H. Chudnovsky et al. [14] answered it for $K_{2,t}$ (extending a result of Myers [62]), and Kostochka and Prince [52] did it for $K_{3,t}$ (and the $K_{1,t}$ result is obvious):

Theorem 3.11. *Let G be an n-vertex graph with no H minor.*

- *If $H = K_{1,t}$ then $|E(G)| \leq \frac{1}{2}(t-1)n$;*
- *if $H = K_{2,t}$ then $|E(G)| \leq \frac{1}{2}(t+1)(n-1)$; and*
- *if $H = K_{3,t}$ and $t \geq 6300$ and $n \geq t+3$ then $|E(G)| \leq \frac{1}{2}(t+3)(n-2)+1$.*

All three results are exact for infinitely many values of n. (By the way, when $H = K_{1,t}$, if we restrict to connected graphs G then the answer is quite different, namely $|E(G)| \leq n + (t+1)(t-2)/2$ if $n \geq t+2$; see [20].)

What about $H = K_{s,t}$ in general, if $t \geq s$? For fixed s and large t, the value of $\phi(K_{s,t})$ is not determined by Theorem 3.8, so this is an interesting case. It turns out to be more natural to exclude $K^*_{s,t}$ instead; this is the graph obtained from $K_{s,t}$ by adding edges joining all pairs of vertices in the side of cardinality s. Extrapolating from Theorem 3.11, one might hope that if an n-vertex graph has no $K_{s,t}$ minor then

$$|E(G)| \leq \frac{1}{2}(2s+t-3)n - \frac{1}{2}(s-1)(s+t-1),$$

because again this can be attained with equality for infinitely many n (take many disjoint copies of K_t and add $s-1$ extra vertices adjacent to everything). But this is not true, at least for $s > 18$. Kostochka and Prince [53, 54] proved (and see also [55] for a related result) that with the function $\phi(H)$ as before (here logarithms are binary):

Theorem 3.12. *Let s, t be positive integers with $t > (180s\log s)^{1+6s\log s}$. Then*

$$3s - 5s^{1/2} + t \leq \phi(K_{s,t}) \leq \phi(K^*_{s,t}) < 3s + t.$$

All these results tell us that the graphs with a certain minor H excluded have average degree at most some constant, and therefore have minimum degree at most the same constant; and that gives us a bound on their degeneracy. In particular, from Theorem 3.6, every graph with no K_t minor has degeneracy at most $O(t(\log t)^{1/2})$, and therefore chromatic number at most the same. For large t, this is the best bound known on the chromatic number of graphs excluding K_t.

Incidentally, bounding minimum degree by average degree is natural, but it might not give the right answer. For instance, graphs with no K_4 minor can have average degree > 3; and yet they always have minimum degree at most 2. When we exclude K_5, average degree gives the true bound for minimum degree; but what happens with K_6? Graphs with no K_6 minor can have average degree more than 7, but must they have minimum degree at most 6? I think this is open.

4 Stability Number

One possible cause of the intractability of Hadwiger's conjecture is that we need to use the fact that the chromatic number is large, and graphs can have large chromatic number for obscure reasons. What if we make our lives easier, and look at graphs that have large chromatic number for obvious reasons? The *stability number* $\alpha(G)$ of a graph G is the size of the largest stable set (a set of vertices is *stable* if no two of its members are adjacent). (This is different from Thomason's α, which we do not need any more.) Every n-vertex graph G has chromatic number at least $\lceil n/\alpha(G)\rceil$, and should contain a clique minor of this size if Hadwiger's conjecture is true. Can we prove this at least?

The signs are not good; the only known proof that every n-vertex planar graph has stability number at least $n/4$ is via the 4CT. Nevertheless, there are some results. There is an elegant argument by Duchet and Meyniel [23] proving:

Theorem 4.1. *Every n-vertex graph G has a K_t minor where $t \geq n/(2\alpha(G)-1)$.*

Their argument can also be used to show a result that seems to have been overlooked:

Theorem 4.2. *For every graph G with no K_{t+1} minor, there exists a t-colourable induced subgraph containing at least half the vertices of G.*

Theorem 4.1 is within a factor of 2 of what should be true, and there have been subsequent improvements, notably by Fox [29] (who proved a factor slightly less than 2) and then Balogh and Kostochka [6], who reduced Fox's factor a little further and currently have the record. They showed the following:

Theorem 4.3. *Every n-vertex graph G has a K_t minor where $t \geq 0.51338n/\alpha(G)$.*

A different strengthening, better than Theorem 4.3 when α is small, was proved by Kawarabayashi and Song [47]:

Theorem 4.4. *Every n-vertex graph G with $\alpha(G) \geq 3$ has a K_t minor where $t \geq n/(2\alpha(G) - 2)$.*

Returning to Theorem 4.1: it implies that if G has no K_{t+1} minor then some stable set has cardinality at least $n/(2t)$. Suppose we give each vertex of G a nonnegative real weight. Hadwiger's conjecture would imply that there is a stable set such that the total weight of its members is at least $1/t$ times the sum of all weights. One might hope to prove a weighted version of Theorem 4.1 (without the -1 in the denominator) and this turns out to be true, though more difficult to prove. Say the *fractional chromatic number* of a graph G is the minimum real number k such that for some integer $s > 0$, there is a list of ks stable sets of G such that every vertex is in s of them. Via linear programming duality, the weighted Duchet-Meyniel statement is equivalent to the following, proved by Reed and myself [65]:

Theorem 4.5. *Every graph with no K_{t+1} minor has fractional chromatic number at most $2t$.*

The proof also gives a corresponding extension of Theorem 4.2:

Theorem 4.6. *In every graph G with no K_{t+1} minor, there is a non-null list of t-colourable subsets of $V(G)$, such that every vertex is in exactly half of the sets in the list.*

Graphs G with $\alpha(G) = 2$ are particularly interesting, because these graphs are more tractable for colouring; for instance, there is a polynomial-time algorithm to find the chromatic number of such a graph (just find the largest matching in the complement graph). Here is another nice feature of them: say a *seagull* in a graph G is an induced 3-vertex path. If $\alpha(G) = 2$ and S is a seagull in G then every other vertex of G has a neighbour in S, and so finding many disjoint seagulls is a way to find a large clique minor. In [15], Chudnovsky and I proved there is a min-max formula for the maximum number of disjoint seagulls in a graph G with $\alpha(G) = 2$.

For an n-vertex graph G with $\alpha(G) = 2$, the Duchet-Meyniel theorem implies that there is a K_t minor with $t \geq n/3$. This was strengthened by Böhme et al. [9], who proved (for graphs with arbitrary stability number):

Theorem 4.7. *Every n-vertex graph with chromatic number k has a K_t minor where $t \geq (4k - n)/3$.*

But Hadwiger's conjecture implies that if $\alpha(G) = 2$ then there should be a K_t minor with $t \geq n/2$. This seems to me to be an excellent place to look for a counterexample. My own belief is, if it is true for graphs with stability number two then it is probably true in general, so it would be very nice to decide this case. Despite some intensive effort the following remains open:

Open Question 4.8. *Does there exists $c > \frac{1}{3}$ such that every graph G with $\alpha(G) = 2$ has a K_t minor where $t \geq c|V(G)|$?*

A graph is *claw-free* if no vertex has three pairwise nonadjacent neighbours. Thus graphs with stability number two are claw-free. Fradkin [31] proved:

Theorem 4.9. *Every n-vertex connected claw-free graph G with $\alpha(G) \geq 3$ has a K_t minor where $t \geq n/\alpha(G)$.*

Chudnovsky and Fradkin [13] proved:

Theorem 4.10. *Every claw-free graph G with no K_{t+1} minor is $\lfloor 3t/2 \rfloor$-colourable.*

Line graphs are claw-free, so these last two results are related to a theorem of Reed and myself; we proved [66] that Hadwiger's conjecture is true for line graphs (of multigraphs).

5 Weakenings

The statement of Hadwiger's conjecture is:

For all $t \geq 0$ and every graph G, either G has a K_{t+1} minor or $V(G)$ can be partitioned into t stable sets.

How can we weaken this and still have something non-trivial? Section 2 covered changing "For all $t \geq 0$" to "For a few $t > 0$"; Sect. 3 did changing "t stable sets" to "$f(t)$ stable sets"; and Sect. 4 covered changing "partitioned into stable sets" to "fractional chromatic number"; but there are several other ways to weaken the statement. Here are some.

Change "every graph G" to "almost every graph G" (The meaning of "almost every" here is that the proportion of n-vertex graphs that satisfy the statement tends to 1 as $n \to \infty$.) This weakening is true. It follows from a combination of a theorem of Bollobás et al. [10] and a theorem of Grimmett and McDiarmid [35]:

Theorem 5.1. *For all $d > 2$, almost every n-vertex graph has a K_t minor where $t \geq n/((\log n)^{1/2} + 4)$, and has chromatic number at most $2n/\log n$.*

Change "K_{t+1}" to something else If we hope to prove that every graph with no H minor has chromatic number at most t, then H had better have at most $t+1$ vertices, or else taking $G = K_{t+1}$ is a counterexample. So, which subgraphs H of K_{t+1} work? Kostochka [48, 50] proved the following.

Theorem 5.2. *For all s there exists t_0 such that for all $t \geq t_0$, every graph with no $K^*_{s,t}$ minor is $(s + t - 1)$-colourable.*

Change "stable sets" to something else Here is a recent theorem of Edwards et al. [26]:

Theorem 5.3. *For all t there exists k such that if G has no K_{t+1} minor then $V(G)$ can be partitioned into t sets $X_1, \ldots, X_t$, such that for $1 \leq i \leq t$, $G[X_i]$ has maximum degree at most k.*

(For $X \subseteq V(G)$, $G[X]$ denotes the subgraph induced on X.) This result is quite easy, but it has two attractive features; first, it is best possible in that if we ask for a partition into $t - 1$ sets there is no such k; and second, it and Theorem 6.8 below are the only results known that derive a partition into t sets with *any* non-trivial property from the absence of a K_{t+1} minor.

There are more weakenings to describe yet, but they deserve a new section.

6 Bounded Component-Size

What if we try to improve Theorem 5.3? Let us say $X \subseteq V(G)$ has *component-size* k if the largest component of $G[X]$ has k vertices. Thus having bounded component-size is more restrictive than have bounded maximum degree, though less than what we really want, being stable. Instead of just saying that each $G[X_i]$ has bounded maximum degree, what if we ask that each of them has bounded component-size? It has not been proved that for graphs G with no K_{t+1} minor, we can partition into t sets with this property, but there has been a series of papers proving that $V(G)$ can be partitioned into a linear number of parts each with bounded component-size. Initially Kawarabayashi and Mohar [43] proved:

Theorem 6.1. *For all $t \geq 0$ there exists k such that if G has no K_t minor, then $V(G)$ can be partitioned into at most $f(t)$ parts each with component-size at most k, where $f(t) = \lceil 31t/2 \rceil$.*

Wood [84] proved the same with $f(t) = \lceil 7t/2 - 3/2 \rceil$ (using Theorem 10.6, an unpublished theorem of Norin and Thomas which we discuss later), and there have been further improvements which we describe below, culminating in Norin's result that the same holds with $f(t) = 2(t - 1)$.

There is a set of lemmas here that can be combined in various ways. DeVos et al. [16] proved:

Theorem 6.2. *For all t there exists w such that for every graph G with no K_t minor, there is a partition of $V(G)$ into two parts, such that the subgraph induced on each part has tree-width at most w.*

Alon et al. [3] showed:

Theorem 6.3. *For all w, d and for every graph G with tree-width at most $w \geq 3$ and maximum degree at most $d \geq 1$, there is a partition of $V(G)$ into two parts each with component-size at most $24wd$.*

Wood [85] improved this, replacing $24kd$ with $5(k+1)(7d-2)/4$. Liu (unpublished) has recently proved a list-colouring version:

Theorem 6.4. *For all w, d there exists k such that for every graph G with tree-width at most w and maximum degree at most d, and every assignment of a set L_v with $|L_v| \geq 2$ to each vertex v, there is a choice of $c(v) \in L_v$ for each v such that for each x, the set of all vertices v with $c(v) = x$ has component-size at most k.*

Incidentally, an interesting asymmetric version was proved by Ding and Dziobiak [19]:

Theorem 6.5. *For all $t \geq 0$ there exists $w \geq 0$ such that for every graph G with no K_t minor, $V(G)$ can be partitioned into two sets X, Y, where $G[X]$ has tree-width at most w, and $G[Y]$ is $(t+1)$-degenerate.*

By combining Theorems 6.2 and 6.3, Alon et al. deduced:

Theorem 6.6. *For all t, d there exists k such that for every graph G with no K_t minor and maximum degree at most d, there is a partition of $V(G)$ into four parts each with component-size at most k.*

Recently, Liu and Oum [58] improved this, replacing "four" by "three". If we then combine their result with Theorem 5.3 we deduce an improvement of Theorem 6.1 with $f(t) = 3(t-1)$. Even more recently, Norin used a different approach to do better. He proved the following lemma [64]:

Theorem 6.7. *For all $t, w \geq 0$ there exists N with the following property. Let G be a graph with $|V(G)| \geq N$, with tree-width at most w and with no K_t minor. Then for every $S \subseteq V(G)$ with $|S| \leq 2w$, there exists $I \subseteq V(G) \setminus S$, nonempty, such that at most $2w$ vertices in $V(G) \setminus I$ have a neighbour in I, and every vertex in I has at most $t-2$ neighbours in $V(G) \setminus I$.*

With the aid of this lemma, an easy inductive argument yields:

Theorem 6.8. *For all $t, w \geq 0$ there exists k such that for every graph G with tree-width at most w and no K_t minor, there is a partition of $V(G)$ into $t-1$ parts such that each part has component-size at most k.*

Then this, combined with Theorem 6.2, yields an improvement of Theorem 6.1 with $f(t) = 2(t-1)$.

Dvořák and Norin have now announced a proof of Theorem 6.1 with $f(t) = t-1$, the best possible.

7 Odd Minors

We have finished with weakenings of Hadwiger's conjecture now; time to turn to strengthenings.

Graphs that are not 2-colourable not only have a K_3 minor (or equivalently, a cycle); they have an odd cycle. It is tempting to try to make some corresponding strengthening of Hadwiger's conjecture. Here is what seems to be the most natural way to do it. If G is a graph and $X \subseteq V(G)$, $\delta(X)$ denotes the set of edges of G with one end in X and the other in $V(G) \setminus X$. We say that $F \subseteq E(G)$ is a *cut* of G if $F = \delta(X)$ for some $X \subseteq V(G)$. Now let G, H be graphs. We say that H is an *odd minor* of G if H can be obtained from a subgraph G' of G by contracting a set of edges that is a cut of G'. (Note that $\emptyset$ is a cut.) Thus a graph is not 2-colourable if and only if it contains K_3 as an odd minor. In 1979, Catlin [12] proved:

Theorem 7.1. *If G has no K_4 odd minor then G is 3-colourable.*

Incidentally, a much stronger statement than this has now been proved. Say a *fully odd K_4* in G is a subgraph of G which is obtained from K_4 by replacing each edge of K_4 by a path of odd length (the *length* of a path is the number of edges in it) in such a way that the interiors of these six paths are disjoint. Toft [77] conjectured in 1975 and Zang [86] proved in 1998 (and, independently, Thomassen [76] proved in 2001) that:

Theorem 7.2. *If G contains no fully odd K_4 then G is 3-colourable.*

Returning to odd minors: there is a result giving a construction for all graphs with no K_4 odd minor, due to Lovász, Schrijver, Truemper and myself. It is rather awkward to state, and not published, although it was proved many years ago (in the early 1980s, and mostly on a riverboat in Bonn, if I remember correctly). We omit its statement here; see [33].

In view of its truth for $t \leq 3$, Gerards and I conjectured the following strengthening of Hadwiger's conjecture (see [38]):

Conjecture 7.3. *For every $t \geq 0$, if G has no K_{t+1} odd minor, then G is t-colourable.*

Several of the results mentioned earlier approaching Hadwiger's conjecture have extensions to odd minors. For instance, Guenin [36] announced at a meeting in Oberwolfach in 2005 that:

Theorem 7.4. *Every graph with no K_5 odd minor is 4-colourable.*

Geelen et al. [32] proved (see also [41] for a simpler proof):

Theorem 7.5. *If G has no K_t odd minor then $\chi(G) \leq O(t(\log t)^{1/2})$.*

Kawarabayashi and Song [47] proved an odd minor version of Theorem 4.1, namely:

Theorem 7.6. *Every n-vertex graph G has a K_t odd minor where $t \geq n/(2\alpha(G)-1)$.*

Kawarabayashi and Reed [45] proved:

Theorem 7.7. *Every graph with no K_t odd minor is fractionally $2t$-colourable.*

Kawarabayashi and Song [47] proved an odd minor relative of Theorem 10.2:

Theorem 7.8. *For every $t \geq 0$, there exists N such that for every $(496t + 13)$-connected graph G with at least N vertices, either G has a K_t odd minor, or there exists $X \subseteq V(G)$ with $|X| \leq 8t$ such that $G \setminus X$ is bipartite.*

Kawarabayashi [40] proved:

Theorem 7.9. *If G has no K_t odd minor, then there is a partition of $V(G)$ into $496t$ parts such that each part induces a subgraph of bounded maximum degree.*

8 Other Strengthenings

There have been some other strengthenings of Hadwiger's conjecture proposed, but they have mostly not fared so well as Conjecture 7.3. For instance, say G is a *subdivision* of H if G can be obtained from H by replacing each edge by a path, where the paths have disjoint interiors. Hajós conjectured in the 1940s (but did not publish it) that

False Conjecture 8.1. *For all $t \geq 0$, if no subgraph of G is a subdivision of K_{t+1} then G is t-colourable.*

This is true for $t \leq 3$, but it is still open for $t = 4, 5$, and Catlin [12] gave a counterexample for all $t \geq 6$. Indeed, Erdős and Fajtlowicz [27] proved that almost all graphs are counterexamples, because of the following:

Theorem 8.2. *There are constants C_1, C_2 such that for almost every n-vertex graph G, no subgraph of G is a subdivision of K_t for $t \geq C_1 n^{1/2}$, and the chromatic number of G is at least $C_2 n/\log(n)$.*

Erdős and Fajtlowicz conjectured and Fox et al. [30] proved that the ratio of the chromatic number over the clique subdivision number over all n-vertex graphs is maximized up to a constant factor by the random graph on n vertices; in other words, the uniform random graph is essentially the strongest counterexample to the Hajós conjecture.

Another conjectured strengthening of Hadwiger's conjecture was proposed by Borowiecki [11]. A graph G is *t-choosable* if for every assignment of a t-element set L_v to each vertex v of G, it is possible to select a member $c(v) \in L_v$ for each v such that $c(u) \neq c(v)$ if u, v are adjacent. Borowiecki asked whether

False Conjecture 8.3. *Every graph with no K_{t+1} minor is t-choosable.*

This is true for $t \leq 3$, but false for $t = 4$; Voigt [81] gave a planar graph that is not 4-choosable. Thomassen [75] proved that all planar graphs are 5-choosable, but an additive constant adjustment is not enough to repair the conjecture in general; Barát et al. [7] showed that for all $t \geq 1$ there is a graph with no K_{3t+2} minor that is not $4t$-choosable. Kawarabayashi and Mohar [43] conjectured:

Conjecture 8.4. *For all t, every graph with no K_t minor is $3t/2$-choosable.*

A third strengthening was proposed by Ding et al. [21]:

Conjecture 8.5. *For all integers $t \geq s \geq 2$, if G has no K_t minor, then there is a partition of $V(G)$ to $t - s + 1$ parts, such that the subgraph induced on each part has no K_s minor.*

For $s = 2$ this is Hadwiger's conjecture, but it has not been disproved for any values of s, t. For $s \geq t - 1$ it is easy, and it was proved for $s = t - 2$ by Gonçalves [34].

Reed and I proposed a fourth variation in [65] (this one is not a strengthening):

Conjecture 8.6. *For every graph, there is a partition of its vertex set, such that each part induces a connected bipartite graph, and contracting each part to a vertex yields a graph with no induced cycle of length more than three.*

This would imply that all graphs with no K_{t+1} minor are $2t$-colourable. It remains open.

A fifth possible extension is to infinite graphs. (Henceforth, graphs may be infinite in this section.) By compactness, if t is an integer and HC(t) holds for finite graphs then it also holds for infinite graphs; but we could try to extend Hadwiger's conjecture to allow infinitely many colours. One might hope that

False Conjecture 8.7. *For every cardinal t, every graph with no K_t minor has chromatic number less than t;*

but this is trivially false (let G be the disjoint union of infinitely many finite cliques, one of each size; then G cannot be coloured with finitely many colours, but has no infinite clique minor). A better formulation is:

Conjecture 8.8. *For every cardinal t, let s be the least cardinal larger than t; every graph with no K_s minor is t-colourable.*

I believe Conjecture 8.8 remains open, but van der Zypen [80] proved the following:

Theorem 8.9. *For every infinite cardinal t, every graph with no subgraph which is a subdivision of K_t is t-colourable.*

Van der Zypen's proof uses the fact that when t is an infinite cardinal, one can give a construction of the graphs that contain no subdivision of K_t, a result due to Robertson et al. [70]. It is also possible [69] to do the same for graphs that contain no K_t minor when t is an infinite cardinal.

9 Immersions

There is an interesting conjecture, parallel to Hadwiger's conjecture, that was proposed by Lescure and Meyniel [57], (and independently, by Abu-Khzam and Langston [1], later). Let G, H be graphs. An *immersion* of H in G is a choice $\eta(v) \in V(G)$ for each $v \in V(H)$, all distinct, and a choice $\eta(e)$ for each $e \in E(G)$, where for $e = uv$, $\eta(e)$ is a path of G between $\eta(u)$ and $\eta(v)$, and all the paths $\eta(e)$ are pairwise edge-disjoint (they may share vertices; and an end-point of one path may be an internal vertex of another). Let us say G *immerses* H if there is an immersion of H in G. Lescure and Meyniel proposed:

Conjecture 9.1. *For every integer $t \geq 0$, every graph that does not immerse K_{t+1} is t-colourable.*

This neither implies nor is implied by Hadwiger's conjecture, since immersing K_{t+1} neither implies nor is implied by having a K_{t+1} minor; but it is in some respects similar. (In one respect it is very different: planar graphs can immerse huge complete graphs.) Conjecture 9.1 was proved for $t \leq 6$ by Lescure and Meyniel (though they did not publish the proof for $t = 6$), and more recently DeVos et al. [18] published a proof for $t = 6$. Both sets of authors used the same approach, proving the stronger statement that for $t \leq 6$, every simple graph with minimum degree at least t immerses K_t. For $t \geq 9$, it is not true that every graph with minimum degree at least t immerses K_t; but DeVos et al. [17] proved the following (in fact they proved it for "strong" immersion, in which the vertices $\eta(v)$ are not permitted to be internal vertices of the paths $\eta(e)$):

Theorem 9.2. *For all $t \geq 0$, every graph of minimum degree at least $200t$ immerses K_t.*

Recently, Dvořák and Yepremyan [25] have improved this:

Theorem 9.3. *For all $t \geq 0$, every graph of minimum degree at least $11t + 7$ immerses K_t.*

It follows that

Theorem 9.4. *Every graph that does not immerse K_t is $11t + 7$-colourable.*

10 Big Graphs

The constructions of Kostochka and Fernandez de la Vega mentioned earlier show that there are graphs with no K_t minor with average degree of the order of $t(\log t)^{1/2}$, and indeed their minimum degree and connectivity are also of this order. But there is a feeling that honest, sensible graphs with no K_t minor are not really like this; they will have vertices of degree about t. How can we make this intuition closer to a true statement?

The intuition comes mostly from the Graph Minors structure theorem of Robertson and myself [68], which says very roughly that to make graphs with no K_{t+1} minor, one takes graphs on surfaces of bounded genus and adds a bounded number of extra vertices; and if these extra vertices are not just attached to small parts of the surface, there had better not be many of them (or else we will get a K_{t+1} minor); in fact at most $t - 4$ of them, and fewer if the surface is not the plane. But vertices in the surface have average degree (in the surface) less than six, so total degree less than $t + 2$. There are have been several attempts to bring this very vague argument closer to reality, and in this section we discuss some of them.

The feeling is that the examples of Kostochka and Fernandez de la Vega have only bounded size (which they do) in some essential way (which remains to be made precise). Of course we can make bigger examples by taking disjoint unions of the little ones, but then the connectivity is lost. What if we impose some connectivity restriction? Can there still be large examples?

Thomas and I conjectured:

Conjecture 10.1. *For all $t \geq 0$ there exists N such that every $(t-2)$-connected graph G with no K_t minor and with $n \geq N$ vertices satisfies*

$$|E(G)| \leq (t-2)n - (t-1)(t-2)/2.$$

This remains open. Böhme et al. [8] proved:

Theorem 10.2. *For all positive integers t, there exists N such that every $3t+2$-connected graph with no K_t minor and with at least N vertices has a vertex of degree less than* $31(t+1)/2 - 3$.

This is very encouraging: everything is linear, the frightening $(\log t)^{1/2}$ term has disappeared.

How can we arrange some decent connectivity? To prove HC(t) it is enough to prove the impossibility of minimal or minimum counterexamples to HC(t) (a counterexample is "minimal" if no proper minor of itself is a counterexample; and "minimum" if no counterexample is smaller.) What about the connectivity of minimal counterexamples? Kawarabayashi [42] proved:

Theorem 10.3. *For $t \geq 0$, every minimal counterexample to HC(t) is $\lceil 2(t+1)/27 \rceil$-connected, and every minimum counterexample to HC(t) is $\lceil (t+1)/3 \rceil$-connected.*

Mader [61] proved that for any value of t, if G is a minimal counterexample to HC(t) then G is 6-connected, and 7-connected if $t \geq 6$. When $t = 5$ this is particularly interesting, because it means that to prove HC(5) we only have to consider 6-connected graphs without K_6 minors. And Jørgensen [39] conjectured the following:

Conjecture 10.4. *Every 6-connected graph with no K_6 minor is apex.*

(We recall that a graph is *apex* if it can be made planar by deleting one vertex, and in particular all apex graphs are 5-colourable.) Thus if only we could prove

Jørgensen's conjecture, we would obtain a much more appealing proof of HC(5). Unfortunately it remains open; but it might point a way to solve Hadwiger's conjecture in general, if we could only figure out an analogue of this conjecture for larger values of t (and then figure out how to prove it).

Kawarabayashi et al. [44] proved that Conjecture 10.4 itself is true in large graphs:

Theorem 10.5. *There exists N such that every 6-connected graph with at least N vertices and with no K_6 minor is apex.*

More recently Norin and Thomas have announced the following analogue of Conjecture 10.5 for general values of t (this is a difficult result with a huge proof, and is still being written at this time):

Theorem 10.6. *For all $t \geq 0$ there exists N such that every $t + 1$-connected graph with at least N vertices and with no K_{t+1} minor can be made planar by deleting $t-4$ vertices.*

So it would be nice to know that minimal counterexamples to HC(t) are $t + 1$-connected; but we do not know this.

But recently it may have been shown that in fact there are no large minimal counterexamples to HC(t), using a feature of them slightly different from connectivity. A *cutset* of G means (in this paper) a partition (A, B, C) of $V(G)$ with $A, B \neq \emptyset$, such that there are no edges between A and B. A *one-way clique cutset* of G means a cutset (A, B, C), and for each $v \in C$ a connected subgraph $X_v \subseteq B \cup C$ containing v, such that X_u, X_v are disjoint and some edge has an end in X_u and an end in X_v, for all distinct $u, v \in C$. In other words, we can turn C into a clique by contracting edges within $B \cup C$. Suppose that G is a minimal counterexample to HC(t), for some value of t. Then it is easy to show that:

- no vertex has degree at most t;
- no vertex of degree $t + 1$ has three nonadjacent neighbours;
- there is no one-way clique cutset; and
- G cannot be made planar by deleting $t - 4$ vertices.

Robertson and I announced (about 1993) that we proved:

Theorem 10.7. *For all $t \geq 0$ and for every graph G with no K_{t+1} minor, if G satisfies the four bullets above then G has bounded tree-width.*

This had all kinds of pleasing consequences, but the proof was very long, and was never written down, and now it is lost. Fortunately, almost the same thing, and with the same desirable consequences, has recently been announced by Kawarabayashi and Reed [46], and their proof seems more manageable, and may get written down. (At the moment the proof sketched in [46] has developed a few cracks, but Kawarabayashi maintains it can be fixed.) They added a fifth bullet to the four above:

- there do not exist a cutset (A, B, C) of G and disjoint connected subgraphs $X_1, \ldots, X_k$ of of $G[A \cup C]$, each including a stable subset of C, such that if B' denotes the graph obtained from $B \cup X_1 \cup \cdots \cup X_t$ by contracting the edges of $X_1, \ldots, X_t$, then every t-colouring of B' extends to one of G.

This evidently also holds in any minimal counterexample to HC(t). They claim:

Theorem 10.8. *For all $t \geq 0$ and for every graph G with no K_{t+1} minor, if G satisfies the five bullets above then G has bounded tree-width.*

This would have several consequences. The most important is probably an explicit function $f(t)$ such that for all t, every minimal counterexample to HC(t) has at most $f(t)$ vertices.

Acknowledgements Thanks to several people who made suggestions for improvement, particularly Jacob Fox, Ken-ichi Kawarabayashi, Sasha Kostochka, Bojan Mohar, Sergey Norin, Sophie Spirkl, Robin Thomas, Bjarne Toft, Paul Wollan and David Wood. Supported by ONR grant N00014-10-1-0680 and NSF grant DMS-1265563.

References

1. F. N. Abu-Khzam and M. A. Langston, "Graph coloring and the immersion order", *Computing and Combinatorics, Lecture Notes in Comput. Sci.* 2697 (2003) (Springer, Berlin) 394–403.
2. B. Albar and D. Gonçalves, "On triangles in K_r-minor free graphs", submitted for publication (manuscript April 2013), http://arXiv.org/abs/1304.5468.
3. N. Alon, G. Ding, B. Oporowski, and D. Vertigan, "Partitioning into graphs with only small components", *J. Combinatorial Theory, Ser. B,* 87 (2003), 231–243.
4. K. Appel and A. Haken, "Every planar map is four colorable. Part I. Discharging", *Illinois J. Math.* 21 (1977), 429–490.
5. K. Appel, A. Haken and J. Koch, "Every planar map is four colorable. Part II. Reducibility", *Illinois J. Math.* 21 (1977), 491–567.
6. J. Balogh and A. V. Kostochka, "Large minors in graphs with given independence number", *Discrete Math.* 311 (2011), 2203–2215.
7. J. Barát, G. Joret and D. R. Wood, "Disproof of the list Hadwiger conjecture", *Electronic Journal of Combinatorics* 18 (2011), p232.
8. T. Böhme, K. Kawarabayashi, J. Maharry and B. Mohar, "Linear connectivity forces large complete bipartite graph minors", *J. Combinatorial Theory, Ser. B,* 99 (2009), 557–582.
9. T. Böhme, A. Kostochka and A. Thomason, "Minors in graphs with high chromatic number", *Combin. Probab. Comput.* 20 (2011), 513–518.
10. B. Bollobás, P. A. Catlin and P. Erdős, "Hadwiger's conjecture is true for almost every graph", *Europ. J. Combinatorics* 1 (1980), 195–199.
11. M. Borowiecki, "Research problem 172", *Discrete Mathematics* 121 (1993), 235–236.
12. P. Catlin, "Hajós' graph-coloring conjecture: Variations and counterexamples," *J. Combinatorial Theory, Ser. B,* 26 (1979), 268–274.
13. M. Chudnovsky and A. Fradkin, "An approximate version of Hadwiger's conjecture for claw-free graphs", *J. Graph Theory* 63 (2010), 259–278.
14. M. Chudnovsky, B. Reed and P. Seymour, "The edge-density for $K_{2,t}$ minors", *J. Combinatorial Theory, Ser. B,* 101 (2011), 18–46.
15. M. Chudnovsky and P. Seymour, "Packing seagulls", *Combinatorica*, 32 (2012), 251–282.

16. M. Devos, G. Ding, B. Oporowski, B. Reed, D. Sanders, P. Seymour and D. Vertigan, "Excluding any graph as a minor allows a low tree-width 2-coloring", *J. Combinatorial Theory, Ser. B*, 91 (2004), 25–41.
17. M. DeVos, Z. Dvořék, J. Fox, J. McDonald, B. Mohar and D. Scheide, "A minimum degree condition forcing complete graph immersion", *Combinatorica* 34 (2014), 279–298.
18. M. DeVos, K. Kawarabayashi, B. Mohar and H. Okamura, "Immersing small complete graphs", *Ars Mathematica Contemporanea* 3 (2010), 139–146.
19. G. Ding and S. Dziobiak, "Vertex-bipartition method for colouring minor-closed classes of graphs", *Combinatorics, Probability and Computing* 19 (2010), 579–591.
20. G. Ding, T. Johnson and P. Seymour, "Spanning trees with many leaves", *J. Graph Theory*, 37 (2001), 189–197.
21. G. Ding, B. Oporowski, D. Sanders, and D. Vertigan, "Surfaces, tree-width, clique-minors, and partitions", *J. Combinatorial Theory, Ser. B*, 79 (2000), 221–246.
22. G. A. Dirac, "A property of 4-chromatic graphs and some remarks on critical graphs", *J. London Math. Soc.* 27 (1952), 85–92.
23. P. Duchet and H. Meyniel, "On Hadwiger's number and the stability number", in *Graph Theory* (Proc. conf. on graph theory, Cambridge, 1981; B. Bollobás, ed.), *Annals of Discrete Math.* 13, North-Holland, Amsterdam, New York, 71–73; *North-Holland Mathematical Studies* 62 (1982), 71–73.
24. R.J. Duffin, "Topology of series–parallel networks", *J. Math. Analys. Appl.* 10 (1965), 303–318.
25. Zdeněk Dvořák and Liana Yepremyan, "Complete graph immersions and minimum degree", http://arxiv.org/abs/1512.00513.
26. K. Edwards, D. Y. Kang, J. Kim, S. Oum and P. Seymour, "A relative of Hadwiger's conjecture", *SIAM J. Discrete Math.* 29 (2015), 2385–2388.
27. P. Erdős and S. Fajtlowicz, "On the conjecture of Hajós", *Combinatorica* 1 (1981), 141–143.
28. W. Fernandez de la Vega, "On the maximum density of graphs which have no subcontraction to K_s", *Discrete Math.* 46 (1983), 109–110.
29. J. Fox, "Complete minors and independence number", *SIAM J. Discrete Math.* 24 (2010), 1313–1321.
30. J. Fox, C. Lee and B. Sudakov, "Chromatic number, clique subdivisions, and the conjectures of Hajós and Erdős-Fajtlowicz", *Combinatorica* 33 (2013), 181–197.
31. A. Fradkin, "Clique minors in claw-free graphs", *J. Combinatorial Theory, Ser. B*, 102 (2012), 71–85.
32. J. Geelen, A. Gerards, B. Reed, P. Seymour and A. Vetta, "On the odd-minor variant of Hadwiger's conjecture", *J. Combinatorial Theory, Ser B*, 99 (2009), 20–29.
33. A. M. H. Gerards, *Graphs and polyhedra. Binary spaces and cutting planes*, volume 73 of CWI Tract. Stichting Mathematisch Centrum voor Wiskunde en Informatica, Amsterdam, 1990; http://oai.cwi.nl/oai/asset/12714/12714A.pdf.
34. D. Gonçalves, "On vertex partitions and some minor-monotone graph parameters", *J. Graph Theory* 66 (2011), 49–56.
35. G. R. Grimmitt and C. J. H. McDiarmid, "On colouring random graphs", *Math. Proc. Cambridge Phil. Soc.* 77 (1975), 313–324.
36. B. Guenin, "Graphs without odd-K5 minors are 4-colourable", in preparation.
37. H. Hadwiger, "Über eine Klassifikation der Streckenkomplexe", *Vierteljschr. Naturforsch. Ges. Zürich* 88 (1943), 133–143.
38. T. Jensen and B. Toft, *Graph Coloring Problems*, Wiley, Chichester UK, 1995, page 115.
39. L. Jørgensen, "Contractions to K_8", *J. Graph Theory* 18 (1994), 431–448.
40. K. Kawarabayashi, "A weakening of the odd Hadwiger's conjecture", *Combinatorics, Probability and Computing* 17 (2008), 815–821.
41. K. Kawarabayashi, "Note on coloring graphs without odd-K_k-minors", *J. Combinatorial Theory, Ser. B*, 99 (2009), 728–731.
42. K. Kawarabayashi, "On the connectivity of minimum and minimal counterexamples to Hadwiger's conjecture", *J. Combinatorial Theory, Ser. B*, 97 (2007), 144–150.

43. K. Kawarabayashi and B. Mohar, "A relaxed Hadwiger's conjecture for list colorings", *J. Combinatorial Theory Ser. B*, 97 (2007), 647–651.
44. K. Kawarabayashi, S. Norin, R. Thomas and P. Wollan, "K_6 minors in large 6-connected graphs", in preparation; http://arXiv.org/abs/1203.2192.
45. K. Kawarabayashi and B. Reed, "Fractional coloring and the odd Hadwiger's conjecture", *European J. Comb.* 29 (2008), 411–417.
46. K. Kawarabayashi and B. Reed, "Hadwiger's conjecture is decidable", *Proc. 41st Annual ACM Symposium on Theory of Computing*, STOC 2009, 445–454.
47. K. Kawarabayashi and Z. Song, "Some remarks on the odd Hadwiger's conjecture", *Combinatorica*, 27 (2007), 429–438.
48. A. V. Kostochka, "$K_{s,t}$ minors in $(s+t)$-chromatic graphs, II", *J. Graph Theory* 75 (2014), 377–386.
49. A. V. Kostochka, "Lower bound on the Hadwiger number of graphs by their average degree", *Combinatorica* 4 (1984), 307–316.
50. A. V. Kostochka, "On $K_{s,t}$ minors in $(s+t)$-chromatic graphs", *J. Graph Theory* 65 (2010), 343–350.
51. A. V. Kostochka, "The minimum Hadwiger number for graphs with a given mean degree of vertices", *Metody Diskret. Analiz.* 38 (1982), 37–58; *AMS Translations* (2), 132 (1986), 15–32.
52. A. V. Kostochka and N. Prince, "Dense graphs have $K_{3,t}$ minors", *Discrete Math.* 310 (2010), 2637–2654.
53. A. V. Kostochka and N. Prince, "On $K_{s,t}$ minors in graphs of given average degree", *Discrete Math.* 308 (2008), 4435–4445.
54. A. V. Kostochka and N. Prince, "On $K_{s,t}$-minors in graphs with given average degree, II", *Discrete Math.* 312 (2012), 3517–3522.
55. D. Kühn and D. Osthus, "Forcing complete unbalanced bipartite minors", *Europ. J. Combinatorics* 26 (2005), 75–81.
56. K. Kuratowski, "Sur le problème des courbes gauches en topologie", *Fund. Math.* 15 (1930), 271–283.
57. F. Lescure and H. Meyniel, "On a problem upon configurations contained in graphs with given chromatic number", *Graph Theory in Memory of G. A. Dirac* (Sandbjerg, 1985), *Ann. Discrete Math.* 41, North-Holland, Amsterdam, 1989, 325–331.
58. C. Liu and S. Oum, "Partitioning H-minor free graphs into three subgraphs with no large components", manuscript March 2015; http://arXiv.org/abs/1503.08371v1.
59. W. Mader, "Homomorphieeigenschaften und mittlere Kantendichte von Graphen" *Math. Ann.* 174 (1967), 265–268.
60. W. Mader, "Homomorphiesätze für Graphen", *Math. Ann.*, 178 (1968), 154–168.
61. W. Mader, "Über trennende Eckenmengen in homomorphiekritischen Graphen", *Math. Ann.* 175 (1968), 245–252.
62. J.S. Myers, "The extremal function for unbalanced bipartite minors", *Discrete Math.* 271 (2003), 209–222.
63. J. S. Myers and A. Thomason, "The extremal function for noncomplete minors", *Combinatorica* 25 (2005), 725–753.
64. S. Norin, "Conquering graphs of bounded treewidth", unpublished manuscript, April 2015.
65. B. Reed and P. Seymour, "Fractional colouring and Hadwiger's conjecture", *J. Combinatorial Theory, Ser. B*, 74 (1998), 147–152.
66. B. Reed and P. Seymour, "Hadwiger's conjecture for line graphs", *European J. Math.*, 25 (2004), 873–876.
67. B. Reed and D. Wood, "Forcing a sparse minor", *Combinatorics, Probability and Computing*, 25 (2016), 300–322.
68. N. Robertson and P. Seymour, "Graph minors. XVI. Excluding a non-planar graph", *J. Combinatorial Theory, Ser. B*, 89 (2003), 43–76.
69. N. Robertson, P. Seymour and R. Thomas, "Excluding infinite clique minors", *Memoirs Amer. Math. Soc.*, no. 566, vol. 118 (1995).

70. N. Robertson, P. Seymour and R. Thomas, "Excluding subdivisions of infinite cliques", *Trans. Amer. Math. Soc.* 332 (1992), 211–223.
71. N. Robertson, P. Seymour and R. Thomas, "Hadwiger's conjecture for K_6-free graphs", *Combinatorica* 13 (1993), 279–361.
72. Z. Song and R. Thomas, "The extremal function for K_9 minors", *J. Combinatorial Theory, Ser. B,* 96 (2006), 240–252.
73. A. Thomason, "An extremal function for contractions of graphs", *Math. Proc. Camb. Phil. Soc.* 95 (1984), 261–265.
74. A. Thomason, "The extremal function for complete minors", *J. Combinatorial Theory Ser. B,* 81 (2001), 318–338.
75. C. Thomassen, "Every planar graph is 5-choosable", *J. Combinatorial Theory, Ser. B,* 62 (1994), 180–181.
76. C. Thomassen, "Totally odd K_4-subdivisions in 4-chromatic graphs", *Combinatorica 21* (2001), 417–443.
77. B. Toft, "Problem 10," in *Recent Advances in Graph Theory*, Proc. Symp. Prague June 1974, M. Fiedler (Ed.), Academia Praha, 1975, 543–544.
78. B. Toft, *A Survey of Hadwiger's Conjecture*, in: *Surveys in Graph Theory* (edited by G. Chartrand and M. Jacobson), Congr. Numer. 115 (1996), 249–283.
79. W. T. Tutte, "On the algebraic theory of graph colorings", *J. Combinatorial Theory*, 1 (1966), 15–50.
80. D. van der Zypen, "A weak form of Hadwiger's conjecture" *SOP Trans. Appl. Math.* 1 (2014), 84–87.
81. M. Voigt, "List colourings of planar graphs", *Discrete Mathematics* 120 (1993), 215–219.
82. K. Wagner, "Beweis einer Abschwächung der Hadwiger-Vermutung", *Math. Ann.* 153 (1964), 139–141.
83. K. Wagner, "Über eine Eigenschaft der ebenen Komplexe", *Math. Ann.* 114 (1937), 570–590.
84. D. R. Wood, "Contractibility and the Hadwiger conjecture", *Europ. J. Combinatorics.*, 31 (2010), 2102–2109.
85. D. R. Wood, "On tree-partition-width", *European J. Combinatorics* 30 (2009), 1245–1253.
86. W. Zang, "Proof of Toft's conjecture: every graph containing no fully odd K_4 is 3-colorable", *J. Combinatorial Optimization* 2 (1998), 117–199.

The Hadwiger–Nelson Problem

Alexander Soifer

Abstract Inspired by the Four-Color Conjecture, the Hadwiger–Nelson Problem became one of the famous open problems of mathematics in its own rights. It has withstood all assaults for 65 years, and attracted many mathematicians from many fields, including Paul Erdős and Ronald L. Graham. John F. Nash admired this problem and chose it for the present book. In this chapter we will discuss this problem, its history and generalizations, several of the many related open problems, and the state of the art results. [1]

1 The Problem

The title problem asks to find the smallest number of colors sufficient for coloring the points of the Euclidean plane E^2 in such a way that no two points of the same color are unit distance apart.

This number is called the *chromatic number of the plane* (CNP) and is denoted by $\chi(E^2)$.

2 The History and the Authorship of the Problem

While the distinguished Swiss geometer Hugo Hadwiger admired this problem and wrote about it in the early 1960s, the problem was posed earlier by one remarkable young person. As documented in [49], the problem was created in October–November 1950 by the 18-year old Edward Nelson (May 4, 1932, Decatur,

[1]Unlike the previous version of this survey [55], more history of the problem and its authorship has been included here.

A. Soifer (✉)
University of Colorado, 1420 Austin Bluffs Parkway, Colorado Springs, CO 80918, USA
e-mail: asoifer@uccs.edu

J.F. Nash, Jr., M.Th. Rassias (eds.), *Open Problems in Mathematics*,
DOI 10.1007/978-3-319-32162-2_14

GA–September 10, 2014, Princeton, NJ), who also determined a lower bound 4; his 20-year old friend John Isbell found an upper bound 7: $4 \leq \chi(E^2) \leq 7$.

During his talk at 25th South Eastern International Conference on Combinatorics, Computing and Graph Theory in Boca Raton, Florida at 9:30–10:30 A.M. on March 10, 1994, Paul Erdős summarized the results of the author's historical research on the authorship of this problem in a characteristically Erdősian style [17]:

> "There is a mathematician called Nelson who in 1950 when he was an epsilon, that is he was 18, discovered the following question. Suppose you join two points in the plane whose distance is 1. It is an infinite graph. What is chromatic number of this graph?
>
> Now, de Bruijn and I showed that if an infinite graph which is chromatic number k, it always has a finite subgraph, which is chromatic number k. So this problem is really [a] finite problem, not an infinite problem. And it was not difficult to prove that the chromatic number of the plane is between 4 and 7. I would bet it is bigger than 4, but I am not sure. And the problem is still open.
>
> If it would be my problem, I would certainly offer money for it. You know, I can't offer money for every nice problem because I would go broke immediately. I was asked once what would happen if all your problems would be solved, could you pay? Perhaps not, but it doesn't matter. What would happen to the strongest bank if all the people who have money there would ask for money back? Or what would happen to the strongest country if they suddenly ask for money? Even Japan or Switzerland would go broke. You see, Hungary would collapse instantly. Even the United States would go broke immediately . . .
>
> Actually it was often attributed to me, this problem. It is certain that I had nothing to do with the problem. I first learned the problem, the chromatic number of the plane, in 1958, in the winter, when I was visiting [Leo] Moser. He did not tell me from where this nor the other problems came from. It was also attributed to Hadwiger but Soifer's careful research showed that the problem is really due to Nelson."

It is therefore fitting to call this problem **The Nelson Problem** or, as it is often called, **The Chromatic Number of the Plane Problem (CNP).**

There are fascinating similarities between the famous Four-Color Problem (4CP) and the Nelson Problem (CNP). 4CP was created during 1850–1852 timeframe by the 18–20 year old Francis Guthrie. CNP was created a century later, in 1950 by the 18-year old Edward Nelson. Augustus De Morgan was instrumental in keeping 4CP alive for decades. Paul Erdős, like De Morgan a century earlier, kept the flaming torch of the problem lit. He made the chromatic number of the plane problem well known by posing it in his countless problem talks and many publications.

In fact, the starting point of Edward Nelson in creating CNP problem was 4CP. In his October 5, 1991, letter [33], he conveyed to the author the Story of Creation:

> "Dear Professor Soifer:
>
> In the autumn of 1950, I was a student at the University of Chicago and among other things was interested in the four-color problem, the problem of coloring graphs topologically embedded in the plane. These graphs are visualizable as nodes connected by wires. I asked myself whether a sufficiently rich class of such graphs might possibly be subgraphs of one big graph whose coloring could be established once and for all, for example, the graph of all points in the plane with the relation of being unit distance apart (so that the wires become rigid, straight, of the same length, but may cross). The idea did not hold up, but the other problem was interesting in its own right and I mentioned it to several people."

Both problems, 4CP and CNP, required a very long time to be conquered. Victor Klee and Stan Wagon [54], observing that solving 4CP took 124 years, suggested that CNP might require as long for its solution: "If a solution of CNP takes as long as 4CC, then we will have a solution by the year 2084." Will we succeed by 2084? Paul Erdős would have said, "We shall see!" The great composer Arnold Schoenberg believed that faith can move mountains. Erdős urged us to believe that the transfinite Book of all theorems and their best proofs exists. A similar belief led Appel and Haken to succeed in solving 4CP at the breaking point of available computing. Such a belief is needed to conquer CNP, the author's favorite open problem of mathematics.

3 Translation of the Nelson Problem into the Language of Graph Theory

Graph Theory was born out of our neglecting geometric considerations of shape and size, and preserving only adjacency. Surprisingly, the past century has witnessed a renewed interest in graphs, where geometrical considerations such as distance matter, in fact, define the adjacency.

We can create a graph $G(E^2)$ out of the Euclidean plane E^2 by taking all of its points as vertices, and joining two vertices by an edge if and only if they are at distance 1 apart. More generally, we call a graph *unit-distance* when any two vertices are adjacent if and only if they are at distance 1 apart. In this language, the Nelson Problem can be translated as follows:

The Nelson Problem 1 Find the chromatic number of the graph $G(E^2)$.

After 65 years of intensive work by many scholars, using tools from graph theory, geometry, abstract algebra, topology and measure theory, we have been unable to improve on the above bounds for $\chi(E^2)$ in the general case. Ronald L. Graham believes that the chromatic number of the plane is 5 or 6 (see [21, 22]). He cites a theorem of Paul O'Donnell (see [34, 49]) showing the existence of 4-chromatic unit-distance graphs of arbitrarily large girth (Theorem 28 below) as "perhaps, the evidence that χ is at least 5." Paul Erdős [14, 15] believed that the chromatic number of the plane was 5, 6, or 7. The author conjectured the answer to be 7.

4 The Polychromatic Number: The Lower Bound

When a problem withstands all assaults, mathematicians create related problems, solutions of which may shed light on the original problem. Let us look at one of them here. We say that a point set S *realizes distance* d if S contains two points at distance d apart. In 1958 Erdős posed the following question.

Fig. 1 The Mosers spindle

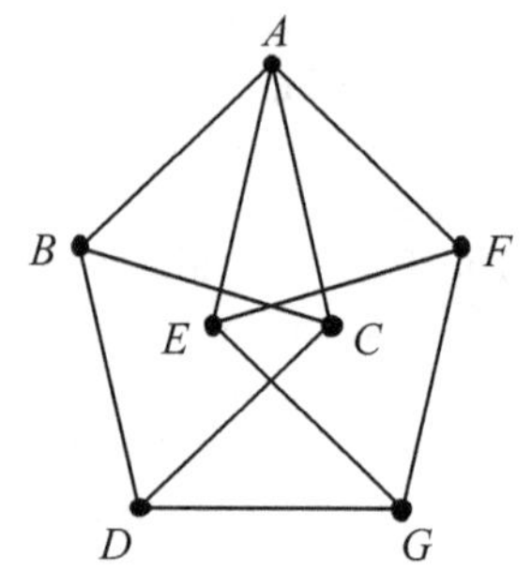

Polychromatic Number of the Plane Problem 2 What is the smallest number of colors needed to color the plane so that no color realizes all distances?

In 1992 Soifer [43] named this invariant the *polychromatic number of the plane*, and denoted by $\chi_p(E^2)$. Clearly $\chi_p(E^2) \leq \chi(E^2) \leq 7$. In 1970 upper and lower bounds for $\chi_p(E^2)$ were published by a Russian high-school student, Dmitry E. Raiskii [40]. We look at the lower bound here and return to the upper bound in Sect. 6. In 1961 the brothers Leo and Willie Moser produced [31] a seven-point plane configuration that we now call *The Mosers Spindle* (Fig. 1): each edge in the spindle has length 1.

Lemma 3 Any three points of the Mosers Spindle contain two points at distance 1 apart. Consequently, in any coloring of a Mosers Spindle that forbids monochromatic distance 1, at most two points can be of the same color.

Raiskii [40] proved the following lower bound.

Theorem 4 $\chi_p(E^2) \geq 4$

Proof The following striking proof was found in 1997 by Alexei Merkov, another Russian high-school student, and appeared in an obscure brochure [37]. It is presented here with modifications by the author. We assume that the plane is colored in three colors, red, white and blue, and that each color forbids a distance r, w and b, respectively. Equip the three-colored plane with Cartesian coordinates with origin O, and construct three 7-point sets S_r, S_w and S_b, each being the Mosers Spindle (Fig. 2), in such a way that all three spindles share O as one of their seven vertices and have edge lengths all equal to r, w and b, respectively. This construction defines 18 vectors: six 'red' ones, $v_1, v_2, \ldots, v_6$ from the origin O to each remaining point of S_r, six 'white' ones, $v_7, v_8, \ldots, v_{12}$ from O to the points of S_w, and six 'blue' ones $v_{13}, v_{14}, \ldots, v_{18}$ from O to the points of S_b.

We now introduce 18-dimensional Euclidean space E^{18} and a function $M: E^{18} \rightarrow E^2$ that maps each vector $(a_1, a_2, \ldots, a_{18})$ to $a_1v_1 + a_2v_2 + \ldots + a_{18}v_{18}$. This function induces a three-coloring of E^{18} by assigning to a point of E^{18} the color of the corresponding point of the plane. In accordance with the colors of the vectors v_i, we call the first six axes of E^{18} 'red', the next six axes 'white', and the last six axes 'blue.'

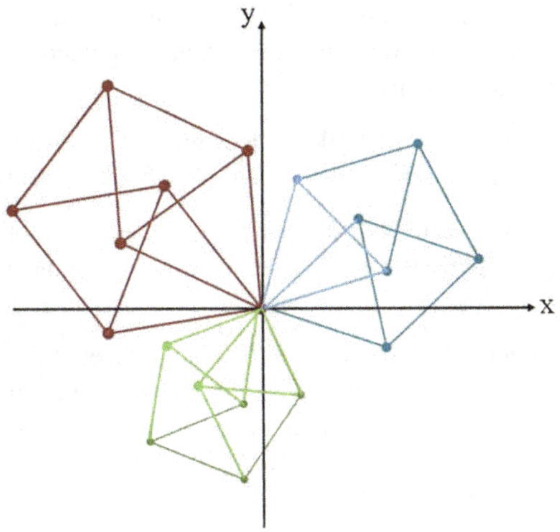

Fig. 2 The Cartesian plane with three Mosers Spindles

Let W be the subset of E^{18} consisting of all points whose coordinates include at most one equal to 1 for each of the three colors of the axes, and the remaining (15 or more) coordinates equal to 0; it is easy to verify that W has $7^3 = 343$ points. For any fixed array of coordinates allowable in W on white and blue axes, we get the 7-element set A of points in W with these fixed coordinates on white and blue axes. The image $M(A)$ forms in the plane a translate of the original 7-point set S_r. If we fix another array of white and blue coordinates, we obtain another 7-element set in E^{18}, whose image under M in the plane would form another translate of S_r. Thus, the set W gets partitioned into 49 subsets, each of which maps onto a translate of S_r.

By Lemma 3, any translate of S_r has at most two red points among its 7 points. Since W has been *partitioned* into translates of S_r, at most $^2/_7$ of the points of W are red. We can now start all over again, and similarly show that at most $^2/_7$ of the points of W are white, and at most $^2/_7$ of the points of W are blue. But $^2/_7 + ^2/_7 + ^2/_7$ is less than 1. This contradiction implies that at least one of the colors realizes all distances, as required. ■

5 The de Bruijn–Erdős Theorem

We can expand the notion of the chromatic number to any subset S of the plane. The *chromatic number* $\chi(S)$ is the smallest number of colors sufficient to color the points of S in such a way that forbids monochromatic pairs of points at distance 1 apart. In other words, we create on S a *unit distance graph*, with vertex set S and two points adjacent if and only if they are distance 1 apart, and denote by $\chi(S)$ the chromatic number of this graph.

In 1951 Nicolaas G. de Bruijn and Paul Erdős [11] published a lemma that implies the following important 'compactness theorem'; its proof assumes the axiom of choice.

The De Bruijn-Erdős Theorem 5 The chromatic number of the plane is equal to the maximum chromatic number of its finite subsets.

Accordingly, Erdős used to say that the problem of finding the chromatic number of the plane is a problem about finite sets in the plane. In 1975 he posed the following problem [13]:

Erdős' $25 Problem 6 Let S be a subset of the plane which contains no equilateral triangle of side length 1. Join two points of S if their distance is 1. Does this graph have chromatic number at most 3? If the answer is no—assume that the graph defined by S contains no C_l [cycles of length l] for $3 \leq l \leq t$ and ask the same question.

Erdős was unsure of the outcome (which was rare for him): he expected triangle-free unit distance graphs to have chromatic number at most 3, or else that chromatic number 3 can be forced by prohibiting all small cycles up to C_t, for sufficiently large t.

In 1979 Nicholas Wormald [53] disproved the first, easier conjecture, contained in Problem 6, and Erdős promptly paid $25 and reported the result in a lecture and later in print [16]:

> In a recent paper (still unpublished), Wormald found a set S for which the unit distance graph $G_1(S)$ has girth 5 and chromatic number 4. His construction involved elaborate computations and is fairly complicated. Indeed, aided by a computer, he had proved in [49] the existence of a set S of 6448 points with chromatic number 4 and without triangles or quadrilaterals with all sides of length 1.

In 1992 the author in a talk posed this Erdős' $25 Problem, in a form of competition:

Problem 7 Find the smallest number σ_4 of points in a plane set with chromatic number 4 without unit equilateral triangles, and classify all such sets S of σ_4 points.

Several young mathematicians entered the race, and the graphs obtained by the record setters were as mathematically significant as they were beautiful (see [49]). The two record-holding graphs, created by roommates-graduate students of Rutgers University Robert Hochberg and Paul O'Donnell [24], are shown in Figs. 3 and 4; both graphs are 4-chromatic unit-distance graphs.

6 The Polychromatic Number: The Upper Bound

An example proving an upper bound $\chi_p(E^2) \leq 6$ was found by Sergei B. Stechkin (see Fig. 5) and published by Raiskii [40].

The 'unit of construction' is a parallelogram consisting of four regular hexagons and eight equilateral triangles, all of side-length 1. We color the hexagons with

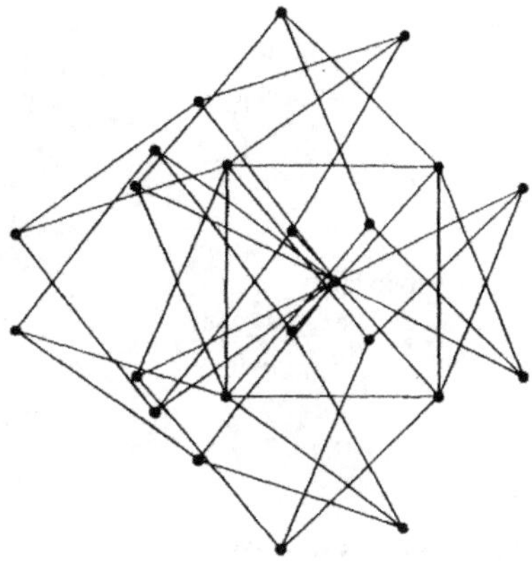

Fig. 3 The Hochberg–O'Donnell fish graph, of girth 4 and order 23

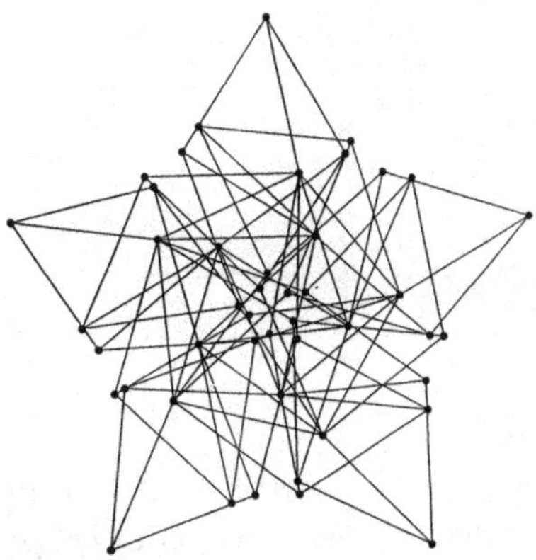

Fig. 4 The Hochberg–O'Donnell star graph, of girth 5 and order 45

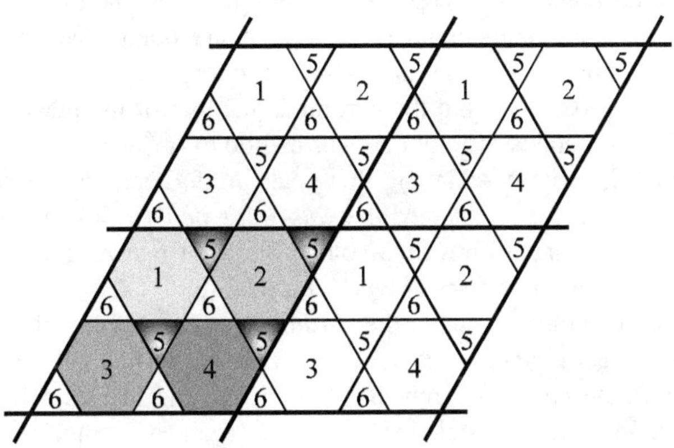

Fig. 5 Stechkin's six-coloring of the plane

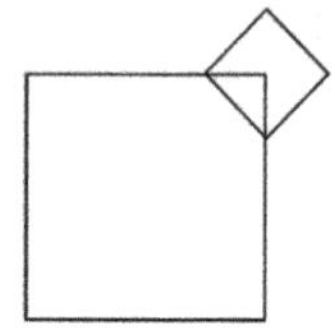

Fig. 6 Two squares that induce the tiling of the plane with octagons and squares

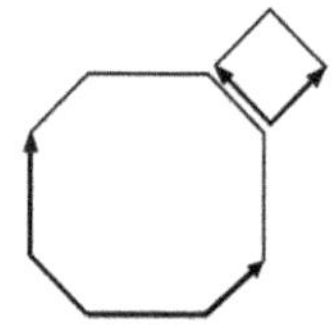

Fig. 7 Boundary points included in the colors of the polygons are shown in bold

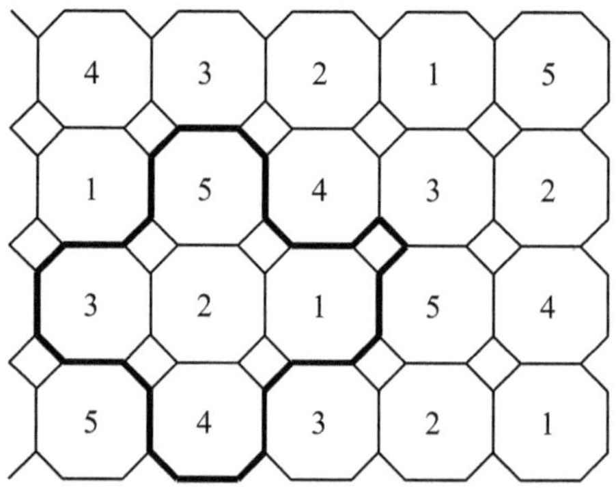

Fig. 8 Soifer's six-coloring of the plane

colors 1, 2, 3 and 4. We then partition the triangles into two types: assign color 5 to the triangles with a vertex below their horizontal base, and color 6 to those with a vertex above their horizontal base. While coloring, assume that every hexagon includes its entire boundary, except for its one right-most and two lowest vertices, and that every triangle includes none of its boundary points. We can now tile the entire plane with translates of the 'unit of construction'.

If our ultimate goal is to find the chromatic number of the plane, or at least to improve its known bounds, it might be worthwhile to 'measure' how close a given coloring of the plane is to achieving this goal. In 1992 such a measurement was introduced by the author and named the *coloring type* (see Soifer [43, 44]): given an n-coloring of the plane for which color i does not realize the distance d_i (for $1 \leq i \leq n$), we say this coloring is of *type* $(d_1, d_2, \ldots, d_n)$.

It would improve our search for the chromatic number of the plane if we could find a six-coloring of type (1, 1, 1, 1, 1, 1), or show that one does not exist. With an appropriate choice of unit, Stechkin's coloring in Fig. 5 has type (1, 1, 1, 1, $^1/_2$, $^1/_2$). In 1973 Woodall [52] found a second six-coloring of the plane with no color realizing all distances; his coloring had a property that each of the six monochromatic sets was closed. However, his example had three 'missing distances': it had type (1, 1, 1, $1/\sqrt{3}$, $1/\sqrt{3}$, $1/2\sqrt{3}$). In 1991 a new six-coloring was found by the author [44], using a tiling of the plane by squares and non-regular octagons; it had type (1,1,1,1,1, $1/\sqrt{5}$). To construct it, we start with two squares, one of side 2 and the other of diagonal 1 (see Fig. 6), and use them to create a tiling of the plane with squares and non-regular octagons (see Fig. 8); colors 1–5 were used for the octagons, and all the squares were colored 6. With each octagon and each square we include half of its boundary (bold lines in Fig. 7) without the endpoints of that half. It is easy to verify that $\sqrt{5}$ is not realized by any of the colors 1–5, and 1 is not realized by color 6. By shrinking all linear sizes by a factor of $\sqrt{5}$, we obtain the six-coloring of type (1,1,1,1,1, $1/\sqrt{5}$). To simplify the verification, we define the unit of construction as the region bounded by the bold line in Fig. 8; its translates tile the plane.

The above six-coloring gave birth to a new definition (see [25]). The *almost chromatic number* $\chi_a(E^2)$ of the plane is the minimal number of colors that are required to color the plane so that all but one of the colors forbid unit distance, and the remaining color forbids a distance, which is not necessarily unit. Note that $4 \leq \chi_a(E^2) \leq 6$; the lower bound follows from Raiskii [40], and the upper bound follows form the above six-coloring. The problem of determining $\chi_a(E^2)$ is still open (see [49]).

7 The Continuum of Six-Colorings

In 1993 another six-coloring was found by the 15-year old violinist Ilya Hoffman and the author (see [25, 26]), with type (1, 1, 1, 1, 1, $\sqrt{2}-1$); the story of its discovery can be found in [49]. To construct it we first tile the plane with squares of diagonals 1 and $\sqrt{2}-1$ (see Fig. 9). We use colors 1–5 for larger squares and color 6 for all small squares. With each square we include the left and lower sides of its boundary without the endpoints of this half (see Fig. 10). To verify that this coloring does the job, define the unit of construction that is bounded by the bold line in Fig. 9; its translates tile the plane.

In 1993 the above two examples prompted the introduction of new terminology and the translation of earlier results and problems into this new language (see [45, 46]). We let X_6 denote the 6-*realizable set* of all positive numbers α for which there exists a six-coloring of the plane of type (1, 1, 1, 1, 1, α). The problem, posed by the author, which is still open and extremely difficult, is to find X_6.

We know, from our above discussion, that $1/\sqrt{5}$ and $\sqrt{2}-1$ both lie in X_6. As shown by the author in [45, 46] (see also [49]), these are extreme examples of the general case, which includes a continuum of 'working' six-colorings: for every α between $\sqrt{2}-1$ and $1/\sqrt{5}$, there is a six-coloring of type (1, 1, 1, 1, 1, α).

Theorem 8 X_6 contains the entire closed interval $[\sqrt{2}-1, 1/\sqrt{5}]$.

Outline of Proof We tile the plane with congruent non-regular octagons and 'small' squares (see Fig. 11). We now color this tiling in six colors. Denote by F the unit of our construction, bounded by a bold line and consisting of five octagons and four 'small' squares. Use colors 1 5 for the octagons inside F and color 6 for all 'small' squares, and include in the colors of octagons and 'small' squares those parts of their boundaries that are shown in bold in Fig. 12. This is followed by a proof that for each α from the given in the theorem interval, we can vary the relative sizes of the hexagons and squares and the angle between them to guarantee that this six-coloring has type $(1, 1, 1, 1, 1, \alpha)$. We have found a continuum of values for α and a continuum of 'working' six-colorings of the plane.

8 The Chromatic Number of the Plane in Special Circumstances

In 1973 Douglas R. Woodall [52] attempted to prove a lower bound for the chromatic number of the plane for the special case of 'map-type colorings'. However, in 1979 Stephen P. Townsend constructed a counter-example that showed that one essential idea of Woodall's proof was incorrect. By that time, Townsend had already proved the same result, and his proof was much more elaborate than Woodall's attempt. For decades Townsend's proof was unavailable until he produced a clear version of it where the definition of the map-type coloring can also be found (see [49]). The following result was thus conjectured by Woodall and proven by Townsend.

The Townsend–Woodall Theorem 9 The chromatic number of the plane under map-type coloring is 6 or 7.

Woodall [52] showed that this result implies another result worth mentioning.

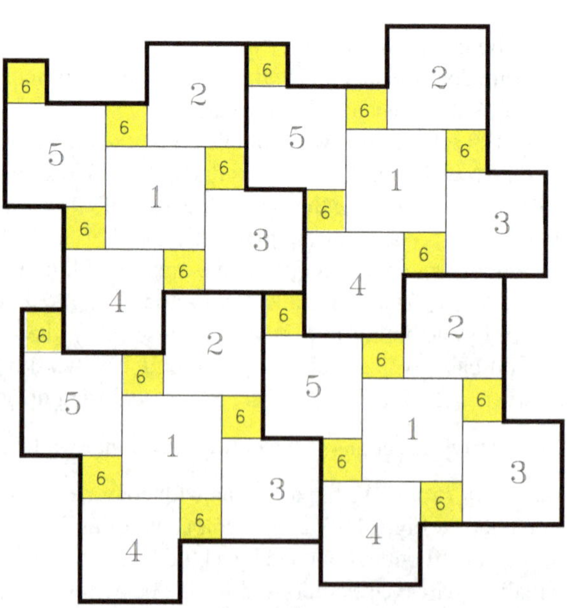

Fig. 9 Hoffman and Soifer's six-coloring of the plane

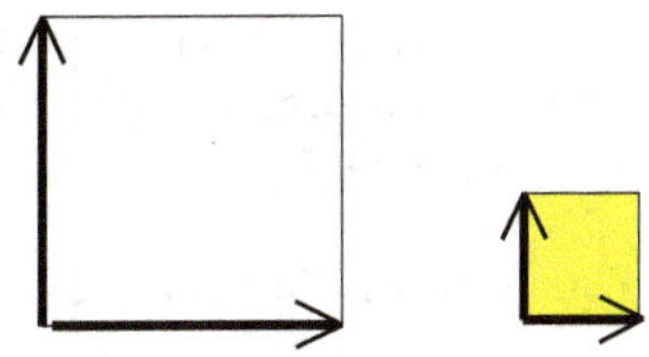

Fig. 10 Boundary points included in the colors of the squares are shown in bold

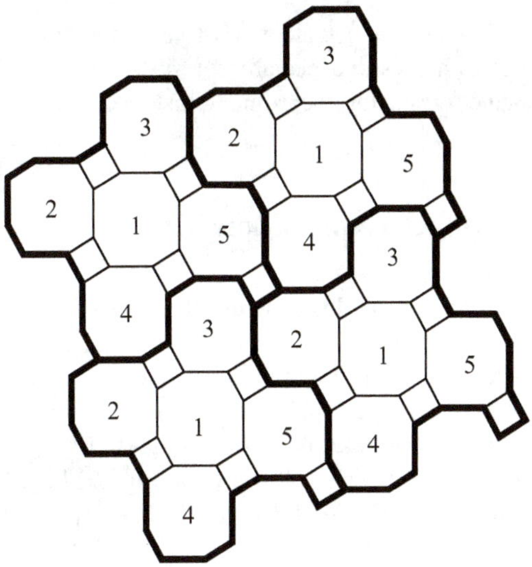

Fig. 11 Continuum of six-colorings of the plane

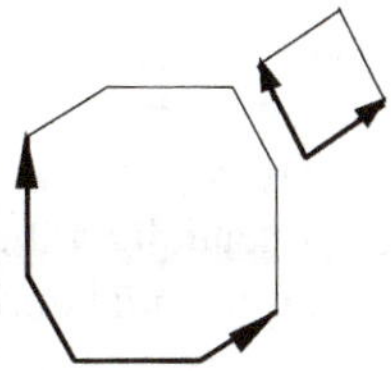

Fig. 12 Boundary points included in the colors of the polygons are shown in bold

Theorem 10 The chromatic number of the plane under coloring with closed monochromatic sets is 6 or 7.

In 1993–94 three American undergraduate students, Nathanial Brown, Nathan Dunfield and Greg Perry, proved that a similar result is true for coloring with open monochromatic sets (see [3–5]).

Theorem 11 The chromatic number of the plane under coloring with open monochromatic sets is 6 or 7.

A related graph-theoretic result was obtained by Carsten Thomassen ([49, 51]). He proved that for a "nice" coloring of a connected graph on a surface, such as the plane, seven colors are needed.

Meanwhile, Kenneth J. Falconer [18], while still a graduate student, proved the following important result, that the *measurable chromatic number* $\chi_m(E^2)$ of the plane is 5, 6 or 7.

The Falconer Theorem 12 If $E^2 = \bigcup_{i=1}^{4} A_i$ is a covering of the Euclidean plane E^2 by four disjoint measurable sets, then one of the sets A_i realizes distance 1.

Decades later, Falconer wrote a much more detailed and self-contained exposition of his result especially for [49]. Theorem 12 means that if a four-coloring of the plane forbids monochromatic distance 1, then one of the classes is non-measurable.

9 Space Explorations

Around 1961 Erdős generalized the problem of finding the chromatic number of the plane to n-dimensional Euclidean space E^n. He was interested both in asymptotic behaviour as n increases, and in exact values of the chromatic number $\chi(E^n)$ for small n (especially $n = 2$ and 3).

In 1970 Raiskii [40] proved that $\chi(E^n) \geq n + 2$, for all $n > 1$; thus, for $n = 3$ we have $\chi(E^3) \geq 5$. This lower bound for E^3 lasted until 2000, when Oren Nechushtan proved that $\chi(E^3) \geq 6$ (see [32]). For upper bounds, David Coulson [9, 10] proved that $\chi(E^3) \leq 15$, by using a face-centred cubic lattice (see Conway and Sloane [8] for more on cubic lattices). For higher dimensions, Kent Cantwell [6] proved in 1996 that $\chi(E^4) \geq 7$ and $\chi(E^5) \geq 9$; these remain the best results known. Then in 2008 the Czech student Josef Cibulka [7] proved that $\chi(E^5) \geq 11$.

Many years ago, Erdős conjectured that the chromatic number $\chi(E^n)$ increases exponentially with n. This conjecture was settled in the affirmative by two results, an exponential upper bound, found in 1972 by D. G. Larman and C. A. Rogers [28], and an exponential lower bound obtained in 1981 by P. Frankl and R. M. Wilson [20]:

For all n, $(1 + o(1))\ 1.2^n \leq \chi(E^n) \leq (3 + o(1))^n$.

Asymptotically, Larman and Rogers's upper bound remains the best known today. In 2000 Frankl and Wilson's asymptotic lower bound was improved by Andrei Raigorodskii [39] to $(1.239\ldots + o(1))^n$. Narrowing the remaining gap is desirable.

The polychromatic number $\chi_p(E^2)$ also generalizes to higher dimensions. Raiskii [40] proved that $\chi_p(E^n) \geq n + 2$, for all $n > 1$. Larman and Rogers's upper bound implied $\chi_p(E^n) \leq (3 + o(1))^n$. Their conjecture that $\chi_p(E^n)$ grows exponentially in n, was proved by Frankl and Wilson [20], who showed $(1 + o(1))\ 1.2^n \leq \chi_p(E^n)$.

10 The Chromatic Number of Rational Spaces

Another approach to the chromatic number of the plane E^2 is to use Cartesian coordinates. Here, E^2 is the set of all ordered pairs (x, y) with real coordinates x and y, with the distance between two points defined in Euclidean way. Since it suffices to deal with finite subsets of E^2, by de Bruijn and Erdős' Theorem 3.1, we can restrict the coordinates to some subset C of E. The problem is: which subset should we choose?

Problem 13 *Find a countable subset C of the set of real numbers E whose chromatic number $\chi(C^2)$ equals that of the plane.*

The set Q of all rational numbers does not work, as shown by Woodall [52]:

Theorem 14 $\chi(Q^2) = 2$.

In 1975 there appeared 'the legendary unpublished manuscript', as P. D. Johnson, Jr. referred to the manuscript by Miro Benda and Micha Perles. This widely circulated manuscript was called *Colorings of Metric Spaces*; Johnson tells its story in the *Geombinatorics*, where in January 2000 the Benda–Perles paper was finally published [2]. It included the following results:

Theorem 15 $\chi(Q^3) = 2$ and $\chi(Q^4) = 4$.

Benda and Perles [3] then posed some important open problems.

Problem 16 Find $\chi(Q^5)$ and, in general, $\chi(Q^n)$.

Problem 17 *Find the chromatic number of* $Q\left(\sqrt{2}\right)^2$ *and, in general, of any algebraic extension of* Q^2.

This direction was developed by P. D. Johnson, Jr., Joseph Zaks, Klaus Fischer, Kiran B. Chilakamarri, Michael Reid, Douglas Jungreis, David Witte, and Timothy Chow. In 2006 Johnson [29] published in *Geombinatorics* "A tentative history and compendium" of this direction of inquiry.

More recently Matthias Mann [29] has proved that $\chi(Q^5) \geq 7$. This jump from $\chi(Q^4) = 4$ explains the difficulty of finding $\chi(Q^5)$, whose exact value is still unknown. Mann [30] then obtained further lower bounds: $\chi(Q^6) \geq 10$; $\chi(Q^7) \geq 13$; $\chi(Q^8) \geq 16$. In 2008 Cibulka [7] obtained new lower bounds for these chromatic numbers, improving some of Mann's results: $\chi(Q^5) \geq 8$ and $\chi(Q^7) \geq 15$

11 One Odd Graph

In 1994 Moshe Rosenfeld [41] defined the *odd-distance graph* E_{odd} to be the graph with vertex-set E^2 in which two vertices are adjacent whenever the distance between them is an odd integer. He showed that E_{odd} does not have a subgraph K_4, and asked whether the chromatic number of E_{odd} is finite. In fact, while the problem was new, the absence of K_4-subgraphs was not, following from a more general result of Graham, Rothschild and Straus [23]:

Theorem 18 *In E^n there exist $n+2$ points for which the distance between any two of them is an odd integer if and only if $n \equiv 14 \pmod{16}$.*

In the necessary part of the proof, the authors used a Victorian result about determinants by Arthur Cayley. The main problem remains wide open:

Problem 19 *Find $\chi(E_{\text{odd}})$.*

We do not even know whether $\chi(E_{\text{odd}})$ is finite. In 2009 Ardal, Manuch, Rosenfeld, Shelah and Stacho [1] improved the lower bound to $\chi(E_{\text{odd}}) \geq 5$. Let me pose here, for the first time, the following conjecture:

Conjecture 20 $\chi(E_{\text{odd}}) \geq \aleph_0$.

In fact, let us denote the *measurable chromatic number* of the odd-distance graph by $\chi_m(E_{\text{odd}})$. In 1986 Falconer and Marstrand [19] proved that plane sets with positive density at infinity contain all large distances. This proves the conjecture in a special case: $\chi_m(E_{\text{odd}}) \geq \aleph_0$.

12 The Influence of Set Theory Axioms on Combinatorial Results

Here we need some ideas from set theory. The standard Zermelo–Fraenkel–Choice system of axioms for set theory will be denoted by **ZFC**; the countable axiom of choice by $\mathbf{AC}_{\aleph_0}$, the principle of dependent choices by **DC**. We will use one further axiom, **LM**: every set of real numbers is Lebesgue measurable. Our first task is to extend the definition of the chromatic number of a graph. Without the axiom of choice, the chromatic number of a graph may not exist. In allowing a system of axioms for set theory to exclude the axiom of choice, we need to create a much broader definition of the chromatic number than the usual one, if we want it to exist. In fact, instead of the chromatic number we ought to talk about the *set of chromatic cardinalities*. There are several meaningful ways to define this. Here is one that the author chose in [49].

Definition 21 Let G be a graph and let $\boldsymbol{A}$ be a system of axioms for set theory. The *set of chromatic cardinalities* $\chi^A(G)$ of G is the set of all cardinal numbers $\tau \leq |V(G)|$ for which there is a proper coloring of the vertices $v(G)$ of G in τ colors, and τ is minimum with respect to this property.

As can be seen, the set of chromatic cardinalities needs not have just one element as with $\boldsymbol{A} = \mathbf{ZFC}$. It can also be empty. The advantage of this definition is its simplicity. Best of all, we can use inequalities on sets of chromatic cardinalities as follows.

Let τ be a cardinal number. The *inequality* $\chi^A(G) > \tau$ means that, for every $\sigma \in \chi^A(G)$, $\sigma > \tau$; the inequalities $<$, $\leq$ and $\geq$ are defined analogously. We also agree that the empty set is greater than or equal to any other set of cardinal numbers. Finally, if τ is a cardinal number and $\chi^A(G) = \{\tau\}$ is a one-element set of chromatic cardinalities (as is the case with the chromatic number when $\mathbf{A} = \mathbf{ZFC}$), then we will simplify our notation by omitting parentheses and writing $\chi^A(G) = \tau$.

An infinite cardinal $\aleph_\alpha$ is *regular* if cf $\omega_\alpha = \omega_\alpha$, and κ is a *strong limit* cardinal if, for every cardinal λ, $\lambda < \kappa$ implies that $2^\lambda < \kappa$. A cardinal κ is called *inaccessible* if $\kappa > \aleph_0$, κ is regular, and κ is a strong limit cardinal. Assuming the existence of an inaccessible cardinal, and using Paul Cohen's forcing, Robert Solovay [50] constructed in 1964 and published in 1970 a model that proved a remarkable theorem. In his honor the author introduced the following definitions [49].

The *Zermelo–Fraenkel–Solovay system of axioms* **ZFS** for set theory is defined by $\mathbf{ZFS} = \mathbf{ZF} + \mathbf{AC}_{\aleph_0} + \mathbf{LM}$, and **ZFS+** stands for $\mathbf{ZF} + \mathbf{DC} + \mathbf{LM}$.

Solovay's theorem can now be formulated very concisely:

Solovay's Theorem 22 **ZFS+** is consistent.

Saharon Shelah and Alexander Soifer [42] constructed the following example. Define a graph G as follows: the vertex-set is the set of real numbers, and the set of edges is $\{(s, t): s - t - \sqrt{2} \in Q\}$. They proved that for this graph $\chi^{\mathbf{ZFC}}(G) = 2$, while $\chi^{\mathbf{ZFS}}(G) > \aleph_0$.

This example and its analogues in 2- and more generally n-dimensional Euclidean spaces prompted Jean-Paul Delahaye [12] to point out philosophical questions of the foundations stemming from these examples:

> "It turns out that knowing if the world of sets satisfies the axiom of choice or a competing axiom is a determining factor in the solution of problems that no one had imagined depended on them. The questions raised by the new results are tied to the fundamental nature of the world of sets. Is it reasonable to believe that the mathematical world of sets is real? If it exists, does the true world of sets—the one in which we think we live—allows the coloring of S. Shelah and A. Soifer in two colors or does it require an infinity of colors? ...
>
> A series of results concerning the theory of graphs, published in 2003 and 2004 by Alexander Soifer of Princeton University and Saharon Shelah, of the University of Jerusalem, should temper our attitude and invite us to greater curiosity for the alternatives offered by the axiom of choice. The observation demonstrated by A. Soifer and S. Shelah should force mathematicians to reflect on the problems of foundations: what axioms must be retained to form the basis of mathematics for physicists and for mathematicians?"

Delahaye observes (ibid): "These [Shelah-Soifer's] results mean, as with the parallels postulate, that several different universes can be considered," and continues:

> "In the case of geometry, the independence of the parallels postulate proved that non-Euclidian geometries deserved to be studied and that they could even be used in physics: Albert Einstein took advantage of these when, between 1907 and 1915, he worked out his general theory of relativity.
>
> Regarding the axiom of choice, a similar logical conclusion was warranted; the universes where the axiom of choice is not satisfied must be explored and could be useful in physics."

Similar examples with the plane E^2, and in general E^n, as the vertex-set, were constructed in [48] and [47] respectively. These examples illuminate the influence of the system of axioms for set theory on combinatorial results. They also suggest that the chromatic number of E^n may not exist 'in the absolute' (that is, in **ZF**), but may depend upon the system of axioms chosen for set theory. An important example later came from the Australian student Michael Payne [35], who started with the unit-distance graph G_1 whose vertex-set is the rational plane Q^2, where, as usual, two vertices are adjacent if and only if they are at distance 1 apart. He then showed that the desired unit-distance graph G on the vertex-set E^2 is obtained by tiling the plane by translates of the graph G_1 – that is, its edge-set is $\{(p_1, p_2): p_1, p_2 \in E^2, p_1 - p_2 \in Q^2$, and $|p_1 - p_2| = 1\}$, and proved that $\chi^{\mathbf{ZFC}}(G) = 2$ and $3 \leq \chi^{\mathbf{ZFS}}(G) \leq 7$. Payne proved first that any measurable set S of positive Lebesgue measure contains the endpoints of a path of length 3 in G. Of course, this rules out a 2-coloring of S. Payne continued: "We can then proceed in a similar fashion to Shelah and Soifer's proof [42]." In 2009 Payne [36] constructed a new class of unit distance graphs on the vertex-set E^n whose chromatic number depends upon the system of axioms for set theory.

13 Predicting the Future

In 2003 Shelah and Soifer [42] observed the following surprising result.

Theorem 23 Assume that any finite unit-distance plane graph has chromatic number not exceeding 4. Then $\chi^{\mathbf{ZFC}}\left(E^2\right) = 4$, but $\chi^{\mathbf{ZFS+}}\left(E^2\right) \geq 5$.

Can we obtain any results unconditionally? Yes, we can [49], but not yet in **ZFC**.

Theorem 24 $\chi^{\mathbf{ZFS+}}\left(E^2\right) \geq 5$.

We conclude with the author's conjectures [49] of the expected value of the chromatic number of the plane, and more generally of E^n—in **ZFC**. Formulating conjectures *is* a form of predicting the future, is it not?

Conjecture 25 $\chi\left(E^2\right) = 4$ or 7.

If the chromatic number of the plane were 4, then Theorem 23 would imply that the chromatic number of the plane does not exist in absolute, but depends upon the choice of the system of axioms for set theory. However, if the author were limited to a single value, he would conjecture the value 7:

The Chromatic Number of the Plane Conjecture 26 $\chi\left(E^2\right) = 7$.

If this conjecture is true, then a unit-distance 7-chromatic finite graph exists in the plane. In 1998 Dan Pritikin [38] published a lower bound for the order of such a graph:

Theorem 27 Any unit-distance 7-chromatic graph G has at least 6198 vertices.

In fact, the order of the smallest such graph may be even larger.

The following important result from the 1999 Ph.D. thesis of Paul O'Donnell makes many of us think that the chromatic number of the place is greater than 4:

The O'Donnell Theorem 28 For any $n \geq 3$, there is a unit distance 4-chromatic graph of girth n.

For the 3-dimensional space the author [49] conjectures as follows:

Conjecture 29 $\chi\left(E^3\right) = 15$.

In general, the author conjectures [49]:

The Main Chromatic Number Conjecture 30 For any positive integer $n > 1$,

$$\chi\left(E^n\right) = 2^{n+1} - 1.$$

To paraphrase Paul Erdős' words about some of his conjectures, we can say, the Main Conjecture will likely withstand centuries, but we shall see!

The author is forever indebted to the late Paul Erdős and Edward Nelson for their inspiration and friendship. A heartfelt gratitude goes to John F. Nash, Jr. and Michael Rassias for their appreciation of this problem and a kind invitation to write this chapter in a group of distinguished colleagues for this spectacular collection of open problems of mathematics.

References

1. H. Ardal, J. Manuch, M. Rosenfeld, S. Shelah and L. Stacho, The odd-distance plane graph, *Discrete Comput. Geom.* 42 (2009), 132–141.
2. M. Benda and M. Perles, Colorings of metric spaces, *Geombinatorics* IX (3) (2000), 113–126.
3. N. Brown, N. Dunfield and G. Perry, Colorings of the plane I, *Geombinatorics* III (2) (1993), 24–31.
4. N. Brown, N. Dunfield and G. Perry, Colorings of the plane II, *Geombinatorics* III (3) (1993), 64–74.
5. N. Brown, N. Dunfield and G. Perry, Colorings of the plane III, *Geombinatorics* III (4) (1993), 110–114.
6. K. Cantwell, All regular polytopes are Ramsey, *J. Combin. Theory (A)* 114 (2007), 555–562.
7. J. Cibulka, On the chromatic number of real and rational spaces, *Geombinatorics* XVIII (2) (2008), 53–65.
8. J. H. Conway and N. J. A. Sloane, *Sphere Packings, Lattices and Groups*, 3rd edn., Springer-Verlag, 1999.
9. D. Coulson, A 15-coloring of 3-space omitting distance one, *Discrete Math.* 256 (2002), 83–90.
10. D. Coulson, Tilings and colorings of 3-space, *Geombinatorics* XII (3) (2003), 102–116.
11. N. G. de Bruijn and P. Erdős, A color problem for infinite graphs and a problem in the theory of relations, *Indag. Math.* 13 (1951), 369–373.
12. J.-P. Delahaye, Coloriages irréals, *Pour la Science*, Février 2005, 88-93. English translation: Imaginary Coloring, *Geombinatorics* XV(3), 2006, 101-119.

13. P. Erdős, Problem (p. 681); Unsolved problems, *Proceedings of the Fifth British Combinatorial Conference 1975, University of Aberdeen, July 14–18, 1975*, (ed. C. St.J. A. Nash-Williams and J. Sheehan), *Congr. Numer.* XV, Utilitas Mathematica, 1976.
14. P. Erdős, Combinatorial problems in geometry and number theory, *Relations between combinatorics and other parts of mathematics*, *Proc. Sympos. Pure Math.*, Ohio State Univ., (1978); *Proc. Sympos. Pure Math.*, XXXIV, Amer. Math. Soc. (1979), 149–162.
15. P. Erdős, Some combinatorial problems in geometry, *Geom. & Diff. Geom. Proc.*, Haifa, Israel, (1979); *Lecture Notes in Math.* 792, Springer (1980), 46–53.
16. P. Erdős, Some new problems and results in graph theory and other branches of combinatorial mathematics, *Combinatorics and Graph Theory, Proc. Symp. Calcutta 1980, Lecture Notes Math.* 885, Springer (1981), 9–17.
17. P. Erdős, Video recording of the talk "Twenty Five Years of Questions and Answers", *25^{th}South-Eastern International Conference On Combinatorics, Graph Theory and Computing*, Florida Atlantic University, Boca Raton, March 10, 1994.
18. K. J. Falconer, The realization of distances in measurable subsets covering R^n, *J. Combin. Theory (A)* 31 (1981), 187–189.
19. K. J. Falconer, and J. M. Marstrand, Plane sets with positive density at infinity contain all large distances. *Bull. London Math. Soc.* 18 (1986), 471–474.
20. P. Frankl and R. M. Wilson, Intersection theorems with geometrical consequences, *Combinatorica* 1 (1981), 357–368.
21. R. L. Graham, Some of My Favourite Problems in Ramsey Theory, B. Landman et al. (ed's), *Proceedings of the 'Integers Conference 2005' in Celebration of the 70th Birthday of Ronald Graham, Carrolton, Georgia, USA, 2005*, Walter de Gruyter (2007), 229–236.
22. R. L. Graham, Old and new problems and results in Ramsey theory, *Horizons of Combinatorics* (Conference and EMS Summer School), Budapest and Lake Balaton, Hungary, 2006, *(Bolyai Society Mathematical Studies)* (ed. E. Gyori, G. Katona and L. Lovasz), (2008), 105–118.
23. R. L. Graham, B. L. Rothschild, and E. G. Straus, Are there $n+2$ points in E^n with odd integral distances?, *Amer. Math. Monthly* 81 (1974) 21–25.
24. R. Hochberg and P. O'Donnell, Some 4-chromatic unit-distance graphs without small cycles, *Geombinatorics* V (4) (1996), 137–141.
25. I. Hoffman and A. Soifer, Almost chromatic number of the plane, *Geombinatorics* III (2) (1993), 38–40.
26. I. Hoffman and A. Soifer, Another six-coloring of the plane, *Discrete Math.* 150 (1996), 427–429.
27. P. D. Johnson, Jr., Coloring the rational points to forbid the distance one – A tentative history and compendium, *Geombinatorics* XVI (1) (2006), 209–218.
28. D. G. Larman, and C. A. Rogers, The realization of distances within sets in Euclidean space, *Mathematika* 19 (1972), 1–24.
29. M. Mann, A new bound for the chromatic number of the rational five-space, *Geombinatorics* XI (2) (2001), 49–53.
30. M. Mann, Hunting unit-distance graphs in rational n-spaces, *Geombinatorics* XIII (2) (2003), 86–97.
31. L. Moser and W. Moser, Solution to problem 10, *Canad. Math. Bull.* 4 (1961), 187–189.
32. O. Nechushtan, On the space chromatic number, *Discrete Math.* 256 (2002), 499–507.
33. E. Nelson, Letter to A. Soifer of October 5, 1991.
34. P. O'Donnell, *High Girth Unit-distance Graphs*, Ph.D. thesis, Rutgers University, 1999.
35. M. S. Payne, A unit distance graph with ambiguous chromatic number, arXiv: 0707.1177v1 [math.CO], 9 July 2007.
36. M. S. Payne, Unit distance graphs with ambiguous chromatic number, *Electronic J. Combin.* 16 (2009).
37. Materialy Konferenzii "Poisk-97", Moscow, 1997 (Russian).
38. D. Pritikin, All unit-distance graphs of order 6197 are 6-colorable, *J. Combin. Theory (B)* (1998), 159–163.

39. A. M. Raigorodskii, On the chromatic number of a space, *Russ. Math. Surv.* 55 (2000), 351–352.
40. D. E. Raiskii, Realizing of all distances in a decomposition of the space R^n into $n+1$ parts, *Mat. Zametki* 7 (1970), 319–323 [Russian]; English transl., *Math. Notes* 7 (1970), 194–196.
41. M. Rosenfeld, Odd integral distances among points in the plane, *Geombinatorics* V (4) (1996), 156–159.
42. S. Shelah and A. Soifer, Axiom of choice and chromatic number of the plane, *J. Combin. Theory (A)* 103 (2003), 387–391.
43. A. Soifer, Relatives of chromatic number of the plane I, *Geombinatorics* I (4) (1992), 13–15.
44. A. Soifer, A six-coloring of the plane, *J. Comb. Theory (A)* 61 (1992), 292–294.
45. A. Soifer, Six-realizable set X_6, *Geombinatorics* III (4), 1994, 140-145.
46. A. Soifer, An infinite class of 6-colorings of the plane, *Congr. Numer.* 101 (1994), 83–86.
47. A. Soifer, Axiom of choice and chromatic number of R^n, *J. Combin. Theory (A)*, 110 (2005), 169–173.
48. A. Soifer and S. Shelah, Axiom of choice and chromatic number: an example on the plane, *J. Combin. Theory (A)* 105 (2004), 359–364.
49. A. Soifer, *The Mathematical Coloring Book: Mathematics of Coloring and the Colorful Life of its Creators*, Springer, New York, 2009.
50. R. M. Solovay, A model of set theory in which every set of reals is Lebesgue measurable, *Ann. of Math.* 92 (1970), 1–56.
51. C. Thomassen, On the Nelson unit distance coloring problem, Amer. Math. Monthly, 106 (1999), 850–853
52. D. R. Woodall, Distances realized by sets covering the plane, *J. Combin. Theory (A)* 14 (1973), 187–200.
53. N. C. Wormald, A 4-chromatic graph with a special plane drawing, *J. Austral. Math. Soc. (A)* 28 (1979), 1–8.
54. V. Klee and S. Wagon, Old and New Unsolved Problems in Plane Geometry and Number Theory, *Math. Assoc. Amer., 2nd edn (with Addendum)*, Washington D.C., 1991.
55. A. Soifer, Geometric graphs, in L.W. Beineke and R.J. Wilson (ed's) Topics in Chromatic Number Theory, Cambridge University Press, 2015, 161–180.

Erdős's Unit Distance Problem

Endre Szemerédi

Abstract We survey some problems and results around one of Paul Erdős's favorite questions, first published 70 years ago: What is the maximum number of times that the unit distance can occur among n points in the plane? This simple and beautiful question has generated a lot of important research in discrete geometry, in extremal combinatorics, in additive number theory, in Fourier analysis, in algebra, and in other fields, but we still do not seem to be close to a satisfactory answer.

1 A Short History of the Problem

Diameters Consider all $\binom{n}{2}$ pairs taken from an n-element point set P in the plane. In 1934, Hopf and Pannwitz [33] discovered that the number of pairs that determine the largest distance, that is, the *diameter* of P is at most n. This bound is attained by the vertex set of the regular n-gon and by many other configurations. Shortly after, Erdős generalized the question in several different ways. In particular, he asked that at most how many times the diameter can occur among n points in 3-dimensional space and in higher dimensions. His childhood friend, Andy Vázsonyi conjectured that in 3-space the answer was $2n - 3$, for every $n \geq 4$. This was proved independently by Grünbaum [28], Heppes [32], and Straszewicz [60]. The bound is sharp for the set of vertices of a tetrahedron with $n - 4$ additional points that lie on a circle C passing through two vertices of the tetrahedron, such that the perpendicular axis of C is the line induced by the two other vertices. This result easily implies that *Borsuk's* famous conjecture [6] is true in 3-space: every d-dimensional point set can be decomposed into $d + 1$ pieces of smaller diameter. Surprisingly, in 1993, Kahn and Kalai [36] disproved Borsuk's conjecture when d is sufficiently large. At present, the smallest known counterexample is 64-dimensional [34]. It is an interesting question to decide what happens in lower dimensions.

E. Szemerédi (✉)
Renyi Alfred Mathematical Institute of the Hungarian Academy of Sciences, Reáltanoda utca 13-15, H-1053 Budapest, Hungary
e-mail: szemered@cs.rutgers.edu

J.F. Nash, Jr., M.Th. Rassias (eds.), *Open Problems in Mathematics*,
DOI 10.1007/978-3-319-32162-2_15

The *diameter graph* of a point set P is the graph on the vertex set P, in which two vertices are connected by an edge if and only if their distance is the diameter of P. A beautiful generalization of the above mentioned result of Hopf and Pannwitz was suggested by Schur [53]. He conjectured that the maximum number of *d-cliques* (complete subgraphs with d vertices) in the diameter graph of a d-dimensional point set is n. The answer does not depend on the dimension! Schur et al. [53] proved this conjecture for $d = 3$. Morić and Pach [48] verified it in the case when any two d-cliques of the diameter graph share $d - 2$ vertices, and conjectured this condition always holds. Finally, Kupavskii and Polyansky [42, 43] proved this last conjecture and, hence, settled Schur's conjecture in the affirmative.

Geometric Graphs Erdős made another fruitful observation concerning the Hopf-Pannwitz result on diameters. He noticed that if we draw the edges of the diameter graph as straight-line segments, any two edges share an endpoint or an interior point. We call a graph whose vertices are points in the plane, no 3 on a line, and whose edges are closed segments, a *geometric graph*. Erdős pointed out that essentially the original proof of the Hopf-Pannwitz result also gives that every geometric graph with n vertices and no 2 disjoint edges has at most n edges. Following this observation, Erdős, Avital and Hanani [3], Kupitz [44], and later Perles and Pach [49], started to systematically explore extremal problems for geometric graphs, where the forbidden configuration is "geometrically defined". What is the maximum number of edges that a geometric graph of n vertices can have if it contains no k disjoint edges, no k pairwise intersecting edges, no non-selfintersecting path of length k, etc.? This has led to the birth of a rich separate area within combinatorial geometry, which is usually called *geometric graph theory*. See [50], for a recent survey.

Unit Distances In 1946, Erdős [20] published a short paper in the *American Mathematical Monthly*, in which he suggested a very natural modification of the Hopf-Pannwitz question. Let P be a set of n points in the plane. What happens if we want to determine or estimate $u(n)$, the largest number of unordered pairs $\{p, q\} \subset P$ such that p and q are at a fixed distance, which is not necessarily the largest distance between two elements of P? Without loss of generality we can assume that this distance is the *unit distance*. This explains why Erdős's question is usually referred to as the *unit distance problem*. That is,

$$u(n) = \max_{P \subset \mathbb{R}^2, |P|=n} |\{\{p, q\} \subset P : |p - q| = 1\}|.$$

Using classical results of Fermat and Lagrange, Erdős showed that one can choose an integer $x \leq n/10$ that can be written as the sum of two squares in at least $n^{c/\log\log n}$ different ways, for a suitable constant $c > 0$. Thus, among the points of the $\sqrt{n} \times \sqrt{n}$ integer lattice, there are at least $(1/2)n^{1+c/\log\log n}$ pairs whose distance is $\sqrt{x}$. Scaling this point set by a factor of $1/\sqrt{x}$, we obtain a set of n points with at least $(1/2)n^{1+c/\log\log n}$, i.e., with a superlinear number of, unit distance pairs.

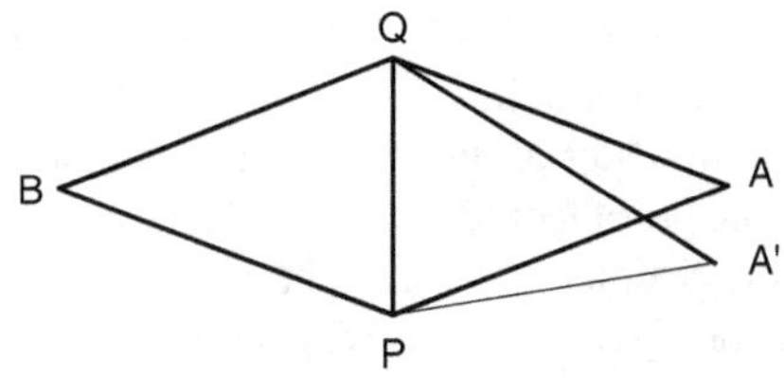

Fig. 1 If the segments PA, QA, PB, QB, QA' are of length 1, then the length of PA' cannot be 1

For any set of n points P, define the *unit distance graph* of P as the geometric graph G_P on the vertex set $V(G_P) = P$, in which two vertices are joined by a segment if and only if their distance is 1. We want to estimate the maximum number of edges that such a graph can have. Erdős noticed that G_P does not contain $K_{2,3}$, a complete bipartite subgraph with 2 and three vertices in its classes (see Fig. 1). He proved that the maximum number of edges of a $K_{2,3}$-free graph on n vertices is $c_2n^{3/2}$, for some $c_2 > 0$; see [19]. This statement and Turán's theorem [68] were the first instances of a result that belongs to the—by now vast—field called (Turán-type) *extremal graph theory*; see Bollobás [5]. Summarizing, Erdős proved that

$$n^{1+c_1/\log\log n} \leq u(n) \leq c_2n^{3/2},$$

for some $c_1, c_2 > 0$, and he conjectured that the order of magnitude of $u(n)$ is roughly $n^{1+c/\log\log n}$. In spite of many efforts to improve on the upper bound, 70 years after the publication of the paper in the *Monthly*, the best known upper bound is still only slightly better than the above estimate. Erdős's upper bound was first improved by Józsa and Szemerédi [35] to $o(n^{3/2})$, and 10 years later by Beck and Spencer [4] to $O(n^{13/9})$. In a joint paper with Spencer and Trotter [59], I proved $u(n) = O(n^{4/3})$, which is the best currently known result.

Erdős's approach has had a tremendous impact on combinatorial geometry. Today, when we try to solve a geometric optimization problem, first we attempt to extract some special combinatorial property of the underlying structure. If we can associate a graph or a hypergraph H with the geometric situation, we check if we can exclude a so-called *forbidden configuration* (subgraph or subhypergraph) H_0. If this is the case, we forget about geometry, and try to establish or apply some known results to bound the number of (hyper)edges of an H_0*-free* graph or hypergraph. This point of view has fertilized extremal combinatorics and motivated many new problems and results.

Incidences Given a curve γ and a point p, we say that there is an *incidence* between them if γ passes through p, that is, $p \in \gamma$. Up to a factor of at most 2, the unit distance problem can now be rephrased as follows. Given a set P of n points in the plane and a set Γ of n unit circles, what is the maximum number of incidences between them? Of course, the same question can be asked for any set of n curves taken from a fixed family of curves other than the family of unit circles. An important special case is when Γ consists of n distinct straight lines. For this case, Erdős conjectured and I proved together with Trotter [64, 65] that the maximum number of incidences is $O(n^{4/3})$. Note that in a *finite projective plane* the number of incidences between n points and n lines can be as large as $cn^{3/2}$. Thus, our result establishes a combinatorial distinction between the Euclidean plane and finite projective planes. The proof technique of the original paper was extended by Tóth [67], who generalize the Szemerédi-Trotter theorem to the complex plane. Later a slightly weaker, but more general result was proved by Solymosi and Tao [57], using the polynomial ham-sandwich theorem of Guth and Katz [29, 30] and methods from algebraic geometry.

Our bounds on the number of incidences between points and unit circles, and points and lines in the plane have a common generalization. A family of curves has *k degrees of freedom* and *multiplicity type s* if

1. for any k points there are at most s curves of Γ passing through all of them, and
2. any pair of curves from Γ intersect in at most s points.

For example, the family of all straight lines in the plane has 2 degrees of freedom and multiplicity type 1. The family of all unit circles has 2 degrees of freedom and multiplicity type 2. The family of all circles of arbitrary radii has 3 degrees of freedom and multiplicity type 2. The family of all polynomial curves defined by an equation $y = p(x)$, where p is a polynomial of degree d, has $d+1$ degrees of freedom and multiplicity type d. According to the Pach-Sharir theorem [51], the number of incidences between n points in the plane and m curves taken from a family Γ of curves with k degrees of freedom and multiplicity type s is at most

$$c(k, s)\left(n^{k/(2k-1)}m^{(2k-2)/(2k-1)} + n + m\right),$$

for a constant $c(k, s)$ depending only on k and s. In the special case $m = n, k = 2$, we obtain the bound $u(n) = O(n^{4/3})$ on the maximum number of unit distances among n points in the plane and the Szemerédi-Trotter theorem on incidences between points and lines.

Computational vs. Combinatorial Geometry In the late 1980s, it became clear that results about incidences play an important role in analyzing many basic algorithms in computational geometry, computer vision, and robotics. Clarkson, Edelsbrunner, Guibas, Sharir, Welzl, and other computer scientists revisited these questions, and made many important contributions. In particular, they gave a new proof of the Szemerédi-Trotter theorem based on space partitions constructed by random sampling methods [16]. They also generalized the incidence results in a natural way, motivated by applications in motion planning. A set of m lines divide the plane into at most $\binom{m+1}{2} + 1$ cells. We pick n of them, and we would like to bound from above the total number of their sides. It was shown in [16] that this quantity is at most $O(n^{2/3}m^{2/3} + n + m)$. By perturbing a little the arrangement of lines, so that a point incident to k lines becomes a cell with $O(k)$ sides, we obtain that the number of incidences between m lines and n points is $O(n^{2/3}m^{2/3} + n + m)$. Similar generalizations have been established for the total number of sides of n cells in arrangements of various kinds of curves. Combinatorial and computational geometry have both benefitted a lot from decades of interaction.

Distinct Distances In the same 1946 paper published in the *American Mathematical Monthly* where Erdős raised the unit distance problem, he also asked the following closely related question. What is the minimum number $f(n)$ of *distinct distances* determined by a set P of n points in the plane? In other words, if we list with multiplicities all $\binom{n}{2}$ distances between the points of P, at least how many of these numbers are necessarily distinct? Erdős conjectured that $f(n) \geq cn/\sqrt{\log n}$ for a suitable constant $c > 0$. Again, this would be asymptotically tight for an integer lattice of size $\sqrt{n} \times \sqrt{n}$. Obviously, if we have an *upper bound* on $u(n)$, the maximum number of times the same distance can occur among n points, this yields a *lower bound* on $f(n)$, the minimum number of distinct distances. This follows from the trivial inequality

$$f(n) \geq \frac{\binom{n}{2}}{u(n)}.$$

Therefore, the best known bound $u(n) = O(n^{4/3})$ only implies that $f(n) \geq cn^{2/3}$. This result was subsequently improved in [14, 15, 39, 58, 63], culminating in the result of Katz and Tardos [40], which states that for any $\varepsilon > 0$, any set of $n \geq n_0(\varepsilon)$ points in the plane has an element from which there are at least $n^{\alpha-\varepsilon}$ distinct distances, where $\alpha = (48-14e)/(55-16e) \approx 0.8641$. Thus, we have $f(n) \geq n^{0.864}$, provided that n is sufficiently large. These developments suggested that it might be easier to settle the problem of distinct distances than the unit distance problem. Indeed, in a sensational breakthrough, Guth and Katz [30] almost completely settled Erdős's question. They proved that $f(n) \geq cn/\log n$. Curiously, no similar bound is known for the stronger version, where we want to find a point from which there are many distinct distances.

2 A Short Proof of the Best Upper Bound on $u(n)$

In the original paper [59], in which we proved the best known upper bound, $u(n) = O(n^{4/3})$, for the unit distance problem, we modified the ideas in [65]. The proof had two main components: a regular subdivision of the plane and a counting argument. Combining them, we found that if the number of unit distances were too large, the total number of intersection points between the unit circles drawn around the n points would exceed $2\binom{n}{2}$, which is impossible. The aim of this section is to present a much more elegant argument, due to Székely [63], which is based on the Crossing Lemma of Ajtai et al. [1] and, independently, Leighton [45]. In retrospect, the idea of this proof does not differ much from the original one, but there is no need to construct a regular subdivision of the plane: it will be automatically created by a random "thinning" of the unit distance graph; see below. As was mentioned earlier, there is a third proof [16], in which the regular subdivision itself is constructed by random sampling. The same proof is "redressed" by Kaplan et al. [38]. Here the only difference is that the partitioning is created using the polynomial ham-sandwich theorem of Guth and Katz.

We start with presenting the Crossing Lemma. We need to extend the notion of geometric graphs. A *topological graph* is a graph drawn in the plane so that its vertices are represented by points and its edges are represented by Jordan arcs connecting the corresponding point pairs, where

1. no arc passes through any point representing a vertex, other than its endpoints;
2. any two arcs intersect in a finite number of points;
3. no two arcs are tangent to each other.

If two arcs have an interior point in common, they must properly cross at this point. We refer to such a point, as a *crossing*, and we say that the two edges (arcs) *cross* each other. For simplicity, in terminology, we make no distinction between vertices and the corresponding points and between edges and the corresponding arcs.

Crossing Lemma ([1, 45]). *Let G be a topological graph with n vertices and $m \geq 4n$ edges. Then $cr(G)$, the number of crossing pairs of edges, satisfies the inequality*

$$cr(G) \geq \frac{1}{64} \cdot \frac{m^3}{n^2}.$$

Proof. The following proof is folkloristic: it was discovered by Lovász, Matoušek, Pach, and others. It follows by induction on m that

$$cr(G) \geq m - 3n + 6 > m - 3n. \tag{1}$$

For any $0 < p \leq 1$, define a random (topological) subgraph G_p of G, by selecting each vertex of G independently with probability p, and letting G_p denote the subgraph induced by the selected vertices. That is, an edge (arc) of G belongs to G_p if and only if both of its endpoints have been selected. Let $n(G_p)$ and $m(G_p)$ denote the number of vertices and the number of edges in G_p, respectively. According to (1), we have

$$cr(G_p) \geq m(G_p) - 3n(G_p).$$

Taking the expected values of both sides, we obtain

$$\mathbf{E}[cr(G_p)] \geq \mathbf{E}[m(G_p)] - 3\mathbf{E}[n(G_p)].$$

Clearly, we have $\mathbf{E}[n(G_p)] = pn$ and $\mathbf{E}[m(G_p)] = p^2m$. On the other hand, the expected number of crossing pairs of edges not incident to the same vertex of G_p is equal to $p^4cr(G)$. Consequently,

$$p^4cr(G) \geq p^2m - 3pn.$$

Setting $p = 4n/m$, we obtain that

$$cr(G) \geq \frac{m}{p^2} - \frac{3n}{p^3} = \frac{m^3}{16n^2} - \frac{3m^3}{64n^2} = \frac{1}{64} \cdot \frac{m^3}{n^2},$$

which completes the proof. □

As before, let $u(n)$ denote the maximum number of times that the unit distance can occur among n points in the plane. The desired inequality $u(n) \leq cn^{4/3}$, for a suitable positive constant c, is an immediate corollary of the $m = n$ special case of the following theorem.

Theorem ([16]). *There is an absolute constant $c > 0$ such that the number of incidences between n points and m unit circles in the plane is at most $c(n^{2/3}m^{2/3} + n + m)$.*

Proof. The following argument is due to Székely [63]. Let P be a set of n points and Γ a set of m unit circles in the plane. Let $I(P, \Gamma)$ denote the number of incidences between the elements of P and Γ.

Define a topological graph G on the vertex set P, as follows. Let any pair of points $p_1, p_2 \in P$ that are consecutive along a circle $\gamma \in \Gamma$ be connected by the arc of γ that passes through no other points in P. We delete those arcs that run along circles incident to at most two points.

We have $|V(G)| = n$ and $|E(G)| \geq I(P, C) - 2m$. Notice that G is a "multigraph," in the sense that the same edge can occur twice, along two arcs belonging to two different circles $\gamma_1, \gamma_2 \in \Gamma$. Keep only one of these arcs. This will reduce the number of edges by a factor of at most 2. With a slight abuse of notation, the resulting topological graph is still denoted by G. Applying the Crossing Lemma, we obtain for the number of crossing pairs of edges

$$cr(G) \geq \frac{1}{64} \frac{((I(P, C) - 2m)/2)^3}{n^2},$$

provided that $|E(G)| \geq 4n$. On the other hand, we clearly have

$$cr(G) \leq m(m - 1),$$

as the total number of intersection points between m circles, with multiplicities, cannot exceed $2\binom{m}{2} = m(m - 1)$. Comparing the last two inequalities, the theorem follows. □

Incidences Between Points and Lines Székely's argument, with some simplifications, can also be used to establish the Szemerédi-Trotter theorem mentioned above: There is an absolute constant $c > 0$ such that the number of incidences between n points and m lines in the plane is at most $c(n^{2/3}m^{2/3} + n + m)$. The order of magnitude of this bound is best possible. In [65], this result was stated in the following (seemingly weaker, but equivalent) form.

Theorem ([65]). *Let t and n be positive integers, $t < \sqrt{n}$. Let P be a set of n points, Γ a set of lines in the plane. Then the number of distinct lines in Γ that pass through at least t points of P is at most cn^2/t^3.*

Sum-Product Estimates Given a set of t non-zero reals A, let $A+A$ and $A \cdot A$ denote the set of all pairwise sums $\{a+b : a, b \in A\}$ and pairwise products $\{ab : a, b \in A\}$ formed by the elements of A. If the elements of A form an arithmetic progression or a geometric progression, then $|A+A| \leq 2t-1$ and $|A \cdot A| \leq 2t-1$, respectively. In a joint paper with Erdős [25], it was shown that it cannot occur that both of these sets have $O(t)$ elements. More precisely, we proved the existence of a small $\varepsilon > 0$ such that $\max\{|A+A|, |A \cdot A|\} \geq t^{1+\varepsilon}$.

György Elekes, whose ingenious ideas contributed a lot to the near-solution of Erdős's problem on distinct distances by Guth and Katz, surprised us by a very elegant improvement on our theorem, based on the above upper bound on the number of lines that pass through at least t elements of an n-element point set in the plane.

Theorem ([18]). *There is a constant $c > 0$ such that for every set of t reals, we have*

$$\max\{|A+A|, |A \cdot A|\} \geq ct^{5/4}.$$

Proof. Let $a_1, \ldots, a_t$ denote the elements of A. For every j and k with $1 \leq j, k \leq t$, define

$$f_{jk}(x) := a_j(x - a_k).$$

Notice that for a fixed k, the function f_{jk} maps the elements $a_k + a_i$ into $a_j a_i$, $i = 1, \ldots, t$. Thus, setting $P := (A+A) \times A \cdot A$, we find that the graph of each function f_{jk} is a line that passes through at least t points of P. We have t^2 such lines. Applying the previous theorem, we obtain that

$$t^2 \leq c\frac{|P|^2}{t^3},$$

which implies that $|P| = |A+A| \cdot |A \cdot A| \geq t^{5/2}/\sqrt{c}$, as required. □

Elekes's bound has been slightly improved by Solymosi [56], who established the inequality

$$\max\{|A+A|, |A \cdot A|\} \geq \frac{t^{14/11}}{\log^{3/11} t}.$$

It is conjectured that the lower bound can be replaced by $t^{2-\varepsilon}$, for every $\varepsilon > 0$, but currently such a result does not seem to be within reach. Today sum-product problems have a huge literature. In particular, there are many deep sum-product estimates for finite and other infinite fields, found by Bourgain et al. [7], Bourgain et al. [8], Konyagin and Rudnev [41], and others. The subject has become an important separate theme within additive combinatorics [66].

3 A Forbidden Submatrix Argument

Erdős's trick of reducing a geometric problem to a question in extremal graph or hypergraph theory has become a standard approach in combinatorial geometry. Unfortunately, it does not always work. Sometimes it might work, but the combinatorial essence of the structure is hidden and cannot be easily extracted. Most of the above mentioned proofs of the bound $u(n) = O(n^{4/3})$ for the maximum number of times the unit distance can occur among n points in the plane, use a *forbidden subgraph* argument, but it has to be applied to several carefully selected subgraphs of the unit distance graph. Applying them directly to the full graph, we obtain Erdős's initial bound $u(n) = O(n^{3/2})$.

The only proof we know that follows by a direct application of a forbidden substructure theorem is due to Pach and Tardos [52]. To state their combinatorial result, we need some definitions.

Given a graph G on n vertices $v_1, \ldots, v_n$, let $A(G)$ denote its $n \times n$ *adjacency matrix*, in which we put a 1 at the position (i,j) if and only if $v_i v_j$ is an edge of G. Otherwise, the entry is 0. Conversely, for any symmetric 0-1 matrix A with an all-zero diagonal, let $G(A)$ denote the graph whose adjacency matrix is A. A sequence $C = (p_0, p_1, \ldots, p_{2k})$ of positions in a zero-one matrix A forms an *orthogonal cycle* if $p_0 = p_{2k}$ and the positions p_{2i} and p_{2i+1} belong to the same row, while the positions p_{2i+1} and p_{2i+2} belong to the same column, for every $0 \le i < k$. If the entry of A in position p_i is 1 for all $0 \le i \le 2k$, then we call C an *orthogonal cycle of* A. For any symmetric 0-1 matrix A, every cycle of $G(A)$ with a fixed starting point and orientation corresponds to an orthogonal cycle of A.

Given a position $p = (i,j)$ of the matrix A and an orthogonal cycle $C = (p_0, p_1, \ldots, p_{2k})$, let $C(i,j)$ be the number of times that the possibly self-intersecting polygon $p_0 p_1 \ldots p_{2k}$ encircles the point $p' = (i + 1/2, j + 1/2)$ of the plane in the counter-clockwise direction. Here the position (i,j) in the matrix is represented by the point (i,j) of the plane. (This convention contradicts the tradition of writing the first row of a matrix on top!) More precisely, let $P(i,j)$ be the set of positions (i',j') with $i' > i$ and $j' > j$, and set

$$C(i,j) = |\{0 < l \le k : p_{2l} \in P(i,j)\}| - |\{0 < l \le k : p_{2l-1} \in P(i,j)\}|.$$

An orthogonal cycle is said to be *positive* if $C(i,j) \ge 0$ for every pair (i,j) and $C(i,j)$ is strictly positive for at least one such pair.

Pach and Tardos proved the following result, from which the bound $u(n) = O(n^{4/3})$ for the number of unit distances can be deduced relatively easily.

Theorem ([52]). *The maximum number of* 1 *entries in an* $n \times n$ 0-1 *matrix that contains no positive orthogonal cycle is* $O(n^{4/3})$. *The order of magnitude of this bound cannot be improved.*

Since the last theorem is tight, this approach cannot lead to any improvement of the $u(n) = O(n^{4/3})$ bound on the number of unit distance pairs in a set of n points in the plane.

Is the $O(n^{4/3})$ Bound on the Number of Unit Distances Tight? It is rather remarkable that none of the known proofs for $u(n) = O(n^{4/3})$ offers any hope to break this barrier, although it is conjectured that $u(n) = O(n^{1+\varepsilon})$, for every $\varepsilon > 0$. How is this possible? All existing proofs easily generalize to the case where the plane is equipped with some other strictly convex metric, i.e., with a metric according to which the unit disk is a centrally symmetric strictly convex region. However, using an idea of Brass [9], Valtr [69] found the following simple metric, with respect to which the number of unit distance pairs among n points in the plane can be as large as $cn^{4/3}$, for some $c > 0$.

Let the unit circle of this metric be the locus of all points (x, y) satisfying the equation $|y| = 1 - x^2$. This is a closed curve γ, centrally symmetric about the origin, and it consists of two parabolic arcs. Assume for simplicity that n is of the form $(2k + 1)(2k^2 + 1)$ for some positive integer k, so that $k \approx (n/4)^{1/3}$. Consider the n-element point set

$$P_n = \left\{ \left(\frac{i}{k}, \frac{j}{k^2} \right) : |i| \leq k, |j| \leq k^2 \right\} \subset [-1, +1]^2.$$

Notice that the unit circle γ passes through $4k$ points of P_n. No matter how we translate γ by a vector belonging to P_n, it will pass through at least k points of P_n. In other words, with respect to this metric, every point of P_n is at unit distance from at least k others. Therefore, the number of unit distance pairs is at least $\frac{1}{2}nk \approx \frac{1}{32^{1/3}}n^{4/3}$.

Algebraically, this metric is quite similar to the Euclidean one! Which special properties of the Euclidean metric do we have to explore in order to break the bound $u(n) = O(n^{4/3})$? Perhaps there is no such property, and the bound $O(n^{4/3})$ is optimal. Frankly, we do not know too many interesting examples of planar point sets, within which there is an exceptionally popular distance. There is no strong evidence supporting the assumption that such an example cannot exist. For convenience, we conjecture that the best constructions are latticelike.

In Almost All Metrics, There Are Few Unit Distances Consider the plane equipped with an arbitrary metric with a strictly convex unit circle γ centered at the origin. Let P_0 be the 1-element set whose only point is the origin. Pick a random element $q_1 \in \gamma$, and set $P_1 := P_0 \cup (P_0 + q_1)$. If the sets $P_0, \ldots, P_{k-1}$ have already been defined, then choose randomly a point $q_k \in \gamma$, and let $P_k := P_{k-1} \cup (P_{k-1} + q_k)$. After k steps, we obtain a set P_k of $n := 2^k$ points in the plane such that from every point there are precisely $k = \log_2 n$ others at unit distance. Thus, the number of unit distance pairs in P_k, with respect to our metric, is at least $\frac{1}{2}n\log n$. In fact, almost surely, exactly this many unit distances will occur in P_k, because there will be no other "accidental" incidences between the points in P_k and the unit circles centered at its elements. Moreover, Matoušek [47] proved that in the Baire category sense, with respect to *almost all* metrics, *every* set of n points in the plane determines only $O(n\log n\log\log n)$ unit distances. We may take it as an indication that Erdős's conjecture that in the Euclidean plane the number unit distances between n points cannot exceed $O(n^{1+c/\log\log n})$. However, this evidence should be taken with a grain of salt. First, $n\log n\log\log n$ is much smaller than $n^{1+c/\log\log n}$. Second, Baire category provides a rather counter-intuitive measure on the class of centrally symmetric convex curves γ. For many curious properties of curves, it is hard to exhibit any explicit example that has that property, yet it can be shown that almost all curves in the Baire category sense possess it. The situation is analogous to the case of random graphs, where for instance we can show that almost all graphs of n vertices have clique number and independence number $O(\log n)$, yet it looks hopelessly difficult to come up with any explicit construction.

Unit Distances on the Sphere In classical geometry, most problems are considered not only in the Euclidean space, but also in spherical and hyperbolic geometry. For our problem, the case of the sphere is particularly interesting. All known planar proofs easily extend to the sphere, so we have that the number of times that the same (angular or Euclidean) distance can occur among n points on the sphere $\mathbb{S}^2$ of radius 1 is $O(n^{4/3})$. In the plane, a $\sqrt{n}\times\sqrt{n}$ piece of the integer lattice is conjectured to provide the maximum number of unit distance pairs, but on the sphere there is no grid with two independent translational (rotational) symmetries.

Leo Moser conjectured that on the unit sphere the same distance δ among n points can occur only $O(n)$ times. However, he overlooked a detail: the answer may depend on δ. As Erdős et al. [22] discovered, for $\delta = \pi/2$, there may be $cn^{4/3}$ point pairs at angular distance δ, that is, the above bound is tight. To see this, suppose that n is even, and take $n/2$ points, $p_1,\ldots,p_{n/2}$, and $n/2$ lines $l_1,\ldots,l_{n/2}$ in the plane $z=-1$ in $\mathbb{R}^3$ with at least $cn^{4/3}$ incidences between them, for some $c>0$. For each l_j, let v_j denote normal vector of the plane spanned by l_j and the origin $\mathbf{0}=(0,0,0)$. For $i=1,\ldots,n/2$, let q_i and r_i denote the intersection points of the line $\mathbf{0}p_i$ and the supporting line of v_i with the unit sphere, respectively. Notice that whenever a line l_j passes through a point p_i, the vectors v_j and $\overrightarrow{\mathbf{0}p_i}$ are perpendicular and, hence, the angular distance between q_i and $r_j\in\mathbb{S}^2$ is equal to $\pi/2$. Thus, the number of point pairs in

$$\{q_1,\ldots,q_{n/2},r_1,\ldots,r_{n/2}\}\subset\mathbb{S}^2$$

at angular distance $\pi/2$ from each other is at least $cn^{4/3}$, the number of incidences between the lines l_j and points v_i.

On the other hand, Erdős et al. [22] disproved Moser's conjecture for any angle δ different from $\pi/2$. They constructed n-element point sets in $\mathbb{S}^2$ with a *superlinear* number of δ-distance pairs, as $n\to\infty$. Their $n\log^* n$ lower bound, where

$\log^*$ stands for the iterated logarithm function, was improved by Swanepoel and Valtr [62] to $cn\sqrt{\log n}$, for a suitable positive constant c. The gap between the lower and upper bounds is still huge! Strangely, on the sphere, even the trivial $\frac{1}{2}n \log n$ lower bound construction described at the beginning of the previous subsection breaks down.

4 Unit Distances with Special Angles

Apart from his conjecture that the unit distance among n points cannot occur more than $n^{1+c/\log\log n}$ times, Erdős also made the stronger conjecture that, with the possible exception of a few points, all extremal examples can be obtained as subsets of a suitable 2-dimensional lattice. It is needless to say that we are very far from being able to verify or disprove this conjecture. Nevertheless, serious computational and other attempts were made to decide whether such a statement may be true at least for small values of n or in some restricted situations.

Brass [11] managed to prove Erdős's stronger conjecture in the special case when we count only unit distance pairs parallel to one of k fixed directions in the plane. To state his result more precisely, we need some notation. For any point set P in the plane and for any set U of k unit vectors, no two of which are opposite each other, let

$$f(P, U) = |\{(p, q) \in P \times P \, : \, p - q \in U\}|.$$

Let $f(n, k)$ be defined as the maximum of $f(P, U)$ over all such sets P of n points and all sets U of k unit vectors with the above property. Obviously, we have $f(n, 1) = n-1$. The unit distance graph of a point set restricted to two directions is the disjoint union of subgraphs of unit distance graphs of square lattices. Therefore, it follows from a theorem of Harary and Harborth [31] that

$$f(n, 2) = \lfloor 2n - 2\sqrt{n} \rfloor.$$

For any two positive functions, g and h, we write that $g(n) = \Theta(h(n))$ if $c_1 h(n) \leq g(n) \leq c_2 h(n)$ for suitable constants $c_1, c_2 > 0$.

Theorem ([11]). *For any fixed $k \geq 3$, we have $f(n, k) = kn - \Theta\left(\sqrt{n}\right)$.*

Furthermore, there are a finite number of lattices $\Lambda_1(k), \Lambda_2(k), \ldots$ and an integer $n_0(k)$ such that for every $n \geq n_0(k)$ the extremal configurations P_k, U_k are subsets of one of the lattices $\Lambda_i(k)$. Moreover, at least one of the extremal n-element point sets P_k, with the exception of $\sqrt{n}$ points, can be obtained by intersecting $\Lambda_i(k)$ with a convex set.

If the number $n_0(k)$ were small, this would prove Erdős's stronger conjecture, because every extremal set for the unit distance problem with no restriction on the number of directions is also an extremal set for some number of directions k. Unfortunately, this is not the case.

The *direction* (or angle) of a straight line in the plane is said to be *rational* if the angle between the line and the x-axis is a rational multiple of π. The special case of Erdős's unit distance problem where we count unit distances in all rational directions, was studied by Schwartz et al. [55]. In this case, they showed that Erdős's weak conjecture is not far from being optimal: *For any $\varepsilon > 0$, there exists $n_0(\varepsilon)$ such that the number of unit distance pairs with rational angles in a set of $n \geq n_0(\varepsilon)$ points in the plane is at most $n^{1+\varepsilon}$.*

Unit vectors in rational directions correspond to roots of unity. The proof proceeds by counting certain paths in the unit distance graph and using a theorem of Mann [46] to bound the number of edges.

Using the Subspace Theorem in place of Mann's Theorem, Ryan Schwartz considered unit distances from a multiplicative group with rank not too large with respect to the number of points [54]. As before, a unit distance in the plane will be considered as a complex number of unit length. So all unit distances can be considered as coming from a subgroup of $\mathbb{C}^*$. Schwartz established the following generalization of the Schwartz-Solymosi-de Zeeuw theorem stated above.

Theorem ([54]). *For any $\varepsilon > 0$, there exist a positive integer $n_0 = n_0(\varepsilon)$ and a constant $c = c(\varepsilon) > 0$ such that given $n > n_0$ points in the plane, the number of unit distances coming from a subgroup $\Gamma \subset \mathbb{C}^*$ with rank $r < c \log n$ is at most $n^{1+\varepsilon}$.*

Proof. Suppose $G = G(V, E)$ is a graph on $v(G) = n$ vertices and $e(G) = m$ edges. We denote the minimum degree in G by $\delta(G)$.

Note that, by removing vertices with degree less than $m/(2n)$, we obtain a subgraph H with at least $e(H) \geq m/2$ edges and $\delta(H) \geq m/(2n)$. The number of vertices in H is at least $v(H) \geq \sqrt{m}$. We will consider such a well behaved subgraph instead of the original graph.

Let G be the unit distance graph on n points with unit distances coming from Γ as edges. We show that there are fewer than $n^{1+\varepsilon}$ such edges, i.e., distances, for any $\varepsilon > 0$. We can assume that $e(G) \geq (1/2)n^{1+\varepsilon}$, $v(G) \geq n^{1/2+\varepsilon/2}$ and $\delta(G) \geq (1/2)n^{\varepsilon}$.

Consider a path in G on k edges $P_k = p_0 p_1 \ldots p_k$. We denote by $u_i(P_k)$ the unit vector between p_i and p_{i+1}. The path is *nondegenerate* if $\sum_{i \in I} u_i(P_k) = 0$ has no solutions where I is a nonempty subset of $\{0, 1, \ldots, k-1\}$. Note that such a sum is a sum of elements of Γ with no vanishing subsums. We will denote by $\mathcal{P}_k(v, w)$ the set of nondegenerate paths of length k between vertices v and w.

The number of nondegenerate paths of length k from any vertex is at least

$$\prod_{\ell=0}^{k-1} (\delta(G) - 2^{\ell} + 1) \geq \frac{n^{k\varepsilon}}{2^{2k}}.$$

The first expression is true since if we consider a path P_ℓ on $\ell < k$ edges then all but $2^{\ell} - 1$ possible continuations give a path $P_{\ell+1}$ with no vanishing subsums. The inequality is true if we assume $2^k \leq (1/2)n^{\varepsilon}$, which is true if $k < (\varepsilon \log n)/\log 2 - 1$.

From this, we get that the number of nondegenerate paths P_k in the graph is at least $n^{1/2+(k+1/2)\varepsilon}/2^{2k+1}$. So there exist vertices v, w in G with

$$|\mathcal{P}_k(v,w)| \geq \frac{n^{(k+1/2)\varepsilon-3/2}}{4^k}.$$

Consider a path $P_k \in \mathcal{P}_k(v,w)$, $P_k = p_0p_1\dots p_k$. Let a be the complex number giving the vector between p_0 and p_k. Since P_k is nondegenerate we get a solution of $(1/a)x_1 + (1/a)x_2 + \cdots + (1/a)x_k = 1$ with no vanishing subsums. Thus, by a corollary of the Subspace Theorem, due to Amoroso and Vieta [2], we obtain

$$|\mathcal{P}_k(v,w)| \leq (8k)^{4k^4(k+kr+1)}.$$

This, with the lower bound, gives

$$\begin{aligned} ((k+1/2)\varepsilon - 3/2)\log n &\leq k\log 4 + 4k^4(k+kr+1)\log(8k) \\ &\leq c'rk^5\log k, \\ \implies \varepsilon &\leq \frac{c'rk^4\log k}{\log n} + \frac{c''}{k}. \end{aligned} \tag{2}$$

Since $r+1 \leq c\log n$, we can choose k an integer satisfying

$$C'((\log n)/r)^{1/5} \leq k \leq C''((\log n)/r)^{1/5}.$$

Then, with this k, the right hand side of (2) goes to zero as n increases. Earlier we assumed that $k \leq (\varepsilon \log n)/\log 2 - 1$. This holds for the value of k given above for n large enough. So the number of unit distances from Γ is less than $cn^{1+\varepsilon}$ for each $\varepsilon > 0$. □

Performing a careful analysis of Erdős' lower bound construction, it is possible to verify that all unit distances come from a group with rank at most $c\log n/\log\log n$ for some $c > 0$. This group is generated by considering solutions of the equation $x^2+y^2 = p$ where p is a prime of the form $4m+1$. Using the prime number theorem for arithmetic progressions, we get the bound on such solutions and thus on the rank. For the details, see [54]. So Erdős' construction satisfies the conditions of the above theorem. A similar approach could be used for other types of lattices. So all the best known lower bounds for the unit distance problem have unit distances coming from a well structured group. It would be interesting to see if every configuration of points with the maximum possible number of unit distances has such a structure.

5 Variations of the Unit Distance Problem

In this short survey, we cannot cover all aspects of the unit distance problems. We only mention those questions that we find most interesting. Our list of references is also somewhat uneven. It reflects our knowledge, our ignorance, and our taste. We refer the interested reader to the monograph *Research Problems in Discrete Geometry* by Brass et al. [12].

However, there are two variants of the problem that we cannot completely ignore even in such a short an subjective review.

Point Sets in Convex Position We say that n points in the plane are in *convex position* if they form the vertex set of a convex polygon. Erdős and Moser [23] conjectured that the number of unit distances, $u_{\mathrm{conv}}(n)$, among n points in convex position in the plane satisfies $u_{\mathrm{conv}}(n) = \frac{5}{3}n+O(1)$. They were wrong: Edelsbrunner and Hajnal [17] exhibited an example with $2n-7$ unit distance pairs, for every $n \geq 7$. It is widely believed that $u_{\mathrm{conv}}(n) = O(n)$, and perhaps even $u_{\mathrm{conv}}(n) = 2n + O(1)$. The best known upper bound is due to Füredi [27], who proved by a forbidden submatrix argument that $u_{\mathrm{conv}}(n) = O(n \log n)$. A very short and elegant inductional argument for the same bound can be found in [13].

Erdős suggested a beautiful approach to prove that $u_{\mathrm{conv}}(n)$ grows at most linearly with n. He conjectured that every convex n-gon in the plane has a vertex from which there are no $k+1$ other vertices at the same distance. Originally, he believed that this is also true with $k = 2$, but Danzer constructed a series of counterexamples. Later, Fishburn and Reeds [26] even found convex polygons whose unit distance graphs are 3-regular, that is, in which for each vertex there are precisely three others at unit distance. If Erdős's latter conjecture is true for some integer k, then this immediately implies by induction that $u_{\mathrm{conv}}(n) < kn$.

Unit Distances in Higher Dimension Erdős's unit distance problem can be asked in any metric space. Let $u_d(n)$ denote the maximum number of unit distance pairs that can occur among n points in $\mathbb{R}^d$. In their landmark paper [16] in which they found alternative proofs and generalizations of the Szemerédi-Trotter theorem on incidences between points and lines, Clarkson, Edelsbrunner, Guibas, Sharir, and Welzl proved that $u_3(n) \leq n^{3/2}\alpha(n)$, where $\alpha(n)$ is an extremely slowly growing function, closely related to the inverse of Ackermann's function. During the last quarter of a century, no improvement was made on this bound, apart from the fact that now we know that the term $\alpha(n)$ can be eliminated; see Zahl [71] and [37], for another proof. By a simple number theoretic argument, Erdős [21] demonstrated that in an $n^{1/3} \times n^{1/3} \times n^{1/3}$ cube lattice, the same (unit) distance occurs at least $cn^{4/3} \log \log n$ times, for a constant $c > 0$. This is the best presently known lower bound, and its order of magnitude is conjectured to be nearly optimal.

The innocent reader may suspect that if it is so difficult to determine the asymptotic behavior of the functions $u_d(n)$ already for $d = 2$ and 3, then the task is even harder for $d > 3$. However, this is not the case, as is shown by the following so-called *Lenz configurations*. For $d > 3$, take $\lfloor d/2 \rfloor$ mutually orthogonal circles of radius $1/\sqrt{2}$ in $\mathbb{R}^d$, centered at the origin. Place n points on these circles, distributed among them as evenly as possible. Notice that the distance between any two points lying on different circles is precisely 1. Thus, the number of unit distances in this example is at least $\frac{1}{2}\left(1 - \frac{1}{\lfloor d/2 \rfloor} + o(1)\right) n^2$.

On the other hand, Erdős [21] proved that the unit distance graph of n points in $\mathbb{R}^d$ contains no complete $(\lfloor d/2 \rfloor + 1)$-partite subgraph $K_{3,3,\ldots,3}$ with three vertices in each of its classes. Using the Erdős-Stone theorem [24], a cornerstone in extremal graph theory, this condition is sufficient to deduce that for $d > 3$ we have

$$u_d(n) = \frac{1}{2}\left(1 - \frac{1}{\lfloor d/2 \rfloor} + o(1)\right) n^2.$$

This is one of the most beautiful early examples of the interplay between discrete geometry and extremal graph theory, from which both fields have richly benefitted.

Swanepoel [61] actually proved that, for every $d > 3$ there exists $n_0(d)$ such that all n-element extremal point sets for the d-dimensional unit distance problem are Lenz configurations, provided that $n \geq n_0(d)$. Moreover, for even dimensions $d \geq 6$, he also succeeded in determining the exact value of $u_d(n)$ for all sufficiently large n. For $d = 4$, completing the work of Brass [10], van Wamelen [70] determined the precise value of $u_4(n)$ for sufficiently large n.

Acknowledgements I would like to thank János Pach and József Solymosi for many interesting discussions on the topic of this survey, and for their help in preparing the manuscript. Work was supported by ERC-AdG. 321104, and OTKA NK 104183 grants.

References

1. M. Ajtai, V. Chvátal, M. Newborn, and E. Szemerédi: Crossing free graphs, *Ann. Discrete Math.* **12** (1982) 9–12.
2. F. Amoroso and E. Viada: Small points on subvarieties of a torus, *Duke Mathematical Journal* **150** (2009), No. 3, 407–442.
3. S. Avital and H. Hanani: Graphs, continuation, *Gilyonot Le'matematika* **3**, issue 2 (1966), 2–8.
4. J. Beck and J. Spencer: Unit distances, *J. Combin. Theory Ser. A* **37** (1984), no. 3, 231–238.
5. B. Bollobás: *Extremal Graph Theory. London Mathematical Society Monographs,* **11**, Academic Press, London-New York, 1978.
6. K. Borsuk: Drei Sätze iiber die n-dimensionale euklidische Sphäre, *Fund. Math.* **20** (1933), 177–190.
7. J. Bourgain, N. Katz, and T. Tao: A sum-product estimate in finite fields, and applications, *Geom. Funct. Anal.* **14** (2004), no. 1, 27–57.
8. J. Bourgain, A. Glibichuk, and S. Konyagin: Estimates for the number of sums and products and for exponential sums in fields of prime order, *J. London Math. Soc. (2)* **73** (2006), no. 2, 380–398.
9. P. Brass: On lattice polyhedra and pseudocircle arrangements, in: *Charlemagne and his Heritage—1200 Years of Civilization and Science in Europe, Vol. 2: Mathematical Arts*, P. L. Butzer et al., eds., Brepols Verlag, 1988, 297–302.

10. P. Brass: On the maximum number of unit distances among n points in dimension four, in: *Intuitive Geometry (I. Bárány et al., eds.), Bolyai Soc. Math. Studies* **4**, Springer, Berlin, 1997, 277–290.
11. P. Brass: On point sets with many unit distances in few directions, *Discrete Comput. Geom.* **19** (1998), no. 3, 355–366.
12. P. Brass, W. Moser, and J. Pach: *Research Problems in Discrete Geometry,* Springer, New York, 2005.
13. P. Brass and J. Pach: The maximum number of times the same distance can occur among the vertices of a convex n-gon, is $O(n \log n)$, *J. Combinatorial Theory Ser. A* **94** (2001), 178–179.
14. F. R. K. Chung: On the number of different distances determined by n points in the plane, *J. Combin. Theory, Ser. A* **36** (1984), 342–354.
15. F. R. K. Chung, E. Szemerédi, and W. T. Trotter: The number of different distances determined by a set of points in the Euclidean plane, *Discrete Comput. Geom.* **7** (1992), 1–11.
16. K. Clarkson, H. Edelsbrunner, L. Guibas, M. Sharir, and E. Welzl: Combinatorial complexity bounds for arrangements of curves and spheres, *Discrete Comput. Geom.* **5** (1990), 99–160.
17. H. Edelsbrunner and P. Hajnal: A lower bound on the number of unit distances between the points of a convex polygon, *J. Combinatorial Theory Ser. A* **56** (1991), 312–316.
18. G. Elekes: On the number of sums and products, *Acta Arith.* **81** (1997), 365–367.
19. P. Erdős: On sequences of integers none of which divides the product of two others, *Mitteilungen des Forschungsinstituts für Mathematik und Mechanik, Tomsk* **2** (1938), 74–82.
20. P. Erdős: On sets of distances of n points, *Amer. Math. Monthly* **53** (1946), 248–250.
21. P. Erdős: On sets of distances of n points in Euclidean space, *Magyar Tudom. Akad. Matem. Kut. Int. Közl. (Publ. Math. Inst. Hung. Acad. Sci.)* **5** (1960), 165–169.
22. P. Erdős, D. Hickerson, and J. Pach: A problem of Leo Moser about repeated distances on the sphere, *Amer. Math. Monthly* **96** (1989), 569–575.
23. P. Erdős and L. Moser: Problem 11, *Canad. Math. Bulletin* **2** (1959), 43.
24. P. Erdős and A. H. Stone: On the structure of linear graphs, *Bull. Amer. Math. Soc.* **52** (1946), 1087–1091.
25. P. Erdős and E. Szemerédi: On sums and products of integers, in: *Studies in Pure Mathematics. To the Memory of Paul Turán* (Erdős et al., eds.), Akadémiai Kiadó–Birkhäuser, Budapest–Basel, 1983, 213–218.
26. P. Fishburn and J. A. Reeds: Unit distances between vertices of a convex polygon, *Comput. Geom. Theory Appl.* **2** (1992), 81–91.
27. Z. Füredi: The maximum number of unit distances in a convex n-gon, *J. Combinatorial Theory Ser. A* **55** (1990) 316–320.
28. B. Grünbaum: A proof of Vázsonyi's conjecture, *Bull. Res. Council Israel, Sect. A* **6** (1956), 77–78.
29. L. Guth and N. H. Katz: Algebraic methods in discrete analogs of the Kakeya problem, *Adv. Math.* **225** (2010), no. 5, 2828–2839.
30. L. Guth and N. H. Katz: On the Erdős distinct distance problem in the plane, *Ann. of Math. (2)* **181** (2015), no. 1, 155–190.
31. F. Harary and H. Harborth: Extremal animals, *J. Combinatorics, Information & System Sciences* **1** (1976), 1–8.
32. A. Heppes: Beweis einer Vermutung von A. Vázsonyi, *Acta Math. Acad. Sci. Hungar.* **7** (1956), 463–466.
33. H. Hopf and E. Pannwitz: Aufgabe Nr. 167, *Jahresbericht d. Deutsch. Math.-Verein.* **43** (1934), 114.
34. T. Jenrich: A 64-dimensional two-distance counterexample to Borsuk's conjecture, arXiv:1308.0206.
35. S. Józsa and E. Szemerédi: The number of unit distance on the plane, in: *Infinite and Finite Sets (Colloq., Keszthely, 1973; dedicated to P. Erdős on his 60th birthday), Colloq. Math. Soc. János Bolyai* **10**, North-Holland, Amsterdam, 1975, 939–950.
36. J. Kahn and G. Kalai: A counterexample to Borsuk's conjecture, *Bull. Amer. Math. Soc. (N.S.)* **29** (1993), 60–62.

37. H. Kaplan, J. Matouěk, Z. Safernová, and M. Sharir: Unit distances in three dimensions, *Combin. Probab. Comput.* **21** (2012), no. 4, 597–610.
38. H. Kaplan, J. Matoušek, and M. Sharir: Simple proofs of classical theorems in discrete geometry via the Guth-Katz polynomial partitioning technique, *Discrete Comput. Geom.* **48** (2012), no. 3, 499–517.
39. N. H. Katz: On arithmetic combinatorics and finite groups, *Illinois J. Math.* **49** (2005), 33–43.
40. N. H. Katz and G. Tardos: A new entropy inequality for the Erdős distance problem, in: *Towards a Theory of Geometric Graphs, Contemp. Math.* **342**, Amer. Math. Soc., Providence, 2004, 119–126.
41. S. Konyagin and M. Rudnev: On new sum-product-type estimates, *SIAM J. Discrete Math.* **27** (2013), no. 2, 973–990.
42. A. Kupavskii: Diameter graphs in $\mathbf{R}^4$, *Discrete Comput. Geom.* **51** (2014), no. 4, 842–858.
43. A. B. Kupavskii and A. Polyanskii: Proof of Schur's conjecture in $\mathbf{R}^d$, arXiv:1402.3694v1.
44. Y. S. Kupitz: *Extremal Problems of Combinatorial Geometry, Lecture Notes Series* **53**, Aarhus University, Denmark, 1979.
45. T. Leighton; *Complexity Issues in VLSI. Foundations of Computing Series*, MIT Press, Cambridge, MA, 1983.
46. H. B. Mann: On linear relations between roots of unity, *Mathematika* **12** (1965), 107–117.
47. J. Matoušek: The number of unit distances is almost linear for most norms, *Adv. Math.* **226** (2011), no. 3, 2618–2628.
48. F. Morić and J. Pach: Remarks on Schur's conjecture, *Comput. Geom.* **48** (2015), no. 7, 520–527.
49. J. Pach: Geometric graph theory, in: *Surveys in Combinatorics, 1999 (J. D. Lamb and D. A. Preece, eds.), London Mathematical Society Lecture Notes* **267**, Cambridge University Press, Cambridge, 1999, 167–200.
50. J. Pach: Geometric intersection patterns and the theory of geometric graphs, in: *Proceedings of the International Congress of Mathematicians 2014 (ICM 2014, Seoul, Korea)*, 455–474.
51. J. Pach and M. Sharir: On the number of incidences between points and curves, *Combin. Probab. Comput.* **7** (1998), 121–127.
52. J. Pach and G. Tardos: Forbidden paths and cycles in ordered graphs and matrices, *Israel J. Math.* **155** (2006), 359–380.
53. Z. Schur, M. A. Perles, H. Martini, and Y. S. Kupitz: On the number of maximal regular simplices determined by n points in $\mathbb{R}^d$, in: *Discrete and Computational Geometry, The Goodman-Pollack Festschrift (Aronov et al., eds.), Algorithms Combin.* **25**, Springer, Berlin, 2003, 767–787.
54. R. Schwartz: Using the subspace theorem to bound unit distances, *Moscow Journal of Combinatorics and Number Theory* **3** (2013), No. 1, 108–117.
55. R. Schwartz, J. Solymosi, and F. de Zeeuw, Rational distances with rational angles, *Mathematika* **58** (2012), no. 2, 409–418.
56. J. Solymosi: On the number of sums and products, *Bull. London Math. Soc.* **37** (2005), no. 4, 491–494.
57. J. Solymosi and T. Tao: An incidence theorem in higher dimensions, *Discrete Comput. Geom.* **48** (2012), no. 2, 255–280.
58. J. Solymosi and Cs. Tóth: Distinct distances in the plane, *Discrete Comput. Geom.* **25** (2001), 629–634.
59. J. Spencer, E. Szemerédi, and W. T. Trotter: Unit distances in the Euclidean plane, in: *Graph Theory and Combinatorics* (B. Bollobás, ed.), Academic Press, London, 1984, 293–303.
60. S. Straszewicz: Sur un problème géométrique de P. Erdős, *Bull. Acad. Pol. Sci., Cl. III* **5** (1957), 39–40.
61. K. J. Swanepoel: Unit distances and diameters in Euclidean spaces, *Discrete Comput. Geom.* **41** (2009), 1–27.
62. K, J. Swanepoel and P. Valtr: The unit distance problem on spheres, in: *Towards a Theory of Geometric Graphs, J. Pach, ed., Contemporary Mathematics* **342**, American Mathematical Society, Providence, 2004, 273–279.

63. L. A. Székely: Crossing numbers and hard Erdős problems in discrete geometry, *Combin. Probab. Comput.* **6** (1997), 353–358.
64. E. Szemerédi and W. T. Trotter, Jr.: A combinatorial distinction between the Euclidean and projective planes, *European J. Combin.* **4** (1983), no. 4, 385–394.
65. E. Szemerédi and W. T. Trotter, Jr.: Extremal problems in discrete geometry, *Combinatorica* **3** (1983), no. 3-4, 381–392.
66. T. Tao and V. H. Vu: *Additive Combinatorics. Cambridge Studies in Advanced Mathematics* **105**, Cambridge University Press, Cambridge, 2010.
67. C. D. Tóth: The Szemerédi-Trotter theorem in the complex plane, *Combinatorica* **35** (2015), no. 1, 95–126.
68. P. Turán: Egy gráfelméleti szélsőértékfeladatról, *Matematikai és Fizikai Lapok* **48** (1941), 436–452.
69. P. Valtr, Strictly convex norms allowing many unit distances and related touching questions, manuscript, Charles University, Prague, 2005.
70. P. van Wamelen: The maximum number of unit distances among *n* points in dimension four, *Beiträge Algebra Geom.* **40** (1999), no. 2, 475–477.
71. J. Zahl: An improved bound on the number of point-surface incidences in three dimensions, *Contrib. Discrete Math.* **8** (2013), no. 1, 100–121.

Goldbach's Conjectures: A Historical Perspective

Robert C. Vaughan

Abstract In 1742, Goldbach and Euler in conversation and in an exchange of letters discussed the representation of numbers as sums of at most three primes. Although the question as to whether every even number is the sum of one or two primes (the binary Goldbach conjecture) is still unresolved, this and associated questions have attracted many mathematicians over the years, and have lead to a range of powerful techniques with many applications. This article is a commentary on the historical developments, the underlying key ideas and their widespread influence on a variety of central questions.

1 Introduction

Christian Goldbach was a German mathematician who was a professor of mathematics and history in St Petersburg. On 7 June 1742, from Moscow, he wrote a letter to Leonhard Euler (letter XLIII) in which he proposed the following conjecture:

> Every integer which can be written as the sum of two primes, can also be written as the sum of as many primes as one wishes, until all terms are units.

In the margin he than added a second conjecture.

> Every integer greater than 2 can be written as the sum of three primes.

Of course he took 1 to be prime. One can also observe that since any representation of an even number would have to include a 2 it would follow that

> Every even number is the sum of two primes.

In a letter dated 30 June 1742 Euler reminded Goldbach of an earlier conversation they had in which Goldbach had pointed out that his original conjecture would follow from this last statement.
Of course today we would state the Goldbach binary and ternary conjectures as follows.

R.C. Vaughan (✉)
Department of Mathematics, Pennsylvania State University, University Park, State College, PA 16802, USA
e-mail: rvaughan@math.psu.edu

J.F. Nash, Jr., M.Th. Rassias (eds.), *Open Problems in Mathematics*,
DOI 10.1007/978-3-319-32162-2_16

Every even integer greater than 2 can be written as the sum of two primes.
Every odd integer greater than 5 can be written as the sum of three primes.
[Include facsimile of letter.]

The above is well known, of course. What is perhaps somewhat less well known is the following assertion of Descartes.

Quod tamen nondum demonftraui. Sed & omnis numerus par fit ex vno vel duobus vel tribis primis.

Descartes[1596–1650], Opuscula Posthuma

It is not yet proved, but all numbers are made from one or two or three primes.

Descartes[1596–1650], posthumous small works

2 Sylvester

There is nothing of consequence in the literature until 1871. Spottiswoode, then President of the London Mathematical Society, in his account [82] of communications received during the meeting of 9th November 1871 describes at some length researches that Sylvester had been undertaking on the behaviour of

$$R(n) = \operatorname{card}\{p_1, p_2 : p_1 + p_2 = n\} \tag{1}$$

when n is even. In modern notation Sylvester asserts that probabilistic arguments suggest that

$$S(n) = \pi(n) \prod_{\substack{p \le \sqrt{n} \\ p \nmid n}} \frac{p-2}{p-1} \tag{2}$$

should be a good approximation to $R(n)$, and further asserts that this is confirmed by the known calculations. The product here is quite interesting. A simple argument based on the observation that a number x is divisible by p with probability $\frac{1}{p}$ leads instead to the expression

$$n \left(\prod_{\substack{p \le \sqrt{n} \\ p \nmid n}} \frac{p-2}{p} \right) \prod_{p \mid n} \frac{p-1}{p}. \tag{3}$$

It turns out that (2) is bad and (3) is worse. However, as we shall see, it is interesting and curious that Sylvester should find the product

$$\prod_{\substack{p \le \sqrt{n} \\ p \nmid n}} \frac{p-2}{p-1}.$$

Of course, the prime number theorem was 20 odd years in the future, and it would be another 3 years before Mertens would prove that

$$\sum_{p \le x} \frac{1}{p} \sim \log \log x$$

and

$$\prod_{p \le x} \frac{p-1}{p} \sim \frac{e^{-\gamma}}{\log x}$$

where γ is Euler's constant. The first of these expressions illustrates the underlying difficulty with probabilistic methods in that the series

$$\sum_{p} \frac{1}{p}$$

diverges. Armed with the above and the prime number theorem it is not hard to show that for even n

$$S(n) \sim 2e^{-\gamma} C \frac{\pi(n)}{\log n} \prod_{\substack{p|n \\ p>2}} \frac{p-1}{p-2}$$

where C is the, so called, *twin prime constant*

$$C = 2 \prod_{p>2} \left(1 - \frac{1}{(p-1)^2}\right)$$

and to deduce as Landau [55] did in reference to similar work of Stäckel [83] that

$$\sum_{n \le x} R(n) \sim \frac{n^2}{2(\log n)^2}$$

whereas

$$\sum_{n \le x} S(n) \sim 2e^{-\gamma} \frac{n^2}{2(\log n)^2}.$$

We have

$$2e^{-\gamma} = 1.1229\ldots,$$

and we now believe that for even n

$$R(n) \sim C\mathrm{li}_2(n) \prod_{\substack{p|n \\ p>2}} \frac{p-1}{p-2} \tag{4}$$

where

$$\mathrm{li}_2(n) = \int_2^{n-2} \frac{dx}{(\log x)\log(n-x)}. \tag{5}$$

Approximately

$$\mathrm{li}_2(n) = \frac{n}{(\log n)^2} + \frac{2n}{(\log n)^3} + \cdots \tag{6}$$

whereas the factor $\pi(n)/\log n$ satisfies

$$\frac{\pi(n)}{\log n} = \frac{n}{(\log n)^2} + \frac{n}{(\log n)^3} + \cdots. \tag{7}$$

These discrepancies will somewhat cancel each other for smaller n and perhaps account for Sylvester's belief that the known calculations supported his formula.

3 Hardy and Ramanujan

The first two substantial lines of progress on Goldbach's conjectures arose almost simultaneously. One was a consequence of a paper which nowhere mentions Goldbach, namely the seminal paper of Hardy and Ramanujan [38] which, as is widely known, is mostly concerned with a formula for the partition function. The fundamental idea is that there is some arithmetical function $R(n)$ of interest and the generating function

$$f(z) = \sum_{n=0}^{\infty} R(n) z^n$$

converges in the unit disc but has singularities on the circle centred at 0 of radius 1, and perhaps even has that circle as a natural boundary. Then by the Cauchy integral formula

$$R(n) = \frac{1}{2\pi i} \int_{\mathscr{C}} f(z) z^{-n-1} dz$$

where $\mathscr{C}$ is a circle of radius r, $0 < r < 1$ about 0 and described in the positive sense. It is also supposed that r is close to 1. How close will depend on n and the behaviour of $f(z)$ on $\mathscr{C}$ as r approaches 1. Typically for functions of arithmetical interest the behaviour of $f(z)$ can be ascertained quite precisely if

$$z = re(a/q + \beta) \quad e(\alpha) = e^{2\pi i\alpha}$$

with r "close" to 1 but β "small" and q is "not too large". Let me illustrate this with an example which relates to Waring's problem. You will recall that Waring had asserted that every positive integer is the sum of four squares, nine cubes, nineteen biquadrates, "and so on". Suppose that $R(n)$ is the number of ways of writing n as the sum of s non-negative cubes. Then it is readily seen that

$$f(z) = \sum_{n=0}^{\infty} R(n)z^n = g(z)^s$$

where

$$g(z) = \sum_{n=0}^{\infty} z^{n^k}.$$

Let z be as above and let us sort the terms according to their residue class modulo q. Then

$$g(z) = \sum_{m=1}^{q} e(am^k/q) \sum_{n\equiv m \bmod q} r^{n^k} e(\beta n^k).$$

When r is close to 1 and β is small we can approximate the sum over n by an integral and after several changes of variables we get

$$g(z) \sim q^{-1}S(q,a)\Gamma(1 + 1/k)(1 - re(\beta))^{-1/k}.$$

Here $S(q, a)$ is the generalized Gauss sum

$$S(q,a) = \sum_{m=1}^{q} e(am^k/q)$$

and was shown by Hardy and Littlewood [36] to satisfy

$$S(q,a) = O\left(q^{1-1/k}\right)$$

provided that $(q, a) = 1$. When $k = 2$ this estimate can be made sufficiently precise that it is of utility on the whole of $\mathscr{C}$, and this is described briefly by Hardy and Ramanujan. For $s \geq 5$ (and when $s = 4$ *via* the Kloosterman refinement) it leads to an approximation to $R(n)$, the number of representations of n as the sum of s squares, of the form

$$R(n) \sim \frac{\Gamma(3/2)^s}{\Gamma(s/2)} n^{\frac{s}{2}-1} \mathfrak{S}(n)$$

where

$$\mathfrak{S}(n) = \sum_{q=1}^{\infty} q^{-s} \sum_{\substack{a=1 \\ (a,q)=1}}^{q} S(q,a)^s e(-an/q).$$

Modern technology (see Vaughan [86]) would enable this to be carried through in a routine way for $k = 3$ and $s \geq 13$ also. When $k > 3$, and at the time when $k = 3$, the approximation for $g(z)$ obtained was only good for a rather small subset of $\mathscr{C}$ and another substantial idea was required. This difficulty Hardy and Littlewood were able to overcome using ideas from Weyl's seminal paper on uniform distribution. However pursuing this would take me too far from our path.

4 Hardy and Littlewood

When Littlewood returned from active duty in the First World war, he and Hardy set themselves the task of developing the fundamental ideas of the Hardy and Ramanujan paper. This eventually saw the light of day as the series of eight papers with the generic title "On some problems of partitio numerorum", the first two of which are on Waring's problem, and the seventh of which never appeared in print.

I understand that the first questions they examined were in fact the Goldbach conjectures. Eventually they realised that they could only make progress by assuming the generalised Riemann hypothesism and they preferred that the first papers using what has come known as the Hardy–Littlewood method established unconditional theorems. However III [33] and V [35] are concerned with the Goldbach conjectures.

To describe these results I am going to make an innovation that only came later, due to Vinogradov. If we were to follow Hardy, Littlewood and Ramanujan the generating function for sums of s primes would be

$$f(z) = g(z)^s, \quad g(z) = \sum_p (\log p) z^p.$$

However if we are only interested in the representation of n we can suppose that the primes p satisfy $p \leq n$. But now since the sum is finite we can work on the unit circle. Thus we will work with

$$F(\alpha) = \sum_{p \leq n} (\log p) e(\alpha p)$$

and now we find that if

$$r_s(n) = \sum_{\substack{p_1,\dots,p_s \\ p_1+\dots+p_s=n}} (\log p_1)\dots(\log p_s) \tag{8}$$

then the substitution $z = e(\alpha)$ in Cauchy's formula gives

$$r_s(n) = \int_0^1 F(\alpha)^s e(-\alpha n)d\alpha.$$

We can also view this as a consequence of the fundamental orthogonality relationship of harmonic analysis.

If we apply the recipe used previously of sorting the terms of F according to the residue classes in which the prime lie modulo q, and we apply results that follow from the generalised Riemann hypothesis, then we find that, provided that also $(q, a) = 1$,

$$F(a/q + \beta) = \frac{\mu(q)}{\phi(q)} \sum_{m=2}^{n} e(\beta m) + O\left((q + qn|\beta|)^{\frac{1}{2}} n^{\frac{1}{2}+\varepsilon}\right).$$

Here μ is the Möbius function and ϕ is Euler's function. There is a famous theorem of Dirichlet on diophantine approximation that tells us, given a real parameter $Q \geq 1$ and any real number α, there are q and a with $(q, a) = 1$ such that $|\alpha - a/q| \leq q^{-1}Q^{-1}$. Now if we take $\beta = \alpha - a/q$ we see that the error term above is always

$$O\left((Q + n/Q)^{\frac{1}{2}} n^{\frac{1}{2}+\varepsilon}\right)$$

and if we make the optimal choice $Q = n^{\frac{1}{2}}$ this is

$$O\left(n^{\frac{3}{4}+\varepsilon}\right).$$

Thus Hardy and Littlewood were able to treat $r_s(n)$ whenever $s \geq 3$ and they established, on the generalised Riemann hypothesis, that, when $n \equiv s \pmod 2$,

$$r_s(n) \sim \frac{n^{s-1}}{(s-1)!}\mathfrak{S}_s(n)$$

where

$$\mathfrak{S}_s(n) = \left(\prod_{p \nmid n} \left(1 + \frac{(-1)^{s+1}}{(p-1)^s}\right)\right) \prod_{p|n} \left(1 + \frac{(-1)^s}{(p-1)^{s-1}}\right).$$

When $s = 2$ they were unable to prove anything quite as precise but they could establish that

$$\sum_{\substack{m=1\\2\nmid m}}^{n} \left(r_2(m) - m\mathfrak{S}_2(m)\right)^2 = O\left(n^{\frac{5}{2}+\varepsilon}\right). \tag{9}$$

This is the strongest evidence we have that for even n,

$$r_2(n) \sim n\mathfrak{S}_2(n).$$

Of course, why stop there. How good an error term should one expect? Well, one could speculate that

$$r_2(n) = n\mathfrak{S}_2(n) + O\big(n^{\frac{1}{2}+\varepsilon}\big)$$

and if true, then this would be essentially best possible as Montgomery and Vaughan [68] have proved unconditionally that

$$\sum_{\substack{m=1\\2\nmid m}}^{n} \left(r_2(m) - m\mathfrak{S}_2(m)\right)^2 = \Omega\left(n^2(\log n)^2\right).$$

Returning to (9) it follows quite easily that the exceptional set

$$E(x) = \mathrm{card}\{m : 2|m, m \le x, r_2(m) \ne 0\} \tag{10}$$

satisfies

$$E(x) = O\left(x^{\frac{1}{2}+\varepsilon}\right). \tag{11}$$

Of course, this assumes the generalised Riemann hypothesis.

5 Vinogradov

In [92] Vinogradov introduced a fundamental new method which enabled theorems of the following kind to be established.

Theorem 1. *Let*

$$F(\alpha) = \sum_{p\le n} e(\alpha p).$$

Suppose that $q \in \mathbb{N}$, $a \in \mathbb{Z}$, $(q, a) = 1$ *and* $|\alpha - a/q| \le q^{-2}$. *Then*

$$F(\alpha) = O\left(n(\log n)^{\frac{9}{2}}\left(q^{-\frac{1}{2}} + (n/q)^{-\frac{1}{2}}\right) + n\exp\left(-\frac{1}{2}\sqrt{\log n}\right)\right).$$

This gives non-trivial estimates when

$$(\log n)^A < q \le n(\log n)^{-A}, \quad |\alpha - a/q| \le (\log n)^A n^{-1} q^{-1}$$

and complements what can be deduced from the more classical theory of Dirichlet L-functions. Thus Vinogradov was able to establish that if n is odd, then

$$R_3(n) \sim \frac{n^2}{2(\log n)^3}\mathfrak{S}_3(n).$$

Of course the method also gives

$$\sum_{\substack{m=1\\ 2|m}}^{n} (R_2(n) - \mathrm{li}_2(n)\mathfrak{S}_2(n))^2 = O\left(n^3(\log n)^{-A}\right)$$

for any fixed positive number A, and consequently

$$E(x) = \left(x(\log x)^{-A}\right),$$

and this was established independently by Chudakov [14], van der Corput [15] and Estermann [21].

Later Montgomery and Vaughan [69] pushed these ideas further and obtained

$$E(x) = O\left(x^{1-\delta}\right)$$

for some positive number δ. A number of authors have computed values for δ. The first was Chen [13] who obtained $\delta = \frac{1}{100}$. More recently Lu [61] has obtained $\delta = 0.121$ and Pintz has claimed $\delta = \frac{1}{3}$.

Let me briefly describe Vinogradov's idea. In principle it works well for sums of the kind

$$\sum_{p \le x} f(p)$$

where the function f is oscillatory, so one expects some cancellation, but is *not* multiplicative. Thus $e(\alpha p)$ or $\chi(p + c)$ are fine, but $\chi(p)$ is not. Actually this rule of thumb can be broken in applications when we average over a large class of χ. See Sect. 11 for an example.

Of course in many ways the primes are rather randomly distributed so such sums are hard to deal with. Vinogradov is able to apply the sieve of Eratosthenes in a rather non-standard way to relate the original sum to two types of sums which are more tractable. These are as follows.

Type I sums. These are sums of the form

$$\sum_{l\le z} a_l \sum_{m\le n/l} f(lm)$$

where the a_l are complex numbers which satisfy, at least for the sake of this exposition,

$$|a_l| = O(\log^C 2l)$$

and are "good" when z is, say, some power of n smaller than n.

Type II sums. These are sums of the form

$$\sum_{l>z} a_l \sum_{z<m\le n/l} b_m f(lm)$$

where the a_l and b_m are complex numbers which satisfy

$$\sum_{l\le x} |a_l|^2 = O(x\log^C x), \quad \sum_{m\le x} |b_m|^2 = O(x\log^C x).$$

These can be considered "good" if z is not too small by comparison with n.

Why have I used the terms "good"? Well, consider the Type I sum

$$\sum_{l\le z} a_l \sum_{m\le n/l} e(\alpha lm).$$

Here the inner sum is just the sum of the terms of a geometric progression, so provided that $\alpha m \notin \mathbb{Z}$ the inner sum is bounded by

$$\frac{1}{|\sin \pi\alpha l|}$$

and so the double sum is bounded by

$$(\log x)^C \sum_{l\le z} \min\left(\frac{n}{l}, \frac{1}{|\sin \pi\alpha l|}\right)$$

and this is something which is quite nice to deal will as long as z is somewhat smaller than n.

How about the Type II sums? In the original treatment Vinogradov makes multiple applications of the Cauchy–Schwarz inequality which look a bit mysterious. Here is what is really happening. We can suppose the support of both the a_l and b_m lies in the interval $(z, x/z)$ and we can write the double sum as

$$\mathbf{a}\mathcal{M}\mathbf{b}^T$$

where $\mathcal{M}$ is (essentially) an $(x/z - z) \times (x/z - z)$ matrix with general entries $f(lm)$. Applying Cauchy-Schwarz once replaces this by

$$\sqrt{(\mathbf{a}.\mathbf{a}^*)\mathbf{b}\mathcal{M}\mathcal{M}^*\mathbf{b}^*}.$$

Now

$$\mathcal{M}\mathcal{M}^* = \left(\sum_l f(kl)\overline{f(lm)}\right)_{km}$$

is a Hermitian matrix and it turns out we just need a bound for the largest eigenvalue. One such bound is

$$\max_k \sum_m \left|\sum_l f(kl)\overline{flm}\right|.$$

In the case $f(x) = e(\alpha x)$ this is also quite nice to deal with.

Note that I have oversimplified this explanation somewhat, but in principle it is the way one proceeds. In particular there is an important technicality I did not mention in order to avoid obscuring the underlying ideas. When one applies the Cauchy–Schwarz inequality, in order to get the best out of it is a good idea to ensure that the order of magnitude of the terms does not vary too much. Thus in the Type II sums it is usual to divide the sum into dyadic ranges, that is into subsums of the kind

$$\sum_{2^r < l \le 2^{r+1}} a_l \sum_{z < m \le n/l} b_m f(lm).$$

Another wrinkle that is useful if one is trying to squeeze as much as possible out of the method is to split some Type I sums into two ranges

$$\sum_{l \le u} a_l \sum_{m \le n/l} f(lm) + \sum_{u < l \le z} \sum_{m \le n/l} f(lm)$$

and treat the second sum here as a Type II sum.

To obtain good Type I and Type II sums Vinogradov applies the sieve of Erathosthenes, but then has to make various combinatorial rearrangements, which for the most powerful results were not well understood. Here is a much simpler way of proceeding [84].

Consider the trivial identity

$$-\frac{\zeta'}{\zeta}(s) = F(s) + G(s)\big(-\zeta'(s)\big)$$
$$- F(s)G(s)\zeta(s) + \big((-\zeta'(s)) - F(s)\zeta(s)\big)\left(\frac{1}{\zeta(s)} - G(s)\right) \quad (12)$$

where

$$F(s) = \sum_{k\le u} \frac{\Lambda(k)}{k^s}, \quad G(s) = \sum_{l\le v} \frac{\mu(l)}{l^s}.$$

We can apply the identity theorem for Dirichlet series to give a partition of the von Mangoldt function

$$\Lambda(m) = c_0(m) + c_1(m) - c_2(m) + c_3(m).$$

Now multiplying by $f(m)$ and summing over m we obtain

$$\sum_{m\le n} \Lambda(m)f(m) = S_0 + S_1 - S_2 + S_3. \quad (13)$$

The sum

$$S_0 = \sum_{k\le u} \Lambda(k)f(k)$$

will be small for u small so we can treat that trivially. The sums

$$S_1 = \sum_{l\le v} \mu(l) \sum_{m\le n/l} (\log m)f(lm)$$

and

$$S_2 = \sum_{l\le uv} a_l \sum_{m\le n/l} f(lm)$$

with

$$a_l = \sum_{\substack{j\le u\\ k\le v\\ jk=l}} \Lambda(j)\mu(k)$$

are good Type I sums, and

$$S_3 = \sum_{l>u} c_l \sum_{v<m\le n/l} \mu(m)f(lm)$$

with

$$c_l = \log l - \sum_{\substack{k\le u\\ k|l}} \Lambda(k)$$

is a good Type II sum.

When applied to the generating function

$$F(\alpha) = \sum_{p\le n} (\log p)e(\alpha p)$$

the method gives

Theorem 2. *Let*

$$F(\alpha) = \sum_{p\le n} (\log p)e(\alpha p).$$

Suppose that $(q,a) = 1$ *and* $|\alpha - a/q| \le q^{-2}$. *Then*

$$F(\alpha) = O\left(n(\log n)^{\frac{5}{2}}\left(q^{-1/2} + n^{-1/5} + (n/q)^{-1/2}\right)\right).$$

This method has had many other applications. Vinogradov and Hua [49] have extended the method to sums of the kind

$$\sum_{p\le X} e(\alpha p^k)$$

and consequently there is a large body of work on the Goldbach–Waring problem. That is, on the solubility of equations

$$p_1^k + \cdots + p_s^k = n$$

in primes $p_1, \ldots, p_s$.

There are also a number of rather different applications. One example, due to Vinagradov [93] himself concerns non-trivial bounds for sums

$$\sum_{p\le x} \chi(p+a)$$

where χ is a non-trivial character modulo q. There are many papers on this and related topics. For a recent incarnation see Friedlander et al. [24] Another due to Piatetski–Shapiro [71] states that when $1 \le c \le \frac{12}{11}$,

$$\mathrm{card}\{n \le x : \lfloor n^c \rfloor \mathrm{prime}\} \sim \frac{x}{c \log x}.$$

There are quite a number of papers increasing the upper bound for c. The best that I have seen is

$$1 \le c < \frac{243}{205} = 1.18\ldots$$

due to Rivat and Wu [77]

6 The Goldbach Ternary Problem

Now let me give a brief outline of the proof of Vinogradov's three primes theorem. To ease expository detail we consider $r_3(n)$ given by (8), rather than

$$R_3(n) = \mathrm{card}\{p_1, p_2, p_3 : p_1 + p_2 + p_3 = n\}$$

considered by Vinogradov.

Theorem 3. *Suppose that A is any fixed positive real number. Then*

$$r_3(n) = \frac{1}{2} n^2 \mathfrak{S}_3(n) + O\left(n^2 (\log n)^{-A}\right)$$

where

$$\mathfrak{S}_3(n) = \left(\prod_{p \nmid n} \left(1 + \frac{1}{(p-1)^3} \right) \right) \prod_{p \mid n} \left(1 - \frac{1}{(p-1)^2} \right).$$

From this it follows that every large odd integer is the sum of three primes. Some of Vinogradov's exposition of this are somewhat complicated by his desire to avoid results in which implicit constants are not computable, with the aim in mind of computing an n_0 beyond which all odd n are the sum of three primes. Recently Helfgott has spoken about showing that *every* odd integer greater than 5 is the sum of three primes, and it has been written up [42], although I am not sure whether it has been accepted for publication. Helfgott refines the method by taking a generating function

$$F(\alpha) = \sum_{p \le n} w(p) e(\alpha p)$$

with more complicated weights $w(p)$ than $\log p$. These have the effect of easing the size of some of the constants which occur in various estimates.

Proof. We have

$$r_3(n) = \int_{\mathfrak{U}} F(\alpha)^3 e(-\alpha n) d\alpha$$

where $\mathfrak{U}$ is a unit interval, and we expect that $r_3(n)$ is roughly n^2 in size. A crude bound for the integrand is

$$\left(\sum_{p \le} n(\log p) \right)^3 \sim n^3.$$

Thus we have to "save" at least n.

Let B be given and assume n is sufficiently large in terms of B. For convenience let $L = (\log n)^B$ and consider the intervals ("major arcs")

$$\mathfrak{M}(q, a) = \{\alpha : |\alpha - a/q| \le Lq^{-1}n^{-1}\}$$

for

$$1 \le a \le q \le L, \quad (a, q) = 1.$$

They are disjoint and contained in the unit interval

$$\mathfrak{U} = (Ln^{-1}, 1 + Ln^{-1}].$$

Let $\mathfrak{M}$ denote their union and define their complement with respect to $\mathfrak{U}$, the "minor arcs", by

$$\mathfrak{m} = \mathfrak{U} \backslash \mathfrak{M}.$$

We are hoping and expecting that the contribution from $\mathfrak{m}$ is $o(n^2)$, so we have to save more than n. For an $\alpha \in \mathfrak{m}$ we use Dirichlet's theorem on diophantine approximation to provide us with q, a so that $1 \le a \le q \le nL^{-1}$, $(q, a) = 1$ and $|\alpha - a/q| \le q^{-1}Ln^{-1}$. Since $\alpha \notin \mathfrak{M}$ it follows that $q > L$. Applying Theorem 5.2 we find that

$$\sup_{\alpha \in \mathfrak{m}} |F(\alpha)| \ll n(\log n)^{\frac{5}{2} - \frac{B}{2}}.$$

But this only "saves" a power of a logarithm. However we can use Parseval's identity

$$\int_{\mathfrak{U}} |F(\alpha)|^2 d\alpha = \sum_{p \le n} (\log p)^2 \sim n \log n$$

which "saves" $n(\log n)^{-1}$. Thus combining the two

$$\int_{\mathfrak{m}} |F(\alpha)|^3 d\alpha = O\left(n^2 (\log n)^{\frac{7}{2}-\frac{B}{2}}\right)$$

and this is

$$O\left(n^2 (\log n)^{-A}\right)$$

provided that $B \ge 2A + 7$. This argument clearly works with any exponent greater than 2 in place of the 3 but fails when it is 2 and this is one reason why binary problems are so much harder than ternary ones.

By the way, although

$$\int_{\mathfrak{U}} |F(\alpha)|^2 d\alpha = \sum_{p \le n} (\log p)^2 \sim n \log n$$

it is probably true that for any fixed $k \neq 0$

$$\int_{\mathfrak{U}} |F(\alpha)|^2 e(2k\alpha) d\alpha \sim n\mathfrak{S}_2(2k)$$

as $n \to \infty$. With $\mathfrak{m}$ defined as here this is equivalent to

$$\int_{\mathfrak{m}} |F(\alpha)|^2 d\alpha \sim n \log n$$

but

$$\int_{\mathfrak{m}} |F(\alpha)|^2 e(2k\alpha) d\alpha = o(n).$$

Having shown that the contribution from the minor arcs is small compared with the expected main term it remains to deal with the contribution from the major arcs. By invoking a standard theorem (the Siegel-Walfisz theorem, [70] Corollary 11.19) on the distribution of primes in arithmetic progressions it follows that whenever $1 \le a \le q \le L$, $(a, q) = 1$ and $\alpha \in \mathfrak{M}(q, a)$ we have

$$F(\alpha) = \frac{\mu(q)}{\phi(q)} \sum_{m=1}^{n} e\big((\alpha - a/q)m\big) + O\left(n \exp\left(-c\sqrt{\log n}\right)\right).$$

On each major arc we replace $F(\alpha)$ by the main term here. The total contribution from the error term to the integral over $\mathfrak{M}$ is

$$O\left(n^2(\log n)^{-A}\right)$$

The main term contributes

$$\sum_{q\le L}\sum_{\substack{a=1\\(a,q)=1}}\frac{\mu(q)}{\phi(q)^3}e(-an/q)\int_{-L/(qn)}^{L/(qn)}T(\beta)^3e(-\beta n)d\beta$$

where

$$T(\beta)=\sum_{m=1}^{n}e(\beta m).$$

The function $T(\beta)$ is bounded by $\|\beta\|^{-1}$ and so the interval $[-L/(qn), L(qn)]$ can be replaced by $\left[-\frac{1}{2},\frac{1}{2}\right]$ with an acceptable error term. Then the integral is simply the number of solutions of $m_1+m_2+m_3=n$ in positive integers m_j, and so is $\frac{1}{2}(n-1)(n-2)$. The sum over q can then be completed to infinity. Thus one finds that for $B\ge B_0(A)$

$$\int_{\mathfrak{M}}F(\alpha)^3e(-\alpha n)d\alpha=\frac{(n-2)(n-1)}{2}\mathfrak{S}_3(n)+O\left(n^2(\log n)^{-A}\right)$$

where

$$\mathfrak{S}_3(n)=\sum_{q=1}^{\infty}\sum_{\substack{a=1\\(a,q)=1}}^{q}\frac{\mu(q)^3}{\phi(q)^3}e(-an/q).$$

7 Twin Primes and Prime k-Tuples

We have already observed that the ideas I have described so far are quite effective for ternary problems, but much less so for binary problems. However they have been quite successful in creating conjectures. The procedure is quite simple. One assumes without proof that the minor arcs give a negligible contribution! In this way Hardy and Littlewood were able to write down a large class of plausible conjectures. One such is the following. Let

$$\psi_{2k}(x)=\sum_{2k<n\le x}\Lambda(n)\Lambda(n-2k).$$

Then

$$\psi_{2k}(x) \sim x\mathfrak{S}_2(2k)$$

as $x \to \infty$. Whilst this seems well out of reach, Lavrik [56] (van der Corput [15] had earlier shown something similar but with the sum over n replaced by a sum restricted to primes p for which $p - 2k$ is also prime) has shown that

$$\sum_{k \le x/2} (\psi_{2k}(x) - (x-2k)\mathfrak{S}_2(2k))^2 \ll x^3(\log x)^{-A}.$$

We shall see later that this has its uses.

Another of the Hardy and Littlewood conjectures concerns k-tuples of primes. Suppose that $\mathbf{h} = h_1, \dots, h_k$ and let $\nu(q) = \nu(q; \mathbf{h})$ be the number of distinct residue classes amongst the $\mathbf{h}$. Let

$$\pi_k(x; \mathbf{h}) = \text{card}\{n \le x : n + h_1, \dots, n + h_k \text{ all prime}\}$$

and suppose that for every $q \in \mathbb{N}$ we have $\nu(q) < q$. Then

$$\pi_k(x; \mathbf{h}) \sim \frac{x}{(\log x)^k}\mathfrak{S}_k(\mathbf{h})$$

where

$$\mathfrak{S}_k(\mathbf{h}) = \left(\prod_{p \nmid \Delta} \frac{p^k - kp^{k-1}}{(p-1)^k}\right) \prod_{p \mid \Delta} \frac{p^k - p^{k-1}\nu(p)}{(p-1)^k}$$

and

$$\Delta = \prod_{1 \le i < j \le k} |h_i - h_j|.$$

Of course, this is still well beyond our compass. However considerations closely connected with this conjecture underly the recent exciting work by Zhang, Maynard and Tao on bounded gaps in the primes, and I will say more about this later.

8 Brun and Selberg

There is another approach to the Goldbach and similar problems. This is the second of the two methods which arose almost simultaneously, namely the advent of a useful sieve method. Of course the idea of the sieve by Erathosthenes is over 2200 years old, but it was not until Brun [11] that a concerted attempt was made to apply the ideas in a more sophisticated way.

Generally given a sequence $\mathscr{A} = \{a(m) : m \in \mathbb{N}\}$ of non-negative real numbers we are interested in

$$S(\mathscr{A}, z) = \sum_{\substack{m \in \mathbb{N} \\ (m, P(z))=1}} a(m)$$

where

$$P(z) = \prod_{p \le z} p$$

and we make some basic assumptions about divisibility and growth, such as

$$A_d := \sum_{\substack{m \in \mathbb{N} \\ d|m}} a(m) = \frac{\rho(d)}{d} X + R_d \tag{14}$$

where ρ is a multiplicative function, X is some measure of A_1 and, we hope, R_d is somewhat smaller than X, at least on average over squarefree d. For example, if we are interested in the twin prime conjecture we might take

$$a(m) = \begin{cases} 1 & m = l(l+2) \text{ for some } l \le X. \\ 0 & \text{otherwise.} \end{cases}$$

Then $\rho(d)$ is the number of solutions of the congruence

$$l(l+2) \equiv 0 \pmod{d}$$

and it is readily seen that ρ is multiplicative, $\rho(2) = 1$ and $\rho(p) = 2$ for odd primes p. Moreover the above holds with

$$|R_d| \le \rho(d).$$

The fundamental problem with this is that if we proceed in the obvious way by using the Möbius function in the form

$$\sum_{d|(m, P(z))} \mu(d)$$

to pick out the condition $(m, P(z)) = 1$ and obtain an expression of the form

$$X \sum_{d|P(z)} \frac{\mu(d)\rho(d)}{d} + \sum_{d|P(z)} R_d$$

the sums over d will have far too many terms. Thus the primary object of sieve theory is to find functions $\lambda_d^{\pm}$ such that

$$\sum_{d|l} \lambda_d^- \leq \sum_{d|l} \mu(d) \leq \sum_{d|l} \lambda_d^+$$

which give reasonable results but whose support is contained in, for example, $[1, D]$ for some parameter D. Brun was the first to do this successfully by adapting Sylvester's principle of inclusion–exclusion for sets to construct $\lambda_d^{\pm}$. The methods are very combinatorial and result in considerable complexities. However there was an immediate success for the method. The simplest of Brun's upper bound methods leads to the inequality

$$\sum_{\substack{p \leq x \\ p+2\text{prime}}} 1 = O\left(\frac{x(\log\log x)^2}{(\log x)^2}\right).$$

It follows that

$$\sum_{\substack{p \\ p+2\text{prime}}} \frac{1}{p} < \infty$$

and so dashes any hopes of proving that there are infinitely many twin primes by imitating Euler's proof that there are infinitely many primes.

Lower bound methods lead to less clear cut conclusions. Let $\mathscr{P}_k$ denote the subset of $\mathbb{N}$ consisting of those $m > 1$ with at most k prime factors. Brun was able to establish that given any large even n there are l, m in $\mathscr{P}_9$ such that $n = l + m$. The ultimate theorem in this direction, after many technical developments, and including the advent of the large sieve and the Bombieri–Vinogradov theorem for which one can see Sects. 10 and 11 below, is the celebrated result of Jing-Run Chen [12] that every large even number can be written as the sum of a prime and an element of $\mathscr{P}_2$.

There is a particularly effective upper bound method due to Selberg [81]. Selberg observes that if we suppose that the sequence $\{\lambda(n)\}$ has $\lambda(1) = 1$ and its support is a subset of the divisors of $P(z)$, then

$$S(\mathscr{A}, z) \leq \sum_m a(m) \left(\sum_{d|m} \lambda(d)\right)^2.$$

On multiplying out, interchanging the order of summation and inserting the assumption (14) we obtain

$$X \sum_{d,e} \frac{\lambda(d)\lambda(e)\rho([d,e])}{[d,e]} + \sum_{d,e} \lambda(d)\lambda(e)R_{[d,e]}.$$

Now it is supposed that the second term here can be taken care of by an appropriate choice for the support of the $\lambda(d)$ and will then be suitably bounded. Thus the initial task is to minimise the quadratic form

$$\sum_{d,e} \frac{\lambda(d)\lambda(e)\rho([d,e])}{[d,e]}$$

subject only to the constraint $\lambda(1) = 1$. It is useful at this stage to suppose further that the support $\mathscr{D}$ of the $\lambda(d)$ is divisor closed, i.e if $d \in \mathscr{D}$, then $e \in \mathscr{D}$ whenever $e|d$, and commonly this is done by taking

$$\mathscr{D} = \{d : d|P(z), d \leq D\}$$

where D is a parameter at our disposal. It turns out that this minimisation process is quite easy and the minimising choice gives

$$S(A,z) \leq \frac{X}{G(D,z)} + \sum_{d\leq D}\sum_{e\leq D} \mu(d)^2\mu(e)^2|R_{[d,e]}|$$

where

$$G(D,z) = \sum_{\substack{d\leq D \\ d|P(z)}} \mu(d)^2 \prod_{p|d} \frac{\rho(p)}{p-\rho(p)}.$$

In the case of the Goldbach conjecture, given an even natural number n we could take $a(m)$ to be the number of l with $l(n-l) = m$ and $1 \leq l \leq n-1$, so that $a(m) = 0$, 1 or 2. Then

$$A_d = \frac{\rho(d)}{d}(n-1) + R_d$$

where $\rho(d)$ is the number of solutions to the congruence $l(n-l) \equiv 0 \pmod{d}$. Thus ρ is multiplicative, $\rho(p) = 1$ when $p|n$ and $\rho(p) = 2$ otherwise. It turns out that the choice $z = D = x^{\frac{1}{2}}(\log x)^{-4}$ leads to

$$R_2(n) \leq \frac{8n}{(\log n)^2}\mathfrak{S}_2(n) + O\left(\frac{n(\log\log n)}{(\log n)^3}\right).$$

This misses the conjectured result by only a factor of 8, and it is remarkable that such a simple and elementary method will do so well. There are various wrinkles that can be applied to this. For example one can take $a(m)$ to be 1 when $m < n$ and $n - m$ is prime and 0 otherwise. Then Selberg's method combined with Bombieri's theorem (see Sect. 11) enables one to replace the 8 by 4. Further small improvements can be made by using more complicated weights $a(m)$ which take account of information from lower bound sieve methods as well. The best that is know is a constant a little bit smaller than 3.5.

The recent sensational work on small gaps between primes (see Sect. 15) shows that there is further information which can be teased out of these methods.

9 Schnirelmann

There is a third attack on the Goldbach problem, due to Schnirelmann [80] which uses just an upper bound sieve method, and then not directly. It appeared before Vinograd's method, and so for a short while was the only unconditional game in town. It also had the merit that it stimulated a whole new area.

Schnirelmann's idea is to consider densities of sets of integers. Given subsets $\mathscr{A}$ and $\mathscr{B}$ of $\mathbb{Z}$ it is natural to consider an new set $\mathscr{C}$,

$$\mathscr{C} = \mathscr{A} + \mathscr{B},$$

the set of integers of the form $a + b$ with $a \in \mathscr{A}$ and $b \in \mathscr{B}$. It is also handy to define iteratively for $s \in \mathbb{N}$,

$$s\mathscr{A} = \mathscr{A} + (s-1)\mathscr{A}$$

and to take

$$A(n) = \sum_{\substack{m=1 \\ m\in\mathscr{A}}}^{n} 1$$

Now suppose we have some idea of density. Then we would hope that the density of $\mathscr{C}$ is greater than the density of either $\mathscr{A}$ or $\mathscr{B}$. If we are being really optimistic, then we might hope that the density of $\mathscr{C}$ is equal to the sum of the densities of $\mathscr{A}$ and $\mathscr{B}$. Unfortunately this fails for any of the more natural definitions of density, such as lower asymptotic density

$$\underline{d}\mathscr{A} = \liminf_{n\to\infty} \frac{A(n)}{n}.$$

For example, let Let $\mathscr{A}$ and $\mathscr{B}$ both consist of the zero residue class modulo q. Then $\mathscr{A} + \mathscr{B}$ gives nothing new.

Schnirelmann's clever idea is to consider

$$\sigma(\mathscr{A}) = \inf_{n\in\mathbb{N}} \frac{A(n)}{n},$$

the Schnirelmann density of $\mathscr{A}$. The nice thing about Schnirelmann density is that if

$$\sigma(\mathscr{A}) = 1$$

then

$$\mathbb{N} \subset \mathscr{A}.$$

Note that if $\mathscr{A}$ contains 0 or any negative integers, then they are not counted by $A(n)$. However 0 plays an important rôle. Schnielmann conjectured the following, which was eventually proved by Mann [63].

Theorem 4. *Suppose that* $0 \in \mathscr{A}$ *and* $0 \in \mathscr{B}$, *then*

$$\sigma(\mathscr{A} + \mathscr{B}) \geq \min\big(1, \sigma(\mathscr{A}) + \sigma(\mathscr{B})\big).$$

Obviously the primes have density 0, and generally there is a parity problem when considering primes. However, consider the set $\mathscr{A}$ consisting of 0 and the n such that $2n$ is prime or the sum of two primes, and suppose this has positive density, say

$$\sigma(\mathscr{A}) \geq \delta$$

for some positive real number δ. Then choose s so that $s\delta \geq 1$. Then by repeated application of Mann's theorem it follows that

$$\sigma(s\mathscr{A}) = 1$$

and so

$$\mathbb{N} \subset s\mathscr{A}.$$

Thus every even number is the sum of at most $2s$ primes.

How might one show unconditionally, prior to Vinogradov, that $\mathscr{A}$ has positive density? The answer lay with the Brun sieve. By the prime number theorem

$$\frac{2x^2}{(\log x)^2} \sim \sum_{\substack{p_1, p_2 \\ p_1+p_2 \leq 2x}} \leq \sum_{n \leq 2x} r_2(n) = \sum_{m \leq x} r_2(2m) + 2\pi(2x).$$

Thus

$$\frac{(2+o(1))x^2}{(\log x)^2} \leq \sum_{m \leq x} r_2(2m)$$

and the Cauchy-Schwarz inequality gives

$$\frac{(4+o(1))x^4}{(\log x)^4} \le A(x)\sum_{m\le x} r_2(2m)^2.$$

Thus the upper bound sieve can be applied to $r_2(m)$ to obtain for $m \ge 2$

$$r_2(2m) \le \frac{(C+o(1))m}{(\log m)^2}\mathfrak{S}_2(2m)$$

and s straightforward calculation then gives

$$A(x) \ge \delta x.$$

for $x \ge 1$. In the early days the sieve gave a poor value for C, and only results somewhat weaker than Mann's were available, so the lower bounds for s were rather large. Nevertheless the methodology is not without interest, especially since it provided a result for *all* even n. It has been refined by a succession of authors. The ultimate result is due to Ramaré, [74] who proved that every even number is the sum of at most six primes.

Let me return to lower asymptotic density. As already pointed out there are problems with developing a theory comparable with that of Schnirelmann when the sets $\mathscr{A}$ and $\mathscr{B}$ are unions of residue classes. However, there is a remarkable theorem of Kneser [53] (see also Halberstam and Roth [30, Chap. 3]) which tells that this is essentially the only bad situation that happens.

From Kneser's theorem it is possible to deduce the following very general form of the local to global principle (Banks et al. [2]) which has many potential applications in additive number theory.

Theorem 5. *Suppose that there are numbers s_1, s_2 such that*

(i) For all $s \ge s_1$ and $m, n \in \mathbb{N}$, the sumset $s\mathscr{A}$ has at least one element in the arithmetic progression n mod m;
(ii) The sumset $s_2\mathscr{A}$ has positive lower asymptotic density, i.e., $\underline{d}(s_2\mathscr{A}) > 0$.

Then, there is a number s_0 with the property that for any $s \ge s_0$ the sumset $s\mathscr{A}$ contains all but finitely many natural numbers.

Theorem 5 has several interesting consequences, such as the next theorem.

Theorem 6. *Let $\mathscr{P}$ be a set of prime numbers with*

$$\liminf_{X\to\infty} \frac{P(X)}{X/\log X} > 0.$$

Suppose that there is a number s_1 such that for all $s \geq s_1$ and $m, n \in \mathbb{N}$, the congruence

$$p_1 + \cdots + p_s \equiv n \pmod{m}$$

has a solution with $p_1, \ldots, p_s \in \mathscr{P}$. Then, there is a number s_0 with the property that for any $s \geq s_0$ the equation

$$p_1 + \cdots + p_s = N$$

has a solution with $p_1, \ldots, p_s \in \mathscr{P}$ for all but finitely many natural numbers N.

10 The Large Sieve

There is another sieve, the large sieve, which apparently, and certainly in its original manifestation, has nothing to do with the Goldbach problems. It was created by Linnik who had in mind a very important application to the least quadratic non-residue.

Let the a_n $(M + 1 \leq n \leq M + N)$ denote arbitrary complex numbers. In the application they may well be specialised to be the indicator function of some set. Let

$$Z(q; h) = \sum_{\substack{n=M+1 \\ n \equiv h \bmod q}}^{M+N} a_n$$

and for brevity write $Z = Z(1; 1)$. Let $\mathscr{H}(p)$ be a set of residue classes modulo p which have the property that if $h \notin \mathscr{H}(p)$, then $Z(p; h) = 0$, let $\rho(p) = p - \mathrm{card}\mathscr{H}(p)$ and consider

$$\sum_{h \in \mathscr{H}(p)} \left| Z(p; h) - \frac{Z}{p - \rho(p)} \right|^2 .$$

By multiplying out this becomes

$$\sum_{h=1}^{p} |Z(p; h)|^2 - \frac{|Z|^2}{p - \rho(p)} = \frac{1}{p} \sum_{a=1}^{p} \left| S\left(\frac{a}{p}\right) \right|^2 - \frac{|Z|^2}{p - \rho(p)}$$

where

$$S(\alpha) = \sum_{n=M+1}^{M+N} a_n e(\alpha n)$$

and we have used the orthogonality of the additive characters modulo p. Note that $S(1) = Z$. Thus on rearranging this and summing over primes $p \le Q$ we find that

$$\sum_{p\le Q} p \sum_{h\in\mathscr{H}(p)} \left| Z(p;h) - \frac{Z}{p-\rho(p)} \right|^2 + |Z|^2 \sum_{p\le Q} \frac{\rho(p)}{p-\rho(p)} = \sum_{p\le Q} \sum_{a=1}^{p-1} \left| S\left(\frac{a}{p}\right)\right|^2 .$$

The sum on the right is bounded by

$$\Delta_0(N,Q) \sum_{n=M+1}^{M+N} |a_n|^2$$

for some choice of $\Delta_0(N,Q)$ independent of the a_n. If we take, as suggested above, the a_n to be the indicator function of some set omitting $\rho(p)$ residues classes modulo p for each prime $p \le Q$ we find that

$$Z \le \frac{\Delta_0(N,Q)}{\sum_{p\le Q} \frac{\rho(p)}{p-\rho(p)}}.$$

There are no restrictions on the size of $\rho(p)$, which is the reason why any decent estimate for $\Delta_0(N,Q)$ is called the large sieve. In Linnik's case $\rho(p) = \frac{p+1}{2}$ when p is odd. In this way he was able to show that for any fixed $\delta > 0$ the number of primes $p \le x$ for which $n_2(p) > p^\delta$ is $O(\log\log x)$.

More generally an inequality of the kind

$$\sum_{r=1}^{R} |S(\alpha_r)|^2 \le \Delta(N,\delta) \sum_{n=M+1}^{M+N} |a_n|^2$$

is called the large sieve. Here

$$\delta = \min_{r\ne s} \|\alpha_r - \alpha_s\|.$$

There was a considerable amount of work on this, beginning with seminal papers of Roth [79] and Bombieri [4] in the mid 1960s, and we now know that we can take

$$\Delta(N,\delta) = N - 1 + \delta^{-1}$$

and that this is best possible. There are two proofs, Montgomery and Vaughan [69] combined with an idea of Cohen, and Selberg [81]. Thus

$$\sum_{q\le Q} \sum_{\substack{a=1\\(a,q)=1}}^{q} \left| S\left(\frac{a}{q}\right)\right|^2 \le (N - 1 + Q^2) \sum_{n=M+1}^{M+N} |a_n|^2.$$

A further development was that Montgomery [31] was able to make use of the composite q, and obtained, again supposing that the a_n are 1 or 0,

$$Z \leq \frac{N - 1 + Q^2}{\sum_{q \leq Q} \mu(q)^2 \prod_{p|q} \frac{\rho(p)}{p-\rho(p)}}.$$

Wait a minute! The sum in the denominator is the same as occurs in the Selberg sieve, and that Q^2 in the numerator looks cleaner to deal with than the remainders in Selberg's sieve. The main drawback is that one has to work on an interval. In fact, it turns out that, in some sense, this is the dual of Selberg's sieve. Anyway, this can certainly be applied to the Goldbach problem. For example, let N be an even natural number and take $a_n = 0$ if $p|n$ or $p|N - n$ for any prime $p \leq Q = \sqrt{N}$.

Moreover there is a lot more information to be extracted from the large sieve inequality. Let

$$S(\chi) = \sum_{n=M+1}^{M+N} a_n \chi(n)$$

where χ is a Dirichlet character modulo q. Then for primitive characters the relationship

$$\sum_{x=1}^{q} \overline{\chi}(x) e(nx/q) = \chi(n)\tau(\overline{\chi}),$$

where the Gauss sum $\tau(\overline{\chi})$ satisfies $\tau(\overline{\chi})$, gives

$$\sum_{q \leq Q} \frac{q}{\phi(q)} \sum_{\chi \bmod q}^{*} |S(\chi)|^2 \leq (N - 1 + Q^2) \sum_{n=M+1}^{M+N} |a_n|^2$$

where $\sum^{*}$ indicates that the sum is restricted to primitive characters.

11 Bombieri

In [4], Bombieri was able to use the large sieve to establish the following. Let

$$\vartheta(y; q, a) = \sum_{\substack{p \leq y \\ p \equiv a \bmod q}} (\log p).$$

Theorem 7. *Given any fixed* $A > 0$ *there is a* $B = B(A)$ *such that if* $Q \le x^{\frac{1}{2}}(\log x)^{-B}$, *then*

$$\sum_{q \le Q} \max_{\substack{a \\ (a,q)=1}} \sup_{y \le x} \left| \vartheta(y; q, a) - \frac{y}{\phi(q)} \right| \ll x(\log x)^{-A}.$$

This is a very powerful result and can substitute for the Generalised Riemann Hypothesis in many applications. Independently and simultaneously a slightly weaker result was established by Vinogradov [91].

Apart from the application to Goldbach *via* the sieve mentioned in Sect. 8, there is another connection with Goldbach. Bombieri's proof is quite technical, and first uses the large sieve to obtain bounds for Dirichlet polynomials. These in turn are used to obtain zero density estimates for the zeros of Dirichlet L-functions. Finally the zero density estimates are applied *via* the explicit formula for

$$\psi(y; \chi) = \sum_{n \le y} \Lambda(n)\chi(n).$$

We now have simpler proofs, and perhaps the simplest is via the identity of Sect. 5 which gives a partition of sums involving the von Mangoldt function into good Type I and Type II sums [85]. Thus things have come full circle. A technique developed to treat the Goldbach problems is then used on a cognate question which then has implications back to Goldbach.

12 Montgomery and Hooley

We are not finished with the large sieve. Barban [3] states without proof a precise estimate for

$$M_2(x; Q) = \sum_{q \le x} \sum_{\substack{a=1 \\ (a,q)=1}}^{q} \left(\psi(x; q, a) - \frac{x}{\phi(q)} \right)^2$$

where

$$\psi(x; q, a) = \sum_{\substack{n \le x \\ n \equiv a \bmod q}} \Lambda(n).$$

Davenport and Halberstam [17] then used the large sieve in a more sophisticated way to show that if $x(\log x)^{-A} \le Q \le x$, then

$$M_2(x; Q) = O\left(Qx(\log x)^5\right).$$

This was refined further by Gallagher who showed that

$$M_2(x;Q) = O\left(Qx(\log x)\right).$$

Gallagher, *inter alia*, had given a very simple proof of a large sieve bound

$$\Delta(N,\delta) \le \pi N + Q^2$$

in which he compares

$$|S(\alpha_r)|^2 \text{with} \int_{\alpha_r-\delta}^{\alpha_r+\delta} |S(\beta)|^2 d\beta.$$

When I first saw this, it was in Roth's inaugural lecture at Imperial College in 1968, I got very excited because, if you think of the specialisation to the case when the α_r are the rationals $\frac{a}{q}$, this looks awfully like the Hardy–Littlewood method.

In [66] Montgomery was able to give a more precise version,

$$M_2(x;Q) = Qx\log x + O\big(Qx\log Q + x^2(\log x)^{-A}\big).$$

The most remarkable thing about these results is that they give something that is beyond what can be deduced from the Generalised Riemann hypothesis. They are saying that on average

$$\psi(x;q,a) - \frac{x}{\phi(q)} = O\left(\left(\frac{x}{q}\right)^{\frac{1}{2}}(\log q)^{\frac{1}{2}}\right).$$

Montgomery's proof begins by multiplying out the left hand side. The terms

$$-2\sum_{q\le Q}\sum_{\substack{a=1\\(a,q)=1}}^{q} \psi(x;q,a)\frac{x}{\phi(q)}$$

and

$$\sum_{q\le Q}\sum_{\substack{a=1\\(a,q)=1}}^{q} \frac{x^2}{\phi(q)^2}$$

are easy to deal with. That leaves

$$\sum_{q\le Q}\sum_{\substack{a=1\\(a,q)=1}}^{q} \psi(x;q,a)^2.$$

Since the residue classes with $(a, q) > 1$ contain very few prime powers this is essentially

$$\sum_{q\le Q}\sum_{m\le x}\sum_{\substack{n\le x\\ n\equiv m \bmod q}} \Lambda(m)\Lambda(n)$$

and this can be rewritten as

$$Q\sum_{n\le x}\Lambda(n)^2 + 2\sum_{q\le Q}\sum_{\substack{l\le x\\ q|l}}\sum_{l<n\le x}\Lambda(n-l)\Lambda(n).$$

Since the vast majority of prime power differences are even this is essentially

$$Q\sum_{n\le x}\Lambda(n)^2 + 2\sum_{k\le x/2}\sum_{\substack{q|2k\\ q\le Q}}\sum_{2k<n\le x}\Lambda(n-2k)\Lambda(n). \tag{15}$$

Then Montgomery appeals to the result by Lavrik

$$\sum_{k\le \frac{1}{2}x}\left(\sum_{n\le x}\Lambda(n)\Lambda(n-2k) - (x-2k)\mathfrak{S}_2(2k)\right)^2 = O\left(x^2(\log x)^{-A}\right),$$

that mentioned in Sect. 7, to show that the sum

$$\sum_{2k<n\le x}\Lambda(n-2k)\Lambda(n)$$

in (12) can be replaced by

$$(x-2k)\mathfrak{S}_2(2k)$$

with an acceptable error. The proof is then completed in a routine way.

Thus Vinogradov's method for dealing with the ternary Goldbach problem has been used to say something quite deep about the distribution of primes in arithmetic progressions.

Soon afterwards Hooley [45] found a more elementary argument which gives even more precise results. However this is not the end of the Hardy–Littlewood–Vinogradov method. In 1996, Goldston and Vaughan have adapted that method to take things even further, and on the assumption of the Generalised Riemann hypothesis have obtained an essentially best possible error term.

13 Hooley and Third Moments

Hooley [44] also considered the third moment

$$M_3^*(x;Q) = \sum_{q\le Q} q \sum_{\substack{a=1\\(a,q)=1}}^{q} \left(\vartheta(x;q,a) - \frac{x}{\phi(q)}\right)^3$$

and has shown for that, any positive constant A, if $Q = o(x/\log x)$ as $x \to \infty$, then

$$M_3^*(x,Q) = o\left(Q^{\frac{3}{2}} x^{\frac{3}{2}} \log^{\frac{3}{2}} x\right) + O\left(\frac{x^3}{\log^A x}\right)$$

and has also shown that if $x/\log x \le Q \le x$, then

$$M_3(x,Q,w_0,\rho_0) = \frac{1}{2\zeta(2)} Q^2 x \log^2 x + O\left(Q^2 x \log x \log^2 \frac{2x}{Q}\right).$$

Later I was able to simplify the proof and obtain somewhat stronger results by using a variation on the main term which takes better account of the fact that when q is close to x there are relatively few primes in each residue class modulo q.

Let me briefly describe the underlying method. We can concentrate on the terms with $R < q \le Q$ for some suitable $R =$, say, $x(\log x)^{-A}$. Then by considering the difference of the intervals $Q < q \le x$ and $R < q \le x$ and cubing out the general term gives four expression

$$-\sum_{R<q\le x} \frac{x^3}{\phi(q)},$$

$$3 \sum_{R<q\le x} \sum_{\substack{a=1\\(a,q)=1}}^{q} \vartheta(x;q,a)\frac{x^2}{\phi(q)},$$

$$-3 \sum_{R<q\le x} \sum_{\substack{a=1\\(a,q)=1}}^{q} \vartheta(x;q,a)^2 x,$$

$$\sum_{R<q\le x} \phi(q) \sum_{\substack{a=1\\(a,q)=1}}^{q} \vartheta(x;q,a)^3$$

for various choices of R. The first two are easily dealt and the third much as in the case of the second moment. The fourth can be rewritten as essentially

$$\sum_{R<q\le Q} \phi(q) \sum_{\substack{p_1\le x, p_2\le x, p_3\le x\\ l\equiv m\equiv n \,\mathrm{card}\, q}} (\log p_1)(\log p_2)(\log p_3).$$

Terms with $p_j = p_k$ for some $j \neq k$ can be dealt with similarly to the second and third sums. This leaves the case when the p_j are distinct. This contributes

$$6 \sum_{R<q\le x} \sum_{\substack{j<x,\, k<x\\ q|j,\, q|k}} \sum_{\substack{p_1\le x, p_2\le x, p_3\le x\\ p_3-p_2=k, p_2-p_1=l}} (\log p_1)(\log p_2)(\log p_3).$$

Thus we need to examine the system

$$p_3 - p_2 = qs,\ p_2 - p_1 = qr.$$

These differences do not exceed x, whereas $q > R$. Hence $s \le x/R$ and $r \le x/R$, and $x/R = (\log x)^A$. Moreover, solving for q, we have

$$\frac{p_3 - p_2}{s} = \frac{p_2 - p_1}{r},$$

in other words

$$rp_3 - (s + r)p_2 + sp_1 = 0.$$

Again, this is a situation which can be treated successfully by the Hardy-Littlewood-Vinogradov method.

14 Hooley's Conjecture

Let

$$V(x;q) = \sum_{\substack{a=1\\ (a,q)=1}}^{q} \left(\psi(x;q,a) - \frac{x}{\phi(q)}\right)^2.$$

In view of the nature of the second moments, Hooley [43], has suggested that, for some unspecified range of q,

$$V(x,q) \sim x \log q$$

and in 1975 he showed that if

$$x(\log x)^{-A} < Q \le x$$

then (12) holds for almost all q in the range $\frac{Q}{2} < q \leq Q$ and that on the generalised Riemann hypothesis the range (14) may be extended to

$$x^{\frac{4}{5}+\varepsilon} < Q \leq x.$$

Friedlander and Goldston [25] have shown that a better approximation to $V(x;q)$ should be

$$U(x,q) = x\log q - x\left(\gamma + \log 2\pi + \sum_{p|q}\frac{\log p}{p-1}\right)$$

and have obtained various estimates for

$$\sum_{\frac{Q}{2}<q\leq Q} |V(x,q) - U(x,q)|\,.$$

In Vaughan [87] the subject was developed further by treating the general moment

$$M_k(x,Q) = \sum_{\frac{Q}{2}<q<Q} |V(x,q) - U(x,q)|^k.$$

Once more the Hardy–Littlewood method plays a crucial rôle.

15 Small Gaps Between Primes

The modern work on small gaps between primes starts with Hardy and Littlewood. In Sect. 4 it was mentioned that the seventh paper in their "Partitio Numerorum" series was never published. This deals with small gaps between prime numbers. Let $2 = p_1 < p_2 < \ldots$ be the sequence of primes in their natural order. The prime number theorem tells us that the average size of $p_{n+1} - p_n$ is $\log p_n$, so at once

$$\liminf_{n\to\infty}\frac{p_{n+1}-p_n}{\log p_n} \leq 1.$$

Hardy and Littlewood had the idea of considering

$$\sum_{h=-H}^{H}(H-|h|)\pi_2(x;h) = \int_0^1 |F(\alpha)|^2|\sum_{h=1}^{H} e(\alpha)|^2 d\alpha$$

where

$$\pi_2(x;h) = \sum_{\substack{p_1\le x,\, p_2\le x\\ p_2-p_1=h}} (\log p_1)(\log p_2)$$

and

$$F(\alpha) = \sum_{p\le x} (\log p)e(\alpha p).$$

On the generalised Riemann hypothesis they were able to take

$$H = \left(\frac{2}{3}+\delta\right)\log x$$

and deduce that the right hand side is so large that the left hand side contains non-zero terms with $h \ne 0$. Thus

$$\liminf_{n\to\infty} \frac{p_{n+1}-p_n}{\log p_n} \le \frac{2}{3}.$$

Erdős obtained an unconditional result and Rankin [75] and Ricci [76] both made contributions, but the first paper to make significant progress is Bombieri and Davenport [5] who used the large sieve *and* the Selberg sieve in the above method to obtain

$$\liminf_{n\to\infty} \frac{p_{n+1}-p_n}{\log p_n} \le \frac{2+\sqrt{3}}{8} = 0.466.$$

This was successively reduced by Pilt'jai [72], Huxley [47, 48], and Maier [62] to 0.248.

The recent spectacular advances on small gaps by Goldston et al. [27], Zhang [94], Maynard [64], Tao are based around a rather different idea, that of seeking primes in "admissible" k-tuples of integers. Let $\mathbf{1}_{\mathbb{P}}$ be the indicator function of the primes $\mathbb{P}$. Typically an expression of the kind

$$\sum_{\substack{N<n\le 2N\\ n\equiv a \pmod q}} \left(\sum_{j=1}^{k} \mathbf{1}_{\mathbb{P}}(n+h_j) - \rho\right)\Bigg(\sum_{\substack{d\le R\\ \mathbf{d}|n+\mathbf{h}\\ (d,q)=1}} \lambda(\mathbf{d})\Bigg)^2 \tag{16}$$

is considered where, perhaps, $\mathbf{d}$ denotes the k-tuple $d_1,\dots,d_k$, $d_j|n+h_j$ $(j=1,\dots,k)$ and $d = d_1\dots d_k$. Other configurations are possible. Maybe $\mathbf{d} = d|(n+h_1)\dots(n+h_j)$. The idea is to use Selberg λ which are related closely to those which would arise in applying his upper bound method to the prime k-tuples conjecture. Since it is only an upper bound sieve method this does not ensure that each $n + h_j$ is prime, but it does make it more likely that primes occur amongst the $n + h_1, \ldots, n + h_k$. If it can be show that the expression in (16) is positive then it will follow that there are n such that

$$\sum_{j=1}^{k} \mathbf{1}_{\mathbb{P}}(n + h_j) > \rho.$$

As of 25th May 2015, the best results achieved by these methods are that unconditionally

$$\liminf_{n\to\infty} p_{n+1} - p_n \le 246,$$

and if one assumes the Elliott–Halberstam conjecture [19], then

$$\liminf_{n\to\infty} p_{n+1} - p_n \le 12.$$

This method has very great flexibility and many potential applications. One is to Dickson's conjecture [1904] which states that if the g_i, h_i are integers and $\prod_{i=1}^{k}(g_i n + h_i)$ has no fixed prime divisor, then there are infinitely many n such that the $g_i n + h_i$ are simultaneously prime. Perhaps the most exciting application has been to large gaps, where Ford et al. [22] have solved a \$5000.00 problem of Erdős by showing that there is a positive constant c such that for infinitely many n

$$p_{n+1} - p_n > \frac{c(\log n)(\log\log n)(\log\log\log\log n)}{\log\log\log n}.$$

16 Goldbach Generalised: Partitions into Primes

There are a variety of generalisations of Goldbach type problems which have been considered recently and to all of which one can apply the Hardy–Littlewood–Ramanujan–Vinogradov method. One such concerns the asymptotic formula for the number $\mathfrak{p}(n)$ of partitions of n into primes. Let

$$\Phi(z) = \sum_{n=0}^{\infty} \mathfrak{p}(n) z^n = \prod_{p} (1 - z^p)^{-1}$$

and

$$\Psi(z) = \sum_{k=1}^{\infty} \sum_{p} \frac{z^{kp}}{k}$$

which are readily seen to converge when $|z| < 1$ and satisfy

$$\Phi(z) = \exp\big(\Psi(z)\big).$$

Moreover, for every $k \in \mathbb{N}^*$ and $\rho \in (0, 1)$,

$$\Psi^{(k)}(\rho) > 0$$

and

$$\Psi^{(k)}(\rho) \to \infty \quad \text{as} \quad \rho \to 1^-.$$

Thus, for every $x \geq 0$ the equation

$$\rho \Psi'(\rho) = x$$

has a unique solution $\rho = \rho(x)$ with $\rho \in [0, 1)$, and $\rho(x) \to 1^-$ as $x \to \infty$. Thus as $x \to \infty$,

$$\begin{aligned} x \log \frac{1}{\rho(x)} &= \pi \sqrt{\frac{x}{3 \log x}} \left(1 + \frac{\log \log x}{\log x} + O(1/\log x)\right), \\ \Psi\big(\rho(x)\big) &= \pi \sqrt{\frac{x}{3 \log x}} \left(1 + \frac{\log \log x}{\log x} + O(1/\log x)\right) \end{aligned}$$

and the prime number theorem gives

$$\Psi_2(\rho) = \left(\rho \frac{\mathrm{d}}{\mathrm{d}\rho}\right)^2 \Psi(\rho) = 2\pi^{-1}(3 \log x)^{\frac{1}{2}} x^{\frac{3}{2}} \big(1 + O(\log \log x / \log x)\big).$$

Then a variant of the method [88] will give the following theorem.

Theorem 8. *Suppose that n is sufficiently large. Then*

$$\mathfrak{p}(n) = \frac{\rho(n)^{-n} \Phi\big(\rho(n)\big)}{\sqrt{2\pi \Psi_2\big(\rho(n)\big)}} \left(1 + O\left(n^{-1/5}\right)\right).$$

Theorem 9. *We have*

$$\mathfrak{p}(n+1) - \mathfrak{p}(n) \sim \frac{\pi \mathfrak{p}(n)}{\sqrt{3n \log n}}$$

as $n \to \infty$.

17 Goldbach Generalised: Beatty Primes

Another collection of recent applications concerns representations by Beatty primes. A Beatty prime is a prime in the set

$$\mathscr{B}(\alpha,\beta) = \{\lfloor \alpha n + \beta \rfloor : n \in \mathbb{N}\} \tag{17}$$

where $\alpha, \beta \in \mathbb{R}$ and $\alpha > 1$. Thus, given $\alpha_1, \dots, \alpha_s, \beta_1, \dots, \beta_s \in \mathbb{R}$ with $\alpha_j > 1$ let

$$R(n; \boldsymbol{\alpha}, \boldsymbol{\beta}) = \sum_{\mathbf{p}}^{*} (\log p_1) \dots (\log p_s) \tag{18}$$

where $\sum_{\mathbf{p}}^{*}$ indicates that the sum is over s-tuples of primes $p_j \in \mathscr{B}(\alpha_j, \beta_j)$ with $p_1 + \cdots + p_s = n$.

See Banks et al. [1], Kumchev [54] and Vaughan [90]. Here is a sample theorem taken from the latter paper.

Theorem 10. *Suppose that* $s \geq 3$, $\alpha_j > 1$ $(1 \leq j \leq s)$, $\alpha_1, \dots, \alpha_s$ *are irrational and there is a pair* i, j *such that* $\alpha_i, \alpha_j, 1$ *are linearly independent over* $\mathbb{Q}$, *and* $n \equiv s \pmod 2$. *Then*

$$R_s(n; \boldsymbol{\alpha}, \boldsymbol{\beta}) \sim \frac{n^{k-1}\mathfrak{S}_k(n)}{\alpha_1 \dots \alpha_k (k-1)!} \tag{19}$$

as $n \to \infty$

18 Goldbach Generalised: Mixed Powers

In addition to the Goldbach–Waring problem mentioned in Sect. 5, there is a considerable body on mixed powers. One modern development has been to combine the Hardy–Littlewood method with sieve techniques. This was first done by Heath-Brown [41] when he showed that there are infinitely many configurations of three primes and an almost prime in arithmetic progression. Of course, this has been overtaken by the recent work of Green and Tao [29] (see also Green [28]), which itself is partially dependent on techniques which can be considered to descend from the Hardy–Littlewood–Ramanujan method.

Let $\mathscr{P}_k$ denote the set of natural numbers having at most k prime factors. Then the elements of $\mathscr{P}_k$ are called almost primes. In the context of the Goldbach–Waring problem the combination of the Hardy–Littlewood method with sieve techniques was first exploited systematically by Brüdern in his Habilitationsschrift [6] and then by Brüdern and Fouvry [9]. They proved that every sufficiently large integer $n \equiv 4 \pmod{24}$ can be written as the sum of four squares of P_{34}-numbers.

This is an area which has really mushroomed and I am going to be very selective.

To set the scene, consider Roth [78] who proved that if n is sufficiently large, then the equation

$$x^3 + p_1^3 + \cdots + p_7^3 = n \tag{20}$$

has solutions in primes $p_1, \ldots, p_7$ and a positive integer x. By comparison, if we insist the all the variables be prime, then the best that can be achieved is nine terms rather than eight. Brüdern showed that, when $n \equiv 4 \pmod{18}$, x can be restricted to be a P_4-number. Then Kawada [50] replaced the P_4 by a P_3. There is a good account of this and similar questions in Brüdern and Kawada [10], Kawada [51] and Kawada [52].

Another class of questions that Brüdern and Kawada consider is to show that in each case for almost all n such that

$$n \not\equiv 0 \pmod{2}, \quad n = x^2 + p_1^3 + p_2^5 \quad x \in \mathscr{P}_{15} \tag{21}$$

$$n \not\equiv 2 \pmod{3}, \quad n = x^2 + p_1^3 + p_2^4 \quad x \in \mathscr{P}_6 \tag{22}$$

$$n \not\equiv 2 \pmod{3}, \quad n = p_1^2 + y^3 + p_2^4 \quad y \in \mathscr{P}_4 \tag{23}$$

$$n \not\equiv 5 \pmod{7}, \quad n = x^2 + p_1^3 + p_2^3 \quad x \in \mathscr{P}_3 \tag{24}$$

$$n \not\equiv 5 \pmod{7}, \quad n = p_1^2 + y^3 + p_2^3 \quad y \in \mathscr{P}_3 \tag{25}$$

there is a solution. Yet another is to show in each case that for

$$n > n_0,\ n \text{ even}, \quad n = p_1 + x^2 + p_2^3 + p_3^4 \quad x \in \mathscr{P}_3 \tag{26}$$

$$n > n_0,\ n \text{ even}, \quad n = p_1 + x^2 + p_2^3 + p_3^5 \quad x \in \mathscr{P}_4 \tag{27}$$

$$n > n_0,\ n \text{ even}, \quad n = x + p_1^2 + p_2^3 + p_3^k \quad x \in \mathscr{P}_2 \tag{28}$$

there is a solution.

Recently Liu [60] has shown in each case if

$$n > n_0,\ n \text{ odd}, \quad n = x + p_1^3 + p_2^3 + p_3^3 + p_4^3 \quad x \in \mathscr{P}_2 \tag{29}$$

$$n > n_0,\ n \text{ odd}, \quad n = p_1 + p_2^3 + p_3^3 + p_4^3 + y^3 \quad y \in \mathscr{P}_3 \tag{30}$$

$$n > n_0,\ n \text{ odd}, \quad n = p_1 + p_2^3 + p_3^3 + y_1^3 + y_2^3 \quad y_j \in \mathscr{P}_2 \tag{31}$$

then there is a solution.

A different approach which works in some cases is to assume the Generalised Riemann Hypothesis. Thus in Vaughan [89] the following were obtained.

Theorem 11. *Assume the Generalised Riemann Hypothesis. Then every sufficiently large even integer is the sum of a prime, the square of a prime and two cubes of primes.*

Theorem 12. *Assume the Generalised Riemann Hypothesis. Then every sufficiently large odd integer is the sum of a prime and four cubes of primes.*

19 Goldbach Generalised: Other Rings

Another rather obvious generalisation of the Goldbach questions is to other rings than $\mathbb{Z}$. The irreducible polynomials play the rôle of prime numbers. There might be some difficulties if one does not have uniqueness of factorisation. This is not a problem for polynomials over a field. It can be a lot easier than the classical case, but can nevertheless give rise to some interesting techniques. Thus Hayes [40] established a Goldbach theorem for polynomials over $\mathbb{Z}$ and there are many generalisations. See Pollack [73] for a generalization and a useful review of the literature.

For polynomials over finite fields Effinger and Hayes [20] give a systematic treatment of the subject. There are many similarities with the classical situation. Thus it is possible to imitate the Hardy–Littlewoo–Ramanujan–Vinogradov method by introducing the Pontryagin dual. This gives an analogue of the fourier transform on the torus which is at the heart of the classical approach.

References

1. W. D. Banks, A. M. Güloğlu & C. W. Nevans, Representation of integers as sums of primes from a Beatty sequence, Acta Arith. 130(2007), 255–275.
2. W. D. Banks, Ahmet M. Guloglu & R. C. Vaughan, On Waring's problem for dense sequences, Journal de Théorie des Nombres de Bordeaux, to appear.
3. M. B. Barban, The "large sieve method" and its application to number theory (Russian), Uspehi Akad. Nauk SSSR 21(1966), 51–102.
4. E. Bombieri, On the large sieve, Mathematika, 12(1965), 201–225.
5. E. Bombieri & H. Davenport, Small differences between prime numbers, Proc. Roy. Soc. A, 293(1966), 1–18.
6. J. Brüdern, Sieves, the circle method, and Waring's problem for cubes, Habilitationsschrift, Mathematica Gottingensis, Vol. 51.
7. J. Brüdern, A sieve approach to the Waring-Goldbach problem I: Sums of four cubes, Ann. scient. Ec. Norm. Sup. (4) 28, 461–476.
8. J. Brüdern, A sieve approach to the Waring-Goldbach problem II: On the seven cubes theorem, Acta Arith. 72, 211–227.
9. J. Brüdern & E. Fouvry, The four square theorem with almost prime variables, J. reine angew. Math. 454, 59–96.
10. J. Brüdern & K. Kawada, Ternary problems in additive prime number theory, Analytic Number Theory, edited by Chaohua Jia and Kohji Matsumoto, Kluwer 2002, 39–91.

11. V. Brun, Über das Goldbachsche Gesetz und die Anzahl der Primzahlpaare, Archiv for Mathematik og Naturvidenskab B34(1915), no. 8, 19pp.
12. Jing-Run Chen, On the representation of a large even integer as the sum of a prime and the product of at most two primes, J. Kexue Tongbao 17(1966), 385–386.
13. Jing-Run Chen & Pan Cheng Dong, The exceptional set of Goldbach numnbers, J. of Shandong Univ, (1979), 1–27.
14. N. G. Chudakov, On the Goldbach problem, C. R. Acad. Sci. URSS, (2)17(1937), 335–338.
15. J. G. van der Corput, Sur l'hypothèse de Goldbach pour presque tous les nombres pairs, Acta Arith. 2(1937), 266–290.
16. H. Davenport, The Collected Works of Harold Davenport IV, edited by B. J. Birch, H. Halberstam, & C. A. Rogers, Academic Press, 1977.
17. H. Davenport & H. Halberstam, Primes in arithmetic progressions, Michigan Math. J. 13(1966), 485–489.
18. Davenport, H.; Halberstam, H. Corrigendum: "Primes in arithmetic progression". Michigan Math. J. 15 1968 505.
19. P. D. T. A. Elliott & H. Halberstam, A conjecture in prime number theory, Symp. Math. 4(1968), 59–72.
20. G. W. Effinger & D. R. Hayes, Additive numbe theory of polynomials over a finite field, Oxford University Press, 1991, 176pp.
21. T. Estermann, On Goldbach's problem: Proof that almost all even positive integers are sums of two primes, Proc. London Math. Soc.(2)44(1938), 307–314.
22. Ford, Kevin; Green, Ben; Konyagin, Sergei; Maynard, James; Tao, Terence, Long gaps between primes, arXiv:1412.5029
23. J. B. Friedlander & H. Iwaniec, Opera de Cribo, A.M.S. Colloquium Publications, vol 57.
24. J. B. Friedlander, K. Gong and I. E. Shparlinski, Character sums over shifted primes, Mat. Zametki 88(2010), 605–619.
25. J. B. Friedlander & D. A. Goldston, Variance of distribution of primes in residue classes, Quart. J. Math. Oxford (2) 47(1996), 313–336.
26. P. X. Gallagher, The large sieve, Mathematika 14(1967), 14–20.
27. D. A. Goldston, J. Pintz, & C. Y. Yıldırım, Primes in tuples I, Ann. of Math. 170(2009), 819–862.
28. B. J. Green, *Generalizing the Hardy–Littlewood method for primes*, Proc. ICM (Madrid 2006), vol. 2, pp. 373–399.
29. B. J. Green & T. Tao, *The primes contain arbitrarily long arithmetic progressions*, Ann. Math. 167(2008), 481–547.
30. H. Halberstam & K. F. Roth, Sequences, Oxford University Press, 1966.
31. G. H. Hardy & J. E. Littlewood, Some problems of partitio numerorum, I: A new solution to Waring's problem, Göttinger Nachrichten (1920), 33–54.
32. G. H. Hardy & J. E. Littlewood, Some problems of partitio numerorum, II: Proof that every large number is the sum of 21 biquadrates, Mat. Z. 9(1921), 14–27.
33. G. H. Hardy & J. E. Littlewood, Some problems of partitio numerorum, III: On the expression of a number as a sum of primes, Acta Math, 44(1922), 1–70.
34. G. H. Hardy & J. E. Littlewood, Some problems of partitio numerorum, IV: The singular series in Waring's problem and the value of the number $G(K)$, Mat. Z. 12(1922), 161–188.
35. G. H. Hardy & J. E. Littlewood, Some problems of partitio numerorum, V: A further contribution to the study of Goldbach's problem, Proc. London Math. Soc. (2)22(1924), 254–269.
36. G. H. Hardy & J. E. Littlewood, Some problems of partitio numerorum, VI: Further researches in Waring's problem, Mat. Z. 23(1925), 1–37.
37. G. H. Hardy & J. E. Littlewood, Some problems of partitio numerorum, VII: The number $\Gamma(k)$ in Waring's problem, Proc. London Math. Soc. (2)28(1928), 518–542.
38. G. H. Hardy & S. Ramanujan, Asymptotic Formulae in Combinatory Analysis, Proc. London Math. Soc. 17(1918), 75–115.

39. G. H. Hardy & E. M. Wright, An Introduction to the Theory of Numbers, Oxford University Proess, fifth edition, 1979.
40. D. R. Hayes, A Goldbach theorem for polynomials with integral coefficients, Amer. Math Monthly 72(1965), 45–46.
41. D. R. Heath-Brown, Three primes and an almost-prime in arithmetic progression, J. London Math. Soc. (2)23(1981), 396–414.
42. H. Helfgott, The ternary Goldbach conjecture is true, arXiv:1312.7748, (2013).
43. C. Hooley, The distribution of sequences in arithmetic progression, Proc. ICM publ. Vancouver 1974.
44. C. Hooley, On the Barban-Davenport-Halberstam theorem: I, J. reine angew. Math. 274/275(1975), 206–223.
45. C. Hooley, On the Barban-Davenport-Halberstam theorem III, J. London Math. Soc. (2) 11(1975), 399–407.
46. C. Hooley, On the Barban-Davenport-Halberstam theorem. VIII. J. Reine Angew. Math. 499 (1998), 1–46.
47. M. N. Huxley, Small differences between consecutive primes, Mathematika 20(1973), 229–232.
48. M. N. Huxley, Small differences between consecutive primes, II, Mathematika 24(1977), 142–152.
49. L. K. Hua, Additive theory of prime numbers, Translations of Mathematical Monographs, 13, American Mathematical Society, Providence, R.I. 1965 xiii+190 pp.
50. K. Kawada, *Note on the sum of cubes of primes and an almost prime*, Arch. Math. (Basel) 69(1997), 13–19.
51. K. Kawada, *On several additive problems that regard variables as prime numbers*, RIMS Kokyuroku 1274(2002), 219–229.
52. K. Kawada, *On sums of seven cubes of almost primes*, Acta Arith. 117(2005), 213–245.
53. M. Kneser, Abschätzungen der asymptotischen Dichte vin Summengen, Math. Z. 58(1953), 459–484.
54. A. V. Kumchev, On sums of primes from Beatty sequences, Integers 8, A8 (2008). 12 pp.
55. E. Landau, Ueber die zahlentheoretische Function $\varphi(n)$ und ihre Bezeihung zum Goldbachschen Satz, Nachrichten von der Königliche Gesellschaft der Wissenschaften zu Göttingen, Mathematisch–Physikalische Klasse, 1900, 177–186.
56. Lavrik, A. F., On the twin prime hypothesis of the theory of primes by the method of I. M. Vinogradov. Dokl. Akad. Nauk SSSR 132(1960) 1013–1015 (Russian); translated as Soviet Math. Dokl. 1 1960 700–702.
57. Yu. V. Linnik, The large sieve, C. R. (Doklady) Acad. Sci. URSS (N.S.) 30(1941), 292–294.
58. Yu. V. Linnik, A remark on the least quadratic non-residue, C. R. (Doklady) Acad. Sci. URSS (N.S.) 36(1942), 119–120.
59. J. E. Littlewood, On the zeros of the Riemann zeta-function, Proc. Cam. Phil. Soc, 22(1924), 295–318.
60. Zhixin Liu, Cubes of primes and almost prime, Journal of Number Theory, (6) 132(2012), 1284–1294.
61. Wen Chao Lu, Exceptional set of Goldbach number, J, Number Theory 130(2010), 2359–2392.
62. H. Maier, Small differences between prime numbers, Michigan Math. J. 35(1988), 323–344.
63. H. B. Mann, A proof of the fundamental theorem on the density of sums of sets of positive integers, Annals of Mathematics, 43(1942), 523–527.
64. J. Maynard, Small gaps between primes, 181(2015), 383–413.
65. H. L. Montgomery, A note on the large sieve, J. London Math. Soc. 43(1968), 93–98.
66. H. L. Montgomery, Primes in arithmetic progressions. Michigan Math. J. 17(1970), 33–39.
67. H. L. Montgomery & R. C. Vaughan, The large sieve, Mathematika, 20(1973), 119–134.
68. H. L. Montgomery & R. C. Vaughan, Error terms in additive prime number theory, Q. J. Math. 24(1973), 207–216.
69. H. L. Montgomery & R. C. Vaughan, The Exceptional Set in Goldbach's Problem, Acta. Arith. 27(1975), 353–370.

70. H. L. Montgomery & R. C. Vaughan, Multiplicative Number Theory I. Classical Theory, Cambridge University Press, 2007.
71. I. Piatetski–Shapiro, On the distribution of prime numbers in sequences of the form [f(n)], Mat. Sbornik N.S. 33(75)(1953), 559–566.
72. G. Z. Pil'tai, Studies in the theory of numbers (Saratov), No. 4, pp. 73–79, Izdat. Saratov. Univ., Saratov, 1972.
73. P. Pollack, On polynomial rings with a Goldbach property, Amer. Math. Monthly 118(2011), 71–77.
74. O. Ramaré, On S'nirel'man's constant, Annali dela Scuola Superiore di Pisa 21(1995), 645–705.
75. R. A. Rankin, The difference between consecutive prime numbers, V, Proc. Edinburgh Math. Soc. 13(1963), 331–332.
76. G. Ricci, Sull'andamento della differenza di numeri primi consecutivi, Riv. Mat. Univ. Parma 5(1954), 3–54.
77. J. Rivat & J. Wu, Prime numbers of the form $\lfloor n^c \rfloor$, Glasg. Math. J. 43(2001), 237–254.
78. K. F. Roth, On Waring's problem for cubes, Proc. London Math. Soc. (2)53(1951), 268–279.
79. K. F. Roth, On the large sieves of Linnik and Rényi, Mathematika 12(1965), 1–9.
80. Schnirelmann, On additive properties of numbers, Ann. Inst. polytechn. Novoerkassk 14(1931), 3–28.
81. A. Selberg, On an elementary method in the theory of primes, Norske Vid. Selsk. Forh. Trondheim 19(1947), 64–67.
82. W. H. Spottiswoode, Description of a communication in the account of the Annual General Meeting, Proc. London Math. Soc. (1871), 3–6.
83. P. G. S. Stäckel, Ueber Goldbachs empirisches Theorem, Nachrichten von der Königliche Gesellschaft der Wissenschaften zu Göttingen, Mathematisch-Physikalische Klasse (1896), 292–299.
84. R. C. Vaughan, Sommes trigonométriques sur les nombres premiers, C.R. Acad. Sci. Paris Sér. A 285(1977), 981–983.
85. R. C. Vaughan, An elementary method in prime number theory, Acta Arithmetica 37(1980), 111–115.
86. R. C. Vaughan, Some remarks on Weyl sums, Colloquia Math. Soc. János Bolyai, 34 Topics in classical number theory, Budapest, 1981, North Holland (1984), 1585–1602.
87. R. C. Vaughan, On a variance associated with the distribution of primes in arithmetic progressions, Proc. London Math. Soc, 82(2001), 533–553.
88. R. C. Vaughan, On the number of partitions into primes, The Ramanujan Journal, 15(2008), 109–121.
89. R. C. Vaughan, Some problems of 'Partitio Numerorum': Hybrid expressions, The Legacy of Srinivasa Ramanujan, RMS-Lecture Notes Series No. 20, 2013, pp. 363–385.
90. R. C. Vaughan, The general Goldbach problem with Beatty primes, The Ramanujan Journal, 34 (2014), 347–359.
91. A. I. Vinogradov, On the density hypothesis for Dirichlet L–functions, Izv. Akad. Nauk SSSR Ser. Mat. 29(1965), 903–934, and Correction to the paper of A. I. Vinogradov, On the desity hypothesis for Dirichlet L–functions, Izv. Akad. Nauk SSSR Ser. Mat. 30(1966), 719–720.
92. I. M. Vinogradov, Representation of an odd number as the sum of three primes, Dokl. Akad. Nauk SSSR, 16 (1937), 179–195.
93. I. M. Vinogradov, The distribution of quadratic residues and non–residues of the kind $p + k$ to a prime modulus, Mat. Sbornik (2)3(45)(1938), 311–320.
94. Yitang Zhang, Bounded gaps between primes, Ann. of Math. 179(2014), 1121–1174.

The Hodge Conjecture

Claire Voisin

Abstract This is an introduction to the Hodge conjecture, which, although intended to a general mathematical audience, assumes some knowledge of topology and complex geometry. The emphasis will be put on the importance of the notion of Hodge structure in complex algebraic geometry.

1 Introduction

The Hodge conjecture stands between algebraic geometry and complex geometry. It relates data coming from topology (a Betti cohomology class), complex geometry (the Hodge decomposition or filtration) and algebraic geometry (the algebraic subvarieties of a complex algebraic variety). We can state it very quickly by saying that it provides a conjectural characterization of *algebraic classes*, that is cohomology classes generated over $\mathbb{Q}$ by classes of algebraic subvarieties of a given dimension of a complex projective manifold X, as *Hodge classes*, that is those rational cohomology classes of degree $2k$ which admit de Rham representatives which are closed forms of type (k,k) for the complex structure on X. The geometry behind this condition is the fact that the integration current defined by a complex submanifold of dimension $n-k$ annihilates forms of type (p,q) with $(p,q) \neq (n-k, n-k)$.

Not much is known about the Hodge conjecture, apart from the Lefschetz theorem on $(1,1)$-classes (Theorem 2) and a beautiful evidence (Theorem 6) provided by Cattani, Deligne and Kaplan, which says roughly that Hodge classes behave in family as if they were algebraic, that is, satisfied the Hodge conjecture. What we plan to do is to explain the basic notions in Hodge theory (Hodge structure, coniveau) giving a strong motivation for the Hodge conjecture (and still more for its generalization, the generalized Hodge conjecture, see Conjecture 3). The Hodge structures on rational cohomology groups are very rich objects associated to a smooth projective complex variety, and the belief is that they carry a lot of qualitative information on the variety: Torelli theorems state that under some

C. Voisin (✉)
CNRS, Institut de Mathématiques de Jussieu, Paris, France
e-mail: claire.voisin@imj-prg.fr

J.F. Nash, Jr., M.Th. Rassias (eds.), *Open Problems in Mathematics*,
DOI 10.1007/978-3-319-32162-2_17

assumptions, the isomorphism class of these Hodge structures determine the variety itself. In another direction, the Hodge conjecture is part of a general picture predicting that these Hodge structures determine the "motive" or at least the Chow groups of the variety. We also wish to present some of the most important facts from Hodge theory allowing to prove some conditional results or implications between various subconjectures. Some very important cases of the Hodge conjecture are summarized under the name of standard conjectures (see [21]), the main one being the Lefschetz standard conjecture (Conjecture 2). These instances of the Hodge conjecture concern Hodge classes of a very special type, which satisfy extra arithmetic conditions (being absolute Hodge, see Definition 5) satisfied by algebraic classes but not known to be satisfied by all Hodge classes (see Conjecture 7). An example of such conditional statement underlining the importance of the Lefschetz standard conjecture is Theorem 7 concerning the variational form of the Hodge conjecture which asks whether, starting from a variety X with a Hodge class α which is algebraic, and deforming X in a family $(X_t)_{t\in B}$ in such a way that the class α remains Hodge along the deformation, the class α_t also remains algebraic on X_t.

This paper is organized as follows: in Sect. 2, we will define Hodge structures, polarizations on them and Hodge classes. In Sect. 3, we will present the Hodge conjecture, its generalized version, and the few cases in which it is known. We will also discuss the standard conjectures. Finally we will turn in Sect. 4 to variational aspects of the Hodge conjecture. Sections 3 and 4 use in an essential way the theory of mixed Hodge structures which is summarized in Sect. 3.3.

This quick presentation of the Hodge conjecture does not contain many examples. It is an invitation to read the book [23] where many specific known cases of the Hodge conjecture are presented.

2 Hodge Structures, Hodge Classes

2.1 Hodge Decomposition

Let X be a complex manifold. The complex structure on X allows to decompose the vector bundle of complex differential 1-forms on X as

$$\Omega_{X,\mathbb{C}} = \Omega_X^{1,0} \oplus \Omega_X^{0,1}, \tag{1}$$

where $\Omega_X^{1,0}$ is the vector bundle of 1-forms which are $\mathbb{C}$-linear for the complex structure on T_X, locally generated by dz_i, the z_i's being local holomorphic coordinates, and $\Omega_X^{0,1} = \overline{\Omega_X^{1,0}}$ is its complex conjugate, locally generated by $d\overline{z_i}$. From (1), we deduce a decomposition of the sheaf of C^∞ complex differential forms of degree k:

$$\mathscr{A}_{X,\mathbb{C}}^k = \oplus_{p+q=k}\mathscr{A}_X^{p,q}, \tag{2}$$

where $\mathscr{A}_X^{p,q}$ is the sheaf of differential forms of type (p,q), which can be written in local holomorphic coordinates z_i as

$$\alpha = \sum_{|I|=p,|J|=q} \alpha_{I,J} dz_I \wedge d\overline{z_J}. \tag{3}$$

It is clear from (3) that the exterior differential $d : \mathscr{A}_{X,\mathbb{C}}^k \to \mathscr{A}_{X,\mathbb{C}}^{k+1}$ satisfies $d\mathscr{A}_X^{p,q} \subset \mathscr{A}_X^{p+1,q} \oplus \mathscr{A}_X^{p,q+1}$. There is thus no reason that the decomposition (2) induces a decomposition on the level of de Rham cohomology, that is on the space

$$H^k(X,\mathbb{C}) = \frac{\mathrm{Ker}\,(d : A^k(X) \to A^{k+1}(X))}{\mathrm{Im}\,(d : A^{k-1}(X) \to A^k(X))}.$$

Here $A^k(X) := \Gamma(X, \mathscr{A}_{X,\mathbb{C}}^k)$ is the space of C^∞ complex differential k-forms on X. However, when X is compact Kähler (and a fortiori projective), the Hodge decomposition theorem says the following:

Theorem 1 (Hodge [17]). *If X is a compact Kähler manifold, one has a canonical decomposition*

$$H^k(X,\mathbb{C}) = \oplus_{p+q=k} H^{p,q}(X), \tag{4}$$

where $H^{p,q}(X)$ is the set of de Rham cohomology classes of closed differential forms on X which are of type (p,q).

The simplest consequence of this statement is the following restriction on the topology of compact Kähler manifolds:

Corollary 1. *If k is odd, and X is a compact Kähler manifold, $b_k(X)$ is even.*

Indeed, the definition we gave of $H^{p,q}(X)$ clearly shows that the Hodge decomposition (4) satisfies the *Hodge symmetry* property:

$$\overline{H^{p,q}(X)} = H^{q,p}(X), \tag{5}$$

where complex conjugation acts naturally on $H^k(X,\mathbb{C}) = H^k(X,\mathbb{R}) \otimes \mathbb{C}$. The conclusion of Corollary 1 is not satisfied by the simplest example of non-Kähler compact complex surface, namely the Hopf surface S, which is the quotient of $\mathbb{C}^2 \setminus \{0\}$ by the action of $\mathbb{Z}$ given by multiplication by $\lambda \neq 0$, where $|\lambda| \neq 1$. Indeed, $\pi_1(T) = \mathbb{Z}$ hence $b_1(T) = 1$.

Note on the other hand that by the change of coefficients theorem, we have

$$H^k(X,\mathbb{C}) = H^k(X,\mathbb{Q}) \otimes \mathbb{C}.$$

This leads us to introduce the basic definition of a *Hodge structure of weight k*:

Definition 1. A rational Hodge structure of weight k is the data of a finite rank $\mathbb{Q}$-vector space L, together with a decomposition

$$L_{\mathbb{C}} := L \otimes \mathbb{C} = \oplus_{p+q=k} L^{p,q}, \tag{6}$$

where the $L^{p,q} \subset L_{\mathbb{C}}$ are complex vector subspaces satisfying the Hodge symmetry condition $\overline{L^{p,q}} = L^{q,p}$.

The data of the Hodge decomposition (6) is equivalent to that of the Hodge filtration (which is a decreasing filtration on $L_{\mathbb{C}}$)

$$F^r L_{\mathbb{C}} := \oplus_{p+q=k, p\geq r} L^{p,q}, \tag{7}$$

since $L^{p,q} = F^p L_{\mathbb{C}} \cap \overline{F^q L_{\mathbb{C}}}$.

Hodge structures coming from geometry are "effective", meaning that $L^{p,q} = 0$ for $p < 0$ or $q < 0$. However it is natural to introduce the dual $(L^*, (L^{p,q})^*)$ of a Hodge structure $(L, L^{p,q})$ of weight k and to give it weight $-k$, so the effectivity condition should not be part of the definition.

Morphisms of Hodge structures $(L, L^{p,q})$ of weight k and $(L', L'^{p,q})$ of weight k' are defined only when $k' = k + 2r$, as the set of morphisms $\phi : L \to L'$ of $\mathbb{Q}$-vector spaces satisfying

$$\phi_{\mathbb{C}}(L^{p,q}) \subset L'^{p+r,q+r}.$$

The Tate twist $L(r)$ of a Hodge structure of weight k is the Hodge structure L' of weight $k - 2r$ which has the same underlying vector space $L' = L$ and Hodge decomposition $L'^{p,q} = L^{p+r,q+r}$. If X is a compact Kähler manifold, Poincaré duality provides an isomorphism of weight $-k$ Hodge structures

$$H^k(X, \mathbb{Q})^* \cong H^{2n-k}(X, \mathbb{Q})(n).$$

If X and Y are compact Kähler manifolds and $\phi : X \to Y$ is a holomorphic map,

$$\phi^* : H^k(Y, \mathbb{Q}) \to H^k(X, \mathbb{Q})$$

is a morphism of Hodge structures since the pull-back by ϕ of a closed form of type (p, q) on X is a closed form of type (p, q) on Y; by Poincaré duality, the Gysin morphism

$$\phi_* : H^k(X, \mathbb{Q}) \to H^{k+2r}(Y, \mathbb{Q}), \ \ r = \dim Y - \dim X$$

is also a morphism of Hodge structures.

2.2 Hodge Structures and Polarizations

Given a morphism of Hodge structures $\phi : L \to L'$, it is obvious how to define a Hodge structure on $\operatorname{Ker}\phi$ and on $\operatorname{Im}\phi$ since morphisms of Hodge structures are those which are bigraded after tensoring by $\mathbb{C}$. Hence rational Hodge structures form an abelian category. However this category is not semi-simple. This phenomenon already appears for weight 1 Hodge structures. An effective weight 1 Hodge structure on L is determined by the choice of vector subspace $L^{1,0} \subset L_{\mathbb{C}}$ which has to be in direct sum with its complex conjugate. Suppose now that $(L, L^{1,0})$ contains a Hodge substructure $L' \subset L$, $L'^{1,0} = L'_{\mathbb{C}} \cap L^{1,0}$. The only condition on the space $L^{1,0}$ determining the Hodge structure on L is that its intersection with $L'_{\mathbb{C}}$ has dimension $\frac{1}{2}\dim L'$. We claim that for a general pair (L', L) of Hodge structures as above, there is no splitting $L = L' \oplus L''$ as Hodge structures. Indeed, there are countably many choices of such splitting over $\mathbb{Q}$, and for a given splitting, the condition that $L'' \subset L$ is also a Hodge structure means that $L^{1,0} \cap L''_{\mathbb{C}}$ has dimension $\frac{1}{2}\dim L''$. The complex dimension of the algebraic subset of the Grassmannian $\operatorname{Grass}(k, 2k)$ parameterizing the Hodge structures on L for which $L' \subset L$ is a Hodge substructure is thus equal to $k'^2 + (k - k')k$ while the algebraic subset of the Grassmannian $\operatorname{Grass}(k, 2k)$ parameterizing the Hodge structures on L for which L is a direct sum $L' \oplus L''$ of Hodge structures is a countable union of algebraic subsets of dimension $k'^2 + (k - k')^2$. As $k'^2 + (k - k')k > k'^2 + (k - k')^2$, the claim is proved. The phenomenon described above does not appear in algebraic geometry where the Hodge structures we get are polarized.

Definition 2. A polarization on a rational Hodge structure L of weight k is a nondegenerate intersection form $(\,,\,)$ on L which is symmetric if k is even, skew-symmetric if k is odd and satisfies the Hodge-Riemann bilinear relations:

(1) $(\alpha, \overline{\beta}) = 0$ for $\alpha \in L^{p,q}$, $\beta \in L^{p',q'}$ and $(p,q) \neq (p',q')$.
(2) $\iota^k(-1)^p(\alpha, \overline{\alpha}) > 0$ for $\alpha \in L^{p,q}$, $\alpha \neq 0$.

Admittedly, the sign rules in (2) are complicated, but they are imposed on us by geometry. The importance of the notion comes from the following:

Lemma 1. *Let $L' \subset L$ be a Hodge substructure of a polarized rational Hodge structure. Then there exists a Hodge substructure $L'' \subset L$ such that L is isomorphic to $L' \oplus L''$ as Hodge structure.*

Proof. Indeed, let q be the intersection form giving the polarization on L. It suffices to prove that the restricted form $q_{|L'}$ is nondegenerate since then the orthogonal complement $L'' := L'^{\perp_q}$ is defined over $\mathbb{Q}$, is a Hodge substructure of L by property (1) above and satisfies $L' \oplus L'' = L$. Let $h(u, v) = \iota^k q(u, \overline{v})$ be the Hermitian bilinear form on $L_{\mathbb{C}}$ associated to q. It suffices to show that $h_{|L'_{\mathbb{C}}}$ is nondegenerate. But the Hodge decomposition of $L'_{\mathbb{C}}$ is orthogonal for h by (1) above and each $h_{|L'^{p,q}}$ is nondegenerate by (2) above. Hence $h_{|L'_{\mathbb{C}}}$ is nondegenerate. □

The construction of a polarization on the Hodge structure on $H^k(X,\mathbb{Q})$ when X is a smooth complex projective manifold goes as follows: Let $l \in H^2(X,\mathbb{Q})$ be the Chern class of an ample line bundle on X. Then the hard Lefschetz theorem [31, 6.2.3] gives an isomorphism of Hodge structures

$$l^{n-k}\smile: H^k(X,\mathbb{Q}) \to H^{2n-k}(X,\mathbb{Q}),\ n = \dim X.$$

We can thus assume $k \le n$. We then consider the nondegenerate intersection pairing

$$(\alpha,\beta)_l := \int_X l^{n-k}\smile\alpha\smile\beta,\ \alpha,\ \beta\in H^k(X,\mathbb{Q}).$$

It is nondegenerate by the hard Lefschetz theorem, does not satisfy property (2) above, but satisfies property (1) above. We finally modify it as follows: the Hodge structure $H^k(X,\mathbb{Q})$ admits the Lefschetz decomposition as a direct sum of Hodge substructures

$$H^k(X,\mathbb{Q}) = \oplus_{2r\le k} l^r \smile H^{k-2r}(X,\mathbb{Q})_{prim}, \tag{8}$$

where the primitive cohomology is defined by $H^{k-2r}(X,\mathbb{Q})_{prim} := \mathrm{Ker}\,(l^{n-k+2r+1}\smile: H^{k-2r}(X,\mathbb{Q}) \to H^{2n-k+2r+2}(X,\mathbb{Q}))$. This decomposition is orthogonal for $(\,,\,)_l$. The polarization $(\,,\,)$ on $H^k(X,\mathbb{Q})$ is the unique intersection pairing for which the Lefschetz decomposition is orthogonal, and which is equal to $(-1)^r(\,,\,)_l$ on $l^r\smile H^{k-2r}(X,\mathbb{Q})_{prim}$. The fact that this polarizes (up to a sign) the Hodge structure on $H^k(X,\mathbb{Q})$ is exactly the contents of the Hodge-Riemann bilinear relations (see [31, 6.3.2]).

2.3 *Hodge Classes and Cycle Classes*

2.3.1 Hodge Classes

Let H be a Hodge structure of even weight $2k$, with Hodge decomposition $H_{\mathbb{C}} = \oplus_{p+q=2k} H^{p,q}$.

Definition 3. The Hodge classes in H are the classes in H (hence rational) which, via the inclusion $H \subset H_{\mathbb{C}}$, belong to $H^{k,k}$.

We will denote $\mathrm{Hdg}^{2k}(H)$ the space $H\cap H^{k,k}$ of Hodge classes. Note that this space can be reduced to 0 and will be 0 for a general Hodge structure with given Hodge numbers $h^{p,q} = \dim H^{p,q}$ unless $h^{p,q} = 0$ for $p \neq q$, since the space $H^{k,k}\subset H_{\mathbb{C}}$ needs not be defined over $\mathbb{Q}$, but only over $\mathbb{R}$ (as implied by the Hodge symmetry property, that is condition (5)). If X is a smooth projective variety, we will denote $\mathrm{Hdg}^{2k}(X)$ the space $\mathrm{Hdg}^{2k}(H^{2k}(X,\mathbb{Q}))$.

2.3.2 Cycle Classes

Let X be a smooth complex projective or compact Kähler variety of (complex) dimension n, and let $Z \stackrel{j}{\hookrightarrow} X$ be a closed analytic subset (which in the projective case is the same thing according to Chow as a closed algebraic subset) of codimension k. If Z is smooth, then Z is a codimension $2k$ real submanifold endowed with the complex orientation, so it has a fundamental homology class $[Z]_{fund} \in H_{2n-2k}(Z, \mathbb{Z})$ which gives a homology class

$$j_*[Z]_{fund} \in H_{2n-2k}(X, \mathbb{Z}) \cong H^{2k}(X, \mathbb{Z}),$$

where the last isomorphism is the Poincaré duality isomorphism. If Z is not smooth, then according to Hironaka, one can construct a smooth projective variety $\widetilde{Z}$ with a morphism $\tau : \widetilde{Z} \to Z$ of degree 1. Letting $\tilde{j} := j \circ \tau : \widetilde{Z} \to X$, we can define the class $[Z]$ of Z by

$$[Z] = \tilde{j}_*[\widetilde{Z}]_{fund} \in H^{2k}(X, \mathbb{Z}).$$

Lemma 2. *The class of a closed analytic subset Z in a compact Kähler manifold X is a Hodge class.*

Proof. Let n be the dimension of X. Then we have Poincaré duality

$$H^{2k}(X, \mathbb{C}) = H^{2n-2k}(X, \mathbb{C})^*$$

identifying the space $H^{k,k}(X)$ with the subspace of $H^{2k}(X, \mathbb{C})$ which is orthogonal to $\oplus_{p+q=2n-2k,(p,q)\neq(n-k,n-k)} H^{p,q}(X)$. It thus suffices to show that for $\beta \in H^{p,q}(X)$, $p+q = 2n-2k$, $(p,q) \neq (n-k, n-k)$, one has $\langle [Z], \beta \rangle_X = 0$. Recall from Sect. 2.1 that $H^{p,q}(X)$ consists of classes of closed forms of type (p,q). The class β is thus represented by a closed form $\tilde{\beta}$ which is closed of type (p,q) and introducing a desingularization $\tilde{j} : \widetilde{Z} \to X$ of Z, we have, by definition of the Gysin morphism,

$$\langle [Z], \beta \rangle_X = \langle \tilde{j}_*[\widetilde{Z}]_{fund}, \tilde{\beta} \rangle_X = \langle [\widetilde{Z}]_{fund}, \tilde{j}^* \tilde{\beta} \rangle_{\widetilde{Z}} = \int_{\widetilde{Z}} \tilde{j}^* \tilde{\beta}.$$

The last expression vanishes since the form $\tilde{j}^* \tilde{\beta}$ vanishes on $\widetilde{Z}$ for type reasons.

Important examples of Hodge classes are provided by the following Lemma 3.

Lemma 3. *Let H, H' be two Hodge classes of weights k, $k' = k + 2r$. Then the Hodge classes of the weight $2r$ Hodge structure* Hom (H, H') *are exactly the morphisms of Hodge structures $H \to H'$.*

Here the Hodge structure on H^* has been introduced previously, and the Hodge structure on the tensor product $H^* \otimes H' = \mathrm{Hom}\,(H, H')$ is given by

$$(H^* \otimes H')^{p,q} = \oplus_{t+t'=p, s+s'=q} (H^*)^{t,s} \otimes (H')^{t',s'}. \tag{9}$$

Proof. Indeed a morphism $\phi \in \mathrm{Hom}(H, H') = H^* \otimes H'$ is of type (r, r) for the tensor product Hodge structure if and only if it satisfies $\phi_{\mathbb{C}} \in \oplus_{(t,s)}(H^*)^{t,s} \otimes (H')^{r-t,r-s}$. As we have $(H^*)^{t,s} = (H^{-t,-s})^*$, this is equivalent to

$$\phi_{\mathbb{C}} \in \oplus_{(t,s)}(H^{t,s})^* \otimes (H')^{r+t,r+s} = \oplus_{(t,s)}\mathrm{Hom}(H^{t,s}, (H')^{r+t,r+s}),$$

that is, to the fact that $\phi_{\mathbb{C}}$ shifts the Hodge decomposition by (r, r).

3 The Hodge and Generalized Hodge Conjectures

3.1 *The Hodge Conjecture*

Conjecture 1 (Hodge 1951). *Let X be a projective complex manifold. Then for any k, the space* $\mathrm{Hdg}^{2k}(X)$ *is generated over* $\mathbb{Q}$ *by classes* $[Z]$ *of codimension k closed algebraic subsets of X.*

A codimension k cycle on X is a formal combination $Z = \sum_i \alpha_i Z_i$, $\alpha_i \in \mathbb{Q}$. We will call cycle classes $[Z] := \sum_i \alpha_i [Z_i]$ *algebraic classes*, and will use the notation $H^{2k}(X, \mathbb{Q})_{alg}$ for the space of algebraic classes. We have $H^{2k}(X, \mathbb{Q})_{alg} \subset \mathrm{Hdg}^{2k}(X)$ and the Hodge conjecture states that $H^{2k}(X, \mathbb{Q})_{alg} = \mathrm{Hdg}^{2k}(X)$.

3.1.1 Why Is the Conjecture Important?

There are very few morphisms in algebraic geometry so it is important to consider multivalued morphisms which are given by their graphs $\Gamma \subset X \times Y$. This leads to consider the group $\mathcal{Z}^m(X \times Y)$ of codimension m cycles in $X \times Y$, or better cycles modulo an adequate equivalence relation $\sim$, like rational equivalence, which provides Chow groups, or homological equivalence. When X and Y are smooth and projective, cycles in $X \times Y$ act on many objects, like Chow groups or cohomology. Given an adequate equivalence relation $\sim$ on cycles, the action of $\Gamma \in \mathcal{Z}^m(X \times Y)$ takes the general form

$$\Gamma^*(\alpha) = pr_{1*}(\Gamma \cdot pr_2^* \alpha) \in \mathcal{Z}^{k+m-\dim Y}(X)/\sim, \ \forall \alpha \in \mathcal{Z}^k(X)/\sim,$$

where pr_{1*} is pushforward by the first projection, pr_2^* is pull-back by the second projection and "$\cdot$" is the intersection product. When the equivalence relation is homological equivalence, cycles Z mod. $\sim$ are cohomology classes and the push-forward map is the Gysin map, the intersection product is the cup-product.

The importance of the Hodge conjecture in this context is that, combined with Lemma 3, it predicts exactly which morphisms $Z^* : H^*(Y, \mathbb{Q}) \to H^*(X, \mathbb{Q})$ can be constructed from cycle classes in $X \times Y$. Namely, one should get exactly the morphisms of Hodge structures. The geometric importance of this prediction is obvious: we mentioned in the introduction Torelli type questions, asking whether a variety is determined by its Hodge structures. The Hodge conjecture predicts that if two smooth projective varieties X, Y have isomorphic Hodge structures, they are related by algebraic cycles in $X \times Y$ inducing isomorphisms in cohomology. In a more motivic direction, the Hodge conjecture can thus pedantically rephrased by saying that the category of polarizable Hodge structures contains the category of cohomological motives as a *full* subcategory, so that structure results for the category of polarizable Hodge structures (like semisimplicity, see Lemma 1) also should hold for the category of cohomological motives. This adequation of Hodge theory and algebraic geometry fits also very well with conjectures of Bloch and Beilinson (see [6, 19, 33]) predicting that to a large extent, Hodge structures control Chow groups. In our mind however, the generalized Hodge conjecture which will be explained in Sect. 3.3 is much more important in this context than the Hodge conjecture itself as it says much more, qualitatively, on the relationship between Hodge structures and algebraic cycles than the Hodge conjecture does.

A more technical but important justification of the interest of the Hodge conjecture concerns the Hodge classes which appear in the standard conjecture. Roughly speaking, these Hodge classes are those which can be produced by linear algebra starting from classes of algebraic cycles. The classes so obtained, which will be described in Sect. 3.2, are still Hodge classes for linear algebra reasons, but it is not known if they are algebraic. The importance of these classes also comes from the consideration of the theory of motives.

3.1.2 Positive Evidences

The only instances of the Hodge conjecture which are known for any smooth complex projective n-fold X are first of all the two trivial cases $H^0(X, \mathbb{Q}) = \mathrm{Hdg}^0(X, \mathbb{Q}) = \mathbb{Q}[X]_{fund}$, (where X is assumed to be connected), and $H^{2n}(X, \mathbb{Q}) = \mathrm{Hdg}^{2n}(X, \mathbb{Q}) = \mathbb{Q}[\text{point}]$, and secondly the Lefschetz theorem on $(1,1)$-classes (Theorem 2) which concerns divisor (that is degree 2) classes and its corollary which concerns curve (that is degree $2n-2$) classes.

Theorem 2 (Degree 2**).** *Let X be a complex projective manifold and let $\alpha \in \mathrm{Hdg}^2(X, \mathbb{Z})$ be an integral Hodge class. Then α is a combination with integral coefficients of classes $[D] \in H^2(X, \mathbb{Z})$ of hypersurfaces $D \subset X$.*

Corollary 2 (Degree $2n-2$**).** *Let X be a complex projective n-fold and let $\alpha \in \mathrm{Hdg}^{2n-2}(X)$ be a Hodge class. Then α is a combination with rational coefficients of classes $[C] \in H^{2n-2}(X, \mathbb{Z})$ of curves $C \subset X$.*

Remark 1. The first three cases mentioned above (degrees 0, 2 or $2n$) are the only cases where the Hodge conjecture is true for *integral* Hodge classes, that is integral cohomology classes whose image in rational cohomology is a Hodge class. This follows from Atiyah-Hirzebruch and Kollár counterexamples [3, 20] for integral Hodge classes.

Proof (Proof of Theorem 2). There is a beautiful description in [13] of the original Lefschetz proof. It relies on the notion of normal function associated to a Hodge class. Given a Hodge class $\alpha \in \mathrm{Hdg}^2(X, \mathbb{Z})$, we choose a pencil of hyperplane sections $(X_t)_{t \in \mathbb{P}^1}$ of X and assume that $\alpha_{|X_t} = 0$. The Hodge class α lifts to a class $\tilde{\alpha}$ in the Deligne cohomology group $H^2_{\mathscr{D}}(X, \mathbb{Z}(1))$ (see [31, 12.3.1]). Then $\tilde{\alpha}_{|X_t}$ belongs to

$$\mathrm{Ker}\,(H^2_{\mathscr{D}}(X_t, \mathbb{Z}(1)) \to H^2(X_t, \mathbb{Z})) = J^1(X_t) = \mathrm{Pic}^0(X_t).$$

Associated to α we thus found a family of divisors $t \mapsto \tilde{\alpha}_{|X_t} \in \mathrm{Pic}^0(X_t)$. A large part of this argument works as well for any Hodge class on a smooth projective variety X vanishing on the fibers X_t of a pencil on X. Indeed, the Deligne cohomology group $H^{2k}_{\mathscr{D}}(X, \mathbb{Z}(k))$ fits in the exact sequence

$$0 \to J^k(X) \to H^{2k}_{\mathscr{D}}(X, \mathbb{Z}(k)) \to \mathrm{Hdg}^{2k}(X, \mathbb{Z}) \to 0$$

and similarly for X_t. We can thus lift a Hodge class on X to a Deligne cohomology class and restrict it to the fibers X_t. The problem is that the normal function one shall get this way will be a holomorphic section of the family of intermediate Jacobians $J^k(X_t)_{t \in \mathbb{P}^1}$, and one does not know for $k \geq 2$ what is the image of the Abel-Jacobi map $\mathscr{Z}^k(X_t)_{hom} \to J^k(X_t)$.

The modern proof of Theorem 2 uses the exponential exact sequence and goes as follows:

1) The Picard group of holomorphic line bundles of an analytic space X identifies to $H^1(X, \mathscr{O}_X^*)$, where $\mathscr{O}_X^*$ is the sheaf of invertible holomorphic functions. The exponential exact sequence

$$0 \to \mathbb{Z} \stackrel{2\iota\pi}{\to} \mathscr{O}_X \stackrel{\exp}{\to} \mathscr{O}_X^* \to 1$$

provides the associated cohomology long exact sequence

$$\ldots H^1(X, \mathscr{O}_X^*) \stackrel{c_1}{\to} H^2(X, \mathbb{Z}) \to H^2(X, \mathscr{O}_X) \ldots$$

defining c_1.

2) If X is compact Kähler, the kernel of the natural map $H^2(X, \mathbb{Z}) \to H^2(X, \mathscr{O}_X)$ appearing above is exactly the set of integral Hodge classes. This follows from the fact that this map identifies using Hodge theory with the composite

$$H^2(X, \mathbb{Z}) \stackrel{2\iota\pi}{\to} H^2(X, \mathbb{C}) \to H^{0,2}(X) \cong H^2(X, \mathscr{O}_X),$$

where all maps are natural and the map $H^2(X, \mathbb{C}) \to H^{0,2}(X)$ is the projection given by Hodge decomposition. It thus follows that a class $\alpha \in H^2(X, \mathbb{Z})$ which maps to 0 in $H^2(X, \mathscr{O}_X)$ has $\alpha^{0,2} = 0$ in the Hodge decomposition. But then it also has $\alpha^{2,0} = 0$ since it is real, and thus it is of type $(1, 1)$ hence a Hodge class.

3) At this point we proved that if X is compact Kähler, the set of Hodge classes of degree 2 is equal to the set of classes $c_1(L)$ where L runs through the set of holomorphic line bundles on X. Assume now that X is projective. By Serre GAGA principle [27], holomorphic line bundles and algebraic line bundles are the same objects on X : equivalently, any holomorphic line bundle has a nonzero meromorphic section. Choosing a nonzero meromorphic section σ of L, we introduce its divisor D_σ which is a codimension 1 cycle on X and the final step is Lelong's formula [31, Theorem 11.33] which says that the class $[D_\sigma]$ is equal to $c_1(L)$.

Proof (Proof of Corollary 2). We use for this the Lefschetz isomorphism

$$l^{n-2} \smile : H^2(X, \mathbb{Q}) \to H^{2n-2}(X, \mathbb{Q})$$

given by the choice of a very ample line bundle $\mathcal{L}$ on X with first Chern class l, which is obviously an isomorphism of Hodge structures. A Hodge class β of degree $2n - 2$ can thus be written as $\beta = l^{n-2} \smile \alpha$, where α is a Hodge class of degree 2. The class α is the class of a divisor $D = \sum_i \alpha_i D_i$, where the D_i's are hypersurfaces in X, and thus $\beta = \sum_i \alpha_i [C_i]$ where the curve C_i is the intersection of D_i with a surface $L_1 \cap \ldots \cap L_{n-2}$ complete intersection of hypersurfaces L_i in the linear system $|\mathcal{L}|$ (hence of class l) in general position.

Apart from these four known cases, the best positive evidence in favour of the Hodge conjecture is the fact that Hodge classes behave geometrically as if they were algebraic as predicted by the Hodge conjecture. The precise statement will be explained in Sect. 4.2.

3.1.3 Negative Evidences

Many complex geometry results have been proved in the past by analytic methods working as well in the compact Kähler setting, for example the Hodge decomposition itself, or the study of positivity of divisors by curvature and currents methods [12], or the proof of the existence of Hermite-Einstein metrics on stable vector bundles [29]. In the case of the Hodge conjecture, it has been known for a long time (see [36]) that in the compact Kähler setting, there are not enough closed analytic cycles to generate the Hodge classes: the example, due to Mumford, is a very general complex torus of dimension at least 2 admitting a holomorphic line bundle $\mathcal{L}$ with nontrivial Chern class which is neither positive not negative: such a torus does not contain any hypersurface, while $c_1(\mathcal{L})$ is a nontrivial Hodge class. However, in this example, one can argue that the problem is a lack of effectivity (or positivity), and that we still have a complex geometric object which is a good substitute for the hypersurfaces, namely the line bundle itself (in the projective case, by the existence of rational sections of line bundles, Chern classes of line bundles are combinations of classes of hypersurfaces).

In the paper [30], I constructed examples of Hodge classes on complex tori T, which do not belong to the $\mathbb{Q}$-vector space generated by Chern classes of coherent sheaves on T. It seems that in these cases, there is no way of extending the Hodge conjecture: there is no holomorphic object on T explaining the presence of a Hodge class on T.

The second point which makes not very plausible a solution of the Hodge conjecture by analytic methods is the lack of uniform solutions to the Hodge conjecture, assuming they exist, that is the lack of bound on the cycles (supposed minimal in some way) representing a given Hodge class. This follows from the analysis of some of the known counterexamples to the integral Hodge conjecture. In the case of Kollár counterexamples [20], which are just hypersurfaces X of degree d in projective space $\mathbb{P}^{n+1}$ with the generator α of $H^{2n-2}(X,\mathbb{Z})$ not being algebraic while $d\alpha$ is algebraic, it was observed in [28] that the following phenomenon holds: Let U be the Zariski open set in the space of homogeneous polynomials of degree d such that the corresponding hypersurface is smooth. Then the (locally constant) class $\alpha_t \in H^{2n-2}(X_t,\mathbb{Z})$ is Hodge on X_t for any $t \in U$, the set of points $t \in U$ such that the class α_t is algebraic on X_t is dense in U for the usual topology, while Kollár proves that this set is not the whole of U. This means that for a very general point $0 \in U$, there is a sequence of points $t_n \in U$ converging to 0 and for which the class α_{t_n} is the class of an algebraic cycle Z_n on X_{t_n}. Thus the cycle Z_n is of the form $Z_n^+ - Z_n^-$, but the degrees of the positive part Z_n^+ and the negative part Z_n^- of Z_n cannot be bounded, although the difference Z_n has class α_{t_n} which is locally constant hence bounded. Indeed, if these degrees were bounded, we could use compactness results to make the cycles Z_n^+ and Z_n^- converge respectively to cycles Z^+ and Z^- on X_0 with $[Z^+] - [Z^-] = \alpha$, which is not true.

3.2 The Standard Conjectures

The main source of construction of Hodge classes is Lemma 3. Let X be a complex projective n-fold, and consider $X \times X$. For any integer k, we have

$$\operatorname{End} H^k(X,\mathbb{Q}) \cong H^{2n-k}(X,\mathbb{Q}) \otimes H^k(X,\mathbb{Q}) \subset H^{2n}(X \times X,\mathbb{Q})$$

and Lemma 3 tells us that a morphism $\phi \in \operatorname{End} H^k(X,\mathbb{Q})$ provides a Hodge class on $X \times X$ by the composite map above if and only if ϕ is a morphism of Hodge structures. In particular, the identity of $H^k(X,\mathbb{Q})$ is a morphism of Hodge structures, hence provides a Hodge class $\delta_k \in \mathrm{Hdg}^{2n}(X\times X,\mathbb{Q})$. The sum $\sum_k \delta_k$ is the identity of $H^*(X,\mathbb{Q})$, hence is the class of the diagonal $\Delta_X \subset X \times X$. Hence $\sum_k \delta_k$ is algebraic but it is not known if individually each class δ_k is algebraic, that is, satisfies the Hodge conjecture. The classes δ_k are called the Künneth components of the diagonal of X. The varieties for which it is known that the Künneth components of the diagonal are algebraic include the abelian varieties (that is, projective complex tori) and smooth complete intersections in projective space, for which the non-algebraic cohomology is concentrated in degree n. If A is an abelian variety (or complex torus), A is an abelian group, hence we have for each l the multiplication map

$$\mu_l : A \to A, \; a \mapsto la.$$

We have $\mu_l^* = l^k Id$ on $H^k(A, \mathbb{Q})$ and it easily follows that we can write the Künneth components of A as linear combinations of the classes of the graph Γ_l of μ_l for various l (note that $\mu_l^* = [\Gamma_l]^* : H^*(A, \mathbb{Q}) \to H^*(A, \mathbb{Q})$).

A more subtle construction involves the properties of the Lefschetz operator. Recall from Sect. 2.2 that if l is the first Chern class of an ample line bundle $\mathscr{L}$ on X, the cup-product map

$$l^{n-k} \smile: H^k(X, \mathbb{Q}) \to H^{2n-k}(X, \mathbb{Q}), \; n = \dim X \tag{10}$$

is an isomorphism for any k. It is clear that $l^{n-k} \smile$ acting on $H^*(X, \mathbb{Q})$ is the action of the following cycle on $X \times X$: let $L_1, \ldots, L_{n-k}$ be general hypersurfaces in the linear system $|\mathscr{L}|$ (we may assume $\mathscr{L}$ very ample), and let $Z = L_1 \cap \ldots \cap L_{n-k}$. Then $[Z] = l^{n-k}$ by Lelong's theorem, and $[i_{\Delta *} Z] \in H^{4n-2k}(X \times X, \mathbb{Q})$ acts on $H^*(X, \mathbb{Q})$ by $l^{n-k} \smile$, where $i_{\Delta *} Z$ is the cycle Z supported on the diagonal $\Delta_X \cong X \subset X \times X$. Next we can consider the inverse $\lambda_{n-k} : H^{2n-k}(X, \mathbb{Q}) \to H^k(X, \mathbb{Q})$ of the Lefschetz isomorphism (10). This is a morphism of Hodge structures, hence this provides a Hodge class on $X \times X$.

Conjecture 2 (Lefschetz Standard Conjecture). *There exists a codimension k cycle Z on $X \times X$ such that $[Z]^* : H^{2n-k}(X, \mathbb{Q}) \to H^k(X, \mathbb{Q})$ is equals to λ_{n-k}.*

Again the answer is positive in the case of an abelian variety A, and this is due to the existence of an interesting line bundle $\mathscr{P}$ on $A \times A$, defined as $\mu^* \mathscr{L}$ where $\mu : A \times A \to A$ is the sum map. The line bundle $\mathscr{P}$ is called the Poincaré divisor and its class $p := c_1(\mathscr{P}) \in \mathrm{Hdg}^2(A \times A)$ and its powers $p^k \in \mathrm{Hdg}^{2k}(A \times A)$ are algebraic classes on $A \times A$ which allow to solve the Lefschetz conjecture in this case (see [24]).

The Lefschetz standard conjecture is very important in the theory of motives (see [1]), because of the semisimplicity Lemma 1. This lemma uses the polarization to construct, given a polarized Hodge structure L and a Hodge substructure $L' \subset L$, a decomposition

$$L = L' \oplus L''. \tag{11}$$

The construction of these polarizations when $L = H^k(X, \mathbb{Q})$ for some smooth projective variety X is quite involved, as it uses the Lefschetz decomposition in order to modify the natural pairing into one which satisfies the polarization axioms. If now $L = H^k(X, \mathbb{Q})$ and $L' \subset L$ is defined as the image of a morphism $[Z]^*$ for some algebraic cycle Z on $X \times X$, the Lefschetz standard conjecture is exactly what would be needed in order to construct the orthogonal complement L'' via the action of an algebraic cycle on $X \times X$.

The most concrete consequence of the Lefschetz standard conjecture is the following (cf. [21]):

Lemma 4. *Let X be a smooth complex projective variety of dimension n. Assume the Lefschetz standard conjecture holds for X and some ample class $l \in \mathrm{Hdg}^2(X)$ in all even degrees $2k$. Then for any k, the intersection pairing between $H^{2k}(X,\mathbb{Q})_{alg}$ and $H^{2n-2k}(X,\mathbb{Q})_{alg}$ is nondegenerate.*

Proof. Indeed, if the Lefschetz conjecture holds for X in any even degree, then the Lefschetz isomorphism (10) induces an isomorphism $l^{n-2k} \smile: H^{2k}(X,\mathbb{Q})_{alg} \cong H^{2n-2k}(X,\mathbb{Q})_{alg}$ for all $k \leq n/2$, because the inverse λ_{n-k} preserves algebraic classes. It follows that the space $H^{2k}(X,\mathbb{Q})_{alg}$ is stable under the Lefschetz decomposition (8). It suffices to prove that for $k \leq n/2$ the pairing $(\,,\,)_l$ on $H^{2k}(X,\mathbb{Q})$ defined by $(\alpha,\beta)_l = \langle l^{n-2k} \smile \alpha, \beta\rangle_X$, is nondegenerate on $H^{2k}(X,\mathbb{Q})_{alg} \subset H^{2k}(X,\mathbb{Q})$. By the Hodge-Riemann bilinear relations, the Lefschetz decomposition is orthogonal for this pairing and on each piece $l^r \smile H^{2k-2r}(X,\mathbb{R})_{prim}$, the pairing $(\,,\,)_l$ restricted to the subspace $H^{k-r,k-r}(X)_{\mathbb{R},prim} \subset H^{2k-2r}(X,\mathbb{Q})_{prim}$ of real classes of Hodge type $(k-r,k-r)$ is definite of a sign which depends only on $k-r$. As $l^r \smile H^{2k-2r}(X,\mathbb{Q})_{alg,prim}$ is contained in $H^{k-r,k-r}(X)_{\mathbb{R},prim}$, it follows that the pairing $(\,,\,)_l$ restricted to $l^r \smile H^{2k-2r}(X,\mathbb{Q})_{alg,prim}$ remains definite, and in particular nondegenerate. Hence $(\,,\,)_l$ is nondegenerate on $H^{2k}(X,\mathbb{Q})_{alg}$ which is the orthogoanl direct sum of the spaces $l^r \smile H^{2k-2r}(X,\mathbb{Q})_{alg,prim}$.

Let us give two corollaries:

Corollary 3. *(i) Let $j : Y \to X$ be a morphism, where X, Y are smooth complex projective varieties. Assume X and Y satisfy the Lefschetz standard conjecture. Then if Z is an algebraic cycle on Y whose class $[Z] \in H^{2k}(Y,\mathbb{Q})$ is equal to $j^*\beta$ for some class $\beta \in H^{2k}(X,\mathbb{Q})$, there exists a codimension k cycle Z' on X such that*

$$j^*[Z'] = [Z] \text{ in } H^{2k}(Y,\mathbb{Q}). \tag{12}$$

(ii) If Z is an algebraic cycle on X such that the class $[Z] \in H^{2k}(X,\mathbb{Q})$ is equal to $j_*\beta$ for some class $\beta \in H^{2k-2r}(Y,\mathbb{Q})$, $r = \dim X - \dim Y$, there exists a codimension $k-r$ cycle Z' on Y such that $j_*[Z'] = [Z]$ in $H^{2k}(X,\mathbb{Q})$.

Proof. (i) The class β gives by the Poincaré pairing on X a linear form on $H^{2n-2k}(X,\mathbb{Q})_{alg}$, $n = \dim X$, which by Lemma 4 applied to X is of the form $\langle [Z'],\,\rangle_X$ for some codimension k cycle Z' on X. We now prove that the class $[Z']$ satisfies (12). By Lemma 4, it suffices to show that for any cycle W on Y,

$$\langle j^*[Z'],[W]\rangle_Y = \langle [Z],[W]\rangle_Y. \tag{13}$$

The left hand side is equal to $\langle [Z'], j_*[W]\rangle_X$ where j is the inclusion morphism of Y in X, and by definition of $[Z']$, this is equal to $\langle \beta, j_*[W]\rangle_X$. Finally, by definition of the Gysin morphism j_*, we have $\langle \beta, j_*[W]\rangle_X = \langle j^*\beta,[W]\rangle_Y = \langle [Z],[W]\rangle_Y$.

(ii) is proved exactly in the same way.

The following corollary appears in [33] where it is proved that the conclusion (for all X and Y) is essentially equivalent to the Lefschetz conjecture:

Corollary 4 (See [33]). *Assume the Lefschetz conjecture. Let X be a smooth projective variety and let $Y \subset X$ be a closed algebraic subset. Let Z be a codimension k cycle on X whose cohomology class $[Z]$ vanishes in $H^{2k}(X \setminus Y, \mathbb{Q})$. Then there exists an algebraic cycle Z' supported on Y such that $[Z] = [Z']$ in $H^{2k}(X, \mathbb{Q})$.*

Proof. Our assumption is that there is a homology class $\beta \in H_{2n-2k}(Y, \mathbb{Q})$ such that the image of $j_*\beta \in H_{2n-2k}(X, \mathbb{Q}) \cong H^{2k}(X, \mathbb{Q})$ is equal to $[Z]$. We now apply Lemma 6, which says that if $\tilde{j} : \widetilde{Y} \to X$ is a desingularization of Y, there exists a class $\beta' \in H^{2k-2r}(\widetilde{Y}, \mathbb{Q})$ such that $\tilde{j}_*\beta' = [Z]$, where $r = \dim X - \dim Y$. We then conclude with Corollary 3, (ii).

3.3 *Mixed Hodge Structures and the Generalized Hodge Conjecture*

In [8], Deligne discovered a very important generalization of Hodge structures, namely mixed Hodge structures, see [25]. The definition is as follows:

Definition 4. A mixed Hodge structure is the data of a finite dimensional $\mathbb{Q}$-vector space L equipped with an increasing exhaustive filtration W (the weight filtration), together with a decreasing exhaustive filtration F on $L_\mathbb{C}$ with the property that the induced filtration on Gr_W^i, defined by $F^pGr_W^i = F^p \cap W_iL_\mathbb{C}/F^p \cap W_{i+1}L_\mathbb{C}$, comes from a Hodge structure [see (7)] of weight i on Gr_W^i.

Morphisms of mixed Hodge structures are morphisms of $\mathbb{Q}$-vector spaces preserving both filtrations. The following result is crucial for geometric and topological applications of this notion.

Lemma 5 (Deligne [8]). *Morphisms of mixed Hodge structures are strict for both filtrations.*

Denoting by $\phi : L \to M$ such a morphism, this means that

$$(\operatorname{Im}\phi_\mathbb{C}) \cap F^pM_\mathbb{C} = \phi_\mathbb{C}(F^pL_\mathbb{C}), \ \ (\operatorname{Im}\phi) \cap W_iM_\mathbb{C} = \phi(W_iL).$$

We will call the pure Hodge substructure of a mixed Hodge structure the smallest nonzero piece $W_iL \subset L$ and the pure quotient the quotient L/W_iL where i is maximal such that $W_iL \neq L$. they both carry a Hodge structure.

Deligne proves the following result:

Theorem 3. *For any quasiprojective variety X, its homology groups and cohomology groups carry mixed Hodge structures, which are functorial under pull-back on cohomology and functorial under pushforward on homology.*

If X is smooth, the pure Hodge substructure on $H^k(X,\mathbb{Q})$ has weight k (so all weights are $\geq k$) and is equal to $\mathrm{Im}\,(H^k(\overline{X},\mathbb{Q}) \to H^k(X,\mathbb{Q}))$ *for any smooth projective compactification $\overline{X}$ of X.*

If X is projective, the pure quotient Hodge structure of $H^k(X,\mathbb{Q})$ has weight k (so all weights are $\leq k$) and is equal to $\mathrm{Im}\,(H^k(X,\mathbb{Q}) \to H^k(\widetilde{X},\mathbb{Q}))$ *for any smooth projective desingularization $\widetilde{X}$ of X. The dual statement is that the pure Hodge substructure of $H_k(X,\mathbb{Q})$ is the image* $\mathrm{Im}\,(H_k(\widetilde{X},\mathbb{Q}) \to H_k(X,\mathbb{Q}))$ *for any smooth projective desingularization $\widetilde{X}$ of X.*

Let now X be a smooth projective variety, and $Y \subset X$ be a closed algebraic subset of X. Assume for simplicity that all the irreducible components of Y are of codimension r.

Theorem 4. *Let $U := X \setminus Y$. Then the kernel*

$$\mathrm{Ker}\,(H^k(X,\mathbb{Q}) \to H^k(U,\mathbb{Q}))$$

is a Hodge substructure L_Y of $H^k(X,\mathbb{Q})$ which is of Hodge coniveau $\geq r$, meaning that $L_Y^{p,q} = 0$ for $p < r$ or $q < r$.

Proof. We will use the following consequence of Theorem 3 and Lemma 5 which is of independent interest:

Lemma 6. *In the situation of Theorem 4, the kernel* $\mathrm{Ker}\,(H^k(X,\mathbb{Q}) \to H^k(U,\mathbb{Q}))$ *is equal to the image of the composite map*

$$\tilde{j}_* : H_{2n-k}(\widetilde{Y},\mathbb{Q}) \to H_{2n-k}(X,\mathbb{Q}) \overset{PD}{\cong} H^k(X,\mathbb{Q}), \tag{14}$$

where $\tilde{j} : \widetilde{Y} \to X$ is a desingularization of Y.

Proof. This kernel is the image of the composite map

$$H_{2n-k}(Y,\mathbb{Q}) \to H_{2n-k}(X,\mathbb{Q}) \overset{PD}{\cong} H^k(X,\mathbb{Q}).$$

This map is a morphism of mixed Hodge structures, the right hand side being a pure Hodge structure of weight k. Comparing weights and applying Lemma 5 and Theorem 3, the image of this map is the same as the image of the pure Hodge substructure of $H_{2n-k}(Y,\mathbb{Q})$, that is $\mathrm{Im}\,(H_{2n-k}(\widetilde{Y},\mathbb{Q}) \to H_{2n-k}(Y,\mathbb{Q}))$, which concludes the proof.

Of course, as $\widetilde{Y}$ is smooth and projective, the composite in (14) is the same as the Gysin morphism $\tilde{j}_* : H^{k-2r}(\widetilde{Y},\mathbb{Q}) \to H^k(X,\mathbb{Q})$. As $\tilde{j}_*$ is a morphism of Hodge structures of bidegree (r,r), its image is a substructure of $H^k(X,\mathbb{Q})$ which is of Hodge coniveau $\geq r$.

The generalized Hodge conjecture due to Grothendieck [15] states the following:

Conjecture 3. *Let X be a smooth complex projective variety and let $L \subset H^k(X, \mathbb{Q})$ be a Hodge substructure of Hodge coniveau $\geq r$. Then there exists a closed algebraic subset $Y \subset X$ of codimension $\geq r$ such that $L \subset \mathrm{Ker}\,(H^k(X, \mathbb{Q}) \to H^k(U, \mathbb{Q}))$, $U := X \setminus Y$.*

The Hodge Conjecture 1 is the particular case of Conjecture 3 where $k = 2r$. Indeed, a Hodge substructure of $H^{2r}(X, \mathbb{Q})$ which is of Hodge coniveau $\geq r$ is made of Hodge classes. Conjecture 3 predicts in this case that L vanishes away from a closed algebraic subset $Y \subset X$ of codimension r, which is the same as saying that L is generated by classes of irreducible components of Y (see [31, 11.1.2]). Conjecture 3 corrects an overoptimistic formulation of the Hodge conjecture (see [18]), where any rational cohomology class α of degree k with Hodge decomposition

$$\alpha_{\mathbb{C}} = \alpha^{k-r,r} + \ldots + \alpha^{r,k-r},$$

that is, satisfying $\alpha^{p,q} = 0$ for $p < r$ or $q < r$, is conjectured to be supported on a codimension r closed algebraic subset. This is wrong by Theorem 4 which says that if α is supported on a codimension r closed algebraic subset, then the minimal Hodge substructure $L \subset H^k(X, \mathbb{Q})$ containing α also satisfies $L^{p,q} = 0$ for $p < r$ or $q < r$ (see [15], [32, Exercise 1 p 184]).

The generalized Hodge Conjecture 3 cannot be deduced from the Hodge conjecture, unless the following conjecture is answered affirmatively:

Conjecture 4. *Let X be a smooth projective complex variety and let $L \subset H^k(X, \mathbb{Q})$ be a Hodge substructure of Hodge coniveau $\geq r$ (thus $L(r)$ is effective of weight $k-2r$). Then there exists a smooth projective variety Y, such that $L(r)$ is isomorphic to a Hodge substructure of $H^{k-2r}(Y, \mathbb{Q})$.*

We now have:

Proposition 1. *Conjecture 4 combined with the Hodge conjecture implies Conjecture 3.*

Proof. Note that by the hard Lefschetz theorem, it suffices to prove Conjecture 3 for $L \subset H^k(X, \mathbb{Q})$ with $k \leq n$. Next assume Conjecture 4. Then since $k \leq n$ we can assume by the Lefschetz theorem on hyperplane section that $\dim Y = n - r$. Now $L(r)$ is a direct summand of $H^{k-2r}(Y, \mathbb{Q})$ and the Hodge structure isomorphism $L(r) \cong L \subset H^k(X, \mathbb{Q})$ provides by Lemma 3 a Hodge class α of degree $2n$ on $Y \times X$. Assuming the Hodge conjecture, α is algebraic, which provides a cycle $Z = \sum_i \alpha_i Z_i$, $Z_i \subset Y \times X$, $\dim Z_i = n - r$, such that $L = \mathrm{Im}\,([Z]_* : H^{k-2r}(Y, \mathbb{Q}) \to H^k(X, \mathbb{Q}))$. But then L vanishes away from the codimension $\geq r$ closed algebraic subset $Y' := \cup_i pr_2(Z_i)$ of X.

4 Variational Hodge Conjecture

4.1 The Global Invariant Cycles Theorem

The following result is due to Deligne [8]. Let $\phi : X \to B$ be a holomorphic map from a smooth projective variety X to a connected complex manifold, and let $\phi^0 : X^0 \to B^0$ be the restriction of ϕ over the open subset B^0 of B of regular values of ϕ. By definition, $\phi^0 : X^0 \to B^0$ is proper with smooth fibers, hence is a topological fibration. There is thus a monodromy representation $\rho : \pi_1(B^0, b) \to \operatorname{Aut} H^k(X_b, \mathbb{Q})$, where $b \in B^0$ is a regular value.

Theorem 5. *The image of the restriction map $H^k(X, \mathbb{Q}) \to H^k(X_b, \mathbb{Q})$ is equal to the subspace $H^k(X_b, \mathbb{Q})^\rho$ of monodromy invariant cohomology classes.*

Proof (Sketch of Proof). The proof of this theorem splits into two parts. First of all, Deligne proves in [11] that the Leray spectral sequence for ϕ^0 degenerates at E_2, a result which was also known to Blanchard [4]. This implies that the space $H^k(X_b, \mathbb{Q})^\rho$, which is also the image of $H^0(B^0, R^k\phi^0_*\mathbb{Q})$ in $H^k(X_b, \mathbb{Q})$, is equal to the image of the restriction map

$$H^k(X^0, \mathbb{Q}) \to H^k(X_b, \mathbb{Q}). \tag{15}$$

The second step uses the full strength of Theorem 3. The morphism (15) is a morphism of mixed Hodge structures, the Hodge structure on the right being pure, that is, equal to its minimal Hodge substructure. The mixed Hodge structure on the left has for minimal Hodge substructure (or pure part) the image of the restriction map $H^k(X, \mathbb{Q}) \to H^k(X^0, \mathbb{Q})$. Comparing weights, it then follows from Lemma 5 that the two restriction maps $H^k(X^0, \mathbb{Q}) \to H^k(X_b, \mathbb{Q})$ and $H^k(X, \mathbb{Q}) \to H^k(X_b, \mathbb{Q})$ have the same image.

4.2 The Algebraicity Theorem and Application to the Variational Hodge Conjecture

The following theorem proved in [7] is the best known evidence for the Hodge conjecture. It says that Hodge classes behave geometrically as if they were algebraic. Let $\phi : \mathscr{X} \to B$ be a projective everywhere submersive morphism, with $\mathscr{X}$, B smooth quasi-projective. For any $b \in B$, denote by $\mathscr{X}_b$ the fiber $\phi^{-1}(b)$. Let $\alpha \in \mathrm{Hdg}^{2k}(\mathscr{X}_b)$ be a Hodge class. The Hodge locus of α is defined as the set of points $t \in B$, such that for some path $\gamma : [0, 1] \to B$ with $\gamma(0) = b$, $\gamma(1) = t$, the class $\alpha_s \in H^{2k}(\mathscr{X}_{\gamma(s)}, \mathbb{Q})$ remains a Hodge class for any $s \in [0, 1]$. Here α_s is the class α transported to $\mathscr{X}_{\gamma(s)}$ using the natural isomorphism $H^{2k}(\mathscr{X}_b, \mathbb{Q}) \cong H^{2k}(\mathscr{X}_{\gamma(s)}, \mathbb{Q})$ given by topological trivialization of the pulled-back family $\mathscr{X}_\gamma \to [0, 1]$.

Theorem 6 (Cattani, Deligne, Kaplan 1995). *The Hodge locus of α is a countable union of closed algebraic subsets of B.*

Note that the local structure of this locus, say in an open ball $B' \subset B$, as a countable union of closed analytic subsets of B' was understood since the developments of the theory of variations of Hodge structures due to Griffiths [14]. The difficulty here lies in the comparison between the analytic and the algebraic category (the basis B is almost never projective in the above theorem).

That this is indeed the structure predicted by the Hodge conjecture for the Hodge locus of α follows from the existence of relative Hilbert schemes (or Chow varieties) which are projective over B and parameterize subschemes (or effective cycles) $Z_t \subset X_t$ of a given cohomology class. Using these relative Hilbert schemes M_i, we can construct a countable union of varieties M_{ij} projective over B, defined by $M_{ij} = M_i \times_B M_j$ and parameterizing cycles $Z_t = Z_t^+ - Z_t^-$ in the fibers $\mathscr{X}_t$. For any point $t \in B$, if the class α_t on $\mathscr{X}_t$ is algebraic, α_t is the class of a cycle $Z_t^+ - Z_t^-$ parameterized by a point in the fiber of at least one of these varieties M_{ij}. Hence the Hodge locus is the union of the images of M_{ij} in B over the pairs (i, j) such that the cycles parameterized by M_{ij} are of class α.

Let us explain the importance of this theorem in the context of the "variational Hodge conjecture". Here the situation is the following: $\mathscr{X}$ is a complex manifold, Δ is a complex ball centered at 0, $\mathscr{X} \to \Delta$ is a proper submersive holomorphic map with projective fibers $\mathscr{X}_t$, $t \in \Delta$, and $\alpha \in H^{2k}(\mathscr{X}, \mathbb{Q})$ is a cohomology class which has the property that $\alpha_t := \alpha_{|\mathscr{X}_t}$ is a degree $2k$ Hodge class on $\mathscr{X}_t$ for any $t \in B$.

Conjecture 5 (Variational Hodge Conjecture). *Assume that α_0 satisfies the Hodge conjecture, that is, is algebraic on $\mathscr{X}_0$. Does it follow that α_t is also algebraic?*

Theorem 7. *The variational Hodge conjecture is implied by the Lefschetz conjecture.*

Proof. The family of projective varieties $(\mathscr{X}_b)_{b \in \Delta}$ is the pullback of an algebraic family $\mathscr{X}^{alg} \to B$ via a holomorphic map $f : \Delta \to B$. Our assumption is that $f(\Delta)$ is contained in the Hodge locus B_α of the Hodge class α_0 on $\mathscr{X}_0$. By Theorem 6, this Hodge locus is algebraic, and we can thus replace Δ by an irreducible component B'_α of B_α passing through 0 and containing $f(\Delta)$. We can assume that B'_α is smooth by desingularization. By definition of B_α, the class α_t deduced by parallel transport from the class α_0 is Hodge on all fibers $\mathscr{X}_t$ of the family $\mathscr{X}_\alpha^{alg} \to B'_\alpha$. The monodromy has finite orbits on the set of cohomology classes in fibers which are Hodge everywhere (see [35, Theorem 4.1]). Replacing B'_α by a finite étale cover, we can thus assume that the class α_0 is monodromy invariant on B'_α. Let us introduce a smooth projective completion $\overline{\mathscr{X}_\alpha^{alg}}$ of $\mathscr{X}_\alpha^{alg}$. By Theorem 5, there exists a class $\beta \in H^{2k}(\overline{\mathscr{X}_\alpha^{alg}}, \mathbb{Q})$ such that $\beta_{|\mathscr{X}_0} = \alpha_0$. We now apply Corollary 3 (i) to $X = \overline{\mathscr{X}_\alpha^{alg}}$, $Y = \mathscr{X}_0$. As the class $\alpha_0 = \beta_{|\mathscr{X}_0}$ is algebraic, there exists assuming the Lefschetz standard conjecture a cycle Z on $\overline{\mathscr{X}_\alpha^{alg}}$ such that $[Z]_{|\mathscr{X}_0} = \alpha_0$, hence $[Z]_{|\mathscr{X}_t} = \alpha_t$, $\forall t \in \Delta \subset B'_\alpha$, and thus α_t is also algebraic.

4.3 Algebraic de Rham Cohomology and Absolute Hodge Classes

The following arithmetic counterpart of Theorem 6 is completely open except for abelian varieties [9] (see also [26, 34] for some partial results) :

Conjecture 6. *In the situation of Theorem 6, assume the family $\mathscr{X} \to B$ is defined over a field K (in fact, we can always assume K to be a number field). Then the Hodge locus of α is a countable union of closed algebraic subsets of B which are defined over a finite extension of K.*

Using the global invariant cycle theorem, this conjecture would allow to reduce the Hodge conjecture to the case of varieties X defined over a number field (see [34]). It would be disproved by the existence of a variety X not defined over a number field, with a Hodge class α such that the pair (X, α) is rigid (meaning that under a nontrivial deformation of X, the class α does not remain Hodge).

We next introduce the notion of *absolute Hodge class*. Let X be a smooth projective variety defined over $\mathbb{C}$. In the following, we will write X^{an} for the complex manifold associated with X and cohomology on X will be coherent cohomology with respect to the Zariski topology on X. We have a chain of isomorphisms whose combination gives the Grothendieck comparison isomorphism [16]:

$$\mathbb{H}^k(X, \Omega^\bullet_{X/\mathbb{C}}) \cong \mathbb{H}^k(X^{an}, \Omega^\bullet_{X^{an}}) \cong H^k(X^{an}, \mathbb{C}).$$

The first term is algebraic de Rham cohomology of X over $\mathbb{C}$. The second term is holomorphic de Rham cohomology of X^{an} and the first isomorphism comes from Serre's GAGA theorem [27]. The second isomorphism comes from the fact that the holomorphic de Rham complex is a resolution of the constant sheaf $\mathbb{C}$ on X^{an}. Note that the Grothendieck isomorphism gives an algebraic definition of the Hodge filtration, namely, it induces for any p an isomorphism

$$\mathbb{H}^k(X, \Omega^{\bullet \geq p}_{X/\mathbb{C}}) \cong F^p H^k(X^{an}, \mathbb{C}). \tag{16}$$

Let now $\tau : \mathbb{C} \to \mathbb{C}$ be a field automorphism. Clearly τ induces an isomorphism (which is not $\mathbb{C}$-linear)

$$\tau_* : \mathbb{H}^k(X, \Omega^\bullet_{X/\mathbb{C}}) \cong \mathbb{H}^k(X_\tau, \Omega^\bullet_{X_\tau/\mathbb{C}}), \tag{17}$$

where X_τ is the complex algebraic variety whose equations are obtained by applying τ to the coefficients of the defining equations of X. Composing this automorphism with the Grothendieck isomorphisms

$$\mathbb{H}^k(X, \Omega^\bullet_{X/\mathbb{C}}) \cong H^k(X^{an}, \mathbb{C}) \tag{18}$$

for X and X_τ, we get an isomorphism $H^k(X^{an}, \mathbb{C}) \cong H^k(X^{an}_\tau, \mathbb{C})$, $\alpha \mapsto \alpha_\tau$. This isomorphism is compatible with the Hodge filtrations by (16).

Definition 5. Let α be a degree $2k$ Hodge class on X. We say that α is an absolute Hodge class if the class $(2\iota\pi)^k\alpha =: \alpha'$ has the property that for any field automorphism τ of $\mathbb{C}$, α'_τ belongs to $(2\iota\pi)^k H^{2k}(X_\tau^{an}, \mathbb{Q})$.

Remark 2. The class α'_τ is then $(2\iota\pi)^k$ times a Hodge class on X_τ, as it belongs to $F^kH^{2k}(X_\tau^{an}, \mathbb{C})$ since α' belongs to $F^kH^{2k}(X^{an}, \mathbb{C})$.

We now use the existence of an algebraic cycle class $Z \mapsto [Z_{dR}]$ with value in algebraic de Rham cohomology (see [5] for an explicit construction). It is clear that if τ is a field automorphism of $\mathbb{C}$, and Z is a codimension k algebraic cycle on X,

$$\tau_*[Z]_{dR} = [Z_\tau]_{dR} \text{ in } \mathbb{H}^{2k}(X_\tau, \Omega^\bullet_{X_\tau/\mathbb{C}}),$$

where Z_τ is the cycle of X_τ obtained by applying τ to the defining equations of the components Z_i of Z. Finally we use the comparison formula saying that, via the Grothendieck isomorphism (18), $[Z]_{dR} = (2\iota\pi)^k[Z]$. We then get:

Proposition 2. *Cycle classes on smooth projective varieties are absolute Hodge.*

Conjecture 6 is a weak form (see [34]) of the following Conjecture 7 (which by Proposition 2 is part of the Hodge conjecture).

Conjecture 7. *Hodge classes are absolute Hodge.*

Deligne [9] proves Conjecture 7 for abelian varieties. It follows from the compatibility properties of the Kuga-Satake construction [22] (see [10]) that it is true as for Hodge classes on (powers of) hyper-Kähler manifolds lying in the subalgebra generated by H^2.

In general, one can say from the above discussion that the Hodge conjecture has two independent parts, each of which might be true or wrong, namely Conjecture 7 on the one hand and on the other hand the conjecture that absolute Hodge classes are algebraic, which is in the same spirit as the Lefschetz standard Conjecture 2 but also concerns more mysterious classes, like Weil classes on abelian varieties with complex multiplication.

Let us conclude with an example of an absolute Hodge class which is not known to be "motivated" in the sense of André [2]. André defines the set of motivated classes as the smallest set of classes on smooth projective algebraic varieties containing algebraic classes, and stable under the operators λ_{n-k} inverse of the Lefschetz operators and under any other algebraic correspondence. Motivated classes include classes $\alpha_t \in \mathrm{Hdg}^{2k}(X_t)$, for some Hodge class α on a smooth projective variety $X \to B$ (where B is connected), such that for some regular value $0 \in B$, $\alpha_0 \in \mathrm{Hdg}^{2k}(X_0)$ is algebraic.

Example 1. Let X be smooth complex projective, and let $b_{2k} := \dim H^{2k}(X, \mathbb{Q})$. Then the space

$$\bigwedge^{b_{2k}} H^{2k}(X, \mathbb{Q}) \subset H^{2k}(X, \mathbb{Q})^{\otimes b_{2k}} \subset H^{2kb_{2k}}(X^{b_{2k}}, \mathbb{Q})$$

is clearly a Hodge substructure which is of rank 1, hence generated by a Hodge class on $X^{b_{2k}}$. This class is clearly an absolute Hodge class. Note that one can make the same construction with odd degree cohomology, but in this case the existence of a polarization easily implies that the classes one gets are algebraic, or at least motivated. For this reason, by specializing to Fermat hypersurfaces, the class constructed above is motivated for all smooth hypersurfaces.

References

1. Y. André. Une introduction aux motifs (motifs purs, motifs mixtes, périodes). Panoramas et Synthèses, 17. Société Mathématique de France, Paris, (2004).
2. Y. André. Pour une théorie inconditionnelle des motifs, Inst. Hautes Études Sci. Publ. Math. No. 83 (1996), 5–49.
3. M. Atiyah, F. Hirzebruch. Analytic cycles on complex manifolds, *Topology* 1, 25–45 (1962).
4. A. Blanchard. Sur les variétés analytiques complexes, Ann. Sci. École Norm. Sup. (3) 73 (1956), 157–202.
5. S. Bloch. Semi-regularity and de Rham cohomology. Invent. Math. 17 (1972), 51–66.
6. S. Bloch. *Lectures on algebraic cycles*. Duke University Mathematics Series, IV. Duke University, Mathematics Department, Durham, N.C., (1980).
7. E. Cattani, P. Deligne, A. Kaplan. On the locus of Hodge classes, J. Amer. Math. Soc. 8 (1995), 2, 483–506.
8. P. Deligne. Théorie de Hodge. II, Inst. Hautes Études Sci. Publ. Math. No. 40 , 5–57 (1971).
9. P. Deligne. Hodge cycles on abelian varieties (notes by JS Milne), in Springer LNM, 900 (1982), 9–100.
10. P. Deligne. La conjecture de Weil pour les surfaces K3, Invent. Math. 15 (1972), 206–226.
11. P. Deligne. Théorème de Lefschetz et critères de dégénérescence de suites spectrales, Inst. Hautes Études Sci. Publ. Math. No. 35 1968 259–278.
12. J.-P. Demailly. Regularization of closed positive currents and intersection theory, J. Algebraic Geom. 1 (1992), no. 3, 361–409.
13. Ph. Griffiths. A theorem concerning the differential equations satisfied by normal functions associated to algebraic cycles. Amer. J. Math. 101 (1979), no. 1, 94–131.
14. Ph. Griffiths. Periods of integrals on algebraic manifolds. II. Local study of the period mapping. Amer. J. Math. 90 (1968) 805–865.
15. A. Grothendieck. Hodge's general conjecture is false for trivial reasons, Topology **8** 299–303 (1969).
16. A. Grothendieck. On the de Rham cohomology of algebraic varieties. Pub. math. IHÉS 29, 95–103 (1966).
17. W. Hodge. Differential forms on a Kähler manifold. Proc. Cambridge Philos. Soc. 47, (1951), 504–517.
18. W. Hodge. The topological invariants of algebraic varieties. Proceedings of the International Congress of Mathematicians, Cambridge, Mass., 1950, vol. 1, pp. 182–192. Amer. Math. Soc., Providence, R. I., (1952).
19. U. Jannsen. *Mixed motives and algebraic K-theory*. With appendices by S. Bloch and C. Schoen. Lecture Notes in Mathematics, 1400. Springer-Verlag, Berlin, (1990).
20. J. Kollár. Lemma p. 134 in *Classification of irregular varieties*, edited by E. Ballico, F. Catanese, C. Ciliberto, Lecture Notes in Math. 1515, Springer.
21. S. Kleiman. Algebraic cycles and the Weil conjectures in *Dix exposés sur la cohomologie des schémas*, pp. 359–386. North-Holland, Amsterdam; Masson, Paris, 1968.
22. M. Kuga, I. Satake. Abelian varieties attached to polarized K3-surfaces, Math. Ann. 169 (1967) 239–242.
23. J. Lewis. *A survey of the Hodge conjecture*, second edition with an appendix B by B. Brent Gordon, CRM Monograph Series, 10. American Mathematical Society, Providence, RI, (1999).

24. D. Lieberman, Numerical and homological equivalence of algebraic cycles on Hodge manifolds, *Amer. J. Math.* 90 (1968), 366–374
25. C. Peters, J. Steenbrink. *Mixed Hodge structures*, Ergebnisse der Mathematik und ihrer Grenzgebiete 52. Springer-Verlag, Berlin, (2008).
26. M. Saito, Ch. Schnell. Fields of definition of Hodge loci, arXiv:1408.2488.
27. J.-P. Serre. Géométrie algébrique et géométrie analytique. Ann. Inst. Fourier, Grenoble 6 (1955–1956), 1–42.
28. C. Soulé, C. Voisin. Torsion cohomology classes and algebraic cycles on complex projective manifolds, Advances in Mathematics, Vol 198/1 pp 107–127 (2005).
29. Uhlenbeck, Yau. On the existence of Hermitian-Yang-Mills connections in stable vector bundles, Comm. Pure Appl. Math. 39 (1986), no. S, suppl., S257–S293.
30. C. Voisin. A counterexample to the Hodge conjecture extended to Kähler varieties, IMRN (2002), no. 20, 1057–1075.
31. C. Voisin. Hodge theory and complex algebraic geometry I, Cambridge studies in advanced mathematics 76, Cambridge University Press (2002).
32. C. Voisin. Hodge theory and complex algebraic geometry II, Cambridge studies in advanced mathematics 77, Cambridge University Press (2003).
33. C. Voisin. The generalized Hodge and Bloch conjectures are equivalent for general complete intersections, Annales scientifiques de l'ENS 46, fascicule 3 (2013), 449–475.
34. C. Voisin. Hodge loci and absolute Hodge classes, Compositio Mathematica, Vol. 143 Part 4, 945–958, (2007).
35. C. Voisin. Hodge loci, in Handbook of moduli (Eds G. Farkas and I. Morrison), Advanced Lectures in Mathematics 25, Volume III, International Press, 507–547 (2013).
36. S. Zucker. The Hodge conjecture for cubic fourfolds. Compositio Math. 34 (1977), no. 2, 199–209.

CPSIA information can be obtained
at www.ICGtesting.com
Printed in the USA
LVHW082324080722
723057LV00012B/11

9 783319 321608